$126.95 per copy, (In United States).
Price subject to change without prior notice.

0055

RSMeans

Square Foot Costs

26th Annual Edition

2005

RS**Means**
Construction Publishers & Consultants
63 Smiths Lane
Kingston, MA 02364-0800
(781) 585-7880

Copyright© 2004 by Reed Construction Data, Inc.
All rights reserved.

Printed in the United States of America
ISSN 1540-6326
ISBN 0-87629-759-9

Senior Editor
Barbara Balboni

Contributing Editors
Robert A. Bastoni
John H. Chiang, PE
Robert J. Kuchta
Robert C. McNichols
Robert W. Mewis, CCC
Melville J. Mossman, PE
John J. Moylan
Jeannene D. Murphy
Stephen C. Plotner
Michael J. Regan
Eugene R. Spencer
Marshall J. Stetson
Phillip R. Waier, PE

**Senior Engineering
Operations Manager**
John H. Ferguson, PE

Editorial Advisory Board
James E. Armstrong, CPE, CEM
Program Manager,
Energy Conservation
NSTAR

William R. Barry, CCC
Cost Consultant

Robert F. Cox, PhD
Assistant Professor
ME Rinker Sr. School of Bldg. Constr.
University of Florida

Roy F. Gilley, AIA
Principal
Gilley-Hinkel Architects

Kenneth K. Humphreys, PhD,
PE, CCE
Secretary-Treasurer
International Cost
Engineering Council

Patricia L. Jackson, PE
President
Jackson A&E Assoc., Inc.

Martin F. Joyce
Executive Vice President
Bond Brothers, Inc.

**Senior Vice President
& General Manager**
John Ware

Product Manager
Jane Crudden

**Vice President
of Sales**
John M. Shea

Production Manager
Michael Kokernak

Production Coordinator
Marion E. Schofield

Technical Support
Thomas J. Dion
Jonathan Forgit
Mary Lou Geary
Gary L. Hoitt
Alice McSharry
Paula Reale-Camelio
Robin Richardson
Kathryn S. Rodriguez
Sheryl A. Rose

Book & Cover Design
Norman R. Forgit

 This book is recyclable.

 This book is printed on recycled
stock.

 Reed Construction Data

First Printing

Foreword

RSMeans is a product line of Reed Construction Data, Inc., a leading provider of construction information, products, and services in North America and globally. Reed Construction Data's project information products include more than 100 regional editions, national construction data, sales leads, and local plan rooms in major business centers. Reed Construction Data's PlansDirect provides surveys, plans and specifications. The First Source suite of products consists of *First Source for Products,* SPEC-DATA™, MANU-SPEC™, CADBlocks, Manufacturer Catalogs and First Source Exchange (www.firstsourceexchange.com) for the selection of nationally available building products. Reed Construction Data also publishes ProFile, a database of more than 20,000 U.S. architectural firms. RSMeans provides construction cost data, training, and consulting services in print, CD-ROM and online. Reed Construction Data, headquartered in Atlanta, is owned by Reed Business Information (www.reedconstructiondata.com), a leading provider of critical information and marketing solutions to business professionals in the media, manufacturing, electronics, construction and retail industries. Its market-leading properties include more than 135 business-to-business publications, over 125 Webzines and Web portals, as well as online services, custom publishing, directories, research and direct-marketing lists. Reed Business Information is a member of the Reed Elsevier plc group (NYSE: RUK and ENL)—a world-leading publisher and information provider operating in the science and medical, legal, education and business-to-business industry sectors.

Our Mission

Since 1942, RSMeans has been actively engaged in construction cost publishing and consulting throughout North America.

Today, over 60 years after RSMeans began, our primary objective remains the same: to provide you, the construction and facilities professional, with the most current and comprehensive construction cost data possible.

Whether you are a contractor, an owner, an architect, an engineer, a facilities manager, or anyone else who needs a fast and reliable construction cost estimate, you'll find this publication to be a highly useful and necessary tool.

Today, with the constant flow of new construction methods and materials, it's difficult to find the time to look at and evaluate all the different construction cost possibilities. In addition, because labor and material costs keep changing, last year's cost information is not a reliable basis for today's estimate or budget.

That's why so many construction professionals turn to RSMeans. We keep track of the costs for you, along with a wide range of other key information, from city cost indexes . . . to productivity rates . . . to crew composition . . . to contractor's overhead and profit rates.

RSMeans performs these functions by collecting data from all facets of the industry, and organizing it in a format that is instantly accessible to you. From the preliminary budget to the detailed unit price estimate, you'll find the data in this book useful for all phases of construction cost determination.

The Staff, the Organization, and Our Services

When you purchase one of RSMeans' publications, you are in effect hiring the services of a full-time staff of construction and engineering professionals.

Our thoroughly experienced and highly qualified staff works daily at collecting, analyzing, and disseminating comprehensive cost information for your needs. These staff members have years of practical construction experience and engineering training prior to joining the firm. As a result, you can count on them not only for the cost figures, but also for additional background reference information that will help you create a realistic estimate.

The RSMeans organization is always prepared to help you solve construction problems through its five major divisions: Construction and Cost Data Publishing, Electronic Products and Services, Consulting Services, Insurance Services, and Educational Services.

Besides a full array of construction cost estimating books, Means also publishes a number of other reference works for the construction industry. Subjects include construction estimating and project and business management; special topics such as HVAC, roofing, plumbing, and hazardous waste remediation; and a library of facility management references.

In addition, you can access all of our construction cost data through your computer with RSMeans *CostWorks 2005* CD-ROM, an electronic tool that offers over 50,000 lines of Means detailed construction cost data, along with assembly and whole building cost data. You can also access Means cost information from our Web site at www.rsmeans.com

What's more, you can increase your knowledge and improve your construction estimating and management performance with a Means Construction Seminar or In-House Training Program. These two-day seminar programs offer unparalleled opportunities for everyone in your organization to get updated on a wide variety of construction-related issues.

RSMeans also is a worldwide provider of construction cost management and analysis services for commercial and government owners and of claims and valuation services for insurers.

In short, RSMeans can provide you with the tools and expertise for constructing accurate and dependable construction estimates and budgets in a variety of ways.

Robert Snow Means Established a Tradition of Quality That Continues Today

Robert Snow Means spent years building RSMeans, making certain he always delivered a quality product.

Today, at RSMeans, we do more than talk about the quality of our data and the usefulness of our books. We stand behind all of our data, from historical cost indexes to construction materials and techniques to current costs.

If you have any questions about our products or services, please call us toll-free at 1-800-334-3509. Our customer service representatives will be happy to assist you or visit our Web site at www.rsmeans.com

Table of Contents

Foreword	ii
How the Book Is Built: An Overview	iv
Residential Section	1
Introduction	3
Work Sheet	13
Economy	18
Average	26
Custom	38
Luxury	48
Adjustments	58
Commercial/Industrial/Institutional Section	65
Introduction	67
Examples	73
Building Types	78
Depreciation	224
Assemblies Section	225
Introduction	227
Substructure	239
Shell	247
Interiors	329
Services	353
Equipment & Furnishings	423
Special Construction	433
Building Sitework	441
Reference Section	449
General Conditions	450
Location Factors	451
Historical Cost Indexes	457
Glossary	465
Abbreviations	469
Other Means Publications and Reference Works	Yellow Pages
Installing Contractor's Overhead & Profit	Inside Back Cover

RESIDENTIAL
- INTRODUCTION
- WORKSHEET
- ECONOMY
- AVERAGE
- CUSTOM
- LUXURY
- ADJUSTMENTS

COMM./INDUST./INSTIT.
- INTRODUCTION
- EXAMPLES
- BUILDING TYPES
- DEPRECIATION

ASSEMBLIES
- SUBSTRUCTURE — A
- SHELL — B
- INTERIORS — C
- SERVICES — D
- EQUIPMENT & FURNISHINGS — E
- SPECIAL CONSTRUCTION — F
- BUILDING SITEWORK — G

REFERENCE INFORMATION
- GENERAL CONDITIONS
- LOCATION FACTORS
- COST INDEXES
- GLOSSARY

Related Means Products and Services

Barbara Balboni, Senior Editor of this cost data book, suggests the following Means products and services as **companion** information resources to *Square Foot Costs*:

Construction Cost Data Books

Building Construction Cost Data 2005
Assemblies Cost Data 2005
Light Commercial Cost Data 2005
Facilities Construction Cost Data 2005

Reference Books

Square Foot & Assemblies Estimating Methods, 3rd Ed.
Estimating Handbook, 2nd Ed.
Historic Preservation: Project Planning & Estimating
Successful Estimating Methods: From Concept to Bid
Managing Construction Purchasing

Seminars and In-House Training

Square Foot and Assemblies Costs Estimating

Means on the Internet

Visit Means at **http://www.rsmeans.com.** The site contains useful *interactive* cost and reference material. Request or download *FREE* estimating software demos. Visit our bookstore for convenient ordering, plus learn more about new publications and companion products.

Means Data on CD-ROM

Get the information found in Means traditional cost books on Means new, easy to use CD-ROM, Means *CostWorks 2005*. Means *CostWorks* users can now enhance their Costlist with the Means *CostWorks Estimator*. For more information see the special full color brochure inserted in this book.

Means Consulting Services

Means Consulting Group provides a number of construction cost-related services including: Cost Estimates, Customized Databases, Benchmark Studies, Estimating Audits, Feasibility Studies, Litigation Support, Staff Assessments and Customized Training.

Means Data for Computer Estimating

Construction costs for software applications.
Over 25 unit price and assemblies cost databases are available through a number of leading estimating and facilities management software providers (listed below). For more information see the "yellow" pages at the back of this publication.

- 3D International
- 4Clicks-Solutions, LLC
- Aepco, Inc.
- Applied Flow Technology
- ArenaSoft Estimating
- ARES Corporation
- Barchan Associates
- BSD – Building Systems Design, Inc.
- CMS – Computerized Micro Solutions
- Corecon Technologies, Inc.
- CorVet Systems
- Discover Software
- Estimating Systems, Inc.
- Entech Engineering
- FAMIS Software, Inc.
- ISES Corporation
- Magellan K-12
- Maximus Asset Solutions
- MC^2 – Management Computer Controls
- Neddam Software Technologies
- Prime Time
- Quest Solutions, Inc.
- RIB Software (Americas), Inc.
- Shaw Beneco Enterprises, Inc.
- Schwab Technology Solutions, Inc.
- Timberline Software Corp.
- TMA Systems, LLC
- US Cost, Inc.
- Vanderweil Facility Advisors
- Vertigraph, Inc.

How the Book Is Built: An Overview

A Powerful Construction Tool

You have in your hands one of the most powerful construction tools available today. A successful project is built on the foundation of an accurate and dependable estimate. This book will enable you to construct just such an estimate.

For the casual user the book is designed to be:

- quickly and easily understood so you can get right to your estimate
- filled with valuable information so you can understand the necessary factors that go into the cost estimate

For the regular user, the book is designed to be:

- a handy desk reference that can be quickly referred to for key costs
- a comprehensive, fully reliable source of current construction costs for typical building structures when only minimum dimensions, specifications and technical data are available. It is specifically useful in the conceptual or planning stage for preparing preliminary estimates which can be supported and developed into complete engineering estimates for construction.
- a source book for rapid estimating of replacement costs and preparing capital expenditure budgets

To meet all of these requirements we have organized the book into the following clearly defined sections.

How To Use the Book: The Details

This section contains an in-depth explanation of how the book is arranged . . . and how you can use it to determine a reliable construction cost estimate. It includes information about how we develop our cost figures and how to completely prepare your estimate.

Residential Section

Model buildings for four classes of construction—economy, average, custom and luxury—are developed and shown with complete costs per square foot.

Commercial, Industrial, Institutional Section

This section contains complete costs for 70 typical model buildings expressed as costs per square foot.

Assemblies Section

The cost data in this section has been organized in an "Assemblies" format. These assemblies are the functional elements of a building and are arranged according to the 7 divisions of the UNIFORMAT II classification system. The costs in this section include standard mark-ups for the overhead and profit of the installing contractor. The assemblies costs are those used to calculate the total building costs in the Commercial, Industrial, Institutional section.

Reference Section

This section includes information on General Conditions, Location Factors, Historical Cost Indexes, a Glossary, and a listing of Abbreviations used in this book.

Location Factors: Costs vary depending upon regional economy. You can adjust the "national average" costs in this book to 930 locations throughout the U.S. and Canada by using the data in this section.

Historical Cost Indexes: These indexes provide you with data to adjust construction costs over time. If you know costs for a project completed in the past, you can use these indexes to calculate a rough estimate of what it would cost to construct the same project today.

Glossary: A listing of common terms related to construction and their definitions is included.

Abbreviations: A listing of the abbreviations used throughout this book, along with the terms they represent, is included.

The Scope of This Book

This book is designed to be as comprehensive and as easy to use as possible. To that end we have made certain assumptions and limited its scope in two key ways:

1. We have established material prices based on a "national average."
2. We have computed labor costs based on a 30-city "national average" of union wage rates.

All costs and adjustments in this manual were developed from the *Means Building Construction Cost Data, Means Assemblies Cost Data, Means Residential Cost Data,* and *Means Light Commercial Cost Data.* Costs include labor, material, overhead, and profit, and general conditions and architect's fees where specified. No allowance is made for sales tax, financing, premiums for material and labor, or unusual business and weather conditions.

What's Behind the Numbers? The Development of Cost Data

The staff at RSMeans continuously monitors developments in the construction industry in order to ensure reliable, thorough and up-to-date cost information.

While *overall* construction costs may vary relative to general economic conditions, price fluctuations within the industry are dependent upon many factors. Individual price variations may, in fact, be opposite to overall economic trends. Therefore, costs are continually monitored and complete updates are published yearly. Also, new items are frequently added in response to changes in materials and methods.

Costs—$ (U.S.)

All costs represent U.S. national averages and are given in U.S. dollars. The Means Location Factors can be used to adjust costs to a particular location. The Location Factors for Canada can be used to adjust U.S. national averages to local costs in Canadian dollars.

Factors Affecting Costs

Due to the almost limitless variation of building designs and combinations of construction methods and materials, costs in this book must be used with discretion. The editors have made every effort to develop costs for "typical" building types. By definition, a square foot estimate, due to the short time and limited amount of detail required, involves a relatively lower degree of accuracy than a detailed unit cost estimate, for example. The accuracy of an estimate will increase as the project is defined in more detailed components. The user should examine closely and compare the specific requirements of the project being estimated with the components included in the models in this book. Any differences should be accounted for in the estimate.

In many building projects, there may be factors that increase or decrease the cost beyond the range shown in this manual. Some of these factors are:

- Substitution of building materials or systems for those used in the model.
- Custom designed finishes, fixtures and/or equipment.
- Special structural qualities (allowance for earthquake, future expansion, high winds, long spans, unusual shape).
- Special foundations and substructures to compensate for unusual soil conditions.
- Inclusion of typical "owner supplied" equipment in the building cost (special testing and monitoring equipment for hospitals, kitchen equipment for hotels, offices that have cafeteria accommodations for employees, etc.).
- An abnormal shortage of labor or materials.
- Isolated building site or rough terrain that would affect the transportation of personnel, material or equipment.
- Unusual climatic conditions during the construction process.

We strongly urge that, for maximum accuracy, these factors be considered each time a structure is being evaluated.

If users require greater accuracy than this manual can provide, the editors recommend that the *Means Building Construction Cost Data* or the *Means Assemblies Cost Data* be consulted.

Final Checklist

Estimating can be a straightforward process provided you remember the basics. Here's a checklist of some of the items you should remember to do before completing your estimate.

Did you remember to . . .

- include the Location Factor for your city
- mark up the entire estimate sufficiently for your purposes
- include all components of your project in the final estimate
- double check your figures to be sure of your accuracy
- call RSMeans if you have any questions about your estimate or the data you've found in our publications

Remember, RSMeans stands behind its publications. If you have any questions about your estimate . . . about the costs you've used from our books . . . or even about the technical aspects of the job that may affect your estimate, feel free to call the RSMeans editors at 1-800-334-3509.

Residential Section

Table of Contents

	Page
Introduction	
General	3
How to Use	4
Building Classes	6
Configurations	7
Building Types	8
Garage Types	10
Building Components	11
Exterior Wall Construction	12
Residential Cost Estimate Worksheets	
Instructions	13
Model Residence Example	15

	Page
Economy	
Illustrations	18
Specifications	19
1 Story	20
1-1/2 Story	21
2 Story	22
Bi-Level	23
Tri-Level	24
Wings & Ells	25
Average	
Illustrations	26
Specifications	27
1 Story	28
1-1/2 Story	29
2 Story	30
2-1/2 Story	31
3 Story	32
Bi-Level	33
Tri-Level	34
Solid Wall (log home) 1 Story	35
Solid Wall (log home) 2 Story	36
Wings & Ells	37
Custom	
Illustrations	38
Specifications	39
1 Story	40
1-1/2 Story	41
2 Story	42
2-1/2 Story	43
3 Story	44
Bi-Level	45
Tri-Level	46
Wings & Ells	47

	Page
Luxury	
Illustrations	48
Specifications	49
1 Story	50
1-1/2 Story	51
2 Story	52
2-1/2 Story	53
3 Story	54
Bi-Level	55
Tri-Level	56
Wings & Ells	57
Residential Modifications	
Kitchen Cabinets	58
Kitchen Counter Tops	58
Vanity Bases	59
Solid Surface Vanity Tops	59
Fireplaces & Chimneys	59
Windows & Skylights	59
Dormers	59
Appliances	60
Breezeway	60
Porches	60
Finished Attic	60
Alarm System	60
Sauna, Prefabricated	60
Garages	61
Swimming Pools	61
Wood & Coal Stoves	61
Sidewalks	61
Fencing	61
Carport	61
Insurance Exclusions	62
Architect/Designer Fees	62
Depreciation	63

Introduction to Residential Section

The residential section of this manual contains costs per square foot for four classes of construction in seven building types. Costs are listed for various exterior wall materials which are typical of the class and building type. There are cost tables for Wings and Ells with modification tables to adjust the base cost of each class of building. Non-standard items can easily be added to the standard structures.

Cost estimating for a residence is a three-step process:
- (1) Identification
- (2) Listing dimensions
- (3) Calculations

Guidelines and a sample cost estimating procedure are shown on the following pages.

Identification

To properly identify a residential building, the class of construction, type, and exterior wall material must be determined. Page 6 has drawings and guidelines for determining the class of construction. There are also detailed specifications and additional drawings at the beginning of each set of tables to further aid in proper building class and type identification.

Sketches for eight types of residential buildings and their configurations are shown on pages 8 and 9. Definitions of living area are next to each sketch. Sketches and definitions of garage types are on page 10.

Living Area

Base cost tables are prepared as costs per square foot of living area. The living area of a residence is that area which is suitable and normally designed for full time living. It does not include basement recreation rooms or finished attics, although these areas are often considered full time living areas by the owners.

Living area is calculated from the exterior dimensions without the need to adjust for exterior wall thickness. When calculating the living area of a 1-1/2 story, two story, three story or tri-level residence, overhangs and other differences in size and shape between floors must be considered.

Only the floor area with a ceiling height of six feet or more in a 1-1/2 story residence is considered living area. In bi-levels and tri-levels, the areas that are below grade are considered living area, even when these areas may not be completely finished.

Base Tables and Modifications

Base cost tables show the base cost per square foot without a basement, with one full bath and one full kitchen for economy and average homes and an additional half bath for custom and luxury models. Adjustments for finished and unfinished basements are part of the base cost tables. Adjustments for multi family residences, additional bathrooms, townhouses, alternate roofs, and air conditioning and heating systems are listed in modification tables below the base cost tables.

Costs for other modifications, including garages, breezeways and site improvements, are in the modification tables on pages 58 to 61.

Listing of Dimensions

To use this section of the manual only the dimensions used to calculate the horizontal area of the building and additions and modifications are needed. The dimensions, normally the length and width, can come from drawings or field measurements. For ease in calculation, consider measuring in tenths of feet, i.e., 9 ft. 6 in. = 9.5 ft., 9 ft. 4 in. = 9.3 ft.

In all cases, make a sketch of the building. Any protrusions or other variations in shape should be noted on the sketch with dimensions.

Calculations

The calculations portion of the estimate is a two-step activity:
- (1) The selection of appropriate costs from the tables
- (2) Computations

Selection of Appropriate Costs

To select the appropriate cost from the base tables, the following information is needed:
- (1) Class of construction
    - (a) Economy
    - (b) Average
    - (c) Custom
    - (d) Luxury
- (2) Type of residence
    - (a) One story
    - (b) 1-1/2 story
    - (c) 2 story
    - (d) 3 story
    - (e) Bi-level
    - (f) Tri-level
- (3) Occupancy
    - (a) One family
    - (b) Two family
    - (c) Three family
- (4) Building configuration
    - (a) Detached
    - (b) Town/Row house
    - (c) Semi-detached
- (5) Exterior wall construction
    - (a) Wood frame
    - (b) Brick veneer
    - (c) Solid masonry
- (6) Living areas

Modifications are classified by class, type and size.

Computations

The computation process should take the following sequence:
- (1) Multiply the base cost by the area
- (2) Add or subtract the modifications
- (3) Apply the location modifier

When selecting costs, interpolate or use the cost that most nearly matches the structure under study. This applies to size, exterior wall construction and class.

How to Use the Residential Square Foot Cost Pages

The following is a detailed explanation of a sample entry in the Residential Square Foot Cost Section. Each bold number below corresponds to the item being described on the facing page with the appropriate component of the sample entry following in parenthesis.

Prices listed are costs that include overhead and profit of the installing contractor. Total model costs include an additional mark-up for General Contractors' overhead and profit, and fees specific to class of construction.

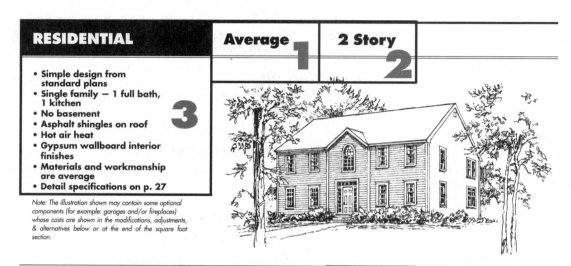

RESIDENTIAL   Average **1**   2 Story **2**

- Simple design from standard plans
- Single family — 1 full bath, 1 kitchen
- No basement
- Asphalt shingles on roof
- Hot air heat
- Gypsum wallboard interior finishes
- Materials and workmanship are average
- Detail specifications on p. 27

3

Note: The illustration shown may contain some optional components (for example: garages and/or fireplaces) whose costs are shown in the modifications, adjustments, & alternatives below or at the end of the square foot section.

Base cost per square foot of living area

5

Exterior Wall **4**	1000	1200	1400	1600	1800	Living Area 2000	2200	2600	3000	3400	3800
Wood Siding - Wood Frame	99.15	90.00	85.80	82.95	80.00	76.75	74.65	70.65	66.55	64.85	63.25
Brick Veneer - Wood Frame	106.65	97.00	92.25	89.15	85.80	82.40	**6**	75.45	71.00	69.05	67.30
Stucco on Wood Frame	98.40	89.35	85.20	82.35	79.40	76.20	74.15	70.15	66.15	64.45	62.90
Solid Masonry **7**	118.35	107.80	102.45	98.85	94.90	91.15	88.30	83.05	78.00	75.70	73.60
Finished Basement, Add	16.05	15.85	15.35	15.00	14.55	14.35	**8**	13.55	13.20	12.95	12.75
Unfinished Basement, Add	6.40	6.00	5.65	5.40	5.15	5.05	4.85	4.50	4.35	4.15	4.05

Modifications

Add to the total cost

Upgrade Kitchen Cabinets	$	+ 3157
Solid Surface Countertops (Included)		
Full Bath - including plumbing, wall and floor finishes		+ 4261
Half Bath - including plumbing, wall and floor finishes		+ 2660
One Car Attached Garage		+ 9490
One Car Detached Garage		+ 12,424
Fireplace & Chimney		+ 4670

9

Adjustments

For multi family - add to total cost

Additional Kitchen	$	+ 5105
Additional Bath		+ 4261
Additional Entry & Exit		+ 1276
Separate Heating		+ 1248
Separate Electric		+ 1575

For Townhouse/Rowhouse - Multiply cost per square foot by **10**

Inner Unit	.90
End Unit	.95

Alternatives

Add to or deduct from the cost per square foot of living area

Cedar Shake Roof	+ 1.17
Clay Tile Roof	+ 2.04
Slate Roof	+ 3.89
Upgrade Walls to Skim Coat Plaster	+ .42
Upgrade Ceilings to Textured Finish	+ .40
Air Conditioning, in Heating Ductwork	+ 1.70
In Separate Ductwork	+ 4.35
Heating Systems, Hot Water	+ 1.39
Heat Pump	+ 2.15
Electric Heat	– .51
Not Heated	– 2.46

11

Additional upgrades or components

Kitchen Cabinets & Countertops	Page 58
Bathroom Vanities	59
Fireplaces & Chimneys	59
Windows, Skylights & Dormers	59
Appliances	60
Breezeways & Porches	60
Finished Attic	60
Garages	61
Site Improvements	61
Wings & Ells	37

12

1. Class of Construction (Average)

The class of construction depends upon the design and specifications of the plan. The four classes are economy, average, custom and luxury.

2. Type of Residence (2 Story)

The building type describes the number of stories or levels in the model. The seven building types are 1 story, 1-1/2 story, 2 story, 2-1/2 story, 3 story, bi-level and tri-level.

3. Specification Highlights (Hot Air Heat)

These specifications include information concerning the components of the model, including the number of baths, roofing types, HVAC systems, and materials and workmanship. If the components listed are not appropriate, modifications can be made by consulting the information shown lower on the page or the Assemblies Section.

4. Exterior Wall System (Wood Siding - Wood Frame)

This section includes the types of exterior wall systems and the structural frame used. The exterior wall systems shown are typical of the class of construction and the building type shown.

5. Living Areas (2000 SF)

The living area is that area of the residence which is suitable and normally designed for full time living. It does not include basement recreation rooms or finished attics. Living area is calculated from the exterior dimensions without the need to adjust for exterior wall thickness. When calculating the living area of a 1-1/2 story, 2 story, 3 story or tri-level residence, overhangs and other differences in size and shape between floors must be considered. Only the floor area with a ceiling height of six feet or more in a 1-1/2 story residence is considered living area. In bi-levels and tri-levels, the areas that are below grade are considered living area, even when these areas may not be completely finished. A range of various living areas for the residential model are shown to aid in selection of values from the matrix.

6. Base Costs per Square Foot of Living Area ($76.75)

Base cost tables show the cost per square foot of living area without a basement, with one full bath and one full kitchen for economy and average homes and an additional half bath for custom and luxury models. When selecting costs, interpolate or use the cost that most nearly matches the

residence under consideration for size, exterior wall system and class of construction. Prices listed are costs that include overhead and profit of the installing contractor, a general contractor markup and an allowance for plans that vary by class of construction. For additional information on contractor overhead and architectural fees, see the Reference Section.

7. Basement Types (Finished)

The two types of basements are finished or unfinished. The specifications and components for both are shown on the Building Classes page in the Introduction to this section.

8. Additional Costs for Basements ($14.35)

These values indicate the additional cost per square foot of living area for either a finished or an unfinished basement.

9. Modifications and Adjustments (Full Bath $4,261)

Modifications and Adjustments are costs added to or subtracted from the total cost of the residence. The total cost of the residence is equal to the cost per square foot of living area times the living area. Typical modifications and adjustments include kitchens, baths, garages and fireplaces.

10. Multiplier for Townhouse/ Rowhouse (Inner Unit .90)

The multipliers shown adjust the base costs per square foot of living area for the common wall condition encountered in townhouses or rowhouses.

11. Alternatives (Skim Coat Plaster $.42)

Alternatives are costs added to or subtracted from the base cost per square foot of living area. Typical alternatives include variations in kitchens, baths, roofing, air conditioning and heating systems.

12. Additional Upgrades or Components (Wings & Ells page 37)

Costs for additional upgrades or components, including wings or ells, breezeways, porches, finished attics and site improvements, are shown in other locations in the Residential Square Foot Cost Section.

Building Classes

Economy Class

An economy class residence is usually built from stock plans. The materials and workmanship are sufficient to satisfy building codes. Low construction cost is more important than distinctive features. The overall shape of the foundation and structure is seldom other than square or rectangular.

An unfinished basement includes 7' high 8" thick foundation wall composed of either concrete block or cast-in-place concrete.

Included in the finished basement cost are inexpensive paneling or drywall as the interior finish on the foundation walls, a low cost sponge backed carpeting adhered to the concrete floor, a drywall ceiling, and overhead lighting.

Custom Class

A custom class residence is usually built from plans and specifications with enough features to give the building a distinction of design. Materials and workmanship are generally above average with obvious attention given to construction details. Construction normally exceeds building code requirements.

An unfinished basement includes a 7'-6" high 10" thick cast-in-place concrete foundation wall or a 7'-6" high 12" thick concrete block foundation wall.

A finished basement includes painted drywall on insulated 2" x 4" wood furring as the interior finish to the concrete walls, a suspended ceiling, carpeting adhered to the concrete floor, overhead lighting and heating.

Average Class

An average class residence is a simple design and built from standard plans. Materials and workmanship are average, but often exceed minimum building codes. There are frequently special features that give the residence some distinctive characteristics.

An unfinished basement includes 7'-6" high 8" thick foundation wall composed of either cast-in-place concrete or concrete block.

Included in the finished basement are plywood paneling or drywall on furring that is fastened to the foundation walls, sponge backed carpeting adhered to the concrete floor, a suspended ceiling, overhead lighting and heating.

Luxury Class

A luxury class residence is built from an architect's plan for a specific owner. It is unique both in design and workmanship. There are many special features, and construction usually exceeds all building codes. It is obvious that primary attention is placed on the owner's comfort and pleasure. Construction is supervised by an architect.

An unfinished basement includes 8' high 12" thick foundation wall that is composed of cast-in-place concrete or concrete block.

A finished basement includes painted drywall on 2" x 4" wood furring as the interior finish, suspended ceiling, tackless carpet on wood subfloor with sleepers, overhead lighting and heating.

Configurations

Detached House

This category of residence is a free-standing separate building with or without an attached garage. It has four complete walls.

Semi-Detached House

This category of residence has two living units side-by-side. The common wall is a fireproof wall. Semi-detached residences can be treated as a row house with two end units. Semi-detached residences can be any of the building types.

Town/Row House

This category of residence has a number of attached units made up of inner units and end units. The units are joined by common walls. The inner units have only two exterior walls. The common walls are fireproof. The end units have three walls and a common wall. Town houses/row houses can be any of the building types.

Building Types

One Story

This is an example of a one-story dwelling. The living area of this type of residence is confined to the ground floor. The headroom in the attic is usually too low for use as a living area.

One-and-a-half Story

The living area in the upper level of this type of residence is 50% to 90% of the ground floor. This is made possible by a combination of this design's high-peaked roof and/or dormers. Only the upper level area with a ceiling height of 6 feet or more is considered living area. The living area of this residence is the sum of the ground floor area plus the area on the second level with a ceiling height of 6 feet or more.

One Story with Finished Attic

The main living area in this type of residence is the ground floor. The upper level or attic area has sufficient headroom for comfortable use as a living area. This is made possible by a high peaked roof. The living area in the attic is less than 50% of the ground floor. The living area of this type of residence is the ground floor area only. The finished attic is considered an adjustment.

Two Story

This type of residence has a second floor or upper level area which is equal or nearly equal to the ground floor area. The upper level of this type of residence can range from 90% to 110% of the ground floor area, depending on setbacks or overhangs. The living area is the sum of the ground floor area and the upper level floor area.

Two-and-one-half Story

This type of residence has two levels of equal or nearly equal area and a third level which has a living area that is 50% to 90% of the ground floor. This is made possible by a high peaked roof, extended wall heights and/or dormers. Only the upper level area with a ceiling height of 6 feet or more is considered living area. The living area of this residence is the sum of the ground floor area, the second floor area and the area on the third level with a ceiling height of 6 feet or more.

Three Story

This type of residence has three levels which are equal or nearly equal. As in the 2 story residence, the second and third floor areas may vary slightly depending on setbacks or overhangs. The living area is the sum of the ground floor area and the two upper level floor areas.

Bi-level

This type of residence has two living areas, one above the other. One area is about 4 feet below grade and the second is about 4 feet above grade. Both areas are equal in size. The lower level in this type of residence is originally designed and built to serve as a living area and not as a basement. Both levels have full ceiling heights. The living area is the sum of the lower level area and the upper level area.

Tri-level

This type of residence has three levels of living area, one at grade level, one about 4 feet below grade and one about 4 feet above grade. All levels are originally designed to serve as living areas. All levels have full ceiling heights. The living area is a sum of the areas of each of the three levels.

Garage Types

Attached Garage

Shares a common wall with the dwelling. Access is typically through a door between dwelling and garage.

Basement Garage

Constructed under the roof of the dwelling but below the living area.

Built-In Garage

Constructed under the second floor living space and above basement level of dwelling. Reduces gross square feet of living area.

Detached Garage

Constructed apart from the main dwelling. Shares no common area or wall with the dwelling.

Building Components

1. Excavation	11. Collar Ties	21. Fascia	31. Backfill	41. Sub-floor
2. Sill Plate	12. Ridge Rafter	22. Downspout	32. Drainage Stone	42. Finish Floor
3. Basement Window	13. Roof Sheathing	23. Shutter	33. Drainage Tile	43. Attic Insulation
4. Floor Joist	14. Roof Felt	24. Window	34. Wall Footing	44. Soffit
5. Shoe Plate	15. Roof Shingles	25. Wall Shingles	35. Gravel	45. Ceiling Strapping
6. Studs	16. Flashing	26. Building Paper	36. Concrete Slab	46. Wall Insulation
7. Drywall	17. Flue Lining	27. Wall Sheathing	37. Column Footing	47. Cross Bridging
8. Plate	18. Chimney	28. Fire Stop	38. Pipe Column	48. Bulkhead Stairs
9. Ceiling Joists	19. Roof Shingles	29. Dampproofing	39. Expansion Joint	
10. Rafters	20. Gutter	30. Foundation Wall	40. Girder	

Exterior Wall Construction

Typical Frame Construction

Typical wood frame construction consists of wood studs with insulation between them. A typical exterior surface is made up of sheathing, building paper and exterior siding consisting of wood, vinyl, aluminum or stucco over the wood sheathing.

Brick Veneer

Typical brick veneer construction consists of wood studs with insulation between them. A typical exterior surface is sheathing, building paper and an exterior of brick tied to the sheathing, with metal strips.

Stone

Typical solid masonry construction consists of a stone or block wall covered on the exterior with brick, stone or other masonry.

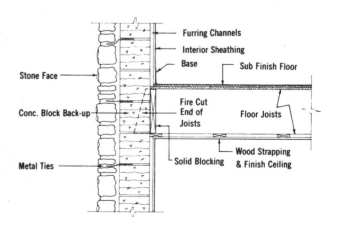

Residential Cost Estimate Worksheets

Worksheet Instructions

The residential cost estimate worksheet can be used as an outline for developing a residential construction or replacement cost. It is also useful for insurance appraisals. The design of the worksheet helps eliminate errors and omissions. To use the worksheet, follow the example below.

1. Fill out the owner's name, residence address, the estimator or appraiser's name, some type of project identifying number or code, and the date.
2. Determine from the plans, specifications, owner's description, photographs or any other means possible the class of construction. The models in this book use economy, average, custom and luxury as classes. Fill in the appropriate box.
3. Fill in the appropriate box for the residence type, configuration, occupancy, and exterior wall. If you require clarification, the pages preceding this worksheet describe each of these.
4. Next, the living area of the residence must be established. The heated or air conditioned space of the residence, not including the basement, should be measured. It is easiest to break the structure up into separate components as shown in the example. The main house (A), a one-and-one-half story wing (B), and a one story wing (C). The breezeway (D), garage (E) and open covered porch (F) will be treated differently. Data entry blocks for the living area are included on the worksheet for your use. Keep each level of each component separate, and fill out the blocks as shown.
5. By using the information on the worksheet, find the model, wing or ell in the following square foot cost pages that best matches the class, type, exterior finish and size of the residence being estimated. Use the *modifications, adjustments, and alternatives* to determine the adjusted cost per square foot of living area for each component.

6. For each component, multiply the cost per square foot by the living area square footage. If the residence is a townhouse/rowhouse, a multiplier should be applied based upon the configuration.
7. The second page of the residential cost estimate worksheet has space for the additional components of a house. The cost for additional bathrooms, finished attic space, breezeways, porches, fireplaces, appliance or cabinet upgrades, and garages should be added on this page. The information for each of these components is found with the model being used, or in the *modifications, adjustments, and alternatives* pages.
8. Add the total from page one of the estimate worksheet and the items listed on page two. The sum is the adjusted total building cost.
9. Depending on the use of the final estimated cost, one of the remaining two boxes should be filled out. Any additional items or exclusions should be added or subtracted at this time. The data contained in this book is a national average. Construction costs are different throughout the country. To allow for this difference, a location factor based upon the first three digits of the residence's zip code must be applied. The location factor is a multiplier that increases or decreases the adjusted total building cost. Find the appropriate location factor and calculate the local cost. If depreciation is a concern, a dollar figure should be subtracted at this point.
10. No residence will match a model exactly. Many differences will be found. At this level of estimating, a variation of plus or minus 10% should be expected.

Adjustments Instructions

No residence matches a model exactly in shape, material, or specifications.

The common differences are:

1. Two or more exterior wall systems
   a. Partial basement
   b. Partly finished basement
2. Specifications or features that are between two classes
3. Crawl space instead of a basement

EXAMPLES

Below are quick examples. See pages 15-17 for complete examples of cost adjustments for these differences:

1. Residence "A" is an average one-story structure with 1,600 S.F. of living area and no basement. Three walls are wood siding on wood frame, and the fourth wall is brick veneer on wood frame. The brick veneer wall is 35% of the exterior wall area.

   Use page 28 to calculate the Base Cost per S.F. of Living Area.
   Wood Siding for 1,600 S.F.   = $78.50 per S.F.
   Brick Veneer for 1,600 S.F.   = $92.15 per S.F.
   .65 ($78.50) + .35 ($92.15) = $83.28 per S.F. of Living Area.

2. a. Residence "B" is the same as Residence "A"; However, it has an unfinished basement under 50% of the building. To adjust the $83.28 per S.F. of living area for this partial basement, use page 28.

   $83.28 + .50 ($8.20) = $87.37 per S.F. of Living Area.

   b. Residence "C" is the same as Residence "A"; However, it has a full basement under the entire building. 640 S.F. or 40% of the basement area is finished.

   Using Page 28:

   $83.28 + .40 ($22.95) + .60 ($8.20) = $97.38 per S.F. of Living Area.

3. When specifications or features of a building are between classes, estimate the percent deviation, and use two tables to calculate the cost per S.F.

   A two-story residence with wood siding and 1,800 S.F. of living area has features 30% better than Average, but 70% less than Custom.

   From pages 30 and 42:

Custom 1,800 S.F. Base Cost	=	$99.35 per S.F.
Average 1,800 S.F. Base Cost	=	$80.00 per S.F.
DIFFERENCE	=	$19.35 per S.F.

   Cost is $80.00 + .30 ($19.35) = $85.80 per S.F. of Living Area.

4. To add the cost of a crawl space, use the cost of an unfinished basement as a maximum. For specific costs of components to be added or deducted, such as vapor barrier, underdrain, and floor, see the "Assemblies" section, pages 237 to 446.

Model Residence Example

First Floor Plan

18'

12'

F

24'

E

8'

D

12'

C

16'

A

28'

12'

20'

24'

12'

12'

46'

18'

B

18'

Second Floor Plan

46'

28'

A

20'

5'

5'

B

5'

5'

13'

A = Main House
B = 1-1/2 Story Wing
C = 1 Story Wing
D = Breezeway
E = Garage
F = Open Covered Porch

RESIDENTIAL
COST ESTIMATE

OWNERS NAME:	**Albert Westenberg**	APPRAISER:	**Nicole Wojtowicz**
RESIDENCE ADDRESS:	**300 Sygiel Road**	PROJECT:	**# 55**
CITY, STATE, ZIP CODE:	**Three Rivers, MA 01080**	DATE:	**Jan. 2, 2005**

CLASS OF CONSTRUCTION	RESIDENCE TYPE	CONFIGURATION	EXTERIOR WALL SYSTEM
☐ ECONOMY	☐ 1 STORY	☑ DETACHED	☑ WOOD SIDING - WOOD FRAME
☑ AVERAGE	☐ 11/2 STORY	☐ TOWN/ROW HOUSE	☐ BRICK VENEER - WOOD FRAME
☐ CUSTOM	☑ 2 STORY	☐ SEMI-DETACHED	☐ STUCCO ON WOOD FRAME
☐ LUXURY	☐ 2 1/2 STORY		☐ PAINTED CONCRETE BLOCK
	☐ 3 STORY	OCCUPANCY	☐ SOLID MASONRY (AVERAGE & CUSTOM)
	☐ BI-LEVEL	☑ ONE FAMILY	☐ STONE VENEER - WOOD FRAME
	☐ TRI-LEVEL	☐ TWO FAMILY	☐ SOLID BRICK (LUXURY)
		☐ THREE FAMILY	☐ SOLID STONE (LUXURY)
		☐ OTHER	

* LIVING AREA (Main Building)			* LIVING AREA (Wing or Ell)	(**B**)		* LIVING AREA (Wing or Ell)	(**C**)
First Level	**1288** S.F.		First Level	**360** S.F.		First Level	**192** S.F.
Second level	**1288** S.F.		Second level	**310** S.F.		Second level	S.F.
Third Level	S.F.		Third Level	S.F.		Third Level	S.F.
Total	**2576** S.F.		Total	**670** S.F.		Total	**192** S.F.

* Basement Area is not part of living area.

MAIN BUILDING			COSTS PER S.F. LIVING AREA	
Cost per Square Foot of Living Area, from Page	**30**		$	**70.65**
Basement Addition: _____ % Finished,	**100**	% Unfinished	+	**4.50**
Roof Cover Adjustment: **Cedar Shake**	Type, Page **42**	(Add or Deduct)	( )	**1.17**
Central Air Conditioning: ☐ Separate Ducts	☑ Heating Ducts, Page	**42**	+	**1.70**
Heating System Adjustment:	Type, Page _____	(Add or Deduct)	( )	
Main Building: Adjusted Cost per S.F. of Living Area			$	**78.02**

MAIN BUILDING TOTAL COST	$ **78.02** /S.F.	x	**2,576** S.F.	x	_____	=	$ **200,980**
	Cost per S.F. Living Area		Living Area		Town/Row House Multiplier (Use 1 for Detached)		TOTAL COST

WING OR ELL (B)	**1 - 1/2** STORY		COSTS PER S.F. LIVING AREA	
Cost per Square Foot of Living Area, from Page	**37**		$	**68.15**
Basement Addition: **100** % Finished,		% Unfinished	+	**20.65**
Roof Cover Adjustment: _____	Type, Page _____	(Add or Deduct)	( )	–
Central Air Conditioning: ☐ Separate Ducts	☑ Heating Ducts, Page	**30 #**	+	**1.7**
Heating System Adjustment: _____	Type, Page _____	(Add or Deduct)	( )	—
Wing or Ell (**B**): Adjusted Cost per S.F. of Living Area			$	**90.50**

WING OR ELL (B) TOTAL COST	$ **90.50** /S.F.	x	**670** S.F.		=	$ **60,635**
	Cost per S.F. Living Area		Living Area			TOTAL COST

WING OR ELL (C)	**1** STORY		COSTS PER S.F. LIVING AREA	
Cost per Square Foot of Living Area, from Page	**37** (WOOD SIDING)		$	**99.50**
Basement Addition: _____ % Finished,		% Unfinished	+	—
Roof Cover Adjustment: _____	Type, Page _____	(Add or Deduct)	( )	—
Central Air Conditioning: ☐ Separate Ducts	☐ Heating Ducts, Page	_____	+	—
Heating System Adjustment: _____	Type, Page _____	(Add or Deduct)	( )	—
Wing or Ell (C) Adjusted Cost per S.F. of Living Area			$	**99.50**

WING OR ELL (C) TOTAL COST	$ **99.50** /S.F.	x	**192** S.F.		=	$ **19,184**
	Cost per S.F. Living Area		Living Area			TOTAL COST

TOTAL THIS PAGE	**280,719**

Page 1 of 2

RESIDENTIAL
COST ESTIMATE

					QUANTITY	UNIT COST		
Total Page 1							$	280,719
Additional Bathrooms:	2	Full,	1 Half	2 @ 4261 1 @ 2660			+	11,182
Finished Attic:	N/A	Ft. x	Ft.			S.F.		
Breezeway: ☑ Open ☐ Enclosed	12	Ft. x	12 Ft.		144	S.F. 18.20	+	2,621
Covered Porch: ☑ Open ☐ Enclosed	18	Ft. x	12 Ft.		216	S.F. 27.00	+	5,832
Fireplace: ☑ Interior Chimney ☐ Exterior Chimney								
☑ No. of Flues (2) ☑ Additional Fireplaces	1 - 2nd Story						+	7880
Appliances:							+	—
Kitchen Cabinets Adjustments:				(±)				—
☑ Garage ☐ Carport: 2 Car(s) Description **Wood, Attached**				(±)				16,130
Miscellaneous:							+	
				ADJUSTED TOTAL BUILDING COST			$	324,364

REPLACEMENT COST		
ADJUSTED TOTAL BUILDING COST	$	324,364
Site Improvements		
(A) Paving & Sidewalks	$	
(B) Landscaping	$	
(C) Fences	$	
(D) Swimming Pools	$	
(E) Miscellaneous	$	
TOTAL	$	324,364
Location Factor	x	1.027
Location Replacement Cost	$	333,122
Depreciation	-$	33,312
LOCAL DEPRECIATED COST	$	299,810

INSURANCE COST		
ADJUSTED TOTAL BUILDING COST	$	
Insurance Exclusions		
(A) Footings, sitework, Underground Piping	-$	
(B) Architects Fees	-$	
Total Building Cost Less Exclusion	$	
Location Factor	x	
LOCAL INSURABLE REPLACEMENT COST	$	

SKETCH AND ADDITIONAL CALCULATIONS

1 Story

© Home Planners, Inc.

1-1/2 Story

2 Story

Bi-Level

Tri-Level

©Design Basics, Inc.

Components	
1 Site Work	Site preparation for slab or excavation for lower level; 4' deep trench excavation for foundation wall.
2 Foundations	Continuous reinforced concrete footing, 8" deep x 18" wide; dampproofed and insulated 8" thick reinforced concrete block foundation wall, 4' deep; 4" concrete slab on 4" crushed stone base and polyethylene vapor barrier, trowel finish.
3 Framing	Exterior walls—2 x 4 wood studs, 16" O.C.; 1/2" insulation board sheathing; wood truss roof 24" O.C. or 2 x 6 rafters 16" O.C. with 1/2" plywood sheathing, 4 in 12 or 8 in 12 roof pitch.
4 Exterior Walls	Beveled wood siding and #15 felt building paper on insulated wood frame walls; sliding sash or double hung wood windows; 2 flush solid core wood exterior doors. **Alternates:** • Brick veneer on wood frame, has 4" veneer of common brick. • Stucco on wood frame has 1" thick colored stucco finish. • Painted concrete block has 8" concrete block sealed and painted on exterior and furring on the interior for the drywall.
5 Roofing	20 year asphalt shingles; #15 felt building paper; aluminum gutters, downspouts, drip edge and flashings.
6 Interiors	Walls and ceilings—1/2" taped and finished drywall, primed and painted with 2 coats; painted baseboard and trim; rubber backed carpeting 80%, asphalt tile 20%; hollow core wood interior doors.
7 Specialties	Economy grade kitchen cabinets—6 L.F. wall and base with plastic laminate counter top and kitchen sink; 30 gallon electric water heater.
8 Mechanical	1 lavatory, white, wall hung; 1 water closet, white; 1 bathtub, enameled steel, white; gas fired warm air heat.
9 Electrical	100 Amp. service; romex wiring; incandescent lighting fixtures, switches, receptacles.
10 Overhead and Profit	General Contractor overhead and profit.

ECONOMY

Adjustments

Unfinished Basement:
7' high 8" concrete block or cast-in-place concrete walls.

Finished Basement:
Includes inexpensive paneling or drywall on foundation walls. Inexpensive sponge backed carpeting on concrete floor, drywall ceiling, and lighting.

- **Mass produced from stock plans**
- **Single family — 1 full bath, 1 kitchen**
- **No basement**
- **Asphalt shingles on roof**
- **Hot air heat**
- **Gypsum wallboard interior finishes**
- **Materials and workmanship are sufficient to meet codes**
- **Detail specifications on page 19**

Note: The illustration shown may contain some optional components (for example: garages and/or fireplaces) whose costs are shown in the modifications, adjustments, & alternatives below or at the end of the square foot section.

©Home Planners, Inc.

Base cost per square foot of living area

Exterior Wall	Living Area										
	600	800	1000	1200	1400	1600	1800	2000	2400	2800	3200
Wood Siding - Wood Frame	88.95	80.90	74.70	69.65	65.20	62.40	61.00	59.15	55.25	52.40	50.50
Brick Veneer - Wood Frame	95.20	86.50	79.80	74.20	69.35	66.25	64.80	62.65	58.50	55.40	53.30
Stucco on Wood Frame	85.80	78.05	72.15	67.35	63.05	60.40	59.10	57.40	53.60	50.90	49.10
Painted Concrete Block	89.10	81.05	74.85	69.80	65.25	62.50	61.10	59.30	55.30	52.50	50.55
Finished Basement, Add	22.70	21.40	20.45	19.60	18.90	18.45	18.20	17.80	17.30	16.95	16.55
Unfinished Basement, Add	10.30	9.20	8.45	7.80	7.25	6.90	6.70	6.40	6.00	5.70	5.40

Modifications

Add to the total cost

Upgrade Kitchen Cabinets	$ + 697
Solid Surface Countertops	+ 684
Full Bath - including plumbing, wall and floor finishes	+ 3395
Half Bath - including plumbing, wall and floor finishes	+ 2120
One Car Attached Garage	+ 8870
One Car Detached Garage	+ 11,433
Fireplace & Chimney	+ 4185

Adjustments

For multi family - add to total cost

Additional Kitchen	$ + 2556
Additional Bath	+ 3395
Additional Entry & Exit	+ 1276
Separate Heating	+ 1248
Separate Electric	+ 842

For Townhouse/Rowhouse -
Multiply cost per square foot by

Inner Unit	.95
End Unit	.97

Alternatives

Add to or deduct from the cost per square foot of living area

Composition Roll Roofing	– .63
Cedar Shake Roof	+ 2.65
Upgrade Walls and Ceilings to Skim Coat Plaster	+ .67
Upgrade Ceilings to Textured Finish	+ .40
Air Conditioning, in Heating Ductwork	+ 2.75
In Separate Ductwork	+ 5.30
Heating Systems, Hot Water	+ 1.46
Heat Pump	+ 1.77
Electric Heat	– 1.24
Not Heated	– 3.16

Additional upgrades or components

Kitchen Cabinets & Countertops	Page 58
Bathroom Vanities	59
Fireplaces & Chimneys	59
Windows, Skylights & Dormers	59
Appliances	60
Breezeways & Porches	60
Finished Attic	60
Garages	61
Site Improvements	61
Wings & Ells	25

- **Mass produced from stock plans**
- **Single family — 1 full bath, 1 kitchen**
- **No basement**
- **Asphalt shingles on roof**
- **Hot air heat**
- **Gypsum wallboard interior finishes**
- **Materials and workmanship are sufficient to meet codes**
- **Detail specifications on page 19**

Note: The illustration shown may contain some optional components (for example: garages and/or fireplaces) whose costs are shown in the modifications, adjustments, & alternatives below or at the end of the square foot section.

Alternatives

Base cost per square foot of living area

Exterior Wall	Living Area										
	600	800	1000	1200	1400	1600	1800	2000	2400	2800	3200
Wood Siding - Wood Frame	102.85	85.90	76.80	72.55	69.55	65.00	62.70	60.40	55.50	53.65	51.70
Brick Veneer - Wood Frame	111.60	92.20	82.60	78.00	74.75	69.75	67.20	64.70	59.30	57.25	54.95
Stucco on Wood Frame	98.45	82.65	73.75	69.75	66.85	62.60	60.45	58.25	53.60	51.85	50.00
Painted Concrete Block	103.05	86.05	76.90	72.70	69.70	65.10	62.85	60.50	55.60	53.75	51.80
Finished Basement, Add	17.50	14.85	14.20	13.70	13.35	12.80	12.50	12.25	11.70	11.45	11.15
Unfinished Basement, Add	9.00	6.90	6.40	6.00	5.70	5.30	5.05	4.85	4.45	4.25	4.00

Modifications

Add to the total cost

Upgrade Kitchen Cabinets	$ + 697
Solid Surface Countertops	+ 684
Full Bath - including plumbing, wall and floor finishes	+ 3395
Half Bath - including plumbing, wall and floor finishes	+ 2120
One Car Attached Garage	+ 8870
One Car Detached Garage	+ 11,433
Fireplace & Chimney	+ 4185

Adjustments

For multi family - add to total cost

Additional Kitchen	$ + 2556
Additional Bath	+ 3395
Additional Entry & Exit	+ 1276
Separate Heating	+ 1248
Separate Electric	+ 842

For Townhouse/Rowhouse -
Multiply cost per square foot by

Inner Unit	.95
End Unit	.97

Alternatives

Add to or deduct from the cost per square foot of living area

Composition Roll Roofing	– .44
Cedar Shake Roof	+ 1.91
Upgrade Walls and Ceilings to Skim Coat Plaster	+ .68
Upgrade Ceilings to Textured Finish	+ .40
Air Conditioning, in Heating Ductwork	+ 2.05
In Separate Ductwork	+ 4.65
Heating Systems, Hot Water	+ 1.40
Heat Pump	+ 1.95
Electric Heat	– .98
Not Heated	– 2.91

Additional upgrades or components

Kitchen Cabinets & Countertops	Page 58
Bathroom Vanities	59
Fireplaces & Chimneys	59
Windows, Skylights & Dormers	59
Appliances	60
Breezeways & Porches	60
Finished Attic	60
Garages	61
Site Improvements	61
Wings & Ells	25

- **Mass produced from stock plans**
- **Single family — 1 full bath, 1 kitchen**
- **No basement**
- **Asphalt shingles on roof**
- **Hot air heat**
- **Gypsum wallboard interior finishes**
- **Materials and workmanship are sufficient to meet codes**
- **Detail specifications on page 19**

Note: The illustration shown may contain some optional components (for example: garages and/or fireplaces) whose costs are shown in the modifications, adjustments, & alternatives below or at the end of the square foot section.

Base cost per square foot of living area

Exterior Wall	Living Area										
	1000	1200	1400	1600	1800	2000	2200	2600	3000	3400	3800
Wood Siding - Wood Frame	81.65	73.70	70.25	67.80	65.35	62.55	60.70	57.25	53.65	52.20	50.85
Brick Veneer - Wood Frame	88.40	79.95	76.05	73.40	70.65	67.55	65.45	61.55	57.65	56.00	54.50
Stucco on Wood Frame	78.20	70.55	67.25	65.00	62.70	59.90	58.25	55.00	51.60	50.25	49.00
Painted Concrete Block	81.85	73.85	70.35	67.95	65.55	62.60	60.85	57.35	53.70	52.25	50.95
Finished Basement, Add	11.90	11.35	11.00	10.75	10.40	10.25	10.05	9.65	9.35	9.20	9.00
Unfinished Basement, Add	5.55	5.15	4.85	4.65	4.35	4.25	4.10	3.75	3.55	3.40	3.30

Modifications

Add to the total cost

Upgrade Kitchen Cabinets	$ + 697
Solid Surface Countertops	+ 684
Full Bath - including plumbing, wall and floor finishes	+ 3395
Half Bath - including plumbing, wall and floor finishes	+ 2120
One Car Attached Garage	+ 8870
One Car Detached Garage	+ 11,433
Fireplace & Chimney	+ 4670

Adjustments

For multi family - add to total cost

Additional Kitchen	$ + 2556
Additional Bath	+ 3395
Additional Entry & Exit	+ 1276
Separate Heating	+ 1248
Separate Electric	+ 842

For Townhouse/Rowhouse -
Multiply cost per square foot by

Inner Unit	.93
End Unit	.96

Alternatives

Add to or deduct from the cost per square foot of living area

Composition Roll Roofing	– .32
Cedar Shake Roof	+ 1.33
Upgrade Walls and Ceilings to Skim Coat Plaster	+ .69
Upgrade Ceilings to Textured Finish	+ .40
Air Conditioning, in Heating Ductwork	+ 1.65
In Separate Ductwork	+ 4.25
Heating Systems, Hot Water	+ 1.36
Heat Pump	+ 2.07
Electric Heat	– .81
Not Heated	– 2.75

Additional upgrades or components

Kitchen Cabinets & Countertops	Page 58
Bathroom Vanities	59
Fireplaces & Chimneys	59
Windows, Skylights & Dormers	59
Appliances	60
Breezeways & Porches	60
Finished Attic	60
Garages	61
Site Improvements	61
Wings & Ells	25

- **Mass produced from stock plans**
- **Single family — 1 full bath, 1 kitchen**
- **No basement**
- **Asphalt shingles on roof**
- **Hot air heat**
- **Gypsum wallboard interior finishes**
- **Materials and workmanship are sufficient to meet codes**
- **Detail specifications on page 19**

Note: The illustration shown may contain some optional components (for example: garages and/or fireplaces) whose costs are shown in the modifications, adjustments, & alternatives below or at the end of the square foot section.

Base cost per square foot of living area

Exterior Wall	Living Area										
	1000	1200	1400	1600	1800	2000	2200	2600	3000	3400	3800
Wood Siding - Wood Frame	76.10	68.55	65.40	63.20	61.00	58.30	56.75	53.60	50.35	49.00	47.85
Brick Veneer - Wood Frame	81.15	73.30	69.80	67.40	64.95	62.10	60.30	56.85	53.35	51.90	50.55
Stucco on Wood Frame	73.45	66.20	63.15	61.10	59.05	56.35	54.90	51.95	48.80	47.60	46.45
Painted Concrete Block	76.20	68.70	65.50	63.30	61.10	58.45	56.85	53.70	50.45	49.10	47.95
Finished Basement, Add	11.90	11.35	11.00	10.75	10.40	10.25	10.05	9.65	9.35	9.20	9.00
Unfinished Basement, Add	5.55	5.15	4.85	4.65	4.35	4.25	4.10	3.75	3.55	3.40	3.30

Modifications

Add to the total cost

Upgrade Kitchen Cabinets	$ + 697
Solid Surface Countertops	+ 684
Full Bath - including plumbing, wall and floor finishes	+ 3395
Half Bath - including plumbing, wall and floor finishes	+ 2120
One Car Attached Garage	+ 8870
One Car Detached Garage	+ 11,433
Fireplace & Chimney	+ 4185

Adjustments

For multi family - add to total cost

Additional Kitchen	$ + 2556
Additional Bath	+ 3395
Additional Entry & Exit	+ 1276
Separate Heating	+ 1248
Separate Electric	+ 842

For Townhouse/Rowhouse -
Multiply cost per square foot by

Inner Unit	.94
End Unit	.97

Alternatives

Add to or deduct from the cost per square foot of living area

Composition Roll Roofing	– .32
Cedar Shake Roof	+ 2.65
Upgrade Walls and Ceilings to Skim Coat Plaster	+ .65
Upgrade Ceilings to Textured Finish	+ .40
Air Conditioning, in Heating Ductwork	+ 1.65
In Separate Ductwork	+ 4.25
Heating Systems, Hot Water	+ 1.36
Heat Pump	+ 2.07
Electric Heat	– .81
Not Heated	– 2.75

Additional upgrades or components

Kitchen Cabinets & Countertops	Page 58
Bathroom Vanities	59
Fireplaces & Chimneys	59
Windows, Skylights & Dormers	59
Appliances	60
Breezeways & Porches	60
Finished Attic	60
Garages	61
Site Improvements	61
Wings & Ells	25

ECONOMY

- Mass produced from stock plans
- Single family — 1 full bath, 1 kitchen
- No basement
- Asphalt shingles on roof
- Hot air heat
- Gypsum wallboard interior finishes
- Materials and workmanship are sufficient to meet codes
- Detail specifications on page 19

Note: The illustration shown may contain some optional components (for example: garages and/or fireplaces) whose costs are shown in the modifications, adjustments, & alternatives below or at the end of the square foot section.

©Design Basics, Inc.

Base cost per square foot of living area

Exterior Wall	Living Area										
	1200	1500	1800	2000	2200	2400	2800	3200	3600	4000	4400
Wood Siding - Wood Frame	70.00	64.40	60.15	58.50	55.95	54.00	52.45	50.35	47.80	46.95	45.00
Brick Veneer - Wood Frame	74.65	68.60	63.95	62.15	59.40	57.25	55.60	53.25	50.50	49.55	47.45
Stucco on Wood Frame	67.60	62.25	58.15	56.65	54.20	52.35	50.85	48.85	46.45	45.60	43.80
Solid Masonry	70.15	64.50	60.25	58.60	56.05	54.10	52.50	50.40	47.90	47.00	45.10
Finished Basement, Add*	14.30	13.65	13.10	12.80	12.60	12.30	12.15	11.80	11.55	11.45	11.25
Unfinished Basement, Add*	6.15	5.65	5.15	5.00	4.80	4.55	4.45	4.20	4.00	3.90	3.75

*Basement under middle level only.

Modifications

Add to the total cost

Upgrade Kitchen Cabinets	$ + 697
Solid Surface Countertops	+ 684
Full Bath - including plumbing, wall and floor finishes	+ 3395
Half Bath - including plumbing, wall and floor finishes	+ 2120
One Car Attached Garage	+ 8870
One Car Detached Garage	+ 11,433
Fireplace & Chimney	+ 4185

Adjustments

For multi family - add to total cost

Additional Kitchen	$ + 2556
Additional Bath	+ 3395
Additional Entry & Exit	+ 1276
Separate Heating	+ 1248
Separate Electric	+ 842

*For Townhouse/Rowhouse -
Multiply cost per square foot by*

Inner Unit	.93
End Unit	.96

Alternatives

Add to or deduct from the cost per square foot of living area

Composition Roll Roofing	– .44
Cedar Shake Roof	+ 1.91
Upgrade Walls and Ceilings to Skim Coat Plaster	+ .60
Upgrade Ceilings to Textured Finish	+ .40
Air Conditioning, in Heating Ductwork	+ 1.40
In Separate Ductwork	+ 3.97
Heating Systems, Hot Water	+ 1.30
Heat Pump	+ 2.15
Electric Heat	– .72
Not Heated	– 2.66

Additional upgrades or components

Kitchen Cabinets & Countertops	Page 58
Bathroom Vanities	59
Fireplaces & Chimneys	59
Windows, Skylights & Dormers	59
Appliances	60
Breezeways & Porches	60
Finished Attic	60
Garages	61
Site Improvements	61
Wings & Ells	25

1 Story — Base cost per square foot of living area

Exterior Wall	Living Area							
	50	100	200	300	400	500	600	700
Wood Siding - Wood Frame	124.55	95.20	82.45	69.10	65.00	62.45	60.80	61.35
Brick Veneer - Wood Frame	140.30	106.40	91.75	75.35	70.60	67.70	65.80	66.15
Stucco on Wood Frame	116.55	89.45	77.65	65.95	62.15	59.75	58.25	58.90
Painted Concrete Block	124.95	95.50	82.70	69.25	65.10	62.60	60.90	61.45
Finished Basement, Add	34.50	28.20	25.55	21.15	20.30	19.80	19.45	19.20
Unfinished Basement, Add	19.10	14.30	12.30	8.90	8.25	7.80	7.60	7.35

1-1/2 Story — Base cost per square foot of living area

Exterior Wall	Living Area							
	100	200	300	400	500	600	700	800
Wood Siding - Wood Frame	97.35	78.20	66.50	59.80	56.25	54.55	52.40	51.75
Brick Veneer - Wood Frame	111.40	89.45	75.90	67.10	63.00	60.95	58.45	57.70
Stucco on Wood Frame	90.25	72.45	61.70	56.05	52.80	51.30	49.35	48.70
Painted Concrete Block	97.70	78.45	66.75	59.95	56.40	54.75	52.55	51.90
Finished Basement, Add	23.05	20.40	18.70	16.75	16.20	15.85	15.55	15.50
Unfinished Basement, Add	11.80	9.80	8.45	7.00	6.60	6.30	6.10	6.00

2 Story — Base cost per square foot of living area

Exterior Wall	Living Area							
	100	200	400	600	800	1000	1200	1400
Wood Siding - Wood Frame	99.35	73.95	62.80	52.20	48.65	46.40	44.95	45.70
Brick Veneer - Wood Frame	115.10	85.20	72.20	58.45	54.25	51.65	49.95	50.50
Stucco on Wood Frame	91.35	68.25	58.00	49.00	45.80	43.75	42.40	43.25
Painted Concrete Block	99.75	74.25	63.05	52.35	48.75	46.60	45.10	45.80
Finished Basement, Add	17.25	14.10	12.75	10.60	10.20	9.90	9.75	9.60
Unfinished Basement, Add	9.55	7.15	6.10	4.45	4.15	3.90	3.80	3.70

Base costs do not include bathroom or kitchen facilities. Use Modifications/Adjustments/Alternatives on pages 58-61 where appropriate.

Important: See the Reference Section for Location Factors (to adjust for your city) and Estimating Forms.

1 Story

1 - 1/2 Story

2 Story

2 - 1/2 Story

Bi-Level

Tri-Level

	Components
1 Site Work	Site preparation for slab or excavation for lower level; 4' deep trench excavation for foundation wall.
2 Foundations	Continuous reinforced concrete footing, 8" deep x 18" wide; dampproofed and insulated 8" thick reinforced concrete foundation wall, 4' deep; 4" concrete slab on 4" crushed stone base and polyethylene vapor barrier, trowel finish.
3 Framing	Exterior walls—2 x 4 wood studs, 16" O.C.; 1/2" plywood sheathing; 2 x 6 rafters 16" O.C. with 1/2" plywood sheathing, 4 in 12 or 8 in 12 roof pitch, 2 x 6 ceiling joists; 1/2" plywood subfloor on 1 x 2 wood sleepers 16" O.C.; 2 x 8 floor joists with 5/8" plywood subfloor on models with more than one level.
4 Exterior Walls	Beveled wood siding and #15 felt building paper on insulated wood frame walls; double hung windows; 3 flush solid core wood exterior doors with storms. **Alternates:** • Brick veneer on wood frame, has 4" veneer of common brick. • Stucco on wood frame has 1" thick colored stucco finish. • Solid masonry has a 6" concrete block load bearing wall with insulation and a brick or stone exterior. Also solid brick or stone walls.
5 Roofing	25 year asphalt shingles; #15 felt building paper; aluminum gutters, downspouts, drip edge and flashings.
6 Interiors	Walls and ceilings—1/2" taped and finished drywall, primed and painted with 2 coats; painted baseboard and trim, finished hardwood floor 40%, carpet with 1/2" underlayment 40%, vinyl tile with 1/2" underlayment 15%, ceramic tile with 1/2" underlayment 5%; hollow core and louvered interior doors.
7 Specialties	Average grade kitchen cabinets—14 L.F. wall and base with plastic laminate counter top and kitchen sink; 40 gallon electric water heater.
8 Mechanical	1 lavatory, white, wall hung; 1 water closet, white; 1 bathtub, enameled steel, white; gas fired warm air heat.
9 Electrical	100 Amp. service; romex wiring; incandescent lighting fixtures, switches, receptacles.
10 Overhead and Profit	General Contractor overhead and profit.

Adjustments

Unfinished Basement:
7'-6" high cast-in-place concrete walls 8" thick or 8" concrete block.

Finished Basement:
Includes plywood paneling or drywall on furring on foundation walls, sponge backed carpeting on concrete floor, suspended ceiling, lighting and heating.

- **Simple design from standard plans**
- **Single family — 1 full bath, 1 kitchen**
- **No basement**
- **Asphalt shingles on roof**
- **Hot air heat**
- **Gypsum wallboard interior finishes**
- **Materials and workmanship are average**
- **Detail specifications on p. 27**

Note: The illustration shown may contain some optional components (for example: garages and/or fireplaces) whose costs are shown in the modifications, adjustments, & alternatives below or at the end of the square foot section.

©Home Planners, Inc.

Base cost per square foot of living area

Exterior Wall	Living Area										
	600	800	1000	1200	1400	1600	1800	2000	2400	2800	3200
Wood Siding - Wood Frame	110.45	100.55	93.00	87.00	81.75	78.50	76.70	74.50	69.95	66.70	64.45
Brick Veneer - Wood Frame	126.70	116.05	107.95	101.35	95.65	92.15	90.20	87.70	82.85	79.35	76.85
Stucco on Wood Frame	119.10	109.25	101.80	95.80	90.60	87.40	85.60	83.45	78.90	75.70	73.45
Solid Masonry	137.55	125.70	116.70	109.30	102.85	98.85	96.70	93.90	88.45	84.50	81.65
Finished Basement, Add	27.75	26.85	25.60	24.50	23.55	22.95	22.60	22.10	21.45	20.95	20.40
Unfinished Basement, Add	11.65	10.60	9.80	9.15	8.55	8.20	7.95	7.70	7.25	7.00	6.70

Modifications

Add to the total cost

Upgrade Kitchen Cabinets	$	+ 3157
Solid Surface Countertops (Included)		
Full Bath - including plumbing, wall and floor finishes		+ 4261
Half Bath - including plumbing, wall and floor finishes		+ 2660
One Car Attached Garage		+ 9490
One Car Detached Garage		+ 12,424
Fireplace & Chimney		+ 4185

Adjustments

For multi family - add to total cost

Additional Kitchen	$	+ 5105
Additional Bath		+ 4261
Additional Entry & Exit		+ 1276
Separate Heating		+ 1248
Separate Electric		+ 1575

For Townhouse/Rowhouse -
Multiply cost per square foot by

Inner Unit	.92
End Unit	.96

Alternatives

Add to or deduct from the cost per square foot of living area

Cedar Shake Roof	+ 2.34
Clay Tile Roof	+ 4.08
Slate Roof	+ 7.77
Upgrade Walls to Skim Coat Plaster	+ .36
Upgrade Ceilings to Textured Finish	+ .40
Air Conditioning, in Heating Ductwork	+ 2.85
In Separate Ductwork	+ 5.50
Heating Systems, Hot Water	+ 1.50
Heat Pump	+ 1.84
Electric Heat	– .68
Not Heated	– 2.63

Additional upgrades or components

Kitchen Cabinets & Countertops	Page 58
Bathroom Vanities	59
Fireplaces & Chimneys	59
Windows, Skylights & Dormers	59
Appliances	60
Breezeways & Porches	60
Finished Attic	60
Garages	61
Site Improvements	61
Wings & Ells	37

- **Simple design from standard plans**
- **Single family — 1 full bath, 1 kitchen**
- **No basement**
- **Asphalt shingles on roof**
- **Hot air heat**
- **Gypsum wallboard interior finishes**
- **Materials and workmanship are average**
- **Detail specifications on p. 27**

Note: The illustration shown may contain some optional components (for example: garages and/or fireplaces) whose costs are shown in the modifications, adjustments, & alternatives below or at the end of the square foot section.

©By Designer

AVERAGE

Base cost per square foot of living area

Exterior Wall	Living Area										
	600	800	1000	1200	1400	1600	1800	2000	2400	2800	3200
Wood Siding - Wood Frame	124.60	104.85	94.15	89.10	85.55	80.20	77.60	74.80	69.25	67.10	64.75
Brick Veneer - Wood Frame	134.30	111.90	100.65	95.25	91.35	85.55	82.65	79.55	73.45	71.10	68.40
Stucco on Wood Frame	123.70	104.20	93.50	88.55	84.95	79.75	77.10	74.40	68.85	66.75	64.35
Solid Masonry	149.45	122.90	110.80	104.75	100.35	93.75	90.45	87.00	80.00	77.30	74.15
Finished Basement, Add	22.65	19.80	18.95	18.30	17.85	17.15	16.75	16.40	15.65	15.35	14.95
Unfinished Basement, Add	10.00	7.95	7.35	6.95	6.65	6.20	6.00	5.75	5.35	5.15	4.90

Modifications

Add to the total cost

Upgrade Kitchen Cabinets	$ + 3157
Solid Surface Countertops (Included)	
Full Bath - including plumbing, wall and floor finishes	+ 4261
Half Bath - including plumbing, wall and floor finishes	+ 2660
One Car Attached Garage	+ 9490
One Car Detached Garage	+ 12,424
Fireplace & Chimney	+ 4185

Adjustments

For multi family - add to total cost

Additional Kitchen	$ + 5105
Additional Bath	+ 4261
Additional Entry & Exit	+ 1276
Separate Heating	+ 1248
Separate Electric	+ 1575

For Townhouse/Rowhouse -
Multiply cost per square foot by

Inner Unit	.92
End Unit	.96

Alternatives

Add to or deduct from the cost per square foot of living area

Cedar Shake Roof	+ 1.69
Clay Tile Roof	+ 2.95
Slate Roof	+ 5.61
Upgrade Walls to Skim Coat Plaster	+ .41
Upgrade Ceilings to Textured Finish	+ .40
Air Conditioning, in Heating Ductwork	+ 2.15
In Separate Ductwork	+ 4.80
Heating Systems, Hot Water	+ 1.42
Heat Pump	+ 2.04
Electric Heat	– .60
Not Heated	– 2.54

Additional upgrades or components

Kitchen Cabinets & Countertops	Page 58
Bathroom Vanities	59
Fireplaces & Chimneys	59
Windows, Skylights & Dormers	59
Appliances	60
Breezeways & Porches	60
Finished Attic	60
Garages	61
Site Improvements	61
Wings & Ells	37

- **Simple design from standard plans**
- **Single family — 1 full bath, 1 kitchen**
- **No basement**
- **Asphalt shingles on roof**
- **Hot air heat**
- **Gypsum wallboard interior finishes**
- **Materials and workmanship are average**
- **Detail specifications on p. 27**

Note: The illustration shown may contain some optional components (for example: garages and/or fireplaces) whose costs are shown in the modifications, adjustments, & alternatives below or at the end of the square foot section.

AVERAGE

Base cost per square foot of living area

Exterior Wall	Living Area										
	1000	1200	1400	1600	1800	2000	2200	2600	3000	3400	3800
Wood Siding - Wood Frame	99.15	90.00	85.80	82.95	80.00	76.75	74.65	70.65	66.55	64.85	63.25
Brick Veneer - Wood Frame	106.65	97.00	92.25	89.15	85.80	82.40	80.00	75.45	71.00	69.05	67.30
Stucco on Wood Frame	98.40	89.35	85.20	82.35	79.40	76.20	74.15	70.15	66.15	64.45	62.90
Solid Masonry	118.35	107.80	102.45	98.85	94.90	91.15	88.30	83.05	78.00	75.70	73.60
Finished Basement, Add	16.05	15.85	15.35	15.00	14.55	14.35	14.10	13.55	13.20	12.95	12.75
Unfinished Basement, Add	6.40	6.00	5.65	5.40	5.15	5.05	4.85	4.50	4.35	4.15	4.05

Modifications

Add to the total cost

Upgrade Kitchen Cabinets	$ + 3157
Solid Surface Countertops (Included)	
Full Bath - including plumbing, wall and floor finishes	+ 4261
Half Bath - including plumbing, wall and floor finishes	+ 2660
One Car Attached Garage	+ 9490
One Car Detached Garage	+ 12,424
Fireplace & Chimney	+ 4670

Adjustments

For multi family - add to total cost

Additional Kitchen	$ + 5105
Additional Bath	+ 4261
Additional Entry & Exit	+ 1276
Separate Heating	+ 1248
Separate Electric	+ 1575

For Townhouse/Rowhouse -
Multiply cost per square foot by

Inner Unit	.90
End Unit	.95

Alternatives

Add to or deduct from the cost per square foot of living area

Cedar Shake Roof	+ 1.17
Clay Tile Roof	+ 2.04
Slate Roof	+ 3.89
Upgrade Walls to Skim Coat Plaster	+ .42
Upgrade Ceilings to Textured Finish	+ .40
Air Conditioning, in Heating Ductwork	+ 1.70
In Separate Ductwork	+ 4.35
Heating Systems, Hot Water	+ 1.39
Heat Pump	+ 2.15
Electric Heat	– .51
Not Heated	– 2.46

Additional upgrades or components

Kitchen Cabinets & Countertops	Page 58
Bathroom Vanities	59
Fireplaces & Chimneys	59
Windows, Skylights & Dormers	59
Appliances	60
Breezeways & Porches	60
Finished Attic	60
Garages	61
Site Improvements	61
Wings & Ells	37

- **Simple design from standard plans**
- **Single family — 1 full bath, 1 kitchen**
- **No basement**
- **Asphalt shingles on roof**
- **Hot air heat**
- **Gypsum wallboard interior finishes**
- **Materials and workmanship are average**
- **Detail specifications on p. 27**

Note: The illustration shown may contain some optional components (for example: garages and/or fireplaces) whose costs are shown in the modifications, adjustments, & alternatives below or at the end of the square foot section.

AVERAGE

Base cost per square foot of living area

Exterior Wall	Living Area										
	1200	1400	1600	1800	2000	2400	2800	3200	3600	4000	4400
Wood Siding - Wood Frame	99.05	93.15	85.20	83.65	80.80	76.20	72.45	68.80	66.90	63.50	62.35
Brick Veneer - Wood Frame	107.05	100.40	91.85	90.30	87.05	81.85	77.85	73.70	71.60	67.80	66.55
Stucco on Wood Frame	98.25	92.45	84.55	83.00	80.20	75.70	71.95	68.30	66.50	63.15	62.00
Solid Masonry	119.50	111.65	102.25	100.70	96.80	90.60	86.30	81.30	78.80	74.55	73.05
Finished Basement, Add	13.55	13.30	12.80	12.75	12.45	11.90	11.70	11.30	11.10	10.85	10.70
Unfinished Basement, Add	5.35	4.90	4.60	4.55	4.35	4.05	3.85	3.60	3.50	3.35	3.30

Modifications

Add to the total cost

Upgrade Kitchen Cabinets	$ + 3157
Solid Surface Countertops (Included)	
Full Bath - including plumbing, wall and floor finishes	+ 4261
Half Bath - including plumbing, wall and floor finishes	+ 2660
One Car Attached Garage	+ 9490
One Car Detached Garage	+ 12,424
Fireplace & Chimney	+ 5305

Adjustments

For multi family - add to total cost

Additional Kitchen	$ + 5105
Additional Bath	+ 4261
Additional Entry & Exit	+ 1276
Separate Heating	+ 1248
Separate Electric	+ 1575

For Townhouse/Rowhouse -
Multiply cost per square foot by

Inner Unit	.90
End Unit	.95

Alternatives

Add to or deduct from the cost per square foot of living area

Cedar Shake Roof	+ 1.02
Clay Tile Roof	+ 1.77
Slate Roof	+ 3.37
Upgrade Walls to Skim Coat Plaster	+ .40
Upgrade Ceilings to Textured Finish	+ .40
Air Conditioning, in Heating Ductwork	+ 1.55
In Separate Ductwork	+ 4.20
Heating Systems, Hot Water	+ 1.25
Heat Pump	+ 2.20
Electric Heat	– .94
Not Heated	– 2.87

Additional upgrades or components

Kitchen Cabinets & Countertops	Page 58
Bathroom Vanities	59
Fireplaces & Chimneys	59
Windows, Skylights & Dormers	59
Appliances	60
Breezeways & Porches	60
Finished Attic	60
Garages	61
Site Improvements	61
Wings & Ells	37

AVERAGE

- **Simple design from standard plans**
- **Single family — 1 full bath, 1 kitchen**
- **No basement**
- **Asphalt shingles on roof**
- **Hot air heat**
- **Gypsum wallboard interior finishes**
- **Materials and workmanship are average**
- **Detail specifications on p. 27**

Note: The illustration shown may contain some optional components (for example: garages and/or fireplaces) whose costs are shown in the modifications, adjustments, & alternatives below or at the end of the square foot section.

Base cost per square foot of living area

Exterior Wall	Living Area										
	1500	1800	2100	2500	3000	3500	4000	4500	5000	5500	6000
Wood Siding - Wood Frame	92.00	83.65	80.05	77.25	71.80	69.50	66.20	62.60	61.50	60.20	58.85
Brick Veneer - Wood Frame	99.50	90.60	86.55	83.45	77.45	74.80	71.05	67.05	65.85	64.35	62.80
Stucco on Wood Frame	91.30	83.00	79.45	76.65	71.25	69.00	65.70	62.15	61.10	59.75	58.45
Solid Masonry	111.20	101.45	96.70	93.05	86.25	83.15	78.60	74.05	72.55	70.80	68.90
Finished Basement, Add	11.70	11.60	11.25	11.00	10.55	10.35	10.00	9.80	9.65	9.55	9.40
Unfinished Basement, Add	4.30	4.00	3.80	3.70	3.45	3.25	3.10	2.95	2.90	2.80	2.70

Modifications

Add to the total cost

Upgrade Kitchen Cabinets	$ + 3157
Solid Surface Countertops (Included)	
Full Bath - including plumbing, wall and floor finishes	+ 4261
Half Bath - including plumbing, wall and floor finishes	+ 2660
One Car Attached Garage	+ 9490
One Car Detached Garage	+ 12,424
Fireplace & Chimney	+ 5305

Adjustments

For multi family - add to total cost

Additional Kitchen	$ + 5105
Additional Bath	+ 4261
Additional Entry & Exit	+ 1276
Separate Heating	+ 1248
Separate Electric	+ 1575

For Townhouse/Rowhouse - Multiply cost per square foot by

Inner Unit	.88
End Unit	.94

Alternatives

Add to or deduct from the cost per square foot of living area

Cedar Shake Roof	+ .78
Clay Tile Roof	+ 1.36
Slate Roof	+ 2.59
Upgrade Walls to Skim Coat Plaster	+ .42
Upgrade Ceilings to Textured Finish	+ .40
Air Conditioning, in Heating Ductwork	+ 1.55
In Separate Ductwork	+ 4.20
Heating Systems, Hot Water	+ 1.25
Heat Pump	+ 2.20
Electric Heat	- .73
Not Heated	- 2.67

Additional upgrades or components

Kitchen Cabinets & Countertops	Page 58
Bathroom Vanities	59
Fireplaces & Chimneys	59
Windows, Skylights & Dormers	59
Appliances	60
Breezeways & Porches	60
Finished Attic	60
Garages	61
Site Improvements	61
Wings & Ells	37

- **Simple design from standard plans**
- **Single family — 1 full bath, 1 kitchen**
- **No basement**
- **Asphalt shingles on roof**
- **Hot air heat**
- **Gypsum wallboard interior finishes**
- **Materials and workmanship are average**
- **Detail specifications on p. 27**

Note: The illustration shown may contain some optional components (for example: garages and/or fireplaces) whose costs are shown in the modifications, adjustments, & alternatives below or at the end of the square foot section.

Base cost per square foot of living area

Exterior Wall	Living Area										
	1000	1200	1400	1600	1800	2000	2200	2600	3000	3400	3800
Wood Siding - Wood Frame	92.80	84.15	80.30	77.70	75.05	71.95	70.10	66.50	62.80	61.20	59.80
Brick Veneer - Wood Frame	98.45	89.35	85.20	82.35	79.45	76.20	74.15	70.15	66.15	64.40	62.85
Stucco on Wood Frame	92.30	83.70	79.80	77.25	74.65	71.55	69.75	66.15	62.50	60.90	59.50
Solid Masonry	107.20	97.50	92.80	89.60	86.30	82.80	80.40	75.80	71.40	69.35	67.55
Finished Basement, Add	16.05	15.85	15.35	15.00	14.55	14.35	14.10	13.55	13.20	12.95	12.75
Unfinished Basement, Add	6.40	6.00	5.65	5.40	5.15	5.05	4.85	4.50	4.35	4.15	4.05

Modifications

Add to the total cost

Upgrade Kitchen Cabinets	$ + 3157
Solid Surface Countertops (Included)	
Full Bath - including plumbing, wall and floor finishes	+ 4261
Half Bath - including plumbing, wall and floor finishes	+ 2660
One Car Attached Garage	+ 9490
One Car Detached Garage	+ 12,424
Fireplace & Chimney	+ 4185

Adjustments

For multi family - add to total cost

Additional Kitchen	$ + 5105
Additional Bath	+ 4261
Additional Entry & Exit	+ 1276
Separate Heating	+ 1248
Separate Electric	+ 1575

For Townhouse/Rowhouse -
Multiply cost per square foot by

Inner Unit	.91
End Unit	.96

Alternatives

Add to or deduct from the cost per square foot of living area

Cedar Shake Roof	+ 2.34
Clay Tile Roof	+ 4.08
Slate Roof	+ 3.89
Upgrade Walls to Skim Coat Plaster	+ .39
Upgrade Ceilings to Textured Finish	+ .40
Air Conditioning, in Heating Ductwork	+ 1.70
In Separate Ductwork	+ 4.35
Heating Systems, Hot Water	+ 1.39
Heat Pump	+ 2.15
Electric Heat	− .51
Not Heated	− 2.46

Additional upgrades or components

Kitchen Cabinets & Countertops	Page 58
Bathroom Vanities	59
Fireplaces & Chimneys	59
Windows, Skylights & Dormers	59
Appliances	60
Breezeways & Porches	60
Finished Attic	60
Garages	61
Site Improvements	61
Wings & Ells	37

- **Simple design from standard plans**
- **Single family — 1 full bath, 1 kitchen**
- **No basement**
- **Asphalt shingles on roof**
- **Hot air heat**
- **Gypsum wallboard interior finishes**
- **Materials and workmanship are average**
- **Detail specifications on p. 27**

Note: The illustration shown may contain some optional components (for example: garages and/or fireplaces) whose costs are shown in the modifications, adjustments, & alternatives below or at the end of the square foot section.

©Design Basics, Inc.

Base cost per square foot of living area

Exterior Wall	Living Area										
	1200	1500	1800	2100	2400	2700	3000	3400	3800	4200	4600
Wood Siding - Wood Frame	88.05	81.45	76.45	72.20	69.40	67.85	66.05	64.45	61.65	59.40	58.35
Brick Veneer - Wood Frame	93.25	86.15	80.70	76.05	73.05	71.30	69.30	67.60	64.60	62.20	61.00
Stucco on Wood Frame	87.50	81.00	76.00	71.85	69.05	67.55	65.75	64.15	61.40	59.10	58.10
Solid Masonry	101.30	93.45	87.30	82.05	78.65	76.75	74.45	72.50	69.20	66.45	65.15
Finished Basement, Add*	18.40	18.10	17.25	16.70	16.35	16.10	15.75	15.55	15.25	14.95	14.80
Unfinished Basement, Add*	7.15	6.65	6.10	5.75	5.50	5.40	5.20	5.05	4.85	4.70	4.60

*Basement under middle level only.

Modifications

Add to the total cost

Upgrade Kitchen Cabinets	$ + 3157
Solid Surface Countertops (Included)	
Full Bath - including plumbing, wall and floor finishes	+ 4261
Half Bath - including plumbing, wall and floor finishes	+ 2660
One Car Attached Garage	+ 9490
One Car Detached Garage	+ 12,424
Fireplace & Chimney	+ 4670

Adjustments

For multi family - add to total cost

Additional Kitchen	$ + 5105
Additional Bath	+ 4261
Additional Entry & Exit	+ 1276
Separate Heating	+ 1248
Separate Electric	+ 1575

For Townhouse/Rowhouse -
Multiply cost per square foot by

Inner Unit	.90
End Unit	.95

Alternatives

Add to or deduct from the cost per square foot of living area

Cedar Shake Roof	+ 1.69
Clay Tile Roof	+ 2.95
Slate Roof	+ 5.61
Upgrade Walls to Skim Coat Plaster	+ .34
Upgrade Ceilings to Textured Finish	+ .40
Air Conditioning, in Heating Ductwork	+ 1.43
In Separate Ductwork	+ 4.10
Heating Systems, Hot Water	+ 1.34
Heat Pump	+ 2.23
Electric Heat	– .42
Not Heated	– 2.37

Additional upgrades or components

Kitchen Cabinets & Countertops	Page 58
Bathroom Vanities	59
Fireplaces & Chimneys	59
Windows, Skylights & Dormers	59
Appliances	60
Breezeways & Porches	60
Finished Attic	60
Garages	61
Site Improvements	61
Wings & Ells	37

- **Post and beam frame**
- **Log exterior walls**
- **Simple design from standard plans**
- **Single family — 1 full bath, 1 kitchen**
- **No basement**
- **Asphalt shingles on roof**
- **Hot air heat**
- **Gypsum wallboard interior finishes**
- **Materials and workmanship are average**
- **Detail specification on page 27**

Note: The illustration shown may contain some optional components (for example: garages and/or fireplaces) whose costs are shown in the modifications, adjustments, & alternatives below or at the end of the square foot section.

Base cost per square foot of living area

Exterior Wall	Living Area										
	600	800	1000	1200	1400	1600	1800	2000	2400	2800	3200
6" Log - Solid Wall	108.90	98.40	90.20	83.80	78.60	74.20	72.00	70.50	65.70	61.90	59.90
8" Log - Solid Wall	108.00	97.60	89.50	83.10	78.00	73.60	71.50	70.00	65.30	61.30	59.40
Finished Basement, Add	25.00	23.50	22.40	21.50	20.70	20.20	19.90	19.50	18.80	18.40	18.00
Unfinished Basement, Add	10.70	9.70	9.10	8.40	7.90	7.60	7.40	7.10	6.80	6.40	6.20

Modifications

Add to the total cost

Upgrade Kitchen Cabinets	$ + 3157
Solid Surface Countertops (Included)	
Full Bath - including plumbing, wall and floor finishes	+ 4261
Half Bath - including plumbing, wall and floor finishes	+ 2660
One Car Attached Garage	+ 9490
One Car Detached Garage	+ 12,424
Fireplace & Chimney	+ 4185

Adjustments

For multi family - add to total cost

Additional Kitchen	$ + 5105
Additional Bath	+ 4261
Additional Entry & Exit	+ 1276
Separate Heating	+ 1248
Separate Electric	+ 1575

For Townhouse/Rowhouse - Multiply cost per square foot by

Inner Unit	.92
End Unit	.96

Alternatives

Add to or deduct from the cost per square foot of living area

Cedar Shake Roof	+ 2.34
Air Conditioning, in Heating Ductwork	+ 2.85
In Separate Ductwork	+ 5.50
Heating Systems, Hot Water	+ 1.50
Heat Pump	+ 1.83
Electric Heat	– .70
Not Heated	– 2.63

Additional upgrades or components

Kitchen Cabinets & Countertops	Page 58
Bathroom Vanities	59
Fireplaces & Chimneys	59
Windows, Skylights & Dormers	59
Appliances	60
Breezeways & Porches	60
Finished Attic	60
Garages	61
Site Improvements	61
Wings & Ells	37

SOLID WALL

- **Post and beam frame**
- **Log exterior walls**
- **Simple design from standard plans**
- **Single family — 1 full bath, 1 kitchen**
- **No basement**
- **Asphalt shingles on roof**
- **Hot air heat**
- **Gypsum wallborad interior finishes**
- **Materials and workmanship are average**
- **Detail specification on page 27**

Note: The illustration shown may contain some optional components (for example: garages and/or fireplaces) whose costs are shown in the modifications, adjustments, & alternatives below or at the end of the square foot section.

Base cost per square foot of living area

Exterior Wall	Living Area										
	1000	1200	1400	1600	1800	2000	2200	2600	3000	3400	3800
6" Log-Solid	109.00	89.80	85.70	82.70	79.30	76.20	73.90	69.50	64.50	62.80	61.10
8" Log-Solid	104.70	93.90	89.30	86.40	82.90	79.30	77.20	72.50	67.50	65.70	63.70
Finished Basement, Add	14.50	14.00	13.50	13.20	12.90	12.70	12.50	12.00	11.70	11.60	11.30
Unfinished Basement, Add	5.80	5.50	5.20	4.90	4.80	4.50	4.50	4.20	3.90	3.80	3.70

Modifications

Add to the total cost

Upgrade Kitchen Cabinets	$ + 3157
Solid Surface Countertops (Included)	
Full Bath - including plumbing, wall and floor finishes	+ 4261
Half Bath - including plumbing, wall and floor finishes	+ 2660
One Car Attached Garage	+ 9490
One Car Detached Garage	+ 12,424
Fireplace & Chimney	+ 4670

Adjustments

For multi family - add to total cost

Additional Kitchen	$ + 5105
Additional Bath	+ 4261
Additional Entry & Exit	+ 1276
Separate Heating	+ 1248
Separate Electric	+ 1575

For Townhouse/Rowhouse - Multiply cost per square foot by

Inner Unit	.92
End Unit	.96

Alternatives

Add to or deduct from the cost per square foot of living area

Cedar Shake Roof	+ 1.69
Air Conditioning, in Heating Ductwork	+ 1.70
In Separate Ductwork	+ 4.35
Heating Systems, Hot Water	+ 1.39
Heat Pump	+ 2.15
Electric Heat	– .51
Not Heated	– 2.46

Additional upgrades or components

Kitchen Cabinets & Countertops	Page 58
Bathroom Vanities	59
Fireplaces & Chimneys	59
Windows, Skylights & Dormers	59
Appliances	60
Breezeways & Porches	60
Finished Attic	60
Garages	61
Site Improvements	61
Wings & Ells	37

1 Story — Base cost per square foot of living area

Exterior Wall	Living Area							
	50	100	200	300	400	500	600	700
Wood Siding - Wood Frame	146.90	113.85	99.50	84.80	80.15	77.25	75.35	76.10
Brick Veneer - Wood Frame	155.15	117.10	100.60	82.45	77.05	73.75	71.60	72.20
Stucco on Wood Frame	145.05	112.45	98.30	83.95	79.30	76.50	74.65	75.45
Solid Masonry	191.70	145.90	126.20	102.60	96.10	92.15	90.90	90.90
Finished Basement, Add	44.05	36.60	33.00	26.90	25.65	24.95	24.45	24.15
Unfinished Basement, Add	20.75	15.80	13.75	10.30	9.60	9.15	8.90	8.70

1-1/2 Story — Base cost per square foot of living area

Exterior Wall	Living Area							
	100	200	300	400	500	600	700	800
Wood Siding - Wood Frame	118.20	95.40	81.85	74.25	70.10	68.15	65.70	64.85
Brick Veneer - Wood Frame	172.15	127.10	105.05	91.95	85.20	81.70	77.90	76.35
Stucco on Wood Frame	154.95	113.35	93.65	83.05	77.00	73.95	70.60	69.05
Solid Masonry	196.50	146.60	121.35	104.65	96.95	92.70	88.35	86.70
Finished Basement, Add	29.80	27.00	24.55	21.90	21.15	20.65	20.20	20.15
Unfinished Basement, Add	13.05	11.00	9.60	8.10	7.70	7.40	7.20	7.10

2 Story — Base cost per square foot of living area

Exterior Wall	Living Area							
	100	200	400	600	800	1000	1200	1400
Wood Siding - Wood Frame	117.60	88.75	76.05	64.15	60.05	57.55	55.85	56.80
Brick Veneer - Wood Frame	175.15	121.25	96.45	77.85	71.35	67.35	64.80	65.00
Stucco on Wood Frame	155.95	107.50	85.00	70.15	64.50	61.00	58.70	59.10
Solid Masonry	202.40	140.80	112.75	88.65	81.05	76.45	73.45	73.35
Finished Basement, Add	23.55	19.90	18.00	15.05	14.40	14.05	13.80	13.60
Unfinished Basement, Add	10.55	8.05	6.95	5.30	4.90	4.75	4.60	4.50

Base costs do not include bathroom or kitchen facilities. Use Modifications/Adjustments/Alternatives on pages 58-61 where appropriate.

CUSTOM

1 Story

1-1/2 Story

2 Story

2-1/2 Story

Bi-Level

Tri-Level

CUSTOM

	Components
1 Site Work	Site preparation for slab or excavation for lower level; 4' deep trench excavation for foundation wall.
2 Foundations	Continuous reinforced concrete footing, 8" deep x 18" wide; dampproofed and insulated 8" thick reinforced concrete foundation wall, 4' deep; 4" concrete slab on 4" crushed stone base and polyethylene vapor barrier, trowel finish.
3 Framing	Exterior walls—2 x 6 wood studs, 16" O.C.; 1/2" plywood sheathing; 2 x 8 rafters 16" O.C. with 1/2" plywood sheathing, 4 in 12, 6 in 12 or 8 in 12 roof pitch, 2 x 6 or 2 x 8 ceiling joists; 1/2" plywood subfloor on 1 x 3 wood sleepers 16" O.C.; 2 x 10 floor joists with 5/8" plywood subfloor on models with more than one level.
4 Exterior Walls	Horizontal beveled wood siding and #15 felt building paper on insulated wood frame walls; double hung windows; 3 flush solid core wood exterior doors with storms. **Alternates:** • Brick veneer on wood frame, has 4" veneer high quality face brick or select common brick. • Stone veneer on wood frame has exterior veneer of field stone or 2" thick limestone. • Solid masonry has an 8" concrete block wall with insulation and a brick or stone facing. It may be a solid brick or stone structure.
5 Roofing	30 year asphalt shingles; #15 felt building paper; aluminum gutters, downspouts and drip edge; copper flashings.
6 Interiors	Walls and ceilings—5/8" drywall, skim coat plaster, primed and painted with 2 coats; hardwood baseboard and trim; sanded and finished, hardwood floor 70%, ceramic tile with 1/2" underlayment 20%, vinyl tile with 1/2" underlayment 10%; wood panel interior doors, primed and painted with 2 coats.
7 Specialties	Custom grade kitchen cabinets—20 L.F. wall and base with plastic laminate counter top and kitchen sink; 4 L.F. bathroom vanity; 75 gallon electric water heater; medicine cabinet.
8 Mechanical	Gas fired warm air heat/air conditioning; one full bath including bathtub, corner shower, built in lavatory and water closet; one 1/2 bath including built in lavatory and water closet.
9 Electrical	100 Amp. service; romex wiring; incandescent lighting fixtures, switches, receptacles.
10 Overhead and Profit	General Contractor overhead and profit.

Adjustments

Unfinished Basement:
7'-6" high cast-in-place concrete walls 10" thick or 12" concrete block.

Finished Basement:
Includes painted drywall on 2 x 4 wood furring with insulation, suspended ceiling, carpeting on concrete floor, heating and lighting.

- **A distinct residence from designer's plans**
- **Single family — 1 full bath, 1 half bath, 1 kitchen**
- **No basement**
- **Asphalt shingles on roof**
- **Forced hot air heat/air conditioning**
- **Gypsum wallboard interior finishes**
- **Materials and workmanship are above average**
- **Detail specifications on page 39**

Note: The illustration shown may contain some optional components (for example: garages and/or fireplaces) whose costs are shown in the modifications, adjustments, & alternatives below or at the end of the square foot section.

eDesign Basics, Inc.

Base cost per square foot of living area

Exterior Wall	Living Area										
	800	1000	1200	1400	1600	1800	2000	2400	2800	3200	3600
Wood Siding - Wood Frame	134.20	121.95	112.15	104.25	98.90	95.95	92.45	85.90	81.10	77.85	74.20
Brick Veneer - Wood Frame	150.70	137.85	127.45	119.00	113.35	110.20	106.45	99.50	94.45	90.95	87.10
Stone Veneer - Wood Frame	155.20	141.95	131.15	122.30	116.50	113.25	109.25	102.15	96.80	93.15	89.15
Solid Masonry	157.30	143.80	132.85	123.85	117.95	114.70	110.60	103.30	97.90	94.15	90.10
Finished Basement, Add	42.55	42.30	40.45	38.95	37.95	37.45	36.65	35.55	34.70	33.95	33.30
Unfinished Basement, Add	18.00	17.00	16.10	15.40	14.90	14.65	14.25	13.70	13.30	12.95	12.65

Modifications

Add to the total cost

Upgrade Kitchen Cabinets	$ + 910
Solid Surface Countertops (Included)	
Full Bath - including plumbing, wall and floor finishes	+ 5050
Half Bath - including plumbing, wall and floor finishes	+ 3152
Two Car Attached Garage	+ 19,382
Two Car Detached Garage	+ 22,032
Fireplace & Chimney	+ 4355

Adjustments

For multi family - add to total cost

Additional Kitchen	$ + 11,494
Additional Full Bath & Half Bath	+ 8202
Additional Entry & Exit	+ 1276
Separate Heating & Air Conditioning	+ 4978
Separate Electric	+ 1575

For Townhouse/Rowhouse -
Multiply cost per square foot by

Inner Unit	.90
End Unit	.95

Alternatives

Add to or deduct from the cost per square foot of living area

Cedar Shake Roof	+ 1.96
Clay Tile Roof	+ 3.70
Slate Roof	+ 7.43
Upgrade Ceilings to Textured Finish	+ .40
Air Conditioning, in Heating Ductwork	Base System
Heating Systems, Hot Water	+ 1.53
Heat Pump	+ 1.82
Electric Heat	− 2.03
Not Heated	− 3.30

Additional upgrades or components

Kitchen Cabinets & Countertops	Page 58
Bathroom Vanities	59
Fireplaces & Chimneys	59
Windows, Skylights & Dormers	59
Appliances	60
Breezeways & Porches	60
Finished Attic	60
Garages	61
Site Improvements	61
Wings & Ells	47

- **A distinct residence from designer's plans**
- **Single family — 1 full bath, 1 half bath, 1 kitchen**
- **No basement**
- **Asphalt shingles on roof**
- **Forced hot air heat/air conditioning**
- **Gypsum wallboard interior finishes**
- **Materials and workmanship are above average**
- **Detail specifications on page 39**

Note: The illustration shown may contain some optional components (for example: garages and/or fireplaces) whose costs are shown in the modifications, adjustments, & alternatives below or at the end of the square foot section.

●Donald A. Gardner Architects, Inc.

CUSTOM

Base cost per square foot of living area

Exterior Wall	Living Area										
	1000	1200	1400	1600	1800	2000	2400	2800	3200	3600	4000
Wood Siding - Wood Frame	122.40	114.10	108.15	101.05	97.00	93.00	85.30	81.80	78.65	76.10	72.75
Brick Veneer - Wood Frame	129.55	120.80	114.60	106.90	102.50	98.25	89.90	86.20	82.70	80.05	76.35
Stone Veneer - Wood Frame	134.30	125.25	118.80	110.75	106.15	101.65	92.95	89.05	85.40	82.60	78.75
Solid Masonry	136.45	127.30	120.80	112.50	107.85	103.25	94.35	90.40	86.60	83.85	79.90
Finished Basement, Add	28.25	28.45	27.60	26.45	25.80	25.20	24.10	23.55	22.90	22.55	22.10
Unfinished Basement, Add	12.05	11.50	11.20	10.60	10.35	10.00	9.50	9.20	8.90	8.80	8.55

Modifications

Add to the total cost

Upgrade Kitchen Cabinets	$ + 910
Solid Surface Countertops (Included)	
Full Bath - including plumbing, wall and floor finishes	+ 5050
Half Bath - including plumbing, wall and floor finishes	+ 3152
Two Car Attached Garage	+ 19,382
Two Car Detached Garage	+ 22,032
Fireplace & Chimney	+ 4915

Adjustments

For multi family - add to total cost

Additional Kitchen	$ + 11,494
Additional Full Bath & Half Bath	+ 8202
Additional Entry & Exit	+ 1276
Separate Heating & Air Conditioning	+ 4978
Separate Electric	+ 1575

For Townhouse/Rowhouse -
Multiply cost per square foot by

Inner Unit	.90
End Unit	.95

Alternatives

Add to or deduct from the cost per square foot of living area

Cedar Shake Roof	+ 1.42
Clay Tile Roof	+ 2.67
Slate Roof	+ 5.37
Upgrade Ceilings to Textured Finish	+ .40
Air Conditioning, in Heating Ductwork	Base System
Heating Systems, Hot Water	+ 1.45
Heat Pump	+ 1.92
Electric Heat	– 1.76
Not Heated	– 3.04

Additional upgrades or components

Kitchen Cabinets & Countertops	Page 58
Bathroom Vanities	59
Fireplaces & Chimneys	59
Windows, Skylights & Dormers	59
Appliances	60
Breezeways & Porches	60
Finished Attic	60
Garages	61
Site Improvements	61
Wings & Ells	47

- **A distinct residence from designer's plans**
- **Single family — 1 full bath, 1 half bath, 1 kitchen**
- **No basement**
- **Asphalt shingles on roof**
- **Forced hot air heat/air conditioning**
- **Gypsum wallboard interior finishes**
- **Materials and workmanship are above average**
- **Detail specifications on page 39**

Note: The illustration shown may contain some optional components (for example: garages and/or fireplaces) whose costs are shown in the modifications, adjustments, & alternatives below or at the end of the square foot section.

CUSTOM

Base cost per square foot of living area

Exterior Wall	Living Area										
	1200	1400	1600	1800	2000	2400	2800	3200	3600	4000	4400
Wood Siding - Wood Frame	114.90	108.20	103.55	99.35	94.85	88.30	82.75	79.15	77.00	74.75	72.80
Brick Veneer - Wood Frame	122.55	115.40	110.35	105.80	101.05	93.90	87.85	83.90	81.60	79.00	77.00
Stone Veneer - Wood Frame	127.60	120.10	114.85	110.00	105.15	97.55	91.15	87.05	84.65	81.90	79.75
Solid Masonry	129.95	122.30	116.95	112.05	107.05	99.25	92.70	88.50	86.05	83.20	81.00
Finished Basement, Add	22.75	22.90	22.25	21.60	21.20	20.25	19.50	19.00	18.70	18.30	18.10
Unfinished Basement, Add	9.70	9.30	9.00	8.70	8.50	8.10	7.70	7.50	7.35	7.15	7.05

Modifications

Add to the total cost

Upgrade Kitchen Cabinets	$ + 910
Solid Surface Countertops (Included)	
Full Bath - including plumbing, wall and floor finishes	+ 5050
Half Bath - including plumbing, wall and floor finishes	+ 3152
Two Car Attached Garage	+ 19,382
Two Car Detached Garage	+ 22,032
Fireplace & Chimney	+ 4915

Adjustments

For multi family - add to total cost

Additional Kitchen	$ + 11,494
Additional Full Bath & Half Bath	+ 8202
Additional Entry & Exit	+ 1276
Separate Heating & Air Conditioning	+ 4978
Separate Electric	+ 1575

For Townhouse/Rowhouse -
Multiply cost per square foot by

Inner Unit	.87
End Unit	.93

Alternatives

Add to or deduct from the cost per square foot of living area

Cedar Shake Roof	+ .98
Clay Tile Roof	+ 1.85
Slate Roof	+ 3.72
Upgrade Ceilings to Textured Finish	+ .40
Air Conditioning, in Heating Ductwork	Base System
Heating Systems, Hot Water	+ 1.42
Heat Pump	+ 2.14
Electric Heat	- 1.76
Not Heated	- 2.88

Additional upgrades or components

Kitchen Cabinets & Countertops	Page 58
Bathroom Vanities	59
Fireplaces & Chimneys	59
Windows, Skylights & Dormers	59
Appliances	60
Breezeways & Porches	60
Finished Attic	60
Garages	61
Site Improvements	61
Wings & Ells	47

- A distinct residence from designer's plans
- Single family — 1 full bath, 1 half bath, 1 kitchen
- No basement
- Asphalt shingles on roof
- Forced hot air heat/air conditioning
- Gypsum wallboard interior finishes
- Materials and workmanship are above average
- Detail specifications on page 39

Note: The illustration shown may contain some optional components (for example: garages and/or fireplaces) whose costs are shown in the modifications, adjustments, & alternatives below or at the end of the square foot section.

CUSTOM

Base cost per square foot of living area

Exterior Wall	Living Area										
	1500	1800	2100	2400	2800	3200	3600	4000	4500	5000	5500
Wood Siding - Wood Frame	113.80	102.65	96.20	92.25	87.10	82.35	79.65	75.40	73.25	71.20	69.20
Brick Veneer - Wood Frame	121.80	110.00	102.75	98.50	93.10	87.75	84.85	80.15	77.70	75.55	73.30
Stone Veneer - Wood Frame	127.00	114.80	107.10	102.60	97.05	91.30	88.15	83.25	80.65	78.30	75.90
Solid Masonry	129.45	117.10	109.05	104.50	98.80	92.95	89.75	84.75	82.00	79.60	77.20
Finished Basement, Add	18.05	18.00	17.00	16.60	16.15	15.55	15.20	14.75	14.45	14.25	13.95
Unfinished Basement, Add	7.75	7.35	6.90	6.65	6.45	6.15	6.05	5.80	5.70	5.55	5.45

Modifications

Add to the total cost

Upgrade Kitchen Cabinets	$ + 910
Solid Surface Countertops (Included)	
Full Bath - including plumbing, wall and floor finishes	+ 5050
Half Bath - including plumbing, wall and floor finishes	+ 3152
Two Car Attached Garage	+ 19,382
Two Car Detached Garage	+ 22,032
Fireplace & Chimney	+ 5555

Adjustments

For multi family - add to total cost

Additional Kitchen	$ + 11,494
Additional Full Bath & Half Bath	+ 8202
Additional Entry & Exit	+ 1276
Separate Heating & Air Conditioning	+ 4978
Separate Electric	+ 1575

For Townhouse/Rowhouse -
Multiply cost per square foot by

Inner Unit	.87
End Unit	.94

Alternatives

Add to or deduct from the cost per square foot of living area

Cedar Shake Roof	+ .85
Clay Tile Roof	+ 1.60
Slate Roof	+ 3.22
Upgrade Ceilings to Textured Finish	+ .40
Air Conditioning, in Heating Ductwork	Base System
Heating Systems, Hot Water	+ 1.28
Heat Pump	+ 2.20
Electric Heat	− 3.14
Not Heated	− 2.88

Additional upgrades or components

Kitchen Cabinets & Countertops	Page 58
Bathroom Vanities	59
Fireplaces & Chimneys	59
Windows, Skylights & Dormers	59
Appliances	60
Breezeways & Porches	60
Finished Attic	60
Garages	61
Site Improvements	61
Wings & Ells	47

CUSTOM

- **A distinct residence from designer's plans**
- **Single family — 1 full bath, 1 half bath, 1 kitchen**
- **No basement**
- **Asphalt shingles on roof**
- **Forced hot air heat/air conditioning**
- **Gypsum wallboard interior finishes**
- **Materials and workmanship are above average**
- **Detail specifications on page 39**

Note: The illustration shown may contain some optional components (for example: garages and/or fireplaces) whose costs are shown in the modifications, adjustments, & alternatives below or at the end of the square foot section.

Base cost per square foot of living area

Exterior Wall	Living Area										
	1500	1800	2100	2500	3000	3500	4000	4500	5000	5500	6000
Wood Siding - Wood Frame	113.35	102.30	97.05	92.65	85.80	82.45	78.25	73.75	72.20	70.55	68.80
Brick Veneer - Wood Frame	121.65	110.00	104.20	99.45	92.05	88.35	83.55	78.75	76.95	75.15	73.10
Stone Veneer - Wood Frame	127.10	115.05	108.85	103.95	96.10	92.20	87.10	82.00	80.15	78.15	75.90
Solid Masonry	129.65	117.35	111.10	106.05	97.95	93.95	88.70	83.45	81.55	79.55	77.30
Finished Basement, Add	15.90	15.80	15.15	14.75	14.10	13.65	13.20	12.75	12.60	12.45	12.25
Unfinished Basement, Add	6.75	6.40	6.15	5.95	5.65	5.45	5.20	5.05	4.95	4.85	4.75

Modifications

Add to the total cost

Upgrade Kitchen Cabinets	$ + 910
Solid Surface Countertops (Included)	
Full Bath - including plumbing, wall and floor finishes	+ 5050
Half Bath - including plumbing, wall and floor finishes	+ 3152
Two Car Attached Garage	+ 19,382
Two Car Detached Garage	+ 22,032
Fireplace & Chimney	+ 5555

Adjustments

For multi family - add to total cost

Additional Kitchen	$ + 11,494
Additional Full Bath & Half Bath	+ 8202
Additional Entry & Exit	+ 1276
Separate Heating & Air Conditioning	+ 4978
Separate Electric	+ 1575

For Townhouse/Rowhouse -
Multiply cost per square foot by

Inner Unit	.85
End Unit	.93

Alternatives

Add to or deduct from the cost per square foot of living area

Cedar Shake Roof	+ .65
Clay Tile Roof	+ 1.23
Slate Roof	+ 2.48
Upgrade Ceilings to Textured Finish	+ .40
Air Conditioning, in Heating Ductwork	Base System
Heating Systems, Hot Water	+ 1.28
Heat Pump	+ 2.20
Electric Heat	– 3.14
Not Heated	– 2.78

Additional upgrades or components

Kitchen Cabinets & Countertops	Page 58
Bathroom Vanities	59
Fireplaces & Chimneys	59
Windows, Skylights & Dormers	59
Appliances	60
Breezeways & Porches	60
Finished Attic	60
Garages	61
Site Improvements	61
Wings & Ells	47

- **A distinct residence from designer's plans**
- **Single family — 1 full bath, 1 half bath, 1 kitchen**
- **No basement**
- **Asphalt shingles on roof**
- **Forced hot air heat/air conditioning**
- **Gypsum wallboard interior finishes**
- **Materials and workmanship are above average**
- **Detail specifications on page 39**

Note: The illustration shown may contain some optional components (for example: garages and/or fireplaces) whose costs are shown in the modifications, adjustments, & alternatives below or at the end of the square foot section.

CUSTOM

Base cost per square foot of living area

Exterior Wall	Living Area										
	1200	1400	1600	1800	2000	2400	2800	3200	3600	4000	4400
Wood Siding - Wood Frame	108.80	102.50	98.05	94.20	89.80	83.85	78.65	75.30	73.35	71.25	69.45
Brick Veneer - Wood Frame	114.50	107.85	103.15	99.00	94.45	88.00	82.50	78.90	76.80	74.50	72.60
Stone Veneer - Wood Frame	118.30	111.40	106.55	102.20	97.55	90.80	84.95	81.25	79.05	76.65	74.70
Solid Masonry	120.05	113.00	108.15	103.65	98.95	92.05	86.15	82.30	80.10	77.60	75.60
Finished Basement, Add	22.75	22.90	22.25	21.60	21.20	20.25	19.50	19.00	18.70	18.30	18.10
Unfinished Basement, Add	9.70	9.30	9.00	8.70	8.50	8.10	7.70	7.50	7.35	7.15	7.05

Modifications

Add to the total cost

Upgrade Kitchen Cabinets	$ + 910
Solid Surface Countertops (Included)	
Full Bath - including plumbing, wall and floor finishes	+ 5050
Half Bath - including plumbing, wall and floor finishes	+ 3152
Two Car Attached Garage	+ 19,382
Two Car Detached Garage	+ 22,032
Fireplace & Chimney	+ 4355

Adjustments

For multi family - add to total cost

Additional Kitchen	$ + 11,494
Additional Full Bath & Half Bath	+ 8202
Additional Entry & Exit	+ 1276
Separate Heating & Air Conditioning	+ 4978
Separate Electric	+ 1575

For Townhouse/Rowhouse -
Multiply cost per square foot by

Inner Unit	.89
End Unit	.95

Alternatives

Add to or deduct from the cost per square foot of living area

Cedar Shake Roof	+ 1.96
Clay Tile Roof	+ 3.70
Slate Roof	+ 3.72
Upgrade Ceilings to Textured Finish	+ .40
Air Conditioning, in Heating Ductwork	Base System
Heating Systems, Hot Water	+ 1.42
Heat Pump	+ 2.14
Electric Heat	– 1.77
Not Heated	– 2.78

Additional upgrades or components

Kitchen Cabinets & Countertops	Page 58
Bathroom Vanities	59
Fireplaces & Chimneys	59
Windows, Skylights & Dormers	59
Appliances	60
Breezeways & Porches	60
Finished Attic	60
Garages	61
Site Improvements	61
Wings & Ells	47

- **A distinct residence from designer's plans**
- **Single family — 1 full bath, 1 half bath, 1 kitchen**
- **No basement**
- **Asphalt shingles on roof**
- **Forced hot air heat/air conditioning**
- **Gypsum wallboard interior finishes**
- **Materials and workmanship are above average**
- **Detail specifications on page 39**

Note: The illustration shown may contain some optional components (for example: garages and/or fireplaces) whose costs are shown in the modifications, adjustments, & alternatives below or at the end of the square foot section.

©Design Basics, Inc.

Base cost per square foot of living area

Exterior Wall	Living Area										
	1200	1500	1800	2100	2400	2800	3200	3600	4000	4500	5000
Wood Siding - Wood Frame	111.80	101.90	94.35	88.15	84.00	80.95	77.40	73.60	72.05	68.45	66.45
Brick Veneer - Wood Frame	117.50	107.05	99.00	92.35	88.00	84.75	80.95	76.90	75.25	71.35	69.30
Stone Veneer - Wood Frame	121.25	110.45	102.10	95.15	90.60	87.30	83.30	79.00	77.35	73.30	71.10
Solid Masonry	123.00	112.00	103.55	96.45	91.80	88.50	84.35	80.10	78.30	74.20	71.90
Finished Basement, Add*	28.30	28.20	27.00	25.95	25.30	24.85	24.20	23.65	23.40	22.90	22.55
Unfinished Basement, Add*	12.00	11.35	10.75	10.25	9.95	9.75	9.45	9.15	9.05	8.80	8.60

*Basement under middle level only.

Modifications

Add to the total cost

Upgrade Kitchen Cabinets	$ + 910
Solid Surface Countertops (Included)	
Full Bath - including plumbing, wall and floor finishes	+ 5050
Half Bath - including plumbing, wall and floor finishes	+ 3152
Two Car Attached Garage	+ 19,382
Two Car Detached Garage	+ 22,032
Fireplace & Chimney	+ 4915

Adjustments

For multi family - add to total cost

Additional Kitchen	$ + 11,494
Additional Full Bath & Half Bath	+ 8202
Additional Entry & Exit	+ 1276
Separate Heating & Air Conditioning	+ 4978
Separate Electric	+ 1575

For Townhouse/Rowhouse -
Multiply cost per square foot by

Inner Unit	.87
End Unit	.94

Alternatives

Add to or deduct from the cost per square foot of living area

Cedar Shake Roof	+ 1.42
Clay Tile Roof	+ 2.67
Slate Roof	+ 5.37
Upgrade Ceilings to Textured Finish	+ .40
Air Conditioning, in Heating Ductwork	Base System
Heating Systems, Hot Water	+ 1.37
Heat Pump	+ 2.22
Electric Heat	– 1.58
Not Heated	– 2.78

Additional upgrades or components

Kitchen Cabinets & Countertops	Page 58
Bathroom Vanities	59
Fireplaces & Chimneys	59
Windows, Skylights & Dormers	59
Appliances	60
Breezeways & Porches	60
Finished Attic	60
Garages	61
Site Improvements	61
Wings & Ells	47

1 Story — Base cost per square foot of living area

Exterior Wall	Living Area							
	50	100	200	300	400	500	600	700
Wood Siding - Wood Frame	172.05	134.00	117.45	100.10	94.65	91.50	89.30	89.95
Brick Veneer - Wood Frame	191.40	147.80	128.95	107.75	101.60	97.90	95.45	95.90
Stone Veneer - Wood Frame	204.10	156.85	136.50	112.80	106.10	102.10	99.45	99.80
Solid Masonry	210.05	161.15	140.05	115.15	108.20	104.10	101.30	101.55
Finished Basement, Add	66.00	56.60	51.40	42.75	41.00	39.95	39.25	38.75
Unfinished Basement, Add	49.20	33.95	26.90	20.75	19.10	18.10	17.45	17.00

1-1/2 Story — Base cost per square foot of living area

Exterior Wall	Living Area							
	100	200	300	400	500	600	700	800
Wood Siding - Wood Frame	137.20	112.60	97.60	89.05	84.45	82.35	79.55	78.75
Brick Veneer - Wood Frame	154.50	126.40	109.15	98.00	92.70	90.15	87.00	86.10
Stone Veneer - Wood Frame	165.80	135.45	116.65	103.85	98.15	95.30	91.85	90.90
Solid Masonry	171.10	139.70	120.25	106.65	100.75	97.70	94.15	93.10
Finished Basement, Add	44.30	41.10	37.60	33.85	32.80	32.15	31.45	31.40
Unfinished Basement, Add	29.45	22.40	19.05	16.30	15.30	14.65	14.05	13.90

2 Story — Base cost per square foot of living area

Exterior Wall	Living Area							
	100	200	400	600	800	1000	1200	1400
Wood Siding - Wood Frame	136.95	104.75	90.55	77.30	72.65	69.90	68.10	69.00
Brick Veneer - Wood Frame	156.25	118.55	102.10	84.95	79.60	76.35	74.20	74.95
Stone Veneer - Wood Frame	168.95	127.60	109.60	89.95	84.10	80.55	78.25	78.85
Solid Masonry	174.90	131.85	113.20	92.35	86.20	82.60	80.10	80.60
Finished Basement, Add	33.05	28.30	25.75	21.40	20.55	20.00	19.70	19.45
Unfinished Basement, Add	24.60	17.05	13.45	10.40	9.60	9.05	8.75	8.50

Base costs do not include bathroom or kitchen facilities. Use Modifications/Adjustments/Alternatives on pages 58-61 where appropriate.

1 Story

1-1/2 Story

2 Story

2-1/2 Story

Bi-Level

Tri-Level

Components

1 Site Work
Site preparation for slab or excavation for lower level; 4' deep trench excavation for foundation wall.

2 Foundations
Continuous reinforced concrete footing, 8" deep x 18" wide; dampproofed and insulated 12" thick reinforced concrete foundation wall, 4' deep; 4" concrete slab on 4" crushed stone base and polyethylene vapor barrier, trowel finish.

3 Framing
Exterior walls—2 x 6 wood studs, 16" O.C.; 1/2" plywood sheathing; 2 x 8 rafters 16" O.C. with 1/2" plywood sheathing, 4 in 12, 6 in 12 or 8 in 12 roof pitch, 2 x 6 or 2 x 8 ceiling joists; 1/2" plywood subfloor on 1 x 3 wood sleepers 16" O.C.; 2 x 10 floor joists with 5/8" plywood subfloor on models with more than one level.

4 Exterior Walls
Face brick veneer and #15 felt building paper on insulated wood frame walls; double hung windows; 3 flush solid core wood exterior doors with storms.
Alternates:
• Wood siding on wood frame, has top quality cedar or redwood siding or hand split cedar shingles or shakes.
• Solid brick may have solid brick exterior wall or brick on concrete block.
• Solid stone concrete block with selected fieldstone or limestone exterior.

5 Roofing
Red cedar shingles; #15 felt building paper; aluminum gutters, downspouts and drip edge; copper flashings.

6 Interiors
Walls and ceilings—5/8" drywall, skim coat plaster, primed and painted with 2 coats; hardwood baseboard and trim; sanded and finished, hardwood floor 70%, ceramic tile with 1/2" underlayment 20%, vinyl tile with 1/2" underlayment 10%; wood panel interior doors, primed and painted with 2 coats.

7 Specialties
Luxury grade kitchen cabinets—25 L.F. wall and base with plastic laminate counter top and kitchen sink; 6 L.F. bathroom vanity; 75 gallon electric water heater; medicine cabinet.

8 Mechanical
Gas fired warm air heat/air conditioning; one full bath including bathtub, corner shower, built in lavatory and water closet; one 1/2 bath including built in lavatory and water closet.

9 Electrical
100 Amp. service; romex wiring; incandescent lighting fixtures, switches, receptacles.

10 Overhead and Profit
General Contractor overhead and profit.

Adjustments

Unfinished Basement:
8" high cast-in-place concrete walls 12" thick or 12" concrete block.

Finished Basement:
Includes painted drywall on 2 x 4 wood furring with insulation, suspended ceiling, carpeting on subfloor with sleepers, heating and lighting.

LUXURY

- Unique residence built from an architect's plan
- Single family — 1 full bath, 1 half bath, 1 kitchen
- No basement
- Cedar shakes on roof
- Forced hot air heat/air conditioning
- Gypsum wallboard interior finishes
- Many special features
- Extraordinary materials and workmanship
- Detail specifications on page 49

©Home Planners, Inc.

Note: The illustration shown may contain some optional components (for example: garages and/or fireplaces) whose costs are shown in the modifications, adjustments, & alternatives below or at the end of the square foot section.

Base cost per square foot of living area

Exterior Wall	Living Area										
	1000	1200	1400	1600	1800	2000	2400	2800	3200	3600	4000
Wood Siding - Wood Frame	155.40	143.85	134.40	128.10	124.50	120.30	112.55	107.00	103.15	99.10	95.70
Brick Veneer - Wood Frame	162.55	150.25	140.20	133.55	129.80	125.25	117.10	111.20	107.05	102.70	99.15
Solid Brick	173.50	160.10	149.20	141.95	137.85	132.90	124.10	117.60	113.00	108.20	104.30
Solid Stone	173.15	159.85	148.90	141.70	137.65	132.70	123.80	117.45	112.80	108.05	104.10
Finished Basement, Add	42.30	45.45	43.65	42.50	41.75	40.75	39.40	38.40	37.50	36.70	36.05
Unfinished Basement, Add	18.95	17.85	17.00	16.45	16.10	15.60	15.00	14.50	14.00	13.65	13.30

Modifications

Add to the total cost

Upgrade Kitchen Cabinets	$ + 1200
Solid Surface Countertops (Included)	
Full Bath - including plumbing, wall and floor finishes	+ 5858
Half Bath - including plumbing, wall and floor finishes	+ 3657
Two Car Attached Garage	+ 22,219
Two Car Detached Garage	+ 25,057
Fireplace & Chimney	+ 6105

Adjustments

For multi family - add to total cost

Additional Kitchen	$ + 14,855
Additional Full Bath & Half Bath	+ 9515
Additional Entry & Exit	+ 2114
Separate Heating & Air Conditioning	+ 4978
Separate Electric	+ 1575

For Townhouse/Rowhouse -
Multiply cost per square foot by

Inner Unit	.90
End Unit	.95

Alternatives

Add to or deduct from the cost per square foot of living area

Heavyweight Asphalt Shingles	– 1.96
Clay Tile Roof	+ 1.74
Slate Roof	+ 5.65
Upgrade Ceilings to Textured Finish	+ .40
Air Conditioning, in Heating Ductwork	Base System
Heating Systems, Hot Water	+ 1.63
Heat Pump	+ 1.96
Electric Heat	– 1.76
Not Heated	– 3.57

Additional upgrades or components

Kitchen Cabinets & Countertops	Page 58
Bathroom Vanities	59
Fireplaces & Chimneys	59
Windows, Skylights & Dormers	59
Appliances	60
Breezeways & Porches	60
Finished Attic	60
Garages	61
Site Improvements	61
Wings & Ells	57

- Unique residence built from an architect's plan
- Single family — 1 full bath, 1 half bath, 1 kitchen
- No basement
- Cedar shakes on roof
- Forced hot air heat/air conditioning
- Gypsum wallboard interior finishes
- Many special features
- Extraordinary materials and workmanship
- Detail specifications on page 49

Note: The illustration shown may contain some optional components (for example: garages and/or fireplaces) whose costs are shown in the modifications, adjustments, & alternatives below or at the end of the square foot section.

•Larry E. Belk Designs

LUXURY

Base cost per square foot of living area

Exterior Wall	Living Area										
	1000	1200	1400	1600	1800	2000	2400	2800	3200	3600	4000
Wood Siding - Wood Frame	143.80	133.75	126.55	118.15	113.30	108.45	99.45	95.25	91.55	88.65	84.65
Brick Veneer - Wood Frame	152.05	141.55	133.90	124.80	119.60	114.40	104.70	100.30	96.20	93.10	88.80
Solid Brick	164.70	153.35	145.20	135.10	129.35	123.70	112.95	108.10	103.35	100.05	95.25
Solid Stone	164.30	152.95	144.85	134.80	129.05	123.40	112.65	107.80	103.10	99.80	95.05
Finished Basement, Add	29.80	32.30	31.30	29.90	29.15	28.40	26.95	26.30	25.50	25.10	24.55
Unfinished Basement, Add	13.50	12.95	12.50	11.80	11.45	11.10	10.45	10.15	9.75	9.55	9.30

Modifications

Add to the total cost

Upgrade Kitchen Cabinets	$ + 1200
Solid Surface Countertops (Included)	
Full Bath - including plumbing, wall and floor finishes	+ 5858
Half Bath - including plumbing, wall and floor finishes	+ 3657
Two Car Attached Garage	+ 22,219
Two Car Detached Garage	+ 25,057
Fireplace & Chimney	+ 6695

Adjustments

For multi family - add to total cost

Additional Kitchen	$ + 14,855
Additional Full Bath & Half Bath	+ 9515
Additional Entry & Exit	+ 2114
Separate Heating & Air Conditioning	+ 4978
Separate Electric	+ 1575

For Townhouse/Rowhouse -
Multiply cost per square foot by

Inner Unit	.90
End Unit	.95

Alternatives

Add to or deduct from the cost per square foot of living area

Heavyweight Asphalt Shingles	– 1.42
Clay Tile Roof	+ 1.25
Slate Roof	+ 4.14
Upgrade Ceilings to Textured Finish	+ .40
Air Conditioning, in Heating Ductwork	Base System
Heating Systems, Hot Water	+ 1.56
Heat Pump	+ 2.17
Electric Heat	– 1.76
Not Heated	– 3.29

Additional upgrades or components

Kitchen Cabinets & Countertops	Page 58
Bathroom Vanities	59
Fireplaces & Chimneys	59
Windows, Skylights & Dormers	59
Appliances	60
Breezeways & Porches	60
Finished Attic	60
Garages	61
Site Improvements	61
Wings & Ells	57

Important: See the Reference Section for Location Factors (to adjust for your city) and Estimating Forms

- **Unique residence built from an architect's plan**
- **Single family — 1 full bath, 1 half bath, 1 kitchen**
- **No basement**
- **Cedar shakes on roof**
- **Forced hot air heat/air conditioning**
- **Gypsum wallboard interior finishes**
- **Many special features**
- **Extraordinary materials and workmanship**
- **Detail specifications on page 49**

LUXURY

Note: The illustration shown may contain some optional components (for example: garages and/or fireplaces) whose costs are shown in the modifications, adjustments, & alternatives below or at the end of the square foot section.

Base cost per square foot of living area

Exterior Wall	Living Area										
	1200	1400	1600	1800	2000	2400	2800	3200	3600	4000	4400
Wood Siding - Wood Frame	133.40	125.40	119.75	114.85	109.40	101.80	95.25	91.05	88.60	85.90	83.55
Brick Veneer - Wood Frame	142.20	133.55	127.60	122.20	116.55	108.20	101.05	96.55	93.90	90.85	88.35
Solid Brick	155.75	146.20	139.65	133.55	127.50	118.05	110.10	105.00	101.95	98.45	95.75
Solid Stone	155.30	145.80	139.25	133.15	127.15	117.80	109.75	104.75	101.75	98.30	95.50
Finished Basement, Add	23.95	25.95	25.20	24.45	23.90	22.80	21.80	21.25	20.85	20.45	20.20
Unfinished Basement, Add	10.90	10.45	10.05	9.70	9.45	8.95	8.50	8.25	8.05	7.85	7.75

Modifications

Add to the total cost

Upgrade Kitchen Cabinets	$ + 1200
Solid Surface Countertops (Included)	
Full Bath - including plumbing, wall and floor finishes	+ 5858
Half Bath - including plumbing, wall and floor finishes	+ 3657
Two Car Attached Garage	+ 22,219
Two Car Detached Garage	+ 25,057
Fireplace & Chimney	+ 6695

Adjustments

For multi family - add to total cost

Additional Kitchen	$ + 14,855
Additional Full Bath & Half Bath	+ 9515
Additional Entry & Exit	+ 2114
Separate Heating & Air Conditioning	+ 4978
Separate Electric	+ 1575

For Townhouse/Rowhouse -
Multiply cost per square foot by

Inner Unit	.86
End Unit	.93

Alternatives

Add to or deduct from the cost per square foot of living area

Heavyweight Asphalt Shingles	– .98
Clay Tile Roof	+ .87
Slate Roof	+ 2.83
Upgrade Ceilings to Textured Finish	+ .40
Air Conditioning, in Heating Ductwork	Base System
Heating Systems, Hot Water	+ 1.52
Heat Pump	+ 2.29
Electric Heat	– 1.58
Not Heated	– 3.11

Additional upgrades or components

Kitchen Cabinets & Countertops	Page 58
Bathroom Vanities	59
Fireplaces & Chimneys	59
Windows, Skylights & Dormers	59
Appliances	60
Breezeways & Porches	60
Finished Attic	60
Garages	61
Site Improvements	61
Wings & Ells	57

- **Unique residence built from an architect's plan**
- **Single family — 1 full bath, 1 half bath, 1 kitchen**
- **No basement**
- **Cedar shakes on roof**
- **Forced hot air heat/air conditioning**
- **Gypsum wallboard interior finishes**
- **Many special features**
- **Extraordinary materials and workmanship**
- **Detail specifications on page 49**

Note: The illustration shown may contain some optional components (for example: garages and/or fireplaces) whose costs are shown in the modifications, adjustments, & alternatives below or at the end of the square foot section.

©Larry W. Garnett & Associates, Inc

LUXURY

Base cost per square foot of living area

Exterior Wall	Living Area										
	1500	1800	2100	2500	3000	3500	4000	4500	5000	5500	6000
Wood Siding - Wood Frame	130.60	117.60	109.95	104.40	96.70	91.10	85.70	83.15	80.80	78.50	76.00
Brick Veneer - Wood Frame	139.70	126.05	117.50	111.50	103.15	96.95	91.20	88.30	85.70	83.20	80.45
Solid Brick	153.80	138.95	129.10	122.50	113.05	106.00	99.60	96.20	93.25	90.45	87.35
Solid Stone	153.40	138.65	128.70	122.15	112.80	105.70	99.35	95.95	93.00	90.15	87.15
Finished Basement, Add	19.05	20.50	19.35	18.70	17.85	17.05	16.55	16.15	15.90	15.60	15.30
Unfinished Basement, Add	8.75	8.30	7.75	7.45	7.05	6.70	6.45	6.25	6.05	5.95	5.85

Modifications

Add to the total cost

Upgrade Kitchen Cabinets	$ + 1200
Solid Surface Countertops (Included)	
Full Bath - including plumbing, wall and floor finishes	+ 5858
Half Bath - including plumbing, wall and floor finishes	+ 3657
Two Car Attached Garage	+ 22,219
Two Car Detached Garage	+ 25,057
Fireplace & Chimney	+ 7320

Adjustments

For multi family - add to total cost

Additional Kitchen	$ + 14,855
Additional Full Bath & Half Bath	+ 9515
Additional Entry & Exit	+ 2114
Separate Heating & Air Conditioning	+ 4978
Separate Electric	+ 1575

For Townhouse/Rowhouse -
Multiply cost per square foot by

Inner Unit	.86
End Unit	.93

Alternatives

Add to or deduct from the cost per square foot of living area

Heavyweight Asphalt Shingles	– .85
Clay Tile Roof	+ .75
Slate Roof	+ 2.47
Upgrade Ceilings to Textured Finish	+ .40
Air Conditioning, in Heating Ductwork	Base System
Heating Systems, Hot Water	+ 1.37
Heat Pump	+ 2.36
Electric Heat	– 3.14
Not Heated	– 3.11

Additional upgrades or components

Kitchen Cabinets & Countertops	Page 58
Bathroom Vanities	59
Fireplaces & Chimneys	59
Windows, Skylights & Dormers	59
Appliances	60
Breezeways & Porches	60
Finished Attic	60
Garages	61
Site Improvements	61
Wings & Ells	57

- **Unique residence built from an architect's plan**
- **Single family — 1 full bath, 1 half bath, 1 kitchen**
- **No basement**
- **Cedar shakes on roof**
- **Forced hot air heat/air conditioning**
- **Gypsum wallboard interior finishes**
- **Many special features**
- **Extraordinary materials and workmanship**
- **Detail specifications on page 49**

Note: The illustration shown may contain some optional components (for example: garages and/or fireplaces) whose costs are shown in the modifications, adjustments, & alternatives below or at the end of the square foot section.

Base cost per square foot of living area

Exterior Wall	Living Area										
	1500	1800	2100	2500	3000	3500	4000	4500	5000	5500	6000
Wood Siding - Wood Frame	129.80	117.00	110.60	105.35	97.45	93.45	88.60	83.55	81.65	79.65	77.60
Brick Veneer - Wood Frame	139.30	125.80	118.80	113.20	104.60	100.15	94.75	89.15	87.10	84.95	82.55
Solid Brick	153.90	139.30	131.40	125.30	115.60	110.50	104.20	97.90	95.50	93.00	90.20
Solid Stone	153.45	138.85	131.00	124.95	115.20	110.20	103.85	97.60	95.25	92.75	90.00
Finished Basement, Add	16.75	18.05	17.30	16.75	15.95	15.45	14.85	14.40	14.15	13.90	13.60
Unfinished Basement, Add	7.65	7.25	6.90	6.65	6.30	6.10	5.80	5.55	5.45	5.35	5.20

Modifications

Add to the total cost

Upgrade Kitchen Cabinets	$ + 1200
Solid Surface Countertops (Included)	
Full Bath - including plumbing, wall and floor finishes	+ 5858
Half Bath - including plumbing, wall and floor finishes	+ 3657
Two Car Attached Garage	+ 22,219
Two Car Detached Garage	+ 25,057
Fireplace & Chimney	+ 7320

Adjustments

For multi family - add to total cost

Additional Kitchen	$ + 14,855
Additional Full Bath & Half Bath	+ 9515
Additional Entry & Exit	+ 2114
Separate Heating & Air Conditioning	+ 4978
Separate Electric	+ 1575

For Townhouse/Rowhouse -
Multiply cost per square foot by

Inner Unit	.84
End Unit	.92

Alternatives

Add to or deduct from the cost per square foot of living area

Heavyweight Asphalt Shingles	– .65
Clay Tile Roof	+ .58
Slate Roof	+ 1.88
Upgrade Ceilings to Textured Finish	+ .40
Air Conditioning, in Heating Ductwork	Base System
Heating Systems, Hot Water	+ 1.37
Heat Pump	+ 2.36
Electric Heat	– 3.14
Not Heated	– 3.01

Additional upgrades or components

Kitchen Cabinets & Countertops	Page 58
Bathroom Vanities	59
Fireplaces & Chimneys	59
Windows, Skylights & Dormers	59
Appliances	60
Breezeways & Porches	60
Finished Attic	60
Garages	61
Site Improvements	61
Wings & Ells	57

- **Unique residence built from an architect's plan**
- **Single family — 1 full bath, 1 half bath, 1 kitchen**
- **No basement**
- **Cedar shakes on roof**
- **Forced hot air heat/air conditioning**
- **Gypsum wallboard interior finishes**
- **Many special features**
- **Extraordinary materials and workmanship**
- **Detail specifications on page 49**

Note: The illustration shown may contain some optional components (for example: garages and/or fireplaces) whose costs are shown in the modifications, adjustments, & alternatives below or at the end of the square foot section.

Base cost per square foot of living area

Exterior Wall	Living Area										
	1200	1400	1600	1800	2000	2400	2800	3200	3600	4000	4400
Wood Siding - Wood Frame	126.45	118.95	113.55	108.95	103.75	96.70	90.75	86.70	84.50	81.95	79.80
Brick Veneer - Wood Frame	133.05	125.10	119.40	114.50	109.10	101.50	95.10	90.80	88.40	85.70	83.35
Solid Brick	143.15	134.60	128.45	122.95	117.30	108.95	101.85	97.15	94.45	91.35	88.90
Solid Stone	142.85	134.30	128.15	122.75	117.05	108.65	101.65	96.95	94.30	91.25	88.75
Finished Basement, Add	23.95	25.95	25.20	24.45	23.90	22.80	21.80	21.25	20.85	20.45	20.20
Unfinished Basement, Add	10.90	10.45	10.05	9.70	9.45	8.95	8.50	8.25	8.05	7.85	7.75

Modifications

Add to the total cost

Upgrade Kitchen Cabinets	$ + 1200
Solid Surface Countertops (Included)	
Full Bath - including plumbing, wall and floor finishes	+ 5858
Half Bath - including plumbing, wall and floor finishes	+ 3657
Two Car Attached Garage	+ 22,219
Two Car Detached Garage	+ 25,057
Fireplace & Chimney	+ 6105

Adjustments

For multi family - add to total cost

Additional Kitchen	$ + 14,855
Additional Full Bath & Half Bath	+ 9515
Additional Entry & Exit	+ 2114
Separate Heating & Air Conditioning	+ 4978
Separate Electric	+ 1575

**For Townhouse/Rowhouse -
Multiply cost per square foot by**

Inner Unit	.89
End Unit	.94

Alternatives

Add to or deduct from the cost per square foot of living area

Heavyweight Asphalt Shingles	– 1.96
Clay Tile Roof	+ 1.74
Slate Roof	+ 5.65
Upgrade Ceilings to Textured Finish	+ .40
Air Conditioning, in Heating Ductwork	Base System
Heating Systems, Hot Water	+ 1.52
Heat Pump	+ 2.29
Electric Heat	– 1.58
Not Heated	– 3.11

Additional upgrades or components

Kitchen Cabinets & Countertops	Page 58
Bathroom Vanities	59
Fireplaces & Chimneys	59
Windows, Skylights & Dormers	59
Appliances	60
Breezeways & Porches	60
Finished Attic	60
Garages	61
Site Improvements	61
Wings & Ells	57

- **Unique residence built from an architect's plan**
- **Single family — 1 full bath, 1 half bath, 1 kitchen**
- **No basement**
- **Cedar shakes on roof**
- **Forced hot air heat/air conditioning**
- **Gypsum wallboard interior finishes**
- **Many special features**
- **Extraordinary materials and workmanship**
- **Detail specifications on page 49**

Note: The illustration shown may contain some optional components (for example: garages and/or fireplaces) whose costs are shown in the modifications, adjustments, & alternatives below or at the end of the square foot section.

©Home Planners, Inc.

Base cost per square foot of living area

Exterior Wall	Living Area										
	1500	1800	2100	2400	2800	3200	3600	4000	4500	5000	5500
Wood Siding - Wood Frame	119.15	110.20	102.90	98.00	94.45	90.20	85.85	84.00	79.80	77.50	74.95
Brick Veneer - Wood Frame	125.05	115.55	107.75	102.55	98.85	94.30	89.65	87.70	83.15	80.65	78.00
Solid Brick	134.20	123.85	115.20	109.60	105.55	100.55	95.45	93.30	88.40	85.55	82.60
Solid Stone	133.90	123.60	115.00	109.40	105.35	100.35	95.20	93.15	88.20	85.45	82.50
Finished Basement, Add*	28.20	30.40	29.15	28.35	27.80	27.00	26.30	26.00	25.35	24.90	24.45
Unfinished Basement, Add*	12.60	11.85	11.35	10.90	10.70	10.30	9.95	9.80	9.50	9.25	9.10

*Basement under middle level only.

Modifications

Add to the total cost

Upgrade Kitchen Cabinets	$ + 1200
Solid Surface Countertops (Included)	
Full Bath - including plumbing, wall and floor finishes	+ 5858
Half Bath - including plumbing, wall and floor finishes	+ 3657
Two Car Attached Garage	+ 22,219
Two Car Detached Garage	+ 25,057
Fireplace & Chimney	+ 6695

Adjustments

For multi family - add to total cost

Additional Kitchen	$ + 14,855
Additional Full Bath & Half Bath	+ 9515
Additional Entry & Exit	+ 2114
Separate Heating & Air Conditioning	+ 4978
Separate Electric	+ 1575

For Townhouse/Rowhouse -
Multiply cost per square foot by

Inner Unit	.86
End Unit	.93

Alternatives

Add to or deduct from the cost per square foot of living area

Heavyweight Asphalt Shingles	– 1.42
Clay Tile Roof	+ 1.25
Slate Roof	+ 4.14
Upgrade Ceilings to Textured Finish	+ .40
Air Conditioning, in Heating Ductwork	Base System
Heating Systems, Hot Water	+ 1.47
Heat Pump	+ 2.38
Electric Heat	– 1.40
Not Heated	– 3.01

Additional upgrades or components

Kitchen Cabinets & Countertops	Page 58
Bathroom Vanities	59
Fireplaces & Chimneys	59
Windows, Skylights & Dormers	59
Appliances	60
Breezeways & Porches	60
Finished Attic	60
Garages	61
Site Improvements	61
Wings & Ells	57

1 Story — Base cost per square foot of living area

Exterior Wall	Living Area							
	50	100	200	300	400	500	600	700
Wood Siding - Wood Frame	191.90	149.25	130.55	111.00	104.90	101.25	98.85	99.65
Brick Veneer - Wood Frame	214.05	165.05	143.80	119.80	112.80	108.60	105.80	106.40
Solid Brick	248.20	189.40	164.10	133.35	125.00	120.00	116.65	116.90
Solid Stone	247.10	188.60	163.45	132.85	124.60	119.65	116.35	116.45
Finished Basement, Add	72.80	66.40	59.80	48.85	46.65	45.30	44.45	43.80
Unfinished Basement, Add	37.75	29.45	26.00	20.20	19.05	18.35	17.90	17.55

1-1/2 Story — Base cost per square foot of living area

Exterior Wall	Living Area							
	100	200	300	400	500	600	700	800
Wood Siding - Wood Frame	152.20	124.60	107.75	97.95	92.85	90.50	87.30	86.45
Brick Veneer - Wood Frame	172.00	140.40	120.95	108.20	102.35	99.45	95.85	94.85
Solid Brick	202.45	164.80	141.25	124.05	117.00	113.20	108.90	107.75
Solid Stone	201.55	164.00	140.60	123.55	116.55	112.80	108.50	107.35
Finished Basement, Add	48.20	47.90	43.45	38.60	37.30	36.40	35.55	35.50
Unfinished Basement, Add	24.30	20.75	18.45	15.90	15.25	14.80	14.35	14.30

2 Story — Base cost per square foot of living area

Exterior Wall	Living Area							
	100	200	400	600	800	1000	1200	1400
Wood Siding - Wood Frame	149.95	113.70	97.80	82.70	77.50	74.45	72.40	73.45
Brick Veneer - Wood Frame	172.05	129.50	110.95	91.50	85.40	81.85	79.40	80.25
Solid Brick	206.15	153.85	131.30	105.00	97.65	93.20	90.20	90.70
Solid Stone	205.15	153.05	130.60	104.55	97.25	92.80	89.85	90.30
Finished Basement, Add	36.40	33.30	29.95	24.40	23.40	22.70	22.25	21.95
Unfinished Basement, Add	18.90	14.75	13.00	10.10	9.55	9.20	9.00	8.80

Base costs do not include bathroom or kitchen facilities. Use Modifications/Adjustments/Alternatives on pages 58-61 where appropriate.

ADJUSTMENTS

Kitchen cabinets - Base units, hardwood *(Cost per Unit)*

	Economy	Average	Custom	Luxury
24" deep, 35" high,				
One top drawer,				
One door below				
12" wide	$130	$173	$230	$305
15" wide	170	226	300	395
18" wide	183	244	325	425
21" wide	191	254	340	445
24" wide	216	288	385	505
Four drawers				
12" wide	278	370	490	650
15" wide	220	293	390	515
18" wide	244	325	430	570
24" wide	263	350	465	615
Two top drawers,				
Two doors below				
27" wide	236	315	420	550
30" wide	255	340	450	595
33" wide	263	350	465	615
36" wide	270	360	480	630
42" wide	293	390	520	685
48" wide	311	415	550	725
Range or sink base				
(Cost per unit)				
Two doors below				
30" wide	214	285	380	500
33" wide	229	305	405	535
36" wide	236	315	420	550
42" wide	255	340	450	595
48" wide	266	355	470	620
Corner Base Cabinet				
(Cost per unit)				
36" wide	341	455	605	795
Lazy Susan *(Cost per unit)*				
With revolving door	334	445	590	780

Kitchen cabinets - Wall cabinets, hardwood *(Cost per Unit)*

	Economy	Average	Custom	Luxury
12" deep, 2 doors				
12" high				
30" wide	$140	$187	$250	$ 325
36" wide	160	213	285	375
15" high				
30" wide	146	195	260	340
33" wide	161	215	285	375
36" wide	164	218	290	380
24" high				
30" wide	177	236	315	415
36" wide	194	259	345	455
42" wide	212	283	375	495
30" high, 1 door				
12" wide	128	170	225	300
15" wide	143	190	255	335
18" wide	154	205	275	360
24" wide	171	228	305	400
30" high, 2 doors				
27" wide	207	276	365	485
30" wide	201	268	355	470
36" wide	229	305	405	535
42" wide	244	325	430	570
48" wide	274	365	485	640
Corner wall, 30" high				
24" wide	147	196	260	345
30" wide	172	229	305	400
36" wide	185	246	325	430
Broom closet				
84" high, 24" deep				
18" wide	371	495	660	865
Oven Cabinet				
84" high, 24" deep				
27" wide	529	705	940	1235

Kitchen countertops *(Cost per L.F.)*

	Economy	Average	Custom	Luxury
Solid Surface				
24" wide, no backsplash	$ 86	$115	$155	$200
with backsplash	94	125	165	220
Stock plastic laminate, 24" wide				
with backsplash	15	21	25	35
Custom plastic laminate, no splash				
7/8" thick, alum. molding	23	30	40	55
1-1/4" thick, no splash	25	34	45	60
Marble				
1/2" - 3/4" thick w/splash	41	55	70	95
Maple, laminated				
1-1/2" thick w/splash	60	80	105	140
Stainless steel				
(per S.F.)	110	146	195	255
Cutting blocks, recessed				
16" x 20" x 1" (each)	81	108	145	190

Vanity bases *(Cost per Unit)*

2 door, 30" high, 21" deep	Economy	Average	Custom	Luxury
24" wide	$170	$227	$300	$395
30" wide	198	264	350	460
36" wide	259	345	460	605
48" wide	308	410	545	720

Solid surface vanity tops *(Cost Each)*

Center bowl	Economy	Average	Custom	Luxury
22" x 25"	$285	$308	$332	$359
22" x 31"	325	351	379	409
22" x 37"	375	405	437	472
22" x 49"	465	502	542	586

Fireplaces & Chimneys *(Cost per Unit)*

	1-1/2 Story	2 Story	3 Story
Economy (prefab metal)			
Exterior chimney & 1 fireplace	$4200	$4640	$5085
Interior chimney & 1 fireplace	4020	4475	4685
Average (masonry)			
Exterior chimney & 1 fireplace	4185	4670	5305
Interior chimney & 1 fireplace	3920	4400	4790
For more than 1 flue, add	305	515	855
For more than 1 fireplace, add	2965	2965	2965
Custom (masonry)			
Exterior chimney & 1 fireplace	4355	4915	5555
Interior chimney & 1 fireplace	4085	4620	4980
For more than 1 flue, add	340	590	805
For more than 1 fireplace, add	3130	3130	3130
Luxury (masonry)			
Exterior chimney & 1 fireplace	6105	6695	7320
Interior chimney & 1 fireplace	5825	6370	6740
For more than 1 flue, add	505	840	1175
For more than 1 fireplace, add	4810	4810	4810

Windows and Skylights *(Cost Each)*

	Economy	Average	Custom	Luxury
Fixed Picture Windows				
3'-6" x 4'-0"	$ 454	$ 491	$ 530	$ 572
4'-0" x 6'-0"	652	704	760	821
5'-0" x 6'-0"	922	995	1075	1161
6'-0" x 6'-0"	943	1019	1100	1188
Bay/Bow Windows				
8'-0" x 5'-0"	1200	1296	1400	1512
10'-0" x 5'-0"	1329	1435	1550	1674
10'-0" x 6'-0"	2122	2292	2475	2673
12'-0" x 6'-0"	2658	2870	3100	3348
Palladian Windows				
3'-2" x 6'-4"		1667	1800	1944
4'-0" x 6'-0"		1667	1800	1944
5'-5" x 6'-10"		2060	2225	2403
8'-0" x 6'-0"		2477	2675	2889
Skylights				
46" x 21-1/2"	350	378	479	517
46" x 28"	385	416	471	509
57" x 44"	479	517	637	688

Dormers *(Cost/S.F. of plan area)*

Framing and Roofing Only	Economy	Average	Custom	Luxury
Gable dormer, 2" x 6" roof frame	$21	$24	$27	$43
2" x 8" roof frame	23	25	28	45
Shed dormer, 2" x 6" roof frame	14	15	17	28
2" x 8" roof frame	15	16	18	29
2" x 10" roof frame	16	18	20	30

ADJUSTMENTS

ADJUSTMENTS

Appliances (Cost per Unit)

	Economy	Average	Custom	Luxury
Range				
30" free standing, 1 oven	$ 310	$1030	$1390	$1750
2 oven	1675	1713	1731	1750
30" built-in, 1 oven	550	1163	1469	1775
2 oven	1200	1575	1763	1950
21" free standing				
1 oven	320	370	395	420
Counter Top Ranges				
4 burner standard	264	477	584	690
As above with griddle	570	698	761	825
Microwave Oven	180	395	503	610
Combination Range,				
Refrigerator, Sink				
30" wide	1025	1538	1794	2050
60" wide	2400	2760	2940	3120
72" wide	2725	3134	3339	3543
Comb. Range, Refrig., Sink,				
Microwave Oven & Ice Maker	4168	4793	5106	5418
Compactor				
4 to 1 compaction	520	573	599	625
Deep Freeze				
15 to 23 C.F.	475	568	614	660
30 C.F.	885	993	1046	1100
Dehumidifier, portable, auto.				
15 pint	164	189	201	213
30 pint	183	211	224	238
Washing Machine, automatic	425	950	1213	1475
Water Heater				
Electric, glass lined				
30 gal.	365	450	493	535
80 gal.	750	963	1069	1175
Water Heater, Gas, glass lined				
30 gal.	520	643	704	765
50 gal.	815	1008	1104	1200
Water Softener, automatic				
30 grains/gal.	615	708	754	800
100 grains/gal.	835	961	1023	1086
Dishwasher, built-in				
2 cycles	385	463	501	540
4 or more cycles	400	448	471	495
Dryer, automatic	500	825	988	1150
Garage Door Opener	320	385	418	450
Garbage Disposal	91	148	177	205
Heater, Electric, built-in				
1250 watt ceiling type	169	209	229	249
1250 watt wall type	202	235	252	268
Wall type w/blower				
1500 watt	228	317	361	405
3000 watt	405	446	466	486
Hood For Range, 2 speed				
30" wide	124	460	627	795
42" wide	330	550	660	770
Humidifier, portable				
7 gal. per day	164	189	201	213
15 gal. per day	197	227	241	256
Ice Maker, automatic				
13 lb. per day	445	512	546	579
51 lb. per day	1275	1467	1562	1658
Refrigerator, no frost				
10-12 C.F.	530	680	755	830
14-16 C.F.	540	583	604	625
18-20 C.F.	620	835	943	1050
21-29 C.F.	780	1603	2014	2425
Sump Pump, 1/3 H.P.	214	295	335	375

Breezeway (Cost per S.F.)

Class	Type	Area (S.F.)			
		50	100	150	200
Economy	Open	$ 19.10	$ 16.25	$13.65	$13.40
	Enclosed	91.60	70.80	58.75	51.45
Average	Open	23.65	20.80	18.20	16.55
	Enclosed	100.40	74.80	61.20	53.85
Custom	Open	33.70	29.65	25.80	23.65
	Enclosed	138.45	103.00	84.20	74.00
Luxury	Open	34.80	30.50	27.50	26.65
	Enclosed	140.10	103.90	84.50	75.25

Porches (Cost per S.F.)

Class	Type	Area (S.F.)				
		25	50	100	200	300
Economy	Open	$ 54.75	$ 36.65	$28.60	$24.25	$20.70
	Enclosed	109.40	76.25	57.60	44.95	38.50
Average	Open	66.45	42.30	32.45	27.00	27.00
	Enclosed	131.30	88.95	67.30	52.10	44.15
Custom	Open	86.10	57.10	43.00	37.55	33.65
	Enclosed	171.10	116.90	88.95	69.25	59.65
Luxury	Open	92.15	60.25	44.45	39.95	35.60
	Enclosed	180.30	126.65	94.00	72.95	62.85

Finished attic (Cost per S.F.)

Class	Area (S.F.)				
	400	500	600	800	1000
Economy	$14.65	$14.10	$13.55	$13.30	$12.85
Average	22.50	21.95	21.45	21.15	20.60
Custom	27.45	26.80	26.25	25.80	25.30
Luxury	34.55	33.75	32.95	32.20	31.65

Alarm system (Cost per System)

	Burglar Alarm	Smoke Detector
Economy	$ 370	$ 57
Average	425	70
Custom	720	140
Luxury	1075	172

Sauna, prefabricated
(Cost per unit, including heater and controls—7' high)

Size	Cost
6' x 4'	$ 4275
6' x 5'	4800
6' x 6'	5125
6' x 9'	6350
8' x 10'	8350
8' x 12'	9800
10' x 12'	10400

Garages *

(Costs include exterior wall systems comparable with the quality of the residence. Included in the cost is an allowance for one personnel door, manual overhead door(s) and electrical fixture.)

Class	Type									
	Detached			Attached			Built-in		Basement	
	One Car	Two Car	Three Car	One Car	Two Car	Three Car	One Car	Two Car	One Car	Two Car
Economy										
Wood	$11,433	$17,426	$23,418	$ 8870	$15,261	$21,254	$1600	$3200	$11,680	$1499
Masonry	15,732	22,806	29,879	11,560	19,032	26,106	2128	4256		
Average										
Wood	12,424	18,665	24,906	9490	16,130	22,371	1722	3443	1336	1835
Masonry	15,886	22,998	30,110	11,656	19,167	26,279	2147	3723		
Custom										
Wood	14,438	22,032	29,626	11,248	19,382	26,976	3322	3592	1976	3115
Masonry	17,690	26,102	34,514	13,284	22,235	30,647	3722	4391		
Luxury										
Wood	16,139	25,057	33,976	12,761	22,219	31,138	3384	3716	2687	4208
Masonry	20,612	30,655	40,699	15,560	26,143	36,187	3934	4814		

*See the Introduction to this section for definitions of garage types.

Swimming pools *(Cost per S.F.)*

Residential	(includes equipment)
In-ground	$ 19.35 - 48.50
Deck equipment	1.30
Paint pool, preparation & 3 coats (epoxy)	3.00
Rubber base paint	2.78
Pool Cover	0.72
Swimming Pool Heaters	(Cost per unit)
(not including wiring, external piping, base or pad)	
Gas	
155 MBH	$ 3050
190 MBH	4075
500 MBH	9950
Electric	
15 KW 7200 gallon pool	2250
24 KW 9600 gallon pool	3000
54 KW 24,000 gallon pool	4475

Wood and coal stoves

Wood Only	
Free Standing (minimum)	$ 1275
Fireplace Insert (minimum)	1285
Coal Only	
Free Standing	$ 1452
Fireplace Insert	1590
Wood and Coal	
Free Standing	$ 2987
Fireplace Insert	3060

Sidewalks *(Cost per S.F.)*

Concrete, 3000 psi with wire mesh	4" thick	$ 2.92
	5" thick	3.56
	6" thick	3.99
Precast concrete patio blocks (natural)	2" thick	8.50
Precast concrete patio blocks (colors)	2" thick	9.00
Flagstone, bluestone	1" thick	12.25
Flagstone, bluestone	1-1/2" thick	16.05
Slate (natural, irregular)	3/4" thick	12.00
Slate (random rectangular)	1/2" thick	17.80
Seeding		
Fine grading & seeding includes lime, fertilizer & seed	per S.Y.	1.93
Lawn Sprinkler System	per S.F.	.76

Fencing *(Cost per L.F.)*

Chain Link, 4' high, galvanized	$ 13.75
Gate, 4' high (each)	144.00
Cedar Picket, 3' high, 2 rail	10.25
Gate (each)	146.00
3 Rail, 4' high	12.30
Gate (each)	155.00
Cedar Stockade, 3 Rail, 6' high	12.50
Gate (each)	155.00
Board & Battens, 2 sides 6' high, pine	17.85
6' high, cedar	25.50
No. 1 Cedar, basketweave, 6' high	14.30
Gate, 6' high (each)	178.00

Carport *(Cost per S.F.)*

Economy	$ 6.89
Average	10.44
Custom	15.57
Luxury	17.68

ADJUSTMENTS

Insurance Exclusions

Insurance exclusions are a matter of policy coverage and are not standard or universal. When making an appraisal for insurance purposes, it is recommended that some time be taken in studying the insurance policy to determine the specific items to be excluded. Most homeowners insurance policies have, as part of their provisions, statements that read "the policy permits the insured to exclude from value, in determining whether the amount of the insurance equals or exceeds 80% of its replacement cost, such items as building excavation, basement footings, underground piping and wiring, piers and other supports which are below the undersurface of the lowest floor or where there is no basement below the ground." Loss to any of these items, however, is covered.

Costs for Excavation, Spread and Strip Footings and Underground Piping

This chart shows excluded items expressed as a percentage of total building cost.

Class	Number of Stories			
	1	1-1/2	2	3
Luxury	3.4%	2.7%	2.4%	1.9%
Custom	3.5%	2.7%	2.4%	1.9%
Average	5.0%	4.3%	3.8%	3.3%
Economy	5.0%	4.4%	4.1%	3.5%

Architect/Designer Fees

Architect/designer fees as presented in the following chart are typical ranges for 4 classes of residences. Factors affecting these ranges include economic conditions, size and scope of project, site selection, and standardization. Where superior quality and detail is required or where closer supervision is required, fees may run higher than listed. Lower quality or simplicity may dictate lower fees.

The editors have included architect/designers fees from the "mean" column in calculating costs.

Listed below are average costs expressed as a range and mean of total building cost for 4 classes of residences.

Class		Range	Mean
Luxury	— Architecturally designed and supervised	5%–15%	10%
Custom	— Architecturally modified designer plans	$2000-$3500	$2500
Average	— Designer plans	$800-$1500	$1100
Economy	— Stock plans	$100-$500	$300

Depreciation as generally defined is "the loss of value due to any cause." Specifically, depreciation can be broken down into three categories: **Physical Depreciation, Functional Obsolescence,** and **Economic Obsolescence.**

Physical Depreciation is the loss of value from the wearing out of the building's components. Such causes may be decay, dry rot, cracks, or structural defects.

Functional Obsolescence is the loss of value from features which render the residence less useful or desirable. Such features may be higher than normal ceilings, design and/or style changes, or outdated mechanical systems.

Economic Obsolescence is the loss of value from external features which render the residence less useful or desirable. These features may be changes in neighborhood socio-economic grouping, zoning changes, legislation, etc.

Depreciation as it applies to the residential portion of this manual deals with the observed physical condition of the residence being appraised. It is in essence *"the cost to cure."*

The ultimate method to arrive at *"the cost to cure"* is to analyze each component of the residence.

For example:

    Component, Roof Covering − Cost = $2400
    Average Life = 15 years
    Actual Life = 5 years
    Depreciation = 5 ÷ 15 = .33 or 33%

    $2400 × 33% = $792 = Amount of Depreciation

The following table, however, can be used as a guide in estimating the % depreciation using the actual age and general condition of the residence.

Depreciation Table — Residential

Age in Years	Good	Average	Poor
2	2%	3%	10%
5	4	6	20
10	7	10	25
15	10	15	30
20	15	20	35
25	18	25	40
30	24	30	45
35	28	35	50
40	32	40	55
45	36	45	60
50	40	50	65

ADJUSTMENTS

Commercial/Industrial/Institutional Section

Table of Contents

Table No.   **Page**

Introduction

General 67
How To Use 68

Examples

Example 1 73
Example 2 75

Building Types

M.010 Apartment, 1-3 Story 78
M.020 Apartment, 4-7 Story 80
M.030 Apartment, 8-24 Story 82
M.035 Assisted - Senior Living 84
M.040 Auditorium 86
M.050 Bank 88
M.060 Bowling Alley 90
M.070 Bus Terminal 92
M.080 Car Wash 94
M.090 Church 96
M.100 Club, Country 98
M.110 Club, Social 100
M.120 College, Classroom, 2-3 Story 102
M.130 College, Dorm., 2-3 Story 104
M.140 College, Dorm., 4-8 Story 106
M.150 College, Laboratory 108
M.160 College, Student Union 110
M.170 Community Center 112
M.180 Courthouse, 1 Story 114
M.190 Courthouse, 2-3 Story 116
M.195 Day Care Center 118

Table No.   **Page**

M.200 Factory, 1 Story 120
M.210 Factory, 3 Story 122
M.220 Fire Station, 1 Story 124
M.230 Fire Station, 2 Story 126
M.240 Fraternity/Sorority House 128
M.250 Funeral Home 130
M.260 Garage, Auto Sales 132
M.270 Garage, Parking 134
M.280 Garage, Underground Parking 136
M.290 Garage, Repair 138
M.300 Garage, Service Station 140
M.310 Gymnasium 142
M.320 Hangar, Aircraft 144
M.330 Hospital, 2-3 Story 146
M.340 Hospital, 4-8 Story 148
M.350 Hotel, 4-7 Story 150
M.360 Hotel, 8-24 Story 152
M.370 Jail 154
M.380 Laundromat 156
M.390 Library 158
M.400 Medical Office, 1 Story 160
M.410 Medical Office, 2 Story 162
M.420 Motel, 1 Story 164
M.430 Motel, 2-3 Story 166
M.440 Movie Theatre 168
M.450 Nursing Home 170
M.455 Office, 1 Story 172

Table No.   **Page**

M.460 Office, 2-4 Story 174
M.470 Office, 5-10 Story 176
M.480 Office, 11-20 Story 178
M.490 Police Station 180
M.500 Post Office 182
M.510 Racquetball Court 184
M.520 Religious Education 186
M.530 Restaurant 188
M.540 Restaurant, Fast Food 190
M.550 Rink, Hockey/Indoor Soccer 192
M.560 School, Elementary 194
M.570 School, High, 2-3 Story 196
M.580 School, Jr. High, 2-3 Story 198
M.590 School, Vocational 200
M.600 Store, Convenience 202
M.610 Store, Department, 1 Story 204
M.620 Store, Department, 3 Story 206
M.630 Store, Retail 208
M.640 Supermarket 210
M.650 Swimming Pool, Enclosed 212
M.660 Telephone Exchange 214
M.670 Town Hall, 1 Story 216
M.680 Town Hall, 2-3 Story 218
M.690 Warehouse 220
M.700 Warehouse, Mini 222

Depreciation   224

Introduction to the Commercial/Industrial/Institutional Section

General

The Commercial/Industrial/Institutional section of this manual contains base building costs per square foot of floor area for 72 model buildings. Each model has a table of square foot costs for combinations of exterior wall and framing systems. This table is supplemented by a list of common additives and their unit costs. A breakdown of the component costs used to develop the base cost for the model is on the opposite page. The total cost derived from the base cost and additives must be modified using the appropriate location factor from the Reference Section.

This section may be used directly to estimate the construction cost of most types of buildings knowing only the floor area, exterior wall construction and framing systems. To adjust the base cost for components which are different than the model, use the tables from the Assemblies Section.

Building Identification & Model Selection

The building models in this section represent structures by use. Occupancy, however, does not necessarily identify the building, i.e., a restaurant could be a converted warehouse. In all instances, the building should be described and identified by its own physical characteristics. The model selection should also be guided by comparing specifications with the model. In the case of converted use, data from one model may be used to supplement data from another.

Adjustments

The base cost tables represent the base cost per square foot of floor area for buildings without a basement and without unusual special features. Basement costs and other common additives are listed below the base cost table. Cost adjustments can also be made to the model by using the tables from the Assemblies Section. This table is for example only.

Dimensions

All base cost tables are developed so that measurement can be readily made during the inspection process. Areas are calculated from exterior dimensions and story heights are measured from the top surface of one floor to the top surface of the floor above. Roof areas are measured by horizontal area covered and costs related to inclines are converted with appropriate factors. The precision of measurement is a matter of the user's choice and discretion. For ease in calculation, consideration should be given to measuring in tenths of a foot, i.e., 9 ft. 6 in. = 9.5 ft., 9 ft. 4 in. = 9.3 ft.

Floor Area

The expression "Floor Area" as used in this section includes the sum of floor plate at grade level and above. This dimension is measured from the outside face of the foundation wall. Basement costs are calculated separately. The user must exercise his own judgment, where the lowest level floor is slightly below grade, whether to consider it at grade level or make the basement adjustment.

How to Use the Commercial/Industrial/Institutional Section

The following is a detailed explanation of a sample entry in the Commercial/Industrial/Institutional Square Foot Cost Section. Each bold number below corresponds to the item being described on the following page with the appropriate component or cost of the sample entry following in parenthesis.

Prices listed are costs that include overhead and profit of the installing contractor and additional mark-ups for General Conditions and Architects' Fees.

| COMMERCIAL/INDUSTRIAL/INSTITUTIONAL | M.010 | Apartment, 1-3 Story |

Costs per square foot of floor area

Exterior Wall	S.F. Area	8000	12000	15000	19000	22500	25000	29000	32000	36000
	L.F. Perimeter	213	280	330	350	400	433	442	480	520
Face Brick with Concrete Block Back-up	Wood Joists	150.05	134.20	127.85	119.35	116.20	114.40	110.20	109.00	107.30
	Steel Joists	151.10	136.00	129.95	122.30	119.40	117.65	113.95	112.80	111.25
Stucco on Concrete Block	Wood Joists	133.10	118.80	113.10	106.30	103.45	101.85	.60	97.50	96.10
	Steel Joists	142.25	127.90	122.10	115.20	112.35	110.75	107.50	106.35	104.95
Wood Siding	Wood Frame	132.75	118.50	112.75	106.05	103.20	101.60	98.45	97.30	95.90
Brick Veneer	Wood Frame	139.00	124.50	118.45	110.70	107.70	106.00	5	101.05	99.55
Perimeter Adj., Add or Deduct	Per 100 L.F.	15.90	10.65	8.55	6.70	5.70	5.10	4.40	4.00	3.55
Story Hgt. Adj., Add or Deduct	Per 1 Ft.	3.00	2.60	2.50	2.05	2.00	1.95	1.70	1.65	1.65

For Basement, add $25.60 per square foot of basement area

The above costs were calculated using the basic specifications shown on the facing page. These costs should be adjusted where necessary for design alternatives and owner's requirements. Reported completed project costs, for this type of structure, range from $42.70 to $158.80 per S.F.

Common additives

Description	Unit	$ Cost
Appliances		
Cooking range, 30" free standing		
1 oven	Each	330 - 1800
2 oven	Each	1700 - 1775
30" built-in		
1 oven	Each	570 - 1875
2 oven	Each	1250 - 2050
Counter top cook tops, 4 burner	Each	285 - 730
Microwave oven	Each	212 - 675
Combination range, refrig. & sink, 30" wide	Each	1300 - 2575
72" wide	Each	3825
Combination range, refrigerator, sink,		
microwave oven & icemaker	Each	5650
Compactor, residential, 4-1 compaction	Each	540 - 655
Dishwasher, built-in, 2 cycles	Each	515 - 810
4 cycles	Each	530 - 1075
Garbage disposer, sink type	Each	144 - 258
Hood for range, 2 speed, vented, 30" wide	Each	224 - 960
42" wide	Each	430 - 1150
Refrigerator, no frost 10-12 C.F.	Each	550 - 860
18-20 C.F.	Each	645 - 1100

Description	Unit	$ Cost
Closed Circuit Surveillance, One station		
Camera and monitor	Each	1525
For additional camera stations, add	Each	820
Elevators, Hydraulic passenger, 2 stops		
2000# capacity	Each	49,000
2500# capacity	Each	50,000
3500# capacity	Each	53,700
Additional stop, add	Each	4550
Emergency Lighting, 25 watt, battery operated		
Lead battery	Each	247
Nickel cadmium	Each	695
Laundry Equipment		
Dryer, gas, 16 lb. capacity	Each	785
30 lb. capacity	Each	3000
Washer, 4 cycle	Each	875
Commercial	Each	1325
Smoke Detectors		
Ceiling type	Each	161
Duct type	Each	425

1 Model Number (M.010)

"M" distinguishes this section of the book and stands for model. The number designation is a sequential number.

2 Type of Building (Apartment, 1-3 Story)

There are 43 different types of commercial/industrial/institutional buildings highlighted in this section.

3 Exterior Wall Construction and Building Framing Options (Face Brick with Concrete Block Back-up and Open Web Steel Bar Joists)

Three or more commonly used exterior walls, and in most cases, two typical building framing systems are presented for each type of building. The model selected should be based on the actual characteristics of the building being estimated.

4 Total Square Foot of Floor Area and Base Perimeter Used to Compute Base Costs (22,500 Square Feet and 400 Linear Feet)

Square foot of floor area is the total gross area of all floors at grade, and above, and does not include a basement. The perimeter in linear feet used for the base cost is for a generally rectangular economical building shape.

5 Cost per Square Foot of Floor Area ($119.40)

The highlighted cost is for a building of the selected exterior wall and framing system and floor area. Costs for buildings with floor areas other than those calculated may be interpolated between the costs shown.

6 Building Perimeter and Story Height Adjustments

Square foot costs for a building with a perimeter or floor to floor story height significantly different from the model used to calculate the base cost may be adjusted, add or deduct, to reflect the actual building geometry.

7 Cost per Square Foot of Floor Area for the Perimeter and/or Height Adjustment ($5.70 for Perimeter Difference and $2.00 for Story Height Difference)

Add (or deduct) $5.70 to the base square foot cost for each 100 feet of perimeter difference between the model and the actual building. Add (or deduct) $2.00 to the base square foot cost for each 1 foot of story height difference between the model and the actual building.

8 Optional Cost per Square Foot of Basement Floor Area ($25.60)

The cost of an unfinished basement for the building being estimated is $22.90 times the gross floor area of the basement.

9 Range of Cost per Square Foot of Floor Area for Similar Buildings ($42.70 to $158.80)

Many different buildings of the same type have been built using similar materials and systems. Means historical cost data of actual construction projects indicates a range of $42.70 to $158.80 for this type of building.

10 Common Additives

Common components and/or systems used in this type of building are listed. These costs should be added to the total building cost. Additional selections may be found in the Assemblies Section.

How to Use the Commercial/ Industrial/Institutional Section (Continued)

The following is a detailed explanation of a specification and costs for a model building in the Commercial/Industrial/Institutional Square Foot Cost Section. Each bold number below corresponds to the item being described on the following page with the appropriate component of the sample entry following in parenthesis.

Prices listed are costs that include overhead and profit of the installing contractor.

Model costs calculated for a 3 story building with 10' story height and 22,500 square feet of floor area — **1**

2 — **Apartment, 1-3 Story**

				Unit	Unit Cost	Cost Per S.F.	% Of Sub-Total
A. SUBSTRUCTURE							
1010	Standard Foundations	Poured concrete; strip and spread footings		S.F. Ground	4.89	1.63	
1030	Slab on Grade	4" reinforced concrete with vapor barrier and granular base		S.F. Slab	3.98	1.32	4.7%
2010	Basement Excavation	Site preparation for slab and trench for foundation wall and footing		S.F. Ground	.14	.05	
2020	Basement Walls	4' foundation wall		L.F. Wall	115	1.19	
B. SHELL							
	B10 Superstructure						
1010	Floor Construction	Open web steel joists, slab form, concrete, interior steel columns		S.F. Floor	15.29	10.19	14.3%
1020	Roof Construction	Open web steel joists with rib metal deck, interior steel columns		S.F. Roof	7.32	2.44	
	B20 Exterior Enclosure						
2010	Exterior Walls	Face brick with concrete block backup	88% of wall	S.F. Wall	18.15	8.52	
2020	Exterior Windows	Aluminum horizontal sliding	12% of wall	Each	346	1.48	11.6%
2030	Exterior Doors	Aluminum and glass		Each	1279	.23	
	B30 Roofing						
3010	Roof Coverings	Built-up tar and gravel with flashing; perlite/EPS composite insulation		S.F. Roof	4.41	1.47	1.7%
3020	Roof Openings	N/A		–	–	–	
C. INTERIORS		**3**	**4**	**6**			**9**
1010	Partitions	Gypsum board and sound deadening board on metal studs	10 S.F. of Floor/L.F. Partition	S.F. Partition	5.28	4.69	
1020	Interior Doors	15% solid core wood, 85% hollow core wood	80 S.F. Floor/Door	Each	429	5.37	
1030	Fittings	Kitchen Cabinets		S.F. Floor	2.01	2.01	
2010	Stair Construction	Concrete filled metal pan		Flight	6425	1.43	26.2%
3010	Wall Finishes	70% paint, 25% vinyl wall covering, 5% ceramic tile	**5**	S.F. Surface	2.00		
3020	Floor Finishes	60% carpet, 30% vinyl composition tile, 10% ceramic tile		S.F. Floor		**7** **8**	
3030	Ceiling Finishes	Painted gypsum board on resilient channels		S.F. Ceiling			
D. SERVICES							
	D10 Conveying						
1010	Elevators & Lifts	One hydraulic passenger elevator		Each	74,250	3.30	3.7%
1020	Escalators & Moving Walks	N/A		–	–	–	
	D20 Plumbing						
2010	Plumbing Fixtures	Kitchen, bathroom and service fixtures, supply and drainage	1 Fixture/200 S.F. Floor	Each	1674	8.37	
2020	Domestic Water Distribution	Gas fired water heater		S.F. Floor	2.43	2.43	12.5%
2040	Rain Water Drainage	Roof drains		S.F. Roof	.81	.27	
	D30 HVAC						
3010	Energy Supply	Oil fired hot water, baseboard radiation		S.F. Floor	5.43	5.43	
3020	Heat Generating Systems	N/A		–	–	–	
3030	Cooling Generating Systems	Chilled water, air cooled condenser system		S.F. Floor	6.43	6.43	13.4%
3050	Terminal & Package Units	N/A		–	–	–	
3090	Other HVAC Sys. & Equipment	N/A		–	–	–	
	D40 Fire Protection						
4010	Sprinklers	Wet pipe sprinkler system		S.F. Floor	2.09	2.09	2.4%
4020	Standpipes	N/A		–	–	–	
	D50 Electrical						
5010	Electrical Service/Distribution	600 ampere service, panel board and feeders		S.F. Floor	1.88	1.88	
5020	Lighting & Branch Wiring	Incandescent fixtures, receptacles, switches, A.C. and misc. power		S.F. Floor	5.78	5.78	9.5%
5030	Communications & Security	Alarm systems and emergency lighting		S.F. Floor	.59	.59	
5090	Other Electrical Systems	Emergency Generator, 11.5KW		S.F. Floor	.16	.16	
E. EQUIPMENT & FURNISHINGS							
1010	Commercial Equipment	N/A		–	–	–	
1020	Institutional Equipment	N/A		–	–	–	
1030	Vehicular Equipment	N/A		–	–	–	0.0%
1090	Other Equipment	N/A		–	–	–	
F. SPECIAL CONSTRUCTION					**10**		
1020	Integrated Construction	N/A		–	–	–	
1040	Special Facilities	N/A		–	–	–	0.0%
G. BUILDING SITEWORK	**N/A**						
					Sub-Total	88.42	100%
	CONTRACTOR FEES (General Requirements: 10%, Overhead: 5%, Profit: 10%) — **11**		**12**			25%	22.14
	ARCHITECT FEES					8%	8.84
					Total Building Cost	**119.40**	

1 Building Description
(Model costs are calculated for a three-story apartment building with 10' story height and 22,500 square feet of floor area)

The model highlighted is described in terms of building type, number of stories, typical story height and square footage.

2 Type of Building
(Apartment, 1-3 Story)

3 Division C Interiors
(C1020 Interior Doors)

System costs are presented in divisions according to the 7-division UNIFORMAT II classifications. Each of the component systems are listed.

4 Specification Highlights
(15% solid core wood; 85% hollow core wood)

All systems in each subdivision are described with the material and proportions used.

5 Quality Criteria
(80 S.F. Floor/Door)

The criteria used in determining quantities for the calculations are shown.

6 Unit (Each)

The unit of measure shown in this column is the unit of measure of the particular system shown that corresponds to the unit cost.

7 Unit Cost ($429)

The cost per unit of measure of each system subdivision.

8 Cost per Square Foot ($5.37)

The cost per square foot for each system is the unit cost of the system times the total number of units divided by the total square feet of building area.

9 % of Sub-Total (26.2%)

The percent of sub-total is the total cost per square foot of all systems in the division divided by the sub-total cost per square foot of the building.

10 Sub-Total ($88.42)

The sub-total is the total of all the system costs per square foot.

11 Project Fees
(Contractor Fees) (25%)
(Architect Fees) (8%)

Contractor Fees to cover the general requirements, overhead and profit of the General Contractor are added as a percentage of the sub-total. Architect Fees, also as a percentage of the sub-total, are also added. These values vary with the building type.

12 Total Building Cost ($119.40)

The total building cost per square foot of building area is the sum of the square foot costs of all the systems plus the General Contractor's general requirements, overhead and profit, and the Architect fees. The total building cost is the amount which appears shaded in the Cost per Square Foot of Floor Area table shown previously.

Examples

Example 1

This example illustrates the use of the base cost tables. The base cost is adjusted for different exterior wall systems, different story height and a partial basement.

CII APPRAISAL

1. SUBJECT PROPERTY:	**Westernberg Holdings, Inc.**	5. DATE:	**Jan. 3, 2005**
2. BUILDING:	**# 623**	6. APPRAISER:	**SJS**
3. ADDRESS:	**55 Pinewood Rd., Bondsville, N.J. 07410**		
4. BUILDING USE:	**Apartments**	7. YEAR BUILT:	**1985**

N
170 '

W
64.7'

E

S

North wall - 4" stone Ashlar w/8" block back up (split Face)
East, West & South walls - 12" Decorative concrete block

8. EXTERIOR WALL CONSTRUCTION:			
9. FRAME:	**Steel**		
10. GROUND FLOOR AREA:	**11,000**	11. GROSS FLOOR AREA (EXCL. BASEMENT):	**55,000** S.F.
12. NUMBER OF STORIES:	**5**	13. STORY HEIGHT:	**11 Ft.**
14. PERIMETER:	**469.4**	15. BASEMENT AREA:	None S.F.
16. GENERAL COMMENTS:			

Example 1 (continued)

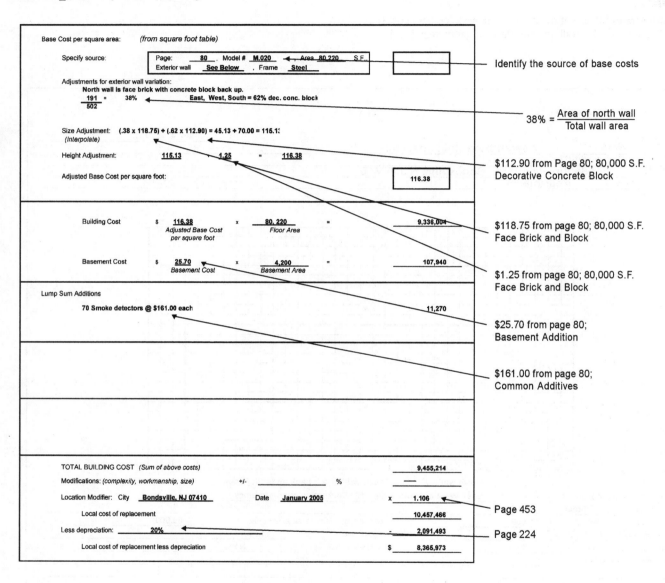

Base Cost per square area: *(from square foot table)*

Specify source:

Page:	**80** , Model # **M.020**	Area **80,220** S.F.
Exterior wall	**See Below** , Frame **Steel**	

→ Identify the source of base costs

Adjustments for exterior wall variation:
North wall is face brick with concrete block back up.

$$\frac{191}{502} = 38\%$$ East, West, South = 62% dec. conc. block

$$38\% = \frac{\text{Area of north wall}}{\text{Total wall area}}$$

Size Adjustment: (.38 x 118.75) + (.62 x 112.90) = 45.13 + 70.00 = 115.1:
(Interpolate)

Height Adjustment: **115.13**   **1.25**   =   **116.38**

$112.90 from Page 80; 80,000 S.F. Decorative Concrete Block

Adjusted Base Cost per square foot: | 116.38 |

Building Cost $ **116.38** x **80,220** = **9,336,004**
  Adjusted Base Cost   Floor Area
  per square foot

$118.75 from page 80; 80,000 S.F. Face Brick and Block

Basement Cost $ **25.70** x **4,200** = **107,940**
  Basement Cost   Basement Area

$1.25 from page 80; 80,000 S.F. Face Brick and Block

Lump Sum Additions

  70 Smoke detectors @ $161.00 each   **11,270**

$25.70 from page 80; Basement Addition

$161.00 from page 80; Common Additives

TOTAL BUILDING COST *(Sum of above costs)*		**9,455,214**
Modifications: *(complexity, workmanship, size)* +/- _____ %		—
Location Modifier: City **Bondsville, NJ 07410** Date **January 2005** x **1.106**		
Local cost of replacement		**10,457,466**
Less depreciation: **20%**		- **2,091,493**
Local cost of replacement less depreciation	$	**8,365,973**

Page 453

Page 224

Example 2

This example shows how to modify a model building.

Model #M.020, 4–7 Story Apartment, page 80, matches the example quite closely. The model specifies a 6-story building with a 10′ story height.

The following adjustments must be made:
- add stone ashlar wall
- adjust model to a 5-story building
- change partitions
- change floor finish

CII APPRAISAL

			5. DATE:	**Jan. 3, 2005**
1. SUBJECT PROPERTY:	**Westernberg Holdings, Inc.**		6. APPRAISER:	**SJS**
2. BUILDING:	**# 623**			
3. ADDRESS:	**55 Pinewood Rd., Bondsville, N.J. 07410**			
4. BUILDING USE:	**Apartments**		7. YEAR BUILT:	**1985**

```
              N
             170'

W
64.7'                                    E


              S
```

Two exterior wall types

North wall - 4" stone Ashlar w/8" block back up (split Face)
East, West & South walls - 12" Decorative concrete block

8. EXTERIOR WALL CONSTRUCTION:

9. FRAME:	**Steel**			
10. GROUND FLOOR AREA:	**11,000**	11. GROSS FLOOR AREA (EXCL. BASEMENT):	**55,000**	S.F.
12. NUMBER OF STORIES:	**5**	13. STORY HEIGHT:	**11 Ft.**	
14. PERIMETER:	**469.4**	15. BASEMENT AREA:	None	S.F.

16. GENERAL COMMENTS:

Example 2 (continued)

Square foot area (excluding basement) from item 11 ___55,000___ S.F.

Perimeter from item 14 ___469.4___ Lin. Ft.

Item 17-Model square foot costs (from model sub-total) ___$92.01___

From page 81, Model #M.020 Sub-total

$.22/S.F. ground area
5 stories

$.22/S.F. ground area
6 stories

Area of Elevated Floors
$18.83/S.F. x 4 x 11,000 S.F.
55,000

Building Floor Area
6.84/S.F. roof
5 stories

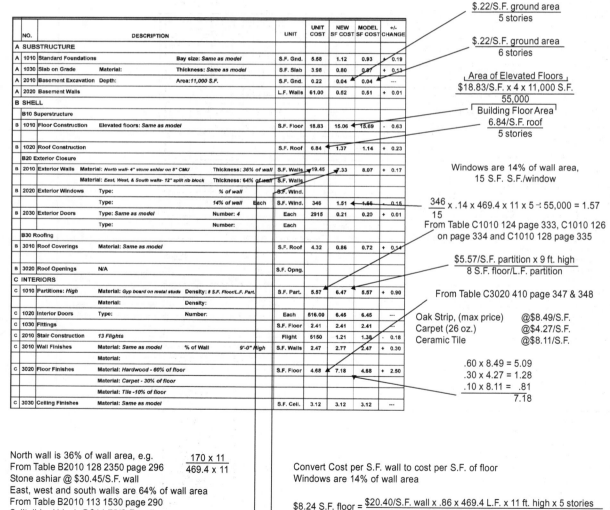

NO.	DESCRIPTION		UNIT	UNIT COST	NEW SF COST	MODEL SF COST	+/- CHANGE
A SUBSTRUCTURE							
A 1010 Standard Foundations	Bay size: Same as model		S.F. Gnd.	5.58	1.12	0.93	+ 0.19
A 1030 Slab on Grade	Material:	Thickness: Same as model	S.F. Slab	3.98	0.80	0.67	+ 0.13
A 2010 Basement Excavation Depth:	Area:11,000 S.F.		S.F. Gnd.	0.22	0.04	0.04	---
A 2020 Basement Walls			L.F. Walls	61.00	0.52	0.51	+ 0.01
B SHELL							
B10 Superstructure							
B 1010 Floor Construction	Elevated floors: Same as model		S.F. Floor	18.83	15.06	15.69	- 0.63
B 1020 Roof Construction			S.F. Roof	6.84	1.37	1.14	+ 0.23
B20 Exterior Closure							
B 2010 Exterior Walls	Material: North wall- 4" stone ashlar on 8" CMU	Thickness: 36% of wall	S.F. Walls	19.45	7.33	8.07	+ 0.17
	Material: East, West, & South walls- 12" split rib block	Thickness: 64% of wall	S.F. Walls				
B 2020 Exterior Windows	Type:	% of wall	S.F. Wind.				
	Type:	14% of wall Each	S.F. Wind.	346	1.51	1.66	- 0.15
B 2030 Exterior Doors	Type: Same as model	Number: 4	Each	2915	0.21	0.20	+ 0.01
	Type:	Number:	Each				
B30 Roofing							
B 3010 Roof Coverings	Material: Same as model		S.F. Roof	4.32	0.86	0.72	+ 0.14
B 3020 Roof Openings	N/A		S.F. Opng.				
C INTERIORS							
C 1010 Partitions: High	Material: Gyp board on metal studs Density: 8 S.F. Floor/L.F. Part.		S.F. Part.	5.57	6.47	5.57	+ 0.90
	Material:	Density:					
C 1020 Interior Doors	Type:	Number:	Each	516.00	6.45	6.45	---
C 1030 Fittings			S.F. Floor	2.41	2.41	2.41	
C 2010 Stair Construction	13 Flights		Flight	5150	1.21	1.39	- 0.18
C 3010 Wall Finishes	Material: Same as model	% of Wall	9'-0" High	S.F. Walls	2.47	2.77	2.47	+ 0.30
	Material:						
C 3020 Floor Finishes	Material: Hardwood - 60% of floor		S.F. Floor	4.68	7.18	4.68	+ 2.50
	Material: Carpet - 30% of floor						
	Material: Tile -10% of floor						
C 3030 Ceiling Finishes	Material: Same as model		S.F. Ceil.	3.12	3.12	3.12	---

Windows are 14% of wall area,
15 S.F. S.F./window

$$\frac{346}{15} \times .14 \times 469.4 \times 11 \times 5 \div 55,000 = 1.57$$

From Table C1010 124 page 333, C1010 126
on page 334 and C1010 128 page 335

$$\frac{\$5.57/S.F. \text{ partition} \times 9 \text{ ft. high}}{8 \text{ S.F. floor/L.F. partition}}$$

From Table C3020 410 page 347 & 348

Oak Strip, (max price) @$8.49/S.F.
Carpet (26 oz.) @$4.27/S.F.
Ceramic Tile @$8.11/S.F.

.60 x 8.49 = 5.09
.30 x 4.27 = 1.28
.10 x 8.11 = .81
7.18

North wall is 36% of wall area, e.g.
From Table B2010 128 2350 page 296
Stone ashlar @ $30.45/S.F. wall
East, west and south walls are 64% of wall area
From Table B2010 113 1530 page 290
Split ribbed block @$14.75/S.F. wall

$$\frac{170 \times 11}{469.4 \times 11}$$

New wall cost
(.36 x $30.45/S.F.) + (.64 x $14.75/S.F.) = $20.45/S.F. wall

Convert Cost per S.F. wall to cost per S.F. of floor
Windows are 14% of wall area

$$\$8.24 \text{ S.F. floor} = \frac{\$20.40/S.F. \text{ wall} \times .86 \times 469.4 \text{ L.F.} \times 11 \text{ ft. high} \times 5 \text{ stories}}{55,000 \text{ S.F. of floor}}$$

Example 2 (Continued)

	NO.	SYSTEM/COMPONENT	DESCRIPTION	UNIT	UNIT COST	NEW S.F. COST	MODEL S.F. COST	+/- CHANGE
D SERVICES								
		D10 Conveying						
D	1010	Elevators & Lifts 2	Type: *Same as model* Capacity: Stops:	Each	147,900	5.38	4.93	+ 0.45
D	1020	Escalators & Moving Walks	Type:	Each				
		D20 Plumbing						
D	2010	Plumbing *Kitchen, bath, + service fixtures*	*1 Fixture/215 S.F. Floor*	Each	1642	7.64	7.64	---
D	2020	Domestic Water Distribution		S.F. Floor	2.34	2.34	2.34	---
D	2040	Rain Water Drainage		S.F. Roof	0.9	0.18	0.15	+ 0.03
		D30 HVAC						
D	3010	Energy Supply		S.F. Floor	4.72	4.72	4.72	---
D	3020	Heat Generating Systems	Type: *Same as model*	S.F. Floor				
D	3030	Cooling Generating Systems	Type: *Same as model*	S.F. Floor	6.52	6.52	6.52	---
D	3090	Other HVAC Sys. & Equipment		Each				
		D40 Fire Protection						
D	4010	Sprinklers	*Same as model*	S.F. Floor	1.73	1.73	1.73	---
D	4020	Standpipes	*Same as model*	S.F. Floor	0.33	0.33	0.33	---
		D50 Electrical						
D	5010	Electrical Service/Distribution	*Same as model*	S.F. Floor	1.78	1.78	1.78	---
D	5020	Lighting & Branch Wiring		S.F. Floor	5.76	5.76	5.76	---
D	5030	Communications & Security		S.F. Floor	0.26	0.26	0.26	---
D	5090	Other Electrical Systems		S.F. Floor	0.14	0.14	0.14	---
E EQUIPMENT & FURNISHINGS								
E	1010	Commercial Equipment		Each				
E	1020	Institutional Equipment		Each				
E	1030	Vehicular Equipment		Each				
E	1090	Other Equipment		Each				
F SPECIAL CONSTRUCTION								
F	1020	Integrated Construction	N/A	S.F.				
F	1040	Special Facilities	N/A	S.F.				
G BUILDING SITEWORK								

						Total Change	+ 4.10
ITEM							
17	Model 020 total	$ 92.01					
18	Adjusted S.F. cost $ 96.11	item 17 +/- changes					
19	Building area - from item 11 55,000	S.F. x adjusted S.F. cost			$	5,286,050	
20	Basement area - from item 15	S.F. x S.F. cost $			$		
21	Base building sub-total - item 20 + item 19				$	5,286,050	
22	Miscellaneous addition (quality, etc.)				$		
23	Sub-total - item 22 + 21				$	5,286,050	
24	General conditions -25 % of item 23				$	1,321,513	
25	Sub-total - item 24 + item 23				$	6,607,563	
26	Architects fees 5.5 % of item 25				$	363,416	
27	Sub-total - item 26 + item 27				$	6,970,979	
28	Location modifier zip code 07410				x	1.106	From page 453
29	Local replacement cost - item 28 x item 27				$	7,709,903	
30	Depreciation 20 % of item 29				$	1,541,981	
31	Depreciated local replacement cost - item 29 less item 30				$	7,167,922	
32	Exclusions				$		
33	Net depreciated replacement cost - item 31 less item 32				$	7,167,922	Use table on page 224

From page 453
Use table on page 224

EXAMPLES

77

BUILDING TYPES

Costs per square foot of floor area

Exterior Wall	S.F. Area	8000	12000	15000	19000	22500	25000	29000	32000	36000
	L.F. Perimeter	213	280	330	350	400	433	442	480	520
Face Brick with Concrete Block Back-up	Wood Joists	150.05	134.20	127.85	119.35	116.20	114.40	110.20	109.00	107.30
	Steel Joists	151.10	136.00	129.95	122.30	119.40	117.65	113.95	112.80	111.25
Stucco on Concrete Block	Wood Joists	133.10	118.80	113.10	106.30	103.45	101.85	98.60	97.50	96.10
	Steel Joists	142.25	127.90	122.10	115.20	112.35	110.75	107.50	106.35	104.95
Wood Siding	Wood Frame	132.75	118.50	112.75	106.05	103.20	101.60	98.45	97.30	95.90
Brick Veneer	Wood Frame	139.60	124.50	118.45	110.70	107.70	106.00	102.25	101.05	99.55
Perimeter Adj., Add or Deduct	Per 100 L.F.	15.90	10.65	8.55	6.70	5.70	5.10	4.40	4.00	3.55
Story Hgt. Adj., Add or Deduct	Per 1 Ft.	3.00	2.60	2.50	2.05	2.00	1.95	1.70	1.65	1.65

For Basement, add $25.60 per square foot of basement area

The above costs were calculated using the basic specifications shown on the facing page. These costs should be adjusted where necessary for design alternatives and owner's requirements. Reported completed project costs, for this type of structure, range from $42.70 to $158.80 per S.F.

Common additives

Description	Unit	$ Cost
Appliances		
Cooking range, 30" free standing		
1 oven	Each	330 - 1800
2 oven	Each	1700 - 1775
30" built-in		
1 oven	Each	570 - 1875
2 oven	Each	1250 - 2050
Counter top cook tops, 4 burner	Each	285 - 730
Microwave oven	Each	212 - 675
Combination range, refrig. & sink, 30" wide	Each	1300 - 2575
72" wide	Each	3825
Combination range, refrigerator, sink,		
microwave oven & icemaker	Each	5650
Compactor, residential, 4-1 compaction	Each	540 - 655
Dishwasher, built-in, 2 cycles	Each	515 - 810
4 cycles	Each	530 - 1075
Garbage disposer, sink type	Each	144 - 258
Hood for range, 2 speed, vented, 30" wide	Each	224 - 960
42" wide	Each	430 - 1150
Refrigerator, no frost 10-12 C.F.	Each	550 - 860
18-20 C.F.	Each	645 - 1100

Description	Unit	$ Cost
Closed Circuit Surveillance, One station		
Camera and monitor	Each	1525
For additional camera stations, add	Each	820
Elevators, Hydraulic passenger, 2 stops		
2000# capacity	Each	49,000
2500# capacity	Each	50,000
3500# capacity	Each	53,700
Additional stop, add	Each	4550
Emergency Lighting, 25 watt, battery operated		
Lead battery	Each	247
Nickel cadmium	Each	695
Laundry Equipment		
Dryer, gas, 16 lb. capacity	Each	785
30 lb. capacity	Each	3000
Washer, 4 cycle	Each	875
Commercial	Each	1325
Smoke Detectors		
Ceiling type	Each	161
Duct type	Each	425

Important: See the Reference Section for Location Factors

Model costs calculated for a 3 story building with 10' story height and 22,500 square feet of floor area

				Unit	Unit Cost	Cost Per S.F.	% Of Sub-Total
A.	**SUBSTRUCTURE**						
1010	Standard Foundations	Poured concrete; strip and spread footings		S.F. Ground	4.89	1.63	
1030	Slab on Grade	4" reinforced concrete with vapor barrier and granular base		S.F. Slab	3.98	1.32	
2010	Basement Excavation	Site preparation for slab and trench for foundation wall and footing		S.F. Ground	.14	.05	4.7%
2020	Basement Walls	4' foundation wall		L.F. Wall	115	1.19	
B.	**SHELL**						
	B10 Superstructure						
1010	Floor Construction	Open web steel joists, slab form, concrete, interior steel columns		S.F. Floor	15.29	10.19	14.3%
1020	Roof Construction	Open web steel joists with rib metal deck, interior steel columns		S.F. Roof	7.32	2.44	
	B20 Exterior Enclosure						
2010	Exterior Walls	Face brick with concrete block backup	88% of wall	S.F. Wall	18.15	8.52	
2020	Exterior Windows	Aluminum horizontal sliding	12% of wall	Each	346	1.48	11.6%
2030	Exterior Doors	Aluminum and glass		Each	1279	.23	
	B30 Roofing						
3010	Roof Coverings	Built-up tar and gravel with flashing; perlite/EPS composite insulation		S.F. Roof	4.41	1.47	1.7%
3020	Roof Openings	N/A		—	—	—	
C.	**INTERIORS**						
1010	Partitions	Gypsum board and sound deadening board on metal studs	10 S.F. of Floor/L.F. Partition	S.F. Partition	5.28	4.69	
1020	Interior Doors	15% solid core wood, 85% hollow core wood	80 S.F. Floor/Door	Each	429	5.37	
1030	Fittings	Kitchen Cabinets		S.F. Floor	2.01	2.01	
2010	Stair Construction	Concrete filled metal pan		Flight	6425	1.43	26.2%
3010	Wall Finishes	70% paint, 25% vinyl wall covering, 5% ceramic tile		S.F. Surface	2.00	2.00	
3020	Floor Finishes	60% carpet, 30% vinyl composition tile, 10% ceramic tile		S.F. Floor	4.68	4.68	
3030	Ceiling Finishes	Painted gypsum board on resilient channels		S.F. Ceiling	3.00	2.99	
D.	**SERVICES**						
	D10 Conveying						
1010	Elevators & Lifts	One hydraulic passenger elevator		Each	74,250	3.30	3.7%
1020	Escalators & Moving Walks	N/A		—	—	—	
	D20 Plumbing						
2010	Plumbing Fixtures	Kitchen, bathroom and service fixtures, supply and drainage	1 Fixture/200 S.F. Floor	Each	1674	8.37	
2020	Domestic Water Distribution	Gas fired water heater		S.F. Floor	2.43	2.43	12.5%
2040	Rain Water Drainage	Roof drains		S.F. Roof	.81	.27	
	D30 HVAC						
3010	Energy Supply	Oil fired hot water, baseboard radiation		S.F. Floor	5.43	5.43	
3020	Heat Generating Systems	N/A		—	—	—	
3030	Cooling Generating Systems	Chilled water, air cooled condenser system		S.F. Floor	6.43	6.43	13.4%
3050	Terminal & Package Units	N/A		—	—	—	
3090	Other HVAC Sys. & Equipment	N/A		—	—	—	
	D40 Fire Protection						
4010	Sprinklers	Wet pipe sprinkler system		S.F. Floor	2.09	2.09	2.4%
4020	Standpipes	N/A		—	—	—	
	D50 Electrical						
5010	Electrical Service/Distribution	600 ampere service, panel board and feeders		S.F. Floor	1.88	1.88	
5020	Lighting & Branch Wiring	Incandescent fixtures, receptacles, switches, A.C. and misc. power		S.F. Floor	5.78	5.78	9.5%
5030	Communications & Security	Alarm systems and emergency lighting		S.F. Floor	.59	.59	
5090	Other Electrical Systems	Emergency Generator, 11.5KW		S.F. Floor	.16	.16	
E.	**EQUIPMENT & FURNISHINGS**						
1010	Commercial Equipment	N/A		—	—	—	
1020	Institutional Equipment	N/A		—	—	—	
1030	Vehicular Equipment	N/A		—	—	—	0.0%
1090	Other Equipment	N/A		—	—	—	
F.	**SPECIAL CONSTRUCTION**						
1020	Integrated Construction	N/A		—	—	—	
1040	Special Facilities	N/A		—	—	—	0.0%
G.	**BUILDING SITEWORK**	**N/A**					

	Sub-Total	88.42	100%
CONTRACTOR FEES (General Requirements: 10%, Overhead: 5%, Profit: 10%)	25%	22.14	
ARCHITECT FEES	8%	8.84	

Total Building Cost	**119.40**

BUILDING TYPES

BUILDING TYPES

Costs per square foot of floor area

Exterior Wall	S.F. Area	40000	45000	50000	55000	60000	70000	80000	90000	100000
	L.F. Perimeter	366	400	433	466	500	510	530	550	594
Face Brick with Concrete Block Back-up	Steel Frame	126.35	125.25	124.40	123.70	123.05	120.45	118.75	117.45	116.90
	R/Conc. Frame	131.80	130.60	129.65	128.90	128.25	125.25	123.30	121.80	121.15
Decorative Concrete Block	Steel Frame	119.25	118.30	117.50	116.90	116.40	114.20	112.90	111.80	111.30
	R/Conc. Frame	121.10	120.10	119.35	118.70	118.25	116.00	114.60	113.55	113.00
Precast Concrete Panels	Steel Frame	123.85	122.70	121.90	121.10	120.55	117.85	116.20	114.90	114.25
	R/Conc. Frame	125.45	124.35	123.50	122.80	122.15	119.45	117.75	116.45	115.85
Perimeter Adj., Add or Deduct	Per 100 L.F.	5.15	4.60	4.10	3.70	3.50	3.00	2.60	2.35	2.05
Story Hgt. Adj., Add or Deduct	Per 1 Ft.	1.70	1.60	1.60	1.50	1.60	1.35	1.25	1.10	1.15

For Basement, add $25.70 per square foot of basement area

The above costs were calculated using the basic specifications shown on the facing page. These costs should be adjusted where necessary for design alternatives and owner's requirements. Reported completed project costs, for this type of structure, range from $51.15 to $148.40 per S.F.

Common additives

Description	Unit	$ Cost
Appliances		
Cooking range, 30" free standing		
1 oven	Each	330 - 1800
2 oven	Each	1700 - 1775
30" built-in		
1 oven	Each	570 - 1875
2 oven	Each	1250 - 2050
Counter top cook tops, 4 burner	Each	285 - 730
Microwave oven	Each	212 - 675
Combination range, refrig. & sink, 30" wide	Each	1300 - 2575
72" wide	Each	3825
Combination range, refrigerator, sink,		
microwave oven & icemaker	Each	5650
Compactor, residential, 4-1 compaction	Each	540 - 655
Dishwasher, built-in, 2 cycles	Each	515 - 810
4 cycles	Each	530 - 1075
Garbage disposer, sink type	Each	144 - 258
Hood for range, 2 speed, vented, 30" wide	Each	224 - 960
42" wide	Each	430 - 1150
Refrigerator, no frost 10-12 C.F.	Each	550 - 860
18-20 C.F.	Each	645 - 1100

Description	Unit	$ Cost
Closed Circuit Surveillance, One station		
Camera and monitor	Each	1525
For additional camera stations, add	Each	820
Elevators, Electric passenger, 5 stops		
2000# capacity	Each	110,300
3500# capacity	Each	116,800
5000# capacity	Each	121,800
Additional stop, add	Each	6550
Emergency Lighting, 25 watt, battery operated		
Lead battery	Each	247
Nickel cadmium	Each	695
Laundry Equipment		
Dryer, gas, 16 lb. capacity	Each	785
30 lb. capacity	Each	3000
Washer, 4 cycle	Each	875
Commercial	Each	1325
Smoke Detectors		
Ceiling type	Each	161
Duct type	Each	425

Model costs calculated for a 6 story building with 10'-4" story height and 60,000 square feet of floor area

				Unit	Unit Cost	Cost Per S.F.	% Of Sub-Total
A. SUBSTRUCTURE							
1010	Standard Foundations	Poured concrete; strip and spread footings and 4' foundation wall		S.F. Ground	5.58	.93	
1030	Slab on Grade	4" reinforced concrete with vapor barrier and granular base		S.F. Slab	3.98	.67	2.3%
2010	Basement Excavation	Site preparation for slab and trench for foundation wall and footing		S.F. Ground	.22	.04	
2020	Basement Walls	4' foundation wall		L.F. Wall	61	.51	
B. SHELL							
	B10 Superstructure						
1010	Floor Construction	Open web steel joists, slab form, concrete, steel columns		S.F. Floor	18.83	15.69	18.3%
1020	Roof Construction	Open web steel joists with rib metal deck, steel columns		S.F. Roof	6.84	1.14	
	B20 Exterior Enclosure						
2010	Exterior Walls	Face brick with concrete block backup	86% of wall	S.F. Wall	18.16	8.07	
2020	Exterior Windows	Aluminum horizontal sliding	14% of wall	Each	346	1.66	10.8%
2030	Exterior Doors	Aluminum and glass		Each	2915	.20	
	B30 Roofing						
3010	Roof Coverings	Built-up tar and gravel with flashing; perlite/EPS composite insulation		S.F. Roof	4.32	.72	0.8%
3020	Roof Openings	N/A		—	—	—	
C. INTERIORS							
1010	Partitions	Gypsum board and sound deadening board on metal studs	8 S.F. Floor/L.F. Partitions	S.F. Partition	5.57	5.57	
1020	Interior Doors	15% solid core wood, 85% hollow core wood	80 S.F Floor/Door	Each	516	6.45	
1030	Fittings	Kitchen cabinets		S.F. Floor	2.41	2.41	
2010	Stair Construction	Concrete filled metal pan		Flight	5150	1.38	28.3%
3010	Wall Finishes	70% paint, 25% vinyl wall covering, 5% ceramic tile		S.F. Surface	2.47	2.47	
3020	Floor Finishes	60% carpet, 30% vinyl composition tile, 10% ceramic tile		S.F. Floor	4.68	4.68	
3030	Ceiling Finishes	Painted gypsum board on resilient channels		S.F. Ceiling	3.12	3.12	
D. SERVICES							
	D10 Conveying						
1010	Elevators & Lifts	Two geared passenger elevators		Each	147,900	4.93	5.4%
1020	Escalators & Moving Walks	N/A		—	—	—	
	D20 Plumbing						
2010	Plumbing Fixtures	Kitchen, bathroom and service fixtures, supply and drainage	1 Fixture/215 S.F. Floor	Each	1642	7.64	
2020	Domestic Water Distribution	Gas fired water heater		S.F. Floor	2.34	2.34	11.0%
2040	Rain Water Drainage	Roof drains		S.F. Roof	.90	.15	
	D30 HVAC						
3010	Energy Supply	Oil fired hot water, baseboard radiation		S.F. Floor	4.72	4.72	
3020	Heat Generating Systems	N/A		—	—	—	
3030	Cooling Generating Systems	Chilled water, air cooled condenser system		S.F. Floor	6.52	6.52	12.2%
3050	Terminal & Package Units	N/A		—	—	—	
3090	Other HVAC Sys. & Equipment	N/A		—	—	—	
	D40 Fire Protection						
4010	Sprinklers	Wet pipe sprinkler system		S.F. Floor	1.73	1.73	2.2%
4020	Standpipes	Standpipe		S.F. Floor	.33	.33	
	D50 Electrical						
5010	Electrical Service/Distribution	1600 ampere service, panel board and feeders		S.F. Floor	1.78	1.78	
5020	Lighting & Branch Wiring	Incandescent fixtures, receptacles, switches, A.C. and misc. power		S.F. Floor	5.76	5.76	8.6%
5030	Communications & Security	Alarm systems, emergency lighting, and intercom		S.F. Floor	.26	.26	
5090	Other Electrical Systems	Emergency generator, 11.5 kW		S.F. Floor	.14	.14	
E. EQUIPMENT & FURNISHINGS							
1010	Commercial Equipment	N/A		—	—	—	
1020	Institutional Equipment	N/A		—	—	—	0.0%
1030	Vehicular Equipment	N/A		—	—	—	
1090	Other Equipment	N/A		—	—	—	
F. SPECIAL CONSTRUCTION							
1020	Integrated Construction	N/A		—	—	—	0.0%
1040	Special Facilities	N/A		—	—	—	
G. BUILDING SITEWORK	**N/A**						

	Sub-Total	92.01	**100%**
CONTRACTOR FEES (General Requirements: 10%, Overhead: 5%, Profit: 10%)	25%	22.99	
ARCHITECT FEES	7%	8.05	
	Total Building Cost	**123.05**	

Costs per square foot of floor area

Exterior Wall	S.F. Area	95000	112000	129000	145000	170000	200000	275000	400000	600000
	L.F. Perimeter	345	386	406	442	480	510	530	570	630
Ribbed Precast Concrete Panel	Steel Frame	148.10	145.20	142.00	140.45	137.85	135.05	129.45	124.90	121.50
	R/Conc. Frame	138.50	135.70	132.65	131.20	128.80	126.10	120.95	116.65	113.45
Face Brick with Concrete Block Back-up	Steel Frame	134.15	131.45	128.50	127.10	124.70	122.20	117.25	113.20	110.15
	R/Conc. Frame	136.80	134.10	131.15	129.80	127.45	124.90	120.00	116.00	112.95
Stucco on Concrete Block	Steel Frame	126.40	124.15	121.80	120.65	118.75	116.80	113.15	110.10	107.90
	R/Conc. Frame	129.05	126.75	124.50	123.30	121.45	119.50	115.95	112.95	110.75
Perimeter Adj., Add or Deduct	Per 100 L.F.	7.05	6.05	5.15	4.65	3.95	3.40	2.45	1.70	1.10
Story Hgt. Adj., Add or Deduct	Per 1 Ft.	2.30	2.20	1.95	1.95	1.75	1.65	1.20	.90	.65

For Basement, add *$26.25 per square foot of basement area*

The above costs were calculated using the basic specifications shown on the facing page. These costs should be adjusted where necessary for design alternatives and owner's requirements. Reported completed project costs, for this type of structure, range from $65.45 to $150.80 per S.F.

Common additives

Description	Unit	$ Cost
Appliances		
Cooking range, 30" free standing		
1 oven	Each	330 - 1800
2 oven	Each	1700 - 1775
30" built-in		
1 oven	Each	570 - 1875
2 oven	Each	1250 - 2050
Counter top cook tops, 4 burner	Each	285 - 730
Microwave oven	Each	212 - 675
Combination range, refrig. & sink, 30" wide	Each	1300 - 2575
72" wide	Each	3825
Combination range, refrigerator, sink,		
microwave oven & icemaker	Each	5650
Compactor, residential, 4-1 compaction	Each	540 - 655
Dishwasher, built-in, 2 cycles	Each	515 - 810
4 cycles	Each	530 - 1075
Garbage disposer, sink type	Each	144 - 258
Hood for range, 2 speed, vented, 30" wide	Each	224 - 960
42" wide	Each	430 - 1150
Refrigerator, no frost 10-12 C.F.	Each	550 - 860
18-20 C.F.	Each	645 - 1100

Description	Unit	$ Cost
Closed Circuit Surveillance, One station		
Camera and monitor	Each	1525
For additional camera stations, add	Each	820
Elevators, Electric passenger, 10 stops		
3000# capacity	Each	239,500
4000# capacity	Each	242,000
5000# capacity	Each	246,500
Additional stop, add	Each	6550
Emergency Lighting, 25 watt, battery operated		
Lead battery	Each	247
Nickel cadmium	Each	695
Laundry Equipment		
Dryer, gas, 16 lb. capacity	Each	785
30 lb. capacity	Each	3000
Washer, 4 cycle	Each	875
Commercial	Each	1325
Smoke Detectors		
Ceiling type	Each	161
Duct type	Each	425

Important: See the Reference Section for Location Factors

BUILDING TYPES

Model costs calculated for a 15 story building with 10'-6" story height and 145,000 square feet of floor area

				Unit	Unit Cost	Cost Per S.F.	% Of Sub-Total
A. SUBSTRUCTURE							
1010	Standard Foundations	Poured concrete; strip and spread footings		S.F. Ground	12.45	.83	
1030	Slab on Grade	4" reinforced concrete with vapor barrier and granular base		S.F. Slab	3.98	.26	
2010	Basement Excavation	Site preparation for slab and trench for foundation wall and footing		S.F. Ground	.22	.01	1.2%
2020	Basement Walls	4' Foundation wall		L.F. Wall	115	.22	
B. SHELL							
	B10 Superstructure						
1010	Floor Construction	Open web steel joists, slab form, concrete, interior steel columns		S.F. Floor	16.79	15.67	15.1%
1020	Roof Construction	Open web steel joists with rib metal deck, interior steel columns		S.F. Roof	5.10	.34	
	B20 Exterior Enclosure						
2010	Exterior Walls	Ribbed precast concrete panel	87% of wall	S.F. Wall	26	11.11	
2020	Exterior Windows	Aluminum horizontal sliding	13% of wall	Each	346	1.44	13.7%
2030	Exterior Doors	Aluminum and glass		Each	2267	1.93	
	B30 Roofing						
3010	Roof Coverings	Built-up tar and gravel with flashing; perlite/EPS composite insulation		S.F. Roof	4.35	.29	0.3%
3020	Roof Openings	N/A		—	—	—	
C. INTERIORS							
1010	Partitions	Gypsum board on concrete block and metal studs	10 S.F. of Floor/L.F. Partition	S.F. Partition	10.70	10.70	
1020	Interior Doors	15% solid core wood, 85% hollow core wood	80 S.F. Floor/Door	Each	525	6.57	
1030	Fittings	Kitchen cabinets		S.F. Floor	2.49	2.49	
2010	Stair Construction	Concrete filled metal pan		Flight	5150	1.53	29.8%
3010	Wall Finishes	70% paint, 25% vinyl wall covering, 5% ceramic tile		S.F. Surface	2.45	2.45	
3020	Floor Finishes	60% carpet, 30% vinyl composition tile, 10% ceramic tile		S.F. Floor	4.68	4.68	
3030	Ceiling Finishes	Painted gypsum board on resilient channels		S.F. Ceiling	3.00	3.12	
D. SERVICES							
	D10 Conveying						
1010	Elevators & Lifts	Four geared passenger elevators		Each	262,812	7.25	6.8%
1020	Escalators & Moving Walks	N/A		—	—	—	
	D20 Plumbing						
2010	Plumbing Fixtures	Kitchen, bathroom and service fixtures, supply and drainage	1 Fixture/210 S.F. Floor	Each	1837	8.75	
2020	Domestic Water Distribution	Gas fired water heater		S.F. Floor	2.69	2.69	10.9%
2040	Rain Water Drainage	Roof drains		S.F Roof	1.80	.12	
	D30 HVAC						
3010	Energy Supply	Oil fired hot water, baseboard radiation		S.F.Floor	4.72	4.72	
3020	Heat Generating Systems	N/A		—	—	—	
3030	Cooling Generating Systems	Chilled water, air cooled condenser system		S.F. Floor	6.52	6.52	10.6%
3050	Terminal & Package Units	N/A		—	—	—	
3090	Other HVAC Sys. & Equipment	N/A		—	—	—	
	D40 Fire Protection						
4010	Sprinklers	Wet pipe sprinkler system		S.F. Floor	1.67	1.67	2.4%
4020	Standpipes	Standpipe		S.F. Floor	.89	.89	
	D50 Electrical						
5010	Electrical Service/Distribution	4000 ampere service, panel board and feeders		S.F. Floor	1.58	1.58	
5020	Lighting & Branch Wiring	Incandescent fixtures, receptacles, switches, A.C. and misc. power		S.F. Floor	6.18	6.18	
5030	Communications & Security	Alarm systems, emergency lighting, antenna, intercom and security television		S.F. Floor	1.82	1.82	9.2%
5090	Other Electrical Systems	Emergency generator, 80KW		S.F. Floor	.16	.16	
E. EQUIPMENT & FURNISHINGS							
1010	Commercial Equipment	N/A		—	—	—	
1020	Institutional Equipment	N/A		—	—	—	
1030	Vehicular Equipment	N/A		—	—	—	0.0%
1090	Other Equipment	N/A		—	—	—	
F. SPECIAL CONSTRUCTION							
1020	Integrated Construction	N/A		—	—	—	
1040	Special Facilities	N/A		—	—	—	0.0%
G. BUILDING SITEWORK	**N/A**						

			Sub-Total	105.99	100%
	CONTRACTOR FEES (General Requirements: 10%, Overhead: 5%, Profit: 10%)		25%	26.51	
	ARCHITECT FEES		6%	7.95	
		Total Building Cost		**140.45**	

Costs per square foot of floor area

Exterior Wall	S.F. Area	2000	3000	4000	5000	7500	10000	15000	25000	50000
	L.F. Perimeter	188	224	260	300	356	400	504	700	1200
Brick Veneer	Wood Frame	188.55	162.95	150.15	142.80	130.40	123.60	117.30	112.05	108.25
	Steel Frame	198.70	173.40	160.70	153.50	141.20	134.45	128.30	123.10	119.30
Brick Veneer on Block	Wood Truss	178.85	157.10	146.20	140.00	128.80	122.65	116.95	112.20	108.75
	Bearing Wall	206.80	180.30	167.00	159.50	146.45	139.30	132.70	127.20	123.15
Wood Siding	Wood Frame	178.15	154.55	142.75	135.95	124.70	118.70	113.00	108.25	104.80
Vinyl Siding	Wood Frame	176.25	153.20	141.65	135.00	124.15	118.40	112.90	108.35	105.10
Perimeter Adj., Add or Deduct	Per 100 L.F.	23.10	15.40	11.55	9.25	6.15	4.65	3.10	1.85	.85
Story Hgt. Adj., Add or Deduct	Per 1 Ft.	2.80	2.30	2.00	1.90	1.40	1.20	1.05	.85	.70

For Basement, add $ 23.05 per square foot of basement area

The above costs were calculated using the basic specifications shown on the facing page. These costs should be adjusted where necessary for design alternatives and owner's requirements. Reported completed project costs, for this type of structure, range from $ 64.30 to $ 161.20 per S.F.

Common additives

Description	Unit	$ Cost
Appliances		
Cooking range, 30" free standing		
1 oven	Each	330 - 1800
2 oven	Each	1700 - 1775
30" built-in		
1 oven	Each	570 - 1875
2 oven	Each	1250 - 2050
Counter top cook tops, 4 burner	Each	285 - 730
Microwave oven	Each	212 - 675
Combination range, refrig. & sink, 30" wide	Each	1300 - 2575
60" wide	Each	3350
72" wide	Each	3825
Combination range, refrigerator, sink,		
microwave oven & icemaker	Each	5650
Compactor, residential, 4-1 compaction	Each	540 - 655
Dishwasher, built-in, 2 cycles	Each	515 - 810
4 cycles	Each	530 - 1075
Garbage disposer, sink type	Each	144 - 258
Hood for range, 2 speed, vented, 30" wide	Each	224 - 960
42" wide	Each	430 - 1150

Description	Unit	$ Cost
Appliances, cont.		
Refrigerator, no frost 10-12 C.F.	Each	550 - 860
14-16 C.F.	Each	560 - 665
18-20 C.F.	Each	645 - 1100
Laundry Equipment		
Dryer, gas, 16 lb. capacity	Each	785
30 lb. capacity	Each	3000
Washer, 4 cycle	Each	875
Commercial	Each	1325
Sound System		
Amplifier, 250 watts	Each	1925
Speaker, ceiling or wall	Each	160
Trumpet	Each	305

BUILDING TYPES

Model costs calculated for a 1 story building with 10' story height and 10,000 square feet of floor area

					Unit	Unit Cost	Cost Per S.F.	% Of Sub-Total
A. SUBSTRUCTURE								
1010	Standard Foundations	Poured concrete strip footings			S.F. Ground	1.53	1.53	
1030	Slab on Grade	4" reinforced concrete with vapor barrier and granular base			S.F. Slab	3.98	3.98	9.4%
2010	Basement Excavation	Site preparation for slab and trench for foundation wall and footing			S.F. Ground	.43	.43	
2020	Basement Walls	4' foundation wall			L.F. Wall	61	2.44	
B. SHELL								
	B10 Superstructure							
1010	Floor Construction	N/A			—	—	—	6.9%
1020	Roof Construction	Plywood on wood trusses (pitched); fiberglass batt insulation			S.F. Roof	6.17	6.17	
	B20 Exterior Enclosure							
2010	Exterior Walls	Brick veneer on wood studs, insulated	80% of wall		S.F. Wall	18.56	5.94	
2020	Exterior Windows	Double hung wood	20% of wall		Each	475	2.38	10.4%
2030	Exterior Doors	Aluminum and glass; 18 ga. steel flush			Each	1630	.98	
	B30 Roofing							
3010	Roof Coverings	Asphalt shingles with flashing (pitched); al. gutters & downspouts			S.F. Roof	1.78	1.78	2.0%
3020	Roof Openings	N/A			—	—	—	
C. INTERIORS								
1010	Partitions	Gypsum board on wood studs	6 S.F. Floor/L.F. Partition		S.F. Partition	4.02	5.36	
1020	Interior Doors	Single leaf wood	80 S.F. Floor/Door		Each	480	5.64	
1030	Fittings	N/A			—	—	—	
2010	Stair Construction	N/A			—	—	—	30.1%
3010	Wall Finishes	Paint on drywall			S.F. Surface	.63	1.68	
3020	Floor Finishes	80% carpet, 20% ceramic tile			S.F. Floor	7.46	7.46	
3030	Ceiling Finishes	Gypsum board on wood furring, painted			S.F. Ceiling	7.00	6.66	
D. SERVICES								
	D10 Conveying							
1010	Elevators & Lifts	N/A			—	—	—	0.0%
1020	Escalators & Moving Walks	N/A			—	—	—	
	D20 Plumbing							
2010	Plumbing Fixtures	Kitchen, toilet and service fixtures, supply and drainage	1 Fixture/200 S.F. Floor		Each	1184	5.92	
2020	Domestic Water Distribution	Included with D2010			S.F. Floor	—	—	6.6%
2040	Rain Water Drainage	N/A			—	—	—	
	D30 HVAC							
3010	Energy Supply	N/A			—	—	—	
3020	Heat Generating Systems	N/A			—	—	—	
3030	Cooling Generating Systems	N/A			—	—	—	3.8%
3050	Terminal & Package Units	Gas fired hot water boiler; terminal package heat/A/C units			S.F. Floor	3.42	3.42	
3090	Other HVAC Sys. & Equipment	N/A			—	—	—	
	D40 Fire Protection							
4010	Sprinklers	Wet pipe sprinkler system			S.F. Floor	2.05	2.05	2.3%
4020	Standpipes	N/A			—	—	—	
	D50 Electrical							
5010	Electrical Service/Distribution	600 ampere service, panel board, feeders and branch wiring			S.F. Floor	7.19	7.19	
5020	Lighting & Branch Wiring	Fluorescent & incandescent fixtures, receptacles, switches, A.C. and misc. power			S.F. Floor	9.08	9.08	27.0%
5030	Communications & Security	Alarm, fire detection, and communication system			S.F. Floor	7.62	7.62	
5090	Other Electrical Systems	Emergency generator, 7.5 kW			S.F. Floor	.15	.15	
E. EQUIPMENT & FURNISHINGS								
1010	Commercial Equipment	Residential washers and dryers			—	.32	.32	
1020	Institutional Equipment	N/A			—	—	—	1.4%
1030	Vehicular Equipment	N/A			—	—	—	
1090	Other Equipment	Residential ranges and refrigerators			—	.90	.90	
F. SPECIAL CONSTRUCTION								
1020	Integrated Construction	N/A			—	—	—	0.0%
1040	Special Facilities	N/A			—	—	—	
G. BUILDING SITEWORK	**N/A**							

						Sub-Total	89.08	100%

CONTRACTOR FEES (General Requirements: 10%, Overhead: 5%, Profit: 10%)	25%	22.27
ARCHITECT FEES	11%	12.25

Total Building Cost | 123.60

BUILDING TYPES

Costs per square foot of floor area

Exterior Wall	S.F. Area	12000	15000	18000	21000	24000	27000	30000	33000	36000
	L.F. Perimeter	440	500	540	590	640	665	700	732	770
Face Brick with Concrete Block Back-up	Steel Frame	142.25	137.80	133.55	131.05	129.20	126.60	125.00	123.50	122.50
	Bearing Wall	137.55	133.45	129.70	127.45	125.75	123.40	121.95	120.65	119.65
Precast Concrete	Steel Frame	136.55	132.60	128.95	126.75	125.10	122.85	121.50	120.15	119.25
Decorative Concrete Block	Bearing Wall	123.85	120.90	118.25	116.65	115.40	113.80	112.80	111.85	111.20
Concrete Block	Steel Frame	130.55	128.25	126.15	124.85	123.90	122.60	121.85	121.10	120.55
	Bearing Wall	120.20	117.60	115.25	113.80	112.75	111.30	110.50	109.65	109.05
Perimeter Adj., Add or Deduct	Per 100 L.F.	9.95	7.90	6.60	5.65	4.95	4.40	3.95	3.60	3.30
Story Hgt. Adj., Add or Deduct	Per 1 Ft.	1.60	1.40	1.30	1.20	1.10	1.05	.95	.90	.85
For Basement, add $20.95 per square foot of basement area										

The above costs were calculated using the basic specifications shown on the facing page. These costs should be adjusted where necessary for design alternatives and owner's requirements. Reported completed project costs, for this type of structure, range from $62.90 to $164.85 per S.F.

Common additives

Description	Unit	$ Cost
Closed Circuit Surveillance, One station		
Camera and monitor	Each	1525
For additional camera stations, add	Each	820
Emergency Lighting, 25 watt, battery operated		
Lead battery	Each	247
Nickel cadmium	Each	695
Seating		
Auditorium chair, all veneer	Each	193
Veneer back, padded seat	Each	233
Upholstered, spring seat	Each	234
Classroom, movable chair & desk	Set	65 - 120
Lecture hall, pedestal type	Each	179 - 545
Smoke Detectors		
Ceiling type	Each	161
Duct type	Each	425
Sound System		
Amplifier, 250 watts	Each	1925
Speaker, ceiling or wall	Each	160
Trumpet	Each	305

BUILDING TYPES

Model costs calculated for a 1 story building with 24' story height and 24,000 square feet of floor area

				Unit	Unit Cost	Cost Per S.F.	% Of Sub-Total
A. SUBSTRUCTURE							
1010	Standard Foundations	Poured concrete; strip and spread footings		S.F. Ground	1.10	1.10	
1030	Slab on Grade	6" reinforced concrete with vapor barrier and granular base		S.F. Slab	4.84	4.84	8.5%
2010	Basement Excavation	Site preparation for slab and trench for foundation wall and footing		S.F. Ground	.14	.14	
2020	Basement Walls	4" foundation wall		L.F. Wall	68	1.83	
B. SHELL							
	B10 Superstructure						
1010	Floor Construction	Open web steel joists, slab form, concrete	(balcony)	S.F. Floor	18.32	2.29	10.6%
1020	Roof Construction	Metal deck on steel truss		S.F. Roof	7.62	7.62	
	B20 Exterior Enclosure						
2010	Exterior Walls	Precast concrete panel	80% of wall (adjusted for end walls)	S.F. Wall	22	11.35	
2020	Exterior Windows	Glass curtain wall	20% of wall	Each	32	4.19	18.1%
2030	Exterior Doors	Double aluminum and glass and hollow metal		Each	2835	1.42	
	B30 Roofing						
3010	Roof Coverings	Built-up tar and gravel with flashing; perlite/EPS composite insulation		S.F. Roof	4.05	4.05	4.5%
3020	Roof Openings	Roof hatches		S.F. Roof	.12	.12	
C. INTERIORS							
1010	Partitions	Concrete Block and toilet partitions	40 S.F. Floor/L.F. Partition	S.F. Partition	6.75	2.70	
1020	Interior Doors	Single leaf hollow metal	400 S.F. Floor/Door	Each	608	1.52	
1030	Fittings	Toilet partitions		S.F. Floor	—	—	
2010	Stair Construction	Concrete filled metal pan		Flight	7700	.96	20.9%
3010	Wall Finishes	70% paint, 30% epoxy coating		S.F. Surface	6.76	3.38	
3020	Floor Finishes	70% vinyl tile, 30% carpet		S.F. Floor	7.96	7.96	
3030	Ceiling Finishes	Fiberglass board, suspended		S.F. Ceiling	3.00	3.05	
D. SERVICES							
	D10 Conveying						
1010	Elevators & Lifts	One hydraulic passenger elevator		Each	65,520	2.73	2.9%
1020	Escalators & Moving Walks	N/A		—	—	—	
	D20 Plumbing						
2010	Plumbing Fixtures	Toilet and service fixtures, supply and drainage	1 Fixture/800 S.F. Floor	Each	2704	3.38	
2020	Domestic Water Distribution	Gas fired water heater		S.F. Floor	.28	.28	4.6%
2040	Rain Water Drainage	Roof drains		S.F. Roof	.67	.67	
	D30 HVAC						
3010	Energy Supply	N/A		—	—	—	
3020	Heat Generating Systems	Included in D3050		—	—	—	
3030	Cooling Generating Systems	N/A		—	—	—	11.5%
3050	Terminal & Package Units	Single zone rooftop unit, gas heating, electric cooling		S.F. Floor	10.75	10.75	
3090	Other HVAC Sys. & Equipment	N/A		—	—	—	
	D40 Fire Protection						
4010	Sprinklers	Wet pipe sprinkler system		S.F. Floor	2.05	2.05	2.2%
4020	Standpipes	N/A		—	—	—	
	D50 Electrical						
5010	Electrical Service/Distribution	800 ampere service, panel board and feeders		S.F. Floor	1.71	1.71	
5020	Lighting & Branch Wiring	Fluorescent fixtures, receptacles, switches, A.C. and misc. power		S.F. Floor	9.67	9.67	16.2%
5030	Communications & Security	Alarm systems and emergency lighting, and public address system		S.F. Floor	2.74	2.74	
5090	Other Electrical Systems	Emergency generator, 100KW		S.F. Floor	1.03	1.03	
E. EQUIPMENT & FURNISHINGS							
1010	Commercial Equipment	N/A		—	—	—	
1020	Institutional Equipment	N/A		—	—	—	0.0%
1030	Vehicular Equipment	N/A		—	—	—	
1090	Other Equipment	N/A		—	—	—	
F. SPECIAL CONSTRUCTION							
1020	Integrated Construction	N/A		—	—	—	0.0%
1040	Special Facilities	N/A		—	—	—	
G. BUILDING SITEWORK	**N/A**						

		Sub-Total	93.53	100%
CONTRACTOR FEES (General Requirements: 10%, Overhead: 5%, Profit: 10%)		25%	23.39	
ARCHITECT FEES		7%	8.18	
	Total Building Cost		**125.10**	

BUILDING TYPES

Costs per square foot of floor area

Exterior Wall	S.F. Area	2000	2700	3400	4100	4800	5500	6200	6900	7600
	L.F. Perimeter	180	208	236	256	280	303	317	337	357
Face Brick with Concrete Block Back-up	Steel Frame	185.10	173.30	166.30	160.30	156.75	153.85	150.60	148.60	147.00
	R/Conc. Frame	198.45	186.55	179.65	173.60	170.05	167.15	163.95	161.95	160.35
Precast Concrete Panel	Steel Frame	177.45	166.75	160.45	155.00	151.75	149.15	146.25	144.50	143.05
	R/Conc. Frame	190.80	180.05	173.75	168.35	165.10	162.50	159.60	157.75	156.35
Limestone with Concrete Block Back-up	Steel Frame	205.85	191.00	182.35	174.75	170.20	166.60	162.40	159.90	157.85
	R/Conc. Frame	219.20	204.35	195.65	188.00	183.55	179.90	175.70	173.20	171.15
Perimeter Adj., Add or Deduct	Per 100 L.F.	36.25	26.85	21.35	17.75	15.10	13.25	11.70	10.55	9.60
Story Hgt. Adj., Add or Deduct	Per 1 Ft.	3.40	2.90	2.70	2.40	2.25	2.15	1.95	1.90	1.80

For Basement, add $24.25 per square foot of basement area

The above costs were calculated using the basic specifications shown on the facing page. These costs should be adjusted where necessary for design alternatives and owner's requirements. Reported completed project costs, for this type of structure, range from $98.80 to $243.20 per S.F.

Common additives

Description	Unit	$ Cost
Bulletproof Teller Window, 44" x 60"	Each	3625
60" x 48"	Each	4400
Closed Circuit Surveillance, One station		
Camera and monitor	Each	1525
For additional camera stations, add	Each	820
Counters, Complete	Station	4925
Door & Frame, 3' x 6'-8", bullet resistant steel		
with vision panel	Each	4675 - 5975
Drive-up Window, Drawer & micr., not incl. glass	Each	5300 - 10,600
Emergency Lighting, 25 watt, battery operated		
Lead battery	Each	247
Nickel cadmium	Each	695
Night Depository	Each	7400 - 11,300
Package Receiver, painted	Each	1575
stainless steel	Each	2450
Partitions, Bullet resistant to 8' high	L.F.	265 - 425
Pneumatic Tube Systems, 2 station	Each	27,100
With TV viewer	Each	49,900

Description	Unit	$ Cost
Service Windows, Pass thru, steel		
24" x 36"	Each	2625
48" x 48"	Each	3300
72" x 40"	Each	4375
Smoke Detectors		
Ceiling type	Each	161
Duct type	Each	425
Twenty-four Hour Teller		
Automatic deposit cash & memo	Each	45,000
Vault Front, Door & frame		
1 hour test, 32"x 78"	Opening	4425
2 hour test, 32" door	Opening	4775
40" door	Opening	5575
4 hour test, 32" door	Opening	5175
40" door	Opening	6200
Time lock, two movement, add	Each	1725

Important: See the Reference Section for Location Factors

BUILDING TYPES

Model costs calculated for a 1 story building with 14' story height and 4,100 square feet of floor area

				Unit	Unit Cost	Cost Per S.F.	% Of Sub-Total
A. SUBSTRUCTURE							
1010	Standard Foundations	Poured concrete; strip and spread footings		S.F. Ground	3.20	3.20	
1030	Slab on Grade	4" reinforced concrete with vapor barrier and granular base		S.F. Slab	3.98	3.98	9.6%
2010	Basement Excavation	Site preparation for slab and trench for foundation wall and footing		S.F. Ground	.23	.23	
2020	Basement Walls	4' Foundation wall		L.F. Wall	61	4.58	
B. SHELL							
	B10 Superstructure						
1010	Floor Construction	Cast-in-place columns		L.F. Column	86	4.19	13.5%
1020	Roof Construction	Cast-in-place concrete flat plate		S.F. Roof	12.70	12.70	
	B20 Exterior Enclosure						
2010	Exterior Walls	Face brick with concrete block backup	80% of wall	S.F. Wall	23	16.61	
2020	Exterior Windows	Horizontal aluminum sliding	20% of wall	Each	436	5.08	18.2%
2030	Exterior Doors	Double aluminum and glass and hollow metal		Each	2291	1.13	
	B30 Roofing						
3010	Roof Coverings	Built-up tar and gravel with flashing; perlite/EPS composite insulation		S.F. Roof	5.07	5.07	4.1%
3020	Roof Openings	N/A		—	—	—	
C. INTERIORS							
1010	Partitions	Gypsum board on metal studs	20 S.F. of Floor/L.F. Partition	S.F. Partition	8.02	4.01	
1020	Interior Doors	Single leaf hollow metal	200 S.F. Floor/Door	Each	608	3.04	
1030	Fittings	N/A		—	—	—	
2010	Stair Construction	N/A		—	—	—	13.5%
3010	Wall Finishes	50% vinyl wall covering, 50% paint		S.F. Surface	2.20	1.10	
3020	Floor Finishes	50% carpet tile, 40% vinyl composition tile, 10% quarry tile		S.F. Floor	4.63	4.63	
3030	Ceiling Finishes	Mineral fiber tile on concealed zee bars		S.F. Ceiling	4.00	4.09	
D. SERVICES							
	D10 Conveying						
1010	Elevators & Lifts	N/A		—	—	—	0.0%
1020	Escalators & Moving Walks	N/A		—	—	—	
	D20 Plumbing						
2010	Plumbing Fixtures	Toilet and service fixtures, supply and drainage	1 Fixture/580 S.F. Floor	Each	3132	5.40	
2020	Domestic Water Distribution	Gas fired water heater		S.F. Floor	.66	.66	5.4%
2040	Rain Water Drainage	Roof drains		S.F. Roof	.73	.73	
	D30 HVAC						
3010	Energy Supply	N/A		—	—	—	
3020	Heat Generating Systems	Included in D3050		—	—	—	
3030	Cooling Generating Systems	N/A		—	—	—	7.5%
3050	Terminal & Package Units	Single zone rooftop unit, gas heating, electric cooling		S.F. Floor	9.37	9.37	
3090	Other HVAC Sys. & Equipment	N/A		—	—	—	
	D40 Fire Protection						
4010	Sprinklers	Wet pipe sprinkler system		S.F. Floor	2.68	2.68	2.1%
4020	Standpipes	N/A		—	—	—	
	D50 Electrical						
5010	Electrical Service/Distribution	200 ampere service, panel board and feeders		S.F. Floor	1.88	1.88	
5020	Lighting & Branch Wiring	Fluorescent fixtures, receptacles, switches, A.C. and misc. power		S.F. Floor	8.65	8.65	10.1%
5030	Communications & Security	Alarm systems, emergency lighting, and security television		S.F. Floor	2.01	2.01	
5090	Other Electrical Systems	Emergency generator, 15KW		S.F. Floor	.16	.16	
E. EQUIPMENT & FURNISHINGS							
1010	Commercial Equipment	Automatic teller, drive up window, night depository		S.F. Floor	6.23	6.23	
1020	Institutional Equipment	Closed circuit TV monitoring system		S.F. Floor	3.58	3.58	7.8%
1030	Vehicular Equipment	N/A		—	—	—	
1090	Other Equipment	N/A		—	—	—	
F. SPECIAL CONSTRUCTION							
1020	Integrated Construction	N/A		—	—	—	8.1%
1040	Special Facilities	Security vault door		S.F. Floor	10.15	10.15	
G. BUILDING SITEWORK	**N/A**						

			Sub-Total	125.14	100%
CONTRACTOR FEES (General Requirements: 10%, Overhead: 5%, Profit: 10%)		25%	31.25		
ARCHITECT FEES		11%	17.21		
Total Building Cost			**173.60**		

Costs per square foot of floor area

Exterior Wall	S.F. Area	12000	14000	16000	18000	20000	22000	24000	26000	28000
	L.F. Perimeter	460	491	520	540	566	593	620	645	670
Concrete Block	Steel Roof Deck	82.50	80.10	78.25	76.55	75.40	74.45	73.65	73.00	72.30
Decorative Concrete Block	Steel Roof Deck	86.30	83.55	81.55	79.55	78.20	77.15	76.20	75.50	74.65
Face Brick on Concrete Block	Steel Roof Deck	92.05	88.80	86.40	84.05	82.45	81.20	80.05	79.20	78.25
Jumbo Brick on Concrete Block	Steel Roof Deck	89.05	86.10	83.85	81.70	80.25	79.10	78.05	77.25	76.40
Stucco on Concrete Block	Steel Roof Deck	85.35	82.70	80.75	78.90	77.50	76.45	75.60	74.90	74.10
Precast Concrete Panel	Steel Roof Deck	85.05	82.40	80.45	78.60	77.25	76.30	75.35	74.65	73.90
Perimeter Adj., Add or Deduct	Per 100 L.F.	3.85	3.25	2.90	2.60	2.30	2.10	1.90	1.75	1.65
Story Hgt. Adj., Add or Deduct	Per 1 Ft.	.80	.75	.80	.70	.60	.60	.60	.55	.55

For Basement, add $23.90 per square foot of basement area

The above costs were calculated using the basic specifications shown on the facing page. These costs should be adjusted where necessary for design alternatives and owner's requirements. Reported completed project costs, for this type of structure, range from $43.75 to $107.35 per S.F.

Common additives

Description	Unit	$ Cost
Bowling Alleys, Incl. alley, pinsetter, scorer, counter & misc. supplies, average	Lane	53,000
For automatic scorer, add (maximum)	Lane	7925
Emergency Lighting, 25 watt, battery operated		
Lead battery	Each	247
Nickel cadmium	Each	695
Lockers, Steel, Single tier, 60" or 72"	Opening	154 - 241
2 tier, 60" or 72" total	Opening	104 - 134
5 tier, box lockers	Opening	55 - 69
Locker bench, lam., maple top only	L.F.	18.80
Pedestals, steel pipe	Each	57
Seating		
Auditorium chair, all veneer	Each	193
Veneer back, padded seat	Each	233
Upholstered, spring seat	Each	234
Sound System		
Amplifier, 250 watts	Each	1925
Speaker, ceiling or wall	Each	160
Trumpet	Each	305

Important: See the Reference Section for Location Factors

Model costs calculated for a 1 story building with 14' story height and 20,000 square feet of floor area

				Unit	Unit Cost	Cost Per S.F.	% Of Sub-Total
A. SUBSTRUCTURE							
1010	Standard Foundations	Poured concrete; strip and spread footings		S.F. Ground	.93	.93	
1030	Slab on Grade	4" reinforced concrete with vapor barrier and granular base		S.F. Slab	3.98	3.98	12.5%
2010	Basement Excavation	Site preparation for slab and trench for foundation wall and footing		S.F. Ground	.43	.43	
2020	Basement Walls	4' Foundation wall		L.F. Wall	61	1.72	
B. SHELL							
	B10 Superstructure						
1010	Floor Construction	N/A		—	—	—	12.8%
1020	Roof Construction	Metal deck on open web steel joists with columns and beams		S.F. Roof	7.23	7.23	
	B20 Exterior Enclosure						
2010	Exterior Walls	Concrete block	90% of wall	S.F. Wall	8.92	3.18	
2020	Exterior Windows	Horizontal pivoted	10% of wall	Each	1565	2.58	10.9%
2030	Exterior Doors	Double aluminum and glass and hollow metal		Each	1640	.41	
	B30 Roofing						
3010	Roof Coverings	Built-up tar and gravel with flashing; perlite/EPS composite insulation		S.F. Roof	3.88	3.88	7.1%
3020	Roof Openings	Roof hatches		S.F. Roof	.10	.10	
C. INTERIORS							
1010	Partitions	Concrete block	50 S.F. Floor/L.F. Partition	S.F. Partition	7.50	1.50	
1020	Interior Doors	Hollow metal single leaf	1000 S.F. Floor/Door	Each	608	.61	
1030	Fittings	N/A		—	—	—	
2010	Stair Construction	N/A		—	—	—	9.4%
3010	Wall Finishes	Paint and block filler		S.F. Surface	9.15	1.83	
3020	Floor Finishes	Vinyl tile	25% of floor	S.F. Floor	2.36	.59	
3030	Ceiling Finishes	Suspended fiberglass board	25% of area	S.F. Ceiling	3.05	.76	
D. SERVICES							
	D10 Conveying						
1010	Elevators & Lifts	N/A		—	—	—	0.0%
1020	Escalators & Moving Walks	N/A		—	—	—	
	D20 Plumbing						
2010	Plumbing Fixtures	Toilet and service fixtures, supply and drainage	1 Fixture/2200 S.F. Floor	Each	2002	.91	
2020	Domestic Water Distribution	Gas fired water heater		S.F. Floor	.13	.13	2.6%
2040	Rain Water Drainage	Roof drains		S.F. Roof	.43	.43	
	D30 HVAC						
3010	Energy Supply	N/A		—	—	—	
3020	Heat Generating Systems	Included in D3050		—	—	—	
3030	Cooling Generating Systems	N/A		—	—	—	21.7%
3050	Terminal & Package Units	Single zone rooftop unit, gas heating, electric cooling		S.F. Floor	12.25	12.25	
3090	Other HVAC Sys. & Equipment	N/A		—	—	—	
	D40 Fire Protection						
4010	Sprinklers	Sprinklers, light hazard		S.F. Floor	1.75	1.75	3.1%
4020	Standpipes	N/A		—	—	—	
	D50 Electrical						
5010	Electrical Service/Distribution	800 ampere service, panel board and feeders		S.F. Floor	2.07	2.07	
5020	Lighting & Branch Wiring	Fluorescent fixtures, receptacles, switches, A.C. and misc. power		S.F. Floor	8.76	8.76	19.8%
5030	Communications & Security	Alarm systems and emergency lighting		S.F. Floor	.27	.27	
5090	Other Electrical Systems	Emergency generator, 7.5KW		S.F. Floor	.06	.06	
E. EQUIPMENT & FURNISHINGS							
1010	Commercial Equipment	N/A		—	—	—	
1020	Institutional Equipment	N/A		—	—	—	0.0%
1030	Vehicular Equipment	N/A		—	—	—	
1090	Other Equipment	N/A		—	—	—	
F. SPECIAL CONSTRUCTION							
1020	Integrated Construction	N/A		—	—	—	0.0%
1040	Special Facilities	N/A		—	—	—	
G. BUILDING SITEWORK	**N/A**						

		Sub-Total	56.36	100%
CONTRACTOR FEES (General Requirements: 10%, Overhead: 5%, Profit: 10%)		25%	14.11	
ARCHITECT FEES		7%	4.93	
	Total Building Cost		**75.40**	

BUILDING TYPES

91

BUILDING TYPES

Costs per square foot of floor area

Exterior Wall	S.F. Area	6000	8000	10000	12000	14000	16000	18000	20000	22000
	L.F. Perimeter	320	386	453	520	540	597	610	660	710
Face Brick with Concrete Block Back-up	Bearing Walls	116.75	112.10	109.40	107.55	104.25	103.15	100.80	100.00	99.40
	Steel Frame	118.85	114.25	111.65	109.80	106.55	105.45	103.15	102.35	101.75
Decorative Concrete Block	Bearing Walls	110.50	106.45	104.15	102.55	99.80	98.80	96.85	96.20	95.65
	Steel Frame	112.65	108.65	106.40	104.75	102.05	101.10	99.15	98.55	98.00
Precast Concrete Panels	Bearing Walls	110.90	106.80	104.40	102.80	100.05	99.05	97.05	96.40	95.85
	Steel Frame	113.00	108.95	106.70	105.05	102.30	101.40	99.40	98.75	98.25
Perimeter Adj., Add or Deduct	Per 100 L.F.	10.20	7.65	6.10	5.15	4.35	3.80	3.40	3.10	2.80
Story Hgt. Adj., Add or Deduct	Per 1 Ft.	1.65	1.55	1.45	1.40	1.25	1.20	1.00	1.05	1.05
Basement—Not Applicable										

The above costs were calculated using the basic specifications shown on the facing page. These costs should be adjusted where necessary for design alternatives and owner's requirements. Reported completed project costs, for this type of structure, range from $52.80 to $127.30 per S.F.

Common additives

Description	Unit	$ Cost
Directory Boards, Plastic, glass covered		
30" x 20"	Each	575
36" x 48"	Each	1275
Aluminum, 24" x 18"	Each	525
36" x 24"	Each	600
48" x 32"	Each	810
48" x 60"	Each	1875
Emergency Lighting, 25 watt, battery operated		
Lead battery	Each	247
Nickel cadmium	Each	695
Benches, Hardwood	L.F.	92 - 169
Ticket Printer	Each	6850
Turnstiles, One way		
4 arm, 46" dia., manual	Each	465
Electric	Each	1700
High security, 3 arm		
65" dia., manual	Each	3425
Electric	Each	4750
Gate with horizontal bars		
65" dia., 7' high transit type	Each	4400

Important: See the Reference Section for Location Factors

Model costs calculated for a 1 story building with 14' story height and 12,000 square feet of floor area

				Unit	Unit Cost	Cost Per S.F.	% Of Sub-Total
A. SUBSTRUCTURE							
1010	Standard Foundations	Poured concrete; strip and spread footings		S.F. Ground	1.97	1.97	
1030	Slab on Grade	6" reinforced concrete with vapor barrier and granular base		S.F. Slab	6.00	6.00	
2010	Basement Excavation	Site preparation for slab and trench for foundation wall and footing		S.F. Ground	.23	.23	13.6%
2020	Basement Walls	4' Foundation wall		L.F. Wall	63	2.74	
B. SHELL							
	B10 Superstructure						
1010	Floor Construction	N/A		—	—	—	
1020	Roof Construction	Metal deck on open web steel joists with columns and beams		S.F. Roof	5.10	5.10	6.3%
	B20 Exterior Enclosure						
2010	Exterior Walls	Face brick with concrete block backup	70% of wall	S.F. Wall	23	10.08	
2020	Exterior Windows	Store front	30% of wall	Each	21	3.92	18.9%
2030	Exterior Doors	Double aluminum and glass		Each	3625	1.21	
	B30 Roofing						
3010	Roof Coverings	Built-up tar and gravel with flashing; perlite/EPS composite insulation		S.F. Roof	4.53	4.53	5.6%
3020	Roof Openings	N/A		—	—	—	
C. INTERIORS							
1010	Partitions	Lightweight concrete block	15 S.F. Floor/L.F. Partition	S.F. Partition	7.49	4.99	
1020	Interior Doors	Hollow metal	150 S.F. Floor/Door	Each	608	4.05	
1030	Fittings	N/A		—	—	—	
2010	Stair Construction	N/A		—	—	—	26.6%
3010	Wall Finishes	Glazed coating		S.F. Surface	2.07	2.07	
3020	Floor Finishes	Quarry tile and vinyl composition tile		S.F. Floor	6.18	6.18	
3030	Ceiling Finishes	Mineral fiber tile on concealed zee bars		S.F. Ceiling	4.00	4.09	
D. SERVICES							
	D10 Conveying						
1010	Elevators & Lifts	N/A		—	—	—	
1020	Escalators & Moving Walks	N/A		—	—	—	0.0%
	D20 Plumbing						
2010	Plumbing Fixtures	Toilet and service fixtures, supply and drainage	1 Fixture/850 S.F. Floor	Each	2040	2.40	
2020	Domestic Water Distribution	Electric water heater		S.F. Floor	.48	.48	4.2%
2040	Rain Water Drainage	Roof drains		S.F. Roof	.48	.48	
	D30 HVAC						
3010	Energy Supply	N/A		—	—	—	
3020	Heat Generating Systems	Included in D3050		—	—	—	
3030	Cooling Generating Systems	N/A		—	—	—	11.7%
3050	Terminal & Package Units	Single zone rooftop unit, gas heating, electric cooling		S.F. Floor	9.40	9.40	
3090	Other HVAC Sys. & Equipment	N/A		—	—	—	
	D40 Fire Protection						
4010	Sprinklers	Wet pipe sprinkler system		S.F. Floor	2.71	2.71	3.4%
4020	Standpipes	N/A		—	—	—	
	D50 Electrical						
5010	Electrical Service/Distribution	400 ampere service, panel board and feeders		S.F. Floor	1.29	1.29	
5020	Lighting & Branch Wiring	Fluorescent fixtures, receptacles, switches, A.C. and misc. power		S.F. Floor	5.96	5.96	9.7%
5030	Communications & Security	Alarm systems and emergency lighting		S.F. Floor	.22	.22	
5090	Other Electrical Systems	Emergency generator, 7.5 KW		S.F. Floor	.32	.32	
E. EQUIPMENT & FURNISHINGS							
1010	Commercial Equipment	N/A		—	—	—	
1020	Institutional Equipment	N/A		—	—	—	
1030	Vehicular Equipment	N/A		—	—	—	0.0%
1090	Other Equipment	N/A		—	—	—	
F. SPECIAL CONSTRUCTION							
1020	Integrated Construction	N/A		—	—	—	
1040	Special Facilities	N/A		—	—	—	0.0%
G. BUILDING SITEWORK	**N/A**						

		Sub-Total	80.42	100%
CONTRACTOR FEES (General Requirements: 10%, Overhead: 5%, Profit: 10%)		25%	20.09	
ARCHITECT FEES		7%	7.04	
	Total Building Cost		**107.55**	

BUILDING TYPES

Costs per square foot of floor area

Exterior Wall	S.F. Area	600	800	1000	1200	1600	2000	2400	3000	4000
	L.F. Perimeter	100	114	128	139	164	189	214	250	314
Brick with Concrete Block Back-up	Steel Frame	266.85	256.25	249.85	244.45	238.65	235.10	232.75	230.20	228.10
	Bearing Walls	262.75	252.10	245.70	240.35	234.50	231.00	228.65	226.05	223.95
Concrete Block	Steel Frame	235.90	229.75	226.10	222.95	219.60	217.55	216.20	214.70	213.45
	Bearing Walls	231.80	225.60	221.95	218.85	215.45	213.45	212.05	210.55	209.35
Galvanized Steel Siding	Steel Frame	236.45	231.15	227.95	225.25	222.35	220.55	219.40	218.10	217.05
Metal Sandwich Panel	Steel Frame	233.45	226.95	223.05	219.80	216.25	214.05	212.65	211.05	209.80
Perimeter Adj., Add or Deduct	Per 100 L.F.	73.30	54.90	43.95	36.75	27.50	22.00	18.35	14.65	10.95
Story Hgt. Adj., Add or Deduct	Per 1 Ft.	4.50	3.75	3.40	3.15	2.75	2.50	2.40	2.20	2.10
Basement—Not Applicable										

The above costs were calculated using the basic specifications shown on the facing page. These costs should be adjusted where necessary for design alternatives and owner's requirements. Reported completed project costs, for this type of structure, range from $94.75 to $269.10 per S.F.

Common additives

Description	Unit	$ Cost
Air Compressors, Electric		
1-1/2 H.P., standard controls	Each	1125
Dual controls	Each	1325
5 H.P., 115/230 volt, standard control	Each	3050
Dual controls	Each	3175
Emergency Lighting, 25 watt, battery operated		
Lead battery	Each	247
Nickel cadmium	Each	695
Fence, Chain link, 6' high		
9 ga. wire, galvanized	L.F.	17.75
6 ga. wire	L.F.	25
Gate	Each	258
Product Dispenser with vapor recovery		
for 6 nozzles	Each	19,800
Laundry Equipment		
Dryer, gas, 16 lb. capacity	Each	785
30 lb. capacity	Each	3000
Washer, 4 cycle	Each	875
Commercial	Each	1325

Description	Unit	$ Cost
Lockers, Steel, single tier, 60" or 72"	Opening	154 - 241
2 tier, 60" or 72" total	Opening	104 - 134
5 tier, box lockers	Opening	55 - 69
Locker bench, lam. maple top only	L.F.	18.80
Pedestals, steel pipe	Each	57
Paving, Bituminous		
Wearing course plus base course	Sq. Yard	6.80
Safe, Office type, 4 hour rating		
30" x 18" x 18"	Each	3425
62" x 33" x 20"	Each	7450
Sidewalks, Concrete 4" thick	S.F.	3.40
Yard Lighting,		
20' aluminum pole		
with 400 watt		
high pressure sodium		
fixture	Each	2375

Important: See the Reference Section for Location Factors

BUILDING TYPES

Model costs calculated for a 1 story building with 12' story height and 800 square feet of floor area

				Unit	Unit Cost	Cost Per S.F.	% Of Sub-Total
A. SUBSTRUCTURE							
1010	Standard Foundations	Poured concrete; strip and spread footings		S.F. Ground	2.33	2.33	
1030	Slab on Grade	5" reinforced concrete with vapor barrier and granular base		S.F. Slab	5.13	5.13	
2010	Basement Excavation	Site preparation for slab and trench for foundation wall and footing		S.F. Ground	.78	.78	8.3%
2020	Basement Walls	4' Foundation wall		L.F. Wall	54	7.72	
B. SHELL							
	B10 Superstructure						
1010	Floor Construction	N/A		—	—	—	
1020	Roof Construction	Metal deck, open web steel joists, beams, columns		S.F. Roof	6.41	6.41	3.3%
	B20 Exterior Enclosure						
2010	Exterior Walls	Face brick with concrete block backup	70% of wall	S.F. Wall	23	28.43	
2020	Exterior Windows	Horizontal pivoted steel	5% of wall	Each	615	5.85	24.1%
2030	Exterior Doors	Steel overhead hollow metal		Each	2370	11.85	
	B30 Roofing						
3010	Roof Coverings	Built-up tar and gravel with flashing; perlite/EPS composite insulation		S.F. Roof	5.81	5.81	3.0%
3020	Roof Openings	N/A		—	—	—	
C. INTERIORS							
1010	Partitions	Concrete block	20 S.F. Floor/S.F. Partition	S.F. Partition	7.22	3.61	
1020	Interior Doors	Hollow metal	600 S.F. Floor/Door	Each	608	1.01	
1030	Fittings	N/A		—	—	—	
2010	Stair Construction	N/A		—	—	—	2.9%
3010	Wall Finishes	Paint		S.F. Surface	1.92	.96	
3020	Floor Finishes	N/A		—	—	—	
3030	Ceiling Finishes	N/A		—	—	—	
D. SERVICES							
	D10 Conveying						
1010	Elevators & Lifts	N/A		—	—	—	0.0%
1020	Escalators & Moving Walks	N/A		—	—	—	
	D20 Plumbing						
2010	Plumbing Fixtures	Toilet and service fixtures, supply and drainage	1 Fixture/160 S.F. Floor	Each	2305	14.41	
2020	Domestic Water Distribution	Gas fired water heater		S.F. Floor	26	26.20	22.6%
2040	Rain Water Drainage	Roof drains		S.F. Roof	2.66	2.66	
	D30 HVAC						
3010	Energy Supply	N/A		—	—	—	
3020	Heat Generating Systems	Unit heaters		Each	13.95	13.95	
3030	Cooling Generating Systems	N/A		—	—	—	7.3%
3050	Terminal & Package Units	N/A		—	—	—	
3090	Other HVAC Sys. & Equipment	N/A		—	—	—	
	D40 Fire Protection						
4010	Sprinklers	N/A		—	—	—	0.0%
4020	Standpipes	N/A		—	—	—	
	D50 Electrical						
5010	Electrical Service/Distribution	400 ampere service, panel board and feeders		S.F. Floor	17.27	17.27	
5020	Lighting & Branch Wiring	Fluorescent fixtures, receptacles, switches and misc. power		S.F. Floor	35	35.82	28.4%
5030	Communications & Security	N/A		—	—	—	
5090	Other Electrical Systems	Emergency generator, 7.5KW		—	1.37	1.37	
E. EQUIPMENT & FURNISHINGS							
1010	Commercial Equipment	N/A		—	—	—	
1020	Institutional Equipment	N/A		—	—	—	
1030	Vehicular Equipment	N/A		—	—	—	0.0%
1090	Other Equipment	N/A		—	—	—	
F. SPECIAL CONSTRUCTION							
1020	Integrated Construction	N/A		—	—	—	0.0%
1040	Special Facilities	N/A		—	—	—	
G. BUILDING SITEWORK	**N/A**						

			Sub-Total	191.57	100%
CONTRACTOR FEES (General Requirements: 10%, Overhead: 5%, Profit: 10%)		25%	47.92	
ARCHITECT FEES		7%	16.76	

Total Building Cost   256.25

Costs per square foot of floor area

Exterior Wall	S.F. Area	2000	7000	12000	17000	22000	27000	32000	37000	42000
	L.F. Perimeter	180	340	470	540	640	740	762	793	860
Decorative Concrete Block	Wood Arch	189.85	142.55	132.20	124.45	121.55	119.80	116.15	113.70	112.70
	Steel Truss	188.40	141.10	130.75	123.00	120.10	118.35	114.65	112.25	111.25
Stone with Concrete Block Back-up	Wood Arch	230.30	164.35	149.80	138.70	134.60	132.10	126.80	123.30	121.90
	Steel Truss	223.15	157.20	142.65	131.55	127.50	124.95	119.65	116.15	114.75
Face Brick with Concrete Block Back-up	Wood Arch	211.20	154.05	141.50	132.00	128.45	126.30	121.75	118.80	117.55
	Steel Truss	204.05	146.90	134.35	124.85	121.30	119.15	114.60	111.65	110.40
Perimeter Adj., Add or Deduct	Per 100 L.F.	49.35	14.10	8.20	5.85	4.50	3.65	3.05	2.65	2.35
Story Hgt. Adj., Add or Deduct	Per 1 Ft.	2.90	1.55	1.25	.95	.95	.90	.75	.70	.65
For Basement, add $24.95 per square foot of basement area										

The above costs were calculated using the basic specifications shown on the facing page. These costs should be adjusted where necessary for design alternatives and owner's requirements. Reported completed project costs, for this type of structure, range from $60.50 to $247.30 per S.F.

Common additives

Description	Unit	$ Cost	Description	Unit	$ Cost
Altar, Wood, custom design, plain	Each	2300	Pews/Benches, Hardwood	L.F.	92 - 169
Deluxe	Each	11,800	Pulpits, Prefabricated, hardwood	Each	1500 - 9000
Granite or marble, average	Each	9400	Railing, Hardwood	L.F.	175
Deluxe	Each	29,600	Steeples, translucent fiberglass		
Ark, Prefabricated, plain	Each	8700	30" square, 15' high	Each	5350
Deluxe	Each	100,000	25' high	Each	6200
Baptistry, Fiberglass, incl. plumbing	Each	4150 - 7525	Painted fiberglass, 24" square, 14' high	Each	4700
Bells & Carillons, 48 bells	Each	693,000	28' high	Each	5375
24 bells	Each	289,000	Aluminum		
Confessional, Prefabricated wood			20' high, 3'- 6" base	Each	5200
Single, plain	Each	3250	35' high, 8'- 0" base	Each	19,700
Deluxe	Each	8075	60' high, 14'- 0" base	Each	41,300
Double, plain	Each	6175			
Deluxe	Each	18,000			
Emergency Lighting, 25 watt, battery operated					
Lead battery	Each	247			
Nickel cadmium	Each	695			
Lecterns, Wood, plain	Each	805			
Deluxe	Each	7425			

Important: See the Reference Section for Location Factors

BUILDING TYPES

Model costs calculated for a 1 story building with 24' story height and 17,000 square feet of floor area

				Unit	Unit Cost	Cost Per S.F.	% Of Sub-Total
A. SUBSTRUCTURE							
1010	Standard Foundations	Poured concrete; strip and spread footings		S.F. Ground	3.08	3.08	
1030	Slab on Grade	4" reinforced concrete with vapor barrier and granular base		S.F. Slab	3.98	3.98	10.6%
2010	Basement Excavation	Site preparation for slab and trench for foundation wall and footing		S.F. Ground	.14	.14	
2020	Basement Walls	4' Foundation wall		L.F. Wall	61	2.91	
B. SHELL							
	B10 Superstructure						
1010	Floor Construction	N/A		—	—	—	17.3%
1020	Roof Construction	Wood deck on laminated wood arches		S.F. Roof	15.40	16.43	
	B20 Exterior Enclosure						
2010	Exterior Walls	Face brick with concrete block backup	80% of wall (adjusted for end walls)	S.F. Wall	28	17.20	
2020	Exterior Windows	Aluminum, top hinged, in-swinging and curtain wall panels	20% of wall	S.F. Window	—	3.46	22.2%
2030	Exterior Doors	Double hollow metal swinging, single hollow metal		Each	1218	.43	
	B30 Roofing						
3010	Roof Coverings	Asphalt shingles with flashing; polystyrene insulation		S.F. Roof	3.51	3.51	3.7%
3020	Roof Openings	N/A		—	—	—	
C. INTERIORS							
1010	Partitions	Plaster on metal studs	40 S.F. Floor/L.F. Partitions	S.F. Partition	12.07	7.24	
1020	Interior Doors	Hollow metal	400 S.F. Floor/Door	Each	608	1.52	
1030	Fittings	N/A		—	—	—	
2010	Stair Construction	N/A		—	—	—	17.1%
3010	Wall Finishes	Paint		S.F. Surface	1.08	1.08	
3020	Floor Finishes	Carpet		S.F. Floor	6.46	6.46	
3030	Ceiling Finishes	N/A		—	—	—	
D. SERVICES							
	D10 Conveying						
1010	Elevators & Lifts	N/A		—	—	—	0.0%
1020	Escalators & Moving Walks	N/A		—	—	—	
	D20 Plumbing						
2010	Plumbing Fixtures	Kitchen, toilet and service fixtures, supply and drainage	1 Fixture/2430 S.F. Floor	Each	2551	1.05	
2020	Domestic Water Distribution	Gas fired hot water heater		S.F. Floor	.17	.17	1.3%
2040	Rain Water Drainage	N/A		—	—	—	
	D30 HVAC						
3010	Energy Supply	Oil fired hot water, wall fin radiation		S.F. Floor	6.74	6.74	
3020	Heat Generating Systems	N/A		—	—	—	
3030	Cooling Generating Systems	N/A		—	—	—	15.0%
3050	Terminal & Package Units	Split systems with air cooled condensing units		S.F. Floor	7.55	7.55	
3090	Other HVAC Sys. & Equipment	N/A		—	—	—	
	D40 Fire Protection						
4010	Sprinklers	Wet pipe sprinkler system		S.F. Floor	2.05	2.05	2.2%
4020	Standpipes	N/A		—	—	—	
	D50 Electrical						
5010	Electrical Service/Distribution	400 ampere service, panel board and feeders		S.F. Floor	.60	.60	
5020	Lighting & Branch Wiring	Fluorescent fixtures, receptacles, switches, A.C. and misc. power		S.F. Floor	7.09	7.09	10.6%
5030	Communications & Security	Alarm systems, sound system and emergency lighting		S.F. Floor	2.34	2.34	
5090	Other Electrical Systems	Emergency generator, 7.5KW		S.F. Floor	.09	.09	
E. EQUIPMENT & FURNISHINGS							
1010	Commercial Equipment	N/A		—	—	—	
1020	Institutional Equipment	N/A		—	—	—	0.0%
1030	Vehicular Equipment	N/A		—	—	—	
1090	Other Equipment	N/A		—	—	—	
F. SPECIAL CONSTRUCTION							
1020	Integrated Construction	N/A		—	—	—	0.0%
1040	Special Facilities	N/A		—	—	—	
G. BUILDING SITEWORK	**N/A**						

		Sub-Total	95.12	100%
CONTRACTOR FEES (General Requirements: 10%, Overhead: 5%, Profit: 10%)		25%	23.80	
ARCHITECT FEES		11%	13.08	
	Total Building Cost		**132.00**	

Costs per square foot of floor area

Exterior Wall	S.F. Area	2000	4000	6000	8000	12000	15000	18000	20000	22000
	L.F. Perimeter	180	280	340	386	460	535	560	600	640
Stone Ashlar Veneer On Concrete Block	Wood Truss	203.95	177.45	164.10	156.25	147.40	144.65	141.00	139.95	139.10
	Steel Joists	201.25	174.80	161.45	153.60	144.70	142.05	138.35	137.35	136.50
Stucco on Concrete Block	Wood Truss	185.00	162.75	152.15	146.05	139.30	137.20	134.45	133.65	133.05
	Steel Joists	182.65	160.35	149.85	143.75	136.95	134.80	132.10	131.30	130.65
Brick Veneer	Wood Frame	192.55	168.50	156.80	149.90	142.25	139.90	136.80	135.90	135.15
Wood Shingles	Wood Frame	183.25	161.25	150.90	144.95	138.35	136.25	133.60	132.85	132.15
Perimeter Adj., Add or Deduct	Per 100 L.F.	33.95	17.00	11.35	8.45	5.55	4.55	3.75	3.40	3.10
Story Hgt. Adj., Add or Deduct	Per 1 Ft.	3.75	2.95	2.35	2.00	1.55	1.50	1.25	1.30	1.25

For Basement, add $21.90 per square foot of basement area

The above costs were calculated using the basic specifications shown on the facing page. These costs should be adjusted where necessary for design alternatives and owner's requirements. Reported completed project costs, for this type of structure, range from $59.85 to $199.00 per S.F.

Common additives

Description	Unit	$ Cost
Bar, Front bar	L.F.	320
Back bar	L.F.	257
Booth, Upholstered, custom, straight	L.F.	166 - 305
"L" or "U" shaped	L.F.	172 - 290
Fireplaces, Brick not incl. chimney or foundation, 30" x 24" opening	Each	2325
Chimney, standard brick		
Single flue 16" x 20"	V.L.F.	75
20" x 20"	V.L.F.	86
2 flue, 20" x 32"	V.L.F.	129
Lockers, Steel, single tier, 60" or 72"	Opening	154 - 241
2 tier, 60" or 72" total	Opening	104 - 134
5 tier, box lockers	Opening	55 - 69
Locker bench, lam. maple top only	L.F.	18.80
Pedestals, steel pipe	Each	57
Refrigerators, Prefabricated, walk-in		
7'-6" High, 6' x 6'	S.F.	129
10' x 10'	S.F.	101
12' x 14'	S.F.	89
12' x 20'	S.F.	79

Description	Unit	$ Cost
Sauna, Prefabricated, complete, 6' x 4'	Each	4525
6' x 9'	Each	6700
8' x 8'	Each	7875
10' x 12'	Each	11,000
Smoke Detectors		
Ceiling type	Each	161
Duct type	Each	425
Sound System		
Amplifier, 250 watts	Each	1925
Speaker, ceiling or wall	Each	160
Trumpet	Each	305
Steam Bath, Complete, to 140 C.F.	Each	1475
To 300 C.F.	Each	1650
To 800 C.F.	Each	4725
To 2500 C.F.	Each	7200
Swimming Pool Complete, gunite	S.F.	51 - 62
Tennis Court, Complete with fence		
Bituminous	Each	25,500 - 31,700
Clay	Each	28,000 - 34,100

Important: See the Reference Section for Location Factors

Model costs calculated for a 1 story building with 12' story height and 6,000 square feet of floor area

				Unit	Unit Cost	Cost Per S.F.	% Of Sub-Total
A.	**SUBSTRUCTURE**						
1010	Standard Foundations	Poured concrete; strip and spread footings		S.F. Ground	2.17	2.17	
1030	Slab on Grade	4" reinforced concrete with vapor barrier and granular base		S.F. Slab	3.98	3.98	9.2%
2010	Basement Excavation	Site preparation for slab and trench for foundation wall and footing		S.F. Ground	.23	.23	
2020	Basement Walls	4' Foundation wall		L.F. Wall	61	4.50	
B.	**SHELL**						
	B10 Superstructure						
1010	Floor Construction	N/A		—	—	—	4.8%
1020	Roof Construction	Wood truss with plywood sheathing		S.F. Roof	5.64	5.64	
	B20 Exterior Enclosure						
2010	Exterior Walls	Stone ashlar veneer on concrete block	65% of wall	S.F. Wall	30	13.46	
2020	Exterior Windows	Aluminum horizontal sliding	35% of wall	Each	346	5.49	17.6%
2030	Exterior Doors	Double aluminum and glass, hollow metal		Each	1929	1.92	
	B30 Roofing						
3010	Roof Coverings	Asphalt shingles		S.F. Roof	11.98	2.13	1.8%
3020	Roof Openings	N/A		—	—	—	
C.	**INTERIORS**						
1010	Partitions	Gypsum board on metal studs, load bearing	14 S.F. Floor/L.F. Partition	S.F. Partition	6.33	4.52	
1020	Interior Doors	Single leaf wood	140 S.F. Floor/Door	Each	480	3.43	
1030	Fittings	N/A		—	—	—	
2010	Stair Construction	N/A		—	—	—	17.4%
3010	Wall Finishes	40% vinyl wall covering, 40% paint, 20% ceramic tile		S.F. Surface	2.93	2.93	
3020	Floor Finishes	50% carpet, 30% hardwood tile, 20% ceramic tile		S.F. Floor	6.38	6.38	
3030	Ceiling Finishes	Gypsum plaster on wood furring		S.F. Ceiling	3.00	3.30	
D.	**SERVICES**						
	D10 Conveying						
1010	Elevators & Lifts	N/A		—	—	—	0.0%
1020	Escalators & Moving Walks	N/A		—	—	—	
	D20 Plumbing						
2010	Plumbing Fixtures	Kitchen, toilet and service fixtures, supply and drainage	1 Fixture/125 S.F. Floor	Each	1920	15.36	
2020	Domestic Water Distribution	Gas fired hot water		S.F. Floor	3.58	3.58	16.0%
2040	Rain Water Drainage	N/A		—	—	—	
	D30 HVAC						
3010	Energy Supply	N/A		—	—	—	
3020	Heat Generating Systems	Included in D3050		—	—	—	
3030	Cooling Generating Systems	N/A		—	—	—	24.5%
3050	Terminal & Package Units	Multizone rooftop unit, gas heating, electric cooling		S.F. Floor	29	29.00	
3090	Other HVAC Sys. & Equipment	N/A		—	—	—	
	D40 Fire Protection						
4010	Sprinklers	Wet pipe sprinkler system		S.F. Floor	2.68	2.68	2.3%
4020	Standpipes	N/A		—	—	—	
	D50 Electrical						
5010	Electrical Service/Distribution	400 ampere service, panel board and feeders		S.F. Floor	2.62	2.62	
5020	Lighting & Branch Wiring	Fluorescent fixtures, receptacles, switches, A.C. and misc. power		S.F. Floor	4.47	4.47	6.4%
5030	Communications & Security	Alarm systems and emergency lighting		S.F. Floor	.31	.31	
5090	Other Electrical Systems	Emergency generator, 11.5KW		S.F. Floor	.17	.17	
E.	**EQUIPMENT & FURNISHINGS**						
1010	Commercial Equipment	N/A		—	—	—	
1020	Institutional Equipment	N/A		—	—	—	0.0%
1030	Vehicular Equipment	N/A		—	—	—	
1090	Other Equipment	N/A		—	—	—	
F.	**SPECIAL CONSTRUCTION**						
1020	Integrated Construction	N/A		—	—	—	0.0%
1040	Special Facilities	N/A		—	—	—	
G.	**BUILDING SITEWORK**	**N/A**					

		Sub-Total	118.27	100%
	CONTRACTOR FEES (General Requirements: 10%, Overhead: 5%, Profit: 10%)	25%	29.57	
	ARCHITECT FEES	11%	16.26	
	Total Building Cost		**164.10**	

BUILDING TYPES

Costs per square foot of floor area

Exterior Wall	S.F. Area	4000	8000	12000	17000	22000	27000	32000	37000	42000
	L.F. Perimeter	280	386	520	585	640	740	840	940	940
Stone Ashlar on Concrete Block	Steel Joists	151.40	128.85	123.20	115.05	110.25	108.60	107.40	106.65	104.05
	Wood Joists	148.45	126.55	121.10	113.15	108.50	106.90	105.70	104.95	102.50
Face Brick on Concrete Block	Steel Joists	146.50	125.50	120.15	112.70	108.25	106.70	105.55	104.85	102.45
	Wood Joists	144.15	123.55	118.40	111.10	106.80	105.30	104.15	103.50	101.20
Decorative Concrete Block	Steel Joists	140.00	121.00	116.15	109.45	105.55	104.10	103.10	102.50	100.40
	Wood Joists	136.95	118.40	113.70	107.20	103.40	102.05	101.10	100.45	98.45
Perimeter Adj., Add or Deduct	Per 100 L.F.	20.00	10.00	6.70	4.75	3.65	2.95	2.45	2.10	1.90
Story Hgt. Adj., Add or Deduct	Per 1 Ft.	3.30	2.30	2.05	1.65	1.40	1.25	1.25	1.20	1.05
For Basement, add $22.30 per square foot of basement area										

The above costs were calculated using the basic specifications shown on the facing page. These costs should be adjusted where necessary for design alternatives and owner's requirements. Reported completed project costs, for this type of structure, range from $61.20 to $175.80 per S.F.

Common additives

Description	Unit	$ Cost
Bar, Front bar	L.F.	320
Back bar	L.F.	257
Booth, Upholstered, custom, straight	L.F.	166 - 305
"L" or "U" shaped	L.F.	172 - 290
Emergency Lighting, 25 watt, battery operated		
Lead battery	Each	247
Nickel cadmium	Each	695
Flagpoles, Complete		
Aluminum, 20' high	Each	1375
40' High	Each	3025
70' High	Each	9500
Fiberglass, 23' High	Each	1650
39'-5" High	Each	3175
59' High	Each	7950
Kitchen Equipment		
Broiler	Each	3075
Coffee urn, twin 6 gallon	Each	2925
Cooler, 6 ft. long, reach-in	Each	4075
Dishwasher, 10-12 racks per hr.	Each	4300
Food warmer, counter 1.2 kw	Each	665

Description	Unit	$ Cost
Kitchen Equipment, cont.		
Freezer, 44 C.F., reach-in	Each	9175
Ice cube maker, 50 lb. per day	Each	1600
Lockers, Steel, single tier, 60" or 72"	Opening	154 - 241
2 tier, 60" or 72" total	Opening	104 - 134
5 tier, box lockers	Opening	55 - 69
Locker bench, lam. maple top only	L.F.	18.80
Pedestals, steel pipe	Each	57
Refrigerators, Prefabricated, walk-in		
7'-6" High, 6' x 6'	S.F.	129
10' x 10'	S.F.	101
12' x 14'	S.F.	89
12' x 20'	S.F.	79
Smoke Detectors		
Ceiling type	Each	161
Duct type	Each	425
Sound System		
Amplifier, 250 watts	Each	1925
Speaker, ceiling or wall	Each	160
Trumpet	Each	305

Important: See the Reference Section for Location Factors

Model costs calculated for a 1 story building with 12′ story height and 22,000 square feet of floor area

Club, Social

				Unit	Unit Cost	Cost Per S.F.	% Of Sub-Total
A. SUBSTRUCTURE							
1010	Standard Foundations	Poured concrete; strip and spread footings		S.F. Ground	1.36	1.36	
1030	Slab on Grade	4″ reinforced concrete with vapor barrier and granular base		S.F. Slab	3.98	3.98	10.1%
2010	Basement Excavation	Site preparation for slab and trench for foundation wall and footing		S.F. Ground	.14	.14	
2020	Basement Walls	4′ Foundation wall		L.F. Wall	61	2.84	
B. SHELL							
	B10 Superstructure						
1010	Floor Construction	N/A		—	—	—	6.2%
1020	Roof Construction	Metal deck on open web steel joists		S.F. Roof	5.14	5.14	
	B20 Exterior Enclosure						
2010	Exterior Walls	Stone, ashlar veneer on concrete block	65% of wall	S.F. Wall	30	6.91	
2020	Exterior Windows	Window wall	35% of wall	Each	37	4.53	14.5%
2030	Exterior Doors	Double aluminum and glass doors		Each	1929	.53	
	B30 Roofing						
3010	Roof Coverings	Built-up tar and gravel with flashing; perlite/EPS composite insulation		S.F. Roof	4.12	4.12	5.0%
3020	Roof Openings	N/A		—	—	—	
C. INTERIORS							
1010	Partitions	Lightweight concrete block	14 S.F. Floor/L.F. Partition	S.F. Partition	8.46	6.04	
1020	Interior Doors	Single leaf wood	140 S.F. Floor/Door	Each	480	3.43	
1030	Fittings	N/A		—	—	—	
2010	Stair Construction	N/A		—	—	—	26.1%
3010	Wall Finishes	65% paint, 25% vinyl wall covering, 10% ceramic tile		S.F. Surface	1.98	1.98	
3020	Floor Finishes	60% carpet, 35% hardwood, 15% ceramic tile		S.F. Floor	6.00	6.00	
3030	Ceiling Finishes	Mineral fiber tile on concealed zee bars		S.F. Ceiling	4.00	4.09	
D. SERVICES							
	D10 Conveying						
1010	Elevators & Lifts	N/A		—	—	—	0.0%
1020	Escalators & Moving Walks	N/A		—	—	—	
	D20 Plumbing						
2010	Plumbing Fixtures	Kitchen, toilet and service fixtures, supply and drainage	1 Fixture/1050 S.F. Floor	Each	2163	2.06	
2020	Domestic Water Distribution	Gas fired hot water heater		S.F. Floor	.15	.15	3.2%
2040	Rain Water Drainage	Roof drains		S.F. Roof	.39	.39	
	D30 HVAC						
3010	Energy Supply	N/A		—	—	—	
3020	Heat Generating Systems	Included in D3050		—	—	—	
3030	Cooling Generating Systems	N/A		—	—	—	25.5%
3050	Terminal & Package Units	Multizone rooftop unit, gas heating, electric cooling		S.F. Floor	21	21.00	
3090	Other HVAC Sys. & Equipment	N/A		—	—	—	
	D40 Fire Protection						
4010	Sprinklers	Wet pipe sprinkler system		S.F. Floor	2.05	2.05	2.5%
4020	Standpipes	N/A		—	—	—	
	D50 Electrical						
5010	Electrical Service/Distribution	400 ampere service, panel board and feeders		S.F. Floor	.87	.87	
5020	Lighting & Branch Wiring	Fluorescent fixtures, receptacles, switches, A.C. and misc. power		S.F. Floor	4.56	4.56	6.9%
5030	Communications & Security	Alarm systems and emergency lighting		S.F. Floor	.17	.17	
5090	Other Electrical Systems	Emergency generator, 11.5KW		S.F. Floor	.10	.10	
E. EQUIPMENT & FURNISHINGS							
1010	Commercial Equipment	N/A		—	—	—	
1020	Institutional Equipment	N/A		—	—	—	0.0%
1030	Vehicular Equipment	N/A		—	—	—	
1090	Other Equipment	N/A		—	—	—	
F. SPECIAL CONSTRUCTION							
1020	Integrated Construction	N/A		—	—	—	0.0%
1040	Special Facilities	N/A		—	—	—	
G. BUILDING SITEWORK	**N/A**						

		Sub-Total	82.44	100%
CONTRACTOR FEES (General Requirements: 10%, Overhead: 5%, Profit: 10%)		25%	20.60	
ARCHITECT FEES		7%	7.21	
	Total Building Cost		**110.25**	

BUILDING TYPES

Costs per square foot of floor area

Exterior Wall	S.F. Area	15000	20000	28000	38000	50000	65000	85000	100000	150000
	L.F. Perimeter	350	400	480	550	630	660	750	825	1035
Face Brick with Concrete Block Back-up	Steel Frame	163.30	154.35	146.80	140.85	136.80	132.70	130.10	128.90	126.45
	Bearing Walls	162.95	153.15	144.65	137.95	133.45	128.65	125.70	124.30	121.55
Decorative Concrete Block	Steel Frame	157.60	149.45	142.55	137.25	133.70	130.20	128.00	126.90	124.80
	Bearing Walls	157.20	148.15	140.40	134.45	130.35	126.15	123.50	122.30	119.85
Stucco on Concrete Block	Steel Frame	155.10	147.00	140.15	134.95	131.35	127.95	125.75	124.65	122.60
	Bearing Walls	156.85	147.80	140.15	134.20	130.05	125.95	123.40	122.15	119.70
Perimeter Adj., Add or Deduct	Per 100 L.F.	7.15	5.45	3.80	2.85	2.15	1.65	1.30	1.10	.75
Story Hgt. Adj., Add or Deduct	Per 1 Ft.	1.80	1.55	1.25	1.05	.95	.75	.70	.60	.50

For Basement, add $25.95 per square foot of basement area

The above costs were calculated using the basic specifications shown on the facing page. These costs should be adjusted where necessary for design alternatives and owner's requirements. Reported completed project costs, for this type of structure, range from $88.50 to $207.95 per S.F.

Common additives

Description	Unit	$ Cost
Carrels Hardwood	Each	785 - 1000
Clock System		
20 Room	Each	14,200
50 Room	Each	34,600
Elevators, Hydraulic passenger, 2 stops		
1500# capacity	Each	47,900
2500# capacity	Each	50,000
3500# capacity	Each	53,700
Additional stop, add	Each	4550
Emergency Lighting, 25 watt, battery operated		
Lead battery	Each	247
Nickel cadmium	Each	695
Flagpoles, Complete		
Aluminum, 20' high	Each	1375
40' High	Each	3025
70' High	Each	9500
Fiberglass, 23' High		1650
39'-5" High	Each	3175
59' High	Each	7950

Description	Unit	$ Cost
Lockers, Steel, single tier, 60" or 72"	Opening	154 - 241
2 tier, 60" or 72" total	Opening	104 - 134
5 tier, box lockers	Opening	55 - 69
Locker bench, lam. maple top only	L.F.	18.80
Pedestals, steel pipe	Each	57
Seating		
Auditorium chair, all veneer	Each	193
Veneer back, padded seat	Each	233
Upholstered, spring seat	Each	234
Classroom, movable chair & desk	Set	65 - 120
Lecture hall, pedestal type	Each	179 - 545
Smoke Detectors		
Ceiling type	Each	161
Duct type	Each	425
Sound System		
Amplifier, 250 watts	Each	1925
Speaker, ceiling or wall	Each	160
Trumpet	Each	305
TV Antenna, Master system, 12 outlet	Outlet	262
30 outlet	Outlet	168
100 outlet	Outlet	160

Important: See the Reference Section for Location Factors

Model costs calculated for a 2 story building with 12' story height and 50,000 square feet of floor area

				Unit	Unit Cost	Cost Per S.F.	% Of Sub-Total
A.	**SUBSTRUCTURE**						
1010	Standard Foundations	Poured concrete; strip and spread footings		S.F. Ground	.98	.49	
1030	Slab on Grade	4" reinforced concrete with vapor barrier and granular base		S.F. Slab	3.98	1.99	3.9%
2010	Basement Excavation	Site preparation for slab and trench for foundation wall and footing		S.F. Ground	.22	.11	
2020	Basement Walls	4' Foundation wall		L.F. Wall	115	1.18	
B.	**SHELL**						
	B10 Superstructure						
1010	Floor Construction	Open web steel joists, slab form, concrete		S.F. Floor	12.84	6.42	10.3%
1020	Roof Construction	Metal deck on open web steel joists, columns		S.F. Roof	7.28	3.64	
	B20 Exterior Enclosure						
2010	Exterior Walls	Decorative concrete block	65% of wall	S.F. Wall	11.90	2.34	
2020	Exterior Windows	Window wall	35% of wall	Each	31	3.36	6.3%
2030	Exterior Doors	Double glass and aluminum with transom		Each	3625	.44	
	B30 Roofing						
3010	Roof Coverings	Built-up tar and gravel with flashing; perlite/EPS composite insulation		S.F. Roof	4.00	2.00	2.1%
3020	Roof Openings	N/A		—	—	—	
C.	**INTERIORS**						
1010	Partitions	Concrete block	20 S.F. Floor/L.F. Partition	S.F. Partition	11.00	5.50	
1020	Interior Doors	Single leaf hollow metal	200 S.F. Floor/Door	Each	608	3.04	
1030	Fittings	Chalkboards, counters, cabinets		S.F. Floor	3.84	3.84	
2010	Stair Construction	Concrete filled metal pan		Flight	7700	1.54	25.6%
3010	Wall Finishes	95% paint, 5% ceramic tile		S.F. Surface	6.12	3.06	
3020	Floor Finishes	70% vinyl composition tile, 25% carpet, 5% ceramic tile		S.F. Floor	3.91	3.91	
3030	Ceiling Finishes	Mineral fiber tile on concealed zee bars		S.F. Ceiling	4.00	4.09	
D.	**SERVICES**						
	D10 Conveying						
1010	Elevators & Lifts	Two hydraulic passenger elevators		Each	59,250	2.37	2.4%
1020	Escalators & Moving Walks	N/A		—	—	—	
	D20 Plumbing						
2010	Plumbing Fixtures	Toilet and service fixtures, supply and drainage	1 Fixture/455 S.F. Floor	Each	4804	10.56	
2020	Domestic Water Distribution	Oil fired hot water heater		S.F. Floor	.72	.72	12.1%
2040	Rain Water Drainage	Roof drains		S.F. Roof	1.10	.55	
	D30 HVAC						
3010	Energy Supply	N/A		—	—	—	
3020	Heat Generating Systems	Included in D3050		—	—	—	
3030	Cooling Generating Systems	N/A		—	—	—	17.0%
3050	Terminal & Package Units	Multizone unit, gas heating, electric cooling		S.F. Floor	16.60	16.60	
3090	Other HVAC Sys. & Equipment	N/A		—	—	—	
	D40 Fire Protection						
4010	Sprinklers	Sprinklers, light hazard		S.F. Floor	1.54	1.54	1.6%
4020	Standpipes	N/A		—	—	—	
	D50 Electrical						
5010	Electrical Service/Distribution	2000 ampere service, panel board and feeders		S.F. Floor	3.42	3.42	
5020	Lighting & Branch Wiring	Fluorescent fixtures, receptacles, switches, A.C. and misc. power		S.F. Floor	10.13	10.13	18.6%
5030	Communications & Security	Alarm systems, communications systems and emergency lighting		S.F. Floor	4.04	4.04	
5090	Other Electrical Systems	Emergency generator, 100KW		S.F. Floor	.55	.55	
E.	**EQUIPMENT & FURNISHINGS**						
1010	Commercial Equipment	N/A		—	—	—	
1020	Institutional Equipment	N/A		—	—	—	0.0%
1030	Vehicular Equipment	N/A		—	—	—	
1090	Other Equipment	N/A		—	—	—	
F.	**SPECIAL CONSTRUCTION**						
1020	Integrated Construction	N/A		—	—	—	0.0%
1040	Special Facilities	N/A		—	—	—	
G.	**BUILDING SITEWORK**	**N/A**					

	Sub-Total	97.43	100%
CONTRACTOR FEES (General Requirements: 10%, Overhead: 5%, Profit: 10%)	25%	24.39	
ARCHITECT FEES	7%	8.53	
Total Building Cost		**130.35**	

Costs per square foot of floor area

Exterior Wall	S.F. Area	10000	15000	25000	40000	55000	70000	80000	90000	100000
	L.F. Perimeter	260	320	400	476	575	628	684	721	772
Face Brick with Concrete Block Back-up	R/Conc. Frame	161.40	149.30	137.35	129.15	125.95	123.20	122.35	121.40	120.70
	Steel Frame	168.30	156.25	144.30	136.05	132.90	130.20	129.30	128.35	127.70
Decorative Concrete Block	R/Conc. Frame	150.15	140.25	130.70	124.30	121.85	119.75	119.05	118.40	117.85
	Steel Frame	156.45	146.55	137.00	130.65	128.15	126.10	125.40	124.70	124.15
Precast Concrete Panels	R/Conc. Frame	157.50	146.15	135.00	127.40	124.50	121.95	121.15	120.30	119.65
	Steel Frame	164.30	152.95	141.80	134.25	131.30	128.80	128.00	127.10	126.50
Perimeter Adj., Add or Deduct	Per 100 L.F.	14.00	9.35	5.60	3.45	2.55	2.00	1.75	1.55	1.40
Story Hgt. Adj., Add or Deduct	Per 1 Ft.	2.65	2.25	1.65	1.20	1.10	.90	.90	.85	.85
For Basement, add $25.30 per square foot of basement area										

The above costs were calculated using the basic specifications shown on the facing page. These costs should be adjusted where necessary for design alternatives and owner's requirements. Reported completed project costs, for this type of structure, range from $57.15 to $168.45 per S.F.

Common additives

Description	Unit	$ Cost
Carrels Hardwood	Each	785 - 1000
Closed Circuit Surveillance, One station		
Camera and monitor	Each	1525
For additional camera stations, add	Each	820
Elevators, Hydraulic passenger, 2 stops		
2000# capacity	Each	49,000
2500# capacity	Each	50,000
3500# capacity	Each	53,700
Additional stop, add	Each	4550
Emergency Lighting, 25 watt, battery operated		
Lead battery	Each	247
Nickel cadmium	Each	695
Furniture	Student	1900 - 3600
Intercom System, 25 station capacity		
Master station	Each	2200
Intercom outlets	Each	142
Handset	Each	390

Description	Unit	$ Cost
Kitchen Equipment		
Broiler	Each	3075
Coffee urn, twin 6 gallon	Each	2925
Cooler, 6 ft. long	Each	4075
Dishwasher, 10-12 racks per hr.	Each	4300
Food warmer	Each	665
Freezer, 44 C.F., reach-in	Each	9175
Ice cube maker, 50 lb. per day	Each	1600
Range with 1 oven	Each	2100
Laundry Equipment		
Dryer, gas, 16 lb. capacity	Each	785
30 lb. capacity	Each	3000
Washer, 4 cycle	Each	875
Commercial	Each	1325
Smoke Detectors		
Ceiling type	Each	161
Duct type	Each	425
TV Antenna, Master system, 12 outlet	Outlet	262
30 outlet	Outlet	168
100 outlet	Outlet	160

Important: See the Reference Section for Location Factors

Model costs calculated for a 3 story building with 12' story height and 25,000 square feet of floor area

				Unit	Unit Cost	Cost Per S.F.	% Of Sub-Total
A. SUBSTRUCTURE							
1010	Standard Foundations	Poured concrete; strip and spread footings		S.F. Ground	4.83	1.61	
1030	Slab on Grade	4" reinforced concrete with vapor barrier and granular base		S.F. Slab	3.98	1.32	
2010	Basement Excavation	Site preparation for slab and trench for foundation wall and footing		S.F. Ground	.14	.05	3.8%
2020	Basement Walls	4' Foundation wall		L.F. Wall	61	.97	
B. SHELL							
	B10 Superstructure						
1010	Floor Construction	Concrete flat plate		S.F. Floor	19.22	12.81	
1020	Roof Construction	Concrete flat plate		S.F. Roof	11.01	3.67	16.0%
	B20 Exterior Enclosure						
2010	Exterior Walls	Face brick with concrete block backup	80% of wall	S.F. Wall	23	10.95	
2020	Exterior Windows	Aluminum horizontal sliding	20% of wall	Each	559	2.80	14.4%
2030	Exterior Doors	Double glass & aluminum doors		Each	4300	1.04	
	B30 Roofing						
3010	Roof Coverings	Single ply membrane, loose laid and ballasted; perlite/EPS composite insulation		S.F. Roof	3.93	1.31	1.3%
3020	Roof Openings	N/A		—	—	—	
C. INTERIORS							
1010	Partitions	Gypsum board on metal studs, concrete block	9 S.F. Floor/L.F. Partition	S.F. Partition	4.82	5.35	
1020	Interior Doors	Single leaf wood	90 S.F. Floor/Door	Each	480	5.34	
1030	Fittings	Closet shelving, mirrors, bathroom accessories		S.F. Floor	1.78	1.78	
2010	Stair Construction	Cast in place concrete		Flight	4375	2.27	24.8%
3010	Wall Finishes	95% paint, 5% ceramic tile		S.F. Surface	—	3.18	
3020	Floor Finishes	80% carpet, 10% vinyl composition tile, 10% ceramic tile		S.F. Floor	6.81	6.81	
3030	Ceiling Finishes	90% paint, 10% suspended fiberglass board		S.F. Ceiling	1.00	.74	
D. SERVICES							
	D10 Conveying						
1010	Elevators & Lifts	One hydraulic passenger elevator		Each	78,000	3.12	3.0%
1020	Escalators & Moving Walks	N/A		—	—	—	
	D20 Plumbing						
2010	Plumbing Fixtures	Toilet and service fixtures, supply and drainage	1 Fixture/455 S.F. Floor	Each	3298	7.25	
2020	Domestic Water Distribution	Electric water heater		S.F. Floor	2.06	2.06	9.3%
2040	Rain Water Drainage	Roof drains		S.F. Roof	.75	.25	
	D30 HVAC						
3010	Energy Supply	N/A		—	—	—	
3020	Heat Generating Systems	Included in D3050		—	—	—	
3030	Cooling Generating Systems	N/A		—	—	—	10.1%
3050	Terminal & Package Units	Rooftop multizone unit system		S.F. Floor	10.38	10.38	
3090	Other HVAC Sys. & Equipment	N/A		—	—	—	
	D40 Fire Protection						
4010	Sprinklers	Wet pipe sprinkler system		S.F. Floor	1.79	1.79	1.7%
4020	Standpipes	N/A		—	—	—	
	D50 Electrical						
5010	Electrical Service/Distribution	800 ampere service, panel board and feeders		S.F. Floor	1.82	1.82	
5020	Lighting & Branch Wiring	Fluorescent fixtures, receptacles, switches, A.C. and misc. power		S.F. Floor	7.79	7.79	12.3%
5030	Communications & Security	Alarm systems, communications systems and emergency lighting		S.F. Floor	2.94	2.94	
5090	Other Electrical Systems	Emergency generator, 7.5KW		S.F. Floor	.06	.06	
E. EQUIPMENT & FURNISHINGS							
1010	Commercial Equipment	N/A		—	—	—	
1020	Institutional Equipment	N/A		—	—	—	
1030	Vehicular Equipment	N/A		—	—	—	3.1%
2020	Other Equipment	Dormitory furniture		S.F. Floor	702	3.23	
F. SPECIAL CONSTRUCTION							
1020	Integrated Construction	N/A		—	—	—	
1040	Special Facilities	N/A		—	—	—	0.0%
G. BUILDING SITEWORK	**N/A**						

	Sub-Total	102.69	**100%**
CONTRACTOR FEES (General Requirements: 10%, Overhead: 5%, Profit: 10%)	25%	25.67	
ARCHITECT FEES	7%	8.99	
Total Building Cost		137.35	

BUILDING TYPES

105

Costs per square foot of floor area

Exterior Wall	S.F. Area	20000	35000	45000	65000	85000	110000	135000	160000	200000
	L.F. Perimeter	260	340	400	440	500	540	560	590	640
Face Brick with Concrete Block Back-up	R/Conc. Frame	151.95	139.15	135.75	128.85	125.75	122.70	120.35	118.90	117.40
	Steel Frame	162.90	150.10	146.70	139.80	136.75	133.65	131.30	129.85	128.30
Decorative Concrete Block	R/Conc. Frame	141.30	131.15	128.45	123.25	120.90	118.60	116.90	115.75	114.65
	Steel Frame	152.40	142.20	139.50	134.25	131.95	129.70	127.95	126.85	125.75
Precast Concrete Panels With Exposed Aggregate	R/Conc. Frame	150.70	138.25	134.95	128.15	125.20	122.20	119.95	118.55	117.05
	Steel Frame	161.60	149.10	145.80	139.10	136.10	133.15	130.90	129.45	128.00
Perimeter Adj., Add or Deduct	Per 100 L.F.	13.30	7.60	5.95	4.10	3.15	2.45	1.95	1.65	1.35
Story Hgt. Adj., Add or Deduct	Per 1 Ft.	2.65	2.00	1.85	1.40	1.20	1.00	.90	.75	.65
For Basement, add $25.30 per square foot of basement area										

The above costs were calculated using the basic specifications shown on the facing page. These costs should be adjusted where necessary for design alternatives and owner's requirements. Reported completed project costs, for this type of structure, range from $81.25 to $195.95 per S.F.

Common additives

Description	Unit	$ Cost	Description	Unit	$ Cost
Carrels Hardwood	Each	785 - 1000	Kitchen Equipment		
Closed Circuit Surveillance, One station			Broiler	Each	3075
Camera and monitor	Each	1525	Coffee urn, twin, 6 gallon	Each	2925
For additional camera stations, add	Each	820	Cooler, 6 ft. long	Each	4075
Elevators, Electric passenger, 5 stops			Dishwasher, 10-12 racks per hr.	Each	4300
2000# capacity	Each	110,300	Food warmer	Each	665
2500# capacity	Each	113,300	Freezer, 44 C.F., reach-in	Each	9175
3500# capacity	Each	114,800	Ice cube maker, 50 lb. per day	Each	1600
Additional stop, add	Each	6550	Range with 1 oven	Each	2100
Emergency Lighting, 25 watt, battery operated			Laundry Equipment		
Lead battery	Each	247	Dryer, gas, 16 lb. capacity	Each	785
Nickel cadmium	Each	695	30 lb. capacity	Each	3000
Furniture	Student	1900 - 3600	Washer, 4 cycle	Each	875
Intercom System, 25 station capacity			Commercial	Each	1325
Master station	Each	2200	Smoke Detectors		
Intercom outlets	Each	142	Ceiling type	Each	161
Handset	Each	390	Duct type	Each	425
			TV Antenna, Master system, 12 outlet	Outlet	262
			30 outlet	Outlet	168
			100 outlet	Outlet	160

Important: See the Reference Section for Location Factors

Model costs calculated for a 6 story building with 12' story height and 85,000 square feet of floor area

College, Dormitory, 4-8 Story

			Unit	Unit Cost	Cost Per S.F.	% Of Sub-Total
A. SUBSTRUCTURE						
1010	Standard Foundations	Poured concrete; strip and spread footings	S.F. Ground	7.14	1.19	
1030	Slab on Grade	4" reinforced concrete with vapor barrier and granular base	S.F. Slab	3.98	.67	
2010	Basement Excavation	Site preparation for slab and trench for foundation wall and footing	S.F. Ground	.22	.04	2.4%
2020	Basement Walls	4' foundation wall	L.F. Wall	115	.50	
B. SHELL						
	B10 Superstructure					
1010	Floor Construction	Concrete slab with metal deck and beams	S.F. Floor	21	17.52	20.5%
1020	Roof Construction	Concrete slab with metal deck and beams	S.F. Roof	16.02	2.67	
	B20 Exterior Enclosure					
2010	Exterior Walls	Decorative concrete block	80% of wall	S.F. Wall	13.22	4.48	
2020	Exterior Windows	Aluminum horizontal sliding	20% of wall	Each	346	1.95	6.7%
2030	Exterior Doors	Double glass & aluminum doors	Each	2198	.21	
	B30 Roofing					
3010	Roof Coverings	Single ply membrane, loose laid and ballasted; perlite/EPS composite insulation	S.F. Roof	3.54	.59	0.6%
3020	Roof Openings	N/A	—	—	—	
C. INTERIORS						
1010	Partitions	Concrete block	9 S.F. Floor/L.F. Partition	S.F. Partition	7.48	8.31	
1020	Interior Doors	Single leaf wood	90 S.F. Floor/Door	Each	480	5.34	
1030	Fittings	Closet shelving, mirrors, bathroom accessories	S.F. Floor	1.46	1.46	
2010	Stair Construction	Concrete filled metal pan	Flight	7700	1.63	29.0%
3010	Wall Finishes	95% paint, 5% ceramic tile	S.F. Surface	—	4.33	
3020	Floor Finishes	80% carpet, 10% vinyl composition tile, 10% ceramic tile	S.F. Floor	6.81	6.81	
3030	Ceiling Finishes	Mineral fiber tile on concealed zee bars, paint	S.F. Ceiling	1.00	.74	
D. SERVICES						
	D10 Conveying					
1010	Elevators & Lifts	Four geared passenger elevators	Each	151,937	7.15	7.2%
1020	Escalators & Moving Walks	N/A	—	—	—	
	D20 Plumbing					
2010	Plumbing Fixtures	Toilet and service fixtures, supply and drainage	1 Fixture/390 S.F. Floor	Each	2710	6.95	
2020	Domestic Water Distribution	Oil fired water heater	S.F. Floor	.53	.53	7.8%
2040	Rain Water Drainage	Roof drains	S.F. Roof	1.32	.22	
	D30 HVAC					
3010	Energy Supply	Oil fired hot water, wall fin radiation	S.F. Floor	2.99	2.99	
3020	Heat Generating Systems	N/A	—	—	—	
3030	Cooling Generating Systems	Chilled water, air cooled condenser system	S.F. Floor	6.52	6.52	9.6%
3050	Terminal & Package Units	N/A	—	—	—	
3090	Other HVAC Sys. & Equipment	N/A	—	—	—	
	D40 Fire Protection					
4010	Sprinklers	Sprinklers, light hazard	S.F. Floor	1.71	1.71	1.7%
4020	Standpipes	N/A	—	—	—	
	D50 Electrical					
5010	Electrical Service/Distribution	1200 ampere service, panel board and feeders	S.F. Floor	.83	.83	
5020	Lighting & Branch Wiring	Fluorescent fixtures, receptacles, switches, A.C. and misc. power	S.F. Floor	7.60	7.60	11.7%
5030	Communications & Security	Alarm systems, communications systems and emergency lighting	S.F. Floor	2.92	2.92	
5090	Other Electrical Systems	Emergency generator, 30 kW	S.F. Floor	.16	.16	
E. EQUIPMENT & FURNISHINGS						
1010	Commercial Equipment	N/A	—	—	—	
1020	Institutional Equipment	N/A	—	—	—	2.7%
1030	Vehicular Equipment	N/A	—	—	—	
2020	Other Equipment	Built-in dormitory furnishings	S.F. Floor	702	2.64	
F. SPECIAL CONSTRUCTION						
1020	Integrated Construction	N/A	—	—	—	0.0%
1040	Special Facilities	N/A	—	—	—	
G. BUILDING SITEWORK	**N/A**					

		Sub-Total	98.66	**100%**
CONTRACTOR FEES (General Requirements: 10%, Overhead: 5%, Profit: 10%)		25%	24.66	
ARCHITECT FEES		7%	8.63	
	Total Building Cost		131.95	

BUILDING TYPES

Costs per square foot of floor area

Exterior Wall	S.F. Area	12000	20000	28000	37000	45000	57000	68000	80000	92000
	L.F. Perimeter	470	600	698	793	900	1060	1127	1200	1320
Face Brick with Concrete Brick Back-up	Steel Frame	208.25	171.65	154.80	144.15	138.80	133.55	129.20	125.85	123.80
	Bearing Walls	204.50	167.95	151.10	140.45	135.10	129.80	125.45	122.10	120.10
Decorative Concrete Block	Steel Frame	203.15	167.75	151.55	141.35	136.15	131.10	127.00	123.90	121.95
	Bearing Walls	199.65	164.20	148.00	137.80	132.65	127.55	123.45	120.35	118.40
Stucco on Concrete Block	Steel Frame	202.15	167.00	150.95	140.80	135.70	130.60	126.60	123.55	121.60
	Bearing Walls	198.55	163.45	147.40	137.25	132.15	127.10	123.05	119.95	118.05
Perimeter Adj., Add or Deduct	Per 100 L.F.	8.15	4.90	3.50	2.65	2.20	1.75	1.45	1.20	1.05
Story Hgt. Adj., Add or Deduct	Per 1 Ft.	1.50	1.20	.95	.85	.75	.70	.60	.60	.60
For Basement, add $24.10 per square foot of basement area										

The above costs were calculated using the basic specifications shown on the facing page. These costs should be adjusted where necessary for design alternatives and owner's requirements. Reported completed project costs, for this type of structure, range from $117.00 to $218.55 per S.F.

Common additives

Description	Unit	$ Cost
Cabinets, Base, door units, metal	L.F.	218
Drawer units	L.F.	415
Tall storage cabinets, open	L.F.	415
With doors	L.F.	500
Wall, metal 12-1/2" deep, open	L.F.	157
With doors	L.F.	281
Carrels Hardwood	Each	785 - 1000
Countertops, not incl. base cabinets, acid proof	S.F.	38 - 50
Stainless steel	S.F.	105
Fume Hood, Not incl. ductwork	L.F.	730 - 1900
Ductwork	Hood	1950 - 6775
Glassware Washer, Distilled water rinse	Each	7375 - 9975
Seating		
Auditorium chair, all veneer	Each	193
Veneer back, padded seat	Each	233
Upholstered, spring seat	Each	234
Classroom, movable chair & desk	Set	65 - 120
Lecture hall, pedestal type	Each	179 - 545

Description	Unit	$ Cost
Safety Equipment, Eye wash, hand held	Each	325
Deluge shower	Each	625
Sink, One piece plastic		
Flask wash, freestanding	Each	1950
Tables, acid resist. top, drawers	L.F.	164
Titration Unit, Four 2000 ml reservoirs	Each	9475

Model costs calculated for a 1 story building with 12′ story height and 45,000 square feet of floor area

				Unit	Unit Cost	Cost Per S.F.	% Of Sub-Total
A.	**SUBSTRUCTURE**						
1010	Standard Foundations	Poured concrete; strip and spread footings		S.F. Ground	2.35	2.35	
1030	Slab on Grade	4″ reinforced concrete with vapor barrier and granular base		S.F. Slab	3.98	3.98	11.4%
2010	Basement Excavation	Site preparation for slab and trench for foundation wall and footing		S.F. Ground	.14	.14	
2020	Basement Walls	4′ foundation wall		L.F. Wall	115	4.75	
B.	**SHELL**						
	B10 Superstructure						
1010	Floor Construction	Metal deck on open web steel joists	(5680 S.F.)	S.F. Floor	12.77	1.66	5.5%
1020	Roof Construction	Metal deck on open web steel joists		S.F. Roof	3.76	3.76	
	B20 Exterior Enclosure						
2010	Exterior Walls	Face brick with concrete block backup	75% of wall	S.F. Wall	23	4.28	
2020	Exterior Windows	Window wall	25% of wall	Each	31	1.91	7.7%
2030	Exterior Doors	Glass and metal doors and entrances with transom		Each	3035	1.35	
	B30 Roofing						
3010	Roof Coverings	Built-up tar and gravel with flashing; perlite/EPS composite insulation		S.F. Roof	3.86	3.86	4.1%
3020	Roof Openings	Skylight		S.F. Roof	.21	.21	
C.	**INTERIORS**						
1010	Partitions	Concrete block	10 S.F. Floor/L.F. Partition	S.F. Partition	8.67	8.67	
1020	Interior Doors	Single leaf-kalamein fire doors	820 S.F. Floor/Door	Each	735	.90	
1030	Fittings	Lockers		S.F. Floor	.04	.04	
2010	Stair Construction	N/A		—	—	—	23.6%
3010	Wall Finishes	60% paint, 40% epoxy coating		S.F. Surface	4.69	4.69	
3020	Floor Finishes	60% epoxy, 20% carpet, 20% vinyl composition tile		S.F. Floor	4.83	4.83	
3030	Ceiling Finishes	Mineral fiber tile on concealed zee runners		S.F. Ceiling	4.00	4.09	
D.	**SERVICES**						
	D10 Conveying						
1010	Elevators & Lifts	N/A		—	—	—	0.0%
1020	Escalators & Moving Walks	N/A		—	—	—	
	D20 Plumbing						
2010	Plumbing Fixtures	Toilet and service fixtures, supply and drainage	1 Fixture/260 S.F. Floor	Each	3923	15.09	
2020	Domestic Water Distribution	Gas fired water heater		S.F. Floor	.77	.77	16.8%
2040	Rain Water Drainage	Roof drains		S.F. Roof	.67	.67	
	D30 HVAC						
3010	Energy Supply	N/A		—	—	—	
3020	Heat Generating Systems	Included in D3050		—	—	—	
3030	Cooling Generating Systems	N/A		—	—	—	16.9%
3050	Terminal & Package Units	Multizone unit, gas heating, electric cooling		S.F. Floor	16.60	16.60	
3090	Other HVAC Sys. & Equipment	N/A		—	—	—	
	D40 Fire Protection						
4010	Sprinklers	Sprinklers, light hazard		S.F. Floor	1.75	1.75	1.8%
4020	Standpipes	N/A		—	—	—	
	D50 Electrical						
5010	Electrical Service/Distribution	1000 ampere service, panel board and feeders		S.F. Floor	1.47	1.47	
5020	Lighting & Branch Wiring	Fluorescent fixtures, receptacles, switches, A.C. and misc. power		S.F. Floor	8.71	8.71	11.0%
5030	Communications & Security	Alarm systems and emergency lighting		S.F. Floor	.59	.59	
5090	Other Electrical Systems	Emergency generator, 11.5 kW		S.F. Floor	.07	.07	
E.	**EQUIPMENT & FURNISHINGS**						
1010	Commercial Equipment	N/A		—	—	—	
1020	Institutional Equipment	Cabinets, fume hoods, lockers, glassware washer		S.F. Floor	1.05	1.05	1.1%
1030	Vehicular Equipment	N/A		—	—	—	
1090	Other Equipment	N/A		—	—	—	
F.	**SPECIAL CONSTRUCTION**						
1020	Integrated Construction	N/A		—	—	—	0.0%
1040	Special Facilities	N/A		—	—	—	
G.	**BUILDING SITEWORK**	**N/A**					

		Sub-Total	98.24	100%
	CONTRACTOR FEES (General Requirements: 10%, Overhead: 5%, Profit: 10%)	25%	24.58	
	ARCHITECT FEES	10%	12.28	
	Total Building Cost		**135.10**	

BUILDING TYPES

Costs per square foot of floor area

Exterior Wall	S.F. Area	15000	20000	25000	30000	35000	40000	45000	50000	55000
	L.F. Perimeter	354	425	457	513	568	583	629	644	683
Brick Face with Concrete Block Back-up	Steel Frame	135.85	131.80	127.70	125.80	124.45	122.35	121.45	120.05	119.45
	R/Conc. Frame	130.65	126.60	122.55	120.65	119.25	117.20	116.30	114.85	114.25
Precast Concrete Panel	Steel Frame	134.50	130.65	126.75	124.95	123.60	121.60	120.80	119.45	118.80
	R/Conc. Frame	129.70	125.75	121.80	119.90	118.60	116.55	115.70	114.40	113.75
Limestone Face Concrete Block Back-up	Steel Frame	144.25	139.40	134.20	131.90	130.25	127.55	126.45	124.65	123.85
	R/Conc. Frame	139.10	134.20	129.05	126.75	125.05	122.40	121.25	119.45	118.70
Perimeter Adj., Add or Deduct	Per 100 L.F.	7.05	5.30	4.25	3.55	3.00	2.70	2.40	2.15	1.95
Story Hgt. Adj., Add or Deduct	Per 1 Ft.	1.75	1.60	1.35	1.25	1.20	1.10	1.05	.95	.95

For Basement, add $27.15 per square foot of basement area

The above costs were calculated using the basic specifications shown on the facing page. These costs should be adjusted where necessary for design alternatives and owner's requirements. Reported completed project costs, for this type of structure, range from $95.85 to $195.60 per S.F.

Common additives

Description	Unit	$ Cost
Carrels Hardwood	Each	785 - 1000
Elevators, Hydraulic passenger, 2 stops		
2000# capacity	Each	49,000
2500# capacity	Each	50,000
3500# capacity	Each	53,700
Emergency Lighting, 25 watt, battery operated		
Lead battery	Each	247
Nickel cadmium	Each	695
Escalators, Metal		
32" wide, 10' story height	Each	100,300
20' story height	Each	117,300
48" wide, 10' Story height	Each	105,300
20' story height	Each	122,300
Glass		
32" wide, 10' story height	Each	98,800
20' story height	Each	117,300
48" wide, 10' story height	Each	105,300
20' story height	Each	122,300

Description	Unit	$ Cost
Lockers, Steel, Single tier, 60" or 72"	Opening	154 - 241
2 tier, 60" or 72" total	Opening	104 - 134
5 tier, box lockers	Opening	55 - 69
Locker bench, lam. maple top only	L.F.	18.80
Pedestals, steel pipe	Each	57
Sound System		
Amplifier, 250 watts	Each	1925
Speaker, ceiling or wall	Each	160
Trumpet	Each	305

Important: See the Reference Section for Location Factors

Model costs calculated for a 2 story building with 12' story height and 25,000 square feet of floor area

College, Student Union

				Unit	Unit Cost	Cost Per S.F.	% Of Sub-Total
A. SUBSTRUCTURE							
1010	Standard Foundations	Poured concrete; strip and spread footings		S.F. Ground	3.58	1.79	
1030	Slab on Grade	4" reinforced concrete with vapor barrier and granular base		S.F. Slab	3.98	1.99	
2010	Basement Excavation	Site preparation for slab and trench for foundation wall and footing		S.F. Ground	.14	.07	5.4%
2020	Basement Walls	4' foundation wall		L.F. Wall	61	1.12	
B. SHELL							
	B10 Superstructure						
1010	Floor Construction	Concrete flat plate		S.F. Floor	21	10.70	
1020	Roof Construction	Concrete flat plate		S.F. Roof	12.26	6.13	18.4%
	B20 Exterior Enclosure						
2010	Exterior Walls	Face brick with concrete block backup	75% of wall	S.F. Wall	23	7.82	
2020	Exterior Windows	Window wall	25% of wall	Each	31	3.49	12.7%
2030	Exterior Doors	Double aluminum and glass		Each	1880	.30	
	B30 Roofing						
3010	Roof Coverings	Built-up tar and gravel with flashing; perlite/EPS composite insulation		S.F. Roof	4.34	2.17	2.4%
3020	Roof Openings	Roof hatches		S.F. Roof	.10	.05	
C. INTERIORS							
1010	Partitions	Gypsum board on metal studs	14 S.F. Floor/L.F. Partition	S.F. Partition	4.62	3.30	
1020	Interior Doors	Single leaf hollow metal	140 S.F. Floor/Door	Each	608	4.34	
1030	Fittings	N/A		—	—	—	
2010	Stair Construction	Cast in place concrete		Flight	6800	1.08	21.5%
3010	Wall Finishes	50% paint, 50% vinyl wall covering		S.F. Surface	2.60	2.60	
3020	Floor Finishes	50% carpet, 50% vinyl composition tile		S.F. Floor	5.30	5.30	
3030	Ceiling Finishes	Suspended fiberglass board		S.F. Ceiling	3.00	3.05	
D. SERVICES							
	D10 Conveying						
1010	Elevators & Lifts	One hydraulic passenger elevator		Each	59,250	2.37	2.6%
1020	Escalators & Moving Walks	N/A		—	—	—	
	D20 Plumbing						
2010	Plumbing Fixtures	Toilet and service fixtures, supply and drainage	1 Fixture/1040 S.F. Floor	Each	1653	1.59	
2020	Domestic Water Distribution	Gas fired water heater		S.F. Floor	.46	.46	2.4%
2040	Rain Water Drainage	Roof drains		S.F. Roof	.38	.19	
	D30 HVAC						
3010	Energy Supply	N/A		—	—	—	
3020	Heat Generating Systems	Included in D3050		—	—	—	
3030	Cooling Generating Systems	N/A		—	—	—	18.1%
3050	Terminal & Package Units	Multizone unit, gas heating, electric cooling		S.F. Floor	16.60	16.60	
3090	Other HVAC Sys. & Equipment	N/A		—	—	—	
	D40 Fire Protection						
4010	Sprinklers	Wet pipe sprinkler system		S.F. Floor	1.87	1.87	2.0%
4020	Standpipes	N/A		—	—	—	
	D50 Electrical						
5010	Electrical Service/Distribution	600 ampere service, panel board and feeders		S.F. Floor	1.63	1.63	
5020	Lighting & Branch Wiring	Fluorescent fixtures, receptacles, switches, A.C. and misc. power		S.F. Floor	10.07	10.07	14.4%
5030	Communications & Security	Alarm systems, communications systems and emergency lighting		S.F. Floor	1.41	1.41	
5090	Other Electrical Systems	Emergency generator, 11.5 kW		S.F. Floor	.12	.12	
E. EQUIPMENT & FURNISHINGS							
1010	Commercial Equipment	N/A		—	—	—	
1020	Institutional Equipment	N/A		—	—	—	
1030	Vehicular Equipment	N/A		—	—	—	0.0%
1090	Other Equipment	N/A		—	—	—	
F. SPECIAL CONSTRUCTION							
1020	Integrated Construction	N/A		—	—	—	
1040	Special Facilities	N/A		—	—	—	0.0%
G. BUILDING SITEWORK	**N/A**						

					Sub-Total	91.61	100%
	CONTRACTOR FEES (General Requirements: 10%, Overhead: 5%, Profit: 10%)				25%	22.92	
	ARCHITECT FEES				7%	8.02	

	Total Building Cost	**122.55**	

BUILDING TYPES

Costs per square foot of floor area

Exterior Wall	S.F. Area	4000	6000	8000	10000	12000	14000	16000	18000	20000
	L.F. Perimeter	260	340	420	453	460	510	560	610	600
Face Brick with Concrete Block Back-up	Bearing Walls	125.30	118.70	115.50	110.25	105.35	103.95	102.85	102.00	99.35
	Steel Frame	121.75	116.00	113.20	108.75	104.55	103.25	102.35	101.65	99.40
Decorative Concrete Block	Bearing Walls	110.45	105.20	102.70	98.55	94.80	93.70	92.85	92.20	90.15
	Steel Frame	111.70	107.20	105.05	101.70	98.60	97.60	96.90	96.40	94.70
Tilt Up Concrete Wall Panels	Bearing Walls	112.20	107.30	104.90	101.15	97.65	96.60	95.80	95.20	93.30
	Steel Frame	108.70	104.55	102.60	99.60	96.80	95.95	95.35	94.80	93.30
Perimeter Adj., Add or Deduct	Per 100 L.F.	17.00	11.35	8.45	6.75	5.70	4.80	4.25	3.80	3.35
Story Hgt. Adj., Add or Deduct	Per 1 Ft.	2.20	1.85	1.70	1.45	1.30	1.20	1.15	1.15	1.00
For Basement, add $23.75 per square foot of basement area										

The above costs were calculated using the basic specifications shown on the facing page. These costs should be adjusted where necessary for design alternatives and owner's requirements. Reported completed project costs, for this type of structure, range from $58.50 to $184.80 per S.F.

Common additives

Description	Unit	$ Cost
Bar, Front bar	L.F.	320
Back bar	L.F.	257
Booth, Upholstered, custom straight	L.F.	166 - 305
"L" or "U" shaped	L.F.	172 - 290
Bowling Alleys, incl. alley, pinsetter		
Scorer, counter & misc. supplies, average	Lane	53,000
For automatic scorer, add	Lane	7925
Emergency Lighting, 25 watt, battery operated		
Lead battery	Each	247
Nickel cadmium	Each	695
Kitchen Equipment		
Broiler	Each	3075
Coffee urn, twin 6 gallon	Each	2925
Cooler, 6 ft. long	Each	4075
Dishwasher, 10-12 racks per hr.	Each	4300
Food warmer	Each	665
Freezer, 44 C.F., reach-in	Each	9175
Ice cube maker, 50 lb. per day	Each	1600
Range with 1 oven	Each	2100

Description	Unit	$ Cost
Movie Equipment		
Projector, 35mm	Each	10,400 - 14,300
Screen, wall or ceiling hung	S.F.	7.20 - 11.05
Partitions, Folding leaf, wood		
Acoustic type	S.F.	60 - 99
Seating		
Auditorium chair, all veneer	Each	193
Veneer back, padded seat	Each	233
Upholstered, spring seat	Each	234
Classroom, movable chair & desk	Set	65 - 120
Lecture hall, pedestal type	Each	179 - 545
Sound System		
Amplifier, 250 watts	Each	1925
Speaker, ceiling or wall	Each	160
Trumpet	Each	305
Stage Curtains, Medium weight	S.F.	8.55 - 315
Curtain Track, Light duty	L.F.	64
Swimming Pools, Complete, gunite	S.F.	51 - 62

Important: See the Reference Section for Location Factors

Model costs calculated for a 1 story building with 12' story height and 10,000 square feet of floor area

				Unit	Unit Cost	Cost Per S.F.	% Of Sub-Total
A.	**SUBSTRUCTURE**						
1010	Standard Foundations	Poured concrete; strip and spread footings		S.F. Ground	1.73	1.73	
1030	Slab on Grade	4" reinforced concrete with vapor barrier and granular base		S.F. Slab	3.98	3.98	
2010	Basement Excavation	Site preparation for slab and trench for foundation wall and footing		S.F. Ground	.23	.23	14.8%
2020	Basement Walls	4' foundation wall		L.F. Wall	115	6.04	
B.	**SHELL**						
	B10 Superstructure						
1010	Floor Construction	N/A		—	—	—	6.3%
1020	Roof Construction	Metal deck on open web steel joists		S.F. Roof	5.13	5.13	
	B20 Exterior Enclosure						
2010	Exterior Walls	Face brick with concrete block backup	80% of wall	S.F. Wall	23	10.33	
2020	Exterior Windows	Aluminum sliding	20% of wall	Each	507	1.72	15.8%
2030	Exterior Doors	Double aluminum and glass and hollow metal		Each	1748	.70	
	B30 Roofing						
3010	Roof Coverings	Built-up tar and gravel with flashing; perlite/EPS composite insulation		S.F. Roof	4.98	4.98	6.2%
3020	Roof Openings	Roof hatches		S.F. Roof	.06	.06	
C.	**INTERIORS**						
1010	Partitions	Gypsum board on metal studs	14 S.F. Floor/L.F. Partition	S.F. Partition	6.24	4.46	
1020	Interior Doors	Single leaf hollow metal	140 S.F. Floor/Door	Each	608	4.34	
1030	Fittings	Toilet partitions, directory board, mailboxes		S.F. Floor	1.71	1.71	
2010	Stair Construction	N/A		—	—	—	26.8%
3010	Wall Finishes	Paint		S.F. Surface	1.95	1.95	
3020	Floor Finishes	50% carpet, 50% vinyl tile		S.F. Floor	5.12	5.12	
3030	Ceiling Finishes	Mineral fiber tile on concealed zee bars		S.F. Ceiling	4.00	4.09	
D.	**SERVICES**						
	D10 Conveying						
1010	Elevators & Lifts	N/A		—	—	—	0.0%
1020	Escalators & Moving Walks	N/A		—	—	—	
	D20 Plumbing						
2010	Plumbing Fixtures	Kitchen, toilet and service fixtures, supply and drainage	1 Fixture/910 S.F. Floor	Each	1956	2.15	
2020	Domestic Water Distribution	Electric water heater		S.F. Floor	3.82	3.82	7.8%
2040	Rain Water Drainage	Roof drains		S.F. Roof	.32	.32	
	D30 HVAC						
3010	Energy Supply	N/A		—	—	—	
3020	Heat Generating Systems	Included in D3050		—	—	—	
3030	Cooling Generating Systems	N/A		—	—	—	10.7%
3050	Terminal & Package Units	Single zone rooftop unit, gas heating, electric cooling		S.F. Floor	8.64	8.64	
3090	Other HVAC Sys. & Equipment	N/A		—	—	—	
	D40 Fire Protection						
4010	Sprinklers	Wet pipe sprinkler system		S.F. Floor	2.05	2.05	2.5%
4020	Standpipes	N/A		—	—	—	
	D50 Electrical						
5010	Electrical Service/Distribution	200 ampere service, panel board and feeders		S.F. Floor	.76	.76	
5020	Lighting & Branch Wiring	Incandescent fixtures, receptacles, switches, A.C. and misc. power		S.F. Floor	4.11	4.11	6.6%
5030	Communications & Security	Alarm systems and emergency lighting		S.F. Floor	.33	.33	
5090	Other Electrical Systems	Emergency generator, 15 kW		S.F. Floor	.13	.13	
E.	**EQUIPMENT & FURNISHINGS**						
1010	Commercial Equipment	Freezer, chest type		S.F. Floor	.64	.64	
1020	Institutional Equipment	N/A		—	—	—	2.4%
1030	Vehicular Equipment	N/A		—	—	—	
1090	Other Equipment	Kitchen equipment, directory board, mailboxes, built-in coat racks		S.F. Floor	1.28	1.28	
F.	**SPECIAL CONSTRUCTION**						
1020	Integrated Construction	N/A		—	—	—	0.0%
1040	Special Facilities	N/A		—	—	—	
G.	**BUILDING SITEWORK**	**N/A**					

		Sub-Total	80.91	**100%**
CONTRACTOR FEES (General Requirements: 10%, Overhead: 5%, Profit: 10%)			25%	20.24
ARCHITECT FEES			9%	9.10
		Total Building Cost	110.25	

BUILDING TYPES

Costs per square foot of floor area

Exterior Wall	S.F. Area	16000	23000	30000	37000	44000	51000	58000	65000	72000
	L.F. Perimeter	597	763	821	968	954	1066	1090	1132	1220
Limestone with Concrete Block Back-up	R/Conc. Frame	173.10	167.05	160.25	158.40	153.45	152.40	150.00	148.55	147.85
	Steel Frame	169.20	163.15	156.35	154.50	149.60	148.50	146.15	144.65	144.00
Face Brick with Concrete Block Back-up	R/Conc. Frame	160.60	155.95	151.05	149.60	146.20	145.45	143.75	142.70	142.25
	Steel Frame	156.75	152.05	147.15	145.70	142.30	141.50	139.85	138.85	138.35
Stone with Concrete Block Back-up	R/Conc. Frame	164.05	159.00	153.55	152.00	148.20	147.30	145.45	144.25	143.70
	Steel Frame	160.25	155.15	149.75	148.20	144.35	143.50	141.60	140.45	139.95
Perimeter Adj., Add or Deduct	Per 100 L.F.	6.25	4.40	3.35	2.75	2.35	1.95	1.75	1.50	1.40
Story Hgt. Adj., Add or Deduct	Per 1 Ft.	2.25	2.00	1.65	1.55	1.35	1.25	1.15	1.05	1.05

For Basement, add $22.45 per square foot of basement area

The above costs were calculated using the basic specifications shown on the facing page. These costs should be adjusted where necessary for design alternatives and owner's requirements. Reported completed project costs, for this type of structure, range from $102.95 to $191.85 per S.F.

Common additives

Description	Unit	$ Cost		Description	Unit	$ Cost
Benches, Hardwood	L.F.	92 - 169		Flagpoles, Complete		
Clock System				Aluminum, 20' high	Each	1375
20 room	Each	14,200		40' high	Each	3025
50 room	Each	34,600		70' high	Each	9500
Closed Circuit Surveillance, One station				Fiberglass, 23' high	Each	1650
Camera and monitor	Each	1525		39'-5" high	Each	3175
For additional camera stations, add	Each	820		59' high	Each	7950
Directory Boards, Plastic, glass covered				Intercom System, 25 station capacity		
30" x 20"	Each	575		Master station	Each	2200
36" x 48"	Each	1275		Intercom outlets	Each	142
Aluminum, 24" x 18"	Each	525		Handset	Each	390
36" x 24"	Each	600		Safe, Office type, 4 hour rating		
48" x 32"	Each	810		30" x 18" x 18"	Each	3425
48" x 60"	Each	1875		62" x 33" x 20"	Each	7450
Emergency Lighting, 25 watt, battery operated				Smoke Detectors		
Lead battery	Each	247		Ceiling type	Each	161
Nickel cadmium	Each	695		Duct type	Each	425

Model costs calculated for a 1 story building with 14' story height and 30,000 square feet of floor area

				Unit	Unit Cost	Cost Per S.F.	% Of Sub-Total
A.	**SUBSTRUCTURE**						
1010	Standard Foundations	Poured concrete; strip and spread footings		S.F. Ground	1.33	1.33	
1030	Slab on Grade	4" reinforced concrete with vapor barrier and granular base		S.F. Slab	3.98	3.98	
2010	Basement Excavation	Site preparation for slab and trench for foundation wall and footing		S.F. Ground	.14	.14	6.0%
2020	Basement Walls	4' foundation wall		L.F. Wall	61	1.79	
B.	**SHELL**						
	B10 Superstructure						
1010	Floor Construction	Cast-in-place columns		L.F. Column	53	1.35	16.3%
1020	Roof Construction	Cast-in-place concrete waffle slab		S.F. Roof	18.20	18.20	
	B20 Exterior Enclosure						
2010	Exterior Walls	Limestone panels with concrete block backup	75% of wall	S.F. Wall	47	13.68	
2020	Exterior Windows	Aluminum with insulated glass	25% of wall	Each	559	2.32	13.7%
2030	Exterior Doors	Double wood		Each	1665	.38	
	B30 Roofing						
3010	Roof Coverings	Built-up tar and gravel with flashing; perlite/EPS composite insulation		S.F. Roof	3.87	3.87	3.3%
3020	Roof Openings	Roof hatches		S.F. Roof	.06	.06	
C.	**INTERIORS**						
1010	Partitions	Plaster on metal studs	10 S.F. Floor/L.F. Partition	S.F. Partition	10.74	12.89	
1020	Interior Doors	Single leaf wood	100 S.F. Floor/Door	Each	480	4.80	
1030	Fittings	Toilet partitions		S.F. Floor	.50	.50	
2010	Stair Construction	N/A		—	—	—	34.6%
3010	Wall Finishes	70% paint, 20% wood paneling, 10% vinyl wall covering		S.F. Surface	4.02	4.88	
3020	Floor Finishes	60% hardwood, 20% carpet, 20% terrazzo		S.F. Floor	10.63	10.63	
3030	Ceiling Finishes	Gypsum plaster on metal lath, suspended		S.F. Ceiling	8.00	7.81	
D.	**SERVICES**						
	D10 Conveying						
1010	Elevators & Lifts	N/A		—	—	—	0.0%
1020	Escalators & Moving Walks	N/A		—	—	—	
	D20 Plumbing						
2010	Plumbing Fixtures	Toilet and service fixtures, supply and drainage	1 Fixture/1110 S.F. Floor	Each	2264	2.04	
2020	Domestic Water Distribution	Electric hot water heater		S.F. Floor	1.37	1.37	3.2%
2040	Rain Water Drainage	Roof drains		S.F. Roof	.40	.40	
	D30 HVAC						
3010	Energy Supply	N/A		—	—	—	
3020	Heat Generating Systems	Included in D3050		—	—	—	
3030	Cooling Generating Systems	N/A		—	—	—	13.9%
3050	Terminal & Package Units	Multizone unit, gas heating, electric cooling		S.F. Floor	16.60	16.60	
3090	Other HVAC Sys. & Equipment	N/A		—	—	—	
	D40 Fire Protection						
4010	Sprinklers	Wet pipe sprinkler system		S.F. Floor	1.75	1.75	1.5%
4020	Standpipes	N/A		—	—	—	
	D50 Electrical						
5010	Electrical Service/Distribution	400 ampere service, panel board and feeders		S.F. Floor	.72	.72	
5020	Lighting & Branch Wiring	Fluorescent fixtures, receptacles, switches, A.C. and misc. power		S.F. Floor	7.85	7.85	7.5%
5030	Communications & Security	Alarm systems and emergency lighting		S.F. Floor	.36	.36	
5090	Other Electrical Systems	Emergency generator, 11.5 kW		S.F. Floor	.10	.10	
E.	**EQUIPMENT & FURNISHINGS**						
1010	Commercial Equipment	N/A		—	—	—	
1020	Institutional Equipment	N/A		—	—	—	
1030	Vehicular Equipment	N/A		—	—	—	0.0%
1090	Other Equipment	N/A		—	—	—	
F.	**SPECIAL CONSTRUCTION**						
1020	Integrated Construction	N/A		—	—	—	0.0%
1040	Special Facilities	N/A		—	—	—	
G.	**BUILDING SITEWORK**	**N/A**					

			Sub-Total	119.80	**100%**
CONTRACTOR FEES (General Requirements: 10%, Overhead: 5%, Profit: 10%)		25%	29.97	
ARCHITECT FEES		7%	10.48	

Total Building Cost	**160.25**

BUILDING TYPES

Costs per square foot of floor area

Exterior Wall	S.F. Area	30000	40000	45000	50000	60000	70000	80000	90000	100000
	L.F. Perimeter	410	493	535	533	600	666	733	800	795
Limestone with Concrete Block Back-up	R/Conc. Frame	184.25	176.95	174.70	170.25	167.10	164.75	163.05	161.70	158.65
	Steel Frame	181.60	174.40	172.10	167.65	164.50	162.15	160.45	159.10	156.05
Face Brick with Concrete Block Back-up	R/Conc. Frame	172.45	166.40	164.40	161.10	158.50	156.55	155.15	154.10	151.80
	Steel Frame	169.85	163.80	161.80	158.50	155.95	153.95	152.55	151.50	149.20
Stone with Concrete Block Back-up	R/Conc. Frame	175.70	169.35	167.35	163.70	160.95	158.90	157.35	156.20	153.70
	Steel Frame	173.10	166.75	164.70	161.10	158.35	156.30	154.75	153.60	151.10
Perimeter Adj., Add or Deduct	Per 100 L.F.	9.30	7.10	6.25	5.70	4.70	4.00	3.50	3.15	2.80
Story Hgt. Adj., Add or Deduct	Per 1 Ft.	2.95	2.80	2.55	2.35	2.20	2.10	2.00	2.00	1.75
For Basement, add $22.90 per square foot of basement area										

The above costs were calculated using the basic specifications shown on the facing page. These costs should be adjusted where necessary for design alternatives and owner's requirements. Reported completed project costs, for this type of structure, range from $103.45 to $193.00 per S.F.

Common additives

Description	Unit	$ Cost	Description	Unit	$ Cost
Benches, Hardwood	L.F.	92 - 169	Emergency Lighting, 25 watt, battery operated		
Clock System			Lead battery	Each	247
20 room	Each	14,200	Nickel cadmium	Each	695
50 room	Each	34,600	Flagpoles, Complete		
Closed Circuit Surveillance, One station			Aluminum, 20' high	Each	1375
Camera and monitor	Each	1525	40' high	Each	3025
For additional camera stations, add	Each	820	70' high	Each	9500
Directory Boards, Plastic, glass covered			Fiberglass, 23' high	Each	1650
30" x 20"	Each	575	39'-5" high	Each	3175
36" x 48"	Each	1275	59' high	Each	7950
Aluminum, 24" x 18"	Each	525	Intercom System, 25 station capacity		
36" x 24"	Each	600	Master station	Each	2200
48" x 32"	Each	810	Intercom outlets	Each	142
48" x 60"	Each	1875	Handset	Each	390
Elevators, Hydraulic passenger, 2 stops			Safe, Office type, 4 hour rating		
1500# capacity	Each	47,900	30" x 18" x 18"	Each	3425
2500# capacity	Each	50,000	62" x 33" x 20"	Each	7450
3500# capacity	Each	53,700	Smoke Detectors		
Additional stop, add	Each	4550	Ceiling type	Each	161
			Duct type	Each	425

Model costs calculated for a 3 story building with 12' story height and 60,000 square feet of floor area

				Unit	Unit Cost	Cost Per S.F.	% Of Sub-Total
A. SUBSTRUCTURE							
1010	Standard Foundations	Poured concrete; strip and spread footings		S.F. Ground	3.06	1.02	
1030	Slab on Grade	4" reinforced concrete with vapor barrier and granular base		S.F. Slab	3.98	1.32	
2010	Basement Excavation	Site preparation for slab and trench for foundation wall and footing		S.F. Ground	.14	.05	2.7%
2020	Basement Walls	4' foundation wall		L.F. Wall	115	.71	
B. SHELL							
	B10 Superstructure						
1010	Floor Construction	Concrete slab with metal deck and beams		S.F. Floor	21	14.52	16.6%
1020	Roof Construction	Concrete slab with metal deck and beams		S.F. Roof	14.64	4.88	
	B20 Exterior Enclosure						
2010	Exterior Walls	Face brick with concrete block backup	75% of wall	S.F. Wall	25	6.85	
2020	Exterior Windows	Horizontal pivoted steel	25% of wall	Each	589	5.89	11.1%
2030	Exterior Doors	Double aluminum and glass and hollow metal		Each	2087	.21	
	B30 Roofing						
3010	Roof Coverings	Built-up tar and gravel with flashing; perlite/EPS composite insulation		S.F. Roof	4.17	1.39	1.2%
3020	Roof Openings	N/A		—	—	—	
C. INTERIORS							
1010	Partitions	Plaster on metal studs	10 S.F. Floor/L.F. Partition	S.F. Partition	10.87	10.87	
1020	Interior Doors	Single leaf wood	100 S.F. Floor/Door	Each	480	4.80	
1030	Fittings	Toilet partitions		S.F. Floor	.26	.26	
2010	Stair Construction	Concrete filled metal pan		Flight	8975	1.50	34.3%
3010	Wall Finishes	70% paint, 20% wood paneling, 10% vinyl wall covering		S.F. Surface	4.07	4.07	
3020	Floor Finishes	60% hardwood, 20% terrazzo, 20% carpet		S.F. Floor	10.63	10.63	
3030	Ceiling Finishes	Gypsum plaster on metal lath, suspended		S.F. Ceiling	8.00	7.81	
D. SERVICES							
	D10 Conveying						
1010	Elevators & Lifts	Five hydraulic passenger elevators		Each	89,520	7.46	6.4%
1020	Escalators & Moving Walks	N/A		—	—	—	
	D20 Plumbing						
2010	Plumbing Fixtures	Toilet and service fixtures, supply and drainage	1 Fixture/665 S.F. Floor	Each	1363	2.05	
2020	Domestic Water Distribution	Electric water heater		S.F. Floor	2.31	2.31	3.9%
2040	Rain Water Drainage	Roof drains		S.F. Roof	.69	.23	
	D30 HVAC						
3010	Energy Supply	N/A		—	—	—	
3020	Heat Generating Systems	Included in D3050		—	—	—	
3030	Cooling Generating Systems	N/A		—	—	—	14.2%
3050	Terminal & Package Units	Multizone unit, gas heating, electric cooling		S.F. Floor	16.60	16.60	
3090	Other HVAC Sys. & Equipment	N/A		—	—	—	
	D40 Fire Protection						
4010	Sprinklers	Wet pipe sprinkler system		S.F. Floor	1.79	1.79	1.5%
4020	Standpipes	N/A		—	—	—	
	D50 Electrical						
5010	Electrical Service/Distribution	800 ampere service, panel board and feeders		S.F. Floor	.86	.86	
5020	Lighting & Branch Wiring	Fluorescent fixtures, receptacles, switches, A.C. and misc. power		S.F. Floor	7.96	7.96	8.0%
5030	Communications & Security	Alarm systems and emergency lighting		S.F. Floor	.36	.36	
5090	Other Electrical Systems	Emergency generator, 15 kW		S.F. Floor	.17	.17	
E. EQUIPMENT & FURNISHINGS							
1010	Commercial Equipment	N/A		—	—	—	
1020	Institutional Equipment	N/A		—	—	—	
1030	Vehicular Equipment	N/A		—	—	—	0.0%
1090	Other Equipment	N/A		—	—	—	
F. SPECIAL CONSTRUCTION							
1020	Integrated Construction	N/A		—	—	—	0.0%
1040	Special Facilities	N/A		—	—	—	
G. BUILDING SITEWORK	**N/A**						

		Sub-Total	116.57	100%
CONTRACTOR FEES (General Requirements: 10%, Overhead: 5%, Profit: 10%)		25%	29.18	
ARCHITECT FEES		7%	10.20	
	Total Building Cost		**155.95**	

Costs per square foot of floor area

Exterior Wall	S.F. Area	2000	5000	7000	10000	12000	15000	18000	21000	25000
	L.F. Perimeter	200	310	360	440	480	520	560	600	660
Tiltup Concrete Panel	Steel Joists	150.60	122.70	115.90	110.95	108.55	105.50	103.50	102.00	100.80
Decorative Concrete Block	Bearing Walls	143.45	117.20	110.85	106.25	104.00	101.15	99.30	97.90	96.80
Brick on Block	Steel Joists	172.10	140.45	132.45	126.65	123.75	120.00	117.60	115.80	114.30
Stucco on Concrete Block	Wood Truss	146.80	120.30	113.90	109.25	107.05	104.15	102.30	100.95	99.80
Brick Veneer	Wood Frame	153.70	125.60	118.60	113.55	111.05	107.85	105.80	104.30	103.00
Wood Siding	Wood Frame	139.25	117.60	112.30	108.50	106.65	104.35	102.80	101.70	100.70
Perimeter Adj., Add or Deduct	Per 100 L.F.	23.10	9.25	6.55	4.60	3.85	3.05	2.55	2.20	1.80
Story Hgt. Adj., Add or Deduct	Per 1 Ft.	2.55	1.55	1.30	1.10	1.00	.85	.80	.75	.60

For Basement, add $22.25 per square foot of basement area

The above costs were calculated using the basic specifications shown on the facing page. These costs should be adjusted where necessary for design alternatives and owner's requirements. Reported completed project costs, for this type of structure, range from $44.65 to $162.55 per S.F.

Common additives

Description	Unit	$ Cost
Emergency Lighting, 25 watt, battery operated		
Lead battery	Each	247
Nickel cadmium	Each	695
Flagpoles, Complete		
Aluminum, 20' high	Each	1375
40' high	Each	3025
70' high	Each	9500
Fiberglass, 23' high	Each	1650
39'-5" high	Each	3175
59' high	Each	7950
Gym Floor, Incl. sleepers and finish, maple	S.F.	11.10
Intercom System, 25 Station capacity		
Master station	Each	2200
Intercom outlets	Each	142
Handset	Each	390

Description	Unit	$ Cost
Lockers, Steel, single tier, 60" to 72"	Opening	154 - 241
2 tier, 60" to 72" total	Opening	104 - 134
5 tier, box lockers	Opening	55 - 69
Locker bench, lam. maple top only	L.F.	18.80
Pedestals, steel pipe	Each	57
Smoke Detectors		
Ceiling type	Each	161
Duct type	Each	425
Sound System		
Amplifier, 250 watts	Each	1925
Speaker, ceiling or wall	Each	160
Trumpet	Each	305

Model costs calculated for a 1 story building with 12' story height and 10,000 square feet of floor area

				Unit	Unit Cost	Cost Per S.F.	% Of Sub-Total
A.	**SUBSTRUCTURE**						
1010	Standard Foundations	Poured concrete; strip and spread footings		S.F. Ground	1.59	1.59	
1030	Slab on Grade	4" concrete with vapor barrier and granular base		S.F. Slab	3.47	3.47	
2010	Basement Excavation	Site preparation for slab and trench for foundation wall and footing		S.F. Ground	.23	.23	9.6%
2020	Basement Walls	4' foundation wall		L.F. Wall	61	2.69	
B.	**SHELL**						
	B10 Superstructure						
1010	Floor Construction	Wood beams on columns		S.F. Floor	.15	.15	8.6%
1020	Roof Construction	Wood trusses		S.F. Roof	7.02	7.02	
	B20 Exterior Enclosure						
2010	Exterior Walls	Brick veneer on wood studs	85% of wall	S.F. Wall	18.85	8.46	
2020	Exterior Windows	Window wall	15% of wall	Each	37	2.95	17.2%
2030	Exterior Doors	Aluminum and glass; steel		Each	1523	2.90	
	B30 Roofing						
3010	Roof Coverings	Asphalt shingles, 9" fiberglass batt insulation, gutters and downspouts		S.F. Roof	2.69	2.69	3.2%
3020	Roof Openings	N/A		—	—	—	
C.	**INTERIORS**						
1010	Partitions	Gypsum board on wood studs	8 S.F. Floor/S.F. Partition	S.F. Partition	5.88	3.92	
1020	Interior Doors	Single leaf hollow metal	700 S.F. Floor/Door	Each	608	1.62	
1030	Fittings	Toilet partitions		S.F. Floor	.45	.45	
2010	Stair Construction	N/A		—	—	—	15.1%
3010	Wall Finishes	Paint		S.F. Surface	1.61	1.61	
3020	Floor Finishes	5% quarry tile, 95% vinyl composition tile		S.F. Floor	1.96	1.96	
3030	Ceiling Finishes	Fiberglass tile on tee grid		S.F. Ceiling	3.00	3.05	
D.	**SERVICES**						
	D10 Conveying						
1010	Elevators & Lifts	N/A		—	—	—	0.0%
1020	Escalators & Moving Walks	N/A		—	—	—	
	D20 Plumbing						
2010	Plumbing Fixtures	Toilet and service fixtures, supply and drainage	1 Fixture/455 S.F. Floor	Each	1091	6.13	
2020	Domestic Water Distribution	Electric water heater		S.F. Floor	1.36	1.36	9.0%
2040	Rain Water Drainage	N/A		—	—	—	
	D30 HVAC						
3010	Energy Supply	Oil fired hot water, wall fin radiation		S.F. Floor	7.42	7.42	
3020	Heat Generating Systems	N/A		—	—	—	
3030	Cooling Generating Systems	N/A		—	—	—	20.6%
3050	Terminal & Package Units	Split systems with air cooled condensing units		S.F. Floor	9.74	9.74	
3090	Other HVAC Sys. & Equipment	N/A		—	—	—	
	D40 Fire Protection						
4010	Sprinklers	Sprinkler, light hazard		S.F. Floor	2.05	2.05	2.5%
4020	Standpipes	N/A		—	—	—	
	D50 Electrical						
5010	Electrical Service/Distribution	200 ampere service, panel board and feeders		S.F. Floor	.76	.76	
5020	Lighting & Branch Wiring	Fluorescent fixtures, receptacles, switches, A.C. and misc. power		S.F. Floor	7.61	7.61	10.5%
5030	Communications & Security	Alarm systems and emergency lighting		S.F. Floor	.34	.34	
5090	Other Electrical Systems	Emergency generator, 15 kW		S.F. Floor	.07	.07	
E.	**EQUIPMENT & FURNISHINGS**						
1010	Commercial Equipment	N/A		—	—	—	
1020	Institutional Equipment	Cabinets and countertop		S.F. Floor	3.10	3.10	3.7%
1030	Vehicular Equipment	N/A		—	—	—	
1090	Other Equipment	N/A		—	—	—	
F.	**SPECIAL CONSTRUCTION**						
1020	Integrated Construction	N/A		—	—	—	0.0%
1040	Special Facilities	N/A		—	—	—	
G.	**BUILDING SITEWORK**	**N/A**					

			Sub-Total	83.34	**100%**
	CONTRACTOR FEES (General Requirements: 10%, Overhead: 5%, Profit: 10%)		25%	20.83	
	ARCHITECT FEES		9%	9.38	

Total Building Cost	**113.55**

BUILDING TYPES

BUILDING TYPES

Costs per square foot of floor area

Exterior Wall	S.F. Area	12000	18000	24000	30000	36000	42000	48000	54000	60000
	L.F. Perimeter	460	580	713	730	826	880	965	1006	1045
Concrete Block	Steel Frame	90.60	85.65	83.30	80.20	79.00	77.75	77.15	76.20	75.55
	Bearing Walls	88.55	83.50	81.25	78.10	76.95	75.70	75.05	74.10	73.45
Precast Concrete Panels	Steel Frame	94.95	89.25	86.75	82.90	81.65	80.15	79.40	78.35	77.50
Insulated Metal Panels	Steel Frame	92.05	86.85	84.45	81.05	79.90	78.55	77.90	76.90	76.20
Face Brick on Common Brick	Steel Frame	105.15	97.85	94.65	89.45	87.75	85.75	84.80	83.30	82.20
Tilt-up Concrete Panel	Steel Frame	92.50	87.20	84.85	81.40	80.20	78.80	78.15	77.15	76.40
Perimeter Adj., Add or Deduct	Per 100 L.F.	4.20	2.70	2.15	1.65	1.40	1.20	1.00	.95	.80
Story Hgt. Adj., Add or Deduct	Per 1 Ft.	.60	.45	.55	.35	.45	.35	.30	.30	.25
For Basement, add $22.70 per square foot of basement area										

The above costs were calculated using the basic specifications shown on the facing page. These costs should be adjusted where necessary for design alternatives and owner's requirements. Reported completed project costs, for this type of structure, range from $33.10 to $128.20 per S.F.

Common additives

Description	Unit	$ Cost		Description	Unit	$ Cost
Clock System				Dock Levelers, Hinged 10 ton cap.		
20 room	Each	14,200		6' x 8'	Each	5650
50 room	Each	34,600		7' x 8'	Each	5500
Dock Bumpers, Rubber blocks				Partitions, Woven wire, 10 ga., 1-1/2" mesh		
4-1/2" thick, 10" high, 14" long	Each	65		4' wide x 7' high	Each	154
24" long	Each	106		8' high	Each	166
36" long	Each	117		10' High	Each	201
12" high, 14" long	Each	109		Platform Lifter, Portable, 6'x 6'		
24" long	Each	123		3000# cap.	Each	8325
36" long	Each	141		4000# cap.	Each	10,300
6" thick, 10" high, 14" long	Each	101		Fixed, 6' x 8', 5000# cap.	Each	10,700
24" long	Each	135				
36" long	Each	164				
20" high, 11" long	Each	167				
Dock Boards, Heavy						
60" x 60" Aluminum, 5,000# cap.	Each	1325				
9000# cap.	Each	1250				
15,000# cap.	Each	1375				

Important: See the Reference Section for Location Factors

Model costs calculated for a 1 story building with 20' story height and 30,000 square feet of floor area

				Unit	Unit Cost	Cost Per S.F.	% Of Sub-Total
A.	**SUBSTRUCTURE**						
1010	Standard Foundations	Poured concrete; strip and spread footings		S.F. Ground	1.09	1.09	
1030	Slab on Grade	4" reinforced concrete with vapor barrier and granular base		S.F. Slab	5.13	5.13	
2010	Basement Excavation	Site preparation for slab and trench for foundation wall and footing		S.F. Ground	.14	.14	13.6%
2020	Basement Walls	4' foundation wall		L.F. Wall	61	1.78	
B.	**SHELL**						
	B10 Superstructure						
1010	Floor Construction	N/A		—	—	—	
1020	Roof Construction	Metal deck, open web steel joists, beams and columns		S.F. Roof	7.15	7.15	11.9%
	B20 Exterior Enclosure						
2010	Exterior Walls	Concrete block	75% of wall	S.F. Wall	5.86	2.14	
2020	Exterior Windows	Industrial horizontal pivoted steel	25% of wall	Each	702	2.67	9.4%
2030	Exterior Doors	Double aluminum and glass, hollow metal, steel overhead		Each	1630	.81	
	B30 Roofing						
3010	Roof Coverings	Built-up tar and gravel with flashing; perlite/EPS composite insulation		S.F. Roof	3.99	3.99	7.0%
3020	Roof Openings	Roof hatches		S.F. Roof	.21	.21	
C.	**INTERIORS**						
1010	Partitions	Concrete block	60 S.F. Floor/L.F. Partition	S.F. Partition	6.40	1.28	
1020	Interior Doors	Single leaf hollow metal and fire doors	600 S.F. Floor/Door	Each	608	1.01	
1030	Fittings	Toilet partitions		S.F. Floor	1.01	1.01	
2010	Stair Construction	N/A		—	—	—	9.1%
3010	Wall Finishes	Paint		S.F. Surface	7.10	1.42	
3020	Floor Finishes	Vinyl composition tile	10% of floor	S.F. Floor	3.00	.30	
3030	Ceiling Finishes	Fiberglass board on exposed grid system	10% of area	S.F. Ceiling	4.09	.41	
D.	**SERVICES**						
	D10 Conveying						
1010	Elevators & Lifts	N/A		—	—	—	0.0%
1020	Escalators & Moving Walks	N/A		—	—	—	
	D20 Plumbing						
2010	Plumbing Fixtures	Toilet and service fixtures, supply and drainage	1 Fixture/1000 S.F. Floor	Each	2670	2.67	
2020	Domestic Water Distribution	Gas fired water heater		S.F. Floor	.32	.32	5.7%
2040	Rain Water Drainage	Roof drains		S.F. Roof	.45	.45	
	D30 HVAC						
3010	Energy Supply	Oil fired hot water, unit heaters		S.F. Floor	6.37	6.37	
3020	Heat Generating Systems	N/A		—	—	—	
3030	Cooling Generating Systems	Chilled water, air cooled condenser system		S.F. Floor	8.24	8.24	24.4%
3050	Terminal & Package Units	N/A		—	—	—	
3090	Other HVAC Sys. & Equipment	N/A		—	—	—	
	D40 Fire Protection						
4010	Sprinklers	Sprinklers, ordinary hazard		S.F. Floor	2.61	2.61	4.4%
4020	Standpipes	N/A		—	—	—	
	D50 Electrical						
5010	Electrical Service/Distribution	600 ampere service, panel board and feeders		S.F. Floor	.99	.99	
5020	Lighting & Branch Wiring	High intensity discharge fixtures, receptacles, switches, A.C. and misc. power		S.F. Floor	7.40	7.40	14.6%
5030	Communications & Security	Alarm systems and emergency lighting		S.F. Floor	.36	.36	
5090	Other Electrical Systems	N/A		—	—	—	
E.	**EQUIPMENT & FURNISHINGS**						
1010	Commercial Equipment	N/A		—	—	—	
1020	Institutional Equipment	N/A		—	—	—	
1030	Vehicular Equipment	N/A		—	—	—	0.0%
1090	Other Equipment	N/A		—	—	—	
F.	**SPECIAL CONSTRUCTION**						
1020	Integrated Construction	N/A		—	—	—	
1040	Special Facilities	N/A		—	—	—	0.0%
G.	**BUILDING SITEWORK**	N/A					

			Sub-Total	59.95	100%
CONTRACTOR FEES (General Requirements: 10%, Overhead: 5%, Profit: 10%)			25%	15.00	
ARCHITECT FEES			7%	5.25	
		Total Building Cost		**80.20**	

BUILDING TYPES

BUILDING TYPES

Costs per square foot of floor area

Exterior Wall	S.F. Area	20000	30000	40000	50000	60000	70000	80000	90000	100000
	L.F. Perimeter	362	410	493	576	600	628	660	700	744
Face Brick Common Brick Back-up	Steel Frame	117.40	106.70	103.05	100.85	97.40	95.15	93.50	92.45	91.60
	Concrete Frame	114.90	104.20	100.50	98.35	94.95	92.60	90.95	89.90	89.10
Face Brick Concrete Block Back-up	Steel Frame	116.35	105.95	102.35	100.25	96.90	94.70	93.10	92.05	91.30
	Concrete Frame	113.55	103.15	99.55	97.50	94.15	91.95	90.30	89.35	88.50
Stucco on Concrete Block	Steel Frame	108.60	100.15	97.10	95.35	92.70	90.90	89.60	88.80	88.15
	Concrete Frame	105.85	97.40	94.35	92.60	89.90	88.10	86.80	86.00	85.35
Perimeter Adj., Add or Deduct	Per 100 L.F.	9.80	6.55	4.90	3.90	3.30	2.75	2.40	2.15	1.95
Story Hgt. Adj., Add or Deduct	Per 1 Ft.	2.75	2.10	1.85	1.75	1.50	1.35	1.20	1.15	1.15

For Basement, add $24.80 per square foot of basement area

The above costs were calculated using the basic specifications shown on the facing page. These costs should be adjusted where necessary for design alternatives and owner's requirements. Reported completed project costs, for this type of structure, range from $33.10 to $128.25 per S.F.

Common additives

Description	Unit	$ Cost	Description	Unit	$ Cost
Clock System			Dock Levelers, Hinged 10 ton cap.		
20 room	Each	14,200	6'x 8'	Each	5650
50 room	Each	34,600	7'x 8'	Each	5500
Dock Bumpers, Rubber blocks			Elevator, Hydraulic freight, 2 stops		
4-1/2" thick, 10" high, 14" long	Each	65	3500# capacity	Each	70,600
24" long	Each	106	4000# capacity	Each	73,100
36" long	Each	117	Additional stop, add	Each	4550
12" high, 14" long	Each	109	Partitions, Woven wire, 10 ga., 1-1/2" mesh		
24" long	Each	123	4' Wide x 7" high	Each	154
36" long	Each	141	8' High	Each	166
6" thick, 10" high, 14" long	Each	101	10' High	Each	201
24" long	Each	135	Platform Lifter, Portable, 6'x 6'		
36" long	Each	164	3000# cap.	Each	8325
20" high, 11" long	Each	167	4000# cap.	Each	10,300
Dock Boards, Heavy			Fixed, 6'x 8', 5000# cap.	Each	10,700
60" x 60" Aluminum, 5,000# cap.	Each	1325			
9000# cap.	Each	1250			
15,000# cap.	Each	1375			

Model costs calculated for a 3 story building with 12' story height and 90,000 square feet of floor area

			Unit	Unit Cost	Cost Per S.F.	% Of Sub-Total
A. SUBSTRUCTURE						
1010	Standard Foundations	Poured concrete; strip and spread footings	S.F. Ground	5.22	1.74	
1030	Slab on Grade	4" reinforced concrete with vapor barrier and granular base	S.F. Slab	7.81	2.60	
2010	Basement Excavation	Site preparation for slab and trench for foundation wall and footing	S.F. Ground	.14	.05	7.2%
2020	Basement Walls	4' foundation wall	L.F. Wall	61	.47	
B. SHELL						
	B10 Superstructure					
1010	Floor Construction	Concrete flat slab	S.F. Floor	19.05	12.70	
1020	Roof Construction	Concrete flat slab	S.F. Roof	12.60	4.20	24.9%
	B20 Exterior Enclosure					
2010	Exterior Walls	Face brick with common brick backup 70% of wall	S.F. Wall	26	5.10	
2020	Exterior Windows	Industrial, horizontal pivoted steel 30% of wall	Each	1049	5.51	16.1%
2030	Exterior Doors	Double aluminum & glass, hollow metal, overhead doors	Each	1565	.29	
	B30 Roofing					
3010	Roof Coverings	Built-up tar and gravel with flashing; perlite/EPS composite insulation	S.F. Roof	3.90	1.30	
3020	Roof Openings	Roof hatches	S.F. Roof	.21	.07	2.0%
C. INTERIORS						
1010	Partitions	Gypsum board on metal studs 50 S.F. Floor/L.F. Partition	S.F. Partition	3.25	.65	
1020	Interior Doors	Single leaf fire doors 500 S.F. Floor/Door	Each	735	1.47	
1030	Fittings	Toilet partitions	S.F. Floor	.50	.50	
2010	Stair Construction	Concrete	Flight	6800	.61	9.3%
3010	Wall Finishes	Paint	S.F. Surface	1.25	.25	
3020	Floor Finishes	90% metallic hardener, 10% vinyl composition tile	S.F. Floor	2.50	2.50	
3030	Ceiling Finishes	Fiberglass board on exposed grid systems 10% of area	S.F. Ceiling	3.05	.31	
D. SERVICES						
	D10 Conveying					
1010	Elevators & Lifts	Two hydraulic freight elevators	Each	103,500	2.30	
1020	Escalators & Moving Walks	N/A	—	—	—	3.4%
	D20 Plumbing					
2010	Plumbing Fixtures	Toilet and service fixtures, supply and drainage 1 Fixture/1345 S.F. Floor	Each	2891	2.15	
2020	Domestic Water Distribution	Gas fired water heater	S.F. Floor	.20	.20	3.7%
2040	Rain Water Drainage	Roof drains	S.F. Roof	.57	.19	
	D30 HVAC					
3010	Energy Supply	Oil fired hot water, unit heaters	S.F. Floor	3.18	3.18	
3020	Heat Generating Systems	N/A	—	—	—	
3030	Cooling Generating Systems	Chilled water, air cooled condenser system	S.F. Floor	8.24	8.24	16.8%
3050	Terminal & Package Units	N/A	—	—	—	
3090	Other HVAC Sys. & Equipment	N/A	—	—	—	
	D40 Fire Protection					
4010	Sprinklers	Wet pipe sprinkler system	S.F. Floor	2.35	2.35	
4020	Standpipes	N/A	—	—	—	3.5%
	D50 Electrical					
5010	Electrical Service/Distribution	800 ampere service, panel board and feeders	S.F. Floor	.58	.58	
5020	Lighting & Branch Wiring	High intensity discharge fixtures, switches, A.C. and misc. power	S.F. Floor	7.83	7.83	
5030	Communications & Security	Alarm systems and emergency lighting	S.F. Floor	.36	.36	13.1%
5090	Other Electrical Systems	Emergency generator, 30 kW	S.F. Floor	.15	.15	
E. EQUIPMENT & FURNISHINGS						
1010	Commercial Equipment	N/A	—	—	—	
1020	Institutional Equipment	N/A	—	—	—	
1030	Vehicular Equipment	N/A	—	—	—	0.0%
1090	Other Equipment	N/A	—	—	—	
F. SPECIAL CONSTRUCTION						
1020	Integrated Construction	N/A	—	—	—	
1040	Special Facilities	N/A	—	—	—	0.0%
G. BUILDING SITEWORK	**N/A**					

		Sub-Total	67.85	100%
CONTRACTOR FEES (General Requirements: 10%, Overhead: 5%, Profit: 10%)		25%	16.96	
ARCHITECT FEES		6%	5.09	
	Total Building Cost		**89.90**	

BUILDING TYPES

BUILDING TYPES

Costs per square foot of floor area

Exterior Wall	S.F. Area	4000	4500	5000	5500	6000	6500	7000	7500	8000
	L.F. Perimeter	260	280	300	310	320	336	353	370	386
Face Brick Concrete Block Back-up	Steel Joists	135.75	133.30	131.40	128.65	126.30	125.05	124.00	123.00	122.15
	Bearing Walls	131.65	129.20	127.30	124.55	122.20	121.00	119.90	118.95	118.05
Decorative Concrete Block	Steel Joists	124.80	122.85	121.30	119.20	117.35	116.35	115.50	114.75	114.05
	Bearing Walls	120.95	119.00	117.45	115.35	113.50	112.50	111.65	110.90	110.15
Limestone with Concrete Block Back-up	Steel Joists	149.40	146.40	144.00	140.50	137.55	135.95	134.60	133.40	132.30
	Bearing Walls	145.55	142.55	140.15	136.65	133.70	132.10	130.75	129.55	128.45
Perimeter Adj., Add or Deduct	Per 100 L.F.	15.50	13.75	12.35	11.20	10.35	9.55	8.80	8.25	7.75
Story Hgt. Adj., Add or Deduct	Per 1 Ft.	2.00	1.90	1.80	1.75	1.70	1.60	1.55	1.55	1.50
For Basement, add $27.75 per square foot of basement area										

The above costs were calculated using the basic specifications shown on the facing page. These costs should be adjusted where necessary for design alternatives and owner's requirements. Reported completed project costs, for this type of structure, range from $53.50 to $158.55 per S.F.

Common additives

Description	Unit	$ Cost		Description	Unit	$ Cost
Appliances				Appliances, cont.		
Cooking range, 30" free standing				Refrigerator, no frost 10-12 C.F.	Each	550 - 860
1 oven	Each	330 - 1800		14-16 C.F.	Each	560 - 665
2 oven	Each	1700 - 1775		18-20 C.F.	Each	645 - 1100
30" built-in				Lockers, Steel, single tier, 60" or 72"	Opening	154 - 241
1 oven	Each	570 - 1875		2 tier, 60" or 72" total	Opening	104 - 134
2 oven	Each	1250 - 2050		5 tier, box lockers	Opening	55 - 69
Counter top cook tops, 4 burner	Each	285 - 730		Locker bench, lam. maple top only	L.F.	18.80
Microwave oven	Each	212 - 675		Pedestals, steel pipe	Each	57
Combination range, refrig. & sink, 30" wide	Each	1300 - 2575		Sound System		
60" wide	Each	3350		Amplifier, 250 watts	Each	1925
72" wide	Each	3825		Speaker, ceiling or wall	Each	160
Combination range refrigerator, sink				Trumpet	Each	305
microwave oven & icemaker	Each	5650				
Compactor, residential, 4-1 compaction	Each	540 - 655				
Dishwasher, built-in, 2 cycles	Each	515 - 810				
4 cycles	Each	530 - 1075				
Garbage disposer, sink type	Each	144 - 258				
Hood for range, 2 speed, vented, 30" wide	Each	224 - 960				
42" wide	Each	430 - 1150				

Important: See the Reference Section for Location Factors

Model costs calculated for a 1 story building with 14' story height and 6,000 square feet of floor area

				Unit	Unit Cost	Cost Per S.F.	% Of Sub-Total
A. SUBSTRUCTURE							
1010	Standard Foundations	Poured concrete; strip and spread footings		S.F. Ground	2.51	2.51	
1030	Slab on Grade	6" reinforced concrete with vapor barrier and granular base		S.F. Slab	5.13	5.13	
2010	Basement Excavation	Site preparation for slab and trench for foundation wall and footing		S.F. Ground	.43	.43	12.1%
2020	Basement Walls	4' foundation wall		L.F. Wall	61	3.26	
B. SHELL							
	B10 Superstructure						
1010	Floor Construction	N/A		—	—	—	
1020	Roof Construction	Metal deck, open web steel joists, beams on columns		S.F. Roof	7.32	7.32	7.8%
	B20 Exterior Enclosure						
2010	Exterior Walls	Face brick with concrete block backup	75% of wall	S.F. Wall	23	13.30	
2020	Exterior Windows	Aluminum insulated glass	10% of wall	Each	702	1.63	19.3%
2030	Exterior Doors	Single aluminum and glass, overhead, hollow metal	15% of wall	S.F. Door	—	3.16	
	B30 Roofing						
3010	Roof Coverings	Built-up tar and gravel with flashing; perlite/EPS composite insulation		S.F. Roof	4.81	4.81	5.2%
3020	Roof Openings	Skylights, roof hatches		S.F. Roof	.09	.09	
C. INTERIORS							
1010	Partitions	Concrete block	17 S.F. Floor/L.F. Partition	S.F. Partition	6.73	3.96	
1020	Interior Doors	Single leaf hollow metal	500 S.F. Floor/Door	Each	608	1.22	
1030	Fittings	Toilet partitions		S.F. Floor	.50	.50	
2010	Stair Construction	N/A		—	—	—	12.6%
3010	Wall Finishes	Paint		S.F. Surface	1.86	1.86	
3020	Floor Finishes	50% vinyl tile, 50% paint		S.F. Floor	2.18	2.18	
3030	Ceiling Finishes	Fiberglass board on exposed grid, suspended	50% of area	S.F. Ceiling	4.09	2.05	
D. SERVICES							
	D10 Conveying						
1010	Elevators & Lifts	N/A		—	—	—	0.0%
1020	Escalators & Moving Walks	N/A		—	—	—	
	D20 Plumbing						
2010	Plumbing Fixtures	Kitchen, toilet and service fixtures, supply and drainage	1 Fixture/375 S.F. Floor	Each	2347	6.26	
2020	Domestic Water Distribution	Gast fired water heater		S.F. Floor	1.46	1.46	8.9%
2040	Rain Water Drainage	Roof drains		S.F. Roof	.60	.60	
	D30 HVAC						
3010	Energy Supply	N/A		—	—	—	
3020	Heat Generating Systems	Included in D3050		—	—	—	
3030	Cooling Generating Systems	N/A		—	—	—	23.8%
3050	Terminal & Package Units	Rooftop multizone unit system		S.F. Floor	22	22.28	
3090	Other HVAC Sys. & Equipment	N/A		—	—	—	
	D40 Fire Protection						
4010	Sprinklers	Wet pipe sprinkler system		S.F. Floor	2.68	2.68	2.9%
4020	Standpipes	N/A		—	—	—	
	D50 Electrical						
5010	Electrical Service/Distribution	200 ampere service, panel board and feeders		S.F. Floor	1.09	1.09	
5020	Lighting & Branch Wiring	Fluorescent fixtures, receptacles, switches, A.C. and misc. power		S.F. Floor	5.41	5.41	7.3%
5030	Communications & Security	Alarm systems		S.F. Floor	.36	.36	
5090	Other Electrical Systems	N/A		—	—	—	
E. EQUIPMENT & FURNISHINGS							
1010	Commercial Equipment	N/A		—	—	—	
1020	Institutional Equipment	N/A		—	—	—	0.0%
1030	Vehicular Equipment	N/A		—	—	—	
1090	Other Equipment	N/A		—	—	—	
F. SPECIAL CONSTRUCTION							
1020	Integrated Construction	N/A		—	—	—	0.0%
1040	Special Facilities	N/A		—	—	—	
G. BUILDING SITEWORK	**N/A**						

			Sub-Total	93.55	100%
CONTRACTOR FEES (General Requirements: 10%, Overhead: 5%, Profit: 10%)			25%	23.39	
ARCHITECT FEES			8%	9.36	
		Total Building Cost		**126.30**	

BUILDING TYPES

Costs per square foot of floor area

Exterior Wall	S.F. Area	6000	7000	8000	9000	10000	11000	12000	13000	14000
	L.F. Perimeter	220	240	260	280	286	303	320	336	353
Face Brick with Concrete Block Back-up	Steel Joists	139.80	135.75	132.75	130.45	127.10	125.40	124.05	122.75	121.80
	Precast Conc.	145.45	141.45	138.45	136.20	132.90	131.20	129.85	128.65	127.60
Decorative Concrete Block	Steel Joists	130.55	127.10	124.50	122.60	119.90	118.45	117.30	116.25	115.40
	Precast Conc.	136.20	132.80	130.25	128.35	125.70	124.25	123.10	122.05	121.30
Limestone with Concrete Block Back-up	Steel Joists	155.25	150.15	146.40	143.60	139.15	137.00	135.30	133.65	132.40
	Precast Conc.	160.80	155.75	152.05	149.20	144.80	142.70	140.95	139.40	138.15
Perimeter Adj., Add or Deduct	Per 100 L.F.	17.75	15.25	13.35	11.85	10.70	9.70	8.85	8.20	7.60
Story Hgt. Adj., Add or Deduct	Per 1 Ft.	2.30	2.15	2.05	2.00	1.85	1.75	1.65	1.60	1.55

For Basement, add $26.90 per square foot of basement area

The above costs were calculated using the basic specifications shown on the facing page. These costs should be adjusted where necessary for design alternatives and owner's requirements. Reported completed project costs, for this type of structure, range from $53.50 to $158.55 per S.F.

Common additives

Description	Unit	$ Cost	Description	Unit	$ Cost
Appliances			Appliances, cont.		
Cooking range, 30" free standing			Refrigerator, no frost 10-12 C.F.	Each	550 - 860
1 oven	Each	330 - 1800	14-16 C.F.	Each	560 - 665
2 oven	Each	1700 - 1775	18-20 C.F.	Each	645 - 1100
30" built-in			Elevators, Hydraulic passenger, 2 stops		
1 oven	Each	570 - 1875	1500# capacity	Each	47,900
2 oven	Each	1250 - 2050	2500# capacity	Each	50,000
Counter top cook tops, 4 burner	Each	285 - 730	3500# capacity	Each	53,700
Microwave oven	Each	212 - 675	Lockers, Steel, single tier, 60" or 72"	Opening	154 - 241
Combination range, refrig. & sink, 30" wide	Each	1300 - 2575	2 tier, 60" or 72" total	Opening	104 - 134
60" wide	Each	3350	5 tier, box lockers	Opening	55 - 69
72" wide	Each	3825	Locker bench, lam. maple top only	L.F.	18.80
Combination range, refrigerator, sink,			Pedestals, steel pipe	Each	57
microwave oven & icemaker	Each	5650	Sound System		
Compactor, residential, 4-1 compaction	Each	540 - 655	Amplifier, 250 watts	Each	1925
Dishwasher, built-in, 2 cycles	Each	515 - 810	Speaker, ceiling or wall	Each	160
4 cycles	Each	530 - 1075	Trumpet	Each	305
Garbage diposer, sink type	Each	144 - 258			
Hood for range, 2 speed, vented, 30" wide	Each	224 - 960			
42" wide	Each	430 - 1150			

Important: See the Reference Section for Location Factors

Model costs calculated for a 2 story building with 14' story height and 10,000 square feet of floor area

Fire Station, 2 Story

				Unit	Unit Cost	Cost Per S.F.	% Of Sub-Total
A. SUBSTRUCTURE							
1010	Standard Foundations	Poured concrete; strip and spread footings		S.F. Ground	2.36	1.18	
1030	Slab on Grade	6" reinforced concrete with vapor barrier and granular base		S.F. Slab	5.13	2.57	
2010	Basement Excavation	Site preparation for slab and trench for foundation wall and footing		S.F. Ground	.43	.22	6.7%
2020	Basement Walls	4' foundation wall		L.F. Wall	68	1.96	
B. SHELL							
	B10 Superstructure						
1010	Floor Construction	Open web steel joists, slab form, concrete		S.F. Floor	10.92	5.46	8.3%
1020	Roof Construction	Metal deck on open web steel joists		S.F. Roof	3.78	1.89	
	B20 Exterior Enclosure						
2010	Exterior Walls	Decorative concrete block	75% of wall	S.F. Wall	14.85	8.92	
2020	Exterior Windows	Aluminum insulated glass	10% of wall	Each	559	1.95	15.2%
2030	Exterior Doors	Single aluminum and glass, steel overhead, hollow metal	15% of wall	S.F. Door	—	2.62	
	B30 Roofing						
3010	Roof Coverings	Built-up tar and gravel with flashing; perlite/EPS composite insulation		S.F. Roof	4.92	2.46	2.8%
3020	Roof Openings	N/A		—	—	—	
C. INTERIORS							
1010	Partitions	Concrete block	10 S.F. Floor/L.F. Partition	S.F. Partition	6.73	3.96	
1020	Interior Doors	Single leaf hollow metal	500 S.F. Floor/Door	Each	735	1.47	
1030	Fittings	Toilet partitions		S.F. Floor	.45	.45	
2010	Stair Construction	Concrete filled metal pan		Flight	8975	1.80	16.3%
3010	Wall Finishes	Paint		S.F. Surface	3.13	3.13	
3020	Floor Finishes	50% vinyl tile, 50% paint		S.F. Floor	2.18	2.18	
3030	Ceiling Finishes	Fiberglass board on exposed grid, suspended	50% of area	S.F. Ceiling	3.05	1.53	
D. SERVICES							
	D10 Conveying						
1010	Elevators & Lifts	One hydraulic passenger elevator		Each	59,100	5.91	6.7%
1020	Escalators & Moving Walks	N/A		—	—	—	
	D20 Plumbing						
2010	Plumbing Fixtures	Kitchen toilet and service fixtures, supply and drainage	1 Fixture/400 S.F. Floor	Each	2300	5.75	
2020	Domestic Water Distribution	Gas fired water heater		S.F. Floor	1.14	1.14	8.2%
2040	Rain Water Drainage	Roof drains		S.F. Roof	.70	.35	
	D30 HVAC						
3010	Energy Supply	N/A		—	—	—	
3020	Heat Generating Systems	Included in D3050		—	—	—	
3030	Cooling Generating Systems	N/A		—	—	—	25.1%
3050	Terminal & Package Units	Rooftop multizone unit system		S.F. Floor	22	22.28	
3090	Other HVAC Sys. & Equipment	N/A		—	—	—	
	D40 Fire Protection						
4010	Sprinklers	Wet pipe sprinkler system		S.F. Floor	2.26	2.26	2.5%
4020	Standpipes	N/A		—	—	—	
	D50 Electrical						
5010	Electrical Service/Distribution	400 ampere service, panel board and feeders		S.F. Floor	1.23	1.23	
5020	Lighting & Branch Wiring	Fluorescent fixtures, receptacles, switches, A.C. and misc. power		S.F. Floor	5.59	5.59	8.3%
5030	Communications & Security	Alarm systems and emergency lighting		S.F. Floor	.36	.36	
5090	Other Electrical Systems	Emergency generator, 15 kW		S.F. Floor	.19	.19	
E. EQUIPMENT & FURNISHINGS							
1010	Commercial Equipment	N/A		—	—	—	
1020	Institutional Equipment	N/A		—	—	—	
1030	Vehicular Equipment	N/A		—	—	—	0.0%
1090	Other Equipment	N/A		—	—	—	
F. SPECIAL CONSTRUCTION							
1020	Integrated Construction	N/A		—	—	—	
1040	Special Facilities	N/A		—	—	—	0.0%
G. BUILDING SITEWORK	**N/A**						

		Sub-Total	88.81	100%
CONTRACTOR FEES (General Requirements: 10%, Overhead: 5%, Profit: 10%)		25%	22.21	
ARCHITECT FEES		8%	8.88	

Total Building Cost **119.90**

BUILDING TYPES

Costs per square foot of floor area

Exterior Wall	S.F. Area	4000	5000	6000	8000	10000	12000	14000	16000	18000
	L.F. Perimeter	180	205	230	260	300	340	353	386	420
Cedar Beveled Siding	Wood Frame	131.95	124.65	119.80	112.30	108.40	105.70	102.90	101.45	100.20
Aluminum Siding	Wood Frame	131.15	123.85	118.90	111.50	107.60	104.90	102.05	100.65	99.40
Board and Batten	Wood Frame	128.45	122.25	118.15	111.75	108.40	106.10	103.55	102.30	101.25
Face Brick on Block	Bearing Wall	153.35	144.85	139.20	130.05	125.45	122.30	118.55	116.75	115.35
Stucco on Block	Bearing Wall	139.95	132.60	127.70	120.20	116.30	113.55	110.75	109.25	108.05
Decorative Block	Wood Joists	134.55	126.95	121.95	114.15	110.10	107.30	104.30	102.80	101.55
Perimeter Adj., Add or Deduct	Per 100 L.F.	12.90	10.30	8.55	6.45	5.15	4.35	3.65	3.20	2.85
Story Hgt. Adj., Add or Deduct	Per 1 Ft.	1.70	1.50	1.35	1.20	1.15	1.10	.90	.85	.90
For Basement, add $17.10 per square foot of basement area										

The above costs were calculated using the basic specifications shown on the facing page. These costs should be adjusted where necessary for design alternatives and owner's requirements. Reported completed project costs, for this type of structure, range from $77.70 to $170.00 per S.F.

Common additives

Description	Unit	$ Cost		Description	Unit	$ Cost
Appliances				Appliances, cont.		
Cooking range, 30" free standing				Refrigerator, no frost 10-12 C.F.	Each	550 - 860
1 oven	Each	330 - 1800		14-16 C.F.	Each	560 - 665
2 oven	Each	1700 - 1775		18-20 C.F.	Each	645 - 1100
30" built-in				Elevators, Hydraulic passenger, 2 stops		
1 oven	Each	570 - 1875		1500# capacity	Each	47,900
2 oven	Each	1250 - 2050		2500# capacity	Each	50,000
Counter top cook tops, 4 burner	Each	285 - 730		3500# capacity	Each	53,700
Microwave oven	Each	212 - 675		Laundry Equipment		
Combination range, refrig. & sink, 30" wide	Each	1300 - 2575		Dryer, gas, 16 lb. capacity	Each	785
60" wide	Each	3350		30 lb. capacity	Each	3000
72" wide	Each	3825		Washer, 4 cycle	Each	875
Combination range, refrigerator, sink,				Commercial	Each	1325
microwave oven & icemaker	Each	5650		Sound System		
Compactor, residential, 4-1 compaction	Each	540 - 655		Amplifier, 250 watts	Each	1925
Dishwasher, built-in, 2 cycles	Each	515 - 810		Speaker, ceiling or wall	Each	160
4 cycles	Each	530 - 1075		Trumpet	Each	305
Garbage disposer, sink type	Each	144 - 258				
Hood for range, 2 speed, vented, 30" wide	Each	224 - 960				
42" wide	Each	430 - 1150				

Model costs calculated for a 2 story building with 10' story height and 10,000 square feet of floor area

Fraternity/Sorority House

				Unit	Unit Cost	Cost Per S.F.	% Of Sub-Total
A. SUBSTRUCTURE							
1010	Standard Foundations	Poured concrete; strip and spread footings		S.F. Ground	2.30	1.15	
1030	Slab on Grade	4" reinforced concrete with vapor barrier and granular base		S.F. Slab	3.98	1.99	
2010	Basement Excavation	Site preparation for slab and trench for foundation wall and footing		S.F. Ground	.43	.22	6.3%
2020	Basement Walls	4' foundation wall		L.F. Wall	54	1.63	
B. SHELL							
	B10 Superstructure						
1010	Floor Construction	Plywood on wood joists		S.F. Floor	3.58	1.79	4.4%
1020	Roof Construction	Plywood on wood rafters (pitched)		S.F. Roof	3.13	1.75	
	B20 Exterior Enclosure						
2010	Exterior Walls	Cedar bevel siding on wood studs, insulated	80% of wall	S.F. Wall	10.63	5.10	
2020	Exterior Windows	Double hung wood	20% of wall	Each	475	2.28	10.7%
2030	Exterior Doors	Solid core wood		Each	1577	1.12	
	B30 Roofing						
3010	Roof Coverings	Asphalt shingles with flashing (pitched); rigid fiberglass insulation		S.F. Roof	3.28	1.64	2.1%
3020	Roof Openings	N/A		—	—	—	
C. INTERIORS							
1010	Partitions	Gypsum board on wood studs	25 S.F. Floor/L.F. Partition	S.F. Partition	6.25	2.00	
1020	Interior Doors	Single leaf wood	200 S.F. Floor/Door	Each	480	2.40	
1030	Fittings	N/A		—	—	—	
2010	Stair Construction	Wood		Flight	1812	.73	20.7%
3010	Wall Finishes	Paint		S.F. Surface	1.23	.41	
3020	Floor Finishes	10% hardwood, 70% carpet, 20% ceramic tile		S.F. Floor	7.59	7.59	
3030	Ceiling Finishes	Gypsum board on wood furring		S.F. Ceiling	3.00	3.30	
D. SERVICES							
	D10 Conveying						
1010	Elevators & Lifts	One hydraulic passenger elevator		Each	59,100	5.91	7.4%
1020	Escalators & Moving Walks	N/A		—	—	—	
	D20 Plumbing						
2010	Plumbing Fixtures	Kitchen toilet and service fixtures, supply and drainage	1 Fixture/150 S.F. Floor	Each	756	5.04	
2020	Domestic Water Distribution	Gas fired water heater		S.F. Floor	1.33	1.33	8.0%
2040	Rain Water Drainage	N/A		—	—	—	
	D30 HVAC						
3010	Energy Supply	Oil fired hot water, baseboard radiation		S.F. Floor	5.02	5.02	
3020	Heat Generating Systems	N/A		—	—	—	
3030	Cooling Generating Systems	N/A		—	—	—	13.5%
3050	Terminal & Package Units	Split system with air cooled condensing unit		S.F. Floor	5.76	5.76	
3090	Other HVAC Sys. & Equipment	N/A		—	—	—	
	D40 Fire Protection						
4010	Sprinklers	Wet pipe sprinkler system		S.F. Floor	2.26	2.26	2.8%
4020	Standpipes	N/A		—	—	—	
	D50 Electrical						
5010	Electrical Service/Distribution	600 ampere service, panel board and feeders		S.F. Floor	3.59	3.59	
5020	Lighting & Branch Wiring	Fluorescent fixtures, receptacles, switches, A.C. and misc. power		S.F. Floor	8.04	8.04	24.1%
5030	Communications & Security	Alarm, communication system and generator set		S.F. Floor	7.36	7.36	
5090	Other Electrical Systems	Emergency generator, 7.5 kW		S.F. Floor	.15	.15	
E. EQUIPMENT & FURNISHINGS							
1010	Commercial Equipment	N/A		—	—	—	
1020	Institutional Equipment	N/A		—	—	—	
1030	Vehicular Equipment	N/A		—	—	—	0.0%
1090	Other Equipment	N/A		—	—	—	
F. SPECIAL CONSTRUCTION							
1020	Integrated Construction	N/A		—	—	—	
1040	Special Facilities	N/A		—	—	—	0.0%
G. BUILDING SITEWORK	**N/A**						

			Sub-Total	79.56	100%
	CONTRACTOR FEES (General Requirements: 10%, Overhead: 5%, Profit: 10%)		25%	19.89	
	ARCHITECT FEES		9%	8.95	

	Total Building Cost	108.40

BUILDING TYPES

Costs per square foot of floor area

Exterior Wall	S.F. Area	4000	6000	8000	10000	12000	14000	16000	18000	20000
	L.F. Perimeter	260	320	384	424	460	484	510	540	576
Vertical Redwood Siding	Wood Frame	120.60	108.40	102.45	98.10	95.00	92.45	90.55	89.25	88.30
Brick Veneer	Wood Frame	129.85	116.10	109.50	104.35	100.75	97.70	95.50	93.95	92.80
Aluminum Siding	Wood Frame	118.35	106.65	101.00	96.90	93.95	91.60	89.85	88.60	87.70
Brick on Block	Wood Truss	136.00	121.40	114.40	108.85	104.95	101.60	99.25	97.50	96.30
Limestone on Block	Wood Truss	157.30	138.85	130.10	122.65	117.35	112.80	109.50	107.20	105.55
Stucco on Block	Wood Truss	123.75	111.35	105.35	100.90	97.65	95.10	93.20	91.85	90.85
Perimeter Adj., Add or Deduct	Per 100 L.F.	9.35	6.25	4.70	3.75	3.10	2.70	2.45	2.05	1.90
Story Hgt. Adj., Add or Deduct	Per 1 Ft.	1.15	.90	.90	.70	.70	.60	.65	.60	.50
For Basement, add $22.90 per square foot of basement area										

The above costs were calculated using the basic specifications shown on the facing page. These costs should be adjusted where necessary for design alternatives and owner's requirements. Reported completed project costs, for this type of structure, range from $76.10 to $233.95 per S.F.

Common additives

Description	Unit	$ Cost
Autopsy Table, Standard	Each	8525
Deluxe	Each	15,600
Directory Boards, Plastic, glass covered		
30" x 20"	Each	575
36" x 48"	Each	1275
Aluminum, 24" x 18"	Each	525
36" x 24"	Each	600
48" x 32"	Each	810
48" x 60"	Each	1875
Emergency Lighting, 25 watt, battery operated		
Lead battery	Each	247
Nickel cadmium	Each	695
Mortuary Refrigerator, End operated		
Two capacity	Each	12,700
Six capacity	Each	23,500

Description	Unit	$ Cost
Planters, Precast concrete		
48" diam., 24" high	Each	655
7" diam., 36" high	Each	1075
Fiberglass, 36" diam., 24" high	Each	470
60" diam., 24" high	Each	780
Smoke Detectors		
Ceiling type	Each	161
Duct type	Each	425

Important: See the Reference Section for Location Factors

Model costs calculated for a 1 story building with 12' story height and 10,000 square feet of floor area

BUILDING TYPES

				Unit	Unit Cost	Cost Per S.F.	% Of Sub-Total
A. SUBSTRUCTURE							
1010	Standard Foundations	Poured concrete; strip and spread footings		S.F. Ground	1.37	1.37	
1030	Slab on Grade	4" reinforced concrete with vapor barrier and granular base		S.F. Slab	3.98	3.98	11.3%
2010	Basement Excavation	Site preparation for slab and trench for foundation wall and footing		S.F. Ground	.23	.23	
2020	Basement Walls	4' foundation wall		L.F. Wall	54	2.53	
B. SHELL							
	B10 Superstructure						
1010	Floor Construction	N/A		—	—	—	6.2%
1020	Roof Construction	Plywood on wood truss		S.F. Roof	4.44	4.44	
	B20 Exterior Enclosure						
2010	Exterior Walls	1" x 4" vertical T & G redwood siding on wood studs	90% of wall	S.F. Wall	9.59	4.40	
2020	Exterior Windows	Double hung wood	10% of wall	Each	383	1.62	9.8%
2030	Exterior Doors	Wood swinging double doors, single leaf hollow metal		Each	1739	1.04	
	B30 Roofing						
3010	Roof Coverings	Single ply membrane, fully adhered; polyisocyanurate sheets		S.F. Roof	3.83	3.83	5.3%
3020	Roof Openings	N/A		—	—	—	
C. INTERIORS							
1010	Partitions	Gypsum board on wood studs with sound deadening board	15 S.F. Floor/L.F. Partition	S.F. Partition	7.09	3.78	
1020	Interior Doors	Single leaf wood	150 S.F. Floor/Door	Each	480	3.20	
1030	Fittings	N/A		—	—	—	
2010	Stair Construction	N/A		—	—	—	29.7%
3010	Wall Finishes	50% wallpaper, 25% wood paneling, 25% paint		S.F. Surface	2.75	2.75	
3020	Floor Finishes	70% carpet, 30% ceramic tile		S.F. Floor	8.58	8.58	
3030	Ceiling Finishes	Fiberglass board on exposed grid, suspended		S.F. Ceiling	3.00	3.05	
D. SERVICES							
	D10 Conveying						
1010	Elevators & Lifts	N/A		—	—	—	0.0%
1020	Escalators & Moving Walks	N/A		—	—	—	
	D20 Plumbing						
2010	Plumbing Fixtures	Toilet and service fixtures, supply and drainage	1 Fixture/770 S.F. Floor	Each	1532	1.99	
2020	Domestic Water Distribution	Electric water heater		S.F. Floor	5.15	5.15	10.4%
2040	Rain Water Drainage	Roof drain		S.F. Roof	.36	.36	
	D30 HVAC						
3010	Energy Supply	N/A		—	—	—	
3020	Heat Generating Systems	Included in D3030		—	—	—	
3030	Cooling Generating Systems	Multizone rooftop unit, gas heating, electric cooling		S.F. Floor	11.55	11.55	16.0%
3050	Terminal & Package Units	N/A		—	—	—	
3090	Other HVAC Sys. & Equipment	N/A		—	—	—	
	D40 Fire Protection						
4010	Sprinklers	Wet pipe sprinkler system		S.F. Floor	2.05	2.05	2.8%
4020	Standpipes	N/A		—	—	—	
	D50 Electrical						
5010	Electrical Service/Distribution	400 ampere service, panel board and feeders		S.F. Floor	1.36	1.36	
5020	Lighting & Branch Wiring	Fluorescent fixtures, receptacles, switches, A.C. and misc. power		S.F. Floor	4.46	4.46	8.4%
5030	Communications & Security	Alarm systems and emergency lighting		S.F. Floor	.19	.19	
5090	Other Electrical Systems	Emergency generator, 15 kW		S.F. Floor	.07	.07	
E. EQUIPMENT & FURNISHINGS							
1010	Commercial Equipment	N/A		—	—	—	
1020	Institutional Equipment	N/A		—	—	—	0.0%
1030	Vehicular Equipment	N/A		—	—	—	
1090	Other Equipment	N/A		—	—	—	
F. SPECIAL CONSTRUCTION							
1020	Integrated Construction	N/A		—	—	—	0.0%
1040	Special Facilities	N/A		—	—	—	
G. BUILDING SITEWORK	**N/A**						

	Sub-Total	71.98	**100%**
CONTRACTOR FEES (General Requirements: 10%, Overhead: 5%, Profit: 10%)	25%	18.02	
ARCHITECT FEES	9%	8.10	
Total Building Cost		98.10	

BUILDING TYPES

Costs per square foot of floor area

Exterior Wall	S.F. Area	12000	14000	16000	19000	21000	23000	26000	28000	30000
	L.F. Perimeter	440	474	510	556	583	607	648	670	695
E.I.F.S. on Concrete Block	Steel Frame	89.20	86.65	84.75	82.40	81.15	79.95	78.65	77.85	77.20
Tilt-up Concrete Wall	Steel Frame	85.40	83.05	81.40	79.35	78.20	77.20	76.05	75.30	74.75
Face Brick with Concrete Block Back-up	Bearing Walls	88.85	86.00	83.95	81.30	79.85	78.50	77.10	76.20	75.45
	Steel Frame	92.80	89.90	87.80	85.15	83.75	82.45	81.00	80.10	79.35
Stucco on Concrete Block	Bearing Walls	82.55	80.10	78.40	76.15	75.00	73.90	72.75	72.00	71.40
	Steel Frame	86.80	84.40	82.70	80.45	79.30	78.20	77.05	76.25	75.65
Perimeter Adj., Add or Deduct	Per 100 L.F.	5.70	4.80	4.30	3.60	3.25	2.95	2.65	2.40	2.30
Story Hgt. Adj., Add or Deduct	Per 1 Ft.	1.40	1.25	1.20	1.10	1.00	.95	.90	.85	.85

For Basement, add $25.90 per square foot of basement area

The above costs were calculated using the basic specifications shown on the facing page. These costs should be adjusted where necessary for design alternatives and owner's requirements. Reported completed project costs, for this type of structure, range from $40.45 to $117.80 per S.F.

Common additives

Description	Unit	$ Cost
Emergency Lighting, 25 watt, battery operated		
Lead battery	Each	247
Nickel cadmium	Each	695
Smoke Detectors		
Ceiling type	Each	161
Duct type	Each	425
Sound System		
Amplifier, 250 watts	Each	1925
Speaker, ceiling or wall	Each	160
Trumpet	Each	305

Model costs calculated for a 1 story building with 14' story height and 21,000 square feet of floor area

				Unit	Unit Cost	Cost Per S.F.	% Of Sub-Total
A. SUBSTRUCTURE							
1010	Standard Foundations	Poured concrete; strip and spread footings		S.F. Ground	1.10	1.10	
1030	Slab on Grade	4" reinforced concrete with vapor barrier and granular base		S.F. Slab	6.00	6.00	
2010	Basement Excavation	Site preparation for slab and trench for foundation wall and footing		S.F. Ground	.23	.23	14.9%
2020	Basement Walls	4' foundation wall		L.F. Wall	61	1.69	
B. SHELL							
	B10 Superstructure						
1010	Floor Construction	N/A		—	—	—	
1020	Roof Construction	Metal deck, open web steel joists, beams, columns		S.F. Roof	7.51	7.51	12.4%
	B20 Exterior Enclosure						
2010	Exterior Walls	E.I.F.S.	70% of wall	S.F. Wall	22	6.05	
2020	Exterior Windows	Window wall	30% of wall	Each	41	4.78	21.6%
2030	Exterior Doors	Double aluminum and glass, hollow metal, steel overhead		Each	2679	2.29	
	B30 Roofing						
3010	Roof Coverings	Built-up tar and gravel with flashing; perlite/EPS composite insulation		S.F. Roof	3.99	3.99	6.6%
3020	Roof Openings	N/A		—	—	—	
C. INTERIORS							
1010	Partitions	Gypsum board on metal studs	28 S.F. Floor/L.F. Partition	S.F. Partition	3.29	1.41	
1020	Interior Doors	Hollow metal	280 S.F. Floor/Door	Each	608	2.17	
1030	Fittings	N/A		—	—	—	
2010	Stair Construction	N/A		—	—	—	12.9%
3010	Wall Finishes	Paint		S.F. Surface	1.08	.54	
3020	Floor Finishes	50% vinyl tile, 50% paint		S.F. Floor	2.18	2.18	
3030	Ceiling Finishes	Fiberglass board on exposed grid, suspended	50% of area	S.F. Ceiling	3.05	1.53	
D. SERVICES							
	D10 Conveying						
1010	Elevators & Lifts	N/A		—	—	—	0.0%
1020	Escalators & Moving Walks	N/A		—	—	—	
	D20 Plumbing						
2010	Plumbing Fixtures	Toilet and service fixtures, supply and drainage	1 Fixture/1500 S.F. Floor	Each	1575	1.05	
2020	Domestic Water Distribution	Gas fired water heater		S.F. Floor	.95	.95	4.5%
2040	Rain Water Drainage	Roof drains		S.F. Roof	.75	.75	
	D30 HVAC						
3010	Energy Supply	Gas fired hot water, unit heaters (service area)		S.F. Floor	2.90	2.90	
3020	Heat Generating Systems	N/A		—	—	—	
3030	Cooling Generating Systems	N/A		—	—	—	11.1%
3050	Terminal & Package Units	Single zone rooftop unit, gas heating, electric (office and showroom)		S.F. Floor	3.61	3.61	
3090	Other HVAC Sys. & Equipment	Underfloor garage exhaust system		S.F. Floor	.25	.25	
	D40 Fire Protection						
4010	Sprinklers	Wet pipe sprinkler system		S.F. Floor	2.71	2.71	4.5%
4020	Standpipes	N/A		—	—	—	
	D50 Electrical						
5010	Electrical Service/Distribution	200 ampere service, panel board and feeders		S.F. Floor	.34	.34	
5020	Lighting & Branch Wiring	Fluorescent fixtures, receptacles, switches, A.C. and misc. power		S.F. Floor	5.33	5.33	9.7%
5030	Communications & Security	Alarm systems and emergency lighting		S.F. Floor	.19	.19	
5090	Other Electrical Systems	Emergency generator, 7.5 kW		S.F. Floor	.05	.05	
E. EQUIPMENT & FURNISHINGS							
1010	Commercial Equipment	N/A		—	—	—	
1020	Institutional Equipment	N/A		—	—	—	
1030	Vehicular Equipment	Hoists, compressor, fuel pump		S.F. Floor	1.05	1.05	1.7%
1090	Other Equipment	N/A		—	—	—	
F. SPECIAL CONSTRUCTION							
1020	Integrated Construction	N/A		—	—	—	0.0%
1040	Special Facilities	N/A		—	—	—	
G. BUILDING SITEWORK	**N/A**						

		Sub-Total	60.65	100%
CONTRACTOR FEES (General Requirements: 10%, Overhead: 5%, Profit: 10%)		25%	15.19	
ARCHITECT FEES		7%	5.31	
	Total Building Cost		**81.15**	

Costs per square foot of floor area

Exterior Wall	S.F. Area	85000	115000	145000	175000	205000	235000	265000	295000	325000
	L.F. Perimeter	529	638	723	823	923	951	1037	1057	1132
Face Brick with Concrete Block Back-up	Steel Frame	47.45	46.40	45.50	45.15	44.75	44.25	44.05	43.65	43.55
	R/Conc. Frame	36.70	35.65	34.70	34.35	34.00	33.45	33.25	32.85	32.70
Precast Concrete	Steel Frame	48.80	47.65	46.65	46.20	45.85	45.25	45.05	44.60	44.45
	R/Conc. Frame	37.50	36.35	35.35	34.95	34.50	33.95	33.75	33.35	33.20
Reinforced Concrete	Steel Frame	47.45	46.50	45.65	45.30	45.00	44.50	44.35	44.00	43.85
	R/Conc. Frame	35.85	34.85	34.05	33.70	33.35	32.95	32.75	32.40	32.30
Perimeter Adj., Add or Deduct	Per 100 L.F.	1.20	.85	.70	.55	.45	.40	.35	.35	.25
Story Hgt. Adj., Add or Deduct	Per 1 Ft.	.45	.35	.30	.25	.30	.25	.20	.25	.20
Basement—Not Applicable										

The above costs were calculated using the basic specifications shown on the facing page. These costs should be adjusted where necessary for design alternatives and owner's requirements. Reported completed project costs, for this type of structure, range from $23.15 to $96.10 per S.F.

Common additives

Description	Unit	$ Cost
Elevators, Electric passenger, 5 stops		
2000# capacity	Each	110,300
3500# capacity	Each	116,800
5000# capacity	Each	121,800
Barrier gate w/programmable controller	Each	3675
Booth for attendant, average	Each	11,400
Fee computer	Each	13,700
Ticket spitter with time/date stamp	Each	6850
Mag strip encoding	Each	19,200
Collection station, pay on foot	Each	115,500
Parking control software	Each	23,100 - 94,500
Painting, Parking stalls	Stall	9.05
Parking Barriers		
Timber with saddles, 4" x 4"	L.F.	5.95
Precast concrete, 6" x 10" x 6'	Each	56
Traffic Signs, directional, 12" x 18", high densit	Each	66

Important: See the Reference Section for Location Factors

Model costs calculated for a 5 story building with 10' story height and 145,000 square feet of floor area

				Unit	Unit Cost	Cost Per S.F.	% Of Sub-Total
A. SUBSTRUCTURE							
1010	Standard Foundations	Poured concrete; strip and spread footings		S.F. Ground	5.40	1.08	
1030	Slab on Grade	6" reinforced concrete with vapor barrier and granular base		S.F. Slab	4.95	.99	12.1%
2010	Basement Excavation	Site preparation for slab and trench for foundation wall and footing		S.F. Ground	.14	.03	
2020	Basement Walls	4' foundation wall		L.F. Wall	54	1.08	
B. SHELL							
	B10 Superstructure						
1010	Floor Construction	Double tee precast concrete slab, precast concrete columns		S.F. Floor	16.21	12.97	49.5%
1020	Roof Construction	N/A		—	—	—	
	B20 Exterior Enclosure						
2010	Exterior Walls	Face brick with concrete block backup	40% of story height	S.F. Wall	23	2.37	
2020	Exterior Windows	N/A		—	—	—	9.0%
2030	Exterior Doors	N/A		—	—	—	
	B30 Roofing						
3010	Roof Coverings	N/A		—	—	—	
3020	Roof Openings	N/A		—	—	—	0.0%
C. INTERIORS							
1010	Partitions	Concrete block		S.F. Partition	23	.90	
1020	Interior Doors	Hollow metal		Each	12,160	.08	
1030	Fittings	N/A		—	—	—	
2010	Stair Construction	Concrete		Flight	3600	.25	5.1%
3010	Wall Finishes	Paint		S.F. Surface	2.60	.10	
3020	Floor Finishes	N/A		—	—	—	
3030	Ceiling Finishes	N/A		—	—	—	
D. SERVICES							
	D10 Conveying						
1010	Elevators & Lifts	Two hydraulic passenger elevators		Each	105,125	1.45	5.5%
1020	Escalators & Moving Walks	N/A		—	—	—	
	D20 Plumbing						
2010	Plumbing Fixtures	Toilet and service fixtures, supply and drainage	1 Fixture/18,125 S.F. Floor	Each	2356	.13	
2020	Domestic Water Distribution	Electric water heater		S.F. Floor	.04	.04	4.0%
2040	Rain Water Drainage	Roof drains		S.F. Roof	4.45	.89	
	D30 HVAC						
3010	Energy Supply	N/A		—	—	—	
3020	Heat Generating Systems	N/A		—	—	—	
3030	Cooling Generating Systems	N/A		—	—	—	0.0%
3050	Terminal & Package Units	N/A		—	—	—	
3090	Other HVAC Sys. & Equipment	N/A		—	—	—	
	D40 Fire Protection						
4010	Sprinklers	N/A		—	—	—	0.2%
4020	Standpipes	Standpipes and hose systems		S.F. Floor	.05	.05	
	D50 Electrical						
5010	Electrical Service/Distribution	400 ampere service, panel board and feeders		S.F. Floor	.17	.17	
5020	Lighting & Branch Wiring	Fluorescent fixtures, receptacles, switches and misc. power		S.F. Floor	2.43	2.43	10.5%
5030	Communications & Security	Alarm systems and emergency lighting		S.F. Floor	.11	.11	
5090	Other Electrical Systems	Emergency generator, 7.5 kW		S.F. Floor	.05	.05	
E. EQUIPMENT & FURNISHINGS							
1010	Commercial Equipment	N/A		—	—	—	
1020	Institutional Equipment	N/A		—	—	—	3.9%
1030	Vehicular Equipment	Ticket dispensers, booths, automatic gates		S.F. Floor	1.03	1.03	
1090	Other Equipment	N/A		—	—	—	
F. SPECIAL CONSTRUCTION							
1020	Integrated Construction	N/A		—	—	—	0.0%
1040	Special Facilities	N/A		—	—	—	
G. BUILDING SITEWORK	**N/A**						

	Sub-Total	26.20	100%
CONTRACTOR FEES (General Requirements: 10%, Overhead: 5%, Profit: 10%)	25%	6.53	
ARCHITECT FEES	6%	1.97	
	Total Building Cost	34.70	

BUILDING TYPES

Costs per square foot of floor area

Exterior Wall	S.F. Area	20000	30000	40000	50000	75000	100000	125000	150000	175000
	L.F. Perimeter	400	500	600	650	775	900	1000	1100	1185
Reinforced Concrete	R/Conc. Frame	71.80	66.75	64.20	61.75	58.50	56.95	55.80	55.05	54.40
Perimeter Adj., Add or Deduct	Per 100 L.F.	4.50	2.95	2.20	1.85	1.20	.85	.70	.60	.50
Story Hgt. Adj., Add or Deduct	Per 1 Ft.	1.70	1.40	1.30	1.10	.90	.70	.65	.55	.55
Basement—Not Applicable										

The above costs were calculated using the basic specifications shown on the facing page. These costs should be adjusted where necessary for design alternatives and owner's requirements. Reported completed project costs, for this type of structure, range from $36.10 to $86.10 per S.F.

Common additives

Description	Unit	$ Cost
Elevators, Hydraulic passenger, 2 stops		
1500# capacity	Each	47,900
2500# capacity	Each	50,000
3500# capacity	Each	53,700
Barrier gate w/programmable controller	Each	3675
Booth for attendant, average	Each	11,400
Fee computer	Each	13,700
Ticket spitter with time/date stamp	Each	6850
Mag strip encoding	Each	19,200
Collection station, pay on foot	Each	115,500
Parking control software	Each	23,100 - 94,500
Painting, Parking stalls	Stall	9.05
Parking Barriers		
Timber with saddles, 4" x 4"	L.F.	5.95
Precast concrete, 6" x 10" x 6'	Each	56
Traffic Signs, directional, 12" x 18"	Each	66

Model costs calculated for a 2 story building with 10' story height and 100,000 square feet of floor area

Garage, Underground Parking

				Unit	Unit Cost	Cost Per S.F.	% Of Sub-Total
A.	**SUBSTRUCTURE**						
1010	Standard Foundations	Poured concrete; strip and spread footings and waterproofing		S.F. Ground	5.76	2.88	
1030	Slab on Grade	5" reinforced concrete with vapor barrier and granular base		S.F. Slab	5.13	2.57	
2010	Basement Excavation	Excavation 24' deep		S.F. Ground	7.75	3.88	22.1%
2020	Basement Walls	N/A					
B.	**SHELL**						
	B10 Superstructure						
1010	Floor Construction	Cast-in-place concrete beam and slab, concrete columns		S.F. Floor	22	11.03	50.4%
1020	Roof Construction	Cast-in-place concrete beam and slab, concrete columns		S.F. Roof	20	10.23	
	B20 Exterior Enclosure						
2010	Exterior Walls	Cast-in place concrete		S.F. Wall	18.89	3.40	
2020	Exterior Windows	N/A		—	—	—	8.4%
2030	Exterior Doors	Steel overhead, hollow metal		Each	3316	.13	
	B30 Roofing						
3010	Roof Coverings	Neoprene membrane traffic deck		S.F. Roof	3.46	1.73	4.1%
3020	Roof Openings	N/A		—	—	—	
C.	**INTERIORS**						
1010	Partitions	Concrete block		S.F. Partition	30	.59	
1020	Interior Doors	Hollow metal		Each	4864	.05	
1030	Fittings	N/A		—	—	—	
2010	Stair Construction	Concrete		Flight	5600	.28	2.3%
3010	Wall Finishes	Paint		S.F. Surface	3.64	.07	
3020	Floor Finishes	N/A		—	—	—	
3030	Ceiling Finishes	N/A		—	—	—	
D.	**SERVICES**						
	D10 Conveying						
1010	Elevators & Lifts	Two hydraulic passenger elevators		Each	59,000	1.18	2.8%
1020	Escalators & Moving Walks	N/A		—	—	—	
	D20 Plumbing						
2010	Plumbing Fixtures	Drainage in parking areas, toilets, & service fixtures	1 Fixture/5000 S.F. Floor	Each	.25	.25	
2020	Domestic Water Distribution	Electric water heater		S.F. Floor	.06	.06	2.4%
2040	Rain Water Drainage	Roof drains		S.F. Roof	1.42	.71	
	D30 HVAC						
3010	Energy Supply	N/A		—	—	—	
3020	Heat Generating Systems	N/A		—	—	—	
3030	Cooling Generating Systems	N/A		—	—	—	0.2%
3050	Terminal & Package Units	Exhaust fans		S.F. Floor	.09	.09	
3090	Other HVAC Sys. & Equipment	N/A		—	—	—	
	D40 Fire Protection						
4010	Sprinklers	N/A		—	—	—	0.2%
4020	Standpipes	Dry standpipe system, class 1		S.F. Floor	.09	.09	
	D50 Electrical						
5010	Electrical Service/Distribution	200 ampere service, panel board and feeders		S.F. Floor	.10	.10	
5020	Lighting & Branch Wiring	Fluorescent fixtures, receptacles, switches and misc. power		S.F. Floor	2.11	2.11	6.2%
5030	Communications & Security	Alarm systems and emergency lighting		S.F. Floor	.36	.36	
5090	Other Electrical Systems	Emergency generator, 11.5 kW		S.F. Floor	.05	.05	
E.	**EQUIPMENT & FURNISHINGS**						
1010	Commercial Equipment	N/A		—	—	—	
1020	Institutional Equipment	N/A		—	—	—	
1030	Vehicular Equipment	Ticket dispensers, booths, automatic gates		S.F. Floor	.32	.32	0.8%
1090	Other Equipment	N/A		—	—	—	
F.	**SPECIAL CONSTRUCTION**						
1020	Integrated Construction	N/A		—	—	—	0.0%
1040	Special Facilities	N/A		—	—	—	
G.	**BUILDING SITEWORK**	**N/A**					

			Sub-Total	42.16	100%
CONTRACTOR FEES (General Requirements: 10%, Overhead: 5%, Profit: 10%)			25%	10.57	
ARCHITECT FEES			8%	4.22	

Total Building Cost   56.95

BUILDING TYPES

Costs per square foot of floor area

Exterior Wall	S.F. Area	2000	4000	6000	8000	10000	12000	14000	16000	18000
	L.F. Perimeter	180	260	340	420	500	580	586	600	610
Concrete Block	Wood Joists	120.50	102.85	96.95	94.05	92.25	91.05	88.15	86.20	84.55
	Steel Joists	120.90	103.80	98.00	95.20	93.50	92.30	89.45	87.50	85.85
Poured Concrete	Wood Joists	132.90	112.05	105.05	101.65	99.50	98.10	94.25	91.70	89.50
	Steel Joists	134.20	113.40	106.30	102.95	100.80	99.40	95.60	92.95	90.75
Stucco	Wood Frame	131.00	110.70	103.85	100.55	98.45	97.10	93.45	90.85	88.85
Insulated Metal Panels	Steel Frame	118.30	102.75	97.50	94.95	93.40	92.35	89.85	88.20	86.80
Perimeter Adj., Add or Deduct	Per 100 L.F.	19.35	9.70	6.45	4.85	3.90	3.20	2.75	2.45	2.15
Story Hgt. Adj., Add or Deduct	Per 1 Ft.	1.45	1.00	.85	.85	.80	.75	.65	.65	.60
For Basement, add $24.40 per square foot of basement area										

The above costs were calculated using the basic specifications shown on the facing page. These costs should be adjusted where necessary for design alternatives and owner's requirements. Reported completed project costs, for this type of structure, range from $51.85 to $155.60 per S.F.

Common additives

Description	Unit	$ Cost
Air Compressors		
Electric 1-1/2 H.P., standard controls	Each	1125
Dual controls	Each	1325
5 H.P. 115/230 Volt, standard controls	Each	3050
Dual controls	Each	3175
Product Dispenser		
with vapor recovery for 6 nozzles	Each	19,800
Hoists, Single post		
8000# cap., swivel arm	Each	7700
Two post, adjustable frames, 11,000# cap.	Each	10,900
24,000# cap.	Each	16,000
7500# Frame support	Each	9150
Four post, roll on ramp	Each	8475
Lockers, Steel, single tier, 60" or 72"	Opening	154 - 241
2 tier, 60" or 72" total	Opening	104 - 134
5 tier, box lockers	Opening	55 - 69
Locker bench, lam. maple top only	L.F.	18.80
Pedestals, steel pipe	Each	57
Lube Equipment		
3 reel type, with pumps, no piping	Each	9875
Spray Painting Booth, 26' long, complete	Each	16,900

Important: See the Reference Section for Location Factors

BUILDING TYPES

Model costs calculated for a 1 story building with 14' story height and 10,000 square feet of floor area

				Unit	Unit Cost	Cost Per S.F.	% Of Sub-Total
A. SUBSTRUCTURE							
1010	Standard Foundations	Poured concrete; strip and spread footings		S.F. Ground	1.69	1.69	
1030	Slab on Grade	6" reinforced concrete with vapor barrier and granular base		S.F. Slab	6.00	6.00	15.9%
2010	Basement Excavation	Site preparation for slab and trench for foundation wall and footing		S.F. Ground	.23	.23	
2020	Basement Walls	4' foundation wall		L.F. Wall	61	3.06	
B. SHELL							
	B10 Superstructure						
1010	Floor Construction	N/A		—	—	—	6.1%
1020	Roof Construction	Metal deck on open web steel joists		S.F. Roof	4.19	4.19	
	B20 Exterior Enclosure						
2010	Exterior Walls	Concrete block	80% of wall	S.F. Wall	8.91	4.99	
2020	Exterior Windows	Hopper type commercial steel	5% of wall	Each	346	.81	10.6%
2030	Exterior Doors	Steel overhead and hollow metal	15% of wall	S.F. Door	—	1.55	
	B30 Roofing						
3010	Roof Coverings	Built-up tar and gravel; perlite/EPS composite insulation		S.F. Roof	4.55	4.55	6.6%
3020	Roof Openings	Skylight		S.F. Roof	.01	.01	
C. INTERIORS							
1010	Partitions	Concrete block	50 S.F. Floor/L.F. Partition	S.F. Partition	5.65	1.13	
1020	Interior Doors	Single leaf hollow metal	3000 S.F. Floor/Door	Each	608	.21	
1030	Fittings	Toilet partitions		S.F. Floor	.15	.15	
2010	Stair Construction	N/A		—	—	—	7.1%
3010	Wall Finishes	Paint		S.F. Surface	11.60	2.32	
3020	Floor Finishes	90% metallic floor hardener, 10% vinyl composition tile		S.F. Floor	.88	.88	
3030	Ceiling Finishes	Gypsum board on wood joists in office and washrooms	10% of area	S.F. Ceiling	2.33	.23	
D. SERVICES							
	D10 Conveying						
1010	Elevators & Lifts	N/A		—	—	—	0.0%
1020	Escalators & Moving Walks	N/A		—	—	—	
	D20 Plumbing						
2010	Plumbing Fixtures	Toilet and service fixtures, supply and drainage	1 Fixture/500 S.F. Floor	Each	860	1.72	
2020	Domestic Water Distribution	Gas fired water heater		S.F. Floor	.27	.27	4.3%
2040	Rain Water Drainage	Roof drains		S.F. Roof	1.02	1.02	
	D30 HVAC						
3010	Energy Supply	Oil fired hot water, unit heaters		S.F. Floor	5.79	5.79	
3020	Heat Generating Systems	N/A		—	—	—	
3030	Cooling Generating Systems	N/A		—	—	—	19.7%
3050	Terminal & Package Units	Split systems with air cooled condensing units		S.F. Floor	7.49	7.49	
3090	Other HVAC Sys. & Equipment	Garage exhaust system		S.F. Floor	.38	.38	
	D40 Fire Protection						
4010	Sprinklers	Sprinklers, ordinary hazard		S.F. Floor	2.91	2.91	4.2%
4020	Standpipes	N/A		—	—	—	
	D50 Electrical						
5010	Electrical Service/Distribution	200 ampere service, panel board and feeders		S.F. Floor	.39	.39	
5020	Lighting & Branch Wiring	Fluorescent fixtures, receptacles, switches, A.C. and misc. power		S.F. Floor	5.51	5.51	8.8%
5030	Communications & Security	Alarm systems and emergency lighting		S.F. Floor	.14	.14	
5090	Other Electrical Systems	Emergency generator, 15 kW		S.F. Floor	.07	.07	
E. EQUIPMENT & FURNISHINGS							
1010	Commercial Equipment	N/A		—	—	—	
1020	Institutional Equipment	N/A		—	—	—	16.7%
1030	Vehicular Equipment	Hoists		S.F. Floor	11.55	11.55	
1090	Other Equipment	N/A		—	—	—	
F. SPECIAL CONSTRUCTION							
1020	Integrated Construction	N/A		—	—	—	0.0%
1040	Special Facilities	N/A		—	—	—	
G. BUILDING SITEWORK	**N/A**						

			Sub-Total	69.24	100%
	CONTRACTOR FEES (General Requirements: 10%, Overhead: 5%, Profit: 10%)		25%	17.34	
	ARCHITECT FEES		8%	6.92	
		Total Building Cost		**93.50**	

BUILDING TYPES

Costs per square foot of floor area

Exterior Wall	S.F. Area	600	800	1000	1200	1400	1600	1800	2000	2200
	L.F. Perimeter	100	120	126	140	153	160	170	180	190
Face Brick with Concrete Block Back-up	Wood Truss	180.65	165.85	149.55	142.30	136.55	130.50	126.45	123.30	120.75
	Steel Joists	174.55	159.70	143.45	136.10	130.45	124.30	120.30	117.15	114.60
Enameled Sandwich Panel	Steel Frame	157.10	144.10	130.45	124.10	119.30	114.15	110.80	108.15	105.95
Tile on Concrete Block	Steel Joists	187.40	171.40	153.50	145.55	139.40	132.65	128.25	124.75	121.95
Aluminum Siding	Wood Frame	153.25	141.15	128.85	123.05	118.60	114.00	110.95	108.50	106.55
Wood Siding	Wood Frame	156.05	143.70	131.00	125.05	120.45	115.70	112.55	110.05	108.00
Perimeter Adj., Add or Deduct	Per 100 L.F.	88.00	66.00	52.85	43.95	37.75	32.95	29.35	26.40	24.00
Story Hgt. Adj., Add or Deduct	Per 1 Ft.	5.80	5.20	4.40	4.00	3.80	3.45	3.25	3.15	3.00
Basement—Not Applicable										

The above costs were calculated using the basic specifications shown on the facing page. These costs should be adjusted where necessary for design alternatives and owner's requirements. Reported completed project costs, for this type of structure, range from $37.35 to $188.50 per S.F.

Common additives

Description	Unit	$ Cost
Air Compressors		
Electric 1-1/2 H.P., standard controls	Each	1125
Dual controls	Each	1325
5 H.P. 115/230 volt, standard controls	Each	3050
Dual controls	Each	3175
Product Dispenser		
with vapor recovery for 6 nozzles	Each	19,800
Hoists, Single post		
8000# cap. swivel arm	Each	7700
Two post, adjustable frames, 11,000# cap.	Each	10,900
24,000# cap.	Each	16,000
7500# cap.	Each	9150
Four post, roll on ramp	Each	8475
Lockers, Steel, single tier, 60" or 72"	Opening	154 - 241
2 tier, 60" or 72" total	Opening	104 - 134
5 tier, box lockers	Each	55 - 69
Locker bench, lam. maple top only	L.F.	18.80
Pedestals, steel pipe	Each	57
Lube Equipment		
3 reel type, with pumps, no piping	Each	9875

Important: See the Reference Section for Location Factors

Model costs calculated for a 1 story building with 12' story height and 1,400 square feet of floor area

BUILDING TYPES

					Unit	Unit Cost	Cost Per S.F.	% Of Sub-Total
A. SUBSTRUCTURE								
1010	Standard Foundations	Poured concrete; strip and spread footings			S.F. Ground	3.21	3.21	
1030	Slab on Grade	6" reinforced concrete with vapor barrier and granular base			S.F. Slab	6.00	6.00	
2010	Basement Excavation	Site preparation for slab and trench for foundation wall and footing			S.F. Ground	.78	.78	15.7%
2020	Basement Walls	4' foundation wall			L.F. Wall	54	5.92	
B. SHELL								
	B10 Superstructure							
1010	Floor Construction	N/A			—	—	—	
1020	Roof Construction	Plywood on wood trusses			S.F. Roof	5.64	5.64	5.6%
	B20 Exterior Enclosure							
2010	Exterior Walls	Face brick with concrete block backup		60% of wall	S.F. Wall	23	18.68	
2020	Exterior Windows	Store front and metal top hinged outswinging		20% of wall	Each	31	8.34	34.7%
2030	Exterior Doors	Steel overhead, aluminum & glass and hollow metal		20% of wall	Each	.02	8.07	
	B30 Roofing							
3010	Roof Coverings	Asphalt shingles with flashing; perlite/EPS composite insulation			S.F. Roof	2.49	2.49	2.5%
3020	Roof Openings	N/A			—	—	—	
C. INTERIORS								
1010	Partitions	Concrete block		25 S.F. Floor/L.F. Partition	S.F. Partition	5.63	1.80	
1020	Interior Doors	Single leaf hollow metal		700 S.F. Floor/Door	Each	608	.87	
1030	Fittings	Toilet partitions			S.F. Floor	2.16	2.16	
2010	Stair Construction	N/A			—	—	—	9.7%
3010	Wall Finishes	Paint			S.F. Surface	8.28	2.76	
3020	Floor Finishes	Vinyl composition tile		35% of floor area	S.F. Floor	2.97	1.04	
3030	Ceiling Finishes	Painted gypsum board on furring in sales area & washrooms		35% of floor area	S.F. Ceiling	3.30	1.16	
D. SERVICES								
	D10 Conveying							
1010	Elevators & Lifts	N/A			—	—	—	
1020	Escalators & Moving Walks	N/A			—	—	—	0.0%
	D20 Plumbing							
2010	Plumbing Fixtures	Toilet and service fixtures, supply and drainage		1 Fixture/235 S.F. Floor	Each	1736	7.39	
2020	Domestic Water Distribution	Gas fired water heater			S.F. Floor	1.94	1.94	9.2%
2040	Rain Water Drainage	N/A			—	—	—	
	D30 HVAC							
3010	Energy Supply	Oil fired hot water, wall fin radiation			S.F. Floor	5.79	5.79	
3020	Heat Generating Systems	N/A			—	—	—	
3030	Cooling Generating Systems	N/A			—	—	—	12.4%
3050	Terminal & Package Units	Split systems with air cooled condensing units			S.F. Floor	6.77	6.77	
3090	Other HVAC Sys. & Equipment	N/A			—	—	—	
	D40 Fire Protection							
4010	Sprinklers	N/A			—	—	—	
4020	Standpipes	N/A			—	—	—	0.0%
	D50 Electrical							
5010	Electrical Service/Distribution	200 ampere service, panel board and feeders			S.F. Floor	3.56	3.56	
5020	Lighting & Branch Wiring	Fluorescent fixtures, receptacles, switches, A.C. and misc. power			S.F. Floor	5.99	5.99	10.2%
5030	Communications & Security	Alarm systems and emergency lighting			S.F. Floor	.34	.34	
5090	Other Electrical Systems	Emergency generator			S.F. Floor	.47	.47	
E. EQUIPMENT & FURNISHINGS								
1010	Commercial Equipment	N/A			—	—	—	
1020	Institutional Equipment	N/A			—	—	—	
1030	Vehicular Equipment	N/A			—	—	—	0.0%
1090	Other Equipment	N/A			—	—	—	
F. SPECIAL CONSTRUCTION								
1020	Integrated Construction	N/A			—	—	—	
1040	Special Facilities	N/A			—	—	—	0.0%
G. BUILDING SITEWORK	**N/A**							

Sub-Total		101.17	**100%**
CONTRACTOR FEES (General Requirements: 10%, Overhead: 5%, Profit: 10%)	25%	25.26	
ARCHITECT FEES	8%	10.12	
Total Building Cost		**136.55**	

BUILDING TYPES

Costs per square foot of floor area

Exterior Wall	S.F. Area	12000	16000	20000	25000	30000	35000	40000	45000	50000
	L.F. Perimeter	440	520	600	700	708	780	841	910	979
Reinforced Concrete Block	Lam. Wood Arches	123.35	118.60	115.80	113.45	110.05	108.75	107.60	106.85	106.15
	Rigid Steel Frame	123.65	118.85	116.00	113.75	110.30	109.00	107.90	107.10	106.45
Face Brick with Concrete Block Back-up	Lam. Wood Arches	142.85	135.80	131.65	128.30	122.55	120.60	118.75	117.50	116.60
	Rigid Steel Frame	143.10	136.10	131.95	128.60	122.80	120.85	119.05	117.80	116.85
Metal Sandwich Panels	Lam. Wood Arches	119.00	114.75	112.20	110.10	107.25	106.10	105.10	104.45	103.85
	Rigid Steel Frame	119.30	115.00	112.45	110.40	107.55	106.40	105.40	104.70	104.10
Perimeter Adj., Add or Deduct	Per 100 L.F.	5.10	3.85	3.00	2.45	2.05	1.75	1.55	1.35	1.25
Story Hgt. Adj., Add or Deduct	Per 1 Ft.	.75	.65	.55	.55	.45	.40	.40	.40	.40
Basement—Not Applicable										

The above costs were calculated using the basic specifications shown on the facing page. These costs should be adjusted where necessary for design alternatives and owner's requirements. Reported completed project costs, for this type of structure, range from $56.60 to $169.05 per S.F.

Common additives

Description	Unit	$ Cost
Bleachers, Telescoping, manual		
To 15 tier	Seat	101 - 141
16-20 tier	Seat	208 - 255
21-30 tier	Seat	221 - 267
For power operation, add	Seat	40 - 63
Gym Divider Curtain, Mesh top		
Manual roll-up	S.F.	10.45
Gym Mats		
2" naugahyde covered	S.F.	3.70
2" nylon	S.F.	5.55
1-1/2" wall pads	S.F.	7.45
1" wrestling mats	S.F.	4.95
Scoreboard		
Basketball, one side	Each	2800 - 19,600
Basketball Backstop		
Wall mtd., 6' extended, fixed	Each	1875 - 2400
Swing up, wall mtd.	Each	2075 - 3125

Description	Unit	$ Cost
Lockers, Steel, single tier, 60" or 72"	Opening	154 - 241
2 tier, 60" or 72" total	Opening	104 - 134
5 tier, box lockers	Opening	55 - 69
Locker bench, lam. maple top only	L.F.	18.80
Pedestals, steel pipe	Each	57
Sound System		
Amplifier, 250 watts	Each	1925
Speaker, ceiling or wall	Each	160
Trumpet	Each	305
Emergency Lighting, 25 watt, battery operated		
Lead battery	Each	247
Nickel cadmium	Each	695

Important: See the Reference Section for Location Factors

Model costs calculated for a 1 story building with 25' story height and 20,000 square feet of floor area

					Unit	Unit Cost	Cost Per S.F.	% Of Sub-Total
A. SUBSTRUCTURE								
1010	Standard Foundations	Poured concrete; strip and spread footings			S.F. Ground	1.32	1.32	
1030	Slab on Grade	4" reinforced concrete with vapor barrier and granular base			S.F. Slab	3.98	3.98	8.5%
2010	Basement Excavation	Site preparation for slab and trench for foundation wall and footing			S.F. Ground	.14	.14	
2020	Basement Walls	4' foundation wall			L.F. Wall	54	1.90	
B. SHELL								
	B10 Superstructure							
1010	Floor Construction	N/A			—	—	—	17.6%
1020	Roof Construction	Wood deck on laminated wood arches			S.F. Roof	15.26	15.26	
	B20 Exterior Enclosure							
2010	Exterior Walls	Reinforced concrete block (end walls included)		90% of wall	S.F. Wall	9.84	6.64	
2020	Exterior Windows	Metal horizontal pivoted		10% of wall	Each	343	2.58	11.1%
2030	Exterior Doors	Aluminum and glass, hollow metal, steel overhead			Each	1426	.43	
	B30 Roofing							
3010	Roof Coverings	EPDM, 60 mils, fully adhered; polyisocyanurate insulation			S.F. Roof	2.96	2.96	3.4%
3020	Roof Openings	N/A			—	—	—	
C. INTERIORS								
1010	Partitions	Concrete block		50 S.F. Floor/L.F. Partition	S.F. Partition	6.75	1.35	
1020	Interior Doors	Single leaf hollow metal		500 S.F. Floor/Door	Each	608	1.22	
1030	Fittings	Toilet partitions			S.F. Floor	.30	.30	
2010	Stair Construction	N/A			—	—	—	20.2%
3010	Wall Finishes	50% paint, 50% ceramic tile			S.F. Surface	15.40	3.08	
3020	Floor Finishes	90% hardwood, 10% ceramic tile			S.F. Floor	10.93	10.93	
3030	Ceiling Finishes	Mineral fiber tile on concealed zee bars		15% of area	S.F. Ceiling	4.09	.62	
D. SERVICES								
	D10 Conveying							
1010	Elevators & Lifts	N/A			—	—	—	0.0%
1020	Escalators & Moving Walks	N/A			—	—	—	
	D20 Plumbing							
2010	Plumbing Fixtures	Toilet and service fixtures, supply and drainage		1 Fixture/515 S.F. Floor	Each	2322	4.51	
2020	Domestic Water Distribution	Electric water heater			S.F. Floor	2.87	2.87	8.5%
2040	Rain Water Drainage	N/A			—	—	—	
	D30 HVAC							
3010	Energy Supply	N/A			—	—	—	
3020	Heat Generating Systems	Included in D3050			—	—	—	
3030	Cooling Generating Systems	N/A			—	—	—	10.9%
3050	Terminal & Package Units	Single zone rooftop unit, gas heating, electric cooling			S.F. Floor	9.40	9.40	
3090	Other HVAC Sys. & Equipment	N/A			—	—	—	
	D40 Fire Protection							
4010	Sprinklers	Wet pipe sprinkler system			S.F. Floor	2.05	2.05	2.4%
4020	Standpipes	N/A			—	—	—	
	D50 Electrical							
5010	Electrical Service/Distribution	400 ampere service, panel board and feeders			S.F. Floor	.78	.78	
5020	Lighting & Branch Wiring	Fluorescent fixtures, receptacles, switches, A.C. and misc. power			S.F. Floor	7.42	7.42	11.5%
5030	Communications & Security	Alarm systems, sound system and emergency lighting			S.F. Floor	1.58	1.58	
5090	Other Electrical Systems	Emergency generator, 7.5 kW			S.F. Floor	.15	.15	
E. EQUIPMENT & FURNISHINGS								
1010	Commercial Equipment	N/A			—	—	—	
1020	Institutional Equipment	N/A			—	—	—	5.9%
1030	Vehicular Equipment	N/A			—	—	—	
1090	Other Equipment	Bleachers, sauna, weight room			S.F. Floor	5.08	5.08	
F. SPECIAL CONSTRUCTION								
1020	Integrated Construction	N/A			—	—	—	0.0%
1040	Special Facilities	N/A			—	—	—	
G. BUILDING SITEWORK	**N/A**							

		Sub-Total	86.55	100%
CONTRACTOR FEES (General Requirements: 10%, Overhead: 5%, Profit: 10%)		25%	21.68	
ARCHITECT FEES		7%	7.57	
	Total Building Cost		**115.80**	

BUILDING TYPES

Costs per square foot of floor area

Exterior Wall	S.F. Area	5000	10000	15000	20000	30000	40000	50000	75000	100000
	L.F. Perimeter	300	410	500	580	710	830	930	1150	1300
Concrete Block Reinforced	Steel Frame	104.20	91.35	86.10	83.20	79.55	77.60	76.10	73.95	72.35
	Bearing Walls	109.00	95.95	90.65	87.75	84.00	82.05	80.55	78.35	76.70
Precast Concrete	Steel Frame	115.50	99.00	92.35	88.60	84.00	81.45	79.60	76.80	74.75
	Bearing Walls	119.90	103.40	96.70	93.00	88.35	85.85	83.95	81.20	79.10
Galv. Steel Siding	Steel Frame	103.85	91.05	85.90	83.00	79.40	77.45	76.00	73.80	72.20
Metal Sandwich Panel	Steel Frame	107.95	93.90	88.20	85.00	81.05	78.90	77.30	74.90	73.15
Perimeter Adj., Add or Deduct	Per 100 L.F.	13.60	6.80	4.55	3.35	2.35	1.70	1.35	.85	.70
Story Hgt. Adj., Add or Deduct	Per 1 Ft.	1.40	.90	.80	.65	.60	.45	.45	.35	.30
Basement—Not Applicable										

The above costs were calculated using the basic specifications shown on the facing page. These costs should be adjusted where necessary for design alternatives and owner's requirements. Reported completed project costs, for this type of structure, range from $32.50 to $148.15 per S.F.

Common additives

Description	Unit	$ Cost
Closed Circuit Surveillance, One station		
Camera and monitor	Each	1525
For additional camera stations, add	Each	820
Emergency Lighting, 25 watt, battery operated		
Lead battery	Each	247
Nickel cadmium	Each	695
Lockers, Steel, single tier, 60" or 72"	Opening	154 - 241
2 tier, 60" or 72" total	Opening	104 - 134
5 tier, box lockers	Opening	55 - 69
Locker bench, lam. maple top only	L.F.	18.80
Pedestals, steel pipe	Each	57
Safe, Office type, 4 hour rating		
30" x 18" x 18"	Each	3425
62" x 33" x 20"	Each	7450
Sound System		
Amplifier, 250 watts	Each	1925
Speaker, ceiling or wall	Each	160
Trumpet	Each	305

BUILDING TYPES

Model costs calculated for a 1 story building with 24' story height and 20,000 square feet of floor area

Hangar, Aircraft

				Unit	Unit Cost	Cost Per S.F.	% Of Sub-Total
A. SUBSTRUCTURE							
1010	Standard Foundations	Poured concrete; strip and spread footings and		S.F. Ground	1.48	1.48	
1030	Slab on Grade	6" reinforced concrete with vapor barrier and granular base		S.F. Slab	6.00	6.00	
2010	Basement Excavation	Site preparation for slab and trench for foundation wall and footing		S.F. Ground	.23	.23	15.3%
2020	Basement Walls	4' foundation wall		L.F. Wall	61	1.77	
B. SHELL							
	B10 Superstructure						
1010	Floor Construction	N/A		—	—	—	12.9%
1020	Roof Construction	Metal deck, open web steel joists, beams, columns		S.F. Roof	8.00	8.00	
	B20 Exterior Enclosure						
2010	Exterior Walls	Galvanized steel siding		S.F. Wall	7.64	2.66	
2020	Exterior Windows	Industrial horizontal pivoted steel	20% of wall	Each	1565	9.08	29.8%
2030	Exterior Doors	Steel overhead and sliding	30% of wall	Each	32	6.73	
	B30 Roofing						
3010	Roof Coverings	Elastomeric membrane; fiberboard insulation		S.F. Roof	3.27	3.27	5.4%
3020	Roof Openings	Roof hatches		S.F. Roof	.08	.08	
C. INTERIORS							
1010	Partitions	Concrete block	200 S.F. Floor/L.F. Partition	S.F. Partition	7.20	.36	
1020	Interior Doors	Single leaf hollow metal	5000 S.F. Floor/Door	Each	608	.12	
1030	Fittings	Toilet partitions		S.F. Floor	.28	.28	
2010	Stair Construction	N/A		—	—	—	1.4%
3010	Wall Finishes	Paint		S.F. Surface	2.00	.10	
3020	Floor Finishes	N/A		—	—	—	
3030	Ceiling Finishes	N/A		—	—	—	
D. SERVICES							
	D10 Conveying						
1010	Elevators & Lifts	N/A		—	—	—	0.0%
1020	Escalators & Moving Walks	N/A		—	—	—	
	D20 Plumbing						
2010	Plumbing Fixtures	Toilet and service fixtures, supply and drainage	1 Fixture/1000 S.F. Floor	Each	2280	2.28	
2020	Domestic Water Distribution	Gas fired water heater		S.F. Floor	.44	.44	4.4%
2040	Rain Water Drainage	N/A		—	—	—	
	D30 HVAC						
3010	Energy Supply	N/A		—	—	—	
3020	Heat Generating Systems	Unit heaters		Each	5.26	5.26	
3030	Cooling Generating Systems	N/A		—	—	—	8.7%
3050	Terminal & Package Units	Exhaust fan		S.F. Floor	.16	.16	
3090	Other HVAC Sys. & Equipment	N/A		—	—	—	
	D40 Fire Protection						
4010	Sprinklers	Sprinklers, extra hazard		S.F. Floor	4.19	4.19	6.8%
4020	Standpipes	N/A		—	—	—	
	D50 Electrical						
5010	Electrical Service/Distribution	200 ampere service, panel board and feeders		S.F. Floor	.67	.67	
5020	Lighting & Branch Wiring	High intensity discharge fixtures, receptacles, switches and misc. power		S.F. Floor	8.44	8.44	15.4%
5030	Communications & Security	Alarm systems and emergency lighting		S.F. Floor	.40	.40	
5090	Other Electrical Systems	Emergency generator, 7.5 kW		S.F. Floor	.05	.05	
E. EQUIPMENT & FURNISHINGS							
1010	Commercial Equipment	N/A		—	—	—	
1020	Institutional Equipment	N/A		—	—	—	0.0%
1030	Vehicular Equipment	N/A		—	—	—	
1090	Other Equipment	N/A		—	—	—	
F. SPECIAL CONSTRUCTION							
1020	Integrated Construction	N/A		—	—	—	0.0%
1040	Special Facilities	N/A		—	—	—	
G. BUILDING SITEWORK	**N/A**						

		Sub-Total	62.05	100%
CONTRACTOR FEES (General Requirements: 10%, Overhead: 5%, Profit: 10%)		25%	15.52	
ARCHITECT FEES		7%	5.43	

Total Building Cost	**83.00**

Important: See the Reference Section for Location Factors

Costs per square foot of floor area

Exterior Wall	S.F. Area	25000	40000	55000	70000	85000	100000	115000	130000	145000
	L.F. Perimeter	388	520	566	666	766	866	878	962	1045
Face Brick with Structural Facing Tile	Steel Frame	224.80	212.95	204.85	201.55	199.45	197.95	195.55	194.60	193.90
	R/Conc. Frame	231.75	219.90	211.85	208.50	206.40	204.90	202.50	201.60	200.85
Face Brick with Concrete Block Back-up	Steel Frame	219.05	208.20	201.10	198.05	196.15	194.80	192.75	191.90	191.20
	R/Conc. Frame	226.05	215.15	208.10	205.05	203.10	201.75	199.70	198.90	198.20
Precast Concrete Panels	Steel Frame	221.30	210.00	202.55	199.45	197.40	196.00	193.85	192.95	192.25
	R/Conc. Frame	228.20	217.00	209.50	206.35	204.35	202.95	200.75	199.90	199.20
Perimeter Adj., Add or Deduct	Per 100 L.F.	6.90	4.30	3.15	2.45	2.00	1.70	1.50	1.35	1.15
Story Hgt. Adj., Add or Deduct	Per 1 Ft.	2.00	1.70	1.35	1.25	1.15	1.10	1.00	1.00	.90
For Basement, add $25.50 per square foot of basement area										

The above costs were calculated using the basic specifications shown on the facing page. These costs should be adjusted where necessary for design alternatives and owner's requirements. Reported completed project costs, for this type of structure, range from $124.90 to $315.50 per S.F.

Common additives

Description	Unit	$ Cost	Description	Unit	$ Cost
Cabinet, Base, door units, metal	L.F.	218	Nurses Call Station		
Drawer units	L.F.	415	Single bedside call station	Each	270
Tall storage cabinets, 7' high, open	L.F.	415	Ceiling speaker station	Each	122
With doors	L.F.	500	Emergency call station	Each	165
Wall, metal 12-1/2" deep, open	L.F.	157	Pillow speaker	Each	253
With doors	L.F.	281	Double bedside call station	Each	320
Closed Circuit TV (Patient monitoring)			Duty station	Each	278
One station camera & monitor	Each	1525	Standard call button	Each	144
For additional camera, add	Each	820	Master control station for 20 stations	Each	5225
For automatic iris for low light, add	Each	2125	Sound System		
Doctors In-Out Register, 200 names	Each	13,600	Amplifier, 250 watts	Each	1925
Comb. control & recall, 200 names	Each	17,000	Speaker, ceiling or wall	Each	160
Recording register	Each	6425	Trumpet	Each	305
Transformers	Each	293	Station, Dietary with ice	Each	14,500
Pocket pages	Each	965	Sterilizers		
Hubbard Tank, with accessories			Single door, steam	Each	143,500
Stainless steel, 125 GPM 45 psi	Each	23,800	Double door, steam	Each	184,500
For electric hoist, add	Each	2600	Portable, countertop, steam	Each	3450 - 5375
Mortuary Refrigerator, End operated			Gas	Each	35,500
2 capacity	Each	12,700	Automatic washer/sterilizer	Each	48,900
6 capacity	Each	23,500			

Important: See the Reference Section for Location Factors

Model costs calculated for a 3 story building with 12' story height and 55,000 square feet of floor area

				Unit	Unit Cost	Cost Per S.F.	% Of Sub-Total
A. SUBSTRUCTURE							
1010	Standard Foundations	Poured concrete; strip and spread footings		S.F. Ground	4.29	1.43	
1030	Slab on Grade	4" reinforced concrete with vapor barrier and granular base		S.F. Slab	3.98	1.32	
2010	Basement Excavation	Site preparation for slab and trench for foundation wall and footing		S.F. Ground	.14	.05	2.2%
2020	Basement Walls	4' foundation wall		L.F. Wall	61	.63	
B. SHELL							
	B10 Superstructure						
1010	Floor Construction	Cast-in-place concrete beam and slab		S.F. Floor	18.77	12.51	11.4%
1020	Roof Construction	Cast-in-place concrete beam and slab		S.F. Roof	15.42	5.14	
	B20 Exterior Enclosure						
2010	Exterior Walls	Face brick and structural facing tile	85% of wall	S.F. Wall	32	10.25	
2020	Exterior Windows	Aluminum sliding	15% of wall	Each	559	1.35	7.6%
2030	Exterior Doors	Double aluminum and glass and sliding doors		Each	2061	.22	
	B30 Roofing						
3010	Roof Coverings	Built-up tar and gravel with flashing; perlite/EPS composite insulation		S.F. Roof	4.23	1.41	0.9%
3020	Roof Openings	Roof hatches		S.F. Roof	.03	.01	
C. INTERIORS							
1010	Partitions	Concrete block, gypsum board on metal studs	9 S.F. Floor/L.F. Partition	S.F. Partition	6.48	7.20	
1020	Interior Doors	Single leaf hollow metal	90 S.F. Floor/Door	Each	629	6.99	
1030	Fittings	Hospital curtains		S.F. Floor	.73	.73	
2010	Stair Construction	Concrete filled metal pan		Flight	5450	1.29	22.5%
3010	Wall Finishes	40% vinyl wall covering, 35% ceramic tile, 25% epoxy coating		S.F. Surface	5.70	6.43	
3020	Floor Finishes	60% vinyl tile, 20% ceramic, 20% terrazzo		S.F. Floor	8.27	8.27	
3030	Ceiling Finishes	Plaster on suspended metal lath		S.F. Ceiling	4.00	4.09	
D. SERVICES							
	D10 Conveying						
1010	Elevators & Lifts	Two hydraulic hospital elevators		Each	83,325	3.03	1.9%
1020	Escalators & Moving Walks	N/A		—	—	—	
	D20 Plumbing						
2010	Plumbing Fixtures	Medical, patient & specialty fixtures, supply and drainage	1 Fixture/224 S.F. Floor	Each	2997	13.38	
2020	Domestic Water Distribution	Electric water heater		S.F. Floor	4.67	4.67	11.8%
2040	Rain Water Drainage	Roof drains		S.F. Roof	.87	.29	
	D30 HVAC						
3010	Energy Supply	Conditioned air with hot water reheat system		S.F. Floor	2.84	2.84	
3020	Heat Generating Systems	Boiler		Each	74,275	1.94	
3030	Cooling Generating Systems	Conditioned air, chillers, cooling towers		S.F. Floor	4.98	4.98	20.0%
3050	Terminal & Package Units	N/A		—	—	—	
3090	Other HVAC Sys. & Equipment	Conditioned air, ductwork, surgery air curtain		S.F. Floor	21	21.27	
	D40 Fire Protection						
4010	Sprinklers	Wet pipe sprinkler system		S.F. Floor	1.79	1.79	1.5%
4020	Standpipes	Standpipe		S.F. Floor	.47	.47	
	D50 Electrical						
5010	Electrical Service/Distribution	1200 ampere service, panel board and feeders		S.F. Floor	1.28	1.28	
5020	Lighting & Branch Wiring	Hospital grade light fixtures, receptacles, switches, A.C. and misc. power		S.F. Floor	13.31	13.31	12.2%
5030	Communications & Security	Alarm systems, communications systems, emergency lighting		S.F. Floor	3.18	3.18	
5090	Other Electrical Systems	Emergency generator, 200 kW with fuel tank		S.F. Floor	1.18	1.18	
E. EQUIPMENT & FURNISHINGS							
1010	Commercial Equipment	N/A		—	—	—	
1020	Institutional Equipment	Medical gas system, kitchen equip, lab equip., curtain partitions		S.F. Floor	7.49	7.49	8.1%
1030	Vehicular Equipment	N/A		—	—	—	
2020	Other Equipment	Patient wall systems		S.F. Floor	1850	5.05	
F. SPECIAL CONSTRUCTION							
1020	Integrated Construction	N/A		—	—	—	0.0%
1040	Special Facilities	N/A		—	—	—	
G. BUILDING SITEWORK	**N/A**						

					Sub-Total	155.47	**100%**
	CONTRACTOR FEES (General Requirements: 10%, Overhead: 5%, Profit: 10%)				25%	38.89	
	ARCHITECT FEES				9%	17.49	

	Total Building Cost	**211.85**

BUILDING TYPES

Costs per square foot of floor area

Exterior Wall	S.F. Area	100000	125000	150000	175000	200000	225000	250000	275000	300000
	L.F. Perimeter	594	705	816	783	866	950	1033	1116	1200
Face Brick with Structural Facing Tile	Steel Frame	206.55	202.30	199.40	194.50	192.85	191.60	190.55	189.70	189.05
	R/Conc. Frame	215.35	211.05	208.15	203.20	201.55	200.30	199.20	198.35	197.65
Face Brick with Concrete Block Back-up	Steel Frame	201.25	197.15	194.35	190.00	188.40	187.25	186.15	185.40	184.80
	R/Conc. Frame	211.75	207.65	204.90	200.50	198.90	197.70	196.65	195.90	195.25
Precast Concrete Panels With Exposed Aggregate	Steel Frame	201.85	197.80	195.05	190.85	189.30	188.10	187.05	186.30	185.70
	R/Conc. Frame	210.75	206.70	203.95	199.75	198.20	197.05	195.95	195.20	194.60
Perimeter Adj., Add or Deduct	Per 100 L.F.	3.40	2.75	2.30	2.00	1.75	1.55	1.40	1.30	1.20
Story Hgt. Adj., Add or Deduct	Per 1 Ft.	1.50	1.40	1.40	1.20	1.20	1.10	1.05	1.05	1.10

For Basement, add $26.35 per square foot of basement area

The above costs were calculated using the basic specifications shown on the facing page. These costs should be adjusted where necessary for design alternatives and owner's requirements. Reported completed project costs, for this type of structure, range from $127.75 to $311.55 per S.F.

Common additives

Description	Unit	$ Cost	Description	Unit	$ Cost
Cabinets, Base, door units, metal	L.F.	218	Nurses Call Station		
Drawer units	L.F.	415	Single bedside call station	Each	270
Tall storage cabinets, 7' high, open	L.F.	415	Ceiling speaker station	Each	122
With doors	L.F.	500	Emergency call station	Each	165
Wall, metal 12-1/2" deep, open	L.F.	157	Pillow speaker	Each	253
With doors	L.F.	281	Double bedside call station	Each	320
Closed Circuit TV (Patient monitoring)			Duty station	Each	278
One station camera & monitor	Each	1525	Standard call button	Each	144
For additional camera add	Each	820	Master control station for 20 stations	Each	5225
For automatic iris for low light add	Each	2125	Sound System		
Doctors In-Out Register, 200 names	Each	13,600	Amplifier, 250 watts	Each	1925
Comb. control & recall, 200 names	Each	17,000	Speaker, ceiling or wall	Each	160
Recording register	Each	6425	Trumpet	Each	305
Transformers	Each	293	Station, Dietary with ice	Each	14,500
Pocket pages	Each	965	Sterilizers		
Hubbard Tank, with accessories			Single door, steam	Each	143,500
Stainless steel, 125 GPM 45 psi	Each	23,800	Double door, steam	Each	184,500
For electric hoist, add	Each	2600	Portable, counter top, steam	Each	3450 - 5375
Mortuary Refrigerator, End operated			Gas	Each	35,500
2 capacity	Each	12,700	Automatic washer/sterilizer	Each	48,900
6 capacity	Each	23,500			

Important: See the Reference Section for Location Factors

Model costs calculated for a 6 story building with 12' story height and 200,000 square feet of floor area

				Unit	Unit Cost	Cost Per S.F.	% Of Sub-Total
A.	**SUBSTRUCTURE**						
1010	Standard Foundations	Poured concrete; strip and spread footings		S.F. Ground	11.28	1.88	
1030	Slab on Grade	4" reinforced concrete with vapor barrier and granular base		S.F. Slab	6.00	1.00	2.2%
2010	Basement Excavation	Site preparation for slab and trench for foundation wall and footing		S.F. Ground	.14	.02	
2020	Basement Walls	4' foundation wall		L.F. Wall	61	.26	
B.	**SHELL**						
	B10 Superstructure						
1010	Floor Construction	Concrete slab with metal deck and beams, steel columns		S.F. Floor	16.15	13.46	10.3%
1020	Roof Construction	Metal deck, open web steel joists, beams, interior columns		S.F. Roof	6.60	1.10	
	B20 Exterior Enclosure						
2010	Exterior Walls	Face brick and structural facing tile	70% of wall	S.F. Wall	32	7.10	
2020	Exterior Windows	Aluminum sliding	30% of wall	Each	436	2.71	7.3%
2030	Exterior Doors	Double aluminum and glass and sliding doors		Each	3345	.48	
	B30 Roofing						
3010	Roof Coverings	Built-up tar and gravel with flashing; perlite/EPS composite insulation		S.F. Roof	5.52	.92	0.7%
3020	Roof Openings	Roof hatches		S.F. Roof	.06	.01	
C.	**INTERIORS**						
1010	Partitions	Gypsum board on metal studs with sound deadening board	9 S.F. Floor/L.F. Partition	S.F. Partition	5.97	6.63	
1020	Interior Doors	Single leaf hollow metal	90 S.F. Floor/Door	Each	629	6.99	
1030	Fittings	Hospital curtains		S.F. Floor	.81	.81	
2010	Stair Construction	Concrete filled metal pan		Flight	5150	.67	25.1%
3010	Wall Finishes	40% vinyl wall covering, 35% ceramic tile, 25% epoxy coating		S.F. Surface	5.75	6.42	
3020	Floor Finishes	60% vinyl tile, 20% ceramic, 20% terrazzo		S.F. Floor	9.03	9.03	
3030	Ceiling Finishes	Plaster on suspended metal lath		S.F. Ceiling	5.00	5.00	
D.	**SERVICES**						
	D10 Conveying						
1010	Elevators & Lifts	Six geared hospital elevators		Each	161,666	4.85	3.4%
1020	Escalators & Moving Walks	N/A		—	—	—	
	D20 Plumbing						
2010	Plumbing Fixtures	Kitchen, toilet and service fixtures, supply and drainage	1 Fixture/416 S.F. Floor	Each	2179	5.24	
2020	Domestic Water Distribution	Electric water heater		S.F. Floor	2.31	2.31	5.6%
2040	Rain Water Drainage	Roof drains		S.F. Floor	2.34	.39	
	D30 HVAC						
3010	Energy Supply	Oil fired hot water, wall fin radiation		S.F. Floor	2.41	2.41	
3020	Heat Generating Systems	Hot water boilers, steam boiler for services		Each	24,475	.30	
3030	Cooling Generating Systems	Chilled water units		S.F. Floor	2.84	2.84	20.2%
3050	Terminal & Package Units	N/A		—	—	—	
3090	Other HVAC Sys. & Equipment	Conditioned air with reheat, operating room air curtains		S.F. Floor	22	22.98	
	D40 Fire Protection						
4010	Sprinklers	Wet pipe sprinkler system		S.F. Floor	1.71	1.71	1.5%
4020	Standpipes	Standpipe		S.F. Floor	.38	.38	
	D50 Electrical						
5010	Electrical Service/Distribution	4000 ampere service, panel board and feeders		S.F. Floor	3.00	3.00	
5020	Lighting & Branch Wiring	Hospital grade light fixtures, receptacles, switches, A.C. and misc. power		S.F. Floor	15.54	15.54	14.6%
5030	Communications & Security	Alarm systems, communications system, emergency lighting		S.F. Floor	1.18	1.18	
5090	Other Electrical Systems	Emergency generator, 800 kW with fuel takn		S.F. Floor	.90	.90	
E.	**EQUIPMENT & FURNISHINGS**						
1010	Commercial Equipment	N/A		—	—	—	
1020	Institutional Equipment	Medical gases, curtain partitions		S.F. Floor	9.70	9.70	9.2%
1030	Vehicular Equipment	N/A		—	—	—	
2020	Other Equipment	Patient wall systems		S.F. Floor	1850	3.33	
F.	**SPECIAL CONSTRUCTION**						
1020	Integrated Construction	N/A		—	—	—	0.0%
1040	Special Facilities	N/A		—	—	—	
G.	**BUILDING SITEWORK**	**N/A**					

			Sub-Total	141.55	100%
CONTRACTOR FEES (General Requirements: 10%, Overhead: 5%, Profit: 10%)			25%	35.38	
ARCHITECT FEES			9%	15.92	
		Total Building Cost		**192.85**	

BUILDING TYPES

Costs per square foot of floor area

Exterior Wall	S.F. Area	35000	55000	75000	95000	115000	135000	155000	175000	195000
	L.F. Perimeter	314	401	497	555	639	722	754	783	850
Face Brick with Concrete Block Back-up	Steel Frame	137.05	128.70	125.15	122.05	120.60	119.55	117.95	116.70	116.10
	R/Conc. Frame	139.35	131.05	127.50	124.35	122.90	121.90	120.25	119.00	118.40
Glass and Metal Curtain Walls	Steel Frame	131.25	124.05	120.95	118.30	117.05	116.10	114.80	113.75	113.30
	R/Conc. Frame	133.75	126.50	123.40	120.80	119.50	118.60	117.30	116.30	115.75
Precast Concrete Panels	Steel Frame	140.75	131.70	127.95	124.45	122.90	121.80	119.95	118.55	117.90
	R/Conc. Frame	144.05	134.85	131.00	127.50	125.90	124.75	122.90	121.40	120.80
Perimeter Adj., Add or Deduct	Per 100 L.F.	7.50	4.80	3.50	2.80	2.25	1.95	1.70	1.50	1.35
Story Hgt. Adj., Add or Deduct	Per 1 Ft.	2.05	1.75	1.60	1.35	1.35	1.30	1.10	1.05	1.00

For Basement, add $25.05 per square foot of basement area

The above costs were calculated using the basic specifications shown on the facing page. These costs should be adjusted where necessary for design alternatives and owner's requirements. Reported completed project costs, for this type of structure, range from $84.30 to $161.75 per S.F.

Common additives

Description	Unit	$ Cost
Bar, Front bar	L.F.	320
Back bar	L.F.	257
Booth, Upholstered, custom, straight	L.F.	166 - 305
"L" or "U" shaped	L.F.	172 - 290
Closed Circuit Surveillance, One station		
Camera and monitor	Each	1525
For additional camera stations, add	Each	820
Directory Boards, Plastic, glass covered		
30" x 20"	Each	575
36" x 48"	Each	1275
Aluminum, 24" x 18"	Each	525
48" x 32"	Each	810
48" x 60"	Each	1875
Elevators, Electric passenger, 5 stops		
3500# capacity	Each	116,800
5000# capacity	Each	121,800
Additional stop, add	Each	6550
Emergency Lighting, 25 watt, battery operated		
Lead battery	Each	247
Nickel cadmium	Each	695

Description	Unit	$ Cost
Laundry Equipment		
Folders, blankets & sheets, king size	Each	57,500
Ironers, 110" single roll	Each	31,200
Combination washer extractor 50#	Each	10,900
125#	Each	29,300
Sauna, Prefabricated, complete		
6' x 4'	Each	4525
6' x 6'	Each	5425
6' x 9'	Each	6700
8' x 8'	Each	7875
10' x 12'	Each	11,000
Smoke Detectors		
Ceiling type	Each	161
Duct type	Each	425
Sound System		
Amplifier, 250 watts	Each	1925
Speaker, ceiling or wall	Each	160
Trumpet	Each	305
TV Antenna, Master system, 12 outlet	Outlet	262
30 outlet	Outlet	168
100 outlet	Outlet	160

Important: See the Reference Section for Location Factors

Model costs calculated for a 6 story building with 10' story height and 135,000 square feet of floor area

BUILDING TYPES

				Unit	Unit Cost	Cost Per S.F.	% Of Sub-Total
A.	**SUBSTRUCTURE**						
1010	Standard Foundations	Poured concrete; strip and spread footings		S.F. Ground	4.74	.79	
1030	Slab on Grade	4" reinforced concrete with vapor barrier and granular base		S.F. Slab	3.98	.67	2.4%
2010	Basement Excavation	Site preparation for slab and trench for foundation wall and footing		S.F. Ground	.14	.02	
2020	Basement Walls	4' foundation wall		L.F. Wall	61	.65	
B.	**SHELL**						
	B10 Superstructure						
1010	Floor Construction	Concrete slab with metal deck, beams, columns		S.F. Floor	15.41	12.84	15.6%
1020	Roof Construction	Metal deck, open web steel joists, beams, columns		S.F. Roof	7.62	1.27	
	B20 Exterior Enclosure						
2010	Exterior Walls	Face brick with concrete block backup	80% of wall	S.F. Wall	23	6.10	
2020	Exterior Windows	Window wall	20% of wall	Each	38	2.49	9.7%
2030	Exterior Doors	Glass amd metal doors and entrances		Each	2507	.19	
	B30 Roofing						
3010	Roof Coverings	Built-up tar and gravel with flashing; perlite/EPS composite insulation		S.F. Roof	4.14	.69	0.8%
3020	Roof Openings	Roof hatches		S.F. Roof	.06	.01	
C.	**INTERIORS**						
1010	Partitions	Gypsum board and sound deadening board, steel studs	9 S.F. Floor/L.F. Partition	S.F. Partition	5.66	5.03	
1020	Interior Doors	Single leaf hollow metal	90 S.F. Floor/Door	Each	608	6.76	
1030	Fittings	N/A		—	—	—	
2010	Stair Construction	Concrete filled metal pan		Flight	6425	.67	24.5%
3010	Wall Finishes	20% paint, 75% vinyl cover, 5% ceramic tile		S.F. Surface	2.82	2.82	
3020	Floor Finishes	80% carpet, 10% vinyl composition tile, 10% ceramic tile		S.F. Floor	4.68	4.68	
3030	Ceiling Finishes	Mineral fiber tile applied with adhesive		S.F. Ceiling	2.00	2.14	
D.	**SERVICES**						
	D10 Conveying						
1010	Elevators & Lifts	Four geared passenger elevators		Each	148,162	4.39	4.9%
1020	Escalators & Moving Walks	N/A		—	—	—	
	D20 Plumbing						
2010	Plumbing Fixtures	Kitchen, toilet and service fixtures, supply and drainage	1 Fixture/155 S.F. Floor	Each	1861	12.01	
2020	Domestic Water Distribution	Oil fired water heater		S.F. Floor	.87	.87	14.4%
2040	Rain Water Drainage	Roof drains		S.F. Roof	.72	.12	
	D30 HVAC						
3010	Energy Supply	Oil fired hot water, wall fin radiation		S.F. Floor	3.29	3.29	
3020	Heat Generating Systems	N/A		—	—	—	
3030	Cooling Generating Systems	Chilled water, fan coil units		S.F. Floor	9.81	9.81	14.5%
3050	Terminal & Package Units	N/A		—	—	—	
3090	Other HVAC Sys. & Equipment	N/A		—	—	—	
	D40 Fire Protection						
4010	Sprinklers	Sprinkler system, light hazard		S.F. Floor	1.71	1.71	2.0%
4020	Standpipes	Standpipes and hose systems		S.F. Floor	.08	.08	
	D50 Electrical						
5010	Electrical Service/Distribution	2000 ampere service, panel board and feeders		S.F. Floor	.96	.96	
5020	Lighting & Branch Wiring	Fluorescent fixtures, receptacles, switches, A.C. and misc. power		S.F. Floor	6.88	6.88	11.2%
5030	Communications & Security	Alarm systems, communications systems and emergency lighting		S.F. Floor	1.89	1.89	
5090	Other Electrical Systems	Emergency generator, 250 kW		S.F. Floor	.41	.41	
E.	**EQUIPMENT & FURNISHINGS**						
1010	Commercial Equipment	N/A		—	—	—	
1020	Institutional Equipment	N/A		—	—	—	0.0%
1030	Vehicular Equipment	N/A		—	—	—	
1090	Other Equipment	N/A		—	—	—	
F.	**SPECIAL CONSTRUCTION**						
1020	Integrated Construction	N/A		—	—	—	0.0%
1040	Special Facilities	N/A		—	—	—	
G.	**BUILDING SITEWORK**	**N/A**					

				Sub-Total	90.24	**100%**
CONTRACTOR FEES (General Requirements: 10%, Overhead: 5%, Profit: 10%)			25%	22.54	
ARCHITECT FEES			6%	6.77	

Total Building Cost	**119.55**

Costs per square foot of floor area

Exterior Wall	S.F. Area	140000	243000	346000	450000	552000	655000	760000	860000	965000
	L.F. Perimeter	403	587	672	800	936	1073	1213	1195	1312
Face Brick with Concrete Block Back-up	Steel Frame	115.90	110.35	106.35	104.75	103.85	103.30	102.80	101.35	101.10
	R/Conc. Frame	116.45	110.85	106.85	105.25	104.30	103.80	103.35	101.90	101.60
Face Brick Veneer On Steel Studs	Steel Frame	113.95	108.75	105.05	103.55	102.70	102.20	101.75	100.45	100.20
	R/Conc. Frame	114.90	109.65	105.95	104.45	103.65	103.10	102.70	101.35	101.10
Glass and Metal Curtain Walls	Steel Frame	129.85	121.25	116.50	114.35	113.10	112.25	111.60	110.30	109.90
	R/Conc. Frame	130.80	122.20	117.45	115.25	113.95	113.15	112.55	111.20	110.80
Perimeter Adj., Add or Deduct	Per 100 L.F.	4.30	2.45	1.80	1.35	1.10	.90	.80	.75	.65
Story Hgt. Adj., Add or Deduct	Per 1 Ft.	1.70	1.35	1.15	1.05	1.05	.95	.95	.85	.85

For Basement, add $25.95 per square foot of basement area

The above costs were calculated using the basic specifications shown on the facing page. These costs should be adjusted where necessary for design alternatives and owner's requirements. Reported completed project costs, for this type of structure, range from $93.50 to $163.55 per S.F.

Common additives

Description	Unit	$ Cost	Description	Unit	$ Cost
Bar, Front bar	L.F.	320	Laundry Equipment		
Back bar	L.F.	257	Folders, blankets & sheets, king size	Each	57,500
Booth, Upholstered, custom, straight	L.F.	166 - 305	Ironers, 110" single roll	Each	31,200
"L" or "U" shaped	L.F.	172 - 290	Combination washer & extractor 50#	Each	10,900
Closed Circuit Surveillance, One station			125#	Each	29,300
Camera and monitor	Each	1525	Sauna, Prefabricated, complete		
For additional camera stations, add	Each	820	6' x 4'	Each	4525
Directory Boards, Plastic, glass covered			6' x 6'	Each	5425
30" x 20"	Each	575	6' x 9'	Each	6700
36" x 48"	Each	1275	8' x 8'	Each	7875
Aluminum, 24" x 18"	Each	525	10' x 12'	Each	11,000
48" x 32"	Each	810	Smoke Detectors		
48" x 60"	Each	1875	Ceiling type	Each	161
Elevators, Electric passenger, 10 stops			Duct type	Each	425
3500# capacity	Each	239,500	Sound System	Each	
5000# capacity	Each	246,500	Amplifier, 250 watts	Each	1925
Additional stop, add	Each	6550	Speaker, ceiling or wall	Each	160
Emergency Lighting, 25 watt, battery operated			Trumpet	Each	305
Lead battery	Each	247	TV Antenna, Master system, 12 outlet	Outlet	262
Nickel cadmium	Each	695	30 outlet	Outlet	168
			100 outlet	Outlet	160

Important: See the Reference Section for Location Factors

Model costs calculated for a 15 story building with 10' story height and 450,000 square feet of floor area

				Unit	Unit Cost	Cost Per S.F.	% Of Sub-Total
A. SUBSTRUCTURE							
1010	Standard Foundations	Poured concrete; strip and spread footings		S.F. Ground	14.10	.94	
1030	Slab on Grade	4" reinforced concrete with vapor barrier and granular base		S.F. Slab	3.98	.26	
2010	Basement Excavation	Site preparation for slab and trench for foundation wall and footing		S.F. Ground	.14	.01	1.6%
2020	Basement Walls	4' foundation wall		L.F. Wall	61	.13	
B. SHELL							
	B10 Superstructure						
1010	Floor Construction	Open web steel joists, slab form, concrete, columns		S.F. Floor	15.47	14.44	17.2%
1020	Roof Construction	Metal deck, open web steel joists, beams, columns		S.F. Roof	6.60	.44	
	B20 Exterior Enclosure						
2010	Exterior Walls	N/A		—	—	—	
2020	Exterior Windows	Glass and metal curtain walls	100% of wall	Each	18.65	4.97	6.0%
2030	Exterior Doors	Glass and metal doors and entrances		Each	2266	.17	
	B30 Roofing						
3010	Roof Coverings	Built-up tar and gravel with flashing; perlite/EPS composite insulation		S.F. Roof	4.20	.28	0.3%
3020	Roof Openings	N/A		—	—	—	
C. INTERIORS							
1010	Partitions	Gypsum board and sound deadening board, steel studs	9 S.F. Floor/L.F. Partition	S.F. Partition	5.49	4.88	
1020	Interior Doors	Single leaf hollow metal	90 S.F. Floor/Door	Each	608	6.76	
1030	Fittings	N/A		—	—	—	
2010	Stair Construction	Concrete filled metal pan		Flight	6425	1.30	27.3%
3010	Wall Finishes	20% paint, 75% vinyl cover, 5% ceramic tile		S.F. Surface	2.80	2.80	
3020	Floor Finishes	80% carpet tile, 10% vinyl composition tile, 10% ceramic tile		S.F. Floor	4.68	4.68	
3030	Ceiling Finishes	Gypsum board on resilient channel		S.F. Ceiling	3.00	3.12	
D. SERVICES							
	D10 Conveying						
1010	Elevators & Lifts	One geared freight, six geared passenger elevators		Each	269,250	3.59	4.2%
1020	Escalators & Moving Walks	N/A		—	—	—	
	D20 Plumbing						
2010	Plumbing Fixtures	Kitchen, toilet and service fixtures, supply and drainage	1 Fixture/165 S.F. Floor	Each	1815	11.00	
2020	Domestic Water Distribution	Electric water heater		S.F. Floor	2.15	2.15	15.3%
2040	Rain Water Drainage	Roof drains		S.F. Roof	1.20	.08	
	D30 HVAC						
3010	Energy Supply	Oil fired hot water, wall fin radiation		S.F. Floor	1.70	1.70	
3020	Heat Generating Systems	N/A		—	—	—	
3030	Cooling Generating Systems	Chilled water, fan coil units		S.F. Floor	9.81	9.81	13.3%
3050	Terminal & Package Units	N/A		—	—	—	
3090	Other HVAC Sys. & Equipment	N/A		—	—	—	
	D40 Fire Protection						
4010	Sprinklers	Sprinkler system, light hazard		S.F. Floor	2.45	2.45	3.2%
4020	Standpipes	Standpipes and hose systems		S.F. Floor	.27	.27	
	D50 Electrical						
5010	Electrical Service/Distribution	6000 ampere service, panel board and feeders		S.F. Floor	1.07	1.07	
5020	Lighting & Branch Wiring	Fluorescent fixtures, receptacles, switches, A.C. and misc. power		S.F. Floor	6.79	6.79	11.7%
5030	Communications & Security	Alarm systems, communications systems and emergency lighting		S.F. Floor	1.90	1.90	
5090	Other Electrical Systems	Emergency generator, 500 kW		S.F. Floor	.31	.31	
E. EQUIPMENT & FURNISHINGS							
1010	Commercial Equipment	N/A		—	—	—	
1020	Institutional Equipment	N/A		—	—	—	
1030	Vehicular Equipment	N/A		—	—	—	0.0%
1090	Other Equipment	N/A		—	—	—	
F. SPECIAL CONSTRUCTION							
1020	Integrated Construction	N/A		—	—	—	0.0%
1040	Special Facilities	N/A		—	—	—	
G. BUILDING SITEWORK	**N/A**						

		Sub-Total	86.30	100%
CONTRACTOR FEES (General Requirements: 10%, Overhead: 5%, Profit: 10%)		25%	21.58	
ARCHITECT FEES		6%	6.47	

Total Building Cost	**114.35**

BUILDING TYPES

Costs per square foot of floor area

Exterior Wall	S.F. Area	5500	12000	20000	30000	40000	60000	80000	100000	145000
	L.F. Perimeter	310	330	400	490	530	590	680	760	920
Face Brick with Concrete Block Back-up	Steel Frame	336.00	248.45	223.60	211.35	202.60	193.10	189.20	186.60	183.15
	R/Conc. Frame	340.65	253.05	228.15	215.85	207.15	197.65	193.80	191.15	187.70
Stucco on Concrete Block	Steel Frame	313.60	237.60	215.70	204.90	197.45	189.25	185.95	183.70	180.75
	R/Conc. Frame	318.00	242.00	220.15	209.30	201.80	193.70	190.35	188.10	185.15
Reinforced Concrete	Steel Frame	324.75	243.05	219.70	208.15	200.05	191.25	187.65	185.15	182.00
	R/Conc. Frame	329.20	247.45	224.10	212.55	204.45	195.65	192.05	189.60	186.45
Perimeter Adj., Add or Deduct	Per 100 L.F.	37.65	17.30	10.40	6.90	5.15	3.50	2.55	2.10	1.40
Story Hgt. Adj., Add or Deduct	Per 1 Ft.	9.00	4.40	3.20	2.60	2.10	1.55	1.40	1.25	1.05

For Basement, add $36.00 per square foot of basement area

The above costs were calculated using the basic specifications shown on the facing page. These costs should be adjusted where necessary for design alternatives and owner's requirements. Reported completed project costs, for this type of structure, range from $136.10 to $361.50 per S.F.

Common additives

Description	Unit	$ Cost
Clock System		
20 room	Each	14,200
50 room	Each	34,600
Closed Circuit Surveillance, One station		
Camera and monitor	Each	1525
For additional camera stations, add	Each	820
Elevators, Hydraulic passenger, 2 stops		
1500# capacity	Each	47,900
2500# capacity	Each	50,000
3500# capacity	Each	53,700
Additional stop, add	Each	4550
Emergency Generators, Complete system, gas		
15 KW	Each	13,200
70 KW	Each	26,000
85 KW	Each	29,000
115 KW	Each	55,500
170 KW	Each	72,500
Diesel, 50 KW	Each	25,000
100 KW	Each	35,800
150 KW	Each	43,100
350 KW	Each	68,500

Description	Unit	$ Cost
Emergency Lighting, 25 watt, battery operated		
Nickel cadmium	Each	695
Flagpoles, Complete		
Aluminum, 20' high	Each	1375
40' High	Each	3025
70' High	Each	9500
Fiberglass, 23' High	Each	1650
39'-5" High	Each	3175
59' High	Each	7950
Laundry Equipment	Each	
Folders, blankets, & sheets	Each	57,500
Ironers, 110" single roll	Each	31,200
Combination washer & extractor, 50#	Each	10,900
125#	Each	29,300
Safe, Office type, 4 hour rating		
30" x 18" x 18"	Each	3425
62" x 33" x 20"	Each	7450
Sound System		
Amplifier, 250 watts	Each	1925
Speaker, ceiling or wall	Each	160
Trumpet	Each	305

Important: See the Reference Section for Location Factors

Model costs calculated for a 3 story building with 12' story height and 40,000 square feet of floor area

					Unit	Unit Cost	Cost Per S.F.	% Of Sub-Total

A. SUBSTRUCTURE

1010	Standard Foundations	Poured concrete; strip and spread footings		S.F. Ground	1.59	.53	
1030	Slab on Grade	4" reinforced concrete with vapor barrier and granular base		S.F. Slab	3.98	1.32	
2010	Basement Excavation	Site preparation for slab and trench for foundation wall and footing		S.F. Ground	.23	.08	
2020	Basement Walls	4' foundation wall		L.F. Wall	54	.72	1.7%

B. SHELL

B10 Superstructure

| 1010 | Floor Construction | Concrete slab with metal deck and beams | | S.F. Floor | 21 | 14.14 | |
| 1020 | Roof Construction | Concrete slab with metal deck and beams | | S.F. Roof | 16.71 | 5.57 | 13.0% |

B20 Exterior Enclosure

2010	Exterior Walls	Face brick with reinforced concrete block backup	85% of wall	S.F. Wall	23	9.63	
2020	Exterior Windows	Bullet resisting and metal horizontal pivoted	15% of wall	Each	.01	8.37	12.0%
2030	Exterior Doors	Metal		Each	4075	.11	

B30 Roofing

| 3010 | Roof Coverings | Built-up tar and gravel with flashing; perlite/EPS composite insulation | | S.F. Roof | 4.32 | 1.44 | |
| 3020 | Roof Openings | Roof hatches | | S.F. Roof | .03 | .01 | 1.0% |

C. INTERIORS

1010	Partitions	Concrete block	45 S.F. of Floor/L.F. Partition	S.F. Partition	7.46	1.99	
1020	Interior Doors	Hollow metal fire doors	930 S.F. Floor/Door	Each	608	.65	
1030	Fittings	N/A		—	—	—	
2010	Stair Construction	Concrete filled metal pan		Flight	7700	.96	6.2%
3010	Wall Finishes	Paint		S.F. Surface	5.10	1.70	
3020	Floor Finishes	70% vinyl composition tile, 20% carpet, 10% ceramic tile		S.F. Floor	2.81	2.81	
3030	Ceiling Finishes	Mineral tile on zee runners	30% of area	S.F. Ceiling	4.09	1.23	

D. SERVICES

D10 Conveying

| 1010 | Elevators & Lifts | One hydraulic passenger elevator | | Each | 114,400 | 2.86 | |
| 1020 | Escalators & Moving Walks | N/A | | — | — | — | 1.9% |

D20 Plumbing

2010	Plumbing Fixtures	Toilet and service fixtures, supply and drainage	1 Fixture/78 S.F. Floor	Each	1763	22.61	
2020	Domestic Water Distribution	Electric water heater		S.F. Floor	3.52	3.52	17.5%
2040	Rain Water Drainage	Roof drains		S.F. Roof	1.20	.40	

D30 HVAC

3010	Energy Supply	N/A		—	—	—	
3020	Heat Generating Systems	Included in D3050		—	—	—	
3030	Cooling Generating Systems	N/A		—	—	—	8.6%
3050	Terminal & Package Units	Rooftop multizone unit systems		S.F. Floor	12.97	12.97	
3090	Other HVAC Sys. & Equipment	N/A		—	—	—	

D40 Fire Protection

| 4010 | Sprinklers | N/A | | — | — | — | |
| 4020 | Standpipes | Standpipes and hose systems | | S.F. Floor | .35 | .35 | 0.2% |

D50 Electrical

5010	Electrical Service/Distribution	600 ampere service, panel board and feeders		S.F. Floor	.90	.90	
5020	Lighting & Branch Wiring	Fluorescent fixtures, receptacles, switches, A.C. and misc. power		S.F. Floor	8.02	8.02	
5030	Communications & Security	Alarm systems, intercom, and emergency lighting		S.F. Floor	3.07	3.07	8.3%
5090	Other Electrical Systems	Emergency generator, 80 kW		S.F. Floor	.57	.57	

E. EQUIPMENT & FURNISHINGS

1010	Commercial Equipment	N/A		—	—	—	
1020	Institutional Equipment	Prefabricated Cells, visitor cubicles		S.F. Floor	44	44.95	
1030	Vehicular Equipment	N/A		—	—	—	29.7%
1090	Other Equipment	N/A		—	—	—	

F. SPECIAL CONSTRUCTION

| 1020 | Integrated Construction | N/A | | — | — | — | |
| 1040 | Special Facilities | N/A | | — | — | — | 0.0% |

G. BUILDING SITEWORK   N/A

			Sub-Total	151.48	100%
CONTRACTOR FEES (General Requirements: 10%, Overhead: 5%, Profit: 10%)			25%	37.87	
ARCHITECT FEES			7%	13.25	

Total Building Cost	202.60

BUILDING TYPES

Costs per square foot of floor area

Exterior Wall	S.F. Area	1000	2000	3000	4000	5000	10000	15000	20000	25000
	L.F. Perimeter	126	179	219	253	283	400	490	568	632
Decorative Concrete Block	Steel Frame	183.15	155.75	144.80	138.80	134.75	125.45	121.60	119.45	117.85
	Bearing Walls	183.85	156.10	145.00	138.90	134.85	125.30	121.35	119.20	117.55
Face Brick with Concrete Block Back-up	Steel Frame	206.90	173.25	159.50	151.80	146.60	134.45	129.30	126.45	124.35
	Bearing Walls	205.40	171.40	157.50	149.65	144.50	132.15	126.95	124.05	121.95
Metal Sandwich Panel	Steel Frame	182.85	156.35	145.80	139.95	136.10	127.15	123.45	121.40	119.95
Precast Concrete Panel	Bearing Walls	188.75	159.60	147.80	141.35	137.00	126.80	122.60	120.30	118.60
Perimeter Adj., Add or Deduct	Per 100 L.F.	41.55	20.80	13.90	10.35	8.35	4.10	2.70	2.05	1.65
Story Hgt. Adj., Add or Deduct	Per 1 Ft.	2.80	2.00	1.65	1.40	1.30	.90	.70	.55	.60
For Basement, add $23.85 per square foot of basement area										

The above costs were calculated using the basic specifications shown on the facing page. These costs should be adjusted where necessary for design alternatives and owner's requirements. Reported completed project costs, for this type of structure, range from $75.80 to $185.25 per S.F.

Common additives

Description	Unit	$ Cost
Closed Circuit Surveillance, One station		
Camera and monitor	Each	1525
For additional camera stations, add	Each	820
Emergency Lighting, 25 watt, battery operated		
Lead battery	Each	247
Nickel cadmium	Each	695
Laundry Equipment		
Dryers, coin operated 30 lb.	Each	3000
Double stacked	Each	6325
50 lb.	Each	3450
Dry cleaner 20 lb. ·	Each	38,100
30 lb.	Each	52,000
Washers, coin operated	Each	1325
Washer/extractor 20 lb.	Each	4600
30 lb.	Each	8850
50 lb.	Each	10,900
75 lb.	Each	22,800
Smoke Detectors		
Ceiling type	Each	161
Duct type	Each	425

Model costs calculated for a 1 story building with 12' story height and 3,000 square feet of floor area

Laundromat

				Unit	Unit Cost	Cost Per S.F.	% Of Sub-Total
A. SUBSTRUCTURE							
1010	Standard Foundations	Poured concrete; strip and spread footings		S.F. Ground	2.51	2.51	
1030	Slab on Grade	5" reinforced concrete with vapor barrier and granular base		S.F. Slab	4.30	4.30	
2010	Basement Excavation	Site preparation for slab and trench for foundation wall and footing		S.F. Ground	.43	.43	10.3%
2020	Basement Walls	4' Foundation wall		L.F. Wall	54	3.96	
B. SHELL							
	B10 Superstructure						
1010	Floor Construction	N/A		—	—	—	
1020	Roof Construction	Metal deck, open web steel joists, beams, columns		S.F. Roof	7.29	7.29	6.7%
	B20 Exterior Enclosure						
2010	Exterior Walls	Decorative concrete block	90% of wall	S.F. Wall	11.57	9.12	
2020	Exterior Windows	Store front	10% of wall	Each	33	2.90	12.1%
2030	Exterior Doors	Double aluminum and glass		Each	3375	1.12	
	B30 Roofing						
3010	Roof Coverings	Built-up tar and gravel with flashing; perlite/EPS composite insulation		S.F. Roof	6.44	6.44	5.9%
3020	Roof Openings	N/A		—	—	—	
C. INTERIORS							
1010	Partitions	Gypsum board on metal studs	60 S.F. Floor/L.F. Partition	S.F. Partition	19.26	3.21	
1020	Interior Doors	Single leaf wood	750 S.F. Floor/Door	Each	480	.64	
1030	Fittings	N/A		—	—	—	
2010	Stair Construction	N/A		—	—	—	7.4%
3010	Wall Finishes	Paint		S.F. Surface	1.26	.21	
3020	Floor Finishes	Vinyl composition tile		S.F. Floor	1.64	1.64	
3030	Ceiling Finishes	Fiberglass board on exposed grid system		S.F. Ceiling	2.00	2.33	
D. SERVICES							
	D10 Conveying						
1010	Elevators & Lifts	N/A		—	—	—	
1020	Escalators & Moving Walks	N/A		—	—	—	0.0%
	D20 Plumbing						
2010	Plumbing Fixtures	Toilet and service fixtures, supply and drainage	1 Fixture/600 S.F. Floor	Each	3654	6.09	
2020	Domestic Water Distribution	Gas fired hot water heater		S.F. Floor	23	23.07	27.6%
2040	Rain Water Drainage	Roof drains		S.F. Roof	.74	.74	
	D30 HVAC						
3010	Energy Supply	N/A		—	—	—	
3020	Heat Generating Systems	Included in D3050		—	—	—	
3030	Cooling Generating Systems	N/A		—	—	—	7.0%
3050	Terminal & Package Units	Rooftop single zone unit systems		S.F. Floor	7.60	7.60	
3090	Other HVAC Sys. & Equipment	N/A		—	—	—	
	D40 Fire Protection						
4010	Sprinklers	Sprinkler, ordinary hazard		S.F. Floor	2.91	2.91	2.7%
4020	Standpipes	N/A		—	—	—	
	D50 Electrical						
5010	Electrical Service/Distribution	200 ampere service, panel board and feeders		S.F. Floor	2.79	2.79	
5020	Lighting & Branch Wiring	Fluorescent fixtures, receptacles, switches, A.C. and misc. power		S.F. Floor	18.62	18.62	20.1%
5030	Communications & Security	Alarm systems and emergency lighting		S.F. Floor	—	—	
5090	Other Electrical Systems	Emergency generator, 7.5 kW		S.F. Floor	.36	.36	
E. EQUIPMENT & FURNISHINGS							
1010	Commercial Equipment	N/A		—	—	—	
1020	Institutional Equipment	N/A		—	—	—	
1030	Vehicular Equipment	N/A		—	—	—	0.0%
1090	Other Equipment	N/A		—	—	—	
F. SPECIAL CONSTRUCTION							
1020	Integrated Construction	N/A		—	—	—	
1040	Special Facilities	N/A		—	—	—	0.0%
G. BUILDING SITEWORK	**N/A**						

			Sub-Total	108.28	100%
CONTRACTOR FEES (General Requirements: 10%, Overhead: 5%, Profit: 10%)		25%	27.05		
ARCHITECT FEES		7%	9.47		
Total Building Cost			**144.80**		

BUILDING TYPES

Costs per square foot of floor area

Exterior Wall	S.F. Area	7000	10000	13000	16000	19000	22000	25000	28000	31000
	L.F. Perimeter	240	300	336	386	411	435	472	510	524
Face Brick with Concrete Block Back-up	R/Conc. Frame	145.50	137.85	131.45	128.50	124.95	122.20	120.75	119.70	117.95
	Steel Frame	139.60	131.90	125.55	122.55	119.00	116.30	114.90	113.80	112.00
Limestone with Concrete Block	R/Conc. Frame	171.90	160.90	151.30	147.00	141.55	137.40	135.30	133.75	130.95
	Steel Frame	165.95	154.95	145.40	141.15	135.65	131.45	129.40	127.80	125.00
Precast Concrete Panels	R/Conc. Frame	140.70	133.65	127.80	125.10	121.90	119.40	118.10	117.15	115.55
	Steel Frame	134.75	127.70	121.90	119.20	116.00	113.50	112.20	111.20	109.60
Perimeter Adj., Add or Deduct	Per 100 L.F.	17.50	12.20	9.40	7.65	6.40	5.55	5.00	4.40	3.95
Story Hgt. Adj., Add or Deduct	Per 1 Ft.	2.65	2.30	1.95	1.85	1.60	1.50	1.50	1.45	1.30

For Basement, add $34.10 per square foot of basement area

The above costs were calculated using the basic specifications shown on the facing page. These costs should be adjusted where necessary for design alternatives and owner's requirements. Reported completed project costs, for this type of structure, range from $72.80 to $186.85 per S.F.

Common additives

Description	Unit	$ Cost
Carrels Hardwood	Each	785 - 1000
Closed Circuit Surveillance, One station		
Camera and monitor	Each	1525
For additional camera stations, add	Each	820
Elevators, Hydraulic passenger, 2 stops		
1500# capacity	Each	47,900
2500# capacity	Each	50,000
3500# capacity	Each	53,700
Emergency Lighting, 25 watt, battery operated		
Lead battery	Each	247
Nickel cadmium	Each	695
Flagpoles, Complete		
Aluminum, 20' high	Each	1375
40' high	Each	3025
70' high	Each	9500
Fiberglass, 23' high	Each	1650
39'-5" high	Each	3175
59' high	Each	7950

Description	Unit	$ Cost
Library Furnishings		
Bookshelf, 90" high, 10" shelf double face	L.F.	197
single face	L.F.	171
Charging desk, built-in with counter		
Plastic laminated top	L.F.	540
Reading table, laminated		
top 60" x 36"	Each	510

Important: See the Reference Section for Location Factors

Model costs calculated for a 2 story building with 14' story height and 22,000 square feet of floor area

Library

				Unit	Unit Cost	Cost Per S.F.	% Of Sub-Total
A. SUBSTRUCTURE							
1010	Standard Foundations	Poured concrete; strip and spread footings		S.F. Ground	2.98	1.49	
1030	Slab on Grade	4" reinforced concrete with vapor barrier and granular base		S.F. Slab	3.98	1.99	
2010	Basement Excavation	Site preparation for slab and trench for foundation wall and footing		S.F. Ground	.23	.12	5.3%
2020	Basement Walls	4' foundation wall		L.F. Wall	61	1.21	
B. SHELL							
	B10 Superstructure						
1010	Floor Construction	Concrete waffle slab		S.F. Floor	23	11.61	22.7%
1020	Roof Construction	Concrete waffle slab		S.F. Roof	17.82	8.91	
	B20 Exterior Enclosure						
2010	Exterior Walls	Face brick with concrete block backup	90% of wall	S.F. Wall	23	11.89	
2020	Exterior Windows	Window wall	10% of wall	Each	40	2.24	16.0%
2030	Exterior Doors	Double aluminum and glass, single leaf hollow metal		Each	3625	.33	
	B30 Roofing						
3010	Roof Coverings	Single ply membrane, EPDM, fully adhered; perlite/EPS composite insulation		S.F. Roof	3.74	1.87	2.1%
3020	Roof Openings	Roof hatches		S.F. Roof	.06	.03	
C. INTERIORS							
1010	Partitions	Gypsum board on metal studs	30 S.F. Floor/L.F. Partition	S.F. Partition	8.85	3.54	
1020	Interior Doors	Single leaf wood	300 S.F. Floor/Door	Each	480	1.60	
1030	Fittings	N/A		—	—	—	
2010	Stair Construction	Concrete filled metal pan		Flight	7975	.73	15.7%
3010	Wall Finishes	Paint		S.F. Surface	1.00	.50	
3020	Floor Finishes	50% carpet, 50% vinyl tile		S.F. Floor	3.74	3.74	
3030	Ceiling Finishes	Mineral fiber on concealed zee bars		S.F. Ceiling	4.00	4.09	
D. SERVICES							
	D10 Conveying						
1010	Elevators & Lifts	One hydraulic passenger elevator		Each	62,040	2.82	3.1%
1020	Escalators & Moving Walks	N/A		—	—	—	
	D20 Plumbing						
2010	Plumbing Fixtures	Toilet and service fixtures, supply and drainage	1 Fixture/1835 S.F. Floor	Each	1853	1.01	
2020	Domestic Water Distribution	Gas fired water heater		S.F. Floor	.70	.70	2.2%
2040	Rain Water Drainage	Roof drains		S.F. Roof	.50	.25	
	D30 HVAC						
3010	Energy Supply	N/A		—	—	—	
3020	Heat Generating Systems	Included in D3050		—	—	—	
3030	Cooling Generating Systems	N/A		—	—	—	19.3%
3050	Terminal & Package Units	Multizone unit, gas heating, electric cooling		S.F. Floor	17.45	17.45	
3090	Other HVAC Sys. & Equipment	N/A		—	—	—	
	D40 Fire Protection						
4010	Sprinklers	Wet pipe sprinkler system		S.F. Floor	1.87	1.87	2.1%
4020	Standpipes	N/A		—	—	—	
	D50 Electrical						
5010	Electrical Service/Distribution	400 ampere service, panel board and feeders		S.F. Floor	.82	.82	
5020	Lighting & Branch Wiring	Fluorescent fixtures, receptacles, switches, A.C. and misc. power		S.F. Floor	9.05	9.05	11.6%
5030	Communications & Security	Alarm systems and emergency lighting		S.F. Floor	.59	.59	
5090	Other Electrical Systems	Emergency generator, 7.5 kW		S.F. Floor	.07	.07	
E. EQUIPMENT & FURNISHINGS							
1010	Commercial Equipment	N/A		—	—	—	
1020	Institutional Equipment	N/A		—	—	—	0.0%
1030	Vehicular Equipment	N/A		—	—	—	
1090	Other Equipment	N/A		—	—	—	
F. SPECIAL CONSTRUCTION							
1020	Integrated Construction	N/A		—	—	—	0.0%
1040	Special Facilities	N/A		—	—	—	
G. BUILDING SITEWORK	**N/A**						

	Sub-Total	90.52	100%
CONTRACTOR FEES (General Requirements: 10%, Overhead: 5%, Profit: 10%)	25%	22.63	
ARCHITECT FEES	8%	9.05	
Total Building Cost		**122.20**	

BUILDING TYPES

159

Costs per square foot of floor area

Exterior Wall	S.F. Area	4000	5500	7000	8500	10000	11500	13000	14500	16000
	L.F. Perimeter	280	320	380	440	453	503	510	522	560
Face Brick with Concrete Block Back-up	Steel Joists	147.00	139.45	136.55	134.70	131.00	129.90	127.45	125.55	124.95
	Wood Truss	152.30	144.75	141.95	140.00	136.30	135.25	132.75	130.90	130.25
Stucco on Concrete Block	Steel Joists	138.40	132.30	129.95	128.35	125.50	124.60	122.65	121.15	120.65
	Wood Truss	143.75	137.60	135.25	133.70	130.75	129.90	127.95	126.45	126.00
Brick Veneer	Wood Truss	148.95	141.90	139.30	137.55	134.15	133.20	130.85	129.15	128.55
Wood Siding	Wood Frame	141.20	135.50	133.30	131.80	129.10	128.30	126.50	125.15	124.65
Perimeter Adj., Add or Deduct	Per 100 L.F.	12.85	9.35	7.40	6.10	5.15	4.55	3.90	3.55	3.15
Story Hgt. Adj., Add or Deduct	Per 1 Ft.	2.60	2.15	2.05	1.95	1.70	1.70	1.40	1.35	1.25

For Basement, add $22.15 per square foot of basement area

The above costs were calculated using the basic specifications shown on the facing page. These costs should be adjusted where necessary for design alternatives and owner's requirements. Reported completed project costs, for this type of structure, range from $61.40 to $158.00 per S.F.

Common additives

Description	Unit	$ Cost	Description	Unit	$ Cost
Cabinets, Hospital, base			Directory Boards, Plastic, glass covered		
Laminated plastic	L.F.	320	30" x 20"	Each	575
Stainless steel	L.F.	405	36" x 48"	Each	1275
Counter top, laminated plastic	L.F.	61	Aluminum, 24" x 18"	Each	525
Stainless steel	L.F.	133	36" x 24"	Each	600
For drop-in sink, add	Each	620	48" x 32"	Each	810
Nurses station, door type			48" x 60"	Each	1875
Laminated plastic	L.F.	355	Heat Therapy Unit		
Enameled steel	L.F.	330	Humidified, 26" x 78" x 28"	Each	3075
Stainless steel	L.F.	425	Smoke Detectors		
Wall cabinets, laminated plastic	L.F.	230	Ceiling type	Each	161
Enameled steel	L.F.	260	Duct type	Each	425
Stainless steel	L.F.	385	Tables, Examining, vinyl top		
			with base cabinets	Each	1450 - 3925
			Utensil Washer, Sanitizer	Each	10,100
			X-Ray, Mobile	Each	12,100 - 68,000

Important: See the Reference Section for Location Factors

Model costs calculated for a 1 story building with 10' story height and 7,000 square feet of floor area

				Unit	Unit Cost	Cost Per S.F.	% Of Sub-Total
A.	**SUBSTRUCTURE**						
1010	Standard Foundations	Poured concrete; strip and spread footings		S.F. Ground	1.60	1.60	
1030	Slab on Grade	4" reinforced concrete with vapor barrier and granular base		S.F. Slab	3.98	3.98	
2010	Basement Excavation	Site preparation for slab and trench for foundation wall and footing		S.F. Ground	.23	.23	8.8%
2020	Basement Walls	4' foundation wall		L.F. Wall	61	3.31	
B.	**SHELL**						
	B10 Superstructure						
1010	Floor Construction	N/A		—	—	—	
1020	Roof Construction	Plywood on wood trusses		S.F. Roof	5.64	5.64	5.4%
	B20 Exterior Enclosure						
2010	Exterior Walls	Face brick with concrete block backup	70% of wall	S.F. Wall	23	9.06	
2020	Exterior Windows	Wood double hung	30% of wall	Each	475	4.55	15.1%
2030	Exterior Doors	Aluminum and glass doors and entrance with transoms		Each	1218	2.09	
	B30 Roofing						
3010	Roof Coverings	Asphalt shingles with flashing (Pitched); rigid fiber glass insulation, gutters		S.F. Roof	3.03	3.03	2.9%
3020	Roof Openings	N/A		—	—	—	
C.	**INTERIORS**						
1010	Partitions	Gypsum bd. & sound deadening bd. on wood studs w/insul.	6 S.F. Floor/L.F. Partition	S.F. Partition	6.86	9.15	
1020	Interior Doors	Single leaf wood	60 S.F. Floor/Door	Each	480	8.00	
1030	Fittings	N/A		—	—	—	
2010	Stair Construction	N/A		—	—	—	27.5%
3010	Wall Finishes	50% paint, 50% vinyl wall covering		S.F. Surface	2.15	2.90	
3020	Floor Finishes	50% carpet, 50% vinyl composition tile		S.F. Floor	5.30	5.30	
3030	Ceiling Finishes	Mineral fiber tile on concealed zee bars		S.F. Ceiling	3.00	3.31	
D.	**SERVICES**						
	D10 Conveying						
1010	Elevators & Lifts	N/A		—	—	—	
1020	Escalators & Moving Walks	N/A		—	—	—	0.0%
	D20 Plumbing						
2010	Plumbing Fixtures	Toilet and service fixtures, supply and drainage	1 Fixture/195 S.F. Floor	Each	2396	12.29	
2020	Domestic Water Distribution	Gas fired water heater		S.F. Floor	1.16	1.16	12.9%
2040	Rain Water Drainage	N/A		—	—	—	
	D30 HVAC						
3010	Energy Supply	N/A		—	—	—	
3020	Heat Generating Systems	Included in D3050		—	—	—	
3030	Cooling Generating Systems	N/A		—	—	—	14.9%
3050	Terminal & Package Units	Multizone unit, gas heating, electric cooling		S.F. Floor	15.48	15.48	
3090	Other HVAC Sys. & Equipment	N/A		—	—	—	
	D40 Fire Protection						
4010	Sprinklers	Wet pipe sprinkler system		S.F. Floor	2.68	2.68	2.6%
4020	Standpipes	N/A		—	—	—	
	D50 Electrical						
5010	Electrical Service/Distribution	200 ampere service, panel board and feeders		S.F. Floor	1.14	1.14	
5020	Lighting & Branch Wiring	Fluorescent fixtures, receptacles, switches, A.C. and misc. power		S.F. Floor	6.12	6.12	10.0%
5030	Communications & Security	Alarm systems, intercom system, and emergency lighting		S.F. Floor	2.47	2.47	
5090	Other Electrical Systems	Emergency generator, 7.5 kW		S.F. Floor	.66	.66	
E.	**EQUIPMENT & FURNISHINGS**						
1010	Commercial Equipment	N/A		—	—	—	
1020	Institutional Equipment	N/A		—	—	—	
1030	Vehicular Equipment	N/A		—	—	—	0.0%
1090	Other Equipment	N/A		—	—	—	
F.	**SPECIAL CONSTRUCTION**						
1020	Integrated Construction	N/A		—	—	—	
1040	Special Facilities	N/A		—	—	—	0.0%
G.	**BUILDING SITEWORK**	**N/A**					

			Sub-Total	104.15	**100%**
CONTRACTOR FEES (General Requirements: 10%, Overhead: 5%, Profit: 10%)			25%	26.08	
ARCHITECT FEES			9%	11.72	
		Total Building Cost		141.95	

BUILDING TYPES

161

BUILDING TYPES

Costs per square foot of floor area

Exterior Wall	S.F. Area	4000	5500	7000	8500	10000	11500	13000	14500	16000
	L.F. Perimeter	180	210	240	270	286	311	336	361	386
Face Brick with Concrete Block Back-up	Steel Joists	164.90	156.70	152.10	149.05	145.75	143.95	142.65	141.55	140.65
	Wood Joists	169.05	160.85	156.25	153.20	149.95	148.15	146.80	145.70	144.80
Stucco on Concrete Block	Steel Joists	156.55	149.60	145.70	143.15	140.40	138.85	137.80	136.85	136.15
	Wood Joists	160.60	153.75	149.85	147.30	144.55	143.05	142.00	141.05	140.30
Brick Veneer	Wood Frame	164.80	157.35	153.05	150.25	147.25	145.60	144.40	143.40	142.60
Wood Siding	Wood Frame	157.85	151.35	147.70	145.30	142.80	141.40	140.35	139.45	138.85
Perimeter Adj., Add or Deduct	Per 100 L.F.	21.80	15.90	12.50	10.30	8.75	7.65	6.75	6.00	5.50
Story Hgt. Adj., Add or Deduct	Per 1 Ft.	3.15	2.70	2.40	2.25	2.05	1.95	1.80	1.75	1.75

For Basement, add $24.60 per square foot of basement area

The above costs were calculated using the basic specifications shown on the facing page. These costs should be adjusted where necessary for design alternatives and owner's requirements. Reported completed project costs, for this type of structure, range from $62.95 to $215.15 per S.F.

Common additives

Description	Unit	$ Cost
Cabinets, Hospital, base		
Laminated plastic	L.F.	320
Stainless steel	L.F.	405
Counter top, laminated plastic	L.F.	61
Stainless steel	L.F.	133
For drop-in sink, add	Each	620
Nurses station, door type		
Laminated plastic	L.F.	355
Enameled steel	L.F.	330
Stainless steel	L.F.	425
Wall cabinets, laminated plastic	L.F.	230
Enameled steel	L.F.	260
Stainless steel	L.F.	385
Elevators, Hydraulic passenger, 2 stops		
1500# capacity	Each	47,900
2500# capacity	Each	50,000
3500# capacity	Each	53,700

Description	Unit	$ Cost
Directory Boards, Plastic, glass covered		
30" x 20"	Each	575
36" x 48"	Each	1275
Aluminum, 24" x 18"	Each	525
36" x 24"	Each	600
48" x 32"	Each	810
48" x 60"	Each	1875
Emergency Lighting, 25 watt, battery operated		
Lead battery	Each	247
Nickel cadmium	Each	695
Heat Therapy Unit		
Humidified, 26" x 78" x 28"	Each	3075
Smoke Detectors		
Ceiling type	Each	161
Duct type	Each	425
Tables, Examining, vinyl top		
with base cabinets	Each	1450 - 3925
Utensil Washer, Sanitizer	Each	10,100
X-Ray, Mobile	Each	12,100 - 68,000

Important: See the Reference Section for Location Factors

Medical Office, 2 Story

Model costs calculated for a 2 story building with 10' story height and 7,000 square feet of floor area

				Unit	Unit Cost	Cost Per S.F.	% Of Sub-Total
A. SUBSTRUCTURE							
1010	Standard Foundations	Poured concrete; strip and spread footings		S.F. Ground	2.56	1.28	
1030	Slab on Grade	4" reinforced concrete with vapor barrier and granular base		S.F. Slab	3.98	1.99	5.1%
2010	Basement Excavation	Site preparation for slab and trench for foundation wall and footing		S.F. Ground	.14	.07	
2020	Basement Walls	4' foundation wall		L.F. Wall	61	2.09	
B. SHELL							
	B10 Superstructure						
1010	Floor Construction	Open web steel joists, slab form, concrete, columns		S.F. Floor	10.74	5.37	7.1%
1020	Roof Construction	Metal deck, open web steel joists, beams, columns		S.F. Roof	4.46	2.23	
	B20 Exterior Enclosure						
2010	Exterior Walls	Stucco on concrete block		S.F. Wall	14.29	6.86	
2020	Exterior Windows	Outward projecting metal	30% of wall	Each	346	4.75	11.8%
2030	Exterior Doors	Aluminum and glass doors with transoms		Each	3625	1.04	
	B30 Roofing						
3010	Roof Coverings	Built-up tar and gravel with flashing; perlite/EPS composite insulation		S.F. Roof	5.02	2.51	2.4%
3020	Roof Openings	Roof hatches		S.F. Roof	.16	.08	
C. INTERIORS							
1010	Partitions	Gypsum bd. & acous. insul. on metal studs	6 S.F. Floor/L.F. Partition	S.F. Partition	5.90	7.86	
1020	Interior Doors	Single leaf wood	60 S.F. Floor/Door	Each	480	8.00	
1030	Fittings	N/A		—	—	—	
2010	Stair Construction	Concrete filled metal pan		Flight	6425	1.84	27.3%
3010	Wall Finishes	45% paint, 50% vinyl wall coating, 5% ceramic tile		S.F. Surface	2.15	2.90	
3020	Floor Finishes	50% carpet, 50% vinyl asbestos tile		S.F. Floor	5.30	5.30	
3030	Ceiling Finishes	Mineral fiber tile on concealed zee bars		S.F. Ceiling	3.00	3.31	
D. SERVICES							
	D10 Conveying						
1010	Elevators & Lifts	One hydraulic hospital elevator		Each	71,400	10.20	9.5%
1020	Escalators & Moving Walks	N/A		—	—	—	
	D20 Plumbing						
2010	Plumbing Fixtures	Toilet and service fixtures, supply and drainage	1 Fixture/160 S.F. Floor	Each	1443	9.02	
2020	Domestic Water Distribution	Gas fired water heater		S.F. Floor	1.26	1.26	9.8%
2040	Rain Water Drainage	Roof drains		S.F. Roof	.38	.19	
	D30 HVAC						
3010	Energy Supply	N/A		—	—	—	
3020	Heat Generating Systems	Included in D3050		—	—	—	
3030	Cooling Generating Systems	N/A		—	—	—	14.5%
3050	Terminal & Package Units	Multizone unit, gas heating, electric cooling		S.F. Floor	15.48	15.48	
3090	Other HVAC Sys. & Equipment	N/A		—	—	—	
	D40 Fire Protection						
4010	Sprinklers	Wet pipe sprinkler system		S.F. Floor	2.26	2.26	2.1%
4020	Standpipes	N/A		—	—	—	
	D50 Electrical						
5010	Electrical Service/Distribution	400 ampere service, panel board and feeders		S.F. Floor	1.80	1.80	
5020	Lighting & Branch Wiring	Fluorescent fixtures, receptacles, switches, A.C. and misc. power		S.F. Floor	6.12	6.12	10.3%
5030	Communications & Security	Alarm system, intercom system, and emergency lighting		S.F. Floor	2.47	2.47	
5090	Other Electrical Systems	Emergency generator, 7.5 kW		S.F. Floor	.66	.66	
E. EQUIPMENT & FURNISHINGS							
1010	Commercial Equipment	N/A		—	—	—	
1020	Institutional Equipment	N/A		—	—	—	0.0%
1030	Vehicular Equipment	N/A		—	—	—	
1090	Other Equipment	N/A		—	—	—	
F. SPECIAL CONSTRUCTION							
1020	Integrated Construction	N/A		—	—	—	0.0%
1040	Special Facilities	N/A		—	—	—	
G. BUILDING SITEWORK	**N/A**						

					Sub-Total	106.94	100%
	CONTRACTOR FEES (General Requirements: 10%, Overhead: 5%, Profit: 10%)				25%	26.73	
	ARCHITECT FEES				9%	12.03	

Total Building Cost	**145.70**

Costs per square foot of floor area

Exterior Wall	S.F. Area	2000	3000	4000	6000	8000	10000	12000	14000	16000
	L.F. Perimeter	240	260	280	380	480	560	580	660	740
Brick Veneer	Wood Frame	143.50	125.25	116.05	111.35	109.00	106.75	102.95	102.20	101.65
Aluminum Siding	Wood Frame	129.60	115.20	107.90	104.00	102.10	100.25	97.35	96.75	96.25
Wood Siding	Wood Frame	130.05	115.50	108.15	104.25	102.30	100.45	97.50	96.90	96.45
Wood Shingles	Wood Frame	132.45	117.25	109.60	105.55	103.50	101.60	98.50	97.85	97.40
Precast Concrete Block	Wood Truss	128.50	114.40	107.25	103.45	101.55	99.70	96.90	96.25	95.80
Brick on Concrete Block	Wood Truss	149.20	129.35	119.35	114.35	111.85	109.40	105.25	104.40	103.80
Perimeter Adj., Add or Deduct	Per 100 L.F.	22.35	14.85	11.15	7.50	5.60	4.45	3.70	3.15	2.75
Story Hgt. Adj., Add or Deduct	Per 1 Ft.	3.70	2.70	2.15	1.95	1.90	1.75	1.50	1.45	1.45
For Basement, add $16.95 per square foot of basement area										

The above costs were calculated using the basic specifications shown on the facing page. These costs should be adjusted where necessary for design alternatives and owner's requirements. Reported completed project costs, for this type of structure, range from $50.85 to $240.65 per S.F.

Common additives

Description	Unit	$ Cost
Closed Circuit Surveillance, One station		
Camera and monitor	Each	1525
For additional camera stations, add	Each	820
Emergency Lighting, 25 watt, battery operated		
Lead battery	Each	247
Nickel cadmium	Each	695
Laundry Equipment		
Dryer, gas, 16 lb. capacity	Each	785
30 lb. capacity	Each	3000
Washer, 4 cycle	Each	875
Commercial	Each	1325
Sauna, Prefabricated, complete		
6' x 4'	Each	4525
6' x 6'	Each	5425
6' x 9'	Each	6700
8' x 8'	Each	7875
8' x 10'	Each	8800
10' x 12'	Each	11,000
Smoke Detectors		
Ceiling type	Each	161
Duct type	Each	425

Description	Unit	$ Cost
Swimming Pools, Complete, gunite	S.F.	51 - 62
TV Antenna, Master system, 12 outlet	Outlet	262
30 outlet	Outlet	168
100 outlet	Outlet	160

Important: See the Reference Section for Location Factors

Model costs calculated for a 1 story building with 9' story height and 8,000 square feet of floor area

				Unit	Unit Cost	Cost Per S.F.	% Of Sub-Total
A.	**SUBSTRUCTURE**						
1010	Standard Foundations	Poured concrete; strip and spread footings		S.F. Ground	2.41	2.41	
1030	Slab on Grade	4" reinforced concrete with vapor barrier and granular base		S.F. Slab	3.98	3.98	
2010	Basement Excavation	Site preparation for slab and trench for foundation wall and footing		S.F. Ground	.23	.23	13.9%
2020	Basement Walls	4' foundation wall		L.F. Wall	57	4.72	
B.	**SHELL**						
	B10 Superstructure						
1010	Floor Construction	N/A		—	—	—	
1020	Roof Construction	Plywood on wood trusses		S.F. Roof	5.64	5.64	6.9%
	B20 Exterior Enclosure						
2010	Exterior Walls	Face brick on wood studs with sheathing, insulation and paper	80% of wall	S.F. Wall	18.87	8.15	
2020	Exterior Windows	Wood double hung	20% of wall	Each	383	2.95	17.7%
2030	Exterior Doors	Wood solid core		Each	1218	3.35	
	B30 Roofing						
3010	Roof Coverings	Asphalt shingles with flashing (pitched); rigid fiberglass insulation		S.F. Roof	2.90	2.90	3.6%
3020	Roof Openings	N/A		—	—	—	
C.	**INTERIORS**						
1010	Partitions	Gypsum bd. and sound deadening bd. on wood studs	9 S.F. Floor/L.F. Partition	S.F. Partition	7.32	6.51	
1020	Interior Doors	Single leaf hollow core wood	300 S.F. Floor/Door	Each	420	1.40	
1030	Fittings	N/A		—	—	—	
2010	Stair Construction	N/A		—	—	—	25.3%
3010	Wall Finishes	90% paint, 10% ceramic tile		S.F. Surface	2.06	2.06	
3020	Floor Finishes	85% carpet, 15% ceramic tile		S.F. Floor	7.36	7.36	
3030	Ceiling Finishes	Painted gypsum board on furring		S.F. Ceiling	3.00	3.30	
D.	**SERVICES**						
	D10 Conveying						
1010	Elevators & Lifts	N/A		—	—	—	0.0%
1020	Escalators & Moving Walks	N/A		—	—	—	
	D20 Plumbing						
2010	Plumbing Fixtures	Toilet and service fixtures, supply and drainage	1 Fixture/90 S.F. Floor	Each	1118	12.43	
2020	Domestic Water Distribution	Gas fired water heater		S.F. Floor	.95	.95	16.4%
2040	Rain Water Drainage	N/A		—	—	—	
	D30 HVAC						
3010	Energy Supply	N/A		—	—	—	
3020	Heat Generating Systems	Included in D3050		—	—	—	
3030	Cooling Generating Systems	N/A		—	—	—	4.3%
3050	Terminal & Package Units	Through the wall electric heating and cooling units		S.F. Floor	3.52	3.52	
3090	Other HVAC Sys. & Equipment	N/A		—	—	—	
	D40 Fire Protection						
4010	Sprinklers	Wet pipe sprinkler system		S.F. Floor	2.68	2.68	3.3%
4020	Standpipes	N/A		—	—	—	
	D50 Electrical						
5010	Electrical Service/Distribution	200 ampere service, panel board and feeders		S.F. Floor	.97	.97	
5020	Lighting & Branch Wiring	Fluorescent fixtures, receptacles, switches and misc. power		S.F. Floor	5.65	5.65	8.6%
5030	Communications & Security	Alarm systems		S.F. Floor	.36	.36	
5090	Other Electrical Systems	N/A		—	—	—	
E.	**EQUIPMENT & FURNISHINGS**						
1010	Commercial Equipment	N/A		—	—	—	
1020	Institutional Equipment	N/A		—	—	—	
1030	Vehicular Equipment	N/A		—	—	—	0.0%
1090	Other Equipment	N/A		—	—	—	
F.	**SPECIAL CONSTRUCTION**						
1020	Integrated Construction	N/A		—	—	—	
1040	Special Facilities	N/A		—	—	—	0.0%
G.	**BUILDING SITEWORK**	**N/A**					

					Sub-Total	81.52	100%
	CONTRACTOR FEES (General Requirements: 10%, Overhead: 5%, Profit: 10%)				25%	20.35	
	ARCHITECT FEES				7%	7.13	

	Total Building Cost	**109.00**

BUILDING TYPES

Costs per square foot of floor area

Exterior Wall	S.F. Area	25000	37000	49000	61000	73000	81000	88000	96000	104000
	L.F. Perimeter	433	593	606	720	835	911	978	1054	1074
Decorative Concrete Block	Wood Joists	127.70	124.35	119.90	118.70	117.95	117.55	117.25	116.90	116.25
	Precast Conc.	135.30	131.90	127.50	126.30	125.55	125.15	124.85	124.55	123.85
Stucco on Concrete Block	Wood Joists	127.40	124.00	119.50	118.30	117.55	117.15	116.85	116.50	115.80
	Precast Conc.	135.65	132.25	127.75	126.55	125.80	125.40	125.10	124.80	124.05
Wood Siding	Wood Frame	125.65	122.40	118.25	117.10	116.40	115.95	115.70	115.40	114.75
Brick Veneer	Wood Frame	130.00	126.40	121.35	120.05	119.25	118.80	118.45	118.15	117.35
Perimeter Adj., Add or Deduct	Per 100 L.F.	3.65	2.40	1.90	1.55	1.20	1.10	1.05	1.00	.85
Story Hgt. Adj., Add or Deduct	Per 1 Ft.	1.30	1.15	.90	.90	.85	.80	.85	.85	.70
For Basement, add $22.20 per square foot of basement area										

The above costs were calculated using the basic specifications shown on the facing page. These costs should be adjusted where necessary for design alternatives and owner's requirements. Reported completed project costs, for this type of structure, range from $46.80 to $240.65 per S.F.

Common additives

Description	Unit	$ Cost	Description	Unit	$ Cost
Closed Circuit Surveillance, One station			Sauna, Prefabricated, complete		
Camera and monitor	Each	1525	6' x 4'	Each	4525
For additional camera station, add	Each	820	6' x 6'	Each	5425
Elevators, Hydraulic passenger, 2 stops			6' x 9'	Each	6700
1500# capacity	Each	47,900	8' x 8'	Each	7875
2500# capacity	Each	50,000	8' x 10'	Each	8800
3500# capacity	Each	53,700	10' x 12'	Each	11,000
Additional stop, add	Each	4550	Smoke Detectors		
Emergency Lighting, 25 watt, battery operated			Ceiling type	Each	161
Lead battery	Each	247	Duct type	Each	425
Nickel cadmium	Each	695	Swimming Pools, Complete, gunite	S.F.	51 - 62
Laundry Equipment			TV Antenna, Master system, 12 outlet	Outlet	262
Dryer, gas, 16 lb. capacity	Each	785	30 outlet	Outlet	168
30 lb. capacity	Each	3000	100 outlet	Outlet	160
Washer, 4 cycle	Each	875			
Commercial	Each	1325			

Important: See the Reference Section for Location Factors

BUILDING TYPES

Model costs calculated for a 3 story building with 9' story height and 49,000 square feet of floor area

				Unit	Unit Cost	Cost Per S.F.	% Of Sub-Total
A. SUBSTRUCTURE							
1010	Standard Foundations	Poured concrete; strip and spread footings		S.F. Ground	1.11	.37	
1030	Slab on Grade	4" reinforced concrete with vapor barrier and granular base		S.F. Slab	3.98	1.32	3.4%
2010	Basement Excavation	Site preparation for slab and trench for foundation wall and footing		S.F. Ground	.14	.05	
2020	Basement Walls	4' foundation wall		L.F. Wall	61	1.51	
B. SHELL							
	B10 Superstructure						
1010	Floor Construction	Precast concrete plank		S.F. Floor	7.95	5.30	8.2%
1020	Roof Construction	Precast concrete plank		S.F. Roof	7.68	2.56	
	B20 Exterior Enclosure						
2010	Exterior Walls	Decorative concrete block	85% of wall	S.F. Wall	13.21	3.75	
2020	Exterior Windows	Aluminum sliding	15% of wall	Each	436	1.45	9.8%
2030	Exterior Doors	Aluminum and glass doors and entrance with transom		Each	1275	4.21	
	B30 Roofing						
3010	Roof Coverings	Built-up tar and gravel with flashing; perlite/EPS composite insulation		S.F. Roof	4.53	1.51	1.6%
3020	Roof Openings	Roof hatches		S.F. Roof	.09	.03	
C. INTERIORS							
1010	Partitions	Concrete block	7 S.F. Floor/L.F. Partition	S.F. Partition	14.32	16.36	
1020	Interior Doors	Wood hollow core	70 S.F. Floor/Door	Each	420	6.00	
1030	Fittings	N/A		—	—	—	
2010	Stair Construction	Concrete filled metal pan		Flight	6425	1.58	38.5%
3010	Wall Finishes	90% paint, 10% ceramic tile		S.F. Surface	2.30	2.65	
3020	Floor Finishes	85% carpet, 5% vinyl composition tile, 10% ceramic tile		S.F. Floor	7.36	7.36	
3030	Ceiling Finishes	Textured finish		S.F. Ceiling	3.00	3.11	
D. SERVICES							
	D10 Conveying						
1010	Elevators & Lifts	Two hydraulic passenger elevators		Each	74,480	3.04	3.2%
1020	Escalators & Moving Walks	N/A		—	—	—	
	D20 Plumbing						
2010	Plumbing Fixtures	Toilet and service fixtures, supply and drainage	1 Fixture/180 S.F. Floor	Each	3211	17.84	
2020	Domestic Water Distribution	Gas fired water heater		S.F. Floor	.73	.73	19.7%
2040	Rain Water Drainage	Roof drains		S.F. Roof	1.14	.38	
	D30 HVAC						
3010	Energy Supply	N/A		—	—	—	
3020	Heat Generating Systems	Included in D3050		—	—	—	
3030	Cooling Generating Systems	N/A		—	—	—	5.7%
3050	Terminal & Package Units	Through the wall electric heating and cooling units		S.F. Floor	5.47	5.47	
3090	Other HVAC Sys. & Equipment	N/A		—	—	—	
	D40 Fire Protection						
4010	Sprinklers	Sprinklers, light hazard		S.F. Floor	1.45	1.45	1.5%
4020	Standpipes	N/A		—	—	—	
	D50 Electrical						
5010	Electrical Service/Distribution	800 ampere service, panel board and feeders		S.F. Floor	1.39	1.39	
5020	Lighting & Branch Wiring	Fluorescent fixtures, receptacles, switches and misc. power		S.F. Floor	6.17	6.17	8.5%
5030	Communications & Security	Alarm systems and emergency lighting		S.F. Floor	.53	.53	
5090	Other Electrical Systems	Emergency generator, 7.5 kW		S.F. Floor	.10	.10	
E. EQUIPMENT & FURNISHINGS							
1010	Commercial Equipment	N/A		—	—	—	
1020	Institutional Equipment	N/A		—	—	—	0.0%
1030	Vehicular Equipment	N/A		—	—	—	
1090	Other Equipment	N/A		—	—	—	
F. SPECIAL CONSTRUCTION							
1020	Integrated Construction	N/A		—	—	—	0.0%
1040	Special Facilities	N/A		—	—	—	
G. BUILDING SITEWORK	**N/A**						

			Sub-Total	96.22	**100%**
CONTRACTOR FEES (General Requirements: 10%, Overhead: 5%, Profit: 10%)			25%	24.06	
ARCHITECT FEES			6%	7.22	
		Total Building Cost		**127.50**	

BUILDING TYPES

Costs per square foot of floor area

Exterior Wall	S.F. Area	9000	10000	12000	13000	14000	15000	16000	18000	20000
	L.F. Perimeter	385	410	440	460	480	500	510	547	583
Decorative Concrete Block	Steel Joists	121.65	118.50	112.70	110.75	109.05	107.60	105.85	103.70	101.85
Painted Concrete Block	Steel Joists	114.90	111.95	106.90	105.15	103.60	102.25	100.80	98.85	97.25
Face Brick on Conc. Block	Steel Joists	131.85	128.25	121.40	119.15	117.15	115.55	113.45	110.90	108.85
Precast Concrete Panels	Steel Joists	129.05	125.55	119.05	116.85	114.90	113.35	111.35	108.85	106.85
Tilt-up Panels	Steel Joists	126.65	123.30	117.00	114.85	113.00	111.45	109.55	107.20	105.25
Metal Sandwich Panels	Steel Joists	117.25	114.25	108.95	107.05	105.45	104.15	102.60	100.50	98.85
Perimeter Adj., Add or Deduct	Per 100 L.F.	6.85	6.15	5.15	4.70	4.40	4.10	3.90	3.35	3.10
Story Hgt. Adj., Add or Deduct	Per 1 Ft.	1.00	.95	.80	.80	.75	.75	.80	.70	.70
Basement—Not Applicable										

The above costs were calculated using the basic specifications shown on the facing page. These costs should be adjusted where necessary for design alternatives and owner's requirements. Reported completed project costs, for this type of structure, range from $59.95 to $156.65 per S.F.

Common additives

Description	Unit	$ Cost
Emergency Lighting, 25 watt, battery operated		
Lead battery	Each	247
Nickel cadmium	Each	695
Seating		
Auditorium chair, all veneer	Each	193
Veneer back, padded seat	Each	233
Upholstered, spring seat	Each	234
Classroom, movable chair & desk	Set	65 - 120
Lecture hall, pedestal type	Each	179 - 545
Smoke Detectors		
Ceiling type	Each	161
Duct type	Each	425
Sound System		
Amplifier, 250 watts	Each	1925
Speaker, ceiling or wall	Each	160
Trumpet	Each	305

Important: See the Reference Section for Location Factors

Model costs calculated for a 1 story building with 20' story height and 12,000 square feet of floor area

				Unit	Unit Cost	Cost Per S.F.	% Of Sub-Total
A.	**SUBSTRUCTURE**						
1010	Standard Foundations	Poured concrete; strip and spread footings		S.F. Ground	1.08	1.08	
1030	Slab on Grade	4" reinforced concrete with vapor barrier and granular base		S.F. Slab	3.98	3.98	
2010	Basement Excavation	Site preparation for slab and trench for foundation wall and footing		S.F. Ground	2.76	2.76	11.9%
2020	Basement Walls	4' foundation wall		L.F. Wall	61	2.24	
B.	**SHELL**						
	B10 Superstructure						
1010	Floor Construction	Open web steel joists, slab form, concrete	mezzanine 2250 S.F.	S.F. Floor	8.66	1.62	12.2%
1020	Roof Construction	Metal deck on open web steel joists		S.F. Roof	8.64	8.64	
	B20 Exterior Enclosure						
2010	Exterior Walls	Decorative concrete block	100% of wall	S.F. Wall	14.85	10.89	
2020	Exterior Windows	N/A		—	—	—	13.9%
2030	Exterior Doors	Sliding mallfront aluminum and glass and hollow metal		Each	1609	.80	
	B30 Roofing						
3010	Roof Coverings	Built-up tar and gravel with flashing; perlite/EPS composite insulation		S.F. Roof	4.21	4.21	5.0%
3020	Roof Openings	N/A		—	—	—	
C.	**INTERIORS**						
1010	Partitions	Concrete block	40 S.F. Floor/L.F. Partition	S.F. Partition	6.75	2.70	
1020	Interior Doors	Single leaf hollow metal	705 S.F. Floor/Door	Each	608	.86	
1030	Fittings	Toilet partitions		S.F. Floor	.62	.62	
2010	Stair Construction	Concrete filled metal pan		Flight	7700	1.28	19.5%
3010	Wall Finishes	Paint		S.F. Surface	5.70	2.85	
3020	Floor Finishes	Carpet 50%, ceramic tile 5%	50% of area	S.F. Floor	8.06	4.03	
3030	Ceiling Finishes	Mineral fiber tile on concealed zee runners suspended		S.F. Ceiling	4.00	4.09	
D.	**SERVICES**						
	D10 Conveying						
1010	Elevators & Lifts	N/A		—	—	—	0.0%
1020	Escalators & Moving Walks	N/A		—	—	—	
	D20 Plumbing						
2010	Plumbing Fixtures	Toilet and service fixtures, supply and drainage	1 Fixture/500 S.F. Floor	Each	1440	2.88	
2020	Domestic Water Distribution	Gas fired water heater		S.F. Floor	.17	.17	3.9%
2040	Rain Water Drainage	Roof drains		S.F. Roof	.27	.27	
	D30 HVAC						
3010	Energy Supply	N/A		—	—	—	
3020	Heat Generating Systems	Included in D3050		—	—	—	
3030	Cooling Generating Systems	N/A		—	—	—	10.3%
3050	Terminal & Package Units	Single zone rooftop unit, gas heating, electric cooling		S.F. Floor	8.64	8.64	
3090	Other HVAC Sys. & Equipment	N/A		—	—	—	
	D40 Fire Protection						
4010	Sprinklers	Wet pipe sprinkler system		S.F. Floor	2.49	2.49	3.0%
4020	Standpipes	N/A		—	—	—	
	D50 Electrical						
5010	Electrical Service/Distribution	400 ampere service, panel board and feeders		S.F. Floor	1.31	1.31	
5020	Lighting & Branch Wiring	Fluorescent fixtures, receptacles, switches, A.C. and misc. power		S.F. Floor	3.96	3.96	6.9%
5030	Communications & Security	Alarm systems, sound system and emergency lighting		S.F. Floor	.44	.44	
5090	Other Electrical Systems	Emergency generator, 7.5 kW		S.F. Floor	.13	.13	
E.	**EQUIPMENT & FURNISHINGS**						
1010	Commercial Equipment	N/A		—	—	—	
1020	Institutional Equipment	Projection equipment, screen		S.F. Floor	7.62	7.62	13.4%
1030	Vehicular Equipment	N/A		—	—	—	
2010	Other Equipment	Movie theater seating		S.F. Floor	148	3.71	
F.	**SPECIAL CONSTRUCTION**						
1020	Integrated Construction	N/A		—	—	—	0.0%
1040	Special Facilities	N/A		—	—	—	
G.	**BUILDING SITEWORK**	**N/A**					

				Sub-Total	84.27	**100%**
	CONTRACTOR FEES (General Requirements: 10%, Overhead: 5%, Profit: 10%)		25%	21.06	
	ARCHITECT FEES		7%	7.37	

Total Building Cost	**112.70**

Costs per square foot of floor area

Exterior Wall	S.F. Area	10000	15000	20000	25000	30000	35000	40000	45000	50000
	L.F. Perimeter	**286**	**370**	**453**	**457**	**513**	**568**	**624**	**680**	**735**
Precast Concrete Panels	Bearing Walls	132.95	126.50	123.25	**118.10**	116.40	115.20	114.25	113.55	113.00
	Steel Frame	137.95	131.55	128.30	123.10	121.45	120.25	119.30	118.60	118.05
Face Brick with Concrete Block Back-up	Bearing Walls	131.40	125.20	122.10	117.15	115.55	114.35	113.45	112.80	112.30
	Steel Joists	136.85	130.70	127.60	122.65	121.00	119.80	118.95	118.20	117.75
Stucco on Concrete Block	Bearing Walls	124.75	119.45	116.80	112.90	111.55	110.55	109.80	109.25	108.80
	Steel Joists	130.25	124.90	122.25	118.35	117.00	116.00	115.30	114.75	114.25
Perimeter Adj., Add or Deduct	Per 100 L.F.	10.00	6.70	4.95	4.00	3.35	2.80	2.55	2.25	2.00
Story Hgt. Adj., Add or Deduct	Per 1 Ft.	2.30	2.05	1.85	1.50	1.45	1.30	1.30	1.20	1.25
For Basement, add $23.05 per square foot of basement area										

The above costs were calculated using the basic specifications shown on the facing page. These costs should be adjusted where necessary for design alternatives and owner's requirements. Reported completed project costs, for this type of structure, range from $64.30 to $161.15 per S.F.

Common additives

Description	Unit	$ Cost	Description	Unit	$ Cost
Beds, Manual	Each	1000 - 2750	Kitchen Equipment, cont.		
Doctors In-Out Register, 200 names	Each	13,600	Ice cube maker, 50 lb. per day	Each	1600
Elevators, Hydraulic passenger, 2 stops			Range with 1 oven	Each	2100
1500# capacity	Each	47,900	Laundry Equipment		
2500# capacity	Each	50,000	Dryer, gas, 16 lb. capacity	Each	785
3500# capacity	Each	53,700	30 lb. capacity	Each	3000
Emergency Lighting, 25 watt, battery operated			Washer, 4 cycle	Each	875
Lead battery	Each	247	Commercial	Each	1325
Nickel cadmium	Each	695	Nurses Call System		
Intercom System, 25 station capacity			Single bedside call station	Each	270
Master station	Each	2200	Pillow speaker	Each	253
Intercom outlets	Each	142	Refrigerator, Prefabricated, walk-in		
Handset	Each	390	7'-6" high, 6' x 6'	S.F.	129
Kitchen Equipment			10' x 10'	S.F.	101
Broiler	Each	3075	12' x 14'	S.F.	89
Coffee urn, twin 6 gallon	Each	2925	12' x 20'	S.F.	79
Cooler, 6 ft. long	Each	4075	TV Antenna, Master system, 12 outlet	Outlet	262
Dishwasher, 10-12 racks per hr.	Each	4300	30 outlet	Outlet	168
Food warmer	Each	665	100 outlet	Outlet	160
Freezer, 44 C.F., reach-in	Each	9175	Whirlpool Bath, Mobile, 18" x 24" x 60"	Each	4150
			X-Ray, Mobile	Each	12,100 - 68,000

Important: See the Reference Section for Location Factors

Model costs calculated for a 2 story building with 10' story height and 25,000 square feet of floor area

				Unit	Unit Cost	Cost Per S.F.	% Of Sub-Total
A. SUBSTRUCTURE							
1010	Standard Foundations	Poured concrete; strip and spread footings		S.F. Ground	1.08	.54	
1030	Slab on Grade	4" reinforced concrete with vapor barrier and granular base		S.F. Slab	3.98	1.99	
2010	Basement Excavation	Site preparation for slab and trench for foundation wall and footing		S.F. Ground	.23	.12	4.6%
2020	Basement Walls	4' foundation wall		L.F. Wall	54	1.29	
B. SHELL							
	B10 Superstructure						
1010	Floor Construction	Pre-cast double tees with concrete topping		S.F. Floor	9.44	4.72	10.1%
1020	Roof Construction	Pre-Cast double tees		S.F. Roof	7.76	3.88	
	B20 Exterior Enclosure						
2010	Exterior Walls	Precast concrete panels		S.F. Wall	26	8.09	
2020	Exterior Windows	Wood double hung	15% of wall	Each	436	1.59	11.8%
2030	Exterior Doors	Double aluminum & glass doors, single leaf hollow metal		Each	1699	.34	
	B30 Roofing						
3010	Roof Coverings	Built-up tar and gravel with flashing; perlite/EPS composite insulation		S.F. Roof	4.54	2.27	2.7%
3020	Roof Openings	Roof hatches		S.F. Roof	.06	.03	
C. INTERIORS							
1010	Partitions	Gypsum board on metal studs	8 S.F. Floor/L.F. Partition	S.F. Partition	5.72	5.72	
1020	Interior Doors	Single leaf wood	80 S.F. Floor/Door	Each	480	6.00	
1030	Fittings	N/A		—	—	—	
2010	Stair Construction	Concrete stair		Flight	5600	1.34	26.0%
3010	Wall Finishes	50% vinyl wall coverings, 45% paint, 5% ceramic tile		S.F. Surface	2.71	2.71	
3020	Floor Finishes	95% vinyl tile, 5% ceramic tile		S.F. Floor	3.24	3.24	
3030	Ceiling Finishes	Painted gypsum board		S.F. Ceiling	3.00	3.12	
D. SERVICES							
	D10 Conveying						
1010	Elevators & Lifts	One hydraulic hospital elevator		Each	71,500	2.86	3.4%
1020	Escalators & Moving Walks	N/A		—	—	—	
	D20 Plumbing						
2010	Plumbing Fixtures	Kitchen, toilet and service fixtures, supply and drainage	1 Fixture/230 S.F. Floor	Each	2272	9.88	
2020	Domestic Water Distribution	Oil fired water heater		S.F. Floor	.33	.33	12.2%
2040	Rain Water Drainage	Roof drains		S.F. Roof	.40	.20	
	D30 HVAC						
3010	Energy Supply	Oil fired hot water, wall fin radiation		S.F. Floor	5.19	5.19	
3020	Heat Generating Systems	N/A		—	—	—	
3030	Cooling Generating Systems	N/A		—	—	—	12.7%
3050	Terminal & Package Units	Split systems with air cooled condensing units		S.F. Floor	5.65	5.65	
3090	Other HVAC Sys. & Equipment	N/A		—	—	—	
	D40 Fire Protection						
4010	Sprinklers	Sprinkler, light hazard		S.F. Floor	3.71	3.71	4.4%
4020	Standpipes	N/A		—	—	—	
	D50 Electrical						
5010	Electrical Service/Distribution	800 ampere service, panel board and feeders		S.F. Floor	1.49	1.49	
5020	Lighting & Branch Wiring	Incandescent fixtures, receptacles, switches, A.C. and misc. power		S.F. Floor	7.83	7.83	12.1%
5030	Communications & Security	Alarm systems and emergency lighting		S.F. Floor	.59	.59	
5090	Other Electrical Systems	Emergency generator, 15 kW		S.F. Floor	.40	.40	
E. EQUIPMENT & FURNISHINGS							
1010	Commercial Equipment	N/A		—	—	—	
1020	Institutional Equipment	N/A		—	—	—	
1030	Vehicular Equipment	N/A		—	—	—	0.0%
1090	Other Equipment	N/A		—	—	—	
F. SPECIAL CONSTRUCTION							
1020	Integrated Construction	N/A		—	—	—	
1040	Special Facilities	N/A		—	—	—	0.0%
G. BUILDING SITEWORK	**N/A**						

			Sub-Total	85.12	100%
CONTRACTOR FEES (General Requirements: 10%, Overhead: 5%, Profit: 10%)		25%	21.28		
ARCHITECT FEES		11%	11.70		

Total Building Cost | **118.10**

BUILDING TYPES

Costs per square foot of floor area

Exterior Wall	S.F. Area	2000	3000	5000	7000	9000	12000	15000	20000	25000
	L.F. Perimeter	220	260	320	360	420	480	520	640	700
Wood Siding	Wood Truss	152.40	135.20	119.95	112.30	108.90	104.95	102.10	100.25	98.25
Brick Veneer	Wood Truss	192.00	170.55	151.50	141.95	137.75	132.85	129.25	126.95	124.45
Brick on Block	Wood Truss	178.90	156.80	136.95	126.90	122.45	117.30	113.40	111.00	108.25
	Steel Roof Deck	185.95	161.00	138.70	127.55	122.55	116.90	112.60	109.95	106.90
EIFS on Metal Studs	Steel Roof Deck	176.65	154.45	134.80	125.05	120.55	115.60	111.85	109.55	106.95
Tiltup Concrete Panel	Steel Roof Deck	168.75	148.45	130.55	121.65	117.60	113.10	109.70	107.60	105.20
Perimeter Adj., Add or Deduct	Per 100 L.F.	19.20	12.85	7.65	5.45	4.25	3.20	2.55	1.95	1.55
Story Hgt. Adj., Add or Deduct	Per 1 Ft.	2.40	1.85	1.40	1.10	.95	.90	.75	.70	.60

For Basement, add $23.40 per square foot of basement area

The above costs were calculated using the basic specifications shown on the facing page. These costs should be adjusted where necessary for design alternatives and owner's requirements. Reported completed project costs, for this type of structure, range from $49.10 to $189.80 per S.F.

Common additives

Description	Unit	$ Cost
Clock System		
20 room	Each	14,200
50 room	Each	34,600
Closed Circuit Surveillance, One station		
Camera and monitor	Each	1525
For additional camera stations, add	Each	820
Directory Boards, Plastic, glass covered		
30" x 20"	Each	575
36" x 48"	Each	1275
Aluminum, 24" x 18"	Each	525
36" x 24"	Each	600
48" x 32"	Each	810
48" x 60"	Each	1875
Emergency Lighting, 25 watt, battery operated		
Lead battery	Each	247
Nickel cadmium	Each	695

Description	Unit	$ Cost
Smoke Detectors		
Ceiling type	Each	161
Duct type	Each	425
Sound System		
Amplifier, 250 watts	Each	1925
Speaker, ceiling or wall	Each	160
Trumpet	Each	305
TV Antenna, Master system, 12 outlet	Outlet	262
30 outlet	Outlet	168
100 outlet	Outlet	160

Model costs calculated for a 1 story building with 12' story height and 7,000 square feet of floor area

BUILDING TYPES

				Unit	Unit Cost	Cost Per S.F.	% Of Sub-Total
A.	**SUBSTRUCTURE**						
1010	Standard Foundations	Poured concrete; strip and spread footings		S.F. Ground	2.13	2.13	
1030	Slab on Grade	4" reinforced concrete with vapor barrier and granular base		S.F. Slab	3.98	3.98	
2010	Basement Excavation	Site preparation for slab and trench for foundation wall and footing		S.F. Ground	.23	.23	10.3%
2020	Basement Walls	4' foundation wall		L.F. Wall	115	3.29	
B.	**SHELL**						
	B10 Superstructure						
1010	Floor Construction	N/A		—	—	—	
1020	Roof Construction	Steel joists, girders & deck on columns		S.F. Roof	7.34	7.34	7.9%
	B20 Exterior Enclosure						
2010	Exterior Walls	E.I.F.S. on metal studs		S.F. Wall	15.94	7.87	
2020	Exterior Windows	Aluminum outward projecting	20% of wall	Each	559	3.00	13.7%
2030	Exterior Doors	Aluminum and glass, hollow metal		Each	2274	1.96	
	B30 Roofing						
3010	Roof Coverings	Single ply membrane, loose laid and ballasted		S.F. Roof	6.62	6.62	7.2%
3020	Roof Openings	Roof hatch		S.F. Roof	.13	.13	
C.	**INTERIORS**						
1010	Partitions	Gypsum board on metal studs	20 S.F. Floor/L.F. Partition	S.F. Partition	6.64	3.32	
1020	Interior Doors	Single leaf hollow metal	200 S.F. Floor/Door	Each	608	3.04	
1030	Fittings	Toilet partitions		S.F. Floor	.49	.49	
2010	Stair Construction	N/A		—	—	—	19.4%
3010	Wall Finishes	60% vinyl wall covering, 40% paint		S.F. Surface	2.36	1.18	
3020	Floor Finishes	60% carpet, 30% vinyl composition tile, 10% ceramic tile		S.F. Floor	6.04	6.04	
3030	Ceiling Finishes	Mineral fiber tile on concealed zee bars		S.F. Ceiling	4.00	4.09	
D.	**SERVICES**						
	D10 Conveying						
1010	Elevators & Lifts	N/A		—	—	—	0.0%
1020	Escalators & Moving Walks	N/A		—	—	—	
	D20 Plumbing						
2010	Plumbing Fixtures	Toilet and service fixtures, supply and drainage	1 Fixture/1320 S.F. Floor	Each	3009	2.28	
2020	Domestic Water Distribution	Gas fired water heater		S.F. Floor	.61	.61	3.7%
2040	Rain Water Drainage	Roof drains		S.F. Roof	.54	.54	
	D30 HVAC						
3010	Energy Supply	N/A		—	—	—	
3020	Heat Generating Systems	Included in D3050		—	—	—	
3030	Cooling Generating Systems	N/A		—	—	—	19.6%
3050	Terminal & Package Units	Multizone unit gas heating, electric cooling		S.F. Floor	18.35	18.35	
3090	Other HVAC Sys. & Equipment	N/A		—	—	—	
	D40 Fire Protection						
4010	Sprinklers	N/A		—	—	—	0.7%
4020	Standpipes	Standpipes and hose systems		S.F. Floor	.69	.69	
	D50 Electrical						
5010	Electrical Service/Distribution	400 ampere service, panel board and feeders		S.F. Floor	4.33	4.33	
5020	Lighting & Branch Wiring	Fluorescent fixtures, receptacles, switches, A.C. and misc. power		S.F. Floor	9.88	9.88	17.4%
5030	Communications & Security	Alarm systems and emergency lighting		S.F. Floor	1.92	1.92	
5090	Other Electrical Systems	Emergency generator, 7.5 kW		S.F. Floor	.16	.16	
E.	**EQUIPMENT & FURNISHINGS**						
1010	Commercial Equipment	N/A		—	—	—	
1020	Institutional Equipment	N/A		—	—	—	
1030	Vehicular Equipment	N/A		—	—	—	0.0%
1090	Other Equipment	N/A		—	—	—	
F.	**SPECIAL CONSTRUCTION**						
1020	Integrated Construction	N/A		—	—	—	0.0%
1040	Special Facilities	N/A		—	—	—	
G.	**BUILDING SITEWORK**	**N/A**					

		Sub-Total	93.47	100%
	CONTRACTOR FEES (General Requirements: 10%, Overhead: 5%, Profit: 10%)	25%	23.40	
	ARCHITECT FEES	7%	8.18	

Total Building Cost	125.05

Costs per square foot of floor area

Exterior Wall	S.F. Area	5000	8000	12000	16000	20000	35000	50000	65000	80000
	L.F. Perimeter	220	260	310	330	360	440	490	548	580
Face Brick with Concrete Block Back-up	Wood Joists	178.05	152.10	137.30	127.15	121.80	111.35	106.25	103.70	101.60
	Steel Joists	179.50	153.50	138.75	128.60	123.30	112.75	107.65	105.10	103.10
Glass and Metal Curtain Wall	Steel Frame	174.85	151.10	137.60	128.50	123.70	114.25	109.75	107.40	105.65
	R/Conc. Frame	178.05	154.35	140.85	131.75	126.95	117.45	113.00	110.70	108.85
Wood Siding	Wood Frame	146.30	126.15	114.65	107.20	103.20	95.45	91.70	89.85	88.40
Brick Veneer	Wood Frame	160.20	136.40	122.85	113.65	108.85	99.35	94.80	92.50	90.65
Perimeter Adj., Add or Deduct	Per 100 L.F.	28.75	17.95	11.90	9.00	7.20	4.00	2.85	2.20	1.85
Story Hgt. Adj., Add or Deduct	Per 1 Ft.	4.70	3.40	2.70	2.20	1.90	1.30	1.05	.85	.80

For Basement, add $26.05 per square foot of basement area

The above costs were calculated using the basic specifications shown on the facing page. These costs should be adjusted where necessary for design alternatives and owner's requirements. Reported completed project costs, for this type of structure, range from $51.35 to $198.95 per S.F.

Common additives

Description	Unit	$ Cost	Description	Unit	$ Cost
Clock System			Smoke Detectors		
20 room	Each	14,200	Ceiling type	Each	161
50 room	Each	34,600	Duct type	Each	425
Closed Circuit Surveillance, One station			Sound System		
Camera and monitor	Each	1525	Amplifier, 250 watts	Each	1925
For additional camera stations, add	Each	820	Speaker, ceiling or wall	Each	160
Directory Boards, Plastic, glass covered			Trumpet	Each	305
30" x 20"	Each	575	TV Antenna, Master system, 12 outlet	Outlet	262
36" x 48"	Each	1275	30 outlet	Outlet	168
Aluminum, 24" x 18"	Each	525	100 outlet	Outlet	160
36" x 24"	Each	600			
48" x 32"	Each	810			
48" x 60"	Each	1875			
Elevators, Hydraulic passenger, 2 stops					
1500# capacity	Each	47,900			
2500# capacity	Each	50,000			
3500# capacity	Each	53,700			
Additional stop, add	Each	4550			
Emergency Lighting, 25 watt, battery operated					
Lead battery	Each	247			
Nickel cadmium	Each	695			

BUILDING TYPES

Important: See the Reference Section for Location Factors

Model costs calculated for a 3 story building with 12' story height and 20,000 square feet of floor area

				Unit	Unit Cost	Cost Per S.F.	% Of Sub-Total
A. SUBSTRUCTURE							
1010	Standard Foundations	Poured concrete; strip and spread footings		S.F. Ground	5.82	1.94	
1030	Slab on Grade	4" reinforced concrete with vapor barrier and granular base		S.F. Slab	3.98	1.32	
2010	Basement Excavation	Site preparation for slab and trench for foundation wall and footing		S.F. Ground	.14	.05	5.0%
2020	Basement Walls	4' foundation wall		L.F. Wall	115	1.29	
B. SHELL							
	B10 Superstructure						
1010	Floor Construction	Open web steel joists, slab form, concrete, columns		S.F. Floor	14.12	9.41	12.1%
1020	Roof Construction	Metal deck, open web steel joists, columns		S.F. Roof	5.25	1.75	
	B20 Exterior Enclosure						
2010	Exterior Walls	Face brick with concrete block backup	80% of wall	S.F. Wall	23	12.32	
2020	Exterior Windows	Aluminum outward projecting	20% of wall	Each	559	3.15	17.5%
2030	Exterior Doors	Aluminum and glass, hollow metal		Each	2274	.69	
	B30 Roofing						
3010	Roof Coverings	Built-up tar and gravel with flashing; perlite/EPS composite		S.F. Roof	4.83	1.61	1.7%
3020	Roof Openings	N/A		—	—	—	
C. INTERIORS							
1010	Partitions	Gypsum board on metal studs	20 S.F. Floor/L.F. Partition	S.F. Partition	7.68	3.07	
1020	Interior Doors	Single leaf hollow metal	200 S.F. Floor/Door	Each	608	3.04	
1030	Fittings	Toilet partitions		S.F. Floor	1.04	1.04	
2010	Stair Construction	Concrete filled metal pan		Flight	6425	2.25	22.2%
3010	Wall Finishes	60% vinyl wall covering, 40% paint		S.F. Surface	1.88	.94	
3020	Floor Finishes	60% carpet, 30% vinyl composition tile, 10% ceramic tile		S.F. Floor	6.04	6.04	
3030	Ceiling Finishes	Mineral fiber tile on concealed zee bars		S.F. Ceiling	4.00	4.09	
D. SERVICES							
	D10 Conveying						
1010	Elevators & Lifts	Two hydraulic passenger elevators		Each	79,000	7.90	8.6%
1020	Escalators & Moving Walks	N/A		—	—	—	
	D20 Plumbing						
2010	Plumbing Fixtures	Toilet and service fixtures, supply and drainage	1 Fixture/1320 S.F. Floor	Each	1927	1.46	
2020	Domestic Water Distribution	Gas fired water heater		S.F. Floor	.22	.22	2.1%
2040	Rain Water Drainage	Roof drains		S.F. Roof	.75	.25	
	D30 HVAC						
3010	Energy Supply	N/A		—	—	—	
3020	Heat Generating Systems	Included in D3050		—	—	—	
3030	Cooling Generating Systems	N/A		—	—	—	15.3%
3050	Terminal & Package Units	Multizone unit gas heating, electric cooling		S.F. Floor	14.10	14.10	
3090	Other HVAC Sys. & Equipment	N/A		—	—	—	
	D40 Fire Protection						
4010	Sprinklers	N/A		—	—	—	0.8%
4020	Standpipes	Standpipes and hose systems		S.F. Floor	.72	.72	
	D50 Electrical						
5010	Electrical Service/Distribution	1000 ampere service, panel board and feeders		S.F. Floor	3.32	3.32	
5020	Lighting & Branch Wiring	Fluorescent fixtures, receptacles, switches, A.C. and misc. power		S.F. Floor	9.48	9.48	14.7%
5030	Communications & Security	Alarm systems and emergency lighting		S.F. Floor	.67	.67	
5090	Other Electrical Systems	Emergency generator, 7.5 kW		S.F. Floor	.05	.05	
E. EQUIPMENT & FURNISHINGS							
1010	Commercial Equipment	N/A		—	—	—	
1020	Institutional Equipment	N/A		—	—	—	
1030	Vehicular Equipment	N/A		—	—	—	0.0%
1090	Other Equipment	N/A		—	—	—	
F. SPECIAL CONSTRUCTION							
1020	Integrated Construction	N/A		—	—	—	0.0%
1040	Special Facilities	N/A		—	—	—	
G. BUILDING SITEWORK	**N/A**						

		Sub-Total	92.17	100%
CONTRACTOR FEES (General Requirements: 10%, Overhead: 5%, Profit: 10%)		25%	23.07	
ARCHITECT FEES		7%	8.06	

Total Building Cost	**123.30**

BUILDING TYPES

Costs per square foot of floor area

Exterior Wall	S.F. Area	20000	40000	60000	80000	100000	150000	200000	250000	300000
	L.F. Perimeter	260	360	400	420	460	520	600	640	700
Precast Concrete Panel	Steel Frame	152.95	131.85	120.80	114.30	111.20	105.95	103.70	101.80	100.75
	R/Conc. Frame	153.45	132.15	121.10	114.45	111.35	106.10	103.85	101.95	100.80
Face Brick with Concrete Block Back-up	Steel Frame	149.20	129.15	118.80	112.65	109.80	104.90	102.80	101.05	100.00
	R/Conc. Frame	149.25	129.25	118.90	112.80	109.85	104.95	102.90	101.10	100.05
Limestone Panel Concrete Block Back-up	Steel Frame	180.70	151.00	134.95	125.40	120.90	113.30	110.05	107.20	105.65
	R/Conc. Frame	180.85	151.05	135.05	125.55	121.00	113.40	110.15	107.35	105.75
Perimeter Adj., Add or Deduct	Per 100 L.F.	20.05	10.05	6.65	5.05	4.00	2.65	2.05	1.65	1.30
Story Hgt. Adj., Add or Deduct	Per 1 Ft.	4.15	2.85	2.15	1.65	1.45	1.10	1.00	.85	.70

For Basement, add $28.20 per square foot of basement area

The above costs were calculated using the basic specifications shown on the facing page. These costs should be adjusted where necessary for design alternatives and owner's requirements. Reported completed project costs, for this type of structure, range from $57.80 to $169.95 per S.F.

Common additives

Description	Unit	$ Cost
Clock System		
20 room	Each	14,200
50 room	Each	34,600
Closed Circuit Surveillance, One station		
Camera and monitor	Each	1525
For additional camera stations, add	Each	820
Directory Boards, Plastic, glass covered		
30" x 20"	Each	575
36" x 48"	Each	1275
Aluminum, 24" x 18"	Each	525
36" x 24"	Each	600
48" x 32"	Each	810
48" x 60"	Each	1875
Elevators, Electric passenger, 5 stops		
2000# capacity	Each	110,300
3500# capacity	Each	116,800
5000# capacity	Each	121,800
Additional stop, add	Each	6550
Emergency Lighting, 25 watt, battery operated		
Lead battery	Each	247
Nickel cadmium	Each	695

Description	Unit	$ Cost
Intercom System, 25 station capacity		
Master station	Each	2200
Intercom outlets	Each	142
Handset	Each	390
Smoke Detectors		
Ceiling type	Each	161
Duct type	Each	425
Sound System		
Amplifier, 250 watts	Each	1925
Speaker, ceiling or wall	Each	160
Trumpet	Each	305
TV Antenna, Master system, 12 oulet	Outlet	262
30 outlet	Outlet	168
100 outlet	Outlet	160

BUILDING TYPES

Model costs calculated for an 8 story building with 12' story height and 80,000 square feet of floor area.

				Unit	Unit Cost	Cost Per S.F.	% Of Sub-Total
A. SUBSTRUCTURE							
1010	Standard Foundations	Poured concrete; strip and spread footings		S.F. Ground	9.60	1.20	
1030	Slab on Grade	4" reinforced concrete with vapor barrier and granular base		S.F. Slab	3.98	.50	
2010	Basement Excavation	Site preparation for slab and trench for foundation wall and footing		S.F. Ground	.23	.03	2.5%
2020	Basement Walls	4' foundation wall		L.F. Wall	61	.41	
B. SHELL							
	B10 Superstructure						
1010	Floor Construction	Concrete slab with metal deck and beams		S.F. Floor	17.70	15.49	18.7%
1020	Roof Construction	Metal deck, open web steel joists, interior columns		S.F. Roof	5.20	.65	
	B20 Exterior Enclosure						
2010	Exterior Walls	Precast concrete panels	80% of wall	S.F. Wall	26	10.86	
2020	Exterior Windows	Vertical pivoted steel	20% of wall	Each	436	2.93	16.2%
2030	Exterior Doors	Double aluminum and glass doors and entrance with transoms		Each	2657	.17	
	B30 Roofing						
3010	Roof Coverings	Built-up tar and gravel with flashing; perlite/EPS composite insulation		S.F. Roof	4.08	.51	0.6%
3020	Roof Openings	N/A		—	—	—	
C. INTERIORS							
1010	Partitions	Gypsum board on metal studs	30 S.F. Floor/L.F. Partition	S.F. Partition	7.38	2.46	
1020	Interior Doors	Single leaf hollow metal	400 S.F. Floor/Door	Each	608	1.52	
1030	Fittings	Toilet Partitions		S.F. Floor	.69	.69	
2010	Stair Construction	Concrete filled metal pan		Flight	6425	1.36	19.7%
3010	Wall Finishes	60% vinyl wall covering, 40% paint		S.F. Surface	2.37	.79	
3020	Floor Finishes	60% carpet, 30% vinyl composition tile, 10% ceramic tile		S.F. Floor	6.04	6.04	
3030	Ceiling Finishes	Mineral fiber tile on concealed zee bars		S.F. Ceiling	4.00	4.09	
D. SERVICES							
	D10 Conveying						
1010	Elevators & Lifts	Four geared passenger elevators		Each	176,600	8.83	10.2%
1020	Escalators & Moving Walks	N/A		—	—	—	
	D20 Plumbing						
2010	Plumbing Fixtures	Toilet and service fixtures, supply and drainage	1 Fixture/1370 S.F. Floor	Each	1918	1.40	
2020	Domestic Water Distribution	Gas fired water heater		S.F. Floor	.22	.22	2.0%
2040	Rain Water Drainage	Roof drains		S.F. Roof	.88	.11	
	D30 HVAC						
3010	Energy Supply	N/A		—	—	—	
3020	Heat Generating Systems	Included in D3050		—	—	—	
3030	Cooling Generating Systems	N/A		—	—	—	16.3%
3050	Terminal & Package Units	Multizone unit gas heating, electric cooling		S.F. Floor	14.10	14.10	
3090	Other HVAC Sys. & Equipment	N/A		—	—	—	
	D40 Fire Protection						
4010	Sprinklers	N/A		—	—	—	0.2%
4020	Standpipes	Standpipes and hose systems		S.F. Floor	.20	.20	
	D50 Electrical						
5010	Electrical Service/Distribution	1600 ampere service, panel board and feeders		S.F. Floor	1.26	1.26	
5020	Lighting & Branch Wiring	Fluorescent fixtures, receptacles, switches, A.C. and misc. power		S.F. Floor	9.41	9.41	13.6%
5030	Communications & Security	Alarm systems and emergency lighting		S.F. Floor	.57	.57	
5090	Other Electrical Systems	Emergency generator, 100 kW		S.F. Floor	.45	.45	
E. EQUIPMENT & FURNISHINGS							
1010	Commercial Equipment	N/A		—	—	—	
1020	Institutional Equipment	N/A		—	—	—	
1030	Vehicular Equipment	N/A		—	—	—	0.0%
1090	Other Equipment	N/A		—	—	—	
F. SPECIAL CONSTRUCTION							
1020	Integrated Construction	N/A		—	—	—	
1040	Special Facilities	N/A		—	—	—	0.0%
G. BUILDING SITEWORK	**N/A**						

			Sub-Total	86.25	100%
CONTRACTOR FEES (General Requirements: 10%, Overhead: 5%, Profit: 10%)			25%	21.58	
ARCHITECT FEES			6%	6.47	

Total Building Cost	**114.30**

BUILDING TYPES

Costs per square foot of floor area

Exterior Wall	S.F. Area	120000	145000	170000	200000	230000	260000	400000	600000	800000
	L.F. Perimeter	420	450	470	490	510	530	600	730	820
Double Glazed Heat Absorbing Tinted Plate Glass Panels	Steel Frame	120.35	116.00	112.35	109.10	106.65	104.80	99.15	96.20	94.30
	R/Conc. Frame	120.90	115.50	111.15	107.25	104.30	102.10	95.55	92.00	89.75
Face Brick with Concrete Block Back-up	Steel Frame	119.85	115.20	111.50	108.15	105.60	103.75	98.20	95.10	93.20
	R/Conc. Frame	113.75	109.10	105.45	102.15	99.70	97.80	92.40	89.35	87.50
Precast Concrete Panel With Exposed Aggregate	Steel Frame	118.95	114.35	110.75	107.45	105.00	103.15	97.75	94.70	92.95
	R/Conc. Frame	113.10	108.55	105.00	101.75	99.40	97.55	92.25	89.25	87.45
Perimeter Adj., Add or Deduct	Per 100 L.F.	6.95	5.75	4.90	4.10	3.55	3.10	2.10	1.40	1.00
Story Hgt. Adj., Add or Deduct	Per 1 Ft.	2.80	2.50	2.25	1.90	1.80	1.55	1.25	.95	.80

For Basement, add $28.20 per square foot of basement area

The above costs were calculated using the basic specifications shown on the facing page. These costs should be adjusted where necessary for design alternatives and owner's requirements. Reported completed project costs, for this type of structure, range from $72.55 to $176.90 per S.F.

Common additives

Description	Unit	$ Cost	Description	Unit	$ Cost
Clock System			Escalators, Metal		
20 room	Each	14,200	32" wide, 10' story height	Each	100,300
50 room	Each	34,600	20' story height	Each	117,300
Directory Boards, Plastic, glass covered			48" wide, 10' story height	Each	105,300
30" x 20"	Each	575	20' story height	Each	122,300
36" x 48"	Each	1275	Glass		
Aluminum, 24" x 18"	Each	525	32" wide, 10' story height	Each	98,800
36" x 24"	Each	600	20' story height	Each	117,300
48" x 32"	Each	810	48" wide, 10' story height	Each	105,300
48" x 60"	Each	1875	20' story height	Each	122,300
Elevators, Electric passenger, 10 stops			Smoke Detectors		
3000# capacity	Each	239,500	Ceiling type	Each	161
4000# capacity	Each	242,000	Duct type	Each	425
5000# capacity	Each	246,500	Sound System		
Additional stop, add	Each	6550	Amplifier, 250 watts	Each	1925
Emergency Lighting, 25 watt, battery operated			Speaker, ceiling or wall	Each	160
Lead battery	Each	247	Trumpet	Each	305
Nickel cadmium	Each	695	TV Antenna, Master system, 12 oulet	Outlet	262
			30 oulet	Outlet	168
			100 outlet	Outlet	160

Important: See the Reference Section for Location Factors

Model costs calculated for a 16 story building with 10' story height and 260,000 square feet of floor area

				Unit	Unit Cost	Cost Per S.F.	% Of Sub-Total
A. SUBSTRUCTURE							
1010	Standard Foundations	Poured concrete; strip and spread footings		S.F. Ground	11.68	.73	
1030	Slab on Grade	4" reinforced concrete with vapor barrier and granular base		S.F. Slab	3.98	.25	1.6%
2010	Basement Excavation	Site preparation for slab and trench for foundation wall and footing		S.F. Ground	.23	.01	
2020	Basement Walls	4' foundation wall		L.F. Wall	61	.30	
B. SHELL							
	B10 Superstructure						
1010	Floor Construction	Concrete slab, metal deck, beams		S.F. Floor	20	18.79	24.2%
1020	Roof Construction	Metal deck, open web steel joists, beams, columns		S.F. Roof	6.08	.38	
	B20 Exterior Enclosure						
2010	Exterior Walls	N/A		—	—	—	
2020	Exterior Windows	Double glazed heat absorbing, tinted plate glass wall panels	100% of wall	Each	35	11.42	15.0%
2030	Exterior Doors	Double aluminum & glass doors		Each	4107	.45	
	B30 Roofing						
3010	Roof Coverings	Single ply membrane, fully adhered; perlite/EPS composite insulation		S.F. Roof	4.32	.27	0.3%
3020	Roof Openings	N/A		—	—	—	
C. INTERIORS							
1010	Partitions	Gypsum board on metal studs	30 S.F. Floor/L.F. Partition	S.F. Partition	6.60	1.76	
1020	Interior Doors	Single leaf hollow metal	400 S.F. Floor/Door	Each	608	1.52	
1030	Fittings	Toilet partitions		S.F. Floor	.40	.40	
2010	Stair Construction	Concrete filled metal pan		Flight	7700	1.04	17.5%
3010	Wall Finishes	60% vinyl wall covering, 40% paint		S.F. Surface	1.86	.62	
3020	Floor Finishes	60% carpet tile, 30% vinyl composition tile, 10% ceramic tile		S.F. Floor	4.38	4.38	
3030	Ceiling Finishes	Mineral fiber tile on concealed zee bars		S.F. Ceiling	4.00	4.09	
D. SERVICES							
	D10 Conveying						
1010	Elevators & Lifts	Four geared passenger elevators		Each	263,250	4.05	5.1%
1020	Escalators & Moving Walks	N/A		—	—	—	
	D20 Plumbing						
2010	Plumbing Fixtures	Toilet and service fixtures, supply and drainage	1 Fixture/1345 S.F. Floor	Each	1129	.84	
2020	Domestic Water Distribution	Oil fired water heater		S.F. Floor	.07	.07	1.2%
2040	Rain Water Drainage	Roof drains		S.F. Roof	.80	.05	
	D30 HVAC						
3010	Energy Supply	Oil fired hot water		S.F. Floor	3.29	3.29	
3020	Heat Generating Systems	N/A		—	—	—	
3030	Cooling Generating Systems	Chilled water, fan coil units		S.F. Floor	10.55	10.55	17.5%
3050	Terminal & Package Units	N/A		—	—	—	
3090	Other HVAC Sys. & Equipment	N/A		—	—	—	
	D40 Fire Protection						
4010	Sprinklers	Sprinkler system, light hazard		S.F. Floor	2.98	2.98	4.2%
4020	Standpipes	Standpipes and hose systems		S.F. Floor	.31	.31	
	D50 Electrical						
5010	Electrical Service/Distribution	2400 ampere service, panel board and feeders		S.F. Floor	.78	.78	
5020	Lighting & Branch Wiring	Fluorescent fixtures, receptacles, switches, A.C. and misc. power		S.F. Floor	9.31	9.31	13.3%
5030	Communications & Security	Alarm systems and emergency lighting		S.F. Floor	.25	.25	
5090	Other Electrical Systems	Emergency generator, 200 kW		S.F. Floor	.18	.18	
E. EQUIPMENT & FURNISHINGS							
1010	Commercial Equipment	N/A		—	—	—	
1020	Institutional Equipment	N/A		—	—	—	0.0%
1030	Vehicular Equipment	N/A		—	—	—	
1090	Other Equipment	N/A		—	—	—	
F. SPECIAL CONSTRUCTION							
1020	Integrated Construction	N/A		—	—	—	0.0%
1040	Special Facilities	N/A		—	—	—	
G. BUILDING SITEWORK	**N/A**						

			Sub-Total	79.07	100%
CONTRACTOR FEES (General Requirements: 10%, Overhead: 5%, Profit: 10%)		25%	19.80		
ARCHITECT FEES		6%	5.93		

Total Building Cost	**104.80**

BUILDING TYPES

BUILDING TYPES

Costs per square foot of floor area

Exterior Wall	S.F. Area	7000	9000	11000	13000	15000	17000	19000	21000	23000
	L.F. Perimeter	240	280	303	325	354	372	397	422	447
Limestone with Concrete Block Back-up	Bearing Walls	185.20	172.95	162.20	154.65	149.90	145.20	142.05	139.60	137.50
	R/Conc. Frame	196.05	184.25	174.15	167.05	162.65	158.20	155.25	152.95	150.95
Face Brick with Concrete Block Back-up	Bearing Walls	157.40	147.65	139.75	134.25	130.70	127.35	125.00	123.20	121.60
	R/Conc. Frame	174.65	164.85	157.00	151.45	147.90	144.60	142.25	140.40	138.85
Decorative Concrete Block	Bearing Walls	149.45	140.40	133.35	128.45	125.25	122.30	120.15	118.50	117.10
	R/Conc. Frame	166.65	157.65	150.55	145.65	142.40	139.45	137.35	135.70	134.30
Perimeter Adj., Add or Deduct	Per 100 L.F.	26.75	20.75	17.00	14.30	12.50	11.00	9.85	8.85	8.10
Story Hgt. Adj., Add or Deduct	Per 1 Ft.	4.80	4.30	3.85	3.40	3.30	3.00	2.90	2.75	2.70

For Basement, add $20.60 per square foot of basement area

The above costs were calculated using the basic specifications shown on the facing page. These costs should be adjusted where necessary for design alternatives and owner's requirements. Reported completed project costs, for this type of structure, range from $83.30 to $217.35 per S.F.

Common additives

Description	Unit	$ Cost
Cells Prefabricated, 5'-6' wide,		
7'-8' high, 7'-8' deep	Each	9900
Elevators, Hydraulic passenger, 2 stops		
1500# capacity	Each	47,900
2500# capacity	Each	50,000
3500# capacity	Each	53,700
Emergency Lighting, 25 watt, battery operated		
Lead battery	Each	247
Nickel cadmium	Each	695
Flagpoles, Complete		
Aluminum, 20' high	Each	1375
40' high	Each	3025
70' high	Each	9500
Fiberglass, 23' high	Each	1650
39'-5" high	Each	3175
59' high	Each	7950

Description	Unit	$ Cost
Lockers, Steel, Single tier, 60" to 72"	Opening	154 - 241
2 tier, 60" or 72" total	Opening	104 - 134
5 tier, box lockers	Opening	55 - 69
Locker bench, lam. maple top only	L.F.	18.80
Pedestals, steel pipe	Each	57
Safe, Office type, 4 hour rating		
30" x 18" x 18"	Each	3425
62" x 33" x 20"	Each	7450
Shooting Range, Incl. bullet traps,		
target provisions, and contols,		
not incl. structural shell	Each	35,000
Smoke Detectors		
Ceiling type	Each	161
Duct type	Each	425
Sound System		
Amplifier, 250 watts	Each	1925
Speaker, ceiling or wall	Each	160
Trumpet	Each	305

Model costs calculated for a 2 story building with 12' story height and 11,000 square feet of floor area

				Unit	Unit Cost	Cost Per S.F.	% Of Sub-Total
A. SUBSTRUCTURE							
1010	Standard Foundations	Poured concrete; strip and spread footings		S.F. Ground	2.14	1.07	
1030	Slab on Grade	4" reinforced concrete with vapor barrier and granular base		S.F. Slab	3.98	1.99	
2010	Basement Excavation	Site preparation for slab and trench for foundation wall and footing		S.F. Ground	.23	.12	4.6%
2020	Basement Walls	4' foundation wall		L.F. Wall	61	2.24	
B. SHELL							
	B10 Superstructure						
1010	Floor Construction	Open web steel joists, slab form, concrete		S.F. Floor	9.16	4.58	5.4%
1020	Roof Construction	Metal deck on open web steel joists		S.F. Roof	3.66	1.83	
	B20 Exterior Enclosure						
2010	Exterior Walls	Limestone with concrete block backup	80% of wall	S.F. Wall	47	25.17	
2020	Exterior Windows	Metal horizontal sliding	20% of wall	Each	821	7.24	28.3%
2030	Exterior Doors	Hollow metal		Each	1774	1.29	
	B30 Roofing						
3010	Roof Coverings	Built-up tar and gravel with flashing; perlite/EPS composite insulation		S.F. Roof	4.86	2.43	2.0%
3020	Roof Openings	N/A		—	—	—	
C. INTERIORS							
1010	Partitions	Concrete block	20 S.F. Floor/L.F. Partition	S.F. Partition	6.74	3.37	
1020	Interior Doors	Single leaf kalamein fire door	200 S.F. Floor/Door	Each	608	3.04	
1030	Fittings	Toilet partitions		S.F. Floor	.82	.82	
2010	Stair Construction	Concrete filled metal pan		Flight	7700	1.40	16.3%
3010	Wall Finishes	90% paint, 10% ceramic tile		S.F. Surface	5.78	2.89	
3020	Floor Finishes	70% vinyl composition tile, 20% carpet tile, 10% ceramic tile		S.F. Floor	3.80	3.80	
3030	Ceiling Finishes	Mineral fiber tile on concealed zee bars		S.F. Ceiling	4.00	4.09	
D. SERVICES							
	D10 Conveying						
1010	Elevators & Lifts	One hydraulic passenger elevator		Each	59,070	5.37	4.5%
1020	Escalators & Moving Walks	N/A		—	—	—	
	D20 Plumbing						
2010	Plumbing Fixtures	Toilet and service fixtures, supply and drainage	1 Fixture/580 S.F. Floor	Each	2285	3.94	
2020	Domestic Water Distribution	Oil fired water heater		S.F. Floor	.35	.35	4.2%
2040	Rain Water Drainage	Roof drains		S.F. Roof	1.50	.75	
	D30 HVAC						
3010	Energy Supply	Oil fired hot water, wall fin radiation		S.F. Floor	7.42	7.42	
3020	Heat Generating Systems	N/A		—	—	—	
3030	Cooling Generating Systems	N/A		—	—	—	12.8%
3050	Terminal & Package Units	Split systems with air cooled condensing units		S.F. Floor	7.84	7.84	
3090	Other HVAC Sys. & Equipment	N/A		—	—	—	
	D40 Fire Protection						
4010	Sprinklers	Wet pipe sprinkler system		S.F. Floor	2.26	2.26	1.9%
4020	Standpipes	N/A		—	—	—	
	D50 Electrical						
5010	Electrical Service/Distribution	400 ampere service, panel board and feeders		S.F. Floor	1.42	1.42	
5020	Lighting & Branch Wiring	Fluorescent fixtures, receptacles, switches, A.C. and misc. power		S.F. Floor	8.86	8.86	9.3%
5030	Communications & Security	Alarm systems and emergency lighting		S.F. Floor	.59	.59	
5090	Other Electrical Systems	Emergency generator, 15 kW		S.F. Floor	.17	.17	
E. EQUIPMENT & FURNISHINGS							
1010	Commercial Equipment	N/A		—	—	—	
1020	Institutional Equipment	Lockers, detention rooms, cells, gasoline dispensers		S.F. Floor	10.90	10.90	10.7%
1030	Vehicular Equipment	Gasoline dispenser system		S.F. Floor	1.80	1.80	
1090	Other Equipment	N/A		—	—	—	
F. SPECIAL CONSTRUCTION							
1020	Integrated Construction	N/A		—	—	—	0.0%
1040	Special Facilities	N/A		—	—	—	
G. BUILDING SITEWORK	**N/A**						

				Sub-Total		119.04	**100%**
	CONTRACTOR FEES (General Requirements: 10%, Overhead: 5%, Profit: 10%)			25%	29.77	
	ARCHITECT FEES			9%	13.39	
			Total Building Cost		**162.20**	

BUILDING TYPES

Costs per square foot of floor area

Exterior Wall	S.F. Area	5000	7000	9000	11000	13000	15000	17000	19000	21000
	L.F. Perimeter	300	380	420	468	486	513	540	580	620
Face Brick with Concrete Block Back-up	Steel Frame	118.70	111.75	104.90	101.00	96.80	94.00	91.95	90.75	89.80
	Bearing Walls	116.80	109.85	103.00	99.15	94.90	92.10	90.05	88.85	87.90
Limestone with Concrete Block Back-up	Steel Frame	140.55	131.45	121.90	116.50	110.35	106.50	103.45	101.85	100.55
	Bearing Walls	138.45	129.45	119.85	114.40	108.30	104.40	101.45	99.85	98.45
Decorative Concrete Block	Steel Frame	110.60	104.35	98.55	95.20	91.70	89.35	87.60	86.60	85.80
	Bearing Walls	108.70	102.50	96.70	93.30	89.90	87.50	85.70	84.70	83.90
Perimeter Adj., Add or Deduct	Per 100 L.F.	13.60	9.70	7.50	6.20	5.20	4.60	3.95	3.55	3.25
Story Hgt. Adj., Add or Deduct	Per 1 Ft.	2.15	1.90	1.65	1.45	1.30	1.20	1.15	1.05	1.05

For Basement, add $21.60 per square foot of basement area

The above costs were calculated using the basic specifications shown on the facing page. These costs should be adjusted where necessary for design alternatives and owner's requirements. Reported completed project costs, for this type of structure, range from $69.10 to $179.35 per S.F.

Common additives

Description	Unit	$ Cost	Description	Unit	$ Cost
Closed Circuit Surveillance, One station			Mail Boxes, Horizontal, key lock, 15" x 6" x 5"	Each	47
Camera and monitor	Each	1525	Double 15" x 12" x 5"	Each	82
For additional camera stations, add	Each	820	Quadruple 15" x 12" x 10"	Each	145
Emergency Lighting, 25 watt, battery operated			Vertical, 6" x 5" x 15", aluminum	Each	41
Lead battery	Each	247	Bronze	Each	61
Nickel cadmium	Each	695	Steel, enameled	Each	47
Flagpoles, Complete			Scales, Dial type, 5 ton cap.		
Aluminum, 20' high	Each	1375	8' x 6' platform	Each	9400
40' high	Each	3025	9' x 7' platform	Each	12,100
70' high	Each	9500	Smoke Detectors		
Fiberglass, 23' high	Each	1650	Ceiling type	Each	161
39'-5" high	Each	3175	Duct type	Each	425
59' high	Each	7950			

Model costs calculated for a 1 story building with 14' story height and 13,000 square feet of floor area

				Unit	Unit Cost	Cost Per S.F.	% Of Sub-Total
A. SUBSTRUCTURE							
1010	Standard Foundations	Poured concrete; strip and spread footings		S.F. Ground	1.53	1.53	
1030	Slab on Grade	4" reinforced concrete with vapor barrier and granular base		S.F. Slab	3.98	3.98	
2010	Basement Excavation	Site preparation for slab and trench for foundation wall and footing		S.F. Ground	.23	.23	12.3%
2020	Basement Walls	4' foundation wall		L.F. Wall	61	2.97	
B. SHELL							
	B10 Superstructure						
1010	Floor Construction	Steel column fireproofing		L.F. Column	26	.08	9.8%
1020	Roof Construction	Metal deck, open web steel joists, columns		S.F. Roof	6.87	6.87	
	B20 Exterior Enclosure						
2010	Exterior Walls	Face brick with concrete block backup	80% of wall	S.F. Wall	23	9.94	
2020	Exterior Windows	Double strength window glass	20% of wall	Each	559	2.55	18.7%
2030	Exterior Doors	Double aluminum & glass, single aluminum, hollow metal, steel overhead		Each	1740	.80	
	B30 Roofing						
3010	Roof Coverings	Built-up tar and gravel with flashing; perlite/EPS composite insulation		S.F. Roof	4.37	4.37	6.2%
3020	Roof Openings	N/A		—	—	—	
C. INTERIORS							
1010	Partitions	Concrete block	15 S.F. Floor/L.F. Partition	S.F. Partition	6.74	5.39	
1020	Interior Doors	Single leaf hollow metal	150 S.F. Floor/Door	Each	608	4.05	
1030	Fittings	Toilet partitions, cabinets, shelving, lockers		S.F. Floor	1.13	1.13	
2010	Stair Construction	N/A		—	—	—	23.3%
3010	Wall Finishes	Paint		S.F. Surface	2.81	2.81	
3020	Floor Finishes	50% vinyl tile, 50% paint		S.F. Floor	2.18	2.18	
3030	Ceiling Finishes	Mineral fiber tile on concealed zee bars	25% of area	S.F. Ceiling	4.09	1.02	
D. SERVICES							
	D10 Conveying						
1010	Elevators & Lifts	N/A		—	—	—	0.0%
1020	Escalators & Moving Walks	N/A		—	—	—	
	D20 Plumbing						
2010	Plumbing Fixtures	Toilet and service fixtures, supply and drainage	1 Fixture/1180 S.F. Floor	Each	1404	1.19	
2020	Domestic Water Distribution	Gas fired water heater		S.F. Floor	.21	.21	2.8%
2040	Rain Water Drainage	Roof drains		S.F. Roof	.57	.57	
	D30 HVAC						
3010	Energy Supply	N/A		—	—	—	
3020	Heat Generating Systems	Included in D3050		—	—	—	
3030	Cooling Generating Systems	N/A		—	—	—	10.1%
3050	Terminal & Package Units	Single zone, gas heating, electric cooling		S.F. Floor	7.20	7.20	
3090	Other HVAC Sys. & Equipment	N/A		—	—	—	
	D40 Fire Protection						
4010	Sprinklers	Wet pipe sprinkler system		S.F. Floor	2.05	2.05	2.9%
4020	Standpipes	N/A		—	—	—	
	D50 Electrical						
5010	Electrical Service/Distribution	600 ampere service, panel board and feeders		S.F. Floor	2.26	2.26	
5020	Lighting & Branch Wiring	Fluorescent fixtures, receptacles, switches, A.C. and misc. power		S.F. Floor	7.18	7.18	13.9%
5030	Communications & Security	Alarm systems and emergency lighting		S.F. Floor	.36	.36	
5090	Other Electrical Systems	Emergency generator, 15 kW		S.F. Floor	.10	.10	
E. EQUIPMENT & FURNISHINGS							
1010	Commercial Equipment	N/A		—	—	—	
1020	Institutional Equipment	N/A		—	—	—	
1030	Vehicular Equipment	N/A		—	—	—	0.0%
1090	Other Equipment	N/A		—	—	—	
F. SPECIAL CONSTRUCTION							
1020	Integrated Construction	N/A		—	—	—	0.0%
1040	Special Facilities	N/A		—	—	—	
G. BUILDING SITEWORK	**N/A**						

		Sub-Total	71.02	100%
CONTRACTOR FEES (General Requirements: 10%, Overhead: 5%, Profit: 10%)		25%	17.79	
ARCHITECT FEES		9%	7.99	

	Total Building Cost	**96.80**

BUILDING TYPES

Costs per square foot of floor area

Exterior Wall	S.F. Area	5000	10000	15000	21000	25000	30000	40000	50000	60000
	L.F. Perimeter	287	400	500	600	660	700	834	900	1000
Face Brick with Concrete Block Back-up	Steel Frame	187.85	158.35	147.55	140.35	137.10	132.95	129.30	125.60	123.75
	Bearing Walls	182.35	154.55	144.35	137.45	134.35	130.25	126.80	123.20	121.35
Concrete Block	Steel Frame	163.30	140.90	132.85	127.60	125.35	122.55	119.95	117.50	116.25
Brick Veneer	Steel Frame	183.05	154.65	144.30	137.45	134.40	130.60	127.15	123.70	122.00
Galvanized Steel Siding	Steel Frame	154.70	134.80	127.65	123.10	121.15	118.70	116.50	114.40	113.35
Metal Sandwich Panel	Steel Frame	156.55	136.20	128.95	124.25	122.25	119.75	117.55	115.40	114.25
Perimeter Adj., Add or Deduct	Per 100 L.F.	22.75	11.40	7.65	5.35	4.55	3.80	2.85	2.30	1.90
Story Hgt. Adj., Add or Deduct	Per 1 Ft.	4.60	3.25	2.70	2.25	2.15	1.90	1.70	1.45	1.35

For Basement, add $20.80 per square foot of basement area

The above costs were calculated using the basic specifications shown on the facing page. These costs should be adjusted where necessary for design alternatives and owner's requirements. Reported completed project costs, for this type of structure, range from $65.50 to $204.20 per S.F.

Common additives

Description	Unit	$ Cost	Description	Unit	$ Cost
Bar, Front Bar	L.F.	320	Lockers, Steel, single tier, 60" or 72"	Opening	154 - 241
Back Bar	L.F.	257	2 tier, 60" or 72" total	Opening	104 - 134
Booth, Upholstered, custom straight	L.F.	166 - 305	5 tier, box lockers	Opening	55 - 69
"L" or "U" shaped	L.F.	172 - 290	Locker bench, lam. maple top only	L.F.	18.80
Bleachers, Telescoping, manual			Pedestals, steel pipe	Each	57
To 15 tier	Seat	101 - 141	Sauna, Prefabricated, complete		
21-30 tier	Seat	221 - 267	6' x 4'	Each	4525
Courts			6' x 9'	Each	6700
Ceiling	Court	6300	8' x 8'	Each	7875
Floor	Court	11,000	8' x 10'	Each	8800
Walls	Court	21,900	10' x 12'	Each	11,000
Emergency Lighting, 25 watt, battery operated			Sound System		
Lead battery	Each	247	Amplifier, 250 watts	Each	1925
Nickel cadmium	Each	695	Speaker, ceiling or wall	Each	160
Kitchen Equipment			Trumpet	Each	305
Broiler	Each	3075	Steam Bath, Complete, to 140 C.F.	Each	1475
Cooler, 6 ft. long, reach-in	Each	4075	To 300 C.F.	Each	1650
Dishwasher, 10-12 racks per hr.	Each	4300	To 800 C.F.	Each	4725
Food warmer, counter 1.2 KW	Each	665	To 2500 C.F.	Each	7200
Freezer, reach-in, 44 C.F.	Each	9175			
Ice cube maker, 50 lb. per day	Each	1600			

Model costs calculated for a 2 story building with 12' story height and 30,000 square feet of floor area

				Unit	Unit Cost	Cost Per S.F.	% Of Sub-Total
A.	**SUBSTRUCTURE**						
1010	Standard Foundations	Poured concrete; strip and spread footings		S.F. Ground	2.64	1.32	
1030	Slab on Grade	5" reinforced concrete with vapor barrier and granular base		S.F. Slab	5.13	3.34	
2010	Basement Excavation	Site preparation for slab and trench for foundation wall and footing		S.F. Ground	.23	.15	6.3%
2020	Basement Walls	4' foundation wall		L.F. Wall	61	1.43	
B.	**SHELL**						
	B10 Superstructure						
1010	Floor Construction	Open web steel joists, slab form, concrete, columns	50% of area	S.F. Floor	13.48	6.74	10.6%
1020	Roof Construction	Metal deck on open web steel joists, columns		S.F. Roof	7.38	3.69	
	B20 Exterior Enclosure						
2010	Exterior Walls	Face brick with concrete block backup	95% of wall	S.F. Wall	23	12.68	
2020	Exterior Windows	Storefront	5% of wall	S.F. Window	.01	2.12	15.3%
2030	Exterior Doors	Aluminum and glass and hollow metal		Each	2198	.29	
	B30 Roofing						
3010	Roof Coverings	Built-up tar and gravel with flashing; perlite/EPS composite insulation		S.F. Roof	5.48	2.74	2.8%
3020	Roof Openings	Roof hatches		S.F. Roof	.08	.04	
C.	**INTERIORS**						
1010	Partitions	Concrete block, gypsum board on metal studs	25 S.F. Floor/L.F. Partition	S.F. Partition	10.65	4.26	
1020	Interior Doors	Single leaf hollow metal	810 S.F. Floor/Door	Each	608	.75	
1030	Fittings	Toilet partitions		S.F. Floor	.41	.41	
2010	Stair Construction	Concrete filled metal pan		Flight	6425	.64	12.1%
3010	Wall Finishes	Paint		S.F. Surface	1.70	.85	
3020	Floor Finishes	80% carpet tile, 20% ceramic tile	50% of floor area	S.F. Floor	5.20	2.60	
3030	Ceiling Finishes	Mineral fiber tile on concealed zee bars	60% of area	S.F. Ceiling	4.09	2.45	
D.	**SERVICES**						
	D10 Conveying						
1010	Elevators & Lifts	N/A		—	—	—	0.0%
1020	Escalators & Moving Walks	N/A		—	—	—	
	D20 Plumbing						
2010	Plumbing Fixtures	Kitchen, bathroom and service fixtures, supply and drainage	1 Fixture/1000 S.F. Floor	Each	2380	2.38	
2020	Domestic Water Distribution	Gas fired water heater		S.F. Floor	1.50	1.50	4.3%
2040	Rain Water Drainage	Roof drains		S.F. Roof	.62	.31	
	D30 HVAC						
3010	Energy Supply	N/A		—	—	—	
3020	Heat Generating Systems	Included in D3050		—	—	—	
3030	Cooling Generating Systems	N/A		—	—	—	24.1%
3050	Terminal & Package Units	Multizone unit, gas heating, electric cooling		S.F. Floor	23	23.75	
3090	Other HVAC Sys. & Equipment	N/A		—	—	—	
	D40 Fire Protection						
4010	Sprinklers	Sprinklers, light hazard		S.F. Floor	.31	.31	0.3%
4020	Standpipes	N/A		—	—	—	
	D50 Electrical						
5010	Electrical Service/Distribution	400 ampere service, panel board and feeders		S.F. Floor	.64	.64	
5020	Lighting & Branch Wiring	Fluorescent and high intensity discharge fixtures, receptacles, switches, A.C. and misc. power		S.F. Floor	10.18	10.18	11.3%
5030	Communications & Security	Alarm systems and emergency lighting		S.F. Floor	.25	.25	
5090	Other Electrical Systems	Emergency generator, 15 kW		S.F. Floor	.05	.05	
E.	**EQUIPMENT & FURNISHINGS**						
1010	Commercial Equipment	N/A		—	—	—	
1020	Institutional Equipment	N/A		—	—	—	12.8%
1030	Vehicular Equipment	N/A		—	—	—	
1090	Other Equipment	Courts, sauna baths		S.F. Floor	12.60	12.60	
F.	**SPECIAL CONSTRUCTION**						
1020	Integrated Construction	N/A		—	—	—	0.0%
1040	Special Facilities	N/A		—	—	—	
G.	**BUILDING SITEWORK**	**N/A**					

				Sub-Total	98.47	**100%**
CONTRACTOR FEES (General Requirements: 10%, Overhead: 5%, Profit: 10%)				25%	24.63	
ARCHITECT FEES				8%	9.85	

Total Building Cost		132.95

BUILDING TYPES

BUILDING TYPES

Costs per square foot of floor area

Exterior Wall	S.F. Area	5000	6000	7000	8000	9000	10000	11000	12000	13000
	L.F. Perimeter	286	320	353	386	397	425	454	460	486
Face Brick with Concrete Block Back-up	Steel Joists	136.80	133.45	130.95	129.05	126.10	124.80	123.75	121.75	121.00
	Wood Joists	132.75	129.25	126.70	124.70	121.70	120.35	119.25	117.15	116.30
Stucco on Concrete Block	Steel Joists	132.25	129.15	126.90	125.20	122.60	121.35	120.50	118.70	118.00
	Wood Joists	124.95	122.00	119.85	118.15	115.65	114.50	113.60	111.90	111.20
Limestone with Concrete Block Back-up	Steel Joists	158.95	154.10	150.50	147.80	143.25	141.25	139.75	136.60	135.50
	Wood Joists	151.70	147.00	143.40	140.75	136.35	134.35	132.95	129.85	128.70
Perimeter Adj., Add or Deduct	Per 100 L.F.	12.45	10.45	9.00	7.85	6.95	6.30	5.75	5.20	4.75
Story Hgt. Adj., Add or Deduct	Per 1 Ft.	2.15	1.95	1.90	1.90	1.65	1.60	1.60	1.45	1.40

For Basement, add $21.90 per square foot of basement area

The above costs were calculated using the basic specifications shown on the facing page. These costs should be adjusted where necessary for design alternatives and owner's requirements. Reported completed project costs, for this type of structure, range from $57.20 to $177.10 per S.F.

Common additives

Description	Unit	$ Cost
Carrels Hardwood	Each	785 - 1000
Emergency Lighting, 25 watt, battery operated		
Lead battery	Each	247
Nickel cadmium	Each	695
Flagpoles, Complete		
Aluminum, 20' high	Each	1375
40' high	Each	3025
70' high	Each	9500
Fiberglass, 23' high	Each	1650
39'-5" high	Each	3175
59' high	Each	7950
Gym Floor, Incl. sleepers and finish, maple	S.F.	11.10
Intercom System, 25 Station capacity		
Master station	Each	2200
Intercom outlets	Each	142
Handset	Each	390

Description	Unit	$ Cost
Lockers, Steel, single tier, 60" to 72"	Opening	154 - 241
2 tier, 60" to 72" total	Opening	104 - 134
5 tier, box lockers	Opening	55 - 69
Locker bench, lam. maple top only	L.F.	18.80
Pedestals, steel pipe	Each	57
Seating		
Auditorium chair, all veneer	Each	193
Veneer back, padded seat	Each	233
Upholstered, spring seat	Each	234
Classroom, movable chair & desk	Set	65 - 120
Lecture hall, pedestal type	Each	179 - 545
Smoke Detectors		
Ceiling type	Each	161
Duct type	Each	425
Sound System		
Amplifier, 250 watts	Each	1925
Speaker, ceiling or wall	Each	160
Trumpet	Each	305
Swimming Pools, Complete, gunite	S.F.	51 - 62

Model costs calculated for a 1 story building with 12' story height and 10,000 square feet of floor area

				Unit	Unit Cost	Cost Per S.F.	% Of Sub-Total
A. SUBSTRUCTURE							
1010	Standard Foundations	Poured concrete; strip and spread footings		S.F. Ground	1.84	1.84	
1030	Slab on Grade	4" reinforced concrete with vapor barrier and granular base		S.F. Slab	3.98	3.98	
2010	Basement Excavation	Site preparation for slab and trench for foundation wall and footing		S.F. Ground	.23	.23	9.4%
2020	Basement Walls	4' foundation wall		L.F. Wall	61	2.59	
B. SHELL							
	B10 Superstructure						
1010	Floor Construction	N/A		—	—	—	
1020	Roof Construction	Metal deck, open web steel joists, beams, interior columns		S.F. Roof	6.68	6.68	7.3%
	B20 Exterior Enclosure						
2010	Exterior Walls	Face brick with concrete block backup	85% of wall	S.F. Wall	23	10.30	
2020	Exterior Windows	Window wall	15% of wall	Each	37	2.84	16.1%
2030	Exterior Doors	Double aluminum and glass		Each	4075	1.65	
	B30 Roofing						
3010	Roof Coverings	Built-up tar and gravel with flashing; perlite/EPS composite insulation		S.F. Roof	4.35	4.35	4.7%
3020	Roof Openings	N/A		—	—	—	
C. INTERIORS							
1010	Partitions	Concrete block	8 S.F. Floor/S.F. Partition	S.F. Partition	6.74	8.43	
1020	Interior Doors	Single leaf hollow metal	700 S.F. Floor/Door	Each	608	.87	
1030	Fittings	Toilet partitions		S.F. Floor	1.21	1.21	
2010	Stair Construction	N/A		—	—	—	26.4%
3010	Wall Finishes	Paint		S.F. Surface	—	4.42	
3020	Floor Finishes	50% vinyl tile, 50% carpet		S.F. Floor	5.12	5.12	
3030	Ceiling Finishes	Mineral fiber tile on concealed zee bars		S.F. Ceiling	4.00	4.09	
D. SERVICES							
	D10 Conveying						
1010	Elevators & Lifts	N/A		—	—	—	0.0%
1020	Escalators & Moving Walks	N/A		—	—	—	
	D20 Plumbing						
2010	Plumbing Fixtures	Toilet and service fixtures, supply and drainage	1 Fixture/455 S.F. Floor	Each	1701	3.74	
2020	Domestic Water Distribution	Gas fired water heater		S.F. Floor	.84	.84	5.4%
2040	Rain Water Drainage	Roof drains		S.F. Roof	.41	.41	
	D30 HVAC						
3010	Energy Supply	Oil fired hot water, wall fin radiation		S.F. Floor	7.42	7.42	
3020	Heat Generating Systems	N/A		—	—	—	
3030	Cooling Generating Systems	N/A		—	—	—	18.7%
3050	Terminal & Package Units	Split systems with air cooled condensing units		S.F. Floor	9.74	9.74	
3090	Other HVAC Sys. & Equipment	N/A		—	—	—	
	D40 Fire Protection						
4010	Sprinklers	Sprinkler, light hazard		S.F. Floor	2.05	2.05	2.2%
4020	Standpipes	N/A		—	—	—	
	D50 Electrical						
5010	Electrical Service/Distribution	200 ampere service, panel board and feeders		S.F. Floor	.76	.76	
5020	Lighting & Branch Wiring	Fluorescent fixtures, receptacles, switches, A.C. and misc. power		S.F. Floor	7.61	7.61	9.6%
5030	Communications & Security	Alarm systems and emergency lighting		S.F. Floor	.34	.34	
5090	Other Electrical Systems	Emergency generator, 15 kW		S.F. Floor	.07	.07	
E. EQUIPMENT & FURNISHINGS							
1010	Commercial Equipment	N/A		—	—	—	
1020	Institutional Equipment	N/A		—	—	—	
1030	Vehicular Equipment	N/A		—	—	—	0.0%
1090	Other Equipment	N/A		—	—	—	
F. SPECIAL CONSTRUCTION							
1020	Integrated Construction	N/A		—	—	—	0.0%
1040	Special Facilities	N/A		—	—	—	
G. BUILDING SITEWORK	**N/A**						

	Sub-Total	91.58	100%
CONTRACTOR FEES (General Requirements: 10%, Overhead: 5%, Profit: 10%)	25%	22.92	
ARCHITECT FEES	9%	10.30	
Total Building Cost		**124.80**	

BUILDING TYPES

Costs per square foot of floor area

Exterior Wall	S.F. Area	2000	2800	3500	4200	5000	5800	6500	7200	8000
	L.F. Perimeter	180	212	240	268	300	314	336	344	368
Wood Siding	Wood Frame	187.35	174.25	167.75	163.40	159.95	156.05	154.05	151.70	150.20
Brick Veneer	Wood Frame	197.35	182.65	175.30	170.40	166.55	161.95	159.70	156.85	155.20
Face Brick with Concrete Block Back-up	Wood Joists	202.60	187.10	179.30	174.20	170.10	165.20	162.80	159.70	158.00
	Steel Joists	199.10	182.25	173.75	168.25	163.75	158.30	155.70	152.25	150.35
Stucco on Concrete Block	Wood Joists	192.70	178.70	171.75	167.15	163.45	159.25	157.10	154.50	152.95
	Steel Joists	182.20	168.25	161.30	156.65	152.95	148.75	146.60	144.00	142.45
Perimeter Adj., Add or Deduct	Per 100 L.F.	21.45	15.30	12.20	10.20	8.50	7.40	6.60	5.90	5.30
Story Hgt. Adj., Add or Deduct	Per 1 Ft.	2.25	1.85	1.65	1.50	1.45	1.30	1.25	1.20	1.15

For Basement, add $25.30 per square foot of basement area

The above costs were calculated using the basic specifications shown on the facing page. These costs should be adjusted where necessary for design alternatives and owner's requirements. Reported completed project costs, for this type of structure, range from $88.65 to $206.60 per S.F.

Common additives

Description	Unit	$ Cost
Bar, Front Bar	L.F.	320
Back bar	L.F.	257
Booth, Upholstered, custom straight	L.F.	166 - 305
"L" or "U" shaped	L.F.	172 - 290
Cupola, Stock unit, redwood		
30" square, 37" high, aluminum roof	Each	510
Copper roof	Each	277
Fiberglass, 5'-0" base, 63" high	Each	3100 - 3750
6'-0" base, 63" high	Each	4625 - 5100
Decorative Wood Beams, Non load bearing		
Rough sawn, 4" x 6"	L.F.	7.85
4" x 8"	L.F.	9.00
4" x 10"	L.F.	10.80
4" x 12"	L.F.	12.30
8" x 8"	L.F.	15.20
Emergency Lighting, 25 watt, battery operated		
Lead battery	Each	247
Nickel cadmium	Each	695

Description	Unit	$ Cost
Fireplace, Brick, not incl. chimney or foundation		
30" x 29" opening	Each	2325
Chimney, standard brick		
Single flue, 16" x 20"	V.L.F.	75
20" x 20"	V.L.F.	86
2 Flue, 20" x 24"	V.L.F.	108
20" x 32"	V.L.F.	129
Kitchen Equipment		
Broiler	Each	3075
Coffee urn, twin 6 gallon	Each	2925
Cooler, 6 ft. long	Each	4075
Dishwasher, 10-12 racks per hr.	Each	4300
Food warmer, counter, 1.2 KW	Each	665
Freezer, 44 C.F., reach-in	Each	9175
Ice cube maker, 50 lb. per day	Each	1600
Range with 1 oven	Each	2100
Refrigerators, Prefabricated, walk-in		
7'-6" high, 6' x 6'	S.F.	129
10' x 10'	S.F.	101
12' x 14'	S.F.	89
12' x 20'	S.F.	79

Important: See the Reference Section for Location Factors

Model costs calculated for a 1 story building with 12' story height and 5,000 square feet of floor area

Restaurant

				Unit	Unit Cost	Cost Per S.F.	% Of Sub-Total
A.	**SUBSTRUCTURE**						
1010	Standard Foundations	Poured concrete; strip and spread footings		S.F. Ground	1.99	1.99	
1030	Slab on Grade	4" reinforced concrete with vapor barrier and granular base		S.F. Slab	3.98	3.98	
2010	Basement Excavation	Site preparation for slab and trench for foundation wall and footing		S.F. Ground	.43	.43	8.4%
2020	Basement Walls	4' foundation wall		L.F. Wall	61	3.67	
B.	**SHELL**						
	B10 Superstructure						
1010	Floor Construction	Wood columns		S.F. Floor	.40	1.01	
1020	Roof Construction	Plywood on wood truss (pitched)		S.F. Roof	5.63	6.31	6.1%
	B20 Exterior Enclosure						
2010	Exterior Walls	Cedar siding on wood studs with insulation	70% of wall	S.F. Wall	8.67	4.37	
2020	Exterior Windows	Storefront windows	30% of wall	Each	33	7.17	12.7%
2030	Exterior Doors	Aluminum and glass doors and entrance with transom		Each	3591	3.61	
	B30 Roofing						
3010	Roof Coverings	Cedar shingles with flashing (pitched); rigid fiberglass insulation		S.F. Roof	13.45	5.88	5.0%
3020	Roof Openings	Skylights		S.F. Roof	.06	.06	
C.	**INTERIORS**						
1010	Partitions	Gypsum board on wood studs	25 S.F. Floor/L.F. Partition	S.F. Partition	7.88	3.15	
1020	Interior Doors	Hollow core wood	250 S.F. Floor/Door	Each	420	1.68	
1030	Fittings	Toilet partitions		S.F. Floor	.69	.69	
2010	Stair Construction	N/A		—	—	—	14.7%
3010	Wall Finishes	75% paint, 25% ceramic tile		S.F. Surface	3.12	1.56	
3020	Floor Finishes	65% carpet, 35% quarry tile		S.F. Floor	6.43	6.43	
3030	Ceiling Finishes	Mineral fiber tile on concealed zee bars		S.F. Ceiling	4.00	4.09	
D.	**SERVICES**						
	D10 Conveying						
1010	Elevators & Lifts	N/A		—	—	—	0.0%
1020	Escalators & Moving Walks	N/A		—	—	—	
	D20 Plumbing						
2010	Plumbing Fixtures	Kitchen, bathroom and service fixtures, supply and drainage	1 Fixture/355 S.F. Floor	Each	2009	5.66	
2020	Domestic Water Distribution	Gas fired water heater		S.F. Floor	4.00	4.00	8.6%
2040	Rain Water Drainage	Roof drains		S.F. Roof	.65	.65	
	D30 HVAC						
3010	Energy Supply	N/A		—	—	—	
3020	Heat Generating Systems	Included in D3050		—	—	—	
3030	Cooling Generating Systems	N/A		—	—	—	32.5%
3050	Terminal & Package Units	Multizone unit, gas heating, electric cooling		S.F. Floor	38	38.90	
3090	Other HVAC Sys. & Equipment	N/A		—	—	—	
	D40 Fire Protection						
4010	Sprinklers	Sprinklers, light hazard		S.F. Floor	2.05	2.05	1.7%
4020	Standpipes	N/A		—	—	—	
	D50 Electrical						
5010	Electrical Service/Distribution	400 ampere service, panel board and feeders		S.F. Floor	3.25	3.25	
5020	Lighting & Branch Wiring	Fluorescent fixtures, receptacles, switches, A.C. and misc. power		S.F. Floor	8.23	8.23	10.2%
5030	Communications & Security	Alarm systems and emergency lighting		S.F. Floor	.48	.48	
5090	Other Electrical Systems	Emergency generator, 15 kW		S.F. Floor	.26	.26	
E.	**EQUIPMENT & FURNISHINGS**						
1010	Commercial Equipment	N/A		—	—	—	
1020	Institutional Equipment	N/A		—	—	—	
1030	Vehicular Equipment	N/A		—	—	—	0.0%
1090	Other Equipment	N/A		—	—	—	
F.	**SPECIAL CONSTRUCTION**						
1020	Integrated Construction	N/A		—	—	—	0.0%
1040	Special Facilities	N/A		—	—	—	
G.	**BUILDING SITEWORK**	**N/A**					

				Sub-Total		119.56	100%
	CONTRACTOR FEES (General Requirements: 10%, Overhead: 5%, Profit: 10%)				25%	29.93	
	ARCHITECT FEES				7%	10.46	

		Total Building Cost	**159.95**

BUILDING TYPES

BUILDING TYPES

Costs per square foot of floor area

Exterior Wall	S.F. Area	2000	2800	3500	4000	5000	5800	6500	7200	8000
	L.F. Perimeter	180	212	240	260	300	314	336	344	368
Face Brick with Concrete Block Back-up	Bearing Walls	156.25	144.90	139.20	136.35	132.40	128.45	126.70	124.10	122.85
	Steel Frame	159.85	148.40	142.75	139.90	135.90	132.00	130.20	127.70	126.35
Concrete Block With Stucco	Bearing Walls	148.25	138.10	133.05	130.55	127.00	123.65	122.10	119.85	118.75
	Steel Frame	150.95	140.95	135.95	133.45	130.00	126.65	125.10	122.95	121.85
Wood Siding	Wood Frame	143.95	134.50	129.75	127.45	124.15	121.05	119.60	117.60	116.50
Brick Veneer	Steel Frame	155.20	144.50	139.15	136.55	132.80	129.15	127.50	125.20	124.00
Perimeter Adj., Add or Deduct	Per 100 L.F.	27.55	19.60	15.70	13.75	10.95	9.50	8.40	7.65	6.80
Story Hgt. Adj., Add or Deduct	Per 1 Ft.	3.60	3.00	2.70	2.60	2.40	2.15	2.05	1.95	1.80
Basement—Not Applicable										

The above costs were calculated using the basic specifications shown on the facing page. These costs should be adjusted where necessary for design alternatives and owner's requirements. Reported completed project costs, for this type of structure, range from $88.45 to $172.50 per S.F.

Common additives

Description	Unit	$ Cost
Bar, Front Bar	L.F.	320
Back bar	L.F.	257
Booth, Upholstered, custom straight	L.F.	166 - 305
"L" or "U" shaped	L.F.	172 - 290
Drive-up Window	Each	5300 - 10,600
Emergency Lighting, 25 watt, battery operated		
Lead battery	Each	247
Nickel cadmium	Each	695
Kitchen Equipment		
Broiler	Each	3075
Coffee urn, twin 6 gallon	Each	2925
Cooler, 6 ft. long	Each	4075
Dishwasher, 10-12 racks per hr.	Each	4300
Food warmer, counter, 1.2 KW	Each	665
Freezer, 44 C.F., reach-in	Each	9175
Ice cube maker, 50 lb. per day	Each	1600
Range with 1 oven	Each	2100

Description	Unit	$ Cost
Refrigerators, Prefabricated, walk-in		
7'-6" High, 6' x 6'	S.F.	129
10' x 10'	S.F.	101
12' x 14'	S.F.	89
12' x 20'	S.F.	79
Serving		
Counter top (Stainless steel)	L.F.	133
Base cabinets	L.F.	320 - 405
Sound System		
Amplifier, 250 watts	Each	1925
Speaker, ceiling or wall	Each	160
Trumpet	Each	305
Storage		
Shelving	S.F.	10.35
Washing		
Stainless steel counter	L.F.	133

Model costs calculated for a 1 story building with 10' story height and 4,000 square feet of floor area

				Unit	Unit Cost	Cost Per S.F.	% Of Sub-Total
A. SUBSTRUCTURE							
1010	Standard Foundations	Poured concrete; strip and spread footings		S.F. Ground	1.91	1.91	
1030	Slab on Grade	4" reinforced concrete with vapor barrier and granular base		S.F. Slab	3.98	3.98	
2010	Basement Excavation	Site preparation for slab and trench for foundation wall and footing		S.F. Ground	.43	.43	10.2%
2020	Basement Walls	4' foundation wall		L.F. Wall	61	3.97	
B. SHELL							
	B10 Superstructure						
1010	Floor Construction	N/A		—	—	—	
1020	Roof Construction	Metal deck on open web steel joists		S.F. Roof	5.58	5.58	5.5%
	B20 Exterior Enclosure						
2010	Exterior Walls	Face brick with concrete block backup	70% of wall	S.F. Wall	20	10.81	
2020	Exterior Windows	Window wall	30% of wall	Each	37	7.27	23.5%
2030	Exterior Doors	Aluminum and glass		Each	2803	5.61	
	B30 Roofing						
3010	Roof Coverings	Built-up tar and gravel with flashing; perlite/EPS composite insulation		S.F. Roof	4.66	4.66	
3020	Roof Openings	Roof hatches		S.F. Roof	.31	.31	4.9%
C. INTERIORS							
1010	Partitions	Gypsum board on metal studs	25 S.F. Floor/L.F. Partition	S.F. Partition	4.61	1.66	
1020	Interior Doors	Hollow core wood	1000 S.F. Floor/Door	Each	420	.42	
1030	Fittings	N/A		—	—	—	
2010	Stair Construction	N/A		—	—	—	17.6%
3010	Wall Finishes	Paint		S.F. Surface	3.12	1.56	
3020	Floor Finishes	Quarry tile		S.F. Floor	10.07	10.07	
3030	Ceiling Finishes	Mineral fiber tile on concealed zee bars		S.F. Ceiling	4.00	4.09	
D. SERVICES							
	D10 Conveying						
1010	Elevators & Lifts	N/A		—	—	—	
1020	Escalators & Moving Walks	N/A		—	—	—	0.0%
	D20 Plumbing						
2010	Plumbing Fixtures	Kitchen, bathroom and service fixtures, supply and drainage	1 Fixture/400 S.F. Floor	Each	2016	5.04	
2020	Domestic Water Distribution	Gas fired water heater		S.F. Floor	2.87	2.87	8.7%
2040	Rain Water Drainage	Roof drains		S.F. Roof	.91	.91	
	D30 HVAC						
3010	Energy Supply	N/A		—	—	—	
3020	Heat Generating Systems	Included in D3050		—	—	—	
3030	Cooling Generating Systems	N/A		—	—	—	9.3%
3050	Terminal & Package Units	Multizone unit, gas heating, electric cooling		S.F. Floor	9.37	9.37	
3090	Other HVAC Sys. & Equipment	N/A		—	—	—	
	D40 Fire Protection						
4010	Sprinklers	Sprinklers, light hazard		S.F. Floor	2.68	2.68	2.7%
4020	Standpipes	N/A		—	—	—	
	D50 Electrical						
5010	Electrical Service/Distribution	400 ampere service, panel board and feeders		S.F. Floor	3.37	3.37	
5020	Lighting & Branch Wiring	Fluorescent fixtures, receptacles, switches, A.C. and misc. power		S.F. Slab	9.00	9.00	12.9%
5030	Communications & Security	Alarm systems and emergency lighting		S.F. Floor	.40	.40	
5090	Other Electrical Systems	Emergency generator, 7.5 kW		S.F. Floor	.30	.30	
E. EQUIPMENT & FURNISHINGS							
1010	Commercial Equipment	N/A		—	—	—	
1020	Institutional Equipment	N/A		—	—	—	
1030	Vehicular Equipment	N/A		—	—	—	4.7%
1090	Other Equipment	Walk-in refrigerator		S.F. Floor	4.73	4.73	
F. SPECIAL CONSTRUCTION							
1020	Integrated Construction	N/A		—	—	—	
1040	Special Facilities	N/A		—	—	—	0.0%
G. BUILDING SITEWORK	N/A						

		Sub-Total	101.00	100%
CONTRACTOR FEES (General Requirements: 10%, Overhead: 5%, Profit: 10%)		25%	25.25	
ARCHITECT FEES		8%	10.10	
	Total Building Cost		**136.35**	

BUILDING TYPES

Costs per square foot of floor area

Exterior Wall	S.F. Area	10000	15000	20000	25000	30000	35000	40000	45000	50000
	L.F. Perimeter	450	500	600	700	740	822	890	920	966
Face Brick with Concrete Block Back-up	Steel Frame	151.65	137.10	132.55	129.75	125.80	124.25	122.70	120.60	119.25
	Lam. Wood Truss	152.70	138.80	134.35	131.65	127.90	126.45	124.90	122.90	121.65
Concrete Block	Steel Frame	151.20	142.50	139.60	137.85	135.55	134.55	133.65	132.45	131.70
	Lam. Wood Truss	134.65	126.00	123.05	121.25	119.00	118.05	117.10	115.90	115.15
Galvanized Steel Siding	Steel Frame	145.40	136.50	133.45	131.65	129.30	128.30	127.30	126.05	125.35
Metal Sandwich Panel	Steel Joists	152.00	141.35	137.85	135.75	132.90	131.75	130.60	129.10	128.10
Perimeter Adj., Add or Deduct	Per 100 L.F.	10.40	6.95	5.20	4.15	3.45	2.95	2.60	2.30	2.10
Story Hgt. Adj., Add or Deduct	Per 1 Ft.	1.60	1.25	1.05	1.00	.90	.80	.80	.70	.70
Basement—Not Applicable										

The above costs were calculated using the basic specifications shown on the facing page. These costs should be adjusted where necessary for design alternatives and owner's requirements. Reported completed project costs, for this type of structure, range from $52.35 to $162.80 per S.F.

Common additives

Description	Unit	$ Cost
Bar, Front Bar	L.F.	320
Back bar	L.F.	257
Booth, Upholstered, custom straight	L.F.	166 - 305
"L" or "U" shaped	L.F.	172 - 290
Bleachers, Telescoping, manual		
To 15 tier	Seat	101 - 141
16-20 tier	Seat	208 - 255
21-30 tier	Seat	221 - 267
For power operation, add	Seat	40 - 63
Emergency Lighting, 25 watt, battery operated		
Lead battery	Each	247
Nickel cadmium	Each	695
Lockers, Steel, single tier, 60" or 72"	Opening	154 - 241
2 tier, 60" or 72" total	Opening	104 - 134
5 tier, box lockers	Opening	55 - 69
Locker bench, lam. maple top only	L.F.	18.80
Pedestals, steel pipe	Each	57

Description	Unit	$ Cost
Rink		
Dasher boards & top guard	Each	155,500
Mats, rubber	S.F.	18.65
Score Board	Each	11,400 - 33,800

Important: See the Reference Section for Location Factors

				Unit	Unit Cost	Cost Per S.F.	% Of Sub-Total
A. SUBSTRUCTURE							
1010	Standard Foundations	Poured concrete; strip and spread footings		S.F. Ground	1.35	1.35	
1030	Slab on Grade	6" reinforced concrete with vapor barrier and granular base		S.F. Slab	4.84	4.84	
2010	Basement Excavation	Site preparation for slab and trench for foundation wall and footing		S.F. Ground	.14	.14	8.2%
2020	Basement Walls	4' foundation wall		L.F. Wall	115	1.95	
B. SHELL							
	B10 Superstructure						
1010	Floor Construction	Wide flange beams and columns		S.F. Floor	24	16.94	29.0%
1020	Roof Construction	Metal deck on steel joist		S.F. Roof	12.47	12.47	
	B20 Exterior Enclosure						
2010	Exterior Walls	Concrete block	95% of wall	S.F. Wall	8.98	5.05	
2020	Exterior Windows	Store front	5% of wall	Each	36	1.08	6.6%
2030	Exterior Doors	Aluminum and glass, hollow metal, overhead		Each	2251	.60	
	B30 Roofing						
3010	Roof Coverings	Elastomeric neoprene membrane with flashing; perlite/EPS composite insulation		S.F. Roof	3.43	3.43	3.5%
3020	Roof Openings	Roof hatches		S.F. Roof	.15	.15	
C. INTERIORS							
1010	Partitions	Concrete block	140 S.F. Floor/L.F. Partition	S.F. Partition	6.77	.58	
1020	Interior Doors	Hollow metal	2500 S.F. Floor/Door	Each	608	.25	
1030	Fittings	N/A		—	—	—	
2010	Stair Construction	N/A		—	—	—	5.4%
3010	Wall Finishes	Paint		S.F. Surface	17.49	1.59	
3020	Floor Finishes	80% rubber mat, 20% paint	50% of floor area	S.F. Floor	5.32	2.66	
3030	Ceiling Finishes	Mineral fiber tile on concealed zee bar	10% of area	S.F. Ceiling	4.09	.41	
D. SERVICES							
	D10 Conveying						
1010	Elevators & Lifts	N/A		—	—	—	0.0%
1020	Escalators & Moving Walks	N/A		—	—	—	
	D20 Plumbing						
2010	Plumbing Fixtures	Toilet and service fixtures, supply and drainage	1 Fixture/1070 S.F. Floor	Each	2589	2.42	
2020	Domestic Water Distribution	Oil fired water heater		S.F. Floor	1.75	1.75	4.1%
2040	Rain Water Drainage	N/A		—	—	—	
	D30 HVAC						
3010	Energy Supply	Oil fired hot water, unit heaters	10% of area	S.F. Floor	.58	.58	
3020	Heat Generating Systems	N/A		—	—	—	
3030	Cooling Generating Systems	N/A		—	—	—	11.5%
3050	Terminal & Package Units	Single zone, electric cooling	90% of area	S.F. Floor	11.03	11.03	
3090	Other HVAC Sys. & Equipment	N/A		—	—	—	
	D40 Fire Protection						
4010	Sprinklers	N/A		—	—	—	0.0%
4020	Standpipes	N/A		—	—	—	
	D50 Electrical						
5010	Electrical Service/Distribution	600 ampere service, panel board and feeders		S.F. Floor	1.35	1.35	
5020	Lighting & Branch Wiring	High intensity discharge and fluorescent fixtures, receptacles, switches, A.C. and misc. power		S.F. Floor	6.31	6.31	8.6%
5030	Communications & Security	Alarm systems, emergency lighting and public address		S.F. Floor	.90	.90	
5090	Other Electrical Systems	Emergency generator		S.F. Floor	.15	.15	
E. EQUIPMENT & FURNISHINGS							
1010	Commercial Equipment	N/A		—	—	—	
1020	Institutional Equipment	N/A		—	—	—	
1030	Vehicular Equipment	N/A		—	—	—	0.0%
1090	Other Equipment	N/A		—	—	—	
F. SPECIAL CONSTRUCTION							
1020	Integrated Construction	N/A		—	—	—	
1040	Special Facilities	Dasher boards and rink (including ice making system)		S.F. Floor	23	23.36	23.1%
G. BUILDING SITEWORK	**N/A**						

				Sub-Total	101.34	**100%**
	CONTRACTOR FEES (General Requirements: 10%, Overhead: 5%, Profit: 10%)			25%	25.34	
	ARCHITECT FEES			7%	8.87	
			Total Building Cost	135.55		

BUILDING TYPES

Costs per square foot of floor area

Exterior Wall	S.F. Area	25000	30000	35000	40000	45000	50000	55000	60000	65000
	L.F. Perimeter	700	740	823	906	922	994	1033	1100	1166
Face Brick with Concrete Block Back-up	Steel Frame	111.90	108.75	107.40	106.40	104.55	103.80	102.85	102.40	102.05
	Bearing Walls	106.90	103.70	102.35	101.40	99.50	98.80	97.80	97.30	96.95
Stucco on Concrete Block	Steel Frame	108.80	106.00	104.85	103.95	102.30	101.70	100.80	100.35	100.05
	Bearing Walls	103.80	101.00	99.75	98.85	97.20	96.60	95.75	95.25	94.95
Decorative Concrete Block	Steel Frame	109.10	106.30	105.10	104.20	102.50	101.90	101.00	100.50	100.20
	Bearing Walls	104.05	101.25	100.05	99.10	97.45	96.80	95.90	95.50	95.20
Perimeter Adj., Add or Deduct	Per 100 L.F.	2.90	2.50	2.10	1.85	1.65	1.50	1.35	1.20	1.00
Story Hgt. Adj., Add or Deduct	Per 1 Ft.	.95	.80	.85	.80	.70	.75	.65	.55	.55

For Basement, add $ 18.70 per square foot of basement area

The above costs were calculated using the basic specifications shown on the facing page. These costs should be adjusted where necessary for design alternatives and owner's requirements. Reported completed project costs, for this type of structure, range from $61.15 to $155.50 per S.F.

Common additives

Description	Unit	$ Cost
Bleachers, Telescoping, manual		
To 15 tier	Seat	101 - 141
16-20 tier	Seat	208 - 255
21-30 tier	Seat	221 - 267
For power operation, add	Seat	40 - 63
Carrels Hardwood	Each	785 - 1000
Clock System		
20 room	Each	14,200
50 room	Each	34,600
Emergency Lighting, 25 watt, battery operated		
Lead battery	Each	247
Nickel cadmium	Each	695
Flagpoles, Complete		
Aluminum, 20' high	Each	1375
40' high	Each	3025
Fiberglass, 23' high	Each	1650
39'-5" high	Each	3175
Kitchen Equipment		
Broiler	Each	3075
Cooler, 6 ft. long, reach-in	Each	4075

Description	Unit	$ Cost
Kitchen Equipment, cont.		
Dishwasher, 10-12 racks per hr.	Each	4300
Food warmer, counter, 1.2 KW	Each	665
Freezer, 44 C.F., reach-in	Each	9175
Ice cube maker, 50 lb. per day	Each	1600
Range with 1 oven	Each	2100
Lockers, Steel, single tier, 60" to 72"	Opening	154 - 241
2 tier, 60" to 72" total	Opening	104 - 134
5 tier, box lockers	Opening	55 - 69
Locker bench, lam. maple top only	L.F.	18.80
Pedestals, steel pipe	Each	57
Seating		
Auditorium chair, all veneer	Each	193
Veneer back, padded seat	Each	233
Upholstered, spring seat	Each	234
Classroom, movable chair & desk	Set	65 - 120
Lecture hall, pedestal type	Each	179 - 545
Sound System		
Amplifier, 250 watts	Each	1925
Speaker, ceiling or wall	Each	160
Trumpet	Each	305

Important: See the Reference Section for Location Factors

BUILDING TYPES

Model costs calculated for a 1 story building with 12' story height and 45,000 square feet of floor area

				Unit	Unit Cost	Cost Per S.F.	% Of Sub-Total

A. SUBSTRUCTURE

				Unit	Unit Cost	Cost Per S.F.	% Of Sub-Total
1010	Standard Foundations	Poured concrete; strip and spread footings		S.F. Ground	2.29	2.29	
1030	Slab on Grade	4" reinforced concrete with vapor barrier and granular base		S.F. Slab	3.98	3.98	11.5%
2010	Basement Excavation	Site preparation for slab and trench for foundation wall and footing		S.F. Ground	.14	.14	
2020	Basement Walls	4' foundation wall		L.F. Wall	61	2.13	

B. SHELL

B10 Superstructure

				Unit	Unit Cost	Cost Per S.F.	% Of Sub-Total
1010	Floor Construction	N/A		—	—	—	4.5%
1020	Roof Construction	Metal deck on open web steel joists		S.F. Roof	3.32	3.32	

B20 Exterior Enclosure

				Unit	Unit Cost	Cost Per S.F.	% Of Sub-Total
2010	Exterior Walls	Face brick with concrete block backup	70% of wall	S.F. Wall	23	4.09	
2020	Exterior Windows	Steel outward projecting	25% of wall	Each	559	1.80	8.5%
2030	Exterior Doors	Metal and glass	5% of wall	Each	2416	.43	

B30 Roofing

				Unit	Unit Cost	Cost Per S.F.	% Of Sub-Total
3010	Roof Coverings	Built-up tar and gravel with flashing; perlite/EPS composite insulation		S.F. Roof	3.87	3.87	5.2%
3020	Roof Openings	N/A		—	—	—	

C. INTERIORS

				Unit	Unit Cost	Cost Per S.F.	% Of Sub-Total
1010	Partitions	Concrete block	20 S.F. Floor/L.F. Partition	S.F. Partition	7.22	3.61	
1020	Interior Doors	Single leaf kalamein fire doors	700 S.F. Floor/Door	Each	608	.87	
1030	Fittings	Toilet partitions		S.F. Floor	1.68	1.68	
2010	Stair Construction	N/A		—	—	—	25.7%
3010	Wall Finishes	75% paint, 15% glazed coating, 10% ceramic tile		S.F. Surface	6.18	3.09	
3020	Floor Finishes	65% vinyl composition tile, 25% carpet, 10% terrazzo		S.F. Floor	5.79	5.79	
3030	Ceiling Finishes	Mineral fiber tile on concealed zee bars		S.F. Ceiling	4.00	4.09	

D. SERVICES

D10 Conveying

				Unit	Unit Cost	Cost Per S.F.	% Of Sub-Total
1010	Elevators & Lifts	N/A		—	—	—	0.0%
1020	Escalators & Moving Walks	N/A		—	—	—	

D20 Plumbing

				Unit	Unit Cost	Cost Per S.F.	% Of Sub-Total
2010	Plumbing Fixtures	Kitchen, bathroom and service fixtures, supply and drainage	1 Fixture/625 S.F. Floor	Each	2412	3.86	
2020	Domestic Water Distribution	Gas fired water heater		S.F. Floor	.34	.34	6.1%
2040	Rain Water Drainage	Roof drains		S.F. Roof	.33	.33	

D30 HVAC

				Unit	Unit Cost	Cost Per S.F.	% Of Sub-Total
3010	Energy Supply	Oil fired hot water, wall fin radiation		S.F. Floor	6.74	6.74	
3020	Heat Generating Systems	N/A		—	—	—	
3030	Cooling Generating Systems	N/A		—	—	—	21.3%
3050	Terminal & Package Units	Split systems with air cooled condensing units		S.F. Floor	9.14	9.14	
3090	Other HVAC Sys. & Equipment	N/A		—	—	—	

D40 Fire Protection

				Unit	Unit Cost	Cost Per S.F.	% Of Sub-Total
4010	Sprinklers	Sprinklers, light hazard		S.F. Floor	1.75	1.75	2.4%
4020	Standpipes	N/A		—	—	—	

D50 Electrical

				Unit	Unit Cost	Cost Per S.F.	% Of Sub-Total
5010	Electrical Service/Distribution	800 ampere service, panel board and feeders		S.F. Floor	1.00	1.00	
5020	Lighting & Branch Wiring	Fluorescent fixtures, receptacles, switches, A.C. and misc. power		S.F. Floor	8.44	8.44	14.7%
5030	Communications & Security	Alarm systems, communications systems and emergency lighting		S.F. Floor	1.42	1.42	
5090	Other Electrical Systems	Emergency generator, 15 kW		S.F. Floor	.07	.07	

E. EQUIPMENT & FURNISHINGS

				Unit	Unit Cost	Cost Per S.F.	% Of Sub-Total
1010	Commercial Equipment	N/A		—	—	—	
1020	Institutional Equipment	Chalkboards		S.F. Floor	.12	.12	0.2%
1030	Vehicular Equipment	N/A		—	—	—	
1090	Other Equipment	N/A		—	—	—	

F. SPECIAL CONSTRUCTION

				Unit	Unit Cost	Cost Per S.F.	% Of Sub-Total
1020	Integrated Construction	N/A		—	—	—	0.0%
1040	Special Facilities	N/A		—	—	—	

G. BUILDING SITEWORK   N/A

	Unit	Cost Per S.F.	%
Sub-Total		74.39	**100%**
CONTRACTOR FEES (General Requirements: 10%, Overhead: 5%, Profit: 10%)	25%	18.60	
ARCHITECT FEES	7%	6.51	
Total Building Cost		**99.50**	

BUILDING TYPES

Costs per square foot of floor area

Exterior Wall	S.F. Area	50000	70000	90000	110000	130000	150000	170000	190000	210000
	L.F. Perimeter	816	1083	1100	1300	1390	1450	1433	1566	1700
Face Brick with Concrete Block Back-up	Steel Frame	115.55	112.50	107.40	106.25	104.30	102.80	101.00	100.50	100.20
	R/Conc. Frame	118.80	115.95	110.95	109.85	108.05	106.50	104.75	104.30	104.00
Decorative Concrete Block	Steel Frame	111.35	108.70	104.40	103.40	101.80	100.50	99.00	98.55	98.30
	R/Conc. Frame	115.40	112.75	108.45	107.45	105.85	104.55	103.05	102.60	102.35
Limestone with Concrete Block Back-up	Steel Frame	124.25	120.85	114.00	112.70	110.25	108.10	105.65	105.10	104.65
	R/Conc. Frame	128.95	125.60	118.75	117.45	114.95	112.80	110.40	109.80	109.40
Perimeter Adj., Add or Deduct	Per 100 L.F.	2.45	1.75	1.35	1.05	1.00	.80	.70	.65	.60
Story Hgt. Adj., Add or Deduct	Per 1 Ft.	1.45	1.35	1.10	1.00	1.00	.85	.75	.75	.70

For Basement, add $24.45 per square foot of basement area

The above costs were calculated using the basic specifications shown on the facing page. These costs should be adjusted where necessary for design alternatives and owner's requirements. Reported completed project costs, for this type of structure, range from $71.05 to $167.30 per S.F.

Common additives

Description	Unit	$ Cost
Bleachers, Telescoping, manual		
To 15 tier	Seat	101 - 141
16-20 tier	Seat	208 - 255
21-30 tier	Seat	221 - 267
For power operation, add	Seat	40 - 63
Carrels Hardwood	Each	785 - 1000
Clock System		
20 room	Each	14,200
50 room	Each	34,600
Elevators, Hydraulic passenger, 2 stops		
1500# capacity	Each	47,900
2500# capacity	Each	50,000
Emergency Lighting, 25 watt, battery operated		
Lead battery	Each	247
Nickel cadmium	Each	695
Flagpoles, Complete		
Aluminum, 20' high	Each	1375
40' high	Each	3025
Fiberglass, 23' high	Each	1650
39'-5" high	Each	3175

Description	Unit	$ Cost
Kitchen Equipment		
Broiler	Each	3075
Cooler, 6 ft. long, reach-in	Each	4075
Dishwasher, 10-12 racks per hr.	Each	4300
Food warmer, counter, 1.2 KW	Each	665
Freezer, 44 C.F., reach-in	Each	9175
Lockers, Steel, single tier, 60" or 72"	Opening	154 - 241
2 tier, 60" or 72" total	Opening	104 - 134
5 tier, box lockers	Opening	55 - 69
Locker bench, lam. maple top only	L.F.	18.80
Pedestals, steel pipe	Each	57
Seating		
Auditorium chair, all veneer	Each	193
Veneer back, padded seat	Each	233
Upholstered, spring seat	Each	234
Classroom, movable chair & desk	Set	65 - 120
Lecture hall, pedestal type	Each	179 - 545
Sound System		
Amplifier, 250 watts	Each	1925
Speaker, ceiling or wall	Each	160
Trumpet	Each	305

Important: See the Reference Section for Location Factors

Model costs calculated for a 2 story building with 12' story height and 130,000 square feet of floor area

				Unit	Unit Cost	Cost Per S.F.	% Of Sub-Total
A. SUBSTRUCTURE							
1010	Standard Foundations	Poured concrete; strip and spread footings		S.F. Ground	1.70	.85	
1030	Slab on Grade	4" reinforced concrete with vapor barrier and granular base		S.F. Slab	3.98	1.99	
2010	Basement Excavation	Site preparation for slab and trench for foundation wall and footing		S.F. Ground	.14	.07	4.4%
2020	Basement Walls	4' foundation wall		L.F. Wall	61	.65	
B. SHELL							
	B10 Superstructure						
1010	Floor Construction	Concrete slab without drop panel, concrete columns		S.F. Floor	17.30	8.65	18.3%
1020	Roof Construction	Concrete slab without drop panel		S.F. Roof	12.26	6.13	
	B20 Exterior Enclosure						
2010	Exterior Walls	Face brick with concrete block backup	75% of wall	S.F. Wall	23	4.57	
2020	Exterior Windows	Window wall	25% of wall	Each	48	3.13	9.9%
2030	Exterior Doors	Metal and glass		Each	1441	.33	
	B30 Roofing						
3010	Roof Coverings	Built-up tar and gravel with flashing; perlite/EPS composite insulation		S.F. Roof	3.82	1.91	2.4%
3020	Roof Openings	Roof hatches		S.F. Roof	.06	.03	
C. INTERIORS							
1010	Partitions	Concrete block	25 S.F. Floor/L.F. Partition	S.F. Partition	8.38	3.35	
1020	Interior Doors	Single leaf kalamein fire doors	700 S.F. Floor/Door	Each	608	.87	
1030	Fittings	Toilet partitions, chalkboards		S.F. Floor	1.19	1.19	
2010	Stair Construction	Concrete filled metal pan		Flight	6425	.30	21.7%
3010	Wall Finishes	75% paint, 15% glazed coating, 10% ceramic tile		S.F. Surface	4.26	2.13	
3020	Floor Finishes	70% vinyl composition tile, 20% carpet, 10% terrazzo		S.F. Floor	5.59	5.59	
3030	Ceiling Finishes	Mineral fiber tile on concealed zee bars		S.F. Ceiling	4.00	4.09	
D. SERVICES							
	D10 Conveying						
1010	Elevators & Lifts	One hydraulic passenger elevator		Each	59,800	.46	0.6%
1020	Escalators & Moving Walks	N/A		—	—	—	
	D20 Plumbing						
2010	Plumbing Fixtures	Kitchen, bathroom and service fixtures, supply and drainage	1 Fixture/860 S.F. Floor	Each	2743	3.19	
2020	Domestic Water Distribution	Gas fired water heater		S.F. Floor	.40	.40	5.0%
2040	Rain Water Drainage	Roof drains		S.F. Roof	.90	.45	
	D30 HVAC						
3010	Energy Supply	Oil fired hot water, wall fin radiation		S.F. Floor	3.44	3.44	
3020	Heat Generating Systems	N/A		—	—	—	
3030	Cooling Generating Systems	Chilled water, cooling tower systems		S.F. Floor	12.77	12.77	20.1%
3050	Terminal & Package Units	N/A		—	—	—	
3090	Other HVAC Sys. & Equipment	N/A		—	—	—	
	D40 Fire Protection						
4010	Sprinklers	Sprinklers, light hazard		S.F. Floor	1.54	1.54	1.9%
4020	Standpipes	N/A		—	—	—	
	D50 Electrical						
5010	Electrical Service/Distribution	2000 ampere service, panel board and feeders		S.F. Floor	1.00	1.00	
5020	Lighting & Branch Wiring	Fluorescent fixtures, receptacles, switches, A.C. and misc. power		S.F. Floor	8.37	8.37	15.5%
5030	Communications & Security	Alarm systems, communications systems and emergency lighting		S.F. Floor	2.72	2.72	
5090	Other Electrical Systems	Emergency generator, 250 kW		S.F. Floor	.43	.43	
E. EQUIPMENT & FURNISHINGS							
1010	Commercial Equipment	N/A		—	—	—	
1020	Institutional Equipment	Laboratory counters		S.F. Floor	.08	.08	0.2%
1030	Vehicular Equipment	N/A		—	—	—	
1090	Other Equipment	Built-in athletic equipment		S.F. Floor	.10	.10	
F. SPECIAL CONSTRUCTION							
1020	Integrated Construction	N/A		—	—	—	0.0%
1040	Special Facilities	N/A		—	—	—	
G. BUILDING SITEWORK	**N/A**						

		Sub-Total	80.78	100%
CONTRACTOR FEES (General Requirements: 10%, Overhead: 5%, Profit: 10%)		25%	20.20	
ARCHITECT FEES		7%	7.07	
	Total Building Cost		**108.05**	

BUILDING TYPES

Costs per square foot of floor area

Exterior Wall	S.F. Area	50000	65000	80000	95000	110000	125000	140000	155000	170000
	L.F. Perimeter	816	1016	1100	1150	1250	1400	1450	1500	1600
Face Brick with Concrete Block Back-up	Steel Frame	115.70	114.50	112.20	110.25	109.30	109.00	108.05	107.20	106.90
	Bearing Walls	110.40	109.20	106.90	104.95	104.05	103.75	102.75	101.95	101.60
Concrete Block Stucco Face	Steel Frame	111.60	110.55	108.75	107.20	106.45	106.20	105.45	104.85	104.50
	Bearing Walls	106.35	105.30	103.50	101.95	101.20	100.95	100.20	99.60	99.25
Decorative Concrete Block	Steel Frame	112.05	111.00	109.15	107.50	106.75	106.50	105.70	105.00	104.75
	Bearing Walls	106.35	105.30	103.40	101.80	101.00	100.80	100.00	99.30	99.05
Perimeter Adj., Add or Deduct	Per 100 L.F.	2.10	1.65	1.35	1.10	.95	.90	.75	.75	.60
Story Hgt. Adj., Add or Deduct	Per 1 Ft.	1.20	1.20	1.10	.90	.85	.90	.80	.80	.70

For Basement, add $25.50 per square foot of basement area

The above costs were calculated using the basic specifications shown on the facing page. These costs should be adjusted where necessary for design alternatives and owner's requirements. Reported completed project costs, for this type of structure, range from $67.10 to $157.55 per S.F.

Common additives

Description	Unit	$ Cost
Bleachers, Telescoping, manual		
To 15 tier	Seat	101 - 141
16-20 tier	Seat	208 - 255
21-30 tier	Seat	221 - 267
For power operation, add	Seat	40 - 63
Carrels Hardwood	Each	785 - 1000
Clock System		
20 room	Each	14,200
50 room	Each	34,600
Elevators, Hydraulic passenger, 2 stops		
1500# capacity	Each	47,900
2500# capacity	Each	50,000
Emergency Lighting, 25 watt, battery operated		
Lead battery	Each	247
Nickel cadmium	Each	695
Flagpoles, Complete		
Aluminum, 20' high	Each	1375
40' high	Each	3025
Fiberglass, 23' high	Each	1650
39'-5" high	Each	3175

Description	Unit	$ Cost
Kitchen Equipment		
Broiler	Each	3075
Cooler, 6 ft. long, reach-in	Each	4075
Dishwasher, 10-12 racks per hr.	Each	4300
Food warmer, counter, 1.2 KW	Each	665
Freezer, 44 C.F., reach-in	Each	9175
Lockers, Steel, single tier, 60" to 72"	Opening	154 - 241
2 tier, 60" to 72" total	Opening	104 - 134
5 tier, box lockers	Opening	55 - 69
Locker bench, lam. maple top only	L.F.	18.80
Pedestals, steel pipe	Each	57
Seating		
Auditorium chair, all veneer	Each	193
Veneer back, padded seat	Each	233
Upholstered, spring seat	Each	234
Classroom, movable chair & desk	Set	65 - 120
Lecture hall, pedestal type	Each	179 - 545
Sound System		
Amplifier, 250 watts	Each	1925
Speaker, ceiling or wall	Each	160
Trumpet	Each	305

Important: See the Reference Section for Location Factors

Model costs calculated for a 2 story building with 12' story height and 110,000 square feet of floor area

				Unit	Unit Cost	Cost Per S.F.	% Of Sub-Total
A. SUBSTRUCTURE							
1010	Standard Foundations	Poured concrete; strip and spread footings		S.F. Ground	1.56	.78	
1030	Slab on Grade	4" reinforced concrete with vapor barrier and granular base		S.F. Slab	3.98	1.99	
2010	Basement Excavation	Site preparation for slab and trench for foundation wall and footing		S.F. Ground	.14	.07	4.3%
2020	Basement Walls	4' foundation wall		L.F. Wall	61	.69	
B. SHELL							
	B10 Superstructure						
1010	Floor Construction	Open web steel joists, slab form, concrete, columns		S.F. Floor	20	10.31	17.8%
1020	Roof Construction	Metal deck, open web steel joists, columns		S.F. Roof	8.50	4.25	
	B20 Exterior Enclosure						
2010	Exterior Walls	Face brick with concrete block backup	75% of wall	S.F. Wall	23	4.86	
2020	Exterior Windows	Window wall	25% of wall	Each	37	2.54	9.6%
2030	Exterior Doors	Double aluminum & glass		Each	1585	.41	
	B30 Roofing						
3010	Roof Coverings	Built-up tar and gravel with flashing; perlite/EPS composite insulation		S.F. Roof	3.96	1.98	2.4%
3020	Roof Openings	Roof hatches		S.F. Roof	.02	.01	
C. INTERIORS							
1010	Partitions	Concrete block	20 S.F. Floor/L.F. Partition	S.F. Partition	6.74	3.37	
1020	Interior Doors	Single leaf kalamein fire doors	750 S.F. Floor/Door	Each	608	.81	
1030	Fittings	Toilet partitions, chalkboards		S.F. Floor	1.27	1.27	
2010	Stair Construction	Concrete filled metal pan		Flight	6425	.35	25.3%
3010	Wall Finishes	50% paint, 40% glazed coatings, 10% ceramic tile		S.F. Surface	6.04	3.02	
3020	Floor Finishes	50% vinyl composition tile, 30% carpet, 20% terrrazzo		S.F. Floor	7.75	7.75	
3030	Ceiling Finishes	Mineral fiberboard on concealed zee bars		S.F. Ceiling	4.00	4.09	
D. SERVICES							
	D10 Conveying						
1010	Elevators & Lifts	One hydraulic passenger elevator		Each	59,400	.54	0.7%
1020	Escalators & Moving Walks	N/A		—	—	—	
	D20 Plumbing						
2010	Plumbing Fixtures	Kitchen, toilet and service fixtures, supply and drainage	1 Fixture/1170 S.F. Floor	Each	2737	2.34	
2020	Domestic Water Distribution	Gas fired water heater		S.F. Floor	.34	.34	3.6%
2040	Rain Water Drainage	Roof drains		S.F. Roof	.52	.26	
	D30 HVAC						
3010	Energy Supply	N/A		—	—	—	
3020	Heat Generating Systems	Included in D3050		—	—	—	
3030	Cooling Generating Systems	N/A		—	—	—	20.3%
3050	Terminal & Package Units	Multizone unit, gas heating, electric cooling		S.F. Floor	16.60	16.60	
3090	Other HVAC Sys. & Equipment	N/A		—	—	—	
	D40 Fire Protection						
4010	Sprinklers	Sprinklers, light hazard	10% of area	S.F. Floor	.31	.31	0.4%
4020	Standpipes	N/A		—	—	—	
	D50 Electrical						
5010	Electrical Service/Distribution	1600 ampere service, panel board and feeders		S.F. Floor	.97	.97	
5020	Lighting & Branch Wiring	Fluorescent fixtures, receptacles, switches, A.C. and misc. power		S.F. Floor	8.47	8.47	15.5%
5030	Communications & Security	Alarm systems, communications systems and emergency lighting		S.F. Floor	2.91	2.91	
5090	Other Electrical Systems	Emergency generator, 100 kW		S.F. Floor	.33	.33	
E. EQUIPMENT & FURNISHINGS							
1010	Commercial Equipment	N/A		—	—	—	
1020	Institutional Equipment	Laboratory counters		S.F. Floor	.10	.10	0.1%
1030	Vehicular Equipment	N/A		—	—	—	
1090	Other Equipment	N/A		—	—	—	
F. SPECIAL CONSTRUCTION							
1020	Integrated Construction	N/A		—	—	—	0.0%
1040	Special Facilities	N/A		—	—	—	
G. BUILDING SITEWORK	**N/A**						

			Sub-Total	81.72	100%
CONTRACTOR FEES (General Requirements: 10%, Overhead: 5%, Profit: 10%)		25%	20.43		
ARCHITECT FEES		7%	7.15		

Total Building Cost	**109.30**

BUILDING TYPES

Costs per square foot of floor area

Exterior Wall	S.F. Area	20000	30000	40000	50000	60000	70000	80000	90000	100000
	L.F. Perimeter	**400**	**500**	**685**	**700**	**800**	**900**	**1000**	**1100**	**1200**
Face Brick with Concrete Block Back-up	Steel Frame	118.80	114.50	114.60	111.10	110.20	109.55	109.15	108.75	108.45
	Bearing Walls	115.10	110.80	110.85	107.35	106.50	105.85	105.40	105.00	104.75
Decorative Concrete Block	Steel Frame	112.40	109.15	109.10	106.55	105.90	105.45	105.10	104.80	104.55
	Bearing Walls	108.65	105.40	105.35	102.80	102.15	101.70	101.35	101.05	100.85
Steel Siding on Steel Studs	Steel Frame	109.60	106.80	106.70	104.65	104.05	103.60	103.35	103.10	102.95
Metal Sandwich Panel	Steel Frame	109.15	106.50	106.35	104.35	103.75	103.40	103.10	102.90	102.70
Perimeter Adj., Add or Deduct	Per 100 L.F.	5.30	3.55	2.65	2.10	1.70	1.50	1.30	1.20	1.05
Story Hgt. Adj., Add or Deduct	Per 1 Ft.	1.50	1.30	1.30	1.05	1.05	.95	.90	.95	.90
For Basement, add $24.75 per square foot of basement area										

The above costs were calculated using the basic specifications shown on the facing page. These costs should be adjusted where necessary for design alternatives and owner's requirements. Reported completed project costs, for this type of structure, range from $60.20 to $171.95 per S.F.

Common additives

Description	Unit	$ Cost		Description	Unit	$ Cost
Carrels Hardwood	Each	785 - 1000		Flagpoles, Complete		
Clock System				Aluminum, 20' high	Each	1375
20 room	Each	14,200		40' high	Each	3025
50 room	Each	34,600		Fiberglass, 23' high	Each	1650
Directory Boards, Plastic, glass covered				39'-5" high	Each	3175
30" x 20"	Each	575		Seating		
36" x 48"	Each	1275		Auditorium chair, all veneer	Each	193
Aluminum, 24" x 18"	Each	525		Veneer back, padded seat	Each	233
36" x 24"	Each	600		Upholstered, spring seat	Each	234
48" x 32"	Each	810		Classroom, movable chair & desk	Set	65 - 120
48" x 60"	Each	1875		Lecture hall, pedestal type	Each	179 - 545
Elevators, Hydraulic passenger, 2 stops				Shops & Workroom:		
1500# capacity	Each	47,900		Benches, metal	Each	575
2500# capacity	Each	50,000		Parts bins 6'-3" high, 3' wide, 12" deep, 72 bins	Each	555
3500# capacity	Each	53,700		Shelving, metal 1' x 3'	S.F.	10.25
Emergency Lighting, 25 watt, battery operated				Wide span 6' wide x 24" deep	S.F.	10.35
Lead battery	Each	247		Sound System		
Nickel cadmium	Each	695		Amplifier, 250 watts	Each	1925
				Speaker, ceiling or wall	Each	160
				Trumpet	Each	305

Model costs calculated for a 2 story building with 12' story height and 40,000 square feet of floor area

					Unit	Unit Cost	Cost Per S.F.	% Of Sub-Total
A. SUBSTRUCTURE								
1010	Standard Foundations	Poured concrete; strip and spread footings			S.F. Ground	2.06	1.03	
1030	Slab on Grade	5" reinforced concrete with vapor barrier and granular base			S.F. Slab	9.45	4.73	
2010	Basement Excavation	Site preparation for slab and trench for foundation wall and footing			S.F. Ground	.14	.07	8.3%
2020	Basement Walls	4' foundation wall			L.F. Wall	61	1.05	
B. SHELL								
	B10 Superstructure							
1010	Floor Construction	Open web steel joists, slab form, concrete, beams, columns			S.F. Floor	13.70	6.85	11.2%
1020	Roof Construction	Metal deck, open web steel joists, beams, columns			S.F. Roof	4.82	2.41	
	B20 Exterior Enclosure							
2010	Exterior Walls	Face brick with concrete block backup	85% of wall		S.F. Wall	23	8.30	
2020	Exterior Windows	Steel outward projecting	15% of wall		Each	37	2.30	13.3%
2030	Exterior Doors	Metal and glass doors			Each	1979	.39	
	B30 Roofing							
3010	Roof Coverings	Built-up tar and gravel with flashing; perlite/EPS composite insulation			S.F. Roof	4.28	2.14	2.7%
3020	Roof Openings	Roof hatches			S.F. Roof	.20	.10	
C. INTERIORS								
1010	Partitions	Concrete block	20 S.F. Floor/L.F. Partition		S.F. Partition	6.74	3.37	
1020	Interior Doors	Single leaf kalamein fire doors	600 S.F. Floor/Door		Each	608	1.01	
1030	Fittings	Toilet partitions, chalkboards			S.F. Floor	1.42	1.42	
2010	Stair Construction	Concrete filled metal pan			Flight	6425	.64	21.5%
3010	Wall Finishes	50% paint, 40% glazed coating, 10% ceramic tile			S.F. Surface	6.72	3.36	
3020	Floor Finishes	70% vinyl composition tile, 20% carpet, 10% terrazzo			S.F. Floor	5.59	5.59	
3030	Ceiling Finishes	Mineral fiber tile on concealed zee bars	60% of area		S.F. Ceiling	4.09	2.45	
D. SERVICES								
	D10 Conveying							
1010	Elevators & Lifts	One hydraulic passenger elevator			Each	59,200	1.48	1.8%
1020	Escalators & Moving Walks	N/A			—	—	—	
	D20 Plumbing							
2010	Plumbing Fixtures	Toilet and service fixtures, supply and drainage	1 Fixture/700 S.F. Floor		Each	2282	3.26	
2020	Domestic Water Distribution	Gas fired water heater			S.F. Floor	.54	.54	5.0%
2040	Rain Water Drainage	Roof drains			S.F. Roof	.72	.36	
	D30 HVAC							
3010	Energy Supply	Oil fired hot water, wall fin radiation			S.F. Floor	6.36	6.36	
3020	Heat Generating Systems	N/A			—	—	—	
3030	Cooling Generating Systems	Chilled water, cooling tower systems			S.F. Floor	11.03	11.03	21.0%
3050	Terminal & Package Units	N/A			—	—	—	
3090	Other HVAC Sys. & Equipment	N/A			—	—	—	
	D40 Fire Protection							
4010	Sprinklers	Sprinklers, light hazard			S.F. Floor	1.04	1.04	1.3%
4020	Standpipes	N/A			—	—	—	
	D50 Electrical							
5010	Electrical Service/Distribution	800 ampere service, panel board and feeders			S.F. Floor	1.14	1.14	
5020	Lighting & Branch Wiring	Fluorescent fixtures, receptacles, switches, A.C. and misc. power			S.F. Floor	8.51	8.51	13.9%
5030	Communications & Security	Alarm systems, communications systems and emergency lighting			S.F. Floor	1.79	1.79	
5090	Other Electrical Systems	Emergency generator, 11/5 kW			S.F. Floor	.08	.08	
E. EQUIPMENT & FURNISHINGS								
1010	Commercial Equipment	N/A			—	—	—	
1020	Institutional Equipment	Stainless steel countertops			S.F. Floor	.10	.10	0.1%
1030	Vehicular Equipment	N/A			—	—	—	
1090	Other Equipment	N/A			—	—	—	
F. SPECIAL CONSTRUCTION								
1020	Integrated Construction	N/A			—	—	—	0.0%
1040	Special Facilities	N/A			—	—	—	
G. BUILDING SITEWORK	**N/A**							

			Sub-Total	82.90	100%
CONTRACTOR FEES (General Requirements: 10%, Overhead: 5%, Profit: 10%)			25%	20.70	
ARCHITECT FEES			7%	7.25	

Total Building Cost	**110.85**

BUILDING TYPES

Costs per square foot of floor area

Exterior Wall	S.F. Area	1000	2000	3000	4000	6000	8000	10000	12000	15000
	L.F. Perimeter	126	179	219	253	310	358	400	438	490
Wood Siding	Wood Frame	102.10	91.40	86.65	83.70	80.35	78.35	76.95	75.90	74.80
Face Brick Veneer	Wood Frame	123.90	106.55	98.75	94.10	88.60	85.35	83.15	81.40	79.65
Stucco on Concrete Block	Steel Frame	119.20	104.45	97.70	93.70	89.00	86.25	84.30	82.90	81.35
	Bearing Walls	116.25	101.45	94.80	90.80	86.10	83.30	81.35	79.95	78.40
Metal Sandwich Panel	Steel Frame	126.65	109.85	102.30	97.70	92.40	89.20	87.05	85.45	83.70
Precast Concrete	Steel Frame	143.65	121.55	111.60	105.65	98.65	94.50	91.60	89.50	87.15
Perimeter Adj., Add or Deduct	Per 100 L.F.	29.25	14.65	9.65	7.35	4.85	3.60	2.90	2.50	1.95
Story Hgt. Adj., Add or Deduct	Per 1 Ft.	2.10	1.55	1.10	1.10	.85	.70	.65	.65	.50
For Basement, add $18.65 per square foot of basement area										

The above costs were calculated using the basic specifications shown on the facing page. These costs should be adjusted where necessary for design alternatives and owner's requirements. Reported completed project costs, for this type of structure, range from $49.85 to $172.80 per S.F.

Common additives

Description	Unit	$ Cost	Description	Unit	$ Cost
Check Out Counter			Refrigerated Food Cases, cont.		
Single belt	Each	2475	Frozen food, chest type, 12' long	Each	6375
Double belt	Each	4125	Glass door reach-in, 5 door	Each	12,000
Emergency Lighting, 25 watt, battery operated			Island case 12' long, single deck	Each	7175
Lead battery	Each	247	Multi deck	Each	14,900
Nickel cadmium	Each	695	Meat cases, 12' long, single deck	Each	5275
Refrigerators, Prefabricated, walk-in			Multi deck	Each	8875
7'-6" high, 6' x 6'	S.F.	129	Produce, 12' long single deck	Each	6900
10' x 10'	S.F.	101	Multi deck	Each	7650
12' x 14'	S.F.	89	Safe, Office type, 4 hour rating		
12' x 20'	S.F.	79	30" x 18" x 18"	Each	3425
Refrigerated Food Cases			62" x 33" x 20"	Each	7450
Dairy, multi deck, 12' long	Each	9300	Smoke Detectors		
Delicatessen case, single deck, 12' long	Each	6275	Ceiling type	Each	161
Multi deck, 18 S.F. shelf display	Each	7725	Duct type	Each	425
Freezer, self-contained chest type, 30 C.F.	Each	4650	Sound System		
Glass door upright, 78 C.F.	Each	8700	Amplifier, 250 watts	Each	1925
			Speaker, ceiling or wall	Each	160
			Trumpet	Each	305

Model costs calculated for a 1 story building with 12' story height and 4,000 square feet of floor area

				Unit	Unit Cost	Cost Per S.F.	% Of Sub-Total
A. SUBSTRUCTURE							
1010	Standard Foundations	Poured concrete; strip and spread footings		S.F. Ground	1.24	1.24	
1030	Slab on Grade	4" reinforced concrete with vapor barrier and granular base		S.F. Slab	3.98	3.98	14.5%
2010	Basement Excavation	Site preparation for slab and trench for foundation wall and footing		S.F. Ground	.43	.43	
2020	Basement Walls	4' foundation wall		L.F. Wall	54	3.43	
B. SHELL							
	B10 Superstructure						
1010	Floor Construction	N/A		—	—	—	9.0%
1020	Roof Construction	Wood truss with plywood sheathing		S.F. Roof	5.64	5.64	
	B20 Exterior Enclosure						
2010	Exterior Walls	Wood siding on wood studs, insulated	80% of wall	S.F. Wall	7.66	4.65	
2020	Exterior Windows	Storefront	20% of wall	Each	35	3.63	16.0%
2030	Exterior Doors	Double aluminum and glass, solid core wood		Each	2279	1.71	
	B30 Roofing						
3010	Roof Coverings	Asphalt shingles; rigid fiberglass insulation		S.F. Roof	3.19	3.19	5.1%
3020	Roof Openings	N/A		—	—	—	
C. INTERIORS							
1010	Partitions	Gypsum board on wood studs	60 S.F. Floor/L.F. Partition	S.F. Partition	10.08	1.68	
1020	Interior Doors	Single leaf wood, hollow metal	1300 S.F. Floor/Door	Each	755	.59	
1030	Fittings	N/A		—	—	—	
2010	Stair Construction	N/A		—	—	—	14.2%
3010	Wall Finishes	Paint		S.F. Surface	1.80	.30	
3020	Floor Finishes	Vinyl composition tile		S.F. Floor	2.98	2.98	
3030	Ceiling Finishes	Mineral fiber tile on wood furring		S.F. Ceiling	3.00	3.31	
D. SERVICES							
	D10 Conveying						
1010	Elevators & Lifts	N/A		—	—	—	0.0%
1020	Escalators & Moving Walks	N/A		—	—	—	
	D20 Plumbing						
2010	Plumbing Fixtures	Toilet and service fixtures, supply and drainage	1 Fixture/1000 S.F. Floor	Each	1630	1.63	
2020	Domestic Water Distribution	Gas fired water heater		S.F. Floor	.79	.79	3.9%
2040	Rain Water Drainage	N/A		—	—	—	
	D30 HVAC						
3010	Energy Supply	N/A		—	—	—	
3020	Heat Generating Systems	Included in D3050		—	—	—	
3030	Cooling Generating Systems	N/A		—	—	—	10.3%
3050	Terminal & Package Units	Single zone rooftop unit, gas heating, electric cooling		S.F. Floor	6.45	6.45	
3090	Other HVAC Sys. & Equipment	N/A		—	—	—	
	D40 Fire Protection						
4010	Sprinklers	Sprinkler, ordinary hazard		S.F. Floor	2.68	2.68	4.3%
4020	Standpipes	N/A		—	—	—	
	D50 Electrical						
5010	Electrical Service/Distribution	200 ampere service, panel board and feeders		S.F. Floor	2.09	2.09	
5020	Lighting & Branch Wiring	Fluorescent fixtures, receptacles, switches, A.C. and misc. power		S.F. Floor	8.56	8.56	22.8%
5030	Communications & Security	Alarm systems and emergency lighting		S.F. Floor	3.36	3.36	
5090	Other Electrical Systems	Emergency generator, 7.5 kW		S.F. Floor	.27	.27	
E. EQUIPMENT & FURNISHINGS							
1010	Commercial Equipment	N/A		—	—	—	
1020	Institutional Equipment	N/A		—	—	—	0.0%
1030	Vehicular Equipment	N/A		—	—	—	
1090	Other Equipment	N/A		—	—	—	
F. SPECIAL CONSTRUCTION							
1020	Integrated Construction	N/A		—	—	—	0.0%
1040	Special Facilities	N/A		—	—	—	
G. BUILDING SITEWORK	**N/A**						

			Sub-Total	62.59	100%
CONTRACTOR FEES (General Requirements: 10%, Overhead: 5%, Profit: 10%)			25%	15.63	
ARCHITECT FEES			7%	5.48	
		Total Building Cost		**83.70**	

BUILDING TYPES

BUILDING TYPES

Costs per square foot of floor area

Exterior Wall	S.F. Area	50000	65000	80000	95000	110000	125000	140000	155000	170000
	L.F. Perimeter	920	1065	1167	1303	1333	1433	1533	1633	1733
Face Brick with Concrete Block Back-up	R/Conc. Frame	91.45	89.35	87.70	86.75	85.45	84.85	84.35	84.00	83.60
	Steel Frame	79.20	77.05	75.45	74.50	73.20	72.55	72.10	71.75	71.35
Decorative Concrete Block	R/Conc. Frame	87.50	85.80	84.55	83.80	82.85	82.35	81.95	81.70	81.40
	Steel Joists	75.95	74.20	72.90	72.10	71.05	70.55	70.15	69.85	69.50
Precast Concrete Panels	R/Conc. Frame	88.50	86.70	85.40	84.65	83.60	83.10	82.70	82.40	82.10
	Steel Joists	75.80	74.05	72.75	71.95	70.95	70.45	70.05	69.70	69.40
Perimeter Adj., Add or Deduct	Per 100 L.F.	1.45	1.05	.90	.75	.65	.55	.45	.40	.45
Story Hgt. Adj., Add or Deduct	Per 1 Ft.	.75	.65	.55	.60	.45	.45	.40	.40	.45

For Basement, add $ 18.40 per square foot of basement area

The above costs were calculated using the basic specifications shown on the facing page. These costs should be adjusted where necessary for design alternatives and owner's requirements. Reported completed project costs, for this type of structure, range from $ 43.00 to $ 107.05 per S.F.

Common additives

Description	Unit	$ Cost
Closed Circuit Surveillance, One station		
Camera and monitor	Each	1525
For additional camera stations, add	Each	820
Directory Boards, Plastic, glass covered		
30" x 20"	Each	575
36" x 48"	Each	1275
Aluminum, 24" x 18"	Each	525
36" x 24"	Each	600
48" x 32"	Each	810
48" x 60"	Each	1875
Emergency Lighting, 25 watt, battery operated		
Lead battery	Each	247
Nickel cadmium	Each	695
Safe, Office type, 4 hour rating		
30" x 18" x 18"	Each	3425
62" x 33" x 20"	Each	7450
Sound System		
Amplifier, 250 watts	Each	1925
Speaker, ceiling or wall	Each	160
Trumpet	Each	305

Model costs calculated for a 1 story building with 14' story height and 110,000 square feet of floor area

				Unit	Unit Cost	Cost Per S.F.	% Of Sub-Total
A. SUBSTRUCTURE							
1010	Standard Foundations	Poured concrete; strip and spread footings		S.F. Ground	.71	.71	
1030	Slab on Grade	4" reinforced concrete with vapor barrier and granular base		S.F. Slab	3.98	3.98	
2010	Basement Excavation	Site preparation for slab and trench for foundation wall and footing		S.F. Ground	.22	.22	8.8%
2020	Basement Walls	4' foundation wall		L.F. Wall	61	.74	
B. SHELL							
	B10 Superstructure						
1010	Floor Construction	Cast-in-place concrete columns		L.F. Column	53	.69	26.7%
1020	Roof Construction	Precast concrete beam and plank, concrete columns		S.F. Roof	16.50	16.50	
	B20 Exterior Enclosure						
2010	Exterior Walls	Face brick with concrete block backup	90% of wall	S.F. Wall	23	3.64	
2020	Exterior Windows	Storefront	10% of wall	Each	48	.81	7.4%
2030	Exterior Doors	Sliding electric operated entrance, hollow metal		Each	5296	.30	
	B30 Roofing						
3010	Roof Coverings	Built-up tar and gravel with flashing; perlite/EPS composite insulation		S.F. Roof	3.58	3.58	5.6%
3020	Roof Openings	Roof hatches		S.F. Roof	.04	.04	
C. INTERIORS							
1010	Partitions	Gypsum board on metal studs	60 S.F. Floor/L.F. Partition	S.F. Partition	7.74	1.29	
1020	Interior Doors	Single leaf hollow metal	600 S.F. Floor/Door	Each	608	1.01	
1030	Fittings	N/A		—	—	—	
2010	Stair Construction	N/A		—	—	—	25.2%
3010	Wall Finishes	Paint		S.F. Surface	1.26	.21	
3020	Floor Finishes	50% ceramic tile, 50% carpet tile		S.F. Floor	11.62	11.62	
3030	Ceiling Finishes	Mineral fiber board on exposed grid system, suspended		S.F. Ceiling	2.00	2.14	
D. SERVICES							
	D10 Conveying						
1010	Elevators & Lifts	N/A		—	—	—	0.0%
1020	Escalators & Moving Walks	N/A		—	—	—	
	D20 Plumbing						
2010	Plumbing Fixtures	Toilet and service fixtures, supply and drainage	1 Fixture/4075 S.F. Floor	Each	1833	.45	
2020	Domestic Water Distribution	Gas fired water heater		S.F. Floor	.08	.08	1.3%
2040	Rain Water Drainage	Roof drains		S.F. Roof	.30	.30	
	D30 HVAC						
3010	Energy Supply	N/A		—	—	—	
3020	Heat Generating Systems	Included in D3050		—	—	—	
3030	Cooling Generating Systems	N/A		—	—	—	10.3%
3050	Terminal & Package Units	Single zone unit gas heating, electric cooling		S.F. Floor	6.62	6.62	
3090	Other HVAC Sys. & Equipment	N/A		—	—	—	
	D40 Fire Protection						
4010	Sprinklers	Sprinklers, light hazard		S.F. Floor	1.75	1.75	2.7%
4020	Standpipes	N/A		—	—	—	
	D50 Electrical						
5010	Electrical Service/Distribution	1200 ampere service, panel board and feeders		S.F. Floor	.64	.64	
5020	Lighting & Branch Wiring	Fluorescent fixtures, receptacles, switches, A.C. and misc. power		S.F. Floor	6.78	6.78	12.1%
5030	Communications & Security	Alarm systems and emergency lighting		S.F. Floor	.36	.36	
5090	Other Electrical Systems	Emergency generator, 7.5 kW		S.F. Floor	.03	.03	
E. EQUIPMENT & FURNISHINGS							
1010	Commercial Equipment	N/A		—	—	—	
1020	Institutional Equipment	N/A		—	—	—	
1030	Vehicular Equipment	N/A		—	—	—	0.0%
1090	Other Equipment	N/A		—	—	—	
F. SPECIAL CONSTRUCTION							
1020	Integrated Construction	N/A		—	—	—	
1040	Special Facilities	N/A		—	—	—	0.0%
G. BUILDING SITEWORK	**N/A**						

			Sub-Total	64.49	100%
CONTRACTOR FEES (General Requirements: 10%, Overhead: 5%, Profit: 10%)			25%	16.12	
ARCHITECT FEES			6%	4.84	
		Total Building Cost		**85.45**	

BUILDING TYPES

Costs per square foot of floor area

Exterior Wall	S.F. Area	50000	65000	80000	95000	110000	125000	140000	155000	170000
	L.F. Perimeter	533	593	670	715	778	840	871	923	976
Face Brick with Concrete Block Back-up	Steel Frame	108.05	103.80	101.60	99.40	98.15	97.15	95.95	95.25	94.70
	R/Conc. Frame	110.15	105.90	103.75	101.50	100.20	99.25	98.05	97.35	96.75
Face Brick on Steel Studs	Steel Frame	104.95	101.20	99.20	97.25	96.05	95.25	94.20	93.50	93.05
	R/Conc. Frame	108.55	104.75	102.75	100.75	99.60	98.75	97.70	97.10	96.55
Precast Concrete Panels Exposed Aggregate	Steel Frame	106.25	102.30	100.25	98.15	97.00	96.05	94.95	94.30	93.75
	R/Conc. Frame	109.35	105.40	103.35	101.25	100.10	99.20	98.10	97.40	96.85
Perimeter Adj., Add or Deduct	Per 100 L.F.	3.80	2.90	2.35	2.00	1.65	1.50	1.40	1.20	1.10
Story Hgt. Adj., Add or Deduct	Per 1 Ft.	1.20	1.00	.90	.80	.75	.75	.70	.70	.55

For Basement, add $31.15 per square foot of basement area

The above costs were calculated using the basic specifications shown on the facing page. These costs should be adjusted where necessary for design alternatives and owner's requirements. Reported completed project costs, for this type of structure, range from $44.45 to $118.95 per S.F.

Common additives

Description	Unit	$ Cost
Closed Circuit Surveillance, One station		
Camera and monitor	Each	1525
For additional camera stations, add	Each	820
Directory Boards, Plastic, glass covered		
30" x 20"	Each	575
36" x 48"	Each	1275
Aluminum, 24" x 18"	Each	525
36" x 24"	Each	600
48" x 32"	Each	810
48" x 60"	Each	1875
Elevators, Hydraulic passenger, 2 stops		
1500# capacity	Each	47,900
2500# capacity	Each	50,000
3500# capacity	Each	53,700
Additional stop, add	Each	4550
Emergency Lighting, 25 watt, battery operated		
Lead battery	Each	247
Nickel cadmium	Each	695

Description	Unit	$ Cost
Escalators, Metal		
32" wide, 10' story height	Each	100,300
20' story height	Each	117,300
48" wide, 10' story height	Each	105,300
20' story height	Each	122,300
Glass		
32" wide, 10' story height	Each	98,800
20' story height	Each	117,300
48" wide, 10' story height	Each	105,300
20' story height	Each	122,300
Safe, Office type, 4 hour rating		
30" x 18" x 18"	Each	3425
62" x 33" x 20"	Each	7450
Sound System		
Amplifier, 250 watts	Each	1925
Speaker, ceiling or wall	Each	160
Trumpet	Each	305

Important: See the Reference Section for Location Factors

Model costs calculated for a 3 story building with 16' story height and 95,000 square feet of floor area

				Unit	Unit Cost	Cost Per S.F.	% Of Sub-Total
A. SUBSTRUCTURE							
1010	Standard Foundations	Poured concrete; strip and spread footings		S.F. Ground	2.88	.96	
1030	Slab on Grade	4" reinforced concrete with vapor barrier and granular base		S.F. Slab	3.98	1.32	
2010	Basement Excavation	Site preparation for slab and trench for foundation wall and footing		S.F. Ground	.14	.05	3.8%
2020	Basement Walls	4' foundation wall		L.F. Wall	115	.49	
B. SHELL							
	B10 Superstructure						
1010	Floor Construction	Concrete slab with metal deck and beams, steel columns		S.F. Floor	21	14.28	
1020	Roof Construction	Metal deck, open web steel joists, beams, columns		S.F. Roof	7.53	2.51	22.4%
	B20 Exterior Enclosure						
2010	Exterior Walls	Face brick with concrete block backup	90% of wall	S.F. Wall	23	7.76	
2020	Exterior Windows	Storefront	10% of wall	Each	35	1.30	13.7%
2030	Exterior Doors	Revolving and sliding panel, mall-front		Each	8357	1.22	
	B30 Roofing						
3010	Roof Coverings	Built-up tar and gravel with flashing; perlite/EPS composite insulation		S.F. Roof	3.96	1.32	1.8%
3020	Roof Openings	Roof hatches		S.F. Roof	.12	.04	
C. INTERIORS							
1010	Partitions	Gypsum board on metal studs	60 S.F. Floor/L.F. Partition	S.F. Partition	4.62	.77	
1020	Interior Doors	Single leaf hollow metal	600 S.F. Floor/Door	Each	608	1.01	
1030	Fittings	N/A		—	—	—	
2010	Stair Construction	Concrete filled metal pan		Flight	8975	.94	26.4%
3010	Wall Finishes	70% paint, 20% vinyl wall covering, 10% ceramic tile		S.F. Surface	7.38	1.23	
3020	Floor Finishes	50% carpet tile, 40% marble tile, 10% terrazzo		S.F. Floor	11.79	11.79	
3030	Ceiling Finishes	Mineral fiber tile on concealed zee bars		S.F. Ceiling	4.00	4.09	
D. SERVICES							
	D10 Conveying						
1010	Elevators & Lifts	One hydraulic passenger, one hydraulic freight		Each	173,850	1.83	8.4%
1020	Escalators & Moving Walks	Four escalators		Each	106,875	4.50	
	D20 Plumbing						
2010	Plumbing Fixtures	Toilet and service fixtures, supply and drainage	1 Fixture/2570 S.F. Floor	Each	1182	.46	
2020	Domestic Water Distribution	Gas fired water heater		S.F. Floor	.07	.07	1.0%
2040	Rain Water Drainage	Roof drains		S.F. Roof	.69	.23	
	D30 HVAC						
3010	Energy Supply	N/A		—	—	—	
3020	Heat Generating Systems	Included in D3050		—	—	—	
3030	Cooling Generating Systems	N/A		—	—	—	8.8%
3050	Terminal & Package Units	Multizone rooftop unit, gas heating, electric cooling		S.F. Floor	6.62	6.62	
3090	Other HVAC Sys. & Equipment	N/A		—	—	—	
	D40 Fire Protection						
4010	Sprinklers	Sprinklers, light hazard		S.F. Floor	1.45	1.45	1.9%
4020	Standpipes	N/A		—	—	—	
	D50 Electrical						
5010	Electrical Service/Distribution	1200 ampere service, panel board and feeders		S.F. Floor	.97	.97	
5020	Lighting & Branch Wiring	Fluorescent fixtures, receptacles, switches, A.C. and misc. power		S.F. Floor	7.19	7.19	11.7%
5030	Communications & Security	Alarm systems and emergency lighting		S.F. Floor	.36	.36	
5090	Other Electrical Systems	Emergency generator, 50 kW		S.F. Floor	.26	.26	
E. EQUIPMENT & FURNISHINGS							
1010	Commercial Equipment	N/A		—	—	—	
1020	Institutional Equipment	N/A		—	—	—	
1030	Vehicular Equipment	N/A		—	—	—	0.0%
1090	Other Equipment	N/A		—	—	—	
F. SPECIAL CONSTRUCTION							
1020	Integrated Construction	N/A		—	—	—	0.0%
1040	Special Facilities	N/A		—	—	—	
G. BUILDING SITEWORK	**N/A**						

		Sub-Total	75.02	100%
CONTRACTOR FEES (General Requirements: 10%, Overhead: 5%, Profit: 10%)		25%	18.75	
ARCHITECT FEES		6%	5.63	
	Total Building Cost		**99.40**	

BUILDING TYPES

Costs per square foot of floor area

Exterior Wall	S.F. Area	4000	6000	8000	10000	12000	15000	18000	20000	22000
	L.F. Perimeter	260	340	360	410	440	490	540	565	594
Split Face Concrete Block	Steel Joists	105.00	95.65	87.30	83.85	80.60	77.60	75.60	74.40	73.50
Stucco on Concrete Block	Steel Joists	103.15	94.05	86.05	82.65	79.60	76.70	74.75	73.65	72.75
Painted Concrete Block	Steel Joists	97.90	89.10	81.60	78.40	75.55	72.75	70.95	69.90	69.05
Face Brick on Concrete Block	Steel Joists	116.70	105.95	95.45	91.25	87.30	83.55	81.05	79.50	78.40
Painted Reinforced Concrete	Steel Joists	113.65	103.25	93.35	89.30	85.60	82.00	79.60	78.20	77.15
Tilt-up Concrete Panels	Steel Joists	105.95	96.50	88.00	84.45	81.20	78.10	76.10	74.85	73.90
Perimeter Adj., Add or Deduct	Per 100 L.F.	12.25	8.20	6.20	4.90	4.15	3.25	2.75	2.45	2.30
Story Hgt. Adj., Add or Deduct	Per 1 Ft.	1.55	1.35	1.10	1.00	.90	.80	.75	.70	.65
For Basement, add $25.80 per square foot of basement area										

The above costs were calculated using the basic specifications shown on the facing page. These costs should be adjusted where necessary for design alternatives and owner's requirements. Reported completed project costs, for this type of structure, range from $43.35 to $150.85 per S.F.

Common additives

Description	Unit	$ Cost
Emergency Lighting, 25 watt, battery operated		
Lead battery	Each	247
Nickel cadmium	Each	695
Safe, Office type, 4 hour rating		
30" x 18" x 18"	Each	3425
62" x 33" x 20"	Each	7450
Smoke Detectors		
Ceiling type	Each	161
Duct type	Each	425
Sound System		
Amplifier, 250 watts	Each	1925
Speaker, ceiling or wall	Each	160
Trumpet	Each	305

Important: See the Reference Section for Location Factors

Model costs calculated for a 1 story building with 14' story height and 8,000 square feet of floor area

Store, Retail

				Unit	Unit Cost	Cost Per S.F.	% Of Sub-Total
A.	**SUBSTRUCTURE**						
1010	Standard Foundations	Poured concrete; strip and spread footings		S.F. Ground	1.43	1.43	
1030	Slab on Grade	4" reinforced concrete with vapor barrier and granular base		S.F. Slab	3.98	3.98	13.0%
2010	Basement Excavation	Site preparation for slab and trench for foundation wall and footing		S.F. Ground	.23	.23	
2020	Basement Walls	4' foundation wall		L.F. Wall	61	2.75	
B.	**SHELL**						
	B10 Superstructure						
1010	Floor Construction	N/A		—	—	—	8.3%
1020	Roof Construction	Metal deck, open web steel joists, beams, interior columns		S.F. Roof	5.37	5.37	
	B20 Exterior Enclosure						
2010	Exterior Walls	Decorative concrete block	90% of wall	S.F. Wall	13.21	7.49	
2020	Exterior Windows	Storefront windows	10% of wall	Each	35	2.26	15.7%
2030	Exterior Doors	Sliding entrance door and hollow metal service doors		Each	1543	.39	
	B30 Roofing						
3010	Roof Coverings	Built-up tar and gravel with flashing; perlite/EPS composite insulation		S.F. Roof	4.42	4.42	6.9%
3020	Roof Openings	Roof hatches		S.F. Roof	.07	.07	
C.	**INTERIORS**						
1010	Partitions	Gypsum board on metal studs	60 S.F. Floor/L.F. Partition	S.F. Partition	4.62	.77	
1020	Interior Doors	Single leaf hollow metal	600 S.F. Floor/Door	Each	608	1.01	
1030	Fittings	N/A		—	—	—	
2010	Stair Construction	N/A		—	—	—	16.1%
3010	Wall Finishes	Paint		S.F. Surface	9.48	1.58	
3020	Floor Finishes	Vinyl tile		S.F. Floor	2.98	2.98	
3030	Ceiling Finishes	Mineral fiber tile on concealed zee bars		S.F. Ceiling	4.00	4.09	
D.	**SERVICES**						
	D10 Conveying						
1010	Elevators & Lifts	N/A		—	—	—	0.0%
1020	Escalators & Moving Walks	N/A		—	—	—	
	D20 Plumbing						
2010	Plumbing Fixtures	Toilet and service fixtures, supply and drainage	1 Fixture/890 S.F. Floor	Each	1815	2.04	
2020	Domestic Water Distribution	Gas fired water heater		S.F. Floor	2.69	2.69	8.7%
2040	Rain Water Drainage	Roof drains		S.F. Roof	.89	.89	
	D30 HVAC						
3010	Energy Supply	N/A		—	—	—	
3020	Heat Generating Systems	Included in D3050		—	—	—	
3030	Cooling Generating Systems	N/A		—	—	—	10.2%
3050	Terminal & Package Units	Single zone unit, gas heating, electric cooling		S.F. Floor	6.62	6.62	
3090	Other HVAC Sys. & Equipment	N/A		—	—	—	
	D40 Fire Protection						
4010	Sprinklers	Wet pipe sprinkler system		S.F. Floor	2.71	2.71	4.2%
4020	Standpipes	N/A		—	—	—	
	D50 Electrical						
5010	Electrical Service/Distribution	400 ampere service, panel board and feeders		S.F. Floor	1.94	1.94	
5020	Lighting & Branch Wiring	Fluorescent fixtures, receptacles, switches, A.C. and misc. power		S.F. Floor	8.44	8.44	16.9%
5030	Communications & Security	Alarm systems and emergency lighting		S.F. Floor	.36	.36	
5090	Other Electrical Systems	Emergency generator, 15 kW		S.F. Floor	.17	.17	
E.	**EQUIPMENT & FURNISHINGS**						
1010	Commercial Equipment	N/A		—	—	—	
1020	Institutional Equipment	N/A		—	—	—	0.0%
1030	Vehicular Equipment	N/A		—	—	—	
1090	Other Equipment	N/A		—	—	—	
F.	**SPECIAL CONSTRUCTION**						
1020	Integrated Construction	N/A		—	—	—	0.0%
1040	Special Facilities	N/A		—	—	—	
G.	**BUILDING SITEWORK**	N/A					

BUILDING TYPES

	Sub-Total	64.68	**100%**
CONTRACTOR FEES (General Requirements: 10%, Overhead: 5%, Profit: 10%)	25%	16.15	
ARCHITECT FEES	8%	6.47	
Total Building Cost		**87.30**	

209

BUILDING TYPES

Costs per square foot of floor area

Exterior Wall	S.F. Area	12000	16000	20000	26000	32000	38000	44000	52000	60000
	L.F. Perimeter	450	510	570	650	716	780	840	920	1020
Face Brick with Concrete Block Back-up	Steel Frame	101.40	94.50	90.40	86.25	83.30	81.20	79.65	78.05	77.20
	Bearing Walls	98.50	91.55	87.55	83.40	80.45	78.30	76.75	75.15	74.30
Stucco on Concrete Block	Steel Frame	93.85	88.10	84.70	81.25	78.80	77.05	75.80	74.50	73.80
	Bearing Walls	94.25	88.45	85.10	81.60	79.25	77.40	76.20	74.90	74.20
Precast Concrete Panels	Steel Frame	99.65	93.05	89.10	85.10	82.30	80.25	78.75	77.25	76.45
Metal Sandwich Panels	Steel Frame	87.10	82.30	79.55	76.70	74.80	73.35	72.35	71.30	70.75
Perimeter Adj., Add or Deduct	Per 100 L.F.	7.30	5.50	4.45	3.40	2.75	2.30	2.00	1.70	1.50
Story Hgt. Adj., Add or Deduct	Per 1 Ft.	1.50	1.25	1.20	1.00	.90	.80	.75	.70	.70

For Basement, add $18.75 per square foot of basement area

The above costs were calculated using the basic specifications shown on the facing page. These costs should be adjusted where necessary for design alternatives and owner's requirements. Reported completed project costs, for this type of structure, range from $50.65 to $145.10 per S.F.

Common additives

Description	Unit	$ Cost
Check Out Counter		
Single belt	Each	2475
Double belt	Each	4125
Scanner, registers, guns & memory 2 lanes	Each	14,700
10 lanes	Each	140,000
Power take away	Each	5125
Emergency Lighting, 25 watt, battery operated		
Lead battery	Each	247
Nickel cadmium	Each	695
Refrigerators, Prefabricated, walk-in		
7'-6" High, 6' x 6'	S.F.	129
10' x 10'	S.F.	101
12' x 14'	S.F.	89
12' x 20'	S.F.	79
Refrigerated Food Cases		
Dairy, multi deck, 12' long	Each	9300
Delicatessen case, single deck, 12' long	Each	6275
Multi deck, 18 S.F. shelf display	Each	7725
Freezer, self-contained chest type, 30 C.F.	Each	4650
Glass door upright, 78 C.F.	Each	8700

Description	Unit	$ Cost
Refrigerated Food Cases, cont.		
Frozen food, chest type, 12' long	Each	6375
Glass door reach in, 5 door	Each	12,000
Island case 12' long, single deck	Each	7175
Multi deck	Each	14,900
Meat case 12' long, single deck	Each	5275
Multi deck	Each	8875
Produce, 12' long single deck	Each	6900
Multi deck	Each	7650
Safe, Office type, 4 hour rating		
30" x 18" x 18"	Each	3425
62" x 33" x 20"	Each	7450
Smoke Detectors		
Ceiling type	Each	161
Duct type	Each	425
Sound System		
Amplifier, 250 watts	Each	1925
Speaker, ceiling or wall	Each	160
Trumpet	Each	305

Important: See the Reference Section for Location Factors

Model costs calculated for a 1 story building with 18' story height and 44,000 square feet of floor area

Supermarket

				Unit	Unit Cost	Cost Per S.F.	% Of Sub-Total
A. SUBSTRUCTURE							
1010	Standard Foundations	Poured concrete; strip and spread footings		S.F. Ground	.86	.86	
1030	Slab on Grade	4" reinforced concrete with vapor barrier and granular base		S.F. Slab	3.98	3.98	10.7%
2010	Basement Excavation	Site preparation for slab and trench for foundation wall and footing		S.F. Ground	.14	.14	
2020	Basement Walls	4' foundation wall		L.F. Wall	61	1.16	
B. SHELL							
	B10 Superstructure						
1010	Floor Construction	N/A		—	—	—	10.2%
1020	Roof Construction	Metal deck, open web steel joists, beams, interior columns		S.F. Roof	5.84	5.84	
	B20 Exterior Enclosure						
2010	Exterior Walls	Face brick with concrete block backup	85% of wall	S.F. Wall	23	6.93	
2020	Exterior Windows	Storefront windows	15% of wall	Each	53	2.75	18.5%
2030	Exterior Doors	Sliding entrance doors with electrical operator, hollow metal		Each	3743	.93	
	B30 Roofing						
3010	Roof Coverings	Built-up tar and gravel with flashing; perlite/EPS composite insulation		S.F. Roof	3.77	3.77	6.7%
3020	Roof Openings	Roof hatches		S.F. Roof	.07	.07	
C. INTERIORS							
1010	Partitions	50% concrete block, 50% gypsum board on metal studs	50 S.F. Floor/L.F. Partition	S.F. Partition	5.71	1.37	
1020	Interior Doors	Single leaf hollow metal	2000 S.F. Floor/Door	Each	608	.30	
1030	Fittings	N/A		—	—	—	
2010	Stair Construction	N/A		—	—	—	18.5%
3010	Wall Finishes	Paint		S.F. Surface	7.48	1.87	
3020	Floor Finishes	Vinyl composition tile		S.F. Floor	2.98	2.98	
3030	Ceiling Finishes	Mineral fiber tile on concealed zee bars		S.F. Ceiling	4.00	4.09	
D. SERVICES							
	D10 Conveying						
1010	Elevators & Lifts	N/A		—	—	—	0.0%
1020	Escalators & Moving Walks	N/A		—	—	—	
	D20 Plumbing						
2010	Plumbing Fixtures	Toilet and service fixtures, supply and drainage	1 Fixture/1820 S.F. Floor	Each	946	.52	
2020	Domestic Water Distribution	Gas fired water heater		S.F. Floor	.23	.23	1.9%
2040	Rain Water Drainage	Roof drains		S.F. Roof	.32	.32	
	D30 HVAC						
3010	Energy Supply	N/A		—	—	—	
3020	Heat Generating Systems	Included in D3050		—	—	—	
3030	Cooling Generating Systems	N/A		—	—	—	11.2%
3050	Terminal & Package Units	Single zone rooftop unit, gas heating, electric cooling		S.F. Floor	6.44	6.44	
3090	Other HVAC Sys. & Equipment	N/A		—	—	—	
	D40 Fire Protection						
4010	Sprinklers	Sprinklers, light hazard		S.F. Floor	2.05	2.05	3.6%
4020	Standpipes	N/A		—	—	—	
	D50 Electrical						
5010	Electrical Service/Distribution	600 ampere service, panel board and feeders		S.F. Floor	.93	.93	
5020	Lighting & Branch Wiring	Fluorescent fixtures, receptacles, switches, A.C. and misc. power		S.F. Floor	9.65	9.65	18.8%
5030	Communications & Security	Alarm systems and emergency lighting		S.F. Floor	.16	.16	
5090	Other Electrical Systems	Emergency generator, 15 kW		S.F. Floor	.03	.03	
E. EQUIPMENT & FURNISHINGS							
1010	Commercial Equipment	N/A		—	—	—	
1020	Institutional Equipment	N/A		—	—	—	0.0%
1030	Vehicular Equipment	N/A		—	—	—	
1090	Other Equipment	N/A		—	—	—	
F. SPECIAL CONSTRUCTION							
1020	Integrated Construction	N/A		—	—	—	0.0%
1040	Special Facilities	N/A		—	—	—	
G. BUILDING SITEWORK	**N/A**						

		Sub-Total	57.37	100%
	CONTRACTOR FEES (General Requirements: 10%, Overhead: 5%, Profit: 10%)		25%	14.36
	ARCHITECT FEES		7%	5.02
		Total Building Cost	**76.75**	

BUILDING TYPES

Costs per square foot of floor area

Exterior Wall	S.F. Area	10000	16000	20000	22000	24000	26000	28000	30000	32000
	L.F. Perimeter	420	510	600	640	660	684	706	740	737
Face Brick with Concrete Block Back-up	Wood Truss	191.75	175.85	172.45	170.95	168.60	166.80	165.15	164.35	162.05
	Precast Conc.	220.15	201.15	197.25	195.45	192.60	190.50	188.55	187.55	184.70
Metal Sandwich Panel	Wood Truss	190.70	178.70	176.05	174.85	173.10	171.75	170.60	169.95	168.25
Precast Concrete Panel	Precast Conc.	200.25	186.00	183.00	181.60	179.50	177.90	176.45	175.70	173.65
Painted Concrete Block	Wood Frame	191.20	179.40	176.80	175.65	173.90	172.65	171.45	170.85	169.20
	Precast Conc.	203.05	188.20	185.10	183.60	181.40	179.75	178.25	177.45	175.35
Perimeter Adj., Add or Deduct	Per 100 L.F.	12.95	8.05	6.40	5.85	5.35	5.00	4.60	4.30	4.05
Story Hgt. Adj., Add or Deduct	Per 1 Ft.	1.60	1.15	1.15	1.05	1.00	1.05	1.00	.85	.80
For Basement, add $27.86 per square foot of basement area										

The above costs were calculated using the basic specifications shown on the facing page. These costs should be adjusted where necessary for design alternatives and owner's requirements. Reported completed project costs, for this type of structure, range from $80.05 to $235.35 per S.F.

Common additives

Description	Unit	$ Cost		Description	Unit	$ Cost
Bleachers, Telescoping, manual				Sauna, Prefabricated, complete		
To 15 tier	Seat	101 - 141		6' x 4'	Each	4525
16-20 tier	Seat	208 - 255		6' x 6'	Each	5425
21-30 tier	Seat	221 - 267		6' x 9'	Each	6700
For power operation, add	Seat	40 - 63		8' x 8'	Each	7875
Emergency Lighting, 25 watt, battery operated				8' x 10'	Each	8800
Lead battery	Each	247		10' x 12'	Each	11,000
Nickel cadmium	Each	695		Sound System		
Lockers, Steel, single tier, 60" or 72"	Opening	154 - 241		Amplifier, 250 watts	Each	1925
2 tier, 60" or 72" total	Opening	104 - 134		Speaker, ceiling or wall	Each	160
5 tier, box lockers	Opening	55 - 69		Trumpet	Each	305
Locker bench, lam. maple top only	L.F.	18.80		Steam Bath, Complete, to 140 C.F.	Each	1475
Pedestal, steel pipe	Each	57		To 300 C.F.	Each	1650
Pool Equipment				To 800 C.F.	Each	4725
Diving stand, 3 meter	Each	7550		To 2500 C.F.	Each	7200
1 meter	Each	4325				
Diving board, 16' aluminum	Each	3100				
Fiberglass	Each	2550				
Lifeguard chair, fixed	Each	1750				
Portable	Each	1150				
Lights, underwater, 12 volt, 300 watt	Each	1375				

BUILDING TYPES

Model costs calculated for a 1 story building with 24' story height and 20,000 square feet of floor area

Swimming Pool, Enclosed

				Unit	Unit Cost	Cost Per S.F.	% Of Sub-Total
A. SUBSTRUCTURE							
1010	Standard Foundations	Poured concrete; strip and spread footings		S.F. Ground	1.23	1.23	
1030	Slab on Grade	4" reinforced concrete with vapor barrier and granular base		S.F. Slab	3.98	3.98	
2010	Basement Excavation	Site preparation for slab and trench for foundation wall and footing		S.F. Ground	7.75	7.75	15.9%
2020	Basement Walls	4' foundation wall		L.F. Wall	122	7.36	
B. SHELL							
	B10 Superstructure						
1010	Floor Construction	N/A		—	—	—	
1020	Roof Construction	Wood deck on laminated wood truss		S.F. Roof	14.08	14.08	11.0%
	B20 Exterior Enclosure						
2010	Exterior Walls	Face brick with concrete block backup	80% of wall	S.F. Wall	23	13.74	
2020	Exterior Windows	Outward projecting steel	20% of wall	Each	35	5.18	15.1%
2030	Exterior Doors	Double aluminum and glass, hollow metal		Each	1690	.42	
	B30 Roofing						
3010	Roof Coverings	Asphalt shingles with flashing; perlite/EPS composite insulation		S.F. Roof	1.47	1.47	1.2%
3020	Roof Openings	N/A		—	—	—	
C. INTERIORS							
1010	Partitions	Concrete block	100 S.F. Floor/L.F. Partition	S.F. Partition	6.70	.67	
1020	Interior Doors	Single leaf hollow metal	1000 S.F. Floor/Door	Each	608	.61	
1030	Fittings	Toilet partitions		S.F. Floor	.17	.17	
2010	Stair Construction	N/A		—	—	—	16.0%
3010	Wall Finishes	Acrylic glazed coating		S.F. Surface	17.50	1.75	
3020	Floor Finishes	70% terrazzo, 30% ceramic tile		S.F. Floor	16.75	16.75	
3030	Ceiling Finishes	Mineral fiber tile on concealed zee bars	20% of area	S.F. Ceiling	2.14	.43	
D. SERVICES							
	D10 Conveying						
1010	Elevators & Lifts	N/A		—	—	—	0.0%
1020	Escalators & Moving Walks	N/A		—	—	—	
	D20 Plumbing						
2010	Plumbing Fixtures	Toilet and service fixtures, supply and drainage	1 Fixture/210 S.F. Floor	Each	2007	9.56	
2020	Domestic Water Distribution	Gas fired water heater		S.F. Floor	2.00	2.00	9.0%
2040	Rain Water Drainage	N/A		—	—	—	
	D30 HVAC						
3010	Energy Supply	Terminal unit heaters	10% of area	S.F. Floor	.58	.58	
3020	Heat Generating Systems	N/A		—	—	—	
3030	Cooling Generating Systems	N/A		—	—	—	9.1%
3050	Terminal & Package Units	Single zone unit gas heating, electric cooling		S.F. Floor	11.03	11.03	
3090	Other HVAC Sys. & Equipment	N/A		—	—	—	
	D40 Fire Protection						
4010	Sprinklers	Sprinklers, light hazard		S.F. Floor	1.03	1.03	0.8%
4020	Standpipes	N/A		—	—	—	
	D50 Electrical						
5010	Electrical Service/Distribution	400 ampere service, panel board and feeders		S.F. Floor	.82	.82	
5020	Lighting & Branch Wiring	Fluorescent fixtures, receptacles, switches, A.C. and misc. power		S.F. Floor	7.49	7.49	6.8%
5030	Communications & Security	Alarm systems and emergency lighting		S.F. Floor	.28	.28	
5090	Other Electrical Systems	Emergency generator, 15 kW		S.F. Floor	.07	.07	
E. EQUIPMENT & FURNISHINGS							
1010	Commercial Equipment	N/A		—	—	—	
1020	Institutional Equipment	N/A		—	—	—	15.1%
1030	Vehicular Equipment	N/A		—	—	—	
1090	Other Equipment	Swimming pool		S.F. Floor	19.30	19.30	
F. SPECIAL CONSTRUCTION							
1020	Integrated Construction	N/A		—	—	—	0.0%
1040	Special Facilities	N/A		—	—	—	
G. BUILDING SITEWORK	**N/A**						

		Sub-Total	127.75	100%
CONTRACTOR FEES (General Requirements: 10%, Overhead: 5%, Profit: 10%)		25%	31.92	
ARCHITECT FEES		8%	12.78	
	Total Building Cost		**172.45**	

BUILDING TYPES

Costs per square foot of floor area

Exterior Wall	S.F. Area	2000	3000	4000	5000	6000	7000	8000	9000	10000
	L.F. Perimeter	180	220	260	286	320	353	368	397	425
Face Brick with Concrete Block Back-up	Steel Frame	155.55	138.95	130.55	123.55	119.75	117.10	113.30	111.55	110.05
	Bearing Walls	152.30	135.70	127.35	120.30	116.50	113.85	110.05	108.30	106.80
Limestone with Concrete Block Back-up	Steel Frame	184.15	162.25	151.20	141.70	136.70	133.10	127.90	125.60	123.55
	Bearing Walls	180.95	159.00	147.95	138.45	133.45	129.90	124.65	122.35	120.25
Decorative Concrete Block	Steel Frame	144.90	130.25	122.85	116.75	113.40	111.15	107.80	106.35	105.00
	Bearing Walls	141.65	127.00	119.60	113.50	110.20	107.85	104.60	103.10	101.80
Perimeter Adj., Add or Deduct	Per 100 L.F.	36.80	24.50	18.40	14.65	12.30	10.50	9.20	8.20	7.35
Story Hgt. Adj., Add or Deduct	Per 1 Ft.	4.35	3.55	3.10	2.70	2.55	2.40	2.25	2.15	2.05
For Basement, add $27.00 per square foot of basement area										

The above costs were calculated using the basic specifications shown on the facing page. These costs should be adjusted where necessary for design alternatives and owner's requirements. Reported completed project costs, for this type of structure, range from $70.80 to $255.30 per S.F.

Common additives

Description	Unit	$ Cost
Emergency Lighting, 25 watt, battery operated		
Lead battery	Each	247
Nickel cadmium	Each	695
Smoke Detectors		
Ceiling type	Each	161
Duct type	Each	425
Emergency Generators, complete system, gas		
15 kw	Each	13,200
85 kw	Each	29,000
170 kw	Each	72,500
Diesel, 50 kw	Each	25,000
150 kw	Each	43,100
350 kw	Each	68,500

Model costs calculated for a 1 story building with 12' story height and 5,000 square feet of floor area

				Unit	Unit Cost	Cost Per S.F.	% Of Sub-Total
A.	**SUBSTRUCTURE**						
1010	Standard Foundations	Poured concrete; strip and spread footings		S.F. Ground	2.10	2.10	
1030	Slab on Grade	4" reinforced concrete with vapor barrier and granular base		S.F. Slab	3.98	3.98	11.2%
2010	Basement Excavation	Site preparation for slab and trench for foundation wall and footing		S.F. Ground	.43	.43	
2020	Basement Walls	4' foundation wall		L.F. Wall	61	3.49	
B.	**SHELL**						
	B10 Superstructure						
1010	Floor Construction	N/A		—	—	—	8.2%
1020	Roof Construction	Metal deck, open web steel joists, beams, columns		S.F. Roof	7.29	7.29	
	B20 Exterior Enclosure						
2010	Exterior Walls	Face brick with concrete block backup	80% of wall	S.F. Wall	23	13.04	
2020	Exterior Windows	Outward projecting steel	20% of wall	Each	1357	9.32	26.4%
2030	Exterior Doors	Single aluminum glass with transom		Each	2753	1.11	
	B30 Roofing						
3010	Roof Coverings	Built-up tar and gravel with flashing; perlite/EPS composite insulation		S.F. Roof	4.74	4.74	5.3%
3020	Roof Openings	—		—			
C.	**INTERIORS**						
1010	Partitions	Double layer gypsum board on metal studs	15 S.F. Floor/L.F. Partition	S.F. Partition	4.61	3.07	
1020	Interior Doors	Single leaf hollow metal	150 S.F. Floor/Door	Each	608	4.05	
1030	Fittings	Toilet partitions		S.F. Floor	1.21	1.21	
2010	Stair Construction	N/A		—	—	—	23.2%
3010	Wall Finishes	Paint		S.F. Surface	2.16	2.16	
3020	Floor Finishes	90% carpet, 10% terrazzo		S.F. Floor	7.12	7.12	
3030	Ceiling Finishes	Fiberglass board on exposed grid system, suspended		S.F. Ceiling	3.00	3.05	
D.	**SERVICES**						
	D10 Conveying						
1010	Elevators & Lifts	N/A		—	—	—	0.0%
1020	Escalators & Moving Walks	N/A		—	—	—	
	D20 Plumbing						
2010	Plumbing Fixtures	Kitchen, toilet and service fixtures, supply and drainage	1 Fixture/715 S.F. Floor	Each	1944	2.72	
2020	Domestic Water Distribution	Gas fired water heater		S.F. Floor	1.08	1.08	5.9%
2040	Rain Water Drainage	Roof drains		S.F. Roof	1.41	1.41	
	D30 HVAC						
3010	Energy Supply	N/A		—	—	—	
3020	Heat Generating Systems	Included in D3050		—	—	—	
3030	Cooling Generating Systems	N/A		—	—	—	8.1%
3050	Terminal & Package Units	Single zone unit, gas heating, electric cooling		S.F. Floor	7.20	7.20	
3090	Other HVAC Sys. & Equipment	N/A		—	—	—	
	D40 Fire Protection						
4010	Sprinklers	Wet pipe sprinkler system		S.F. Floor	2.91	2.91	3.3%
4020	Standpipes	N/A		—	—	—	
	D50 Electrical						
5010	Electrical Service/Distribution	200 ampere service, panel board and feeders		S.F. Floor	1.33	1.33	
5020	Lighting & Branch Wiring	Fluorescent fixtures, receptacles, switches, A.C. and misc. power		S.F. Floor	4.98	4.98	8.5%
5030	Communications & Security	Alarm systems and emergency lighting		S.F. Floor	1.12	1.12	
5090	Other Electrical Systems	Emergency generator, 15 kW		S.F. Floor	.13	.13	
E.	**EQUIPMENT & FURNISHINGS**						
1010	Commercial Equipment	N/A		—	—	—	
1020	Institutional Equipment	N/A		—	—	—	0.0%
1030	Vehicular Equipment	N/A		—	—	—	
1090	Other Equipment	N/A		—	—	—	
F.	**SPECIAL CONSTRUCTION**						
1020	Integrated Construction	N/A		—	—	—	0.0%
1040	Special Facilities	N/A		—	—	—	
G.	**BUILDING SITEWORK**	**N/A**					

					Sub-Total	89.04	100%
	CONTRACTOR FEES (General Requirements: 10%, Overhead: 5%, Profit: 10%)				25%	22.27	
	ARCHITECT FEES				11%	12.24	

Total Building Cost — **123.55**

BUILDING TYPES

Costs per square foot of floor area

Exterior Wall	S.F. Area	5000	6500	8000	9500	11000	14000	17500	21000	24000
	L.F. Perimeter	300	360	386	396	435	510	550	620	680
Face Brick with Concrete Block Back-up	Steel Joists	116.80	112.15	106.70	102.05	100.15	97.50	94.00	92.45	91.55
	Wood Joists	118.00	113.25	107.65	102.85	101.00	98.25	94.65	93.05	92.10
Stone with Concrete Block Back-up	Steel Joists	121.45	116.40	110.45	105.25	103.20	100.30	96.40	94.70	93.65
	Wood Joists	122.60	117.50	111.40	106.05	104.05	101.05	97.05	95.35	94.25
Brick Veneer	Wood Frame	111.45	107.10	102.10	97.75	96.10	93.55	90.35	88.95	88.10
E.I.F.S.	Wood Frame	105.85	101.85	97.55	93.90	92.40	90.20	87.40	86.20	85.45
Perimeter Adj., Add or Deduct	Per 100 L.F.	11.90	9.10	7.45	6.25	5.45	4.30	3.40	2.80	2.45
Story Hgt. Adj., Add or Deduct	Per 1 Ft.	2.15	1.95	1.75	1.45	1.50	1.30	1.15	1.05	1.00
For Basement, add $22.20 per square foot of basement area										

The above costs were calculated using the basic specifications shown on the facing page. These costs should be adjusted where necessary for design alternatives and owner's requirements. Reported completed project costs, for this type of structure, range from $66.90 to $175.90 per S.F.

Common additives

Description	Unit	$ Cost	Description	Unit	$ Cost
Directory Boards, Plastic, glass covered			Smoke Detectors		
30" x 20"	Each	575	Ceiling type	Each	161
36" x 48"	Each	1275	Duct type	Each	425
Aluminum, 24" x 18"	Each	525	Vault Front, Door & frame		
36" x 24"	Each	600	1 Hour test, 32" x 78"	Opening	4425
48" x 32"	Each	810	2 Hour test, 32" door	Opening	4775
48" x 60"	Each	1875	40" door	Opening	5575
Emergency Lighting, 25 watt, battery operated			4 Hour test, 32" door	Opening	5175
Lead battery	Each	247	40" door	Opening	6200
Nickel cadmium	Each	695	Time lock movement; two movement	Each	1725
Flagpoles, Complete					
Aluminum, 20' high	Each	1375			
40' high	Each	3025			
70' high	Each	9500			
Fiberglass, 23' high	Each	1650			
39'-5" high	Each	3175			
59' high	Each	7950			
Safe, Office type, 4 hour rating					
30" x 18" x 18"	Each	3425			
62" x 33" x 20"	Each	7450			

Model costs calculated for a 1 story building with 12' story height and 11,000 square feet of floor area

				Unit	Unit Cost	Cost Per S.F.	% Of Sub-Total
A. SUBSTRUCTURE							
1010	Standard Foundations	Poured concrete; strip and spread footings		S.F. Ground	1.28	1.28	
1030	Slab on Grade	4" reinforced concrete with vapor barrier and granular base		S.F. Slab	3.98	3.98	
2010	Basement Excavation	Site preparation for slab and trench for foundation wall and footing		S.F. Ground	.23	.23	10.8%
2020	Basement Walls	4' foundation wall		L.F. Wall	61	2.42	
B. SHELL							
	B10 Superstructure						
1010	Floor Construction	N/A		—	—	—	
1020	Roof Construction	Metal deck, open web steel joists, beams, interior columns		S.F. Roof	5.58	5.58	7.6%
	B20 Exterior Enclosure						
2010	Exterior Walls	Face brick with concrete block backup	70% of wall	S.F. Wall	23	7.89	
2020	Exterior Windows	Metal outward projecting	30% of wall	Each	559	3.46	16.3%
2030	Exterior Doors	Metal and glass with transom		Each	1811	.66	
	B30 Roofing						
3010	Roof Coverings	Built-up tar and gravel with flashing; perlite/EPS composite insulation		S.F. Roof	4.43	4.43	6.0%
3020	Roof Openings	Roof hatches		S.F. Roof	—	—	
C. INTERIORS							
1010	Partitions	Gypsum board on metal studs	20 S.F. Floor/L.F. Partition	S.F. Partition	6.86	3.43	
1020	Interior Doors	Wood solid core	200 S.F. Floor/Door	Each	480	2.40	
1030	Fittings	Toilet partitions		S.F. Floor	.47	.47	
2010	Stair Construction	N/A		—	—	—	24.7%
3010	Wall Finishes	90% paint, 10% ceramic tile		S.F. Surface	2.32	1.16	
3020	Floor Finishes	70% carpet tile, 15% terrazzo, 15% vinyl composition tile		S.F. Floor	6.65	6.65	
3030	Ceiling Finishes	Mineral fiber tile on concealed zee bars		S.F. Ceiling	4.00	4.09	
D. SERVICES							
	D10 Conveying						
1010	Elevators & Lifts	N/A		—	—	—	0.0%
1020	Escalators & Moving Walks	N/A		—	—	—	
	D20 Plumbing						
2010	Plumbing Fixtures	Kitchen, toilet and service fixtures, supply and drainage	1 Fixture/500 S.F. Floor	Each	2225	4.45	
2020	Domestic Water Distribution	Gas fired water heater		S.F. Floor	.77	.77	7.9%
2040	Rain Water Drainage	Roof drains		S.F. Roof	.57	.57	
	D30 HVAC						
3010	Energy Supply	N/A		—	—	—	
3020	Heat Generating Systems	Included in D3050		—	—	—	
3030	Cooling Generating Systems	N/A		—	—	—	9.8%
3050	Terminal & Package Units	Multizone unit, gas heating, electric cooling		S.F. Floor	7.20	7.20	
3090	Other HVAC Sys. & Equipment	N/A		—	—	—	
	D40 Fire Protection						
4010	Sprinklers	Wet pipe sprinkler system		S.F. Floor	2.05	2.05	2.8%
4020	Standpipes	N/A		—	—	—	
	D50 Electrical						
5010	Electrical Service/Distribution	400 ampere service, panel board and feeders		S.F. Floor	1.42	1.42	
5020	Lighting & Branch Wiring	Fluorescent fixtures, receptacles, switches, A.C. and misc. power		S.F. Floor	8.29	8.29	14.1%
5030	Communications & Security	Alarm systems and emergency lighting		S.F. Floor	.54	.54	
5090	Other Electrical Systems	Emergency generator, 15 kW		S.F. Floor	.12	.12	
E. EQUIPMENT & FURNISHINGS							
1010	Commercial Equipment	N/A		—	—	—	
1020	Institutional Equipment	N/A		—	—	—	
1030	Vehicular Equipment	N/A		—	—	—	0.0%
1090	Other Equipment	N/A		—	—	—	
F. SPECIAL CONSTRUCTION							
1020	Integrated Construction	N/A		—	—	—	0.0%
1040	Special Facilities	N/A		—	—	—	
G. BUILDING SITEWORK	**N/A**						

				Sub-Total		73.54	100%
	CONTRACTOR FEES (General Requirements: 10%, Overhead: 5%, Profit: 10%)				25%	18.34	
	ARCHITECT FEES				9%	8.27	

Total Building Cost 100.15

BUILDING TYPES

Costs per square foot of floor area

Exterior Wall	S.F. Area	8000	10000	12000	15000	18000	24000	28000	35000	40000
	L.F. Perimeter	206	233	260	300	320	360	393	451	493
Face Brick with Concrete Block Back-up	Steel Frame	141.20	134.30	129.70	125.05	120.35	114.45	112.20	109.60	108.25
	R/Conc. Frame	148.80	141.95	137.35	132.70	127.95	122.05	119.85	117.20	115.90
Stone with Concrete Block Back-up	Steel Frame	138.90	132.50	128.20	123.90	119.45	113.75	111.65	109.15	107.90
	R/Conc. Frame	154.75	147.30	142.30	137.30	132.05	125.50	123.10	120.15	118.75
Limestone with Concrete Block Back-up	Steel Frame	162.25	153.40	147.45	141.40	134.90	126.70	123.75	120.10	118.40
	R/Conc. Frame	169.90	161.00	155.05	149.05	142.55	134.35	131.35	127.75	126.05
Perimeter Adj., Add or Deduct	Per 100 L.F.	18.15	14.50	12.10	9.65	8.05	6.05	5.25	4.10	3.70
Story Hgt. Adj., Add or Deduct	Per 1 Ft.	2.75	2.55	2.30	2.15	1.90	1.55	1.55	1.40	1.35

For Basement, add $21.65 per square foot of basement area

The above costs were calculated using the basic specifications shown on the facing page. These costs should be adjusted where necessary for design alternatives and owner's requirements. Reported completed project costs, for this type of structure, range from $62.45 to $175.90 per S.F.

Common additives

Description	Unit	$ Cost	Description	Unit	$ Cost
Directory Boards, Plastic, glass covered			Flagpoles, Complete		
30" x 20"	Each	575	Aluminum, 20' high	Each	1375
36" x 48"	Each	1275	40' high	Each	3025
Aluminum, 24" x 18"	Each	525	70' high	Each	9500
36" x 24"	Each	600	Fiberglass, 23' high	Each	1650
48" x 32"	Each	810	39'-5" high	Each	3175
48" x 60"	Each	1875	59' high	Each	7950
Elevators, Hydraulic passenger, 2 stops			Safe, Office type, 4 hour rating		
1500# capacity	Each	47,900	30" x 18" x 18"	Each	3425
2500# capacity	Each	50,000	62" x 33" x 20"	Each	7450
3500# capacity	Each	53,700	Smoke Detectors		
Additional stop, add	Each	4550	Ceiling type	Each	161
Emergency Lighting, 25 watt, battery operated			Duct type	Each	425
Lead battery	Each	247	Vault Front, Door & frame		
Nickel cadmium	Each	695	1 Hour test, 32" x 78"	Opening	4425
			2 Hour test, 32" door	Opening	4775
			40" door	Opening	5575
			4 Hour test, 32" door	Opening	5175
			40" door	Opening	6200
			Time lock movement; two movement	Each	1725

Model costs calculated for a 3 story building with 12' story height and 18,000 square feet of floor area

				Unit	Unit Cost	Cost Per S.F.	% Of Sub-Total
A. SUBSTRUCTURE							
1010	Standard Foundations	Poured concrete; strip and spread footings		S.F. Ground	3.15	1.05	
1030	Slab on Grade	4" reinforced concrete with vapor barrier and granular base		S.F. Slab	3.98	1.32	
2010	Basement Excavation	Site preparation for slab and trench for foundation wall and footing		S.F. Ground	.14	.05	4.1%
2020	Basement Walls	4' foundation wall		L.F. Wall	115	1.21	
B. SHELL							
	B10 Superstructure						
1010	Floor Construction	Open web steel joists, slab form, concrete, wide flange steel columns		S.F. Floor	19.01	12.67	16.8%
1020	Roof Construction	Metal deck, open web steel joists, beams, interior columns		S.F. Roof	6.15	2.05	
	B20 Exterior Enclosure						
2010	Exterior Walls	Stone with concrete block backup	70% of wall	S.F. Wall	30	13.65	
2020	Exterior Windows	Metal outward projecting	10% of wall	Each	795	1.84	18.2%
2030	Exterior Doors	Metal and glass with transoms		Each	1690	.46	
	B30 Roofing						
3010	Roof Coverings	Built-up tar and gravel with flashing; perlite/EPS composite insulation		S.F. Roof	4.41	1.47	1.7%
3020	Roof Openings	N/A		—	—	—	
C. INTERIORS							
1010	Partitions	Gypsum board on metal studs	20 S.F. Floor/L.F. Partition	S.F. Partition	7.66	3.83	
1020	Interior Doors	Wood solid core	200 S.F. Floor/Door	Each	480	2.40	
1030	Fittings	Toilet partitions		S.F. Floor	.29	.29	
2010	Stair Construction	Concrete filled metal pan		Flight	7700	2.57	24.0%
3010	Wall Finishes	90% paint, 10% ceramic tile		S.F. Surface	2.32	1.16	
3020	Floor Finishes	70% carpet tile, 15% terrazzo, 15% vinyl composition tile		S.F. Floor	6.65	6.65	
3030	Ceiling Finishes	Mineral fiber tile on concealed zee bars		S.F. Ceiling	4.00	4.09	
D. SERVICES							
	D10 Conveying						
1010	Elevators & Lifts	Two hydraulic elevators		Each	78,930	8.77	10.0%
1020	Escalators & Moving Walks	N/A		—	—	—	
	D20 Plumbing						
2010	Plumbing Fixtures	Toilet and service fixtures, supply and drainage	1 Fixture/1385 S.F. Floor	Each	1772	1.28	
2020	Domestic Water Distribution	Gas fired water heater		S.F. Floor	.30	.30	2.1%
2040	Rain Water Drainage	Roof drains		S.F. Roof	.87	.29	
	D30 HVAC						
3010	Energy Supply	N/A		—	—	—	
3020	Heat Generating Systems	Included in D3050		—	—	—	
3030	Cooling Generating Systems	N/A		—	—	—	8.2%
3050	Terminal & Package Units	Multizone unit, gas heating, electric cooling		S.F. Floor	7.20	7.20	
3090	Other HVAC Sys. & Equipment	N/A		—	—	—	
	D40 Fire Protection						
4010	Sprinklers	Sprinklers, light hazard		S.F. Floor	1.97	1.97	2.2%
4020	Standpipes	N/A		—	—	—	
	D50 Electrical						
5010	Electrical Service/Distribution	400 ampere service, panel board and feeders		S.F. Floor	1.21	1.21	
5020	Lighting & Branch Wiring	Fluorescent fixtures, receptacles, switches, A.C. and misc. power		S.F. Floor	9.18	9.18	12.6%
5030	Communications & Security	Alarm systems and emergency lighting		S.F. Floor	.54	.54	
5090	Other Electrical Systems	Emergency generator, 15 kW		S.F. Floor	.14	.14	
E. EQUIPMENT & FURNISHINGS							
1010	Commercial Equipment	N/A		—	—	—	
1020	Institutional Equipment	N/A		—	—	—	
1030	Vehicular Equipment	N/A		—	—	—	0.0%
1090	Other Equipment	N/A		—	—	—	
F. SPECIAL CONSTRUCTION							
1020	Integrated Construction	N/A		—	—	—	0.0%
1040	Special Facilities	N/A		—	—	—	
G. BUILDING SITEWORK	**N/A**						

			Sub-Total	87.64	**100%**
	CONTRACTOR FEES (General Requirements: 10%, Overhead: 5%, Profit: 10%)			25%	21.95
	ARCHITECT FEES			9%	9.86
			Total Building Cost	**119.45**	

Costs per square foot of floor area

Exterior Wall		S.F. Area	10000	15000	20000	25000	30000	35000	40000	50000	60000
		L.F. Perimeter	410	500	600	640	700	766	833	966	1000
Tiltup Concrete Panels	Steel Frame		84.50	77.35	74.15	70.80	68.95	67.70	66.80	65.55	63.80
Brick with Block Backup	Bearing Walls		99.55	88.95	84.20	78.95	76.10	74.25	72.85	70.95	68.00
Concrete Block	Steel Frame		83.50	76.45	73.25	70.05	68.25	67.05	66.15	64.95	63.25
	Bearing Walls		81.05	73.95	70.70	67.45	65.60	64.45	63.50	62.25	60.55
Galvanized Steel Siding	Steel Frame		88.30	80.95	77.60	74.25	72.35	71.15	70.20	68.90	67.10
Metal Sandwich Panels	Steel Frame		89.40	81.30	77.60	73.75	71.65	70.25	69.20	67.75	65.65
Perimeter Adj., Add or Deduct	Per 100 L.F.		5.55	3.80	2.80	2.20	1.85	1.65	1.40	1.10	.90
Story Hgt. Adj., Add or Deduct	Per 1 Ft.		.65	.55	.45	.40	.35	.35	.35	.30	.25

For Basement, add $20.80 per square foot of basement area

The above costs were calculated using the basic specifications shown on the facing page. These costs should be adjusted where necessary for design alternatives and owner's requirements. Reported completed project costs, for this type of structure, range from $26.65 to $106.30 per S.F.

Common additives

Description	Unit	$ Cost
Dock Leveler, 10 ton cap.		
6' x 8'	Each	5650
7' x 8'	Each	5500
Emergency Lighting, 25 watt, battery operated		
Lead battery	Each	247
Nickel cadmium	Each	695
Fence, Chain link, 6' high		
9 ga. wire	L.F.	17.75
6 ga. wire	L.F.	25
Gate	Each	258
Flagpoles, Complete		
Aluminum, 20' high	Each	1375
40' high	Each	3025
70' high	Each	9500
Fiberglass, 23' high	Each	1650
39'-5" high	Each	3175
59' high	Each	7950
Paving, Bituminous		
Wearing course plus base course	S.Y.	6.80
Sidewalks, Concrete 4" thick	S.F.	3.40

Description	Unit	$ Cost
Sound System		
Amplifier, 250 watts	Each	1925
Speaker, ceiling or wall	Each	160
Trumpet	Each	305
Yard Lighting, 20' aluminum pole with 400 watt high pressure sodium fixture.	Each	2375

BUILDING TYPES

Model costs calculated for a 1 story building with 24' story height and 30,000 square feet of floor area

				Unit	Unit Cost	Cost Per S.F.	% Of Sub-Total
A. SUBSTRUCTURE							
1010	Standard Foundations	Poured concrete; strip and spread footings		S.F. Ground	1.09	1.09	
1030	Slab on Grade	5" reinforced concrete with vapor barrier and granular base		S.F. Slab	9.45	9.45	25.5%
2010	Basement Excavation	Site preparation for slab and trench for foundation wall and footing		S.F. Ground	.14	.14	
2020	Basement Walls	4' foundation wall		L.F. Wall	61	1.85	
B. SHELL							
	B10 Superstructure						
1010	Floor Construction	Mezzanine: open web steel joists, slab form, concrete beams, columns	10% of area	S.F. Floor	15.90	1.59	14.0%
1020	Roof Construction	Metal deck, open web steel joists, beams, columns		S.F. Roof	5.27	5.27	
	B20 Exterior Enclosure						
2010	Exterior Walls	Concrete block	95% of wall	S.F. Wall	8.91	4.74	
2020	Exterior Windows	N/A		—	—	—	11.3%
2030	Exterior Doors	Steel overhead, hollow metal	5% of wall	Each	2164	.80	
	B30 Roofing						
3010	Roof Coverings	Built-up tar and gravel with flashing; perlite/EPS composite insulation		S.F. Roof	3.88	3.88	8.3%
3020	Roof Openings	Roof hatches and skylight		S.F. Roof	.19	.19	
C. INTERIORS							
1010	Partitions	Concrete block (office and washrooms)	100 S.F. Floor/L.F. Partition	S.F. Partition	6.75	.54	
1020	Interior Doors	Single leaf hollow metal	5000 S.F. Floor/Door	Each	608	.12	
1030	Fittings	N/A		—	—	—	
2010	Stair Construction	Steel gate with rails		Flight	6200	.42	9.3%
3010	Wall Finishes	Paint		S.F. Surface	20	1.68	
3020	Floor Finishes	90% hardener, 10% vinyl composition tile		S.F. Floor	1.41	1.41	
3030	Ceiling Finishes	Suspended mineral tile on zee channels in office area	10% of area	S.F. Ceiling	4.09	.41	
D. SERVICES							
	D10 Conveying						
1010	Elevators & Lifts	N/A		—	—	—	0.0%
1020	Escalators & Moving Walks	N/A		—	—	—	
	D20 Plumbing						
2010	Plumbing Fixtures	Toilet and service fixtures, supply and drainage	1 Fixture/2500 S.F. Floor	Each	1925	.77	
2020	Domestic Water Distribution	Gas fired water heater		S.F. Floor	.18	.18	4.1%
2040	Rain Water Drainage	Roof drains		S.F. Roof	1.05	1.05	
	D30 HVAC						
3010	Energy Supply	Oil fired hot water, unit heaters	90% of area	S.F. Floor	3.62	3.62	
3020	Heat Generating Systems	N/A		—	—	—	
3030	Cooling Generating Systems	N/A		—	—	—	8.8%
3050	Terminal & Package Units	Single zone unit gas, heating, electric cooling	10% of area	S.F. Floor	.72	.72	
3090	Other HVAC Sys. & Equipment	N/A		—	—	—	
	D40 Fire Protection						
4010	Sprinklers	Sprinklers, ordinary hazard		S.F. Floor	2.20	2.20	4.5%
4020	Standpipes	N/A		—	—	—	
	D50 Electrical						
5010	Electrical Service/Distribution	200 ampere service, panel board and feeders		S.F. Floor	.37	.37	
5020	Lighting & Branch Wiring	Fluorescent fixtures, receptacles, switches, A.C. and misc. power		S.F. Floor	4.01	4.01	9.7%
5030	Communications & Security	Alarm systems		S.F. Floor	.36	.36	
5090	Other Electrical Systems	N/A		—	—	—	
E. EQUIPMENT & FURNISHINGS							
1010	Commercial Equipment	N/A		—	—	—	
1020	Institutional Equipment	N/A		—	—	—	4.5%
1030	Vehicular Equipment	Dock boards, dock levelers		S.F. Floor	2.20	2.20	
1090	Other Equipment	N/A		—	—	—	
F. SPECIAL CONSTRUCTION							
1020	Integrated Construction	N/A		—	—	—	0.0%
1040	Special Facilities	N/A		—	—	—	
G. BUILDING SITEWORK	**N/A**						

		Sub-Total	49.06	100%
CONTRACTOR FEES (General Requirements: 10%, Overhead: 5%, Profit: 10%)		25%	12.25	
ARCHITECT FEES		7%	4.29	

Total Building Cost	**65.60**

BUILDING TYPES

Costs per square foot of floor area

Exterior Wall	S.F. Area	2000	3000	5000	8000	12000	20000	30000	50000	100000
	L.F. Perimeter	180	220	300	420	580	900	1300	2100	4100
Concrete Block	Steel Frame	132.35	115.80	102.50	94.95	90.80	87.50	85.75	84.45	83.45
	R/Conc. Frame	118.50	108.15	99.95	95.35	92.70	90.70	89.65	88.85	88.25
Metal Sandwich Panel	Steel Frame	112.90	101.75	92.90	87.95	85.20	82.95	81.80	80.90	80.30
Tilt-up Concrete Panel	R/Conc. Frame	123.90	113.25	104.70	99.95	97.25	95.15	94.05	93.15	92.55
Precast Concrete Panel	Steel Frame	112.55	101.35	92.40	87.40	84.60	82.40	81.25	80.40	79.70
	R/Conc. Frame	129.70	117.80	108.35	103.00	99.95	97.60	96.40	95.45	94.80
Perimeter Adj., Add or Deduct	Per 100 L.F.	20.70	13.80	8.20	5.15	3.45	2.05	1.45	.80	.45
Story Hgt. Adj., Add or Deduct	Per 1 Ft.	2.15	1.65	1.40	1.25	1.10	1.05	1.05	.95	1.00
Basement—Not Applicable										

The above costs were calculated using the basic specifications shown on the facing page. These costs should be adjusted where necessary for design alternatives and owner's requirements. Reported completed project costs, for this type of structure, range from $25.50 to $146.30 per S.F.

Common additives

Description	Unit	$ Cost
Dock Leveler, 10 ton cap.		
6' x 8'	Each	5650
7' x 8'	Each	5500
Emergency Lighting, 25 watt, battery operated		
Lead battery	Each	247
Nickel cadmium	Each	695
Fence, Chain link, 6' high		
9 ga. wire	L.F.	17.75
6 ga. wire	L.F.	25
Gate	Each	258
Flagpoles, Complete		
Aluminum, 20' high	Each	1375
40' high	Each	3025
70' high	Each	9500
Fiberglass, 23' high	Each	1650
39'-5" high	Each	3175
59' high	Each	7950
Paving, Bituminous		
Wearing course plus base course	S.Y.	6.80
Sidewalks, Concrete 4" thick	S.F.	3.40

Description	Unit	$ Cost
Sound System		
Amplifier, 250 watts	Each	1925
Speaker, ceiling or wall	Each	160
Trumpet	Each	305
Yard Lighting,	Each	2375
20' aluminum pole with 400 watt high pressure sodium fixture		

Important: See the Reference Section for Location Factors

Model costs calculated for a 1 story building with 12' story height and 20,000 square feet of floor area

				Unit	Unit Cost	Cost Per S.F.	% Of Sub-Total
A. SUBSTRUCTURE							
1010	Standard Foundations	Poured concrete; strip and spread footings		S.F. Ground	1.47	1.47	
1030	Slab on Grade	4" reinforced concrete with vapor barrier and granular base		S.F. Slab	7.81	7.81	
2010	Basement Excavation	Site preparation for slab and trench for foundation wall and footing		S.F. Ground	.23	.23	18.3%
2020	Basement Walls	4' foundation wall		L.F. Wall	54	2.44	
B. SHELL							
	B10 Superstructure						
1010	Floor Construction	N/A		—	—	—	
1020	Roof Construction	Metal deck, open web steel joists, beams, columns		S.F. Roof	6.68	6.68	10.2%
	B20 Exterior Enclosure						
2010	Exterior Walls	Concrete block	70% of wall	S.F. Wall	10.13	3.83	
2020	Exterior Windows	Aluminum projecting	5% of wall	Each	618	5.57	18.1%
2030	Exterior Doors	Steel overhead, hollow metal	25% of wall	Each	1098	2.42	
	B30 Roofing						
3010	Roof Coverings	Built-up tar and gravel with flashing; perlite/EPS composite insulation		S.F. Roof	3.54	3.54	5.4%
3020	Roof Openings	N/A		—	—	—	
C. INTERIORS							
1010	Partitions	Concrete block, gypsum board on metal studs	10.65 S.F. Floor/L.F. Partition	S.F. Partition	6.69	11.47	
1020	Interior Doors	Single leaf hollow metal	300 S.F. Floor/Door	Each	1002	3.34	
1030	Fittings	N/A		—	—	—	
2010	Stair Construction	N/A		—	—	—	22.6%
3010	Wall Finishes	N/A		—	—	—	
3020	Floor Finishes	N/A		—	—	—	
3030	Ceiling Finishes	N/A		—	—	—	
D. SERVICES							
	D10 Conveying						
1010	Elevators & Lifts	N/A		—	—	—	
1020	Escalators & Moving Walks	N/A		—	—	—	0.0%
	D20 Plumbing						
2010	Plumbing Fixtures	Toilet and service fixtures, supply and drainage	1 Fixture/5000 S.F. Floor	Each	1650	.33	
2020	Domestic Water Distribution	Gas fired water heater		S.F. Floor	.14	.14	2.0%
2040	Rain Water Drainage	Roof drains		S.F. Roof	.83	.83	
	D30 HVAC						
3010	Energy Supply	N/A		—	—	—	
3020	Heat Generating Systems	Oil fired hot water, unit heaters		Each	6.22	6.22	
3030	Cooling Generating Systems	N/A		—	—	—	9.5%
3050	Terminal & Package Units	N/A		—	—	—	
3090	Other HVAC Sys. & Equipment	N/A		—	—	—	
	D40 Fire Protection						
4010	Sprinklers	Wet pipe sprinkler system		S.F. Floor	2.71	2.71	4.1%
4020	Standpipes	N/A		—	—	—	
	D50 Electrical						
5010	Electrical Service/Distribution	200 ampere service, panel board and feeders		S.F. Floor	1.23	1.23	
5020	Lighting & Branch Wiring	Fluorescent fixtures, receptacles, switches and misc. power		S.F. Floor	4.72	4.72	9.8%
5030	Communications & Security	Alarm systems		S.F. Floor	.34	.34	
5090	Other Electrical Systems	Emergency generator, 7.5 kW		S.F. Floor	.09	.09	
E. EQUIPMENT & FURNISHINGS							
1010	Commercial Equipment	N/A		—	—	—	
1020	Institutional Equipment	N/A		—	—	—	
1030	Vehicular Equipment	N/A		—	—	—	0.0%
1090	Other Equipment	N/A		—	—	—	
F. SPECIAL CONSTRUCTION							
1020	Integrated Construction	N/A		—	—	—	0.0%
1040	Special Facilities	N/A		—	—	—	
G. BUILDING SITEWORK	**N/A**						

Sub-Total		65.41	**100%**
CONTRACTOR FEES (General Requirements: 10%, Overhead: 5%, Profit: 10%)		25%	16.37
ARCHITECT FEES		7%	5.72
Total Building Cost		**87.50**	

BUILDING TYPES

Deterioration occurs within the structure itself and is determined by the observation of both materials and equipment.

Curable deterioration can be remedied either by maintenance, repair or replacement, within prudent economic limits.

Incurable deterioration that has progressed to the point of actually affecting the structural integrity of the structure, making repair or replacement not economically feasible.

Actual Versus Observed Age

The observed age of a structure refers to the age that the structure appears to be. Periodic maintenance, remodeling and renovation all tend to reduce the amount of deterioration that has taken place, thereby decreasing the observed age. Actual age on the other hand relates solely to the year that the structure was built.

The Depreciation Table shown here relates to the observed age.

Obsolescence arises from conditions either occurring within the structure (functional) or caused by factors outside the limits of the structure (economic).

Functional obsolescence is any inadequacy caused by outmoded design, dated construction materials or oversized or undersized areas, all of which cause excessive operation costs.

Incurable is so costly as not to economically justify the capital expenditure required to correct the deficiency.

Economic obsolescence is caused by factors outside the limits of the structure. The prime causes of economic obsolescence are:

- zoning and environmental laws
- government legislation
- negative neighborhood influences
- business climate
- proximity to transportation facilities

Depreciation Table
Commercial/Industrial/Institutional

Building Material			
Observed Age (Years)	Frame	Masonry On Wood	Masonry On Masonry or Steel
1	1%	0%	0%
2	2	1	0
3	3	2	1
4	4	3	2
5	6	5	3
10	20	15	8
15	25	20	15
20	30	25	20
25	35	30	25
30	40	35	30
35	45	40	35
40	50	45	40
45	55	50	45
50	60	55	50
55	65	60	55
60	70	65	60

DEPRECIATION

Assemblies Section

Table of Contents

Table No.	Page
A SUBSTRUCTURE	239
A1010 Standard Foundations	
A1010 100 Wall Foundations	240
A1010 200 Column Foundations & Pile Caps	241
A1010 300 Perimeter Drainage	241
A1030 Slab on Grade	
A1030 100 Standard Slab on Grade	243
A2010 Basement Excavation	
A2010 100 Excavation for Basements	244
A2020 Basement Walls	
A2020 100 Basement Wall Construction	245
B SHELL	247
B1010 Floor Construction	
B1010 200 Upper Floor Construction	248
B1010 700 Fireproofing	275
B1020 Roof Construction	
B1020 100 Flat Roof Construction	276
B2010 Exterior Walls	
B2010 100 Exterior Wall Construction	282
B2020 Exterior Windows	
B2020 100 Windows	314
B2020 200 Curtain Walls	316
B2030 Exterior Doors	
B2030 100 Glazed Doors/Entrances	318
B2030 200 Exterior Doors	320
B3010 Roof Coverings	
B3010 100 Roof Finishes	322
B3010 300 Roof Insulation/Fill	325
B3010 400 Flashing & Trim	325
B3010 600 Gutters & Downspouts	327
B3020 Roof Openings	
B3020 100 Glazed Roof Openings	328
B3020 200 Roof Hatches	328
C INTERIORS	329
C1010 Partitions	
C1010 100 Fixed Partitions	330
C1010 200 Demountable Partitions	339
C1020 Interior Doors	
C1020 100 Interior Doors	340
C1030 Fittings	
C1030 100 Fabricated Toilet Partitions	341
C1030 300 Storage Shelving/Lockers	342

Table No.	Page
C1030 500 Identifying Devices	342
C1030 700 Gen. Fittings & Misc. Metals	342
C1030 800 Fabricated Cabinets/Counters	342
C1030 900 Other Fittings	343
C2010 Stair Construction	
C2010 100 Regular Stairs	344
C3010 Wall Finishes	
C3010 200 Wall Finishes to Interiors	345
C3020 Floor Finishes	
C3020 400 Flooring	347
C3030 Ceiling Finishes	
C3030 100 Ceiling Finishes	349
C3030 200 Suspended Ceilings	350
D SERVICES	353
D1010 Elevators and Lifts	
D1010 100 Passenger Elevators	354
D1090 Other Conveying Systems	
D1090 500 Chutes	356
D2010 Plumbing Fixtures	
D2010 100 Water Closets	359
D2010 200 Urinals	359
D2010 300 Lavatories	360
D2010 400 Sinks	360
D2010 500 Bathtubs	363
D2010 600 Wash Fountains	363
D2010 700 Showers	364
D2010 800 Drinking Fountains/Coolers	364
D2010 900 Bidets, Other Plumb. Fixt.	366
D2020 Domestic Water Distribution	
D2020 200 Hot Water Service	368
D2040 Rain Water Drainage	
D2040 200 Roof Drains	371
D2090 Other Plumbing Systems	
D2090 800 Piping & Fittings	372
D3010 Energy Supply	
D3010 500 Hot Water Supply System	375
D3020 Heat Generating Systems	
D3020 100 Boilers	376
D3030 Cooling Generating Systems	
D3030 100 Chilled Water Systems	381
D3050 Terminal & Pkg. Units	
D3050 100 Terminal Self-Contained Units	385

Table No.	Page
D3090 Other HVAC Sys. & Equip.	
D3090 300 Dust/Fume Collectors	395
D4010 Sprinklers	
D4010 300 Dry Sprinkler System	398
D4010 400 Wet Sprinkler System	406
D4020 Standpipes	
D4020 300 Standpipe Equipment	408
D4020 400 Fire Hose Equipment	410
D4090 Other Fire Protection Systems	
D4090 900 Misc. Other Fire Prot. Sys.	411
D5010 Electrical Service/Dist.	
D5010 100 High Tension Service & Dist.	412
D5010 200 Low Tension Service & Dist.	412
D5020 Lighting & Branch Wiring	
D5020 100 Branch Wiring Devices	414
D5020 200 Lighting Equipment	418
D5030 Comm. & Security	
D5030 900 Other Comm. & Security	421
D5090 Other Electrical Systems	
D5090 200 Emerg. Light & Power Sys.	422
E EQUIPMENT & FURNISHINGS	423
E1010 Commercial Equipment	
E1010 100 Security & Vault Equipment	424
E1010 500 Mercantile Equipment	424
E1010 600 Laund. & Dry Cleaning Equip.	424
E1020 Institutional Equipment	
E1020 100 Ecclesiastical Equipment	425
E1020 200 Library Equipment	425
E1020 300 Theater & Stage Equipment	425
E1020 600 Detention Equipment	425
E1020 700 Laboratory Equipment	426
E1020 800 Medical Equipment	426
E1030 Vehicular Equipment	
E1030 100 Vehicular Service Equipment	427
E1030 200 Parking Control Equipment	427
E1030 300 Loading Dock Equipment	427
E1090 Other Equipment	
E1090 200 Solid Waste Handling Equip.	428
E1090 300 Food Service Equipment	428
E1090 400 Residential Equipment	428

Table of Contents

Table No.   Page

E1090 600 School Equipment...................... 428
E1090 800 Athletic, Rec. & Therap. Eq........ 429

E2010 Fixed Furnishings

E2010 300 Blinds/Window Treatment 430
E2010 400 Fixed Floor Grilles & Mats 430
E2010 500 Fixed Multiple Seating 430

E2020 Moveable Furnishings

E2020 200 Furniture & Accessories............ 431

F SPECIAL CONSTRUCTION   433

F1010 Special Structures

F1010 100 Air-Supported Structures............ 434
F1010 200 Pre-Engineered Structures.......... 434
F1010 300 Other Special Structures............ 434

F1020 Integrated Construction

F1020 100 Integrated Assemblies................ 436
F1020 200 Special Purpose Rooms 436
F1020 300 Other Integrated Construction.. 436

Table No.   Page

F1030 Special Construction Systems

F1030 100 Sound, Vibration &
            Seismic Const. 437
F1030 200 Radiation Protection 437
F1030 900 Other Special Const. Systems.... 437

F1040 Special Facilities

F1040 200 Ice Rinks.................................... 438
F1040 500 Liquid & Gas Storage Tanks 438
F1040 900 Special Construction................. 439

G BUILDING SITEWORK   441

G1030 Site Earthwork

G1030 800 Utilities Trenching..................... 442

G2010 Roadways

G2010 200 Paving & Surfacing 443

G2020 Parking Lots

G2020 200 Paving & Surfacing 444

Table No.   Page

G2040 Site Development

G2040 800 Flagpoles.................................... 445
G2040 900 Miscellaneous Structures 445

G3030 Storm Sewer

G3030 200 Manholes.................................... 446

G4020 Site Lighting

G4020 200 Site Lighting Controls................ 447

Introduction to the Assemblies Section

This section of the manual contains the technical description and installed cost of all components used in the commercial/industrial/institutional section. In most cases, there is an illustration of the component to aid in identification.

All Costs Include:

Materials purchased in lot sizes typical of normal construction projects with discounts appropriate to established contractors.

Installation work performed by union labor working under standard conditions at a normal pace.

Installing Contractor's Overhead & Profit

Standard installing contractor's overhead & profit are included in the assemblies costs.

These assemblies costs contain no provisions for General Contractor's overhead and profit or Architect's fees, nor do they include premiums for labor or materials or savings which may be realized under certain economic situations.

The assemblies tables are arranged by physical size (dimensions) of the component wherever possible.

How to Use the Assemblies Cost Tables

The following is a detailed explanation of a sample Assemblies Cost Table. Most Assembly Tables are separated into two parts: 1) an illustration of the system to be estimated with descriptive notes; and 2) the costs for similar systems with dimensional and/or size variations. Next to each bold number below is the item being described with the appropriate component of the sample entry following in parenthesis. In most cases, if the work is to be subcontracted, the general contractor will need to add an additional markup (R.S. Means suggests using 10%) to the "Total" figures.

System/Line Numbers (A1030 120 2240)

Each Assemblies Cost Line has been assigned a unique identification number based on the UNIFORMAT II classification sstem.

UNIFORMAT II Major Group

A1030. **120.** **2240.**

UNIFORMAT II Level 3

Means Major Classification

Means Individual Line Number

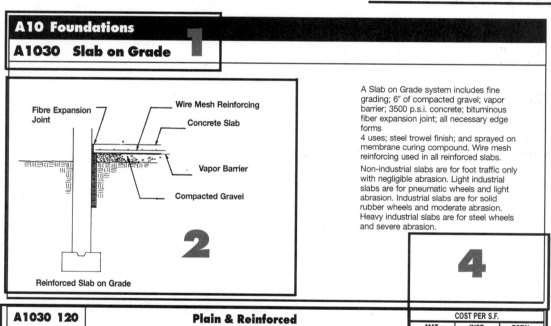

A10 Foundations

A1030 Slab on Grade

A Slab on Grade system includes fine grading; 6" of compacted gravel; vapor barrier; 3500 p.s.i. concrete; bituminous fiber expansion joint; all necessary edge forms 4 uses; steel trowel finish; and sprayed on membrane curing compound. Wire mesh reinforcing used in all reinforced slabs.

Non-industrial slabs are for foot traffic only with negligible abrasion. Light industrial slabs are for pneumatic wheels and light abrasion. Industrial slabs are for solid rubber wheels and moderate abrasion. Heavy industrial slabs are for steel wheels and severe abrasion.

A1030 120	Plain & Reinforced	COST PER S.F.		
		MAT.	INST.	TOTAL
2220	Slab on grade, 4" thick, non industrial, non reinforced	1.44	2.03	3.47
2240	Reinforced	1.66	2.32	3.98
2260	Light industrial, non reinforced	1.80	2.49	4.29
2280	Reinforced	2.02	2.78	4.80
2300	Industrial, non reinforced	2.9	2.0	.9
2320	Reinforced		5.40	.81
3340	5" thick, non industrial, non reinforced	1.71	2.08	3.79
3360	Reinforced	1.93	2.37	4.30
3380	Light industrial, non reinforced	2.08	2.54	4.62
3400	Reinforced	2.30	2.83	5.13
3420	Heavy industrial, non reinforced	2.84	6.10	8.94
3440	Reinforced	3.05	6.40	9.45
4460	6" thick, non industrial, non reinforced	2.07	2.04	4.11
4480	Reinforced	2.41	2.43	4.84
4500	Light industrial, non reinforced	2.45	2.50	4.95
4520	Reinforced	2.98	3.02	6
4540	Heavy industrial, non reinforced	3.22	6.15	9.37
4560	Reinforced	3.56	6.55	10.11
5580	7" thick, non industrial, non reinforced	2.35	2.11	4.46
5600	Reinforced	2.78	2.52	5.30

2 Illustration

At the top of most assembly pages is an illustration, a brief description, and the design criteria used to develop the cost.

3 System Description

The components of a typical system are listed in the description to show what has been included in the development of the total system price. The rest of the table contains prices for other similar systems with dimensional and/or size variations.

4 Unit of Measure for Each System (S.F.)

Costs shown in the three right hand columns have been adjusted by the component quantity and unit of measure for the entire system. In this example, "Cost per S.F." is the unit of measure for this system or "assembly."

5 Materials (1.66)

This column contains the Materials Cost of each component. These cost figures are bare costs plus 10% for profit.

6 Installation (2.32)

Installation includes labor and equipment plus the installing contractor's overhead and profit. Equipment costs are the bare rental costs plus 10% for profit. The labor overhead and profit is defined on the inside back cover of this book.

7 Total (3.98)

The figure in this column is the sum of the material and installation costs.

Material Cost	+	Installation Cost	=	Total
$1.66	+	$2.32	=	$3.98

Example

Assemblies costs can be used to calculate the cost for each component of a building and, accumulated with appropriate markups, produce a cost for the complete structure. Components from a model building with similar characteristics in conjunction with Assemblies components may be used to calculate costs for a complete structure.

Example:

Outline Specifications

General	Building size—60′ x 100′, 8 suspended floors, 12′ floor to floor, 4′ high parapet above roof, full basement 11′-8″ floor to floor, bay size 25′ x 30′, ceiling heights - 9′ in office area and 8′ in core area.
A Substructure	Concrete, spread and strip footings, concrete walls waterproofed, 4″ slab on grade.
B Shell	
B10 Superstructure	Columns, wide flange; 3 hr. fire rated; Floors, composite steel frame & deck with concrete slab; Roof, steel beams, open web joists and deck.
B20 Exterior Enclosure	Walls; North, East & West, brick & lightweight concrete block with 2″ cavity insulation, 25% window; South, 8″ lightweight concrete block insulated, 10% window. Doors, aluminum & glass. Windows, aluminum, 3′-0″ x 5′-4″, insulating glass.
B30 Roofing	Tar & gravel, 2″ rigid insulation, R 12.5.
C Interiors	Core—6″ lightweight concrete block, full height with plaster on exposed faces to ceiling height.
	Corridors—1st & 2nd floor; 3-5/8″ steel studs with F.R. gypsum board, full height.
	Exterior Wall—plaster, ceiling height.
	Fittings—Toilet accessories, directory boards.
	Doors—hollow metal.
	Wall Finishes—lobby, mahogany paneling on furring, remainder plaster & gypsum board, paint.
	Floor Finishes—1st floor lobby, corridors & toilet rooms, terrazzo; remainder, concrete, tenant developed. 2nd thru 8th, toilet rooms, ceramic tile; office & corridor, carpet.
	Ceiling Finishes—24″ x 48″ fiberglass board on Tee grid.
D Services	
D10 Conveying	2-2500 lb. capacity, 200 F.P.M. geared elevators, 9 stops.
D20 Plumbing	Fixtures, see sketch.
	Roof Drains—2-4″ C.I. pipe.
D30 HVAC	Heating—fin tube radiation, forced hot water.
	Air Conditioning—chilled water with water cooled condenser.
D40 Fire Protection	Fire Protection—4″ standpipe, 9 hose cabinets.
D50 Electrical	Lighting, 1st thru 8th, 15 fluorescent fixtures/1000 S.F., 3 Watts/S.F.
	Basement, 10 fluorescent fixtures/1000 S.F., 2 Watts/S.F.
	Receptacles, 1st thru 8th, 16.5/1000 S.F., 2 Watts/S.F.
	Basement, 10 receptacles/1000 S.F., 1.2 Watts/S.F.
	Air Conditioning, 4 Watts/S.F.
	Miscellaneous Connections, 1.2 Watts/S.F.
	Elevator Power, 2-10 H.P., 230 volt motors.
	Wall Switches, 2/1000 S.F.
	Service, panel board & feeder, 2000 Amp.
	Fire Detection System, pull stations, signals, smoke and heat detectors.
	Emergency Lighting Generator, 30 KW.
E Equipment & Furnishings	NA.
F Special Construction	NA.
G Building Sitework	NA.

Front Elevation

Basement Plan

Ground Floor Plan

Typical Floor Plan

Example

If spread footing & column sizes are unknown, develop approximate loads as follows. Enter tables with these loads to determine costs.

Superimposed Load Ranges

Apartments & Residential Structures	65	to	75 psf
Assembly Areas & Retail Stores	110	to	125 psf
Commercial & Manufacturing	150	to	250 psf
Offices	75	to	100 psf

Approximate loads/S.F. for roof & floors.
Roof. Assume 40 psf superimposed load.
Steel joists, beams & deck.
Table B1020 112—Line 3900

B1020 112	Steel Joists, Beams, & Deck on Columns							
	BAY SIZE (FT.)	SUPERIMPOSED LOAD (P.S.F.)	DEPTH (IN.)	TOTAL LOAD (P.S.F.)	COLUMN ADD	COST PER S.F.		
						MAT.	INST.	TOTAL
3500	25x30	20	22	40		2.58	1.23	3.81
3600					columns	.58	.22	.80
3900		40	25	60		3.14	1.45	4.59
4000					columns	.69	.26	.95

B1010 256	Composite Beams, Deck & Slab							
	BAY SIZE (FT.)	SUPERIMPOSED LOAD (P.S.F.)	SLAB THICKNESS (IN.)	TOTAL DEPTH (FT.-IN.)	TOTAL LOAD (P.S.F.)	COST PER S.F.		
						MAT.	INST.	TOTAL
3400	25x30	40	5-1/2	1 - 11-1/2	83	6.50	4.36	10.86
3600		75	5-1/2	1 - 11-1/2	119	7	4.39	11.39
3900		125	5-1/2	1 - 11-1/2	170	8.05	4.95	13
4000		200	6-1/4	2 - 6-1/4	252	9.90	5.60	15.50

Floors–Total load, 119 psf.
Interior foundation load.
Roof

[(25' x 30' x 60 psf) + 8 floors x (25' x 30' x 119 psf)] x 1/1000 lb./Kip =	759 Kips
Approximate Footing Loads, Interior footing =	759 Kips
Exterior footing (1/2 bay) 759 k x .6 =	455 Kips
Corner footing (1/4 bay) 759 k x .45 =	342 Kips
[Factors to convert Interior load to Exterior & Corner loads]	
Approximate average Column load 759 k/2 =	379 Kips

PRELIMINARY ESTIMATE

PROJECT	**Office Building**		TOTAL SITE AREA		
BUILDING TYPE			OWNER		
LOCATION			ARCHITECT		
DATE OF CONSTRUCTION			ESTIMATED CONSTRUCTION PERIOD		

BRIEF DESCRIPTION **Building size: 60' x 100' - 8 structural floors - 12' floor to floor**

4' high parapet above roof, full basement 11' - 8" floor to floor, bay size

25' x 30', ceiling heights - 9' in office area + 8' in core area

TYPE OF PLAN		TYPE OF CONSTRUCTION
QUALITY		BUILDING CAPACITY

Floor			**Wall Area**				
Below Grade Levels			Foundation Walls	L.F.		Ht.	S.F.
Area		S.F.	Frost Walls	L.F.		Ht.	S.F.
Area		S.F.	Exterior Closure		Total		S.F.
Total Area		S.F.	Comment				
Ground Floor			Fenestration		%		S.F.
Area		S.F.			%		S.F.
Area		S.F.	Exterior Wall		%		S.F.
Total Area		S.F.			%		S.F.
Supported Levels			**Site Work**				
Area			Parking		S.F. (For		Cars)
Area		S.F.	Access Roads		L.F. (X		Ft. Wide)
Area		S.F.	Sidewalk		L.F. (X		Ft. Wide)
Area		S.F.	Landscaping		S.F. (		% Unbuilt Site)
Area		S.F.	**Building Codes**				
Total Area		S.F.	City			Country	
Miscellaneous			National			Other	
Area		S.F.	**Loading**				
Area		S.F.	Roof	psf	Ground Floor		psf
Area		S.F.	Supported Floors	psf	Corridor		psf
Area		S.F.	Balcony	psf	Partition, allow		psf
Total Area		S.F.	Miscellaneous				psf
Net Finished Area		S.F.	Live Load Reduction				
Net Floor Area		S.F.	Wind				
Gross Floor Area	**54,000**	S.F.	Earthquake			Zone	
Roof			Comment				
Total Area		S.F.	Soil Type				
Comments			Bearing Capacity				K.S.F.
			Frost Depth				Ft.
Volume			**Frame**				
Depth of Floor System			Type		Bay Spacing		
Minimum		In.	Foundation				
Maximum		In.	Special				
Foundation Wall Height		Ft.	Substructure				
Floor to Floor Height		Ft.	Comment				
Floor to Ceiling Height		Ft.	Superstructure, Vertical				
Subgrade Volume		C.F.	Fireproofing	☐ Columns			Hrs.
Above Grade Volume		C.F.	☐ Girders	Hrs.	☐ Beams		Hrs.
Total Building Volume	**648,000**	C.F.	☐ Floor	Hrs.	☐ None		

NUMBER		QTY.	UNIT	TOTAL COST UNIT	TOTAL	COST PER S.F.
A	**SUBSTRUCTURE**					
A1010 210 7900	Corner Footings 8'-6" SQ. x 27"	4	Ea.	1,495.00	5,980	
8010	Exterior 9'-6" SQ. x 30"	8		1,995.00	15,960	
8300	Interior 12" SQ.	3		3,650.00	10,950	
A1010 110 2700	Strip 2' Wide x 1' Thick					
	320 L.F. [(4 x 8.5) + 8 x 9.5)] =	320	L.F.	29.40	9,408	
A2020 110 7262	Foundation Wall 12' High, 1' Thick	320		188.50	60,320	
A1010 320 2800	Foundation Waterproofing	320		14.22	4,550	
A1030 120 2240	4", non-industrial, reinf. slab on grade	6000	S.F.	3.98	23,880	
A2010 110 3440	Building Excavation + Backfill	6000	S.F.	5.22	31,320	
3500	(Interpolated ; 12' Between					
4620	8' and 16' ; 6,000 Between					
4680	4,000 and 10,000 S.F.					
	Total				162,368	3.01
B	**SHELL**					
B10	**Superstructure**					
B1010 208 5800	Columns: Load 379K Use 400K					
	Exterior 12 x 96' + Interior 3 x 108'	1476	V.L.F.	100.30	148,043	
B1010 720 3650	Column Fireproofing - Interior 4 sides					
3700	Exterior 1/2 (Interpolated for 12")	900	V.L.F	26.65	23,985	
B1010 256 3600	Floors: Composite steel + lt. wt. conc.	48000	S.F.	14.02	672,960	
B1020 112 3900	Roof: Open web joists, beams + decks	6000	S.F.	6.13	36,780	
	Total				881,768	16.33
B20	**Exterior Enclosure**					
B2010 134 1200	4" Brick + 6" Block - Insulated					
	75% NE + W Walls 220' x 100' x .75	16500	S.F.	22.45	370,425	
B2010 110 3410	8" Block - Insulated 90%					
	100' x 100' x .9	9000	S.F.	8.42	75,780	
B2030 110 6950	Double Aluminum + Glass Door	1	Pr.	3,625.00	3,625	
6300	Single	2	Ea.	1,995.00	3,990	
B2020 106 8800	Aluminum windows - Insul. glass					
	(5,500 S.F. + 1,000 S.F.)/(3 x 5.33)	406	Ea.	754.00	306,124	
B2010 105 8480	Precast concrete coping	320	L.F.	28.60	9,152	
	Total				769,096	14.24
B30	**Roofing**					
B3010 320 4020	Insulation: Perlite/Polyisocyanurate	6000	S.F.	1.44	8,640	
B3010 105 1400	Roof: 4 Ply T & G composite	6000		1.88	11,280	
B3010 430 0300	Flashing: Aluminum - fabric backed	640		2.24	1,434	
B3020 210 0500	Roof hatch	1	Ea.	849.00	849	
	Total				22,203	0.41
	Page Total					

234

ASSEMBLY NUMBER		QTY.	UNIT	TOTAL COST		COST PER S.F.
				UNIT	TOTAL	
C	INTERIORS					
C1010 104 5500	Core partitions: 6" Lt. Wt. concrete					
	block 288 L.F. x 11.5' x 8 Floors	26,449	S.F.	6.74	178,266	
C1010 144 0920	Plaster: [(196 L.F. x 8') + (144 L.F. x 9')] 8					
	+ Ext. Wall [(320 x 9 x 8 Floors) - 6500]	39,452	S.F.	2.41	95,079	
C1010 126 5400	Corridor partitions 1st + 2nd floors					
	steel studs + F.R. gypsum board	2,530	S.F.	3.25	8,223	
C1020 102 2600	Interior doors	66	Ea.	608.00	40,128	
	Wall finishes - Lobby: Mahogany w/furring					
C1010 128 0652	Furring: 150 L.F. x 9' + Ht.	1,350	S.F.	1.17	1,580	
C1030 710 0120	Towel Dispenser	16	Ea.	72.00	1,152	
C1030 710 0140	Grab Bar	32	Ea.	38.30	1,226	
C1030 710 0160	Mirror	31	Ea.	146.50	4,542	
C1030 710 0180	Toilet Tissue Dispenser	31	Ea.	26.15	811	
C1030 510 0140	Directory Board	8	Ea.	226.00	1,808	
C2010 110 0780	Stairs: Steel w/conc. fill	18	Flt.	8,975.00	161,550	
C3010 230 1662	Mahogany paneling	1,350	S.F.	4.61	6,224	
	Paint, plaster + gypsum board					
C3010 230 0080	39,452 + [(220' x 9' x 2) - 1350] + 72	42,134	S.F.	0.90	37,921	
	Floor Finishes					
C3020 410 1100	1st floor lobby + terrazzo	2,175	S.F.	9.79	21,293	
	Remainder: Concrete - Tenant finished					
C3020 410 0060	2nd - 8th: Carpet - 4725 S.F. x 7	33,075	S.F.	4.25	140,569	
C3020 410 1720	Ceramic Tile - 300 S.F. x 7	2,100	S.F.	8.11	17,031	
C3030 240 2780	Suspended Ceiling Tiles- 5200 S.F. x 8	41,600	S.F.	1.55	64,480	
C3030 240 3260	Suspended Ceiling Grid- 5200 S.F. x 8	41,600	S.F.	1.07	44,512	
C1030 110 0700	Toilet partitions	31	Ea.	762.00	23,622	
0760	Handicapped addition	16	Ea.	315.00	5,040	
	Total				855,054	15.83
D	SERVICES					
D10	Conveying					
	2 Elevators - 2500# 200'/min. Geared					
D1010 140 1600	5 Floors $113,300					
1800	15 Floors $ 216,000					
	Diff. $102,700/10 = 10,270/Flr.					
	$113,300 + (4 x 10,270) = $154,380/Ea.	2	Ea.	154,380.00	308,760	
	Total				308,760	5.72
D20	Plumbing					
D2010 310 1600	Plumbing - Lavatories	31	Ea.	767.00	23,777	
D2010 440 4340	-Service Sink	8	Ea.	1,845.00	14,760	
D2010 210 2000	-Urinals	8	Ea.	1,030.00	8,240	

ASSEMBLY NUMBER		QTY.	UNIT	TOTAL COST		COST PER S.F.
				UNIT	TOTAL	
D2010 110 2080	-Water Closets	31	Ea.	1,385.00	42,935	
	Water Control, Waste Vent Piping	45%			40,370	
D2040 210 4200	- Roof Drains, 4" C.I.	2	Ea.	1,060.00	2,120	
4240	- Pipe, 9 Flr. x 12' x 12' Ea.	216	L.F.	25.05	5,411	
	Total				137,613	2.55
D30	HVAC					
D3010 520 2000	10,000 S.F. @ $6.74/S.F. Interpolate					
2040	10,000 S.F. @ $2.99/S.F.	48,000	S.F.	4.87	233,760	
	Cooling - Chilled Water, Air Cool, Cond.					
D3030 115 4000	40,000 S.F. @10.32/S.F. Interpolate	48,000	S.F.	10.44	501,120	
4040	60,000 S.F. @10.55/S.F.					
	Total				734,880	13.61
D40	Fire Protection					
D4020 310 0560	Wet Stand Pipe: 4" x 10' 1st floor	12/10	Ea.	4,825.00	5,790	
0580	Additional Floors: (12'/10 x 8)	9.6	Ea.	1,335.00	12,816	
D4020 410 8400	Cabinet Assembly	9	Ea.	753.00	6,777	
	Total				25,383	0.47
D50	Electrical					
D5020 210 0280	Office Lighting, 15/1000 S.F. - 3 Watts/S.F.	48,000	S.F.	6.35	304,800	
0240	Basement Lighting, 10/1000 S.F. - 2 Watts/S.F.	6,000	S.F.	4.22	25,320	
D5020 110 0640	Office Receptacles 16.5/1000 S.F. - 2 Watts/S.F.	48,000	S.F.	3.68	176,640	
0560	Basement Receptacles 10/1000 S.F. - 1.2 Watts/S.F.	6,000	S.F.	2.82	16,920	
D5020 140 0280	Central A.C. - 4 Watts/S.F.	48,000	S.F.	0.48	23,040	
D5020 135 0320	Misc. Connections - 1.2 Watts/S.F.	48,000	S.F.	0.26	12,480	
D5020 145 0680	Elevator Motor Power - 10 H.P.	2	Ea.	2,400.00	4,800	
D5020 130 0280	Wall Switches - 2/1000 S.F.	54,000	S.F.	0.36	19,440	
D5010 120 0560	2000 Amp Service	1	Ea.	31,275.00	31,275	
D5010 240 0400	Switchgear	1	Ea.	42,100.00	42,100	
D5010 230 0560	Feeder	50	L.F.	407.00	20,350	
D5030 910 0400	Fire Detection System - 50 Detectors	1	Ea.	26,200.00	26,200	
D5090 210 0320	Emergency Generator - 30 kW	30	kW	446.00	13,380	
	Total				716,745	13.27
E	EQUIPMENT & FURNISHINGS					
G	BUILDING SITEWORK					
	Miscellaneous					
			Page Total			

Preliminary Estimate Cost Summary

PROJECT: Office Building	TOTAL AREA	54,000 S.F.	SHEET NO.	
LOCATION	TOTAL VOLUME	648,000 C.F.	ESTIMATE NO.	
ARCHITECT	COST PER S.F.		DATE	
OWNER	COST PER C.F.		NO OF STORIES	
QUANTITIES BY	EXTENSIONS BY		CHECKED BY	

DIV	DESCRIPTION	SUBTOTAL COST	COST/S.F.	PERCENTAGE
A	SUBSTRUCTURE	162,368	$ 3.01	
B10	SHELL: SUPERSTRUCTURE	881,768	$ 16.33	
B20	SHELL: EXTERIOR ENCLOSURE	769,096	$ 14.24	
B30	SHELL: ROOFING	22,203	$ 0.41	
C	INTERIORS	855,054	$ 15.83	
D10	SERVICES: CONVEYING	308,760	$ 5.72	
D20	SERVICES: PLUMBING	137,613	$ 2.55	
D30	SERVICES: HVAC	734,880	$ 13.61	
D40	SERVICES: FIRE PROTECTION	25,383	$ 0.47	
D50	SERVICES: ELECTRICAL	716,745	$ 13.27	
E	EQUIPMENT & FURNISHINGS			
F	SPECIAL CONSTRUCTION & DEMO.			
G	BUILDING SITEWORK			
	BUILDING SUBTOTAL	$ 4,613,871		$ 4,613,871

Sales Tax % x Subtotal /2	N/A		$ -
General Conditions 10 % x Subtotal			$ 461,387
		Subtotal "A"	$ 5,075,258
Overhead 5 % x Subtotal "A"			$ 253,763
		Subtotal "B"	$ 5,329,021
Profit 10% x Subtotal "B"			$ 532,902
		Subtotal "C"	$ 5,861,923
Location Factor % x Subtotal "C"	N/A	Localized Cost	
Architects Fee 6.5% x Localized Cost =			$ 381,025
Contingency x Localized Cost =	N/A		$ -
		Project Total Cost	$ 6,242,948
Square Foot Cost $6,242,948/ 54,000 S.F. =		S.F. Cost	$ 115.61
Cubic Foot Cost $6,242,948/ 648,000 C.F. =		C.F. Cost	$ 9.63

237

A SUBSTRUCTURE

A1010 Standard Foundations

The Strip Footing System includes: excavation; hand trim; all forms needed for footing placement; forms for 2″ x 6″ keyway (four uses); dowels; and 3,000 p.s.i. concrete.

The footing size required varies for different soils. Soil bearing capacities are listed for 3 KSF and 6 KSF. Depths of the system range from 8″ to 24″. Widths range from 16″ to 96″. Smaller strip footings may not require reinforcement.

Please see the reference section for further design and cost information.

SUBSTRUCTURE A

A1010 110	Strip Footings	COST PER L.F.		
		MAT.	INST.	TOTAL
2100	Strip footing, load 2.6KLF, soil capacity 3KSF, 16″wide x 8″deep plain	5.45	10.90	16.35
2300	Load 3.9 KLF, soil capacity, 3 KSF, 24″wide x 8″deep, plain	6.95	12.05	19
2500	Load 5.1KLF, soil capacity 3 KSF, 24″wide x 12″deep, reinf.	11.25	18.15	29.40
2700	Load 11.1KLF, soil capacity 6 KSF, 24″wide x 12″deep, reinf.	11.25	18.15	29.40
2900	Load 6.8 KLF, soil capacity 3 KSF, 32″wide x 12″deep, reinf.	13.95	19.85	33.80
3100	Load 14.8 KLF, soil capacity 6 KSF, 32″wide x 12″deep, reinf.	13.95	19.85	33.80
3300	Load 9.3 KLF, soil capacity 3 KSF, 40″wide x 12″deep, reinf.	16.65	21.50	38.15
3500	Load 18.4 KLF, soil capacity 6 KSF, 40″wide x 12″deep, reinf.	16.80	22	38.80
4500	Load 10KLF, soil capacity 3 KSF, 48″wide x 16″deep, reinf.	23.50	27	50.50
4700	Load 22KLF, soil capacity 6 KSF, 48″wide, 16″deep, reinf.	24	27.50	51.50
5700	Load 15KLF, soil capacity 3 KSF, 72″wide x 20″deep, reinf.	42	38.50	80.50
5900	Load 33KLF, soil capacity 6 KSF, 72″wide x 20″deep, reinf.	44.50	41.50	86

A1010 Standard Foundations

Waterproofing

Spread Footing

Strip Footing

A1010 210	Spread Footings	COST EACH		
		MAT.	INST.	TOTAL
7090	Spread footings, 3000 psi concrete, chute delivered			
7100	Load 25K, soil capacity 3 KSF, 3'-0" sq. x 12" deep	46.50	93	139.50
7150	Load 50K, soil capacity 3 KSF, 4'-6" sq. x 12" deep	99	161	260
7200	Load 50K, soil capacity 6 KSF, 3'-0" sq. x 12" deep	46.50	93	139.50
7250	Load 75K, soil capacity 3 KSF, 5'-6" sq. x 13" deep	157	227	384
7300	Load 75K, soil capacity 6 KSF, 4'-0" sq. x 12" deep	80.50	138	218.50
7350	Load 100K, soil capacity 3 KSF, 6'-0" sq. x 14" deep	199	273	472
7410	Load 100K, soil capacity 6 KSF, 4'-6" sq. x 15" deep	122	189	311
7450	Load 125K, soil capacity 3 KSF, 7'-0" sq. x 17" deep	315	390	705
7500	Load 125K, soil capacity 6 KSF, 5'-0" sq. x 16" deep	157	227	384
7550	Load 150K, soil capacity 3 KSF 7'-6" sq. x 18" deep	380	460	840
7610	Load 150K, soil capacity 6 KSF, 5'-6" sq. x 18" deep	209	287	496
7650	Load 200K, soil capacity 3 KSF, 8'-6" sq. x 20" deep	540	605	1,145
7700	Load 200K, soil capacity 6 KSF, 6'-0" sq. x 20" deep	274	355	629
7750	Load 300K, soil capacity 3 KSF, 10'-6" sq. x 25" deep	1,000	985	1,985
7810	Load 300K, soil capacity 6 KSF, 7'-6" sq. x 25" deep	520	595	1,115
7850	Load 400K, soil capacity 3 KSF, 12'-6" sq. x 28" deep	1,575	1,450	3,025
7900	Load 400K, soil capacity 6 KSF, 8'-6" sq. x 27" deep	725	770	1,495
8010	Load 500K, soil capacity 6 KSF, 9'-6" sq. x 30" deep	995	1,000	1,995
8100	Load 600K, soil capacity 6 KSF, 10'-6" sq. x 33" deep	1,350	1,300	2,650
8200	Load 700K, soil capacity 6 KSF, 11'-6" sq. x 36" deep	1,725	1,575	3,300
8300	Load 800K, soil capacity 6 KSF, 12'-0" sq. x 37" deep	1,925	1,725	3,650
8400	Load 900K, soil capacity 6 KSF, 13'-0" sq. x 39" deep	2,375	2,050	4,425
8500	Load 1000K, soil capacity 6 KSF, 13'-6" sq. x 41" deep	2,700	2,300	5,000

A1010 320	Foundation Dampproofing	COST PER L.F.		
		MAT.	INST.	TOTAL
1000	Foundation dampproofing, bituminous, 1 coat, 4' high	.32	3.39	3.71
1400	8' high	.64	6.80	7.44
1800	12' high	.96	10.50	11.46
2000	2 coats, 4' high	.44	4.19	4.63
2400	8' high	.88	8.40	9.28
2800	12' high	1.32	12.90	14.22
3000	Asphalt with fibers, 1/16" thick, 4' high	.72	4.19	4.91
3400	8' high	1.44	8.40	9.84
3800	12' high	2.16	12.90	15.06
4000	1/8" thick, 4' high	1.28	4.99	6.27
4400	8' high	2.56	10	12.56
4800	12' high	3.84	15.30	19.14
5000	Asphalt coated board and mastic, 1/4" thick, 4' high	2.72	4.55	7.27
5400	8' high	5.45	9.10	14.55
5800	12' high	8.15	14	22.15
6000	1/2" thick, 4' high	4.08	6.35	10.43

A1010 Standard Foundations

A1010 320	Foundation Dampproofing	COST PER L.F.		
		MAT.	INST.	TOTAL
6400	8' high	8.15	12.65	20.80
6800	12' high	12.25	19.30	31.55
7000	Cementitious coating, on walls, 1/8" thick coating, 4' high	6.60	4.64	11.24
7400	8' high	13.20	9.25	22.45
7800	12' high	19.80	13.95	33.75
8000	Cementitious/metallic slurry, 2 coat, 1/4" thick, 2' high	38.50	156	194.50
8400	4' high	77	310	387
8800	6' high	116	470	586

Important: See the Reference Section for critical supporting data - Reference Numbers and City Cost Indexes

A1030 Slab on Grade

Fibre Expansion Joint

Wire Mesh Reinforcing

Concrete Slab

Vapor Barrier

Compacted Gravel

Reinforced Slab on Grade

A Slab on Grade system includes fine grading; 6″ of compacted gravel; vapor barrier; 3500 p.s.i. concrete; bituminous fiber expansion joint; all necessary edge forms

4 uses; steel trowel finish; and sprayed on membrane curing compound. Wire mesh reinforcing used in all reinforced slabs.

Non-industrial slabs are for foot traffic only with negligible abrasion. Light industrial slabs are for pneumatic wheels and light abrasion. Industrial slabs are for solid rubber wheels and moderate abrasion. Heavy industrial slabs are for steel wheels and severe abrasion.

A1030 120	Plain & Reinforced	COST PER S.F.		
		MAT.	INST.	TOTAL
2220	Slab on grade, 4″ thick, non industrial, non reinforced	1.44	2.03	3.47
2240	Reinforced	1.66	2.32	3.98
2260	Light industrial, non reinforced	1.80	2.49	4.29
2280	Reinforced	2.02	2.78	4.80
2300	Industrial, non reinforced	2.19	5.10	7.29
2320	Reinforced	2.41	5.40	7.81
3340	5″ thick, non industrial, non reinforced	1.71	2.08	3.79
3360	Reinforced	1.93	2.37	4.30
3380	Light industrial, non reinforced	2.08	2.54	4.62
3400	Reinforced	2.30	2.83	5.13
3420	Heavy industrial, non reinforced	2.84	6.10	8.94
3440	Reinforced	3.05	6.40	9.45
4460	6″ thick, non industrial, non reinforced	2.07	2.04	4.11
4480	Reinforced	2.41	2.43	4.84
4500	Light industrial, non reinforced	2.45	2.50	4.95
4520	Reinforced	2.98	3.02	6
4540	Heavy industrial, non reinforced	3.22	6.15	9.37
4560	Reinforced	3.56	6.55	10.11
5580	7″ thick, non industrial, non reinforced	2.35	2.11	4.46
5600	Reinforced	2.78	2.52	5.30
5620	Light industrial, non reinforced	2.74	2.57	5.31
5640	Reinforced	3.17	2.98	6.15
5660	Heavy industrial, non reinforced	3.50	6.10	9.60
5680	Reinforced	3.84	6.50	10.34
6700	8″ thick, non industrial, non reinforced	2.62	2.14	4.76
6720	Reinforced	2.98	2.49	5.47
6740	Light industrial, non reinforced	3.01	2.60	5.61
6760	Reinforced	3.37	2.95	6.32
6780	Heavy industrial, non reinforced	3.79	6.15	9.94
6800	Reinforced	4.27	6.50	10.77

SUBSTRUCTURE

A

A2010 Basement Excavation

Sand & Gravel 1:1

Clay 1/2:1

Line of Excavation

In general, the following items are accounted for in the table below.

Costs:
1) Excavation for building or other structure to depth and extent indicated.
2) Backfill compacted in place
3) Haul of excavated waste
4) Replacement of unsuitable backfill material with bank run gravel.

A2010 110	Building Excavation & Backfill	COST PER S.F.		
		MAT.	INST.	TOTAL
2220	Excav & fill, 1000 S.F. 4' sand, gravel, or common earth, on site storage		.78	.78
2240	Off site storage		1.16	1.16
2260	Clay excavation, bank run gravel borrow for backfill	1.96	1.73	3.69
2280	8' deep, sand, gravel, or common earth, on site storage		4.62	4.62
2300	Off site storage		9.95	9.95
2320	Clay excavation, bank run gravel borrow for backfill	7.95	8.10	16.05
2340	16' deep, sand, gravel, or common earth, on site storage		13.05	13.05
2350	Off site storage		25	25
2360	Clay excavation, bank run gravel borrow for backfill	22	21	43
3380	4000 S.F.,4' deep, sand, gravel, or common earth, on site storage		.43	.43
3400	Off site storage		.81	.81
3420	Clay excavation, bank run gravel borrow for backfill	1	.88	1.88
3440	8' deep, sand, gravel, or common earth, on site storage		3.20	3.20
3460	Off site storage		5.60	5.60
3480	Clay excavation, bank run gravel borrow for backfill	3.62	4.80	8.42
3500	16' deep, sand, gravel, or common earth, on site storage		7.90	7.90
3520	Off site storage		15.35	15.35
3540	Clay, excavation, bank run gravel borrow for backfill	9.70	11.50	21.20
4560	10,000 S.F., 4' deep, sand gravel, or common earth, on site storage		.23	.23
4580	Off site storage		.46	.46
4600	Clay excavation, bank run gravel borrow for backfill	.61	.55	1.16
4620	8' deep, sand, gravel, or common earth, on site storage		2.76	2.76
4640	Off site storage		4.17	4.17
4660	Clay excavation, bank run gravel borrow for backfill	2.19	3.74	5.93
4680	16' deep, sand, gravel, or common earth, on site storage		6.40	6.40
4700	Off site storage		10.75	10.75
4720	Clay excavation, bank run gravel borrow for backfill	5.80	8.55	14.35
5740	30,000 S.F., 4' deep, sand, gravel, or common earth, on site storage		.14	.14
5760	Off site storage		.28	.28
5780	Clay excavation, bank run gravel borrow for backfill	.35	.30	.65
5860	16' deep, sand, gravel, or common earth, on site storage		5.40	5.40
5880	Off site storage		7.75	7.75
5900	Clay excavation, bank run gravel borrow for backfill	3.23	6.60	9.83
6910	100,000 S.F., 4' deep, sand, gravel, or common earth, on site storage		.07	.07
6940	8' deep, sand, gravel, or common earth, on site storage		2.28	2.28
6970	16' deep, sand, gravel, or common earth, on site storage		4.87	4.87
6980	Off site storage		6.10	6.10
6990	Clay excavation, bank run gravel borrow for backfill	1.74	5.45	7.19

A2020 Basement Walls

The Foundation Bearing Wall System includes: forms up to 16' high (four uses); 3,000 p.s.i. concrete placed and vibrated; and form removal with breaking form ties and patching walls. The wall systems list walls from 6" to 16" thick and are designed with minimum reinforcement.

Excavation and backfill are not included.

Please see the reference section for further design and cost information.

A2020 110		Walls, Cast in Place						
	WALL HEIGHT (FT.)	**PLACING METHOD**	**CONCRETE (C.Y. per L.F.)**	**REINFORCING (LBS. per L.F.)**	**WALL THICKNESS (IN.)**	**COST PER L.F.**		
						MAT.	**INST.**	**TOTAL**
1500	4'	direct chute	.074	3.3	6	11.35	39	50.35
1520			.099	4.8	8	14.20	40	54.20
1540			.123	6.0	10	16.85	40.50	57.35
1561			.148	7.2	12	19.60	41.50	61.10
1580			.173	8.1	14	22	42.50	64.50
1600			.197	9.44	16	25	43.50	68.50
3000	6'	direct chute	.111	4.95	6	17.05	58.50	75.55
3020			.149	7.20	8	21.50	60	81.50
3040			.184	9.00	10	25	61	86
3061			.222	10.8	12	29.50	62.50	92
5000	8'	direct chute	.148	6.6	6	22.50	78	100.50
5020			.199	9.6	8	28.50	80	108.50
5040			.250	12	10	34	81.50	115.50
5061			.296	14.39	12	39	83.50	122.50
6020	10'	direct chute	.248	12	8	35.50	100	135.50
6040			.307	14.99	10	42	102	144
6061			.370	17.99	12	49	104	153
7220	12'	pumped	.298	14.39	8	42.50	125	167.50
7240			.369	17.99	10	50.50	127	177.50
7262			.444	21.59	12	58.50	130	188.50
9220	16'	pumped	.397	19.19	8	57	165	222
9240			.492	23.99	10	67.50	169	236.50
9260			.593	28.79	12	78.50	175	253.50

For information about Means Estimating Seminars, see yellow pages 12 and 13 in back of book

SUBSTRUCTURE

A

B SHELL

General: It is desirable for purposes of consistency and simplicity to maintain constant column sizes throughout building height. To do this, concrete strength may be varied (higher strength concrete at lower stories and lower strength concrete at upper stories), as well as varying the amount of reinforcing.

The table provides probably minimum column sizes with related costs and weight per lineal foot of story height.

	LOAD (KIPS)	STORY HEIGHT (FT.)	COLUMN SIZE (IN.)	COLUMN WEIGHT (P.L.F.)	CONCRETE STRENGTH (PSI)	COST PER V.L.F.		
						MAT.	INST.	TOTAL
0640	100	10	10	96	4000	8.25	35.50	43.75
0680		12	10	97	4000	8	35.50	43.50
0720		14	12	142	4000	10.80	43	53.80
0840	200	10	12	140	4000	10.95	43	53.95
0860		12	12	142	4000	10.85	43	53.85
0900		14	14	196	4000	13.90	49.50	63.40
0920	300	10	14	192	4000	14.30	50	64.30
0960		12	14	194	4000	14.15	50	64.15
0980		14	16	253	4000	16.05	54.50	70.55
1020	400	10	16	248	4000	17.80	57	74.80
1060		12	16	251	4000	17.60	56.50	74.10
1080		14	16	253	4000	17.45	56.50	73.95
1200	500	10	18	315	4000	24.50	68.50	93
1250		12	20	394	4000	24.50	71	95.50
1300		14	20	397	4000	24.50	71	95.50
1350	600	10	20	388	4000	29.50	78	107.50
1400		12	20	394	4000	29	77.50	106.50
1600		14	20	397	4000	28.50	77	105.50
3400	900	10	24	560	4000	42.50	101	143.50
3800		12	24	567	4000	42	100	142
4000		14	24	571	4000	41.50	99.50	141
7300	300	10	14	192	6000	14.15	50	64.15
7500		12	14	194	6000	14	49.50	63.50
7600		14	14	196	6000	13.90	49.50	63.40
8000	500	10	16	248	6000	17.80	57	74.80
8050		12	16	251	6000	17.60	56.50	74.10
8100		14	16	253	6000	17.45	56.50	73.95
8200	600	10	18	315	6000	21.50	66	87.50
8300		12	18	319	6000	21.50	65.50	87
8400		14	18	321	6000	21	65	86
8800	800	10	20	388	6000	25	71.50	96.50
8900		12	20	394	6000	24.50	71	95.50
9000		14	20	397	6000	24.50	71	95.50

B1010 203 — **C.I.P. Column, Square Tied**

Important: See the Reference Section for critical supporting data - Reference Numbers and City Cost Indexes

B1010 Floor Construction

B1010 203			C.I.P. Column, Square Tied					

| | LOAD (KIPS) | STORY HEIGHT (FT.) | COLUMN SIZE (IN.) | COLUMN WEIGHT (P.L.F.) | CONCRETE STRENGTH (PSI) | COST PER V.L.F. | | |
						MAT.	INST.	TOTAL
9100	900	10	20	388	6000	37	89	126
9300		12	20	394	6000	36	88	124
9600		14	20	397	6000	36	87.50	123.50

B1010 204			C.I.P. Column, Square Tied-Minimum Reinforcing					

| | LOAD (KIPS) | STORY HEIGHT (FT.) | COLUMN SIZE (IN.) | COLUMN WEIGHT (P.L.F.) | CONCRETE STRENGTH (PSI) | COST PER V.L.F. | | |
						MAT.	INST.	TOTAL
9913	150	10-14	12	135	4000	10.55	42.50	53.05
9918	300	10-14	16	240	4000	15.90	54.50	70.40
9924	500	10-14	20	375	4000	24	70.50	94.50
9930	700	10-14	24	540	4000	35.50	90.50	126
9936	1000	10-14	28	740	4000	44.50	108	152.50
9942	1400	10-14	32	965	4000	58.50	125	183.50
9948	1800	10-14	36	1220	4000	71	143	214

B SHELL

Concentric Load

Eccentric Load

General: Data presented here is for plant produced members transported 50 miles to 100 miles to the site and erected.

Design and pricing assumptions:
Normal wt. concrete, f'c = 5 KSI

Main reinforcement, fy = 60 KSI
Ties, fy = 40 KSI

Minimum design eccentricity, 0.1t.

Concrete encased structural steel haunches are assumed where practical; otherwise galvanized rebar haunches are assumed.

Base plates are integral with columns.

Foundation anchor bolts, nuts and washers are included in price.

B1010 206 — Tied, Concentric Loaded Precast Concrete Columns

	LOAD (KIPS)	STORY HEIGHT (FT.)	COLUMN SIZE (IN.)	COLUMN WEIGHT (P.L.F.)	LOAD LEVELS	COST PER V.L.F.		
						MAT.	INST.	TOTAL
0560	100	10	12x12	164	2	49.50	9.35	58.85
0570		12	12x12	162	2	46.50	7.80	54.30
0580		14	12x12	161	2	46	7.80	53.80
0590	150	10	12x12	166	3	45.50	7.80	53.30
0600		12	12x12	169	3	45.50	7	52.50
0610		14	12x12	162	3	45.50	7	52.50
0620	200	10	12x12	168	4	46	8.60	54.60
0630		12	12x12	170	4	45.50	7.80	53.30
0640		14	12x12	220	4	46	7.80	53.80

B1010 207 — Tied, Eccentric Loaded Precast Concrete Columns

	LOAD (KIPS)	STORY HEIGHT (FT.)	COLUMN SIZE (IN.)	COLUMN WEIGHT (P.L.F.)	LOAD LEVELS	COST PER V.L.F.		
						MAT.	INST.	TOTAL
1130	100	10	12x12	161	2	45	9.35	54.35
1140		12	12x12	159	2	45.50	7.80	53.30
1150		14	12x12	159	2	45	7.80	52.80
1390	600	10	18x18	385	4	79.50	8.60	88.10
1400		12	18x18	380	4	79.50	7.80	87.30
1410		14	18x18	375	4	79.50	7.80	87.30
1480	800	10	20x20	490	4	76.50	8.60	85.10
1490		12	20x20	480	4	76.50	7.80	84.30
1500		14	20x20	475	4	76	7.80	83.80

SHELL B

MEANS
CostWorks®
2005

Maximize your estimating & budgeting efforts quickly & efficiently . . .

Introducing Means CostWorks Online – your preeminent Cost Data resource is now available in an easy-to-use online format as well as on CD.

RSMeans

Get your Means CostWorks 2005 today!

Fax : 1-800-632-6732, phone: 1-800-334-3509, or at www.rsmeans.com/cwcw.asp
For information on network pricing call 1-800-334-3509.

Catalog No.	Means CostWorks 2005 CD	Price	Cost
65065	Assemblies Cost Data	$202.95	
65015	Building Construction Cost Data	$142.95	
65315	Building Construction Cost Data/Metric	$142.95	
65155	Open Shop Building Construction Cost Data	$142.95	
65115	Concrete & Masonry Cost Data	$142.95	
65035	Electrical Cost Data	$142.95	
65165	Heavy Construction Cost Data	$142.95	
65095	Interior Cost Data	$142.95	
65185	Light Commercial Cost Data	$177.95	
65025	Mechanical Cost Data	$142.95	
65215	Plumbing Cost Data	$142.95	
65285	Site Work & Landscape Cost Data	$142.95	
65055	Square Foot Costs	$202.95	
65605*	CostWorks Estimator	$299.95	
65205	Facilities Construction Cost Data	$329.95	
65045	Repair & Remodeling Cost Data	$142.95	
65365	Heavy Construction Cost Data/Metric	$142.95	
65175	Residential Cost Data	$164.95	
65305	Facilities Maintenance & Repair Cost Data	$329.95	

Packages*

Catalog No.	Title	Price	Cost
65515	Builder's Package	$380.95	
65525	Building Professional's Package	$445.95	
65535	Facility Manager's Package	$572.95	
65545	Design Professional's Package	$445.95	

*(Packages include Means Estimator FREE)

Subtotal	
MA residents please add 5% Sales Tax	
Shipping & Handling	**$6.00**
Grand Total	

Name: _____

Company:_____

Street: _____

City: _____ State: _____ Zip:_____

Phone: (___)_____ Fax: (___)_____

E-Mail: _____

☐ Bill me   ☐ Please charge my credit card

☐ Visa   ☐ MasterCard   ☐ American Express   ☐ Discover

Card # _____ Expiration Date:_____

Cardholder Signature:_____

All items shipped on a single CD with software code to unlock specific selections purchased.
Buy additional keys any time you need to unlock additional cost data books!

To find out more about our exciting new Means CostWorks Online Products, call 1-800-334-3509 or go to www.rsmeans.com/costworks/web/index.asp

CWCW-2005

From the planning stages through the punch list . . .

Maximize your estimating & budgeting efforts quickly & efficiently with *Means CostWorks 2005*

As building requirements grow more and more complex, the estimating and budgeting process has never been more demanding. Tighter deadlines and more stringent cost controls mean you must find ways to work more efficiently. Let *Means CostWorks 2005*, available in CD or Online formats, help you quickly and easily perform in-depth cost research . . . and take your estimating and budgeting to the next level.

Purchase your titles separately or in value priced packages. Plus, with your purchase you receive access to regular cost updates, access to the Means Estimating HOTLINE, and the Wage Rate Utility (with selected titles).

Your 2005 Means CostWorks Subscription now includes, *Free:*

- Access to Means Estimating Hotline
- The *RSMeans Quarterly Update Service*
- Wage Rate Utility – allows you to reference Union and Open Shop rates (included with selected titles)
- Free Means Estimator with your purchase of a *Means CostWorks* package

The same Cost Data you know and trust... delivered electronically in your choice of CD or Online formats!

- Localize costs to your geographic area for better accuracy
- Easily search and organize detailed cost data
- Determine best-cost solutions for each project
- Change and calculate results automatically
- Receive free online cost factor updates with the *RSMeans Quarterly Update Service*

Design custom reports that meet your specifications

Select items and calculate project costs with the CostList

Means CostWorks 2005 in CD or Online formats – Join the thousands of industry professionals who rely on RSMeans Cost Data each year!

Means CostWorks 2005 CD offers 18 detailed cost titles that cover every job costing requirement you can imagine... it turns your desktop or laptop PC into a comprehensive cost library. Each title addresses a specific area of the commercial or residential building and maintenance industry.

Plus you get the productivity-enhancing tools to really put this valuable information to work.

Save hundreds on specially priced CD packages! Choose the one that's right for you...

Builder's Package
Includes:
- Residential Cost Data
- Light Commercial Cost Data
- Site Work & Landscape Cost Data
- Concrete & Masonry Cost Data
- Open Shop Building Construction Cost Data

Just $380.95
Save over $300 off the price of buying each book separately!

Building Professional's Package
Includes:
- Building Construction Cost Data
- Mechanical Cost Data
- Plumbing Cost Data
- Electrical Cost Data
- Square Foot Costs

Just $445.95
Save over $250 off the price of buying each book separately!

Facility Manager's Package
Includes:
- Facilities Construction Cost Data
- Facilities Maintenance & Repair Cost Data
- Repair & Remodeling Cost Data
- Building Construction Cost Data

Just $572.95
Save over $300 off the price of buying each book separately!

Design Professional's Package
Includes:
- Building Construction Cost Data
- Square Foot Costs
- Interior Cost Data
- Assemblies Cost Data

Just $445.95
Save over $200 off the price of buying each book separately!

Go to www.rsmeans.com/costworks/web/index.asp to learn about other packages *available exclusively online* with a subscription to *Means CostWorks Online 2005!*

B10 Superstructure

B1010 Floor Construction

 (A) Wide Flange

 (B) Pipe

(C) Pipe, Concrete Filled

(G) Rectangular Tube, Concrete Filled

B1010 208						Steel Columns		
	LOAD (KIPS)	UNSUPPORTED HEIGHT (FT.)	WEIGHT (P.L.F.)	SIZE (IN.)	TYPE	COST PER V.L.F.		
						MAT.	INST.	TOTAL
1000	25	10	13	4	A	15.10	8.30	23.40
1020			7.58	3	B	8.80	8.30	17.10
1040			15	3-1/2	C	10.75	8.30	19.05
1120			20	4x3	G	12.45	8.30	20.75
1200		16	16	5	A	17.20	6.25	23.45
1220			10.79	4	B	11.60	6.25	17.85
1240			36	5-1/2	C	16.25	6.25	22.50
1320			64	8x6	G	25	6.25	31.25
1600	50	10	16	5	A	18.60	8.30	26.90
1620			14.62	5	B	17	8.30	25.30
1640			24	4-1/2	C	12.85	8.30	21.15
1720			28	6x3	G	16.50	8.30	24.80
1800		16	24	8	A	26	6.25	32.25
1840			36	5-1/2	C	16.25	6.25	22.50
1920			64	8x6	G	25	6.25	31.25
2000	50	20	28	8	A	28.50	6.25	34.75
2040			49	6-5/8	C	20	6.25	26.25
2120			64	8x6	G	24	6.25	30.25
2200	75	10	20	6	A	23	8.30	31.30
2240			36	4-1/2	C	32	8.30	40.30
2320			35	6x4	G	18.65	8.30	26.95
2400		16	31	8	A	33.50	6.25	39.75
2440			49	6-5/8	C	21	6.25	27.25
2520			64	8x6	G	25	6.25	31.25
2600		20	31	8	A	31.50	6.25	37.75
2640			81	8-5/8	C	30.50	6.25	36.75
2720			64	8x6	G	24	6.25	30.25
2800	100	10	24	8	A	28	8.30	36.30
2840			35	4-1/2	C	32	8.30	40.30
2920			46	8x4	G	23	8.30	31.30

SHELL

B

B1010 208					Steel Columns			
	LOAD (KIPS)	UNSUPPORTED HEIGHT (FT.)	WEIGHT (P.L.F.)	SIZE (IN.)	TYPE	COST PER V.L.F.		
						MAT.	INST.	TOTAL
3000		16	31	8	A	33.50	6.25	39.75
3040			56	6-5/8	C	31.50	6.25	37.75
3120			64	8x6	G	25	6.25	31.25
3200		20	40	8	A	40.50	6.25	46.75
3240			81	8-5/8	C	30.50	6.25	36.75
3320			70	8x6	G	34	6.25	40.25
3400	125	10	31	8	A	36	8.30	44.30
3440			81	8	C	34.50	8.30	42.80
3520			64	8x6	G	27	8.30	35.30
3600	125	16	40	8	A	43	6.25	49.25
3640			81	8	C	32	6.25	38.25
3720			64	8x6	G	25	6.25	31.25
3800		20	48	8	A	49	6.25	55.25
3840			81	8	C	30.50	6.25	36.75
3920			60	8x6	G	34	6.25	40.25
4000	150	10	35	8	A	40.50	8.30	48.80
4040			81	8-5/8	C	34.50	8.30	42.80
4120			64	8x6	G	27	8.30	35.30
4200		16	45	10	A	48.50	6.25	54.75
4240			81	8-5/8	C	32	6.25	38.25
4320			70	8x6	G	36	6.25	42.25
4400		20	49	10	A	50	6.25	56.25
4440			123	10-3/4	C	43	6.25	49.25
4520			86	10x6	G	33.50	6.25	39.75
4600	200	10	45	10	A	52.50	8.30	60.80
4640			81	8-5/8	C	34.50	8.30	42.80
4720			70	8x6	G	39	8.30	47.30
4800		16	49	10	A	52.50	6.25	58.75
4840			123	10-3/4	C	45.50	6.25	51.75
4920			85	10x6	G	42	6.25	48.25
5200	300	10	61	14	A	71	8.30	79.30
5240			169	12-3/4	C	60.50	8.30	68.80
5320			86	10x6	G	58	8.30	66.30
5400		16	72	12	A	77.50	6.25	83.75
5440			169	12-3/4	C	56	6.25	62.25
5600		20	79	12	A	80.50	6.25	86.75
5640			169	12-3/4	C	53.50	6.25	59.75
5800	400	10	79	12	A	92	8.30	100.30
5840			178	12-3/4	C	79	8.30	87.30
6000		16	87	12	A	93.50	6.25	99.75
6040			178	12-3/4	C	73	6.25	79.25
6400	500	10	99	14	A	115	8.30	123.30
6600		16	109	14	A	117	6.25	123.25
6800		20	120	12	A	122	6.25	128.25
7000	600	10	120	12	A	139	8.30	147.30
7200		16	132	14	A	142	6.25	148.25
7400		20	132	14	A	134	6.25	140.25
7600	700	10	136	12	A	158	8.30	166.30
7800		16	145	14	A	156	6.25	162.25
8000		20	145	14	A	148	6.25	154.25
8200	800	10	145	14	A	168	8.30	176.30
8300		16	159	14	A	171	6.25	177.25
8400		20	176	14	A	179	6.25	185.25

SHELL B

Important: See the Reference Section for critical supporting data - Reference Numbers and City Cost Indexes

B1010   Floor Construction

B1010 208		Steel Columns						
	LOAD (KIPS)	UNSUPPORTED HEIGHT (FT.)	WEIGHT (P.L.F.)	SIZE (IN.)	TYPE	COST PER V.L.F.		
						MAT.	INST.	TOTAL
8800	900	10	159	14	A	185	8.30	193.30
8900		16	176	14	A	189	6.25	195.25
9000		20	193	14	A	196	6.25	202.25
9100	1000	10	176	14	A	204	8.30	212.30
9200		16	193	14	A	208	6.25	214.25
9300		20	211	14	A	215	6.25	221.25

B1010 Floor Construction

Columns — Interior Bay

Beams

Exterior Bay

Joists

Corner Bay

Description: Table below lists costs per S.F. of bay size for wood columns of various sizes and unsupported heights and the maximum allowable total load per S.F. per bay size.

SHELL B

B1010 210		Wood Columns						
	NOMINAL COLUMN SIZE (IN.)	BAY SIZE (FT.)	UNSUPPORTED HEIGHT (FT.)	MATERIAL (BF per M.S.F.)	TOTAL LOAD (P.S.F.)	COST PER S.F.		
						MAT.	INST.	TOTAL
1000	4 x 4	10 x 8	8	133	100	.25	.19	.44
1050			10	167	60	.32	.24	.56
1200		10 x 10	8	106	80	.20	.15	.35
1250			10	133	50	.25	.19	.44
1400		10 x 15	8	71	50	.13	.10	.23
1450			10	88	30	.17	.13	.30
1600		15 x 15	8	47	30	.09	.07	.16
1650			10	59	15	.11	.08	.19
2000	6 x 6	10 x 15	8	160	230	.37	.21	.58
2050			10	200	210	.47	.27	.74
2200		15 x 15	8	107	150	.25	.14	.39
2250			10	133	140	.31	.18	.49
2400		15 x 20	8	80	110	.19	.11	.30
2450			10	100	100	.23	.13	.36
2600		20 x 20	8	60	80	.14	.08	.22
2650			10	75	70	.17	.10	.27
2800		20 x 25	8	48	60	.11	.06	.17
2850			10	60	50	.14	.08	.22
3400	8 x 8	20 x 20	8	107	160	.27	.13	.40
3450			10	133	160	.34	.16	.50
3600		20 x 25	8	85	130	.16	.12	.28
3650			10	107	130	.20	.15	.35
3800		25 x 25	8	68	100	.18	.08	.26
3850			10	85	100	.22	.10	.32
4200	10 x 10	20 x 25	8	133	210	.35	.15	.50
4250			10	167	210	.43	.19	.62
4400		25 x 25	8	107	160	.28	.12	.40
4450			10	133	160	.35	.15	.50
4700	12 x 12	20 x 25	8	192	310	.50	.21	.71
4750			10	240	310	.62	.26	.88
4900		25 x 25	8	154	240	.40	.17	.57
4950			10	192	240	.50	.21	.71

B1010 Floor Construction

"T" Shaped
Precast Beams

"L" Shaped
Precast Beams

B1010 214		"T" Shaped Precast Beams						
	SPAN (FT.)	SUPERIMPOSED LOAD (K.L.F.)	SIZE W X D (IN.)	BEAM WEIGHT (P.L.F.)	TOTAL LOAD (K.L.F.)	COST PER L.F.		
						MAT.	INST.	TOTAL
2300	15	2.8	12x16	260	3.06	123	16.60	139.60
2500		8.37	12x28	515	8.89	153	16.40	169.40
8900	45	3.34	12x60	1165	4.51	264	15.60	279.60
9900		6.5	24x60	1915	8.42	325	15.60	340.60

B1010 215		"L" Shaped Precast Beams						
	SPAN (FT.)	SUPERIMPOSED LOAD (K.L.F.)	SIZE W X D (IN.)	BEAM WEIGHT (P.L.F.)	TOTAL LOAD (K.L.F.)	COST PER L.F.		
						MAT.	INST.	TOTAL
2250	15	2.58	12x16	230	2.81	91	16.60	107.60
2400		5.92	12x24	370	6.29	108	15.60	123.60
4000	25	2.64	12x28	435	3.08	118	9.35	127.35
4450		6.44	18x36	790	7.23	155	11.70	166.70
5300	30	2.80	12x36	565	3.37	140	9.35	149.35
6400		8.66	24x44	1245	9.90	210	17.15	227.15

SHELL

B

B1010 Floor Construction

Cast in Place Floor Slab, One Way
General: Solid concrete slabs of uniform depth reinforced for flexure in one direction and for temperature and shrinkage in the other direction.

Cast in Place Beam & Slab, One Way
General: Solid concrete one way slab cast monolithically with reinforced concrete beams and girders.

B1010 217		Cast in Place Slabs, One Way				COST PER S.F.		
	SLAB DESIGN & SPAN (FT.)	SUPERIMPOSED LOAD (P.S.F.)	THICKNESS (IN.)	TOTAL LOAD (P.S.F.)		MAT.	INST.	TOTAL
2500	Single 8	40	4	90		2.96	6.20	9.16
2600		75	4	125		2.99	6.75	9.74
2700		125	4-1/2	181		3.13	6.80	9.93
2800		200	5	262		3.51	7.05	10.56
3000	Single 10	40	4	90		3.11	6.70	9.81
3100		75	4	125		3.11	6.70	9.81
3200		125	5	188		3.48	6.85	10.33
3300		200	7-1/2	293		4.85	7.75	12.60
3500	Single 15	40	5-1/2	90		3.77	6.85	10.62
3600		75	6-1/2	156		4.33	7.20	11.53
3700		125	7-1/2	219		4.72	7.35	12.07
3800		200	8-1/2	306		5.15	7.50	12.65
4000	Single 20	40	7-1/2	115		4.67	7.20	11.87
4100		75	9	200		5.05	7.20	12.25
4200		125	10	250		5.60	7.50	13.10
4300		200	10	324		6.05	7.80	13.85

B1010 219		Cast in Place Beam & Slab, One Way				COST PER S.F.		
	BAY SIZE (FT.)	SUPERIMPOSED LOAD (P.S.F.)	MINIMUM COL. SIZE (IN.)	SLAB THICKNESS (IN.)	TOTAL LOAD (P.S.F.)	MAT.	INST.	TOTAL
5000	20x25	40	12	5-1/2	121	4.52	8.55	13.07
5200		125	16	5-1/2	215	5.35	9.80	15.15
5500	25x25	40	12	6	129	4.77	8.40	13.17
5700		125	18	6	227	6.05	10.35	16.40
7500	30x35	40	16	8	158	5.90	9.50	15.40
7600		75	18	8	196	6.30	9.85	16.15
7700		125	22	8	254	7.05	10.90	17.95
7800		200	26	8	332	7.65	11.30	18.95
8000	35x35	40	16	9	169	6.55	9.80	16.35
8200		75	20	9	213	7.10	10.70	17.80
8400		125	24	9	272	7.80	11.15	18.95
8600		200	26	9	355	8.55	11.90	20.45

SHELL B

Important: See the Reference Section for critical supporting data - Reference Numbers and City Cost Indexes

B1010 Floor Construction

General: Solid concrete two way slab cast monolithically with reinforced concrete support beams and girders.

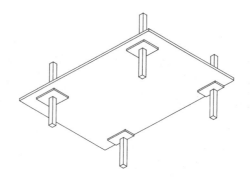

General: Flat Slab: Solid uniform depth concrete two way slabs with drop panels at columns and no column capitals.

B1010 220 — Cast in Place Beam & Slab, Two Way

	BAY SIZE (FT.)	SUPERIMPOSED LOAD (P.S.F.)	MINIMUM COL. SIZE (IN.)	SLAB THICKNESS (IN.)	TOTAL LOAD (P.S.F.)	COST PER S.F. MAT.	COST PER S.F. INST.	COST PER S.F. TOTAL
4000	20 x 25	40	12	7	141	5.20	8.65	13.85
4300		75	14	7	181	5.95	9.45	15.40
4500		125	16	7	236	6.05	9.75	15.80
5100	25 x 25	40	12	7-1/2	149	5.45	8.75	14.20
5200		75	16	7-1/2	185	5.90	9.40	15.30
5300		125	18	7-1/2	250	6.40	10.15	16.55
7600	30 x 35	40	16	10	188	7.10	9.95	17.05
7700		75	18	10	225	7.55	10.30	17.85
8000		125	22	10	282	8.35	11.10	19.45
8500	35 x 35	40	16	10-1/2	193	7.60	10.15	17.75
8600		75	20	10-1/2	233	7.95	10.50	18.45
9000		125	24	10-1/2	287	8.85	11.30	20.15

B1010 222 — Cast in Place Flat Slab with Drop Panels

	BAY SIZE (FT.)	SUPERIMPOSED LOAD (P.S.F.)	MINIMUM COL. SIZE (IN.)	SLAB & DROP (IN.)	TOTAL LOAD (P.S.F.)	COST PER S.F. MAT.	COST PER S.F. INST.	COST PER S.F. TOTAL
1960	20 x 20	40	12	7 - 3	132	4.59	6.90	11.49
1980		75	16	7 - 4	168	4.86	7.05	11.91
2000		125	18	7 - 6	221	5.40	7.30	12.70
3200	25 x 25	40	12	8-1/2 - 5-1/2	154	5.35	7.25	12.60
4000		125	20	8-1/2 - 8-1/2	243	6.05	7.80	13.85
4400		200	24	9 - 8-1/2	329	6.35	7.95	14.30
5000	25 x 30	40	14	9-1/2 - 7	168	5.80	7.50	13.30
5200		75	18	9-1/2 - 7	203	6.20	7.80	14
5600		125	22	9-1/2 - 8	256	6.50	8	14.50
6400	30 x 30	40	14	10-1/2 - 7-1/2	182	6.30	7.75	14.05
6600		75	18	10-1/2 - 7-1/2	217	6.70	8.05	14.75
6800		125	22	10-1/2 - 9	269	7	8.25	15.25
7400	30 x 35	40	16	11-1/2 - 9	196	6.85	8	14.85
7900		75	20	11-1/2 - 9	231	7.30	8.35	15.65
8000		125	24	11-1/2 - 11	284	7.60	8.55	16.15
9000	35 x 35	40	16	12 - 9	202	7	8.10	15.10
9400		75	20	12 - 11	240	7.55	8.45	16
9600		125	24	12 - 11	290	7.80	8.65	16.45

SHELL

B

B1010 Floor Construction

Multispan Joist Slab

General: Flat Plates: Solid uniform depth concrete two way slab without drops or interior beams. Primary design limit is shear at columns.

General: Combination of thin concrete slab and monolithic ribs at uniform spacing to reduce dead weight and increase rigidity.

B1010 223		Cast in Place Flat Plate				COST PER S.F.		
	BAY SIZE (FT.)	SUPERIMPOSED LOAD (P.S.F.)	MINIMUM COL. SIZE (IN.)	SLAB THICKNESS (IN.)	TOTAL LOAD (P.S.F.)	MAT.	INST.	TOTAL
3000	15 x 20	40	14	7	127	4.35	6.65	11
3400		75	16	7-1/2	169	4.62	6.80	11.42
3600		125	22	8-1/2	231	5.05	6.95	12
3800		175	24	8-1/2	281	5.10	6.95	12.05
4200	20 x 20	40	16	7	127	4.36	6.65	11.01
4400		75	20	7-1/2	175	4.67	6.80	11.47
4600		125	24	8-1/2	231	5.10	6.95	12.05
5000		175	24	8-1/2	281	5.10	7	12.10
5600	20 x 25	40	18	8-1/2	146	5.05	6.95	12
6000		75	20	9	188	5.20	7.05	12.25
6400		125	26	9-1/2	244	5.60	7.25	12.85
6600		175	30	10	300	5.85	7.35	13.20
7000	25 x 25	40	20	9	152	5.20	7.05	12.25
7400		75	24	9-1/2	194	5.50	7.20	12.70
7600		125	30	10	250	5.85	7.35	13.20

B1010 226		Cast in Place Multispan Joist Slab				COST PER S.F.		
	BAY SIZE (FT.)	SUPERIMPOSED LOAD (P.S.F.)	MINIMUM COL. SIZE (IN.)	RIB DEPTH (IN.)	TOTAL LOAD (P.S.F.)	MAT.	INST.	TOTAL
2000	15 x 15	40	12	8	115	5.30	8.15	13.45
2100		75	12	8	150	5.30	8.15	13.45
2200		125	12	8	200	5.45	8.25	13.70
2300		200	14	8	275	5.60	8.55	14.15
2600	15 x 20	40	12	8	115	5.40	8.15	13.55
2800		75	12	8	150	5.50	8.30	13.80
3000		125	14	8	200	5.70	8.75	14.45
3300		200	16	8	275	5.95	8.85	14.80
3600	20 x 20	40	12	10	120	5.50	8	13.50
3900		75	14	10	155	5.70	8.45	14.15
4000		125	16	10	205	5.80	8.60	14.40
4100		200	18	10	280	6.05	8.95	15
6200	30 x 30	40	14	14	131	6.10	8.40	14.50
6400		75	18	14	166	6.25	8.65	14.90
6600		125	20	14	216	6.65	9.10	15.75
6700		200	24	16	297	7.10	9.45	16.55

Important: See the Reference Section for critical supporting data - Reference Numbers and City Cost Indexes

Waffle Slab

B1010 227	Cast in Place Waffle Slab							
	BAY SIZE (FT.)	SUPERIMPOSED LOAD (P.S.F.)	MINIMUM COL. SIZE (IN.)	RIB DEPTH (IN.)	TOTAL LOAD (P.S.F.)	COST PER S.F.		
						MAT.	INST.	TOTAL
3900	20 x 20	40	12	8	144	8.65	8.70	17.35
4000		75	12	8	179	8.75	8.80	17.55
4100		125	16	8	229	8.90	8.90	17.80
4200		200	18	8	304	9.30	9.20	18.50
4400	20 x 25	40	12	8	146	8.80	8.75	17.55
4500		75	14	8	181	8.95	8.90	17.85
4600		125	16	8	231	9.10	9	18.10
4700		200	18	8	306	9.45	9.25	18.70
4900	25 x 25	40	12	10	150	8.95	8.85	17.80
5000		75	16	10	185	9.20	9	18.20
5300		125	18	10	235	9.40	9.15	18.55
5500		200	20	10	310	9.60	9.30	18.90
5700	25 x 30	40	14	10	154	9.10	8.90	18
5800		75	16	10	189	9.30	9.05	18.35
5900		125	18	10	239	9.50	9.20	18.70
6000		200	20	12	329	10.25	9.55	19.80
6400	30 x 30	40	14	12	169	9.55	9.05	18.60
6500		75	18	12	204	9.75	9.20	18.95
6600		125	20	12	254	9.85	9.30	19.15
6700		200	24	12	329	10.60	9.85	20.45
6900	30 x 35	40	16	12	169	9.70	9.15	18.85
7000		75	18	12	204	9.70	9.15	18.85
7100		125	22	12	254	10.10	9.50	19.60
7200		200	26	14	334	11	10	21
7400	35 x 35	40	16	14	174	10.15	9.40	19.55
7500		75	20	14	209	10.35	9.50	19.85
7600		125	24	14	259	10.55	9.65	20.20
7700		200	26	16	346	11.15	10.15	21.30
8000	35 x 40	40	18	14	176	10.40	9.55	19.95
8300		75	22	14	211	10.65	9.75	20.40
8500		125	26	16	271	11.10	9.95	21.05
8750		200	30	20	372	11.95	10.50	22.45
9200	40 x 40	40	18	14	176	10.65	9.75	20.40
9400		75	24	14	211	11	10	21
9500		125	26	16	271	11.25	10.10	21.35
9700	40 x 45	40	20	16	186	11	9.90	20.90
9800		75	24	16	221	11.35	10.15	21.50
9900		125	28	16	271	11.50	10.25	21.75

SHELL

B

B1010 Floor Construction

Precast Plank with No Topping

Precast Plank with 2″ Concrete Topping

B1010 229			**Precast Plank with No Topping**					
	SPAN (FT.)	SUPERIMPOSED LOAD (P.S.F.)	TOTAL DEPTH (IN.)	DEAD LOAD (P.S.F.)	TOTAL LOAD (P.S.F.)	COST PER S.F.		
						MAT.	INST.	TOTAL
0720	10	40	4	50	90	4.42	2.60	7.02
0750		75	6	50	125	5.45	2.23	7.68
0770		100	6	50	150	5.45	2.23	7.68
0800	15	40	6	50	90	5.45	2.23	7.68
0820		75	6	50	125	5.45	2.23	7.68
0850		100	6	50	150	5.45	2.23	7.68
0950	25	40	6	50	90	5.45	2.23	7.68
0970		75	8	55	130	6	1.95	7.95
1000		100	8	55	155	6	1.95	7.95
1200	30	40	8	55	95	6	1.95	7.95
1300		75	8	55	130	6	1.95	7.95
1400		100	10	70	170	5.90	1.73	7.63
1500	40	40	10	70	110	5.90	1.73	7.63
1600		75	12	70	145	6.75	1.56	8.31
1700	45	40	12	70	110	6.75	1.56	8.31

B1010 230			**Precast Plank with 2″ Concrete Topping**					
	SPAN (FT.)	SUPERIMPOSED LOAD (P.S.F.)	TOTAL DEPTH (IN.)	DEAD LOAD (P.S.F.)	TOTAL LOAD (P.S.F.)	COST PER S.F.		
						MAT.	INST.	TOTAL
2000	10	40	6	75	115	5.30	4.13	9.43
2100		75	8	75	150	6.30	3.76	10.06
2200		100	8	75	175	6.30	3.76	10.06
2500	15	40	8	75	115	6.30	3.76	10.06
2600		75	8	75	150	6.30	3.76	10.06
2700		100	8	75	175	6.30	3.76	10.06
3100	25	40	8	75	115	6.30	3.76	10.06
3200		75	8	75	150	6.30	3.76	10.06
3300		100	10	80	180	6.85	3.48	10.33
3400	30	40	10	80	120	6.85	3.48	10.33
3500		75	10	80	155	6.85	3.48	10.33
3600		100	10	80	180	6.85	3.48	10.33
4000	40	40	12	95	135	6.75	3.26	10.01
4500		75	14	95	170	7.60	3.09	10.69
5000	45	40	14	95	135	7.60	3.09	10.69

B1010 Floor Construction

Precast Double "T" Beams with No Topping

Precast Double "T" Beams with 2" Topping

Most widely used for moderate span floors and moderate and long span roofs. At shorter spans, they tend to be competitive with hollow core slabs. They are also used as wall panels.

B1010 234	Precast Double "T" Beams with No Topping							
	SPAN (FT.)	SUPERIMPOSED LOAD (P.S.F.)	DBL. "T" SIZE D (IN.) W (FT.)	CONCRETE "T" TYPE	TOTAL LOAD (P.S.F.)	COST PER S.F.		
						MAT.	INST.	TOTAL
4300	50	30	20x8	Lt. Wt.	66	6.15	1.21	7.36
4400		40	20x8	Lt. Wt.	76	6.65	1.10	7.75
4500		50	20x8	Lt. Wt.	86	6.70	1.32	8.02
4600		75	20x8	Lt. Wt.	111	6.75	1.52	8.27
5600	70	30	32x10	Lt. Wt.	78	7.25	.82	8.07
5750		40	32x10	Lt. Wt.	88	7.30	1.01	8.31
5900		50	32x10	Lt. Wt.	98	7.35	1.16	8.51
6000		75	32x10	Lt. Wt.	123	7.40	1.42	8.82
6100		100	32x10	Lt. Wt.	148	7.55	1.93	9.48
6200	80	30	32x10	Lt. Wt.	78	7.35	1.16	8.51
6300		40	32x10	Lt. Wt.	88	7.40	1.42	8.82
6400		50	32x10	Lt. Wt.	98	7.50	1.68	9.18

B1010 235	Precast Double "T" Beams With 2" Topping							
	SPAN (FT.)	SUPERIMPOSED LOAD (P.S.F.)	DBL. "T" SIZE D (IN.) W (FT.)	CONCRETE "T" TYPE	TOTAL LOAD (P.S.F.)	COST PER S.F.		
						MAT.	INST.	TOTAL
7100	40	30	18x8	Reg. Wt.	120	5.75	2.45	8.20
7200		40	20x8	Reg. Wt.	130	5.80	2.33	8.13
7300		50	20x8	Reg. Wt.	140	5.90	2.56	8.46
7400		75	20x8	Reg. Wt.	165	5.90	2.66	8.56
7500		100	20x8	Reg. Wt.	190	6	2.99	8.99
7550	50	30	24x8	Reg. Wt.	120	6.20	2.45	8.65
7600		40	24x8	Reg. Wt.	130	6.25	2.53	8.78
7750		50	24x8	Reg. Wt.	140	6.25	2.54	8.79
7800		75	24x8	Reg. Wt.	165	6.35	2.87	9.22
7900		100	32x10	Reg. Wt.	189	7.10	2.34	9.44

SHELL

B

B1010 Floor Construction

Precast Beam and Plank with No Topping

Precast Beam and Plank with 2" Topping

B1010 236 — Precast Beam & Plank with No Topping

	BAY SIZE (FT.)	SUPERIMPOSED LOAD (P.S.F.)	PLANK THICKNESS (IN.)	TOTAL DEPTH (IN.)	TOTAL LOAD (P.S.F.)	COST PER S.F.		
						MAT.	INST.	TOTAL
6200	25x25	40	8	28	118	13	3.51	16.51
6400		75	8	36	158	13.60	3.51	17.11
6500		100	8	36	183	13.60	3.51	17.11
7000	25x30	40	8	36	110	12.40	3.51	15.91
7200		75	10	36	159	12.60	3.29	15.89
7400		100	12	36	188	14.05	3.12	17.17
7600	30x30	40	8	36	121	12.85	3.51	16.36
8000		75	10	44	140	13.50	3.51	17.01
8250		100	12	52	206	15.25	3.12	18.37
8500	30x35	40	12	44	135	13	3.12	16.12
8750		75	12	52	176	13.85	3.12	16.97
9000	35x35	40	12	52	141	13.85	3.12	16.97
9250		75	12	60	181	14.65	3.12	17.77
9500	35x40	40	12	52	137	13.25	3.90	17.15

B1010 238 — Precast Beam & Plank with 2" Topping

	BAY SIZE (FT.)	SUPERIMPOSED LOAD (P.S.F.)	PLANK THICKNESS (IN.)	TOTAL DEPTH (IN.)	TOTAL LOAD (P.S.F.)	COST PER S.F.		
						MAT.	INST.	TOTAL
4300	20x20	40	6	22	135	13.40	5.30	18.70
4400		75	6	24	173	14.20	5.30	19.50
4500		100	6	28	200	14.60	5.30	19.90
4600	20x25	40	6	26	134	12.55	5.25	17.80
5000		75	8	30	177	13.60	4.99	18.59
5200		100	8	30	202	13.60	4.99	18.59
5400	25x25	40	6	38	143	13.95	5.25	19.20
5600		75	8	38	183	13.95	5.25	19.20
6000		100	8	46	216	15.55	4.96	20.51
6200	25x30	40	8	38	144	13.25	4.94	18.19
6400		75	10	46	200	14.05	4.72	18.77
6600		100	10	46	225	14.05	4.72	18.77
7000	30x30	40	8	46	150	14.35	4.93	19.28
7200		75	10	54	181	15.35	4.93	20.28
7600		100	10	54	231	15.35	4.93	20.28
7800	30x35	40	10	54	166	13.85	4.69	18.54
8000		75	12	54	200	14.45	4.52	18.97

Important: See the Reference Section for critical supporting data - Reference Numbers and City Cost Indexes

B1010 Floor Construction

General: Beams and double tees priced here are for plant produced prestressed members transported to the site and erected.

The 2″ structural topping is applied after the beams and double tees are in place and is reinforced with W.W.F.

Note: Deduct from prices 20% for Southern States. Add to prices 10% for Western States.

With Topping

B1010 239		Precast Double "T" & 2″ Topping on Precast Beams						
	BAY SIZE (FT.)	**SUPERIMPOSED LOAD (P.S.F.)**	**DEPTH (IN.)**		**TOTAL LOAD (P.S.F.)**	**COST PER S.F.**		
						MAT.	**INST.**	**TOTAL**
3000	25x30	40	38		130	12.95	4.29	17.24
3100		75	38		168	12.95	4.29	17.24
3300		100	46		196	13.55	4.29	17.84
3600	30x30	40	46		150	14	4.26	18.26
3750		75	46		174	14	4.26	18.26
4000		100	54		203	14.70	4.26	18.96
4100	30x40	40	46		136	11.45	4.03	15.48
4300		75	54		173	11.95	4.03	15.98
4400		100	62		204	12.60	4.03	16.63
4600	30x50	40	54		138	10.95	3.91	14.86
4800		75	54		181	11.45	3.91	15.36
5000		100	54		219	12.50	3.68	16.18
5200	30x60	40	62		151	11.45	3.69	15.14
5400		75	62		192	12	3.69	15.69
5600		100	62		215	12	3.69	15.69
5800	35x40	40	54		139	12.25	4.02	16.27
6000		75	62		179	13.35	3.92	17.27
6250		100	62		212	13.60	3.92	17.52
6500	35x50	40	62		142	11.70	3.91	15.61
6750		75	62		186	12.20	3.91	16.11
7300		100	62		231	14.05	3.68	17.73
7600	35x60	40	54		154	13	3.58	16.58
7750		75	54		179	13.30	3.58	16.88
8000		100	62		224	14	3.58	17.58
8250	40x40	40	62		145	13.50	4.81	18.31
8400		75	62		187	14.05	4.80	18.85
8750		100	62		223	15.05	4.80	19.85
9000	40x50	40	62		151	12.55	4.66	17.21
9300		75	62		193	13.35	3.96	17.31
9800	40x60	40	62		164	16.10	4.11	20.21

SHELL

B

B1010 Floor Construction

General: The following table is based upon structural wide flange (WF) beam and girder framing. Non-composite action is assumed between WF framing and decking. Deck costs not included.

The deck spans the short direction. The steel beams and girders are fireproofed with sprayed fiber fireproofing.

No columns included in price.

B1010 241		W Shape Beams & Girders						
	BAY SIZE (FT.) BEAM X GIRD	SUPERIMPOSED LOAD (P.S.F.)	STEEL FRAMING DEPTH (IN.)	FIREPROOFING (S.F. PER S.F.)	TOTAL LOAD (P.S.F.)	COST PER S.F.		
						MAT.	INST.	TOTAL
1350	15x20	40	12	.535	50	3.39	1.50	4.89
1400		40	16	.65	90	4.42	1.94	6.36
1450		75	18	.694	125	5.80	2.46	8.26
1500		125	24	.796	175	8	3.35	11.35
2000	20x20	40	12	.55	50	3.78	1.66	5.44
2050		40	14	.579	90	5.15	2.16	7.31
2100		75	16	.672	125	6.20	2.56	8.76
2150		125	16	.714	175	7.40	3.08	10.48
2900	20x25	40	16	.53	50	4.16	1.77	5.93
2950		40	18	.621	96	6.55	2.65	9.20
3000		75	18	.651	131	7.55	3.02	10.57
3050		125	24	.77	200	9.95	4.03	13.98
5200	25x25	40	18	.486	50	4.53	1.87	6.40
5300		40	18	.592	96	6.75	2.71	9.46
5400		75	21	.668	131	8.15	3.24	11.39
5450		125	24	.738	191	10.75	4.29	15.04
6300	25x30	40	21	.568	50	5.15	2.15	7.30
6350		40	21	.694	90	7	2.86	9.86
6400		75	24	.776	125	9	3.59	12.59
6450		125	30	.904	175	10.80	4.41	15.21
7700	30x30	40	21	.52	50	5.55	2.23	7.78
7750		40	24	.629	103	8.50	3.34	11.84
7800		75	30	.715	138	10.15	3.96	14.11
7850		125	36	.822	206	13.30	5.25	18.55
8250	30x35	40	21	.508	50	6.30	2.51	8.81
8300		40	24	.651	109	9.30	3.64	12.94
8350		75	33	.732	150	11.30	4.38	15.68
8400		125	36	.802	225	14.10	5.50	19.60
9550	35x35	40	27	.560	50	6.55	2.60	9.15
9600		40	36	.706	109	11.90	4.56	16.46
9650		75	36	.750	150	12.30	4.74	17.04
9820		125	36	.797	225	16.45	6.35	22.80

SHELL B

B1010 Floor Construction

Description: Table below lists costs for light gage CEE or PUNCHED DOUBLE joists to suit the span and loading with the minimum thickness subfloor required by the joist spacing.

B1010 244		Light Gauge Steel Floor Systems						
	SPAN (FT.)	SUPERIMPOSED LOAD (P.S.F.)	FRAMING DEPTH (IN.)	FRAMING SPAC. (IN.)	TOTAL LOAD (P.S.F.)	COST PER S.F.		
						MAT.	INST.	TOTAL
1500	15	40	8	16	54	2.96	1.66	4.62
1550			8	24	54	3.10	1.69	4.79
1600		65	10	16	80	3.86	2	5.86
1650			10	24	80	3.82	1.95	5.77
1700		75	10	16	90	3.86	2	5.86
1750			10	24	90	3.82	1.95	5.77
1800		100	10	16	116	5.05	2.53	7.58
1850			10	24	116	3.82	1.95	5.77
1900		125	10	16	141	5.05	2.53	7.58
1950			10	24	141	3.82	1.95	5.77
2500	20	40	8	16	55	3.40	1.87	5.27
2550			8	24	55	3.10	1.69	4.79
2600		65	8	16	80	3.94	2.12	6.06
2650			10	24	80	3.83	1.97	5.80
2700		75	10	16	90	3.64	1.90	5.54
2750			12	24	90	4.27	1.73	6
2800		100	10	16	115	3.64	1.90	5.54
2850			10	24	116	4.92	2.20	7.12
2900		125	12	16	142	5.70	2.19	7.89
2950			12	24	141	5.35	2.33	7.68
3500	25	40	10	16	55	3.86	2	5.86
3550			10	24	55	3.82	1.95	5.77
3600		65	10	16	81	5.05	2.53	7.58
3650			12	24	81	4.73	1.88	6.61
3700		75	12	16	92	5.70	2.19	7.89
3750			12	24	91	5.35	2.33	7.68
3800		100	12	16	117	6.40	2.41	8.81
3850		125	12	16	143	7.35	3.10	10.45
4500	30	40	12	16	57	5.70	2.19	7.89
4550			12	24	56	4.71	1.87	6.58
4600		65	12	16	82	6.35	2.40	8.75
4650		75	12	16	92	6.35	2.40	8.75

SHELL

B

Description: The table below lists costs per S.F. for a floor system on bearing walls using open web steel joists, galvanized steel slab form and 2-1/2″ concrete slab reinforced with welded wire fabric. Costs of the bearing walls are not included.

B1010 246		Deck & Joists on Bearing Walls						
	SPAN (FT.)	SUPERIMPOSED LOAD (P.S.F.)	JOIST SPACING FT. or IN.	DEPTH (IN.)	TOTAL LOAD (P.S.F.)	COST PER S.F.		
						MAT.	INST.	TOTAL
1050	20	40	2-0	14-1/2	83	4.29	3	7.29
1070		65	2-0	16-1/2	109	4.65	3.16	7.81
1100		75	2-0	16-1/2	119	4.65	3.16	7.81
1120		100	2-0	18-1/2	145	4.68	3.18	7.86
1150		125	1-9	18-1/2	170	5.65	3.73	9.38
1170	25	40	2-0	18-1/2	84	5.10	3.47	8.57
1200		65	2-0	20-1/2	109	5.30	3.56	8.86
1220		75	2-0	20-1/2	119	5.15	3.39	8.54
1250		100	2-0	22-1/2	145	5.25	3.41	8.66
1270		125	1-9	22-1/2	170	6	3.74	9.74
1300	30	40	2-0	22-1/2	84	5.25	3.05	8.30
1320		65	2-0	24-1/2	110	5.70	3.19	8.89
1350		75	2-0	26-1/2	121	5.90	3.25	9.15
1370		100	2-0	26-1/2	146	6.30	3.37	9.67
1400		125	2-0	24-1/2	172	6.85	4.33	11.18
1420	35	40	2-0	26-1/2	85	6.45	4.16	10.61
1450		65	2-0	28-1/2	111	6.65	4.26	10.91
1470		75	2-0	28-1/2	121	6.65	4.26	10.91
1500		100	1-11	28-1/2	147	7.25	4.53	11.78
1520		125	1-8	28-1/2	172	7.95	4.87	12.82
1550	5/8″ gyp. fireproof.							
1560	On metal furring, add					.55	2.25	2.80

B1010 Floor Construction

The table below lists costs for a floor system on exterior bearing walls and interior columns and beams using open web steel joists, galvanized steel slab form, 2-1/2" concrete slab reinforced with welded wire fabric.

Columns are sized to accommodate one floor plus roof loading but costs for columns are from suspended floor to slab on grade only. Costs of the bearing walls are not included.

Steel Joists and Concrete Slab on Walls and Beams

B1010 248		Steel Joists on Beam & Wall						
	BAY SIZE (FT.)	SUPERIMPOSED LOAD (P.S.F.)	DEPTH (IN.)	TOTAL LOAD (P.S.F.)	COLUMN ADD	COST PER S.F.		
						MAT.	INST.	TOTAL
1720	25x25	40	23	84		6.20	3.72	9.92
1730					columns	.27	.09	.36
1750	25x25	65	29	110		6.50	3.86	10.36
1760					columns	.27	.09	.36
1770	25x25	75	26	120		6.75	3.80	10.55
1780					columns	.32	.11	.43
1800	25x25	100	29	145		7.60	4.16	11.76
1810					columns	.32	.11	.43
1820	25x25	125	29	170		7.95	4.29	12.24
1830					columns	.35	.12	.47
2020	30x30	40	29	84		6.65	3.41	10.06
2030					columns	.25	.08	.33
2050	30x30	65	29	110		7.50	3.68	11.18
2060					columns	.25	.08	.33
2070	30x30	75	32	120		7.70	3.75	11.45
2080					columns	.28	.09	.37
2100	30x30	100	35	145		8.50	4.01	12.51
2110					columns	.33	.12	.45
2120	30x30	125	35	172		9.30	5.05	14.35
2130					columns	.40	.13	.53
2170	30x35	40	29	85		7.50	3.68	11.18
2180					columns	.24	.08	.32
2200	30x35	65	29	111		8.40	4.74	13.14
2210					columns	.27	.09	.36
2220	30x35	75	32	121		8.40	4.74	13.14
2230					columns	.28	.09	.37
2250	30x35	100	35	148		9.05	4.17	13.22
2260					columns	.34	.12	.46
2270	30x35	125	38	173		10.20	4.52	14.72
2280					columns	.35	.12	.47
2320	35x35	40	32	85		7.70	3.74	11.44
2330					columns	.24	.08	.32
2350	35x35	65	35	111		8.85	4.91	13.76
2360					columns	.29	.10	.39
2370	35x35	75	35	121		9.05	4.98	14.03
2380					columns	.29	.10	.39
2400	35x35	100	38	148		9.40	5.15	14.55
2410					columns	.36	.12	.48

SHELL

B

B1010 Floor Construction

B1010 248		Steel Joists on Beam & Wall						
	BAY SIZE (FT.)	SUPERIMPOSED LOAD (P.S.F.)	DEPTH (IN.)	TOTAL LOAD (P.S.F.)	COLUMN ADD	COST PER S.F.		
						MAT.	INST.	TOTAL
2460	5/8 gyp. fireproof.							
2475	On metal furring, add					.55	2.25	2.80

B1010 Floor Construction

The table below lists costs per S.F. for a floor system on steel columns and beams using open web steel joists, galvanized steel slab form, 2-1/2″ concrete slab reinforced with welded wire fabric.

Columns are sized to accommodate one floor plus roof loading but costs for columns are from floor to grade only.

Steel Joists and Concrete Slab on Steel Columns and Beams

B1010 250		Steel Joists, Beams & Slab on Columns						
	BAY SIZE (FT.)	SUPERIMPOSED LOAD (P.S.F.)	DEPTH (IN.)	TOTAL LOAD (P.S.F.)	COLUMN ADD	COST PER S.F.		
						MAT.	INST.	TOTAL
2350	15x20	40	17	83		5.95	3.46	9.41
2400					column	.91	.32	1.23
2450	15x20	65	19	108		6.55	3.67	10.22
2500					column	.91	.32	1.23
2550	15x20	75	19	119		6.85	3.79	10.64
2600					column	.99	.34	1.33
2650	15x20	100	19	144		7.25	3.96	11.21
2700					column	.99	.34	1.33
2750	15x20	125	19	170		8.10	4.41	12.51
2800					column	1.32	.45	1.77
2850	20x20	40	19	83		6.45	3.61	10.06
2900					column	.74	.25	.99
2950	20x20	65	23	109		7.10	3.87	10.97
3000					column	.99	.34	1.33
3100	20x20	75	26	119		7.45	4	11.45
3200					column	.99	.34	1.33
3400	20x20	100	23	144		7.75	4.10	11.85
3450					column	.99	.34	1.33
3500	20x20	125	23	170		8.60	4.43	13.03
3600					column	1.19	.41	1.60
3700	20x25	40	44	83		7.35	4.11	11.46
3800					column	.79	.28	1.07
3900	20x25	65	26	110		8	4.36	12.36
4000					column	.79	.28	1.07
4100	20x25	75	26	120		7.85	4.18	12.03
4200					column	.95	.33	1.28
4300	20x25	100	26	145		8.35	4.34	12.69
4400					column	.95	.33	1.28
4500	20x25	125	29	170		9.25	4.74	13.99
4600					column	1.11	.38	1.49
4700	25x25	40	23	84		7.85	4.28	12.13
4800					column	.76	.26	1.02
4900	25x25	65	29	110		8.30	4.47	12.77
5000					column	.76	.26	1.02
5100	25x25	75	26	120		8.70	4.46	13.16
5200					column	.89	.30	1.19
5300	25x25	100	29	145		9.65	4.85	14.50
5400					column	.89	.30	1.19

SHELL

B

B1010 Floor Construction

B1010 250	Steel Joists, Beams & Slab on Columns

	BAY SIZE (FT.)	SUPERIMPOSED LOAD (P.S.F.)	DEPTH (IN.)	TOTAL LOAD (P.S.F.)	COLUMN ADD	COST PER S.F.		
						MAT.	INST.	TOTAL
5500	25x25	125	32	170		10.20	5.05	15.25
5600					column	.98	.34	1.32
5700	25x30	40	29	84		8.15	4.46	12.61
5800					column	.74	.25	.99
5900	25x30	65	29	110		8.50	4.63	13.13
6000					column	.74	.25	.99
6050	25x30	75	29	120		9.15	4.25	13.40
6100					column	.82	.28	1.10
6150	25x30	100	29	145		9.90	4.49	14.39
6200					column	.82	.28	1.10
6250	25x30	125	32	170		10.65	5.45	16.10
6300					column	.94	.33	1.27
6350	30x30	40	29	84		8.55	4.04	12.59
6400					column	.68	.24	.92
6500	30x30	65	29	110		9.65	4.42	14.07
6600					column	.68	.24	.92
6700	30x30	75	32	120		9.85	4.49	14.34
6800					column	.79	.27	1.06
6900	30x30	100	35	145		10.95	4.83	15.78
7000					column	.92	.32	1.24
7100	30x30	125	35	172		11.95	5.95	17.90
7200					column	1.02	.35	1.37
7300	30x35	40	29	85		9.60	4.40	14
7400					column	.58	.20	.78
7500	30x35	65	29	111		10.70	5.50	16.20
7600					column	.75	.26	1.01
7700	30x35	75	32	121		10.70	5.50	16.20
7800					column	.77	.26	1.03
7900	30x35	100	35	148		11.55	5	16.55
8000					column	.94	.33	1.27
8100	30x35	125	38	173		12.85	5.40	18.25
8200					column	.96	.33	1.29
8300	35x35	40	32	85		9.85	4.47	14.32
8400					column	.67	.24	.91
8500	35x35	65	35	111		11.25	5.70	16.95
8600					column	.81	.28	1.09
9300	35x35	75	38	121		11.50	5.80	17.30
9400					column	.81	.28	1.09
9500	35x35	100	38	148		12.40	6.15	18.55
9600					column	1	.34	1.34
9750	35x35	125	41	173		13.35	5.60	18.95
9800					column	1.02	.34	1.36
9810	5/8 gyp. fireproof.							
9815	On metal furring, add					.55	2.25	2.80

Important: See the Reference Section for critical supporting data - Reference Numbers and City Cost Indexes

B1010   Floor Construction

General: Composite construction of wide flange beams and concrete slabs is most efficiently used when loads are heavy and spans are moderately long. It is stiffer with less deflection than non-composite construction of similar depth and spans.

Composite Beam & Cast in Place Slab

B1010 252	Composite Beam & Cast in Place Slab							
	BAY SIZE (FT.)	SUPERIMPOSED LOAD (P.S.F.)	SLAB THICKNESS (IN.)	TOTAL DEPTH (FT.-IN.)	TOTAL LOAD (P.S.F.)	COST PER S.F.		
						MAT.	INST.	TOTAL
3800	20x25	40	4	1 - 8	94	7.05	8.05	15.10
3900		75	4	1 - 8	130	8.20	8.55	16.75
4000		125	4	1 - 10	181	9.55	9.10	18.65
4100		200	5	2 - 2	272	12.40	10.35	22.75
4200	25x25	40	4-1/2	1 - 8-1/2	99	7.40	8.10	15.50
4300		75	4-1/2	1 - 10-1/2	136	8.90	8.80	17.70
4400		125	5-1/2	2 - 0-1/2	200	10.55	9.55	20.10
4500		200	5-1/2	2 - 2-1/2	278	13.05	10.60	23.65
4600	25x30	40	4-1/2	1 - 8-1/2	100	8.25	8.45	16.70
4700		75	4-1/2	1 - 10-1/2	136	9.65	9.10	18.75
4800		125	5-1/2	2 - 2-1/2	202	11.85	10.05	21.90
4900		200	5-1/2	2 - 5-1/2	279	14.10	11	25.10
5000	30x30	40	4	1 - 8	95	8.25	8.50	16.75
5200		75	4	2 - 1	131	9.55	9	18.55
5400		125	4	2 - 4	183	11.75	9.95	21.70
5600		200	5	2 - 10	274	15	11.35	26.35
5800	30x35	40	4	2 - 1	95	8.70	8.60	17.30
6000		75	4	2 - 4	139	10.50	9.30	19.80
6250		125	4	2 - 4	185	13.60	10.65	24.25
6500		200	5	2 - 11	276	16.60	11.90	28.50
8000	35x35	40	4	2 - 1	96	9.20	8.80	18
8250		75	4	2 - 4	133	10.90	9.55	20.45
8500		125	4	2 - 10	185	13.10	10.45	23.55
8750		200	5	3 - 5	276	17.05	12.10	29.15
9000	35x40	40	4	2 - 4	97	10.20	9.25	19.45
9250		75	4	2 - 4	134	11.90	9.90	21.80
9500		125	4	2 - 7	186	14.50	10.95	25.45
9750		200	5	3 - 5	278	18.80	12.65	31.45

SHELL

B

B1010 Floor Construction

The table below lists costs per S.F. for floors using steel beams and girders, composite steel deck, concrete slab reinforced with W.W.F. and sprayed fiber fireproofing (non- asbestos) on the steel beams and girders and on the steel deck.

Wide Flange, Composite Deck & Slab

B1010 254			W Shape, Composite Deck, & Slab					
	BAY SIZE (FT.) BEAM X GIRD	SUPERIMPOSED LOAD (P.S.F.)	SLAB THICKNESS (IN.)	TOTAL DEPTH (FT.-IN.)	TOTAL LOAD (P.S.F.)	COST PER S.F.		
						MAT.	INST.	TOTAL
0540	15x20	40	5	1-7	89	8.40	5.05	13.45
0560		75	5	1-9	125	9.20	5.35	14.55
0580		125	5	1-11	176	10.70	5.90	16.60
0600		200	5	2-2	254	13.05	6.70	19.75
0700	20x20	40	5	1-9	90	9.15	5.30	14.45
0720		75	5	1-9	126	10.25	5.70	15.95
0740		125	5	1-11	177	11.55	6.25	17.80
0760		200	5	2-5	255	13.90	7	20.90
0780	20x25	40	5	1-9	90	9.30	5.25	14.55
0800		75	5	1-9	127	11.50	6.10	17.60
0820		125	5	1-11	180	13.70	6.65	20.35
0840		200	5	2-5	256	15.05	7.50	22.55
0940	25x25	40	5	1-11	91	10.45	5.65	16.10
0960		75	5	2-5	178	12	6.25	18.25
0980		125	5	2-5	181	14.60	6.95	21.55
1000		200	5-1/2	2-8-1/2	263	16.35	8.05	24.40
1400	25x30	40	5	2-5	91	10.35	5.80	16.15
1500		75	5	2-5	128	12.55	6.50	19.05
1600		125	5	2-8	180	14.15	7.20	21.35
1700		200	5	2-11	259	17.60	8.30	25.90
2200	30x30	40	5	2-2	92	11.45	6.20	17.65
2300		75	5	2-5	129	13.30	6.90	20.20
2400		125	5	2-11	182	15.70	7.75	23.45
2500		200	5	3-2	263	21	9.70	30.70
2600	30x35	40	5	2-5	94	12.45	6.55	19
2700		75	5	2-11	131	14.45	7.20	21.65
2800		125	5	3-2	183	16.80	8.10	24.90
2900		200	5-1/2	3-5-1/2	268	20.50	9.35	29.85
3800	35x35	40	5	2-8	94	12.85	6.55	19.40
3900		75	5	2-11	131	14.80	7.25	22.05
4000		125	5	3-5	184	17.70	8.30	26
4100		200	5-1/2	3-5-1/2	270	22.50	9.85	32.35

Important: See the Reference Section for critical supporting data - Reference Numbers and City Cost Indexes

B1010 Floor Construction

Composite Beam, Deck & Slab

The table below lists costs per S.F. for floors using composite steel beams with welded shear studs, composite steel deck and light weight concrete slab reinforced with W.W.F.

B1010 256		Composite Beams, Deck & Slab						
	BAY SIZE (FT.)	SUPERIMPOSED LOAD (P.S.F.)	SLAB THICKNESS (IN.)	TOTAL DEPTH (FT.-IN.)	TOTAL LOAD (P.S.F.)	COST PER S.F.		
						MAT.	INST.	TOTAL
2400	20x25	40	5-1/2	1 - 5-1/2	80	8.75	4.73	13.48
2500		75	5-1/2	1 - 9-1/2	115	9.10	4.73	13.83
2750		125	5-1/2	1 - 9-1/2	167	11.05	5.55	16.60
2900		200	6-1/4	1 - 11-1/2	251	12.40	6	18.40
3000	25x25	40	5-1/2	1 - 9-1/2	82	8.65	4.49	13.14
3100		75	5-1/2	1 - 11-1/2	118	9.60	4.56	14.16
3200		125	5-1/2	2 - 2-1/2	169	10	4.94	14.94
3300		200	6-1/4	2 - 6-1/4	252	13.45	5.80	19.25
3400	25x30	40	5-1/2	1 - 11-1/2	83	8.85	4.47	13.32
3600		75	5-1/2	1 - 11-1/2	119	9.50	4.52	14.02
3900		125	5-1/2	1 - 11-1/2	170	10.95	5.10	16.05
4000		200	6-1/4	2 - 6-1/4	252	13.50	5.80	19.30
4200	30x30	40	5-1/2	1 - 11-1/2	81	8.80	4.61	13.41
4400		75	5-1/2	2 - 2-1/2	116	9.50	4.83	14.33
4500		125	5-1/2	2 - 5-1/2	168	11.45	5.40	16.85
4700		200	6-1/4	2 - 9-1/4	252	13.65	6.25	19.90
4900	30x35	40	5-1/2	2 - 2-1/2	82	9.20	4.78	13.98
5100		75	5-1/2	2 - 5-1/2	117	10	4.88	14.88
5300		125	5-1/2	2 - 5-1/2	169	11.75	5.55	17.30
5500		200	6-1/4	2 - 9-1/4	254	13.75	6.30	20.05
5750	35x35	40	5-1/2	2 - 5-1/2	84	9.85	4.79	14.64
6000		75	5-1/2	2 - 5-1/2	121	11.20	5.15	16.35
7000		125	5-1/2	2 - 8-1/2	170	13.05	5.85	18.90
7200		200	5-1/2	2 - 11-1/2	254	14.95	6.50	21.45
7400	35x40	40	5-1/2	2 - 5-1/2	85	10.85	5.15	16
7600		75	5-1/2	2 - 5-1/2	121	11.70	5.35	17.05
8000		125	5-1/2	2 - 5-1/2	171	13.35	5.95	19.30
9000		200	5-1/2	2 - 11-1/2	255	16.15	6.75	22.90

SHELL B

B1010 Floor Construction

Description: Table B1010 258 lists S.F. costs for steel deck and concrete slabs for various spans.

Description: Table B1010 261 lists the S.F. costs for wood joists and a minimum thickness plywood subfloor.

Description: Table B1010 264 lists the S.F. costs, total load, and member sizes, for various bay sizes and loading conditions.

Metal Deck/Concrete Fill

Wood Joist

Wood Beam and Joist

B1010 258 — Metal Deck/Concrete Fill

	SUPERIMPOSED LOAD (P.S.F.)	DECK SPAN (FT.)	DECK GAGE DEPTH	SLAB THICKNESS (IN.)	TOTAL LOAD (P.S.F.)	COST PER S.F.		
						MAT.	INST.	TOTAL
0900	125	6	22 1-1/2	4	164	2.31	1.92	4.23
0920		7	20 1-1/2	4	164	2.55	2.03	4.58
0950		8	20 1-1/2	4	165	2.55	2.03	4.58
0970		9	18 1-1/2	4	165	2.97	2.03	5
1000		10	18 2	4	165	2.91	2.13	5.04
1020		11	18 3	5	169	3.54	2.32	5.86

B1010 261 — Wood Joist

		COST PER S.F.		
		MAT.	INST.	TOTAL
2902	Wood joist, 2" x 8",12" O.C.	1.70	1.45	3.15
2950	16" O.C.	1.46	1.24	2.70
3000	24" O.C.	1.58	1.14	2.72
3300	2"x10", 12" O.C.	2.14	1.66	3.80
3350	16" O.C.	1.78	1.39	3.17
3400	24" O.C.	1.79	1.23	3.02
3700	2"x12", 12" O.C.	2.65	1.68	4.33
3750	16" O.C.	2.16	1.41	3.57
3800	24" O.C.	2.05	1.25	3.30
4100	2"x14", 12" O.C.	3.33	1.82	5.15
4150	16" O.C.	2.69	1.52	4.21
4200	24" O.C.	2.40	1.32	3.72
7101	Note: Subfloor cost is included in these prices			

B1010 264 — Wood Beam & Joist

	BAY SIZE (FT.)	SUPERIMPOSED LOAD (P.S.F.)	GIRDER BEAM (IN.)	JOISTS (IN.)	TOTAL LOAD (P.S.F.)	COST PER S.F.		
						MAT.	INST.	TOTAL
2000	15x15	40	8 x 12 4 x 12	2 x 6 @ 16	53	5.65	3.35	9
2050		75	8 x 16 4 x 16	2 x 8 @ 16	90	7.70	3.67	11.37
2100		125	12 x 16 6 x 16	2 x 8 @ 12	144	12.25	4.72	16.97
2150		200	14 x 22 12 x 16	2 x 10 @ 12	227	23	7.20	30.20
3000	20x20	40	10 x 14 10 x 12	2 x 8 @ 16	63	8.65	3.38	12.03
3050		75	12 x 16 8 x 16	2 x 10 @ 16	102	10.80	3.77	14.57
3100		125	14 x 22 12 x 16	2 x 10 @ 12	163	18.30	6.05	24.35

SHELL B

B1010 Floor Construction

Steel Column Fireproofing

"T" Shaped
Precast Beams

"L" Shaped
Precast Beams

B1010 720		Steel Column Fireproofing						
	ENCASEMENT SYSTEM	COLUMN SIZE (IN.)	THICKNESS (IN.)	FIRE RATING (HRS.)	WEIGHT (P.L.F.)	COST PER V.L.F.		
						MAT.	INST.	TOTAL
3000	Concrete	8	1	1	110	5.50	27	32.50
3300		14	1	1	258	9.05	39.50	48.55
3400			2	3	325	10.80	45	55.80
3450	Gypsum board	8	1/2	2	8	2.57	16.55	19.12
3550	1 layer	14	1/2	2	18	2.78	17.65	20.43
3600	Gypsum board	8	1	3	14	3.58	21	24.58
3650	1/2" fire rated	10	1	3	17	3.83	22.50	26.33
3700	2 layers	14	1	3	22	3.97	23	26.97
3750	Gypsum board	8	1-1/2	3	23	4.76	26.50	31.26
3800	1/2" fire rated	10	1-1/2	3	27	5.35	29.50	34.85
3850	3 layers	14	1-1/2	3	35	5.90	32	37.90
3900	Sprayed fiber	8	1-1/2	2	6.3	3.12	5.30	8.42
3950	Direct application		2	3	8.3	4.30	7.30	11.60
4050		10	1-1/2	2	7.9	3.77	6.40	10.17
4200		14	1-1/2	2	10.8	4.68	7.95	12.63

SHELL

B

B1020 Roof Construction

The table below lists prices per S.F. for roof rafters and sheathing by nominal size and spacing. Sheathing is 5/16″ CDX for 12″ and 16″ spacing and 3/8″ CDX for 24″ spacing.

Factors for Converting Inclined to Horizontal

Roof Slope	Approx. Angle	Factor	Roof Slope	Approx. Angle	Factor
Flat	0°	1.000	12 in 12	45.0°	1.414
1 in 12	4.8°	1.003	13 in 12	47.3°	1.474
2 in 12	9.5°	1.014	14 in 12	49.4°	1.537
3 in 12	14.0°	1.031	15 in 12	51.3°	1.601
4 in 12	18.4°	1.054	16 in 12	53.1°	1.667
5 in 12	22.6°	1.083	17 in 12	54.8°	1.734
6 in 12	26.6°	1.118	18 in 12	56.3°	1.803
7 in 12	30.3°	1.158	19 in 12	57.7°	1.873
8 in 12	33.7°	1.202	20 in 12	59.0°	1.943
9 in 12	36.9°	1.250	21 in 12	60.3°	2.015
10 in 12	39.8°	1.302	22 in 12	61.4°	2.088
11 in 12	42.5°	1.357	23 in 12	62.4°	2.162

B1020 102	Wood/Flat or Pitched	COST PER S.F.		
		MAT.	INST.	TOTAL
2500	Flat rafter, 2″x4″, 12″ O.C.	1.13	1.31	2.44
2550	16″ O.C.	1.03	1.12	2.15
2600	24″ O.C.	.97	.97	1.94
2900	2″x6″, 12″ O.C.	1.37	1.31	2.68
2950	16″ O.C.	1.21	1.13	2.34
3000	24″ O.C.	1.08	.97	2.05
3300	2″x8″, 12″ O.C.	1.71	1.41	3.12
3350	16″ O.C.	1.47	1.20	2.67
3400	24″ O.C.	1.26	1.02	2.28
3700	2″x10″, 12″ O.C.	2.15	1.62	3.77
3750	16″ O.C.	1.79	1.35	3.14
3800	24″ O.C.	1.47	1.11	2.58
4100	2″x12″, 12″ O.C.	2.66	1.64	4.30
4150	16″ O.C.	2.17	1.37	3.54
4200	24″ O.C.	1.73	1.13	2.86
4500	2″x14″, 12″ O.C.	3.34	2.38	5.72
4550	16″ O.C.	2.70	1.93	4.63
4600	24″ O.C.	2.08	1.51	3.59
4900	3″x6″, 12″ O.C.	3.03	1.39	4.42
4950	16″ O.C.	2.46	1.18	3.64
5000	24″ O.C.	1.92	1.01	2.93
5300	3″x8″, 12″ O.C.	3.86	1.55	5.41
5350	16″ O.C.	3.07	1.30	4.37
5400	24″ O.C.	2.33	1.09	3.42
5700	3″x10″, 12″ O.C.	4.64	1.77	6.41
5750	16″ O.C.	3.67	1.47	5.14
5800	24″ O.C.	2.73	1.20	3.93
6100	3″x12″, 12″ O.C.	5.50	2.13	7.63
6150	16″ O.C.	4.32	1.74	6.06
6200	24″ O.C.	3.15	1.37	4.52
7001	Wood truss, 4 in 12 slope, 24″ O.C., 24′ to 29′ span	3.45	2	5.45
7100	30′ to 43′ span	3.64	2	5.64
7200	44′ to 60′ span	3.85	2	5.85

SHELL B

Important: See the Reference Section for critical supporting data - Reference Numbers and City Cost Indexes

B1020 Roof Construction

The table below lists the cost per S.F. for a roof system with steel columns, beams and deck using open web steel joists and 1-1/2" galvanized metal deck. Perimeter of system is supported on bearing walls.

Fireproofing is not included. Costs/S.F. are based on a building 4 bays long and 4 bays wide.

Column costs are additive. Costs for the bearing walls are not included.

Steel Joists, Beams and Deck on Bearing Walls

B1020 108		Steel Joists, Beams & Deck on Columns & Walls						
	BAY SIZE (FT.)	SUPERIMPOSED LOAD (P.S.F.)	DEPTH (IN.)	TOTAL LOAD (P.S.F.)	COLUMN ADD	COST PER S.F.		
						MAT.	INST.	TOTAL
3000	25x25	20	18	40		3.01	1.10	4.11
3100					columns	.26	.07	.33
3200		30	22	50		3.39	1.28	4.67
3300					columns	.34	.09	.43
3400		40	20	60		3.57	1.27	4.84
3500					columns	.34	.09	.43
3600	25x30	20	22	40		3.20	1.04	4.24
3700					columns	.29	.08	.37
3800		30	20	50		3.47	1.34	4.81
3900					columns	.29	.08	.37
4000		40	25	60		3.75	1.19	4.94
4100					columns	.34	.09	.43
4200	30x30	20	25	42		3.50	1.11	4.61
4300					columns	.24	.06	.30
4400		30	22	52		3.82	1.21	5.03
4500					columns	.29	.08	.37
4600		40	28	62		3.96	1.25	5.21
4700					columns	.29	.08	.37
4800	30x35	20	22	42		3.64	1.17	4.81
4900					columns	.24	.07	.31
5000		30	28	52		3.84	1.22	5.06
5100					columns	.24	.07	.31
5200		40	25	62		4.15	1.32	5.47
5300					columns	.29	.08	.37
5400	35x35	20	28	42		3.66	1.17	4.83
5500					columns	.21	.05	.26

SHELL

B

Description: The table below lists the cost per S.F. for a roof system with steel columns, beams, and deck, using open web steel joists and 1-1/2″ galvanized metal deck.

Column costs are additive. Fireproofing is not included.

Steel Joists, Beams and Deck on Columns

SHELL B

B1020 112		Steel Joists, Beams, & Deck on Columns						
	BAY SIZE (FT.)	SUPERIMPOSED LOAD (P.S.F.)	DEPTH (IN.)	TOTAL LOAD (P.S.F.)	COLUMN ADD	COST PER S.F.		
						MAT.	INST.	TOTAL
1100	15x20	20	16	40		2.98	1.09	4.07
1200					columns	1.48	.40	1.88
1500		40	18	60		3.32	1.21	4.53
1600					columns	1.48	.40	1.88
2300	20x25	20	18	40		3.38	1.21	4.59
2400					columns	.89	.25	1.14
2700		40	20	60		3.80	1.37	5.17
2800					columns	1.19	.32	1.51
2900	25x25	20	18	40		3.93	1.36	5.29
3000					columns	.71	.19	.90
3100		30	22	50		4.25	1.54	5.79
3200					columns	.95	.26	1.21
3300		40	20	60		4.53	1.55	6.08
3400					columns	.95	.26	1.21
3500	25x30	20	22	40		3.87	1.24	5.11
3600					columns	.79	.22	1.01
3900		40	25	60		4.67	1.46	6.13
4000					columns	.95	.26	1.21
4100	30x30	20	25	42		4.37	1.36	5.73
4200					columns	.66	.18	.84
4300		30	22	52		4.81	1.50	6.31
4400					columns	.79	.22	1.01
4500		40	28	62		5.05	1.55	6.60
4600					columns	.79	.22	1.01
5300	35x35	20	28	42		4.77	1.48	6.25
5400					columns	.58	.16	.74
5500		30	25	52		5.30	1.63	6.93
5600					columns	.68	.18	.86
5700		40	28	62		5.75	1.76	7.51
5800					columns	.75	.21	.96

B1020 Roof Construction

Description: The table below lists the cost per S.F. for a roof system using open web steel joists and 1-1/2" galvanized metal deck. The system is assumed supported on bearing walls or other suitable support. Costs for the supports are not included.

B1020 116		**Steel Joists & Deck on Bearing Walls**						
	BAY SIZE (FT.)	SUPERIMPOSED LOAD (P.S.F.)	DEPTH (IN.)	TOTAL LOAD (P.S.F.)		COST PER S.F.		
						MAT.	INST.	TOTAL
1100	20	20	13-1/2	40		1.97	.80	2.77
1200		30	15-1/2	50		2.01	.81	2.82
1300		40	15-1/2	60		2.15	.88	3.03
1400	25	20	17-1/2	40		2.16	.88	3.04
1500		30	17-1/2	50		2.33	.95	3.28
1600		40	19-1/2	60		2.36	.96	3.32
1700	30	20	19-1/2	40		2.35	.96	3.31
1800		30	21-1/2	50		2.41	.98	3.39
1900		40	23-1/2	60		2.59	1.06	3.65
2000	35	20	23-1/2	40		2.57	.88	3.45
2100		30	25-1/2	50		2.65	.90	3.55
2200		40	25-1/2	60		2.81	.95	3.76
2300	40	20	25-1/2	41		2.85	.96	3.81
2400		30	25-1/2	51		3.03	1.02	4.05
2500		40	25-1/2	61		3.13	1.06	4.19
2600	45	20	27-1/2	41		3.18	1.42	4.60
2700		30	31-1/2	51		3.36	1.50	4.86
2800		40	31-1/2	61		3.54	1.59	5.13
2900	50	20	29-1/2	42		3.54	1.60	5.14
3000		30	31-1/2	52		3.87	1.75	5.62
3100		40	31-1/2	62		4.10	1.87	5.97
3200	60	20	37-1/2	42		4.53	1.63	6.16
3300		30	37-1/2	52		5	1.81	6.81
3400		40	37-1/2	62		5	1.81	6.81
3500	70	20	41-1/2	42		5	1.81	6.81
3600		30	41-1/2	52		5.35	1.92	7.27
3700		40	41-1/2	64		6.50	2.33	8.83
3800	80	20	45-1/2	44		7.30	2.61	9.91
3900		30	45-1/2	54		7.30	2.61	9.91
4000		40	45-1/2	64		8.10	2.89	10.99
4400	100	20	57-1/2	44		6.30	2.15	8.45
4500		30	57-1/2	54		7.55	2.55	10.10
4600		40	57-1/2	65		8.30	2.81	11.11
4700	125	20	69-1/2	44		8.80	2.95	11.75
4800		30	69-1/2	56		10.30	3.43	13.73
4900		40	69-1/2	67		11.75	3.91	15.66

SHELL

B

B1020 Roof Construction

Description: The table below lists costs for a roof system supported on exterior bearing walls and interior columns. Costs include bracing, joist girders, open web steel joists and 1-1/2″ galvanized metal deck.

Column costs are additive. Fireproofing is not included.

Costs/S.F. are based on a building 4 bays long and 4 bays wide. Costs for bearing walls are not included.

SHELL B

B1020 120		Steel Joists, Joist Girders & Deck on Columns & Walls						
	BAY SIZE (FT.) GIRD X JOISTS	SUPERIMPOSED LOAD (P.S.F.)	DEPTH (IN.)	TOTAL LOAD (P.S.F.)	COLUMN ADD	COST PER S.F.		
						MAT.	INST.	TOTAL
2350	30x35	20	32-1/2	40		2.82	1.13	3.95
2400					columns	.25	.07	.32
2550		40	36-1/2	60		3.17	1.25	4.42
2600					columns	.29	.08	.37
3000	35x35	20	36-1/2	40		3.07	1.37	4.44
3050					columns	.21	.05	.26
3200		40	36-1/2	60		3.46	1.52	4.98
3250					columns	.27	.08	.35
3300	35x40	20	36-1/2	40		3.13	1.26	4.39
3350					columns	.21	.06	.27
3500		40	36-1/2	60		3.52	1.41	4.93
3550					columns	.24	.06	.30
3900	40x40	20	40-1/2	41		3.47	1.63	5.10
3950					columns	.21	.05	.26
4100		40	40-1/2	61		3.88	1.79	5.67
4150					columns	.21	.05	.26
5100	45x50	20	52-1/2	41		3.89	1.86	5.75
5150					columns	.15	.04	.19
5300		40	52-1/2	61		4.72	2.24	6.96
5350					columns	.21	.06	.27
5400	50x45	20	56-1/2	41		3.78	1.79	5.57
5450					columns	.15	.04	.19
5600		40	56-1/2	61		4.84	2.28	7.12
5650					columns	.20	.05	.25
5700	50x50	20	56-1/2	42		4.22	2.02	6.24
5750					columns	.15	.04	.19
5900		40	59	64		5.75	2.28	8.03
5950					columns	.21	.05	.26
6300	60x50	20	62-1/2	43		4.27	2.04	6.31
6350					columns	.24	.06	.30
6500		40	71	65		5.90	2.33	8.23
6550					columns	.32	.09	.41

B1020 Roof Construction

Description: The table below lists the cost per S.F. for a roof system supported on columns. Costs include joist girders, open web steel joists and 1-1/2″ galvanized metal deck. Column costs are additive.

Fireproofing is not included. Costs/S.F. are based on a building 4 bays long and 4 bays wide.

Costs for wind columns are not included.

B1020 124		Steel Joists & Joist Girders on Columns						
	BAY SIZE (FT.) GIRD X JOISTS	SUPERIMPOSED LOAD (P.S.F.)	DEPTH (IN.)	TOTAL LOAD (P.S.F.)	COLUMN ADD	COST PER S.F.		
						MAT.	INST.	TOTAL
2000	30x30	20	17-1/2	40		2.89	1.30	4.19
2050					columns	.79	.22	1.01
2100		30	17-1/2	50		3.14	1.40	4.54
2150					columns	.79	.22	1.01
2200		40	21-1/2	60		3.23	1.44	4.67
2250					columns	.79	.22	1.01
2300	30x35	20	32-1/2	40		3.04	1.24	4.28
2350					columns	.68	.18	.86
2500		40	36-1/2	60		3.43	1.38	4.81
2550					columns	.79	.22	1.01
3200	35x40	20	36-1/2	40		3.77	1.91	5.68
3250					columns	.60	.17	.77
3400		40	36-1/2	60		4.25	2.13	6.38
3450					columns	.66	.18	.84
3800	40x40	20	40-1/2	41		3.83	1.91	5.74
3850					columns	.58	.16	.74
4000		40	40-1/2	61		4.29	2.12	6.41
4050					columns	.58	.16	.74
4100	40x45	20	40-1/2	41		4.09	2.16	6.25
4150					columns	.51	.14	.65
4200		30	40-1/2	51		4.45	2.32	6.77
4250					columns	.51	.14	.65
4300		40	40-1/2	61		4.94	2.54	7.48
4350					columns	.58	.16	.74
5000	45x50	20	52-1/2	41		4.31	2.21	6.52
5050					columns	.41	.12	.53
5200		40	52-1/2	61		5.25	2.65	7.90
5250					columns	.53	.14	.67
5300	50x45	20	56-1/2	41		4.20	2.14	6.34
5350					columns	.41	.12	.53
5500		40	56-1/2	61		5.40	2.70	8.10
5550					columns	.53	.14	.67
5600	50x50	20	56-1/2	42		4.69	2.38	7.07
5650					columns	.42	.12	.54
5800		40	59	64		5.10	2.16	7.26
5850					columns	.58	.16	.74

SHELL

B

B2010 Exterior Walls

Cast in Place Concrete Wall

Flat Precast Concrete Wall

Table B2010 101 describes a concrete wall system for exterior closure. There are several types of wall finishes priced from plain finish to a finish with 3/4" rustication strip.

In Table B2010 103, precast concrete wall panels are either solid or insulated. Prices below are based on delivery within 50 miles of a plant. Small jobs can double the prices below. For large, highly repetitive jobs, deduct up to 15%.

SHELL B

B2010 101	Cast In Place Concrete	MAT.	INST.	TOTAL
2100	Conc wall reinforced, 8' high, 6" thick, plain finish, 3000 PSI	3.95	13.85	17.80
2700	Aged wood liner, 3000 PSI	6.05	14.40	20.45
3000	Sand blast light 1 side, 3000 PSI	4.20	14.85	19.05
3700	3/4" bevel rustication strip, 3000 PSI	4.02	16.60	20.62
4000	8" thick, plain finish, 3000 PSI	4.64	14.15	18.79
4100	4000 PSI	4.72	14.15	18.87
4300	Rub concrete 1 side, 3000 PSI	4.67	16.40	21.07
4550	8" thick, aged wood liner, 3000 PSI	6.75	14.75	21.50
4750	Sand blast light 1 side, 3000 PSI	4.89	15.15	20.04
5300	3/4" bevel rustication strip, 3000 PSI	4.71	16.90	21.61
5600	10" thick, plain finish, 3000 PSI	5.30	14.45	19.75
5900	Rub concrete 1 side, 3000 PSI	5.30	16.70	22
6200	Aged wood liner, 3000 PSI	7.35	15.05	22.40
6500	Sand blast light 1 side, 3000 PSI	5.55	15.45	21
7100	3/4" bevel rustication strip, 3000 PSI	5.35	17.20	22.55
7400	12" thick, plain finish, 3000 PSI	6.05	14.85	20.90
7700	Rub concrete 1 side, 3000 PSI	6.10	17.05	23.15
8000	Aged wood liner, 3000 PSI	8.15	15.40	23.55
8300	Sand blast light 1 side, 3000 PSI	6.30	15.85	22.15
8900	3/4" bevel rustication strip, 3000 PSI	6.10	17.55	23.65

B2010 102			Flat Precast Concrete			COST PER S.F.		
	THICKNESS (IN.)	PANEL SIZE (FT.)	FINISHES	RIGID INSULATION (IN)	TYPE	MAT.	INST.	TOTAL
3000	4	5x18	smooth gray	none	low rise	6.25	5.95	12.20
3050		6x18				5.25	4.97	10.22
3100		8x20				9.30	1.87	11.17
3150		12x20				8.80	1.77	10.57
3200	6	5x18	smooth gray	2	low rise	7.10	6.35	13.45
3250		6x18				6.10	5.40	11.50
3300		8x20				10.30	2.37	12.67
3350		12x20				9.35	2.17	11.52
3400	8	5x18	smooth gray	2	low rise	13.25	2.95	16.20
3450		6x18				12.55	2.81	15.36
3500		8x20				11.50	2.60	14.10
3550		12x20				10.50	2.40	12.90

B20 Exterior Enclosure

B2010 Exterior Walls

B2010 102 — Flat Precast Concrete

	THICKNESS (IN.)	PANEL SIZE (FT.)	FINISHES	RIGID INSULATION (IN)	TYPE	COST PER S.F. MAT.	COST PER S.F. INST.	COST PER S.F. TOTAL
3600	4	4x8	white face	none	low rise	22.50	3.44	25.94
3650		8x8				16.85	2.58	19.43
3700		10x10				14.75	2.27	17.02
3750		20x10				13.35	2.06	15.41
3800	5	4x8	white face	none	low rise	23	3.52	26.52
3850		8x8				17.30	2.66	19.96
3900		10x10				15.35	2.35	17.70
3950		20x20				14.05	2.15	16.20
4000	6	4x8	white face	none	low rise	23.50	3.64	27.14
4050		8x8				18.05	2.77	20.82
4100		10x10				15.90	2.43	18.33
4150		20x10				14.60	2.24	16.84
4200	6	4x8	white face	2	low rise	24.50	4.12	28.62
4250		8x8				18.95	3.25	22.20
4300		10x10				16.80	2.91	19.71
4350		20x10				14.60	2.24	16.84
4400	7	4x8	white face	none	low rise	24.50	3.73	28.23
4450		8x8				18.65	2.86	21.51
4500		10x10				16.75	2.57	19.32
4550		20x10				15.35	2.35	17.70
4600	7	4x8	white face	2	low rise	25	4.21	29.21
4650		8x8				19.60	3.34	22.94
4700		10x10				17.65	3.05	20.70
4750		20x10				16.25	2.83	19.08
4800	8	4x8	white face	none	low rise	25	3.82	28.82
4850		8x8				19.15	2.94	22.09
4900		10x10				17.20	2.65	19.85
4950		20x10				15.85	2.43	18.28
5000	8	4x8	white face	2	low rise	26	4.30	30.30
5050		8x8				20	3.42	23.42
5100		10x10				18.15	3.13	21.28
5150		20x10				16.80	2.91	19.71

B2010 103 — Fluted Window or Mullion Precast Concrete

	THICKNESS (IN.)	PANEL SIZE (FT.)	FINISHES	RIGID INSULATION (IN)	TYPE	COST PER S.F. MAT.	COST PER S.F. INST.	COST PER S.F. TOTAL
5200	4	4x8	smooth gray	none	high rise	14.35	13.65	28
5250		8x8				10.25	9.75	20
5300		10x10				17.80	3.57	21.37
5350		20x10				15.60	3.13	18.73
5400	5	4x8	smooth gray	none	high rise	14.60	13.85	28.45
5450		8x8				10.60	10.05	20.65
5500		10x10				18.55	3.72	22.27
5550		20x10				16.40	3.29	19.69
5600	6	4x8	smooth gray	none	high rise	15	14.25	29.25
5650		8x8				11	10.40	21.40
5700		10x10				19.10	3.82	22.92
5750		20x10				16.95	3.39	20.34
5800	6	4x8	smooth gray	2	high rise	15.90	14.70	30.60
5850		8x8				11.90	10.90	22.80
5900		10x10				20	4.30	24.30
5950		20x10				17.85	3.87	21.72

SHELL

B

B2010 Exterior Walls

B2010 103	Fluted Window or Mullion Precast Concrete

	THICKNESS (IN.)	PANEL SIZE (FT.)	FINISHES	RIGID INSULATION (IN)	TYPE	COST PER S.F.		
						MAT.	INST.	TOTAL
6000	7	4x8	smooth gray	none	high rise	15.30	14.50	29.80
6050		8x8				11.30	10.70	22
6100		10x10				19.90	4	23.90
6150		20x10				17.45	3.50	20.95
6200	7	4x8	smooth gray	2	high rise	16.20	15	31.20
6250		8x8				12.20	11.20	23.40
6300		10x10				21	4.48	25.48
6350		20x10				18.40	3.98	22.38
6400	8	4x8	smooth gray	none	high rise	15.55	14.70	30.25
6450		8x8				11.55	10.95	22.50
6500		10x10				20.50	4.12	24.62
6550		20x10				18.35	3.68	22.03
6600	8	4x8	smooth gray	2	high rise	16.45	15.20	31.65
6650		8x8				12.50	11.45	23.95
6700		10x10				21.50	4.60	26.10
6750		20x10				19.25	4.16	23.41

B2010 104	Ribbed Precast Concrete

	THICKNESS (IN.)	PANEL SIZE (FT.)	FINISHES	RIGID INSULATION(IN.)	TYPE	COST PER S.F.		
						MAT.	INST.	TOTAL
6800	4	4x8	aggregate	none	high rise	18.25	13.60	31.85
6850		8x8				13.35	9.95	23.30
6900		10x10				23	3.60	26.60
6950		20x10				20	3.18	23.18
7000	5	4x8	aggregate	none	high rise	18.55	13.85	32.40
7050		8x8				13.70	10.20	23.90
7100		10x10				24	3.74	27.74
7150		20x10				21	3.34	24.34
7200	6	4x8	aggregate	none	high rise	18.95	14.10	33.05
7250		8x8				14.10	10.50	24.60
7300		10x10				24.50	3.86	28.36
7350		20x10				22	3.46	25.46
7400	6	4x8	aggregate	2	high rise	19.85	14.60	34.45
7450		8x8				15	11	26
7500		10x10				25.50	4.34	29.84
7550		20x10				23	3.94	26.94
7600	7	4x8	aggregate	none	high rise	19.35	14.45	33.80
7650		8x8				14.45	10.75	25.20
7700		10x10				25.50	3.98	29.48
7750		20x10				22.50	3.55	26.05
7800	7	4x8	aggregate	2	high rise	20.50	14.95	35.45
7850		8x8				15.35	11.25	26.60
7900		10x10				26	4.46	30.46
7950		20x10				23.50	4.03	27.53
8000	8	4x8	aggregate	none	high rise	19.70	14.70	34.40
8050		8x8				14.85	11.10	25.95
8100		10x10				26	4.12	30.12
8150		20x10				23.50	3.72	27.22
8200	8	4x8	aggregate	2	high rise	20.50	15.15	35.65
8250		8x8				15.75	11.60	27.35
8300		10x10				27	4.60	31.60
8350		20x10				24.50	4.20	28.70

SHELL B

B20 Exterior Enclosure

B2010 Exterior Walls

B2010 105	Precast Concrete Specialties

	TYPE	SIZE				COST PER L.F.		
						MAT.	INST.	TOTAL
8400	Coping, precast	6"wide				16.50	10.10	26.60
8450	Stock units	10" wide				11.55	10.80	22.35
8460		12" wide				22	11.65	33.65
8480		14" wide				15.95	12.65	28.60
8500	Window sills	6" wide				10.35	10.80	21.15
8550	Precast	10" wide				13.75	12.65	26.40
8600		14" wide				14.85	15.15	30
8610								

B2010 Exterior Walls

The advantage of tilt up construction is in the low cost of forms and placing of concrete and reinforcing. Tilt up has been used for several types of buildings, including warehouses, stores, offices, and schools. The panels are cast in forms on the ground, or floor slab. Most jobs use 5-1/2" thick solid reinforced concrete panels.

Tilt Up Concrete Panel

B2010 106	Tilt-Up Concrete Panel	COST PER S.F.		
		MAT.	INST.	TOTAL
3200	Tilt up conc panels, broom finish, 5-1/2" thick, 3000 PSI	3.25	4.60	7.85
3250	5000 PSI	3.34	4.54	7.88
3300	6" thick, 3000 PSI	3.60	4.75	8.35
3350	5000 PSI	3.69	4.66	8.35
3400	7-1/2" thick, 3000 PSI	4.54	4.92	9.46
3450	5000 PSI	4.70	4.84	9.54
3500	8" thick, 3000 PSI	4.89	5.05	9.94
3550	5000 PSI	5.05	4.95	10
3700	Steel trowel finish, 5-1/2" thick, 3000 PSI	3.25	4.72	7.97
3750	5000 PSI	3.34	4.63	7.97
3800	6" thick, 3000 PSI	3.60	4.84	8.44
3850	5000 PSI	3.69	4.75	8.44
3900	7-1/2" thick, 3000 PSI	4.54	5	9.54
3950	5000 PSI	4.70	4.93	9.63
4000	8" thick, 3000 PSI	4.89	5.15	10.04
4050	5000 PSI	5.05	5.05	10.10
4200	Exp. aggregate finish, 5-1/2" thick, 3000 PSI	7.85	4.73	12.58
4250	5000 PSI	7.90	4.64	12.54
4300	6" thick, 3000 PSI	8.15	4.84	12.99
4350	5000 PSI	8.25	4.76	13.01
4400	7-1/2" thick, 3000 PSI	9.10	5	14.10
4450	5000 PSI	9.25	4.93	14.18
4500	8" thick, 3000 PSI	9.45	5.15	14.60
4550	5000 PSI	9.65	5.05	14.70
4600	Exposed aggregate & vert. rustication 5-1/2" thick, 3000 PSI	9.30	7.85	17.15
4650	5000 PSI	9.40	7.75	17.15
4700	6" thick, 3000 PSI	9.65	7.95	17.60
4750	5000 PSI	9.75	7.85	17.60
4800	7-1/2" thick, 3000 PSI	10.60	8.15	18.75
4850	5000 PSI	10.75	8.05	18.80
4900	8" thick, 3000 PSI	10.95	8.25	19.20
4950	5000 PSI	11.10	8.15	19.25
5000	Vertical rib & light sandblast, 5-1/2" thick, 3000 PSI	5.50	5.65	11.15
5100	6" thick, 3000 PSI	5.85	5.75	11.60
5200	7-1/2" thick, 3000 PSI	6.80	5.95	12.75
5300	8" thick, 3000 PSI	7.15	6.05	13.20
6000	Broom finish w/2" polystyrene insulation, 6" thick, 3000 PSI	3.40	5.70	9.10
6100	Broom finish 2" fiberplank insulation, 6" thick, 3000 PSI	3.89	5.65	9.54
6200	Exposed aggregate w/2" polystyrene insulation, 6" thick, 3000 PSI	7.90	5.85	13.75
6300	Exposed aggregate 2" fiberplank insulation, 6" thick, 3000 PSI	8.40	5.80	14.20

SHELL B

Important: See the Reference Section for critical supporting data - Reference Numbers and City Cost Indexes

Exterior concrete block walls are defined in the following terms; structural reinforcement, weight, percent solid, size, strength and insulation. Within each of these categories, two to four variations are shown. No costs are included for brick shelf or relieving angles.

SHELL

B

B2010 109 — Concrete Block Wall - Regular Weight

	TYPE	SIZE (IN.)	STRENGTH (P.S.I.)	CORE FILL		COST PER S.F. MAT.	INST.	TOTAL
1200	Hollow	4x8x16	2,000	none		1.18	4.68	5.86
1250			4,500	none		1.08	4.68	5.76
1400		8x8x16	2,000	perlite		2.29	5.70	7.99
1410				styrofoam		2.76	5.40	8.16
1440				none		1.82	5.40	7.22
1450			4,500	perlite		2.07	5.70	7.77
1460				styrofoam		2.54	5.40	7.94
1490				none		1.60	5.40	7
2000	75% solid	4x8x16	2,000	none		1.15	4.74	5.89
2100	75% solid	6x8x16	2,000	perlite		1.66	5.20	6.86
2500	Solid	4x8x16	2,000	none		1.57	4.85	6.42
2700		8x8x16	2,000	none		2.69	5.60	8.29

B2010 110 — Concrete Block Wall - Lightweight

	TYPE	SIZE (IN.)	WEIGHT (P.C.F.)	CORE FILL		COST PER S.F. MAT.	INST.	TOTAL
3100	Hollow	8x4x16	105	perlite		1.57	5.35	6.92
3110				styrofoam		2.04	5.05	7.09
3200		4x8x16	105	none		1.37	4.58	5.95
3250			85	none		1.36	4.48	5.84
3300		6x8x16	105	perlite		2.16	5.15	7.31
3310				styrofoam		2.78	4.90	7.68
3340				none		1.84	4.90	6.74
3400		8x8x16	105	perlite		2.70	5.55	8.25
3410				styrofoam		3.17	5.25	8.42
3440				none		2.23	5.25	7.48
3450			85	perlite		2.37	5.45	7.82
4000	75% solid	4x8x16	105	none		1.19	4.63	5.82
4050			85	none		2.32	4.53	6.85
4100		6x8x16	105	perlite		1.94	5.10	7.04
4500	Solid	4x8x16	105	none		1.49	4.79	6.28
4700		8x8x16	105	none		2.98	5.55	8.53

B2010 Exterior Walls

| B2010 111 | Reinforced Concrete Block Wall - Regular Weight |

	TYPE	SIZE (IN.)	STRENGTH (P.S.I.)	VERT. REINF & GROUT SPACING		COST PER S.F. MAT.	INST.	TOTAL
5200	Hollow	4x8x16	2,000	#4 @ 48"		1.30	5.20	6.50
5300		6x8x16	2,000	#4 @ 48"		1.89	5.55	7.44
5330				#5 @ 32"		2.06	5.75	7.81
5340				#5 @ 16"		2.41	6.50	8.91
5350			4,500	#4 @ 28"		1.57	5.55	7.12
5390				#5 @ 16"		2.09	6.50	8.59
5400	Hollow	8x8x16	2,000	#4 @ 48"		2.08	5.95	8.03
5430				#5 @ 32"		2.25	6.25	8.50
5440				#5 @ 16"		2.66	7.15	9.81
5450		8x8x16	4,500	#4 @ 48"		1.84	6	7.84
5490				#5 @ 16"		2.44	7.15	9.59
5500		12x8x16	2,000	#4 @ 48"		2.95	7.60	10.55
5540				#5 @ 16"		3.75	8.75	12.50
6100	75% solid	6x8x16	2,000	#4 @ 48"		1.61	5.50	7.11
6140				#5 @ 16"		1.96	6.30	8.26
6150			4,500	#4 @ 48"		1.98	5.50	7.48
6190				#5 @ 16"		2.33	6.30	8.63
6200		8x8x16	2,000	#4 @ 48"		2.35	5.95	8.30
6230				#5 @ 32"		2.48	6.15	8.63
6240				#5 @ 16"		2.72	6.90	9.62
6250			4,500	#4 @ 48"		2.64	5.95	8.59
6280				#5 @ 32"		2.77	6.15	8.92
6290				#5 @ 16"		3.01	6.90	9.91
6500	Solid-double	2-4x8x16	2,000	#4 @ 48" E.W.		3.82	11	14.82
6530	Wythe			#5 @ 16" E.W.		4.41	11.65	16.06
6550			4,500	#4 @ 48" E.W.		3.28	10.85	14.13
6580				#5 @ 16" E.W.		3.87	11.55	15.42

| B2010 112 | Reinforced Concrete Block Wall - Lightweight |

	TYPE	SIZE (IN.)	WEIGHT (P.C.F.)	VERT REINF. & GROUT SPACING		COST PER S.F. MAT.	INST.	TOTAL
7100	Hollow	8x4x16	105	#4 @ 48"		1.34	5.70	7.04
7140				#5 @ 16"		1.94	6.80	8.74
7150			85	#4 @ 48"		3.33	5.55	8.88
7190				#5 @ 16"		3.93	6.70	10.63
7400		8x8x16	105	#4 @ 48"		2.47	5.85	8.32
7440				#5 @ 16"		3.07	7	10.07
7450		8x8x16	85	#4 @ 48"		2.14	5.75	7.89
7490				#5 @ 16"		2.74	6.85	9.59
7800		8x8x24	105	#4 @ 48"		1.87	6.30	8.17
7840				#5 @ 16"		2.47	7.45	9.92
7850			85	#4 @ 48"		3.64	5.40	9.04
7890				#5 @ 16"		4.24	6.50	10.74
8100	75% solid	6x8x16	105	#4 @ 48"		1.89	5.40	7.29
8130				#5 @ 32"		2.02	5.55	7.57
8150			85	#4 @ 48"		2.79	5.30	8.09
8180				#5 @ 32"		2.92	5.45	8.37
8200		8x8x16	105	#4 @ 48"		2.51	5.80	8.31
8230				#5 @ 32"		2.64	6	8.64
8250			85	#4 @ 48"		3.77	5.70	9.47
8280				#5 @ 32"		3.90	5.90	9.80

SHELL B

Important: See the Reference Section for critical supporting data - Reference Numbers and City Cost Indexes

B2010 Exterior Walls

B2010 112	Reinforced Concrete Block Wall - Lightweight

	TYPE	SIZE (IN.)	WEIGHT (P.C.F.)	VERT REINF. & GROUT SPACING		COST PER S.F.		
						MAT.	INST.	TOTAL
8500	Solid-double	2-4x8x16	105	#4 @ 48"		3.66	10.85	14.51
8530	Wythe		105	#5 @ 16"		4.25	11.55	15.80
8600		2-6x8x16	105	#4 @ 48"		5.10	11.60	16.70
8630				#5 @ 16"		5.70	12.30	18
8650			85	#4 @ 48"		6.45	11.10	17.55

B2010 Exterior Walls

Exterior split ribbed block walls are defined in the following terms; structural reinforcement, weight, percent solid, size, number of ribs and insulation. Within each of these categories two to four variations are shown. No costs are included for brick shelf or relieving angles. Costs include control joints every 20' and horizontal reinforcing.

B2010 113	Split Ribbed Block Wall - Regular Weight							
	TYPE	**SIZE (IN.)**	**RIBS**	**CORE FILL**		**COST PER S.F.**		
						MAT.	**INST.**	**TOTAL**
1220	Hollow	4x8x16	4	none		2.23	5.80	8.03
1250			8	none		2.59	5.80	8.39
1280			16	none		2.12	5.85	7.97
1430		8x8x16	8	perlite		4.50	6.90	11.40
1440				styrofoam		4.97	6.60	11.57
1450				none		4.03	6.60	10.63
1530		12x8x16	8	perlite		5.65	9.10	14.75
1540				styrofoam		6.05	8.45	14.50
1550				none		4.90	8.45	13.35
2120	75% solid	4x8x16	4	none		2.79	5.85	8.64
2150			8	none		3.30	5.85	9.15
2180			16	none		2.68	5.95	8.63
2520	Solid	4x8x16	4	none		3.40	5.95	9.35
2550			8	none		3.98	5.95	9.93
2580			16	none		3.21	6.05	9.26

B2010 115	Reinforced Split Ribbed Block Wall - Regular Weight							
	TYPE	**SIZE (IN.)**	**RIBS**	**VERT. REINF. & GROUT SPACING**		**COST PER S.F.**		
						MAT.	**INST.**	**TOTAL**
5200	Hollow	4x8x16	4	#4 @ 48"		2.35	6.35	8.70
5230			8	#4 @ 48"		2.71	6.35	9.06
5260			16	#4 @ 48"		2.24	6.40	8.64
5430		8x8x16	8	#4 @ 48"		4.27	7.20	11.47
5440				#5 @ 32"		4.46	7.45	11.91
5450				#5 @ 16"		4.87	8.35	13.22
5530		12x8x16	8	#4 @ 48"		5.25	9.15	14.40
5540				#5 @ 32"		5.45	9.40	14.85
5550				#5 @ 16"		6.05	10.30	16.35
6230	75% solid	6x8x16	8	#4 @ 48"		4.21	6.70	10.91
6240				#5 @ 32"		4.34	6.85	11.19
6250				#5 @ 16"		4.56	7.50	12.06
6330		8x8x16	8	#4 @ 48"		5.25	7.20	12.45
6340				#5 @ 32"		5.40	7.40	12.80
6350				#5 @ 16"		5.65	8.15	13.80

B20 Exterior Enclosure

B2010 Exterior Walls

Exterior split face block walls are defined in the following terms; structural reinforcement, weight, percent solid, size, scores and insulation. Within each of these categories two to four variations are shown. No costs are included for brick shelf or relieving angles. Costs include control joints every 20′ and horizontal reinforcing.

B2010 117 — Split Face Block Wall - Regular Weight

	TYPE	SIZE (IN.)	SCORES	CORE FILL		COST PER S.F. MAT.	INST.	TOTAL
1200	Hollow	8x4x16	0	perlite		6.55	7.55	14.10
1210				styrofoam		7.05	7.25	14.30
1240				none		6.10	7.25	13.35
1250			1	perlite		7	7.55	14.55
1260				styrofoam		7.50	7.25	14.75
1290				none		6.55	7.25	13.80
1300		12x4x16	0	perlite		8.60	8.65	17.25
1310				styrofoam		9	8	17
1340				none		7.80	8	15.80
1350			1	perlite		9.05	8.65	17.70
1360				styrofoam		9.45	8	17.45
1390				none		8.25	8	16.25
1400		4x8x16	0	none		2.23	5.70	7.93
1450			1	none		2.51	5.80	8.31
1500		6x8x16	0	perlite		3.21	6.55	9.76
1510				styrofoam		3.83	6.30	10.13
1540				none		2.89	6.30	9.19
1550			1	perlite		3.50	6.65	10.15
1560				styrofoam		4.12	6.40	10.52
1590				none		3.18	6.40	9.58
1600		8x8x16	0	perlite		3.89	7.10	10.99
1610				styrofoam		4.36	6.80	11.16
1640				none		3.42	6.80	10.22
1650			1	perlite		4.12	7.20	11.32
1660				styrofoam		4.59	6.90	11.49
1690				none		3.65	6.90	10.55
1700		12x8x16	0	perlite		5	9.25	14.25
1710				styrofoam		5.40	8.60	14
1740				none		4.24	8.60	12.84
1750			1	perlite		5.25	9.45	14.70
1760				styrofoam		5.65	8.80	14.45
1790				none		4.48	8.80	13.28

B2010 Exterior Walls

| B2010 117 | Split Face Block Wall - Regular Weight | | | | | | | |

	TYPE	SIZE (IN.)	SCORES	CORE FILL		COST PER S.F.		
						MAT.	INST.	TOTAL
1800	75% solid	8x4x16	0	perlite		7.95	7.55	15.50
1840				none		7.70	7.40	15.10
1852		8x4x16	1	perlite		8.45	7.55	16
1890				none		8.20	7.40	15.60
2000		4x8x16	0	none		2.79	5.80	8.59
2050			1	none		3.03	5.85	8.88
2400	Solid	8x4x16	0	none		9.40	7.55	16.95
2450			1	none		9.90	7.55	17.45
2900		12x8x16	0	none		6.45	8.95	15.40
2950			1	none		6.65	9.10	15.75

| B2010 119 | Reinforced Split Face Block Wall - Regular Weight | | | | | | | |

	TYPE	SIZE (IN.)	SCORES	VERT. REINF. & GROUT SPACING		COST PER S.F.		
						MAT.	INST.	TOTAL
5200	Hollow	8x4x16	0	#4 @ 48"		6.35	7.85	14.20
5210				#5 @ 32"		6.55	8.10	14.65
5240				#5 @ 16"		6.95	9	15.95
5250			1	#4 @ 48"		6.80	7.85	14.65
5260				#5 @ 32"		7	8.10	15.10
5290				#5 @ 16"		7.40	9	16.40
5700		12x8x16	0	#4 @ 48"		4.58	9.30	13.88
5710				#5 @ 32"		4.81	9.55	14.36
5740				#5 @ 16"		5.40	10.45	15.85
5750			1	#4 @ 48"		4.82	9.50	14.32
5760				#5 @ 32"		5.05	9.75	14.80
5790				#5 @ 16"		5.60	10.65	16.25
6000	75% solid	8x4x16	0	#4 @ 48"		7.80	7.90	15.70
6010				#5 @ 32"		7.95	8.10	16.05
6040				#5 @ 16"		8.20	8.85	17.05
6050			1	#4 @ 48"		8.30	7.90	16.20
6060				#5 @ 32"		8.45	8.10	16.55
6090				#5 @ 16"		8.70	8.85	17.55
6100		12x4x16	0	#4 @ 48"		10.10	8.70	18.80
6110				#5 @ 32"		10.30	8.90	19.20
6140				#5 @ 16"		10.60	9.50	20.10
6150			1	#4 @ 48"		10.50	8.70	19.20
6160				#5 @ 32"		10.70	8.90	19.60
6190				#5 @ 16"		11.05	9.65	20.70
6700	Solid-double	2-4x8x16	0	#4 @ 48" E.W.		7.50	13	20.50
6710	Wythe			#5 @ 32" E.W.		7.70	13.10	20.80
6740				#5 @ 16" E.W.		8.05	13.70	21.75
6750			1	#4 @ 48" E.W.		7.95	13.20	21.15
6760				#5 @ 32" E.W.		8.15	13.30	21.45
6790				#5 @ 16" E.W.		8.55	13.90	22.45
6800		2-6x8x16	0	#4 @ 48" E.W.		9.70	14.45	24.15
6810				#5 @ 32" E.W.		9.90	14.55	24.45
6840				#5 @ 16" E.W.		10.30	15.10	25.40
6850			1	#4 @ 48" E.W.		10.15	14.65	24.80
6860				#5 @ 32" E.W.		10.40	14.75	25.15
6890				#5 @ 16" E.W.		10.75	15.30	26.05

SHELL B

Important: See the Reference Section for critical supporting data - Reference Numbers and City Cost Indexes

B2010 Exterior Walls

Exterior solid brick walls are defined in the following terms; structural reinforcement, type of brick, thickness, and bond. Nine different types of face bricks are presented, with single wythes shown in four different bonds. Shelf angles are included in system components. These walls do not include ties and as such are not tied to a backup wall.

B2010 125 — Solid Brick Walls - Single Wythe

	TYPE	THICKNESS (IN.)	BOND			COST PER S.F. MAT.	INST.	TOTAL
1000	Common	4	running			3.34	10.60	13.94
1010			common			3.98	12.10	16.08
1050			Flemish			4.37	14.70	19.07
1100			English			4.87	15.55	20.42
1150	Standard	4	running			3.98	11	14.98
1160			common			4.66	12.65	17.31
1200			Flemish			5.15	15.10	20.25
1250			English			5.75	16.05	21.80
1300	Glazed	4	running			8.50	11.40	19.90
1310			common			10.10	13.25	23.35
1350			Flemish			11.20	16.05	27.25
1400			English			12.55	17.10	29.65
1600	Economy	4	running			3.97	8.40	12.37
1610			common			4.64	9.65	14.29
1650			Flemish			5.10	11.40	16.50
1700			English			9.60	12.10	21.70
1900	Fire	4-1/2	running			7.75	9.65	17.40
1910			common			9.20	11	20.20
1950			Flemish			10.15	13.25	23.40
2000			English			11.40	13.95	25.35
2050	King	3-1/2	running			3.38	8.65	12.03
2060			common			3.93	9.95	13.88
2100			Flemish			4.74	11.85	16.59
2150			English			5.20	12.40	17.60
2200	Roman	4	running			5.65	9.95	15.60
2210			common			6.70	11.40	18.10
2250			Flemish			7.40	13.60	21
2300			English			8.30	14.70	23
2350	Norman	4	running			4.87	8.25	13.12
2360			common			5.70	9.35	15.05
2400			Flemish			10.15	11.20	21.35
2450			English			7.05	11.85	18.90

SHELL

B

B20 Exterior Enclosure

B2010 Exterior Walls

B2010 125 — Solid Brick Walls - Single Wythe

	TYPE	THICKNESS (IN.)	BOND			COST PER S.F.		
						MAT.	INST.	TOTAL
2500	Norwegian	4	running			3.99	7.35	11.34
2510			common			4.68	8.30	12.98
2550			Flemish			5.15	9.95	15.10
2600			English			5.70	10.45	16.15

B2010 126 — Solid Brick Walls - Double Wythe

	TYPE	THICKNESS (IN.)	COLLAR JOINT THICKNESS (IN.)			COST PER S.F.		
						MAT.	INST.	TOTAL
4100	Common	8	3/4			6.45	17.80	24.25
4150	Standard	8	1/2			7.60	18.40	26
4200	Glazed	8	1/2			16.70	19.05	35.75
4250	Engineer	8	1/2			7.80	16.25	24.05
4300	Economy	8	3/4			7.60	13.95	21.55
4350	Double	8	3/4			9.75	11.30	21.05
4400	Fire	8	3/4			15.20	16.25	31.45
4450	King	7	3/4			6.40	14.30	20.70
4500	Roman	8	1			11	16.75	27.75
4550	Norman	8	3/4			9.35	13.65	23
4600	Norwegian	8	3/4			7.60	12.10	19.70
4650	Utility	8	3/4			7.10	10.50	17.60
4700	Triple	8	3/4			5.55	9.45	15
4750	SCR	12	3/4			10.45	13.95	24.40
4800	Norwegian	12	3/4			9.10	12.35	21.45
4850	Jumbo	12	3/4			9.85	10.80	20.65

B2010 127 — Reinforced Brick Walls

	TYPE	THICKNESS (IN.)	WYTHE	REINF. & SPACING		COST PER S.F.		
						MAT.	INST.	TOTAL
7200	Common	4"	1	#4 @ 48"vert		3.50	11	14.50
7220				#5 @ 32"vert		3.62	11.15	14.77
7230				#5 @ 16"vert		3.83	11.65	15.48
7700	King	9"	2	#4 @ 48" E.W.		6.75	14.95	21.70
7720				#5 @ 32" E.W.		6.95	15.10	22.05
7730				#5 @ 16" E.W.		7.35	15.55	22.90
7800	Common	10"	2	#4 @ 48" E.W.		6.80	18.45	25.25
7820				#5 @ 32" E.W.		7	18.60	25.60
7830				#5 @ 16" E.W.		7.40	19.05	26.45
7900	Standard	10"	2	#4 @ 48" E.W.		7.95	19.05	27
7920				#5 @ 32" E.W.		8.15	19.20	27.35
7930				#5 @ 16" E.W.		8.55	19.65	28.20
8100	Economy	10"	2	#4 @ 48" E.W.		7.95	14.60	22.55
8120				#5 @ 32" E.W.		8.15	14.75	22.90
8130				#5 @ 16" E.W.		8.55	15.20	23.75
8200	Double	10"	2	#4 @ 48" E.W.		10.10	11.95	22.05
8220				#5 @ 32" E.W.		10.30	12.10	22.40
8230				#5 @ 16" E.W.		10.70	12.55	23.25
8300	Fire	10"	2	#4 @ 48" E.W.		15.55	16.90	32.45
8320				#5 @ 32" E.W.		15.75	17.05	32.80
8330				#5 @ 16" E.W.		16.15	17.50	33.65
8400	Roman	10"	2	#4 @ 48" E.W.		11.35	17.40	28.75
8420				#5 @ 32" E.W.		11.55	17.55	29.10
8430				#5 @ 16" E.W.		11.95	18	29.95

SHELL B

Important: See the Reference Section for critical supporting data - Reference Numbers and City Cost Indexes

B2010 Exterior Walls

B2010 127	Reinforced Brick Walls

	TYPE	THICKNESS (IN.)	WYTHE	REINF. & SPACING		COST PER S.F.		
						MAT.	INST.	TOTAL
8500	Norman	10″	2	#4 @ 48″ E.W.		9.70	14.30	24
8520				#5 @ 32″ E.W.		9.90	14.45	24.35
8530				#5 @ 16″ E.W.		10.30	14.90	25.20
8600	Norwegian	10″	2	#4 @ 48″ E.W.		7.95	12.75	20.70
8620				#5 @ 32″ E.W.		8.15	12.90	21.05
8630				#5 @ 16″ E.W.		8.55	13.35	21.90
8800	Triple	10″	2	#4 @ 48″ E.W.		5.90	10.10	16
8820				#5 @ 32″ E.W.		6.10	10.25	16.35
8830				#5 @ 16″ E.W.		6.50	10.70	17.20

SHELL

B

B2010 Exterior Walls

Stone Veneer

The table below lists costs per S.F. for stone veneer walls on various backup using different stone.

B2010 128	Stone Veneer	COST PER S.F.		
		MAT.	INST.	TOTAL
2000	Ashlar veneer,4", 2"x4" stud backup,16" O.C., 8' high, low priced stone	10.90	14.45	25.35
2100	Metal stud backup, 8' high, 16" O.C.	11.90	14.70	26.60
2150	24" O.C.	11.50	14.30	25.80
2200	Conc. block backup, 4" thick	11.60	17.60	29.20
2300	6" thick	12.10	17.80	29.90
2350	8" thick	12.20	18.25	30.45
2400	10" thick	12.90	19.30	32.20
2500	12" thick	13	20.50	33.50
3100	High priced stone, wood stud backup, 10' high, 16" O.C.	17	16.30	33.30
3200	Metal stud backup, 10' high, 16" O.C.	17.90	16.55	34.45
3250	24" O.C.	17.50	16.15	33.65
3300	Conc. block backup, 10' high, 4" thick	17.60	19.45	37.05
3350	6" thick	18.10	19.50	37.60
3400	8" thick	18.20	20	38.20
3450	10" thick	18.90	21	39.90
3500	12" thick	19	22.50	41.50
4000	Indiana limestone 2" thick,sawn finish,wood stud backup,10' high,16" O.C.	22	11.80	33.80
4100	Metal stud backup, 10' high, 16" O.C.	22.50	12.05	34.55
4150	24" O.C.	22.50	11.65	34.15
4200	Conc. block backup, 4" thick	22.50	14.95	37.45
4250	6" thick	23	15.15	38.15
4300	8" thick	23	15.60	38.60
4350	10" thick	24	16.65	40.65
4400	12" thick	24	18.10	42.10
4450	2" thick, smooth finish, wood stud backup, 8' high, 16" O.C.	22	11.80	33.80
4550	Metal stud backup, 8' high, 16" O.C.	22.50	11.90	34.40
4600	24" O.C.	22.50	11.50	34
4650	Conc. block backup, 4" thick	22.50	14.80	37.30
4700	6" thick	23	15	38
4750	8" thick	23	15.45	38.45
4800	10" thick	24	16.65	40.65
4850	12" thick	24	18.10	42.10
5350	4" thick, smooth finish, wood stud backup, 8' high, 16" O.C.	31	11.80	42.80
5450	Metal stud backup, 8' high, 16" O.C.	31.50	12.05	43.55
5500	24" O.C.	31.50	11.65	43.15
5550	Conc. block backup, 4" thick	31.50	14.95	46.45
5600	6" thick	32	15.15	47.15
5650	8" thick	32	15.60	47.60

Important: See the Reference Section for critical supporting data - Reference Numbers and City Cost Indexes

B2010 Exterior Walls

B2010 128	Stone Veneer	COST PER S.F.		
		MAT.	INST.	TOTAL
5700	10" thick	33	16.65	49.65
5750	12" thick	33	18.10	51.10
6000	Granite, grey or pink, 2" thick, wood stud backup, 8' high, 16" O.C.	20.50	20.50	41
6100	Metal studs, 8' high, 16" O.C.	21	20.50	41.50
6150	24" O.C.	21	20.50	41.50
6200	Conc. block backup, 4" thick	21	23.50	44.50
6250	6" thick	21.50	24	45.50
6300	8" thick	21.50	24.50	46
6350	10" thick	22.50	25.50	48
6400	12" thick	22.50	27	49.50
6900	4" thick, wood stud backup, 8' high, 16" O.C.	33	23	56
7000	Metal studs, 8' high, 16" O.C.	33.50	23.50	57
7050	24" O.C.	33.50	23	56.50
7100	Conc. block backup, 4" thick	33.50	26.50	60
7150	6" thick	34	26.50	60.50
7200	8" thick	34	27	61
7250	10" thick	35	28	63
7300	12" thick	35	29.50	64.50

SHELL

B

B2010 Exterior Walls

Exterior brick veneer/stud backup walls are defined in the following terms; type of brick and studs, stud spacing and bond. All systems include a brick shelf, ties to the backup and all necessary dampproofing and insulation.

SHELL B

B2010 129	Brick Veneer/Wood Stud Backup					COST PER S.F.		
	FACE BRICK	STUD BACKUP	STUD SPACING (IN.)	BOND		MAT.	INST.	TOTAL
1100	Standard	2x4-wood	16	running		5.55	13	18.55
1120				common		6.25	14.65	20.90
1140				Flemish		6.70	17.10	23.80
1160				English		7.30	18.05	25.35
1400		2x6-wood	16	running		5.80	13.05	18.85
1420				common		6.45	14.70	21.15
1440				Flemish		6.95	17.15	24.10
1460				English		7.55	18.10	25.65
1700	Glazed	2x4-wood	16	running		10.10	13.40	23.50
1720				common		11.70	15.25	26.95
1740				Flemish		12.80	18.05	30.85
1760				English		14.15	19.10	33.25
2300	Engineer	2x4-wood	16	running		5.65	11.65	17.30
2320				common		6.35	13	19.35
2340				Flemish		6.85	15.25	22.10
2360				English		7.45	15.95	23.40
2900	Roman	2x4-wood	16	running		7.25	11.95	19.20
2920				common		8.30	13.40	21.70
2940				Flemish		9	15.60	24.60
2960				English		9.90	16.70	26.60
4100	Norwegian	2x4-wood	16	running		5.55	9.35	14.90
4120				common		6.25	10.30	16.55
4140				Flemish		6.70	11.95	18.65
4160				English		7.30	12.45	19.75

B2010 130	Brick Veneer/Metal Stud Backup					COST PER S.F.		
	FACE BRICK	STUD BACKUP	STUD SPACING (IN.)	BOND		MAT.	INST.	TOTAL
5100	Standard	25ga.x6"NLB	24	running		5.25	12.70	17.95
5120				common		5.95	14.35	20.30
5140				Flemish		5.95	16.25	22.20
5160				English		7	17.75	24.75
5200		20ga.x3-5/8"NLB	16	running		5.60	13.20	18.80
5220				common		6.30	14.85	21.15
5240				Flemish		6.75	17.30	24.05
5260				English		7.35	18.25	25.60

B2010 Exterior Walls

B2010 130	Brick Veneer/Metal Stud Backup

	FACE BRICK	STUD BACKUP	STUD SPACING (IN.)	BOND		COST PER S.F.		
						MAT.	INST.	TOTAL
5400	Standard	16ga.x3-5/8"LB	16	running		6.15	13.40	19.55
5420				common		6.80	15.05	21.85
5440				Flemish		7.30	17.50	24.80
5460				English		7.90	18.45	26.35
5700	Glazed	25ga.x6"NLB	24	running		9.80	13.10	22.90
5720				common		11.40	14.95	26.35
5740				Flemish		12.50	17.75	30.25
5760				English		13.85	18.80	32.65
5800		20ga.x3-5/8"NLB	24	running		9.90	13.20	23.10
5820				common		11.50	15.05	26.55
5840				Flemish		12.60	17.85	30.45
5860				English		13.95	18.90	32.85
6000		16ga.x3-5/8"LB	16	running		10.65	13.80	24.45
6020				common		12.25	15.65	27.90
6040				Flemish		13.35	18.45	31.80
6060				English		14.70	19.50	34.20
6300	Engineer	25ga.x6"NLB	24	running		5.35	11.35	16.70
6320				common		6.05	12.70	18.75
6340				Flemish		6.55	14.95	21.50
6360				English		7.15	15.65	22.80
6400		20ga.x3-5/8"NLB	16	running		5.70	11.85	17.55
6420				common		6.40	13.20	19.60
6440				Flemish		6.90	15.45	22.35
6460				English		7.50	16.15	23.65
6900	Roman	25ga.x6"NLB	24	running		6.95	11.65	18.60
6920				common		8	13.10	21.10
6940				Flemish		8.70	15.30	24
6960				English		9.60	16.40	26
7000		20ga.x3-5/8"NLB	16	running		7.30	12.15	19.45
7020				common		8.35	13.60	21.95
7040				Flemish		9.05	15.80	24.85
7060				English		9.95	16.90	26.85
7500	Norman	25ga.x6"NLB	24	running		6.15	9.95	16.10
7520				common		7	11.05	18.05
7540				Flemish		11.45	12.90	24.35
7560				English		8.35	13.55	21.90
7600		20ga.x3-5/8"NLB	24	running		6.25	10.05	16.30
7620				common		7.10	11.15	18.25
7640				Flemish		11.55	13	24.55
7660				English		8.45	13.65	22.10
8100	Norwegian	25ga.x6"NLB	24	running		5.25	9.05	14.30
8120				common		5.95	10	15.95
8140				Flemish		6.40	11.65	18.05
8160				English		7	12.15	19.15
8400		16ga.x3-5/8"LB	16	running		6.15	9.75	15.90
8420				common		6.85	10.70	17.55
8440				Flemish		7.30	12.35	19.65
8460				English		7.90	12.85	20.75

SHELL

B

B2010 Exterior Walls

Brick Face Composite Wall

B2010 132		Brick Face Composite Wall - Double Wythe							
	FACE BRICK	**BACKUP MASONRY**	**BACKUP THICKNESS (IN.)**	**BACKUP CORE FILL**		**COST PER S.F.**			
						MAT.	**INST.**	**TOTAL**	
1000	Standard	common brick	4	none		6.90	20	26.90	
1040		SCR brick	6	none		8.95	17.90	26.85	
1080		conc. block	4	none		5.45	16.20	21.65	
1120			6	perlite		6.25	16.75	23	
1160				styrofoam		6.90	16.50	23.40	
1200			8	perlite		6.55	17.20	23.75	
1240				styrofoam		7	16.85	23.85	
1520		glazed block	4	none		11	17.30	28.30	
1560			6	perlite		11.75	17.80	29.55	
1600				styrofoam		12.35	17.55	29.90	
1640			8	perlite		12.85	18.25	31.10	
1680				styrofoam		13.30	17.90	31.20	
1720		clay tile	4	none		9.75	15.55	25.30	
1760			6	none		10.30	16	26.30	
1800			8	none		11.65	16.50	28.15	
1840		glazed tile	4	none		12.15	20.50	32.65	
1880									
2000	Glazed	common brick	4	none		11.40	20.50	31.90	
2040		SCR brick	6	none		13.50	18.30	31.80	
2080		conc. block	4	none		10	16.60	26.60	
2200			8	perlite		11.10	17.60	28.70	
2240				styrofoam		11.55	17.25	28.80	
2280		L.W. block	4	none		10.15	16.50	26.65	
2400			8	perlite		11.50	17.45	28.95	
2520		glazed block	4	none		15.55	17.70	33.25	
2560			6	perlite		16.25	18.20	34.45	
2600				styrofoam		16.90	17.95	34.85	
2640			8	perlite		17.35	18.65	36	
2680				styrofoam		17.85	18.30	36.15	

Important: See the Reference Section for critical supporting data - Reference Numbers and City Cost Indexes

| B2010 132 | Brick Face Composite Wall - Double Wythe | | | | | | | |

	FACE BRICK	BACKUP MASONRY	BACKUP THICKNESS (IN.)	BACKUP CORE FILL		COST PER S.F.		
						MAT.	INST.	TOTAL
2720	Glazed	clay tile	4	none		14.30	15.95	30.25
2760			6	none		14.85	16.40	31.25
2800			8	none		16.20	16.90	33.10
2840		glazed tile	4	none		16.70	21	37.70
3000	Engineer	common brick	4	none		7	18.75	25.75
3040		SCR brick	6	none		9.05	16.55	25.60
3080		conc. block	4	none		5.55	14.85	20.40
3200			8	perlite		6.65	15.85	22.50
3280		L.W. block	4	none		4.57	10.55	15.12
3322		L.W. block	6	perlite		6.50	15.30	21.80
3520		glazed block	4	none		11.10	15.95	27.05
3560			6	perlite		11.85	16.45	28.30
3600				styrofoam		12.45	16.20	28.65
3640			8	perlite		12.95	16.90	29.85
3680				styrofoam		13.40	16.55	29.95
3720		clay tile	4	none		9.85	14.20	24.05
3800			8	none		11.75	15.15	26.90
3840		glazed tile	4	none		12.25	19.15	31.40
4000	Roman	common brick	4	none		8.60	19.05	27.65
4040		SCR brick	6	none		10.65	16.85	27.50
4080		conc. block	4	none		7.15	15.15	22.30
4120			6	perlite		7.95	15.70	23.65
4200			8	perlite		8.25	16.15	24.40
5000	Norman	common brick	4	none		7.80	17.35	25.15
5040		SCR brick	6	none		9.85	15.15	25
5280		L.W. block	4	none		6.55	13.35	19.90
5320			6	perlite		7.35	14.05	21.40
5720		clay tile	4	none		10.65	12.80	23.45
5760			6	none		11.20	13.25	24.45
5840		glazed tile	4	none		13.05	17.75	30.80
6000	Norwegian	common brick	4	none		6.90	16.45	23.35
6040		SCR brick	6	none		8.95	14.25	23.20
6080		conc. block	4	none		5.45	12.55	18
6120			6	perlite		6.30	13.10	19.40
6160				styrofoam		6.90	12.85	19.75
6200			8	perlite		6.55	13.55	20.10
6520		glazed block	4	none		11	13.65	24.65
6560			6	perlite		11.75	14.15	25.90
7000	Utility	common brick	4	none		6.65	15.60	22.25
7040		SCR brick	6	none		8.70	13.40	22.10
7080		conc. block	4	none		5.20	11.65	16.85
7120			6	perlite		6	12.25	18.25
7160				styrofoam		6.65	12	18.65
7200			8	perlite		6.30	12.65	18.95
7240				styrofoam		6.75	12.35	19.10
7280		L.W. block	4	none		5.40	11.55	16.95
7320			6	perlite		6.15	12.10	18.25
7520		glazed block	4	none		10.75	12.80	23.55
7560			6	perlite		11.45	13.30	24.75
7720		clay tile	4	none		9.50	11	20.50
7760			6	none		10.05	11.45	21.50
7840		glazed tile	4	none		11.90	16	27.90

SHELL

B

B2010 Exterior Walls

B2010 133	Brick Face Composite Wall - Triple Wythe

	FACE BRICK	MIDDLE WYTHE	INSIDE MASONRY	TOTAL THICKNESS (IN.)		COST PER S.F.		
						MAT.	INST.	TOTAL
8000	Standard	common brick	standard brick	12		10.40	28.50	38.90
8100		4" conc. brick	standard brick	12		10.15	29.50	39.65
8120		4" conc. brick	common brick	12		9.50	29	38.50
8200	Glazed	common brick	standard brick	12		14.95	29	43.95
8300		4" conc. brick	standard brick	12		14.65	30	44.65
8320		4" conc. brick	glazed brick	12		14.05	29.50	43.55

SHELL B

Important: See the Reference Section for critical supporting data - Reference Numbers and City Cost Indexes

B2010 Exterior Walls

Brick Face Cavity Wall

						COST PER S.F.		
	FACE BRICK	**BACKUP MASONRY**	**TOTAL THICKNESS (IN.)**	**CAVITY INSULATION**		**MAT.**	**INST.**	**TOTAL**
1000	Standard	4"common brick	10	polystyrene		6.80	20	26.80
1020				none		6.60	19.60	26.20
1040		6"SCR brick	12	polystyrene		8.85	17.95	26.80
1060				none		8.70	17.40	26.10
1080		4"conc. block	10	polystyrene		5.35	16.20	21.55
1100				none		5.15	15.70	20.85
1120		6"conc. block	12	polystyrene		5.85	16.55	22.40
1140				none		5.70	16.05	21.75
1160		4"L.W. block	10	polystyrene		5.55	16.10	21.65
1180				none		5.35	15.60	20.95
1200		6"L.W. block	12	polystyrene		6	16.45	22.45
1220				none		5.85	15.90	21.75
1240		4"glazed block	10	polystyrene		10.90	17.35	28.25
1260				none		10.75	16.80	27.55
1280		6"glazed block	12	polystyrene		10.95	17.35	28.30
1300				none		10.75	16.80	27.55
1320		4"clay tile	10	polystyrene		9.65	15.60	25.25
1340				none		9.50	15.05	24.55
1360		4"glazed tile	10	polystyrene		12.05	20.50	32.55
1380				none		11.90	20	31.90
1500	Glazed	4" common brick	10	polystyrene		11.35	20.50	31.85
1520				none		11.15	20	31.15
1580		4" conc. block	10	polystyrene		9.90	16.60	26.50
1600				none		9.70	16.10	25.80
1660		4" L.W. block	10	polystyrene		10.10	16.50	26.60
1680				none		9.90	16	25.90
1740		4" glazed block	10	polystyrene		15.45	17.75	33.20
1760				none		15.25	17.20	32.45
1820		4" clay tile	10	polystyrene		14.20	16	30.20
1840				none		14	15.45	29.45
1860		4" glazed tile	10	polystyrene		14.75	16.40	31.15
1880				none		14.55	15.90	30.45
2000	Engineer	4" common brick	10	polystyrene		6.90	18.80	25.70
2020				none		6.70	18.25	24.95
2080		4" conc. block	10	polystyrene		5.45	14.85	20.30
2100				none		5.25	14.35	19.60

SHELL

B

B2010 Exterior Walls

B2010 134		Brick Face Cavity Wall						
	FACE BRICK	BACKUP MASONRY	TOTAL THICKNESS (IN.)	CAVITY INSULATION		COST PER S.F.		
						MAT.	INST.	TOTAL
2162	Engineer	4" L.W. block	10	polystyrene		5.65	14.75	20.40
2180				none		5.45	14.25	19.70
2240		4" glazed block	10	polystyrene		11	16	27
2260				none		10.85	15.45	26.30
2320		4" clay tile	10	polystyrene		9.75	14.25	24
2340				none		9.60	13.70	23.30
2360		4" glazed tile	10	polystyrene		12.15	19.20	31.35
2380				none		12	18.65	30.65
2500	Roman	4" common brick	10	polystyrene		8.50	19.10	27.60
2520				none		8.30	18.55	26.85
2580		4" conc. block	10	polystyrene		7.05	15.15	22.20
2600				none		6.85	14.65	21.50
2660		4" L.W. block	10	polystyrene		7.25	15.05	22.30
2680				none		7.05	14.55	21.60
2740		4" glazed block	10	polystyrene		12.60	16.30	28.90
2760				none		12.40	15.75	28.15
2820		4"clay tile	10	polystyrene		11.35	14.55	25.90
2840				none		11.15	14	25.15
2860		4"glazed tile	10	polystyrene		13.75	19.50	33.25
2880				none		13.55	18.95	32.50
3000	Norman	4" common brick	10	polystyrene		7.70	17.40	25.10
3020				none		7.50	16.85	24.35
3080		4" conc. block	10	polystyrene		6.25	13.45	19.70
3100				none		6.05	12.95	19
3160		4" L.W. block	10	polystyrene		6.45	13.35	19.80
3180				none		6.25	12.85	19.10
3240		4" glazed block	10	polystyrene		11.80	14.60	26.40
3260				none		11.60	14.05	25.65
3320		4" clay tile	10	polystyrene		10.55	12.85	23.40
3340				none		10.35	12.30	22.65
3360		4" glazed tile	10	polystyrene		12.95	17.80	30.75
3380				none		12.75	17.25	30
3500	Norwegian	4" common brick	10	polystyrene		6.80	16.50	23.30
3520				none		6.60	15.95	22.55
3580		4" conc. block	10	polystyrene		5.35	12.55	17.90
3600				none		5.20	12.05	17.25
3660		4" L.W. block	10	polystyrene		5.55	12.45	18
3680				none		5.35	11.95	17.30
3740		4" glazed block	10	polystyrene		10.95	13.70	24.65
3760				none		10.75	13.15	23.90
3820		4" clay tile	10	polystyrene		9.70	11.95	21.65
3840				none		9.50	11.40	20.90
3860		4" glazed tile	10	polystyrene		12.10	16.90	29
3880				none		11.90	16.35	28.25
4000	Utility	4" common brick	10	polystyrene		6.55	15.60	22.15
4020				none		6.35	15.10	21.45
4080		4" conc. block	10	polystyrene		5.10	11.70	16.80
4100				none		4.91	11.15	16.06
4160		4" L.W. block	10	polystyrene		5.30	11.60	16.90
4180				none		5.10	11.05	16.15
4240		4" glazed block	10	polystyrene		10.65	12.80	23.45
4260				none		10.45	12.30	22.75

SHELL B

Important: See the Reference Section for critical supporting data - Reference Numbers and City Cost Indexes

B2010 Exterior Walls

B2010 134 — Brick Face Cavity Wall

	FACE BRICK	BACKUP MASONRY	TOTAL THICKNESS (IN.)	CAVITY INSULATION		COST PER S.F.		
						MAT.	INST.	TOTAL
4320	Utility	4" clay tile	10	polystyrene		9.40	11.05	20.45
4340				none		9.20	10.55	19.75
4360		4" glazed tile	10	polystyrene		11.80	16	27.80
4380				none		11.60	15.50	27.10

B2010 135 — Brick Face Cavity Wall - Insulated Backup

	FACE BRICK	BACKUP MASONRY	TOTAL THICKNESS (IN.)	BACKUP CORE FILL		COST PER S.F.		
						MAT.	INST.	TOTAL
5100	Standard	6" conc. block	10	perlite		6	16.30	22.30
5120				styrofoam		6.60	16.05	22.65
5180		6" L.W. block	10	perlite		6.15	16.15	22.30
5200				styrofoam		6.75	15.90	22.65
5260		6" glazed block	10	perlite		11.45	17.30	28.75
5280				styrofoam		12.05	17.05	29.10
5340		6" clay tile	10	none		10.05	15.50	25.55
5360		8" clay tile	12	none		11.40	16.05	27.45
5600	Glazed	6" conc. block	10	perlite		10.50	16.70	27.20
5620				styrofoam		11.15	16.45	27.60
5680		6" L.W. block	10	perlite		10.65	16.55	27.20
5700				styrofoam		11.30	16.30	27.60
5760		6" glazed block	10	perlite		16	17.70	33.70
5780				styrofoam		16.60	17.45	34.05
5840		6" clay tile	10	none		14.55	15.90	30.45
5860		8" clay tile	8	none		15.90	16.45	32.35
6100	Engineer	6" conc. block	10	perlite		6.10	14.95	21.05
6120				styrofoam		6.70	14.70	21.40
6180		6" L.W. block	10	perlite		6.25	14.80	21.05
6200				styrofoam		6.85	14.55	21.40
6260		6" glazed block	10	perlite		11.55	15.95	27.50
6280				styrofoam		12.15	15.70	27.85
6340		6" clay tile	10	none		10.15	14.15	24.30
6360		8" clay tile	12	none		11.50	14.70	26.20
6600	Roman	6" conc. block	10	perlite		7.70	15.25	22.95
6620				styrofoam		8.30	15	23.30
6680		6" L.W. block	10	perlite		7.85	15.10	22.95
6700				styrofoam		8.45	14.85	23.30
6760		6" glazed block	10	perlite		13.15	16.25	29.40
6780				styrofoam		13.75	16	29.75
6840		6" clay tile	10	none		11.70	14.45	26.15
6860		8" clay tile	12	none		13.10	15	28.10
7100	Norman	6" conc. block	10	perlite		6.90	13.55	20.45
7120				styrofoam		7.50	13.30	20.80
7180		6" L.W. block	10	perlite		7.05	13.40	20.45
7200				styrofoam		7.65	13.15	20.80
7260		6" glazed block	10	perlite		12.35	14.55	26.90
7280				styrofoam		12.95	14.30	27.25
7340		6" clay tile	10	none		10.90	12.75	23.65
7360		8" clay tile	12	none		12.30	13.30	25.60
7600	Norwegian	6" conc. block	10	perlite		6	12.65	18.65
7620				styrofoam		6.60	12.40	19
7680		6" L.W. block	10	perlite		6.15	12.50	18.65
7700				styrofoam		6.75	12.25	19

SHELL

B

B2010 Exterior Walls

| B2010 135 | Brick Face Cavity Wall - Insulated Backup | | | | | | | |

	FACE BRICK	BACKUP MASONRY	TOTAL THICKNESS (IN.)	BACKUP CORE FILL		COST PER S.F.		
						MAT.	INST.	TOTAL
7760	Norwegian	6" glazed block	10	perlite		11.45	13.65	25.10
7780				styrofoam		12.10	13.40	25.50
7840		6" clay tile	10	none		10.05	11.85	21.90
7860		8" clay tile	12	none		11.40	12.40	23.80
8100	Utility	6" conc. block	10	perlite		5.75	11.75	17.50
8120				styrofoam		6.35	11.50	17.85
8180		6" L.W. block	10	perlite		5.90	11.65	17.55
8200				styrofoam		6.50	11.40	17.90
8260		6" glazed block	10	perlite		11.20	12.80	24
8280				styrofoam		11.80	12.55	24.35
8340		6" clay tile	10	none		9.75	10.95	20.70
8360		8" clay tile	12	none		11.15	11.50	22.65

SHELL B

B2010 Exterior Walls

Glass Block

B2010 140	Glass Block	COST PER S.F.		
		MAT.	INST.	TOTAL
2300	Glass block 4" thick, 6"x6" plain, under 1,000 S.F.	16.85	18.85	35.70
2400	1,000 to 5,000 S.F.	16.55	16.35	32.90
2500	Over 5,000 S.F.	16.05	15.35	31.40
2600	Solar reflective, under 1,000 S.F.	23.50	25.50	49
2700	1,000 to 5,000 S.F.	23	22	45
2800	Over 5,000 S.F.	22.50	20.50	43
3500	8"x8" plain, under 1,000 S.F.	11.30	14.10	25.40
3600	1,000 to 5,000 S.F.	11.05	12.20	23.25
3700	Over 5,000 S.F.	10.75	11	21.75
3800	Solar reflective, under 1,000 S.F.	15.80	18.95	34.75
3900	1,000 to 5,000 S.F.	15.45	16.30	31.75
4000	Over 5,000 S.F.	15	14.60	29.60
5000	12"x12" plain, under 1,000 S.F.	13.15	13.05	26.20
5100	1,000 to 5,000 S.F.	12.90	11	23.90
5200	Over 5,000 S.F.	12.50	10.05	22.55
5300	Solar reflective, under 1,000 S.F.	18.40	17.50	35.90
5400	1,000 to 5,000 S.F.	18.05	14.60	32.65
5600	Over 5,000 S.F.	17.45	13.30	30.75
5800	3" thinline, 6"x6" plain, under 1,000 S.F.	14.45	18.85	33.30
5900	Over 5,000 S.F.	14.45	18.85	33.30
6000	Solar reflective, under 1,000 S.F.	20	25.50	45.50
6100	Over 5,000 S.F.	18.05	20.50	38.55
6200	8"x8" plain, under 1,000 S.F.	8.55	14.10	22.65
6300	Over 5,000 S.F.	7.65	11	18.65
6400	Solar reflective, under 1,000 S.F.	11.95	18.95	30.90
6500	Over 5,000 S.F.	10.70	14.60	25.30

SHELL

B

B2010 Exterior Walls

The table below lists costs for metal siding of various descriptions, not including the steel frame, or the structural steel, of a building. Costs are per S.F. including all accessories and insulation.

Metal Siding

SHELL B

B2010 146	Metal Siding Panel	COST PER S.F.		
		MAT.	INST.	TOTAL
1400	Metal siding aluminum panel, corrugated, .024" thick, natural	1.57	3.04	4.61
1450	Painted	1.79	3.04	4.83
1500	.032" thick, natural	2.06	3.04	5.10
1550	Painted	2.48	3.04	5.52
1600	Ribbed 4" pitch, .032" thick, natural	2.02	3.04	5.06
1650	Painted	2.50	3.04	5.54
1700	.040" thick, natural	2.42	3.04	5.46
1750	Painted	2.77	3.04	5.81
1800	.050" thick, natural	2.75	3.04	5.79
1850	Painted	3.07	3.04	6.11
1900	8" pitch panel, .032" thick, natural	1.93	2.92	4.85
1950	Painted	2.40	2.93	5.33
2000	.040" thick, natural	2.33	2.93	5.26
2050	Painted	2.66	2.96	5.62
2100	.050" thick, natural	2.66	2.95	5.61
2150	Painted	2.99	2.97	5.96
3000	Steel, corrugated or ribbed, 29 Ga. .0135" thick, galvanized	1.41	2.77	4.18
3050	Colored	1.95	2.81	4.76
3100	26 Ga. .0179" thick, galvanized	1.45	2.78	4.23
3150	Colored	2.01	2.82	4.83
3200	24 Ga. .0239" thick, galvanized	1.80	2.79	4.59
3250	Colored	2.24	2.83	5.07
3300	22 Ga. .0299" thick, galvanized	1.81	2.81	4.62
3350	Colored	2.67	2.85	5.52
3400	20 Ga. .0359" thick, galvanized	1.81	2.81	4.62
3450	Colored	2.83	3.01	5.84
4100	Sandwich panels, factory fab., 1" polystyrene, steel core, 26 Ga., galv.	3.38	4.35	7.73
4200	Colored, 1 side	4.13	4.35	8.48
4300	2 sides	5.30	4.35	9.65
4400	2" polystyrene, steel core, 26 Ga., galvanized	4	4.35	8.35
4500	Colored, 1 side	4.75	4.35	9.10
4600	2 sides	5.90	4.35	10.25
4700	22 Ga., baked enamel exterior	7.85	4.60	12.45
4800	Polyvinyl chloride exterior	8.25	4.60	12.85
5100	Textured aluminum, 4' x 8' x 5/16" plywood backing, single face	2.46	2.64	5.10
5200	Double face	3.80	2.64	6.44

B2010 Exterior Walls

The table below lists costs per S.F. for exterior walls with wood siding. A variety of systems are presented using both wood and metal studs at 16″ and 24″ O.C.

Exterior Wall Siding

B2010 148	Panel, Shingle & Lap Siding	MAT.	INST.	TOTAL
1400	Wood siding w/2″x4″studs, 16″O.C., insul. wall, 5/8″text 1-11 fir plywood	2.84	4.35	7.19
1450	5/8″ text 1-11 cedar plywood	4.35	4.20	8.55
1500	1″ x 4″ vert T.&G. redwood	4.66	5.95	10.61
1600	1″ x 8″ vert T.&G. redwood	4.25	5.35	9.60
1650	1″ x 5″ rabbetted cedar bev. siding	5.35	4.80	10.15
1700	1″ x 6″ cedar drop siding	5.40	4.83	10.23
1750	1″ x 12″ rough sawn cedar	2.96	4.39	7.35
1800	1″ x 12″ sawn cedar, 1″ x 4″ battens	3.95	4.73	8.68
1850	1″ x 10″ redwood shiplap siding	4.18	4.59	8.77
1900	18″ no. 1 red cedar shingles, 5-1/2″ exposed	3.54	5.75	9.29
1950	6″ exposed	3.41	5.55	8.96
2000	6-1/2″ exposed	3.28	5.35	8.63
2100	7″ exposed	3.15	5.20	8.35
2150	7-1/2″ exposed	3.02	4.99	8.01
3000	8″ wide aluminum siding	2.93	3.88	6.81
3150	8″ plain vinyl siding	2.34	3.89	6.23
3250	8″ insulated vinyl siding	2.42	3.89	6.31
3300				
3400	2″ x 6″ studs, 16″ O.C., insul. wall, w/ 5/8″ text 1-11 fir plywood	3.14	4.52	7.66
3500	5/8″ text 1-11 cedar plywood	4.65	4.52	9.17
3600	1″ x 4″ vert T.&G. redwood	4.96	6.10	11.06
3700	1″ x 8″ vert T.&G. redwood	4.55	5.55	10.10
3800	1″ x 5″ rabbetted cedar bev siding	5.65	4.97	10.62
3900	1″ x 6″ cedar drop siding	5.70	5	10.70
4000	1″ x 12″ rough sawn cedar	3.26	4.56	7.82
4200	1″ x 12″ sawn cedar, 1″ x 4″ battens	4.25	4.90	9.15
4500	1″ x 10″ redwood shiplap siding	4.48	4.76	9.24
4550	18″ no. 1 red cedar shingles, 5-1/2″ exposed	3.84	5.90	9.74
4600	6″ exposed	3.71	5.75	9.46
4650	6-1/2″ exposed	3.58	5.55	9.13
4700	7″ exposed	3.45	5.35	8.80
4750	7-1/2″ exposed	3.32	5.15	8.47
4800	8″ wide aluminum siding	3.23	4.05	7.28
4850	8″ plain vinyl siding	2.64	4.06	6.70
4900	8″ insulated vinyl siding	2.72	4.06	6.78
4910				
5000	2″ x 6″ studs, 24″ O.C., insul. wall, 5/8″ text 1-11, fir plywood	3	4.26	7.26
5050	5/8″ text 1-11 cedar plywood	4.51	4.26	8.77
5100	1″ x 4″ vert T.&G. redwood	4.82	5.85	10.67
5150	1″ x 8″ vert T.&G. redwood	4.41	5.30	9.71

COST PER S.F.

SHELL

B

B2010 Exterior Walls

B2010 148	Panel, Shingle & Lap Siding	COST PER S.F.		
		MAT.	INST.	TOTAL
5200	1" x 5" rabbetted cedar bev siding	5.50	4.71	10.21
5250	1" x 6" cedar drop siding	5.55	4.74	10.29
5300	1" x 12" rough sawn cedar	3.12	4.30	7.42
5400	1" x 12" sawn cedar, 1" x 4" battens	4.11	4.64	8.75
5450	1" x 10" redwood shiplap siding	4.34	4.50	8.84
5500	18" no. 1 red cedar shingles, 5-1/2" exposed	3.70	5.65	9.35
5550	6" exposed	3.57	5.45	9.02
5650	7" exposed	3.31	5.10	8.41
5700	7-1/2" exposed	3.18	4.90	8.08
5750	8" wide aluminum siding	3.09	3.79	6.88
5800	8" plain vinyl siding	2.50	3.80	6.30
5850	8" insulated vinyl siding	2.58	3.80	6.38
5900	3-5/8" metal studs, 16 Ga., 16" OC insul.wall, 5/8"text 1-11 fir plywood	3.81	4.58	8.39
5950	5/8" text 1-11 cedar plywood	5.30	4.58	9.88
6000	1" x 4" vert T.&G. redwood	5.65	6.15	11.80
6050	1" x 8" vert T.&G. redwood	5.20	5.60	10.80
6100	1" x 5" rabbetted cedar bev siding	6.30	5.05	11.35
6150	1" x 6" cedar drop siding	6.40	5.05	11.45
6200	1" x 12" rough sawn cedar	3.93	4.62	8.55
6250	1" x 12" sawn cedar, 1" x 4" battens	4.92	4.96	9.88
6300	1" x 10" redwood shiplap siding	5.15	4.82	9.97
6350	18" no. 1 red cedar shingles, 5-1/2" exposed	4.51	6	10.51
6500	6" exposed	4.38	5.80	10.18
6550	6-1/2" exposed	4.25	5.60	9.85
6600	7" exposed	4.12	5.40	9.52
6650	7-1/2" exposed	4.12	5.25	9.37
6700	8" wide aluminum siding	3.90	4.11	8.01
6750	8" plain vinyl siding	3.31	3.97	7.28
6800	8" insulated vinyl siding	3.39	3.97	7.36
7000	3-5/8" metal studs, 16 Ga. 24" OC insul wall, 5/8" text 1-11 fir plywood	3.43	4.06	7.49
7050	5/8" text 1-11 cedar plywood	4.94	4.06	9
7100	1" x 4" vert T.&G. redwood	5.25	5.80	11.05
7150	1" x 8" vert T.&G. redwood	4.84	5.25	10.09
7200	1" x 5" rabbetted cedar bev siding	5.95	4.66	10.61
7250	1" x 6" cedar drop siding	6	4.69	10.69
7300	1" x 12" rough sawn cedar	3.55	4.25	7.80
7350	1" x 12" sawn cedar 1" x 4" battens	4.54	4.59	9.13
7400	1" x 10" redwood shiplap siding	4.77	4.45	9.22
7450	18" no. 1 red cedar shingles, 5-1/2" exposed	4.26	5.80	10.06
7500	6" exposed	4.13	5.60	9.73
7550	6-1/2" exposed	4	5.40	9.40
7600	7" exposed	3.87	5.25	9.12
7650	7-1/2" exposed	3.61	4.85	8.46
7700	8" wide aluminum siding	3.52	3.74	7.26
7750	8" plain vinyl siding	2.93	3.75	6.68
7800	8" insul. vinyl siding	3.01	3.75	6.76

SHELL B

Important: See the Reference Section for critical supporting data - Reference Numbers and City Cost Indexes

B2010 Exterior Walls

The table below lists costs for some typical stucco walls including all the components as demonstrated in the component block below. Prices are presented for backup walls using wood studs, metal studs and CMU.

Exterior Stucco Wall

B2010 151	Stucco Wall	COST PER S.F.		
		MAT.	INST.	TOTAL
2100	Cement stucco, 7/8" th., plywood sheathing, stud wall, 2" x 4", 16" O.C.	2.60	7.05	9.65
2200	24" O.C.	2.51	6.85	9.36
2300	2" x 6", 16" O.C.	2.90	7.20	10.10
2400	24" O.C.	2.76	6.95	9.71
2500	No sheathing, metal lath on stud wall, 2" x 4", 16" O.C.	1.70	6.10	7.80
2600	24" O.C.	1.61	5.90	7.51
2700	2" x 6", 16" O.C.	2	6.30	8.30
2800	24" O.C.	1.86	6	7.86
2900	1/2" gypsum sheathing, 3-5/8" metal studs, 16" O.C.	2.80	6.35	9.15
2950	24" O.C.	2.42	6	8.42
3000	Cement stucco, 5/8" th., 2 coats on std. CMU block, 8"x 16", 8" thick	2.13	8.90	11.03
3100	10" Thick	2.83	9.15	11.98
3200	12" Thick	2.90	10.50	13.40
3300	Std. light Wt. block 8" x 16", 8" Thick	2.57	8.55	11.12
3400	10" Thick	3.23	8.80	12.03
3500	12" Thick	3.71	10.05	13.76
3600	3 coat stucco, self furring metal lath 3.4 Lb/SY, on 8" x 16", 8" thick	2.17	9.40	11.57
3700	10" Thick	2.87	9.65	12.52
3800	12" Thick	2.94	11	13.94
3900	Lt. Wt. block, 8" Thick	2.61	9.05	11.66
4000	10" Thick	3.27	9.30	12.57
4100	12" Thick	3.75	10.55	14.30

B2010 152	E.I.F.S.	COST PER S.F.		
		MAT.	INST.	TOTAL
5100	E.I.F.S., plywood sheathing, stud wall, 2" x 4", 16" O.C., 1" EPS	3.88	8.65	12.53
5110	2" EPS	4.16	8.65	12.81
5120	3" EPS	4.27	8.65	12.92
5130	4" EPS	4.62	8.65	13.27
5140	2" x 6", 16" O.C., 1" EPS	4.18	8.80	12.98
5150	2" EPS	4.46	8.80	13.26
5160	3" EPS	4.57	8.80	13.37
5170	4" EPS	4.92	8.80	13.72
5180	Cement board sheathing, 3-5/8" metal studs, 16" O.C., 1" EPS	4.82	10.25	15.07
5190	2" EPS	5.10	10.25	15.35
5200	3" EPS	5.20	10.25	15.45
5210	4" EPS	6.40	10.25	16.65
5220	6" metal studs, 16" O.C., 1" EPS	5.30	10.35	15.65
5230	2" EPS	5.60	10.35	15.95

SHELL

B

B2010 Exterior Walls

B2010 152	E.I.F.S.	COST PER S.F.		
		MAT.	INST.	TOTAL
5240	3″ EPS	5.70	10.35	16.05
5250	4″ EPS	6.90	10.35	17.25
5260	CMU block, 8″ x 8″ x 16″, 1″ EPS	3.89	11.95	15.84
5270	2″ EPS	4.17	11.95	16.12
5280	3″ EPS	4.28	11.95	16.23
5290	4″ EPS	4.63	11.95	16.58
5300	8″ x 10″ x 16″, 1″ EPS	4.59	12.20	16.79
5310	2″ EPS	4.87	12.20	17.07
5320	3″ EPS	4.98	12.20	17.18
5330	4″ EPS	5.35	12.20	17.55
5340	8″ x 12″ x 16″, 1″ EPS	4.66	13.55	18.21
5350	2″ EPS	4.94	13.55	18.49
5360	3″ EPS	5.05	13.55	18.60
5370	4″ EPS	5.40	13.55	18.95

SHELL B

Important: See the Reference Section for critical supporting data - Reference Numbers and City Cost Indexes

Description: The table below lists costs, \$/S.F., for channel girts with sag rods and connector angles top and bottom for various column spacings, building heights and wind loads. Additive costs are shown for wind columns.

B2010 154	Metal Siding Support							

	BLDG. HEIGHT (FT.)	WIND LOAD (P.S.F.)	COL. SPACING (FT.)		INTERMEDIATE COLUMNS	COST PER S.F.		
						MAT.	INST.	TOTAL
3000	18	20	20			1.59	3.03	4.62
3100					wind cols.	.82	.22	1.04
3200		20	25			1.74	3.06	4.80
3300					wind cols.	.66	.18	.84
3400		20	30			1.92	3.12	5.04
3500					wind cols.	.55	.14	.69
3600		20	35			2.11	3.17	5.28
3700					wind cols.	.47	.13	.60
3800		30	20			1.76	3.08	4.84
3900					wind cols.	.82	.22	1.04
4000		30	25			1.92	3.12	5.04
4100					wind cols.	.66	.18	.84
4200		30	30			2.12	3.17	5.29
4300					wind cols.	.74	.20	.94
4600	30	20	20			1.54	2.13	3.67
4700					wind cols.	1.16	.31	1.47
4800		20	25			1.73	2.18	3.91
4900					wind cols.	.93	.25	1.18
5000		20	30			1.95	2.24	4.19
5100					wind cols.	.91	.25	1.16
5200		20	35			2.18	2.30	4.48
5300					wind cols.	.91	.25	1.16
5400		30	20			1.76	2.18	3.94
5500					wind cols.	1.37	.38	1.75
5600		30	25			1.95	2.23	4.18
5700					wind cols.	1.31	.35	1.66
5800		30	30			2.19	2.31	4.50
5900					wind cols.	1.23	.34	1.57

B / SHELL

B2020 Exterior Windows

The table below lists window systems by material, type and size. Prices between sizes listed can be interpolated with reasonable accuracy. Prices include frame, hardware, and casing as illustrated in the component block below.

SHELL B

B2020 102			Wood Windows					
	MATERIAL	**TYPE**	**GLAZING**	**SIZE**	**DETAIL**	**COST PER UNIT**		
						MAT.	**INST.**	**TOTAL**
3000	Wood	double hung	std. glass	2'-8" x 4'-6"		183	187	370
3050				3'-0" x 5'-6"		243	214	457
3100			insul. glass	2'-8" x 4'-6"		196	187	383
3150				3'-0" x 5'-6"		261	214	475
3200		sliding	std. glass	3'-4" x 2'-7"		222	155	377
3250				4'-4" x 3'-3"		253	170	423
3300				5'-4" x 6'-0"		320	204	524
3350			insul. glass	3'-4" x 2'-7"		270	181	451
3400				4'-4" x 3'-3"		310	197	507
3450				5'-4" x 6'-0"		390	233	623
3500		awning	std. glass	2'-10" x 1'-9"		166	93.50	259.50
3600				4'-4" x 2'-8"		270	90.50	360.50
3700			insul. glass	2'-10" x 1'-9"		203	108	311
3800				4'-4" x 2'-8"		330	99	429
3900		casement	std. glass	1'-10" x 3'-2"	1 lite	211	123	334
3950				4'-2" x 4'-2"	2 lite	430	163	593
4000				5'-11" x 5'-2"	3 lite	670	210	880
4050				7'-11" x 6'-3"	4 lite	945	252	1,197
4100				9'-11" x 6'-3"	5 lite	1,250	294	1,544
4150			insul. glass	1'-10" x 3'-2"	1 lite	222	123	345
4200				4'-2" x 4'-2"	2 lite	455	163	618
4250				5'-11" x 5'-2"	3 lite	710	210	920
4300				7'-11" x 6'-3"	4 lite	1,000	252	1,252
4350				9'-11" x 6'-3"	5 lite	1,325	294	1,619
4400		picture	std. glass	4'-6" x 4'-6"		350	211	561
4450				5'-8" x 4'-6"		395	234	629
4500			insul. glass	4'-6" x 4'-6"		430	245	675
4550				5'-8" x 4'-6"		485	272	757
4600		fixed bay	std. glass	8' x 5'		1,225	385	1,610
4650				9'-9" x 5'-4"		870	540	1,410
4700			insul. glass	8' x 5'		1,925	385	2,310
4750				9'-9" x 5'-4"		940	540	1,480
4800		casement bay	std. glass	8' x 5'		1,225	440	1,665
4850			insul. glass	8' x 5'		1,375	440	1,815
4900		vert. bay	std. glass	8' x 5'		1,375	440	1,815
4950			insul. glass	8' x 5'		1,450	440	1,890

B2020 102			Wood Windows					
	MATERIAL	TYPE	GLAZING	SIZE	DETAIL	COST PER UNIT		
						MAT.	INST.	TOTAL
5000	Steel	double hung	1/4" tempered	2'-8" x 4'-6"		670	145	815
5050				3'-4" x 5'-6"		1,025	221	1,246
5100			insul. glass	2'-8" x 4'-6"		680	166	846
5150				3'-4" x 5'-6"		1,050	253	1,303
5202		horiz. pivoted	std. glass	2' x 2'		213	48.50	261.50
5250				3' x 3'		480	109	589
5300				4' x 4'		855	194	1,049
5350				6' x 4'		1,275	290	1,565
5400			insul. glass	2' x 2'		217	55.50	272.50
5450				3' x 3'		490	125	615
5500				4' x 4'		870	222	1,092
5550				6' x 4'		1,300	330	1,630
5600		picture window	std. glass	3' x 3'		291	109	400
5650				6' x 4'		775	290	1,065
5700			insul. glass	3' x 3'		300	125	425
5750				6' x 4'		800	330	1,130
5800		industrial security	std. glass	2'-9" x 4'-1"		635	136	771
5850				4'-1" x 5'-5"		1,250	268	1,518
5900			insul. glass	2'-9" x 4'-1"		650	155	805
5950				4'-1" x 5'-5"		1,275	305	1,580
6000		comm. projected	std. glass	3'-9" x 5'-5"		1,050	246	1,296
6050				6'-9" x 4'-1"		1,425	335	1,760
6100			insul. glass	3'-9" x 5'-5"		1,075	282	1,357
6150				6'-9" x 4'-1"		1,450	380	1,830
6200		casement	std. glass	4'-2" x 4'-2"	2 lite	805	210	1,015
6250			insul. glass	4'-2" x 4'-2"		820	240	1,060
6300			std. glass	5'-11" x 5'-2"	3 lite	1,550	370	1,920
6350			insul. glass	5'-11" x 5'-2"		1,575	425	2,000
6400	Aluminum	projecting	std. glass	3'-1" x 3'-2"		234	109	343
6450				4'-5" x 5'-3"		330	137	467
6500			insul. glass	3'-1" x 3'-2"		281	131	412
6550				4'-5" x 5'-3"		395	164	559
6600		sliding	std. glass	3' x 2'		177	109	286
6650				5' x 3'		225	121	346
6700				8' x 4'		325	182	507
6750				9' x 5'		490	273	763
6800			insul. glass	3' x 2'		197	109	306
6850				5' x 3'		315	121	436
6900				8' x 4'		520	182	702
6950				9' x 5'		785	273	1,058
7000		single hung	std. glass	2' x 3'		158	109	267
7050				2'-8" x 6'-8"		335	137	472
7100				3'-4" x 5'0"		216	121	337
7150			insul. glass	2' x 3'		191	109	300
7200				2'-8" x 6'-8"		430	137	567
7250				3'-4" x 5'		305	121	426
7300		double hung	std. glass	2' x 3'		224	72.50	296.50
7350				2'-8" x 6'-8"		665	215	880
7400				3'-4" x 5'		620	201	821
7450			insul. glass	2' x 3'		230	83	313
7500				2'-8" x 6'-8"		685	247	932
7550				3'-4" x 5'-0"		640	231	871

SHELL

B

B2020 Exterior Windows

B2020 102 — Wood Windows

	MATERIAL	TYPE	GLAZING	SIZE	DETAIL	COST PER UNIT		
						MAT.	INST.	TOTAL
7600	Aluminum	casements	std. glass	3'-1" x 3'-2"		185	118	303
7650				4'-5" x 5'-3"		450	286	736
7700			insul. glass	3'-1" x 3'-2"		195	135	330
7750				4'-5" x 5'-3"		470	325	795
7800		hinged swing	std. glass	3' x 4'		420	145	565
7850				4' x 5'		695	242	937
7900			insul. glass	3' x 4'		430	166	596
7950				4' x 5'		715	277	992
8002		folding type	std. glass	3'-0" x 4'-0"		228	119	347
8050				4'-0" x 5'-0"		320	119	439
8100			insul. glass	3'-0" x 4'-0"		286	119	405
8150				4'-0" x 5'-0"		415	119	534
8200		picture unit	std. glass	2'-0" x 3'-0"		119	72.50	191.50
8250				2'-8" x 6'-8"		355	215	570
8300				3'-4" x 5'-0"		330	201	531
8350			insul. glass	2'-0" x 3'-0"		126	83	209
8400				2'-8" x 6'-8"		375	247	622
8450				3'-4" x 5'-0"		350	231	581
8500		awning type	std. glass	3'-0" x 3'-0"	2 lite	385	78	463
8550				3'-0" x 4'-0"	3 lite	450	109	559
8600				3'-0" x 5'-4"	4 lite	545	109	654
8650				4'-0" x 5'-4"	4 lite	600	121	721
8700			insul. glass	3'-0" x 3'-0"	2 lite	415	78	493
8750				3'-0" x 4'-0"	3 lite	520	109	629
8800				3'-0" x 5'-4"	4 lite	645	109	754
8850				4'-0" x 5'-4"	4 lite	720	121	841
8900		jalousie type	std. glass	1'-7" x 3'-2"		144	109	253
8950				2'-3" x 4'-0"		206	109	315
9000				3'-1" x 2'-0"		162	109	271
9050				3'-1" x 5'-3"		293	109	402
9051								
9052								

B2020 210 — Tubular Aluminum Framing

		COST/S.F. OPNG.		
		MAT.	INST.	TOTAL
1100	Alum flush tube frame, for 1/4"glass, 1-3/4"x4", 5'x6'opng, no inter horizntls	9.85	9.85	19.70
1150	One intermediate horizontal	13.45	11.55	25
1200	Two intermediate horizontals	17	13.25	30.25
1250	5' x 20' opening, three intermediate horizontals	8.85	8.20	17.05
1400	1-3/4" x 4-1/2", 5' x 6' opening, no intermediate horizontals	11.20	9.85	21.05
1450	One intermediate horizontal	14.95	11.55	26.50
1500	Two intermediate horizontals	18.65	13.25	31.90
1550	5' x 20' opening, three intermediate horizontals	10	8.20	18.20
1700	For insulating glass, 2"x4-1/2", 5'x6' opening, no intermediate horizontals	12.25	10.30	22.55
1750	One intermediate horizontal	16.05	12.10	28.15
1800	Two intermediate horizontals	19.80	13.90	33.70
1850	5' x 20' opening, three intermediate horizontals	10.70	8.60	19.30
2000	Thermal break frame, 2-1/4"x4-1/2", 5'x6'opng, no intermediate horizontals	13.20	10.45	23.65
2050	One intermediate horizontal	17.70	12.55	30.25
2100	Two intermediate horizontals	22	14.70	36.70
2150	5' x 20' opening, three intermediate horizontals	12	8.95	20.95

SHELL B

Important: See the Reference Section for critical supporting data - Reference Numbers and City Cost Indexes

Glazing Panel

Plate Glass

Spandrel Glass

Polycarbonate

Sandwich Panel

The table below lists costs of curtain wall and spandrel panels per S.F. Costs do not include structural framing used to hang the panels.

B2020 220	Curtain Wall Panels	COST PER S.F.		
		MAT.	INST.	TOTAL
1000	Glazing panel, insulating, 1/2" thick, 2 lites 1/8" float, clear	7.85	8.40	16.25
1100	Tinted	11.55	8.40	19.95
1200	5/8" thick units, 2 lites 3/16" float, clear	9.55	8.85	18.40
1300	Tinted	9.80	8.85	18.65
1400	1" thick units, 2 lites, 1/4" float, clear	13.30	10.65	23.95
1500	Tinted	16.35	10.65	27
1600	Heat reflective film inside	19.05	9.40	28.45
1700	Light and heat reflective glass, tinted	22	9.40	31.40
2000	Plate glass, 1/4" thick, clear	5.50	6.65	12.15
2050	Tempered	6.80	6.65	13.45
2100	Tinted	5.45	6.65	12.10
2200	3/8" thick, clear	8.70	10.65	19.35
2250	Tempered	13.65	10.65	24.30
2300	Tinted	10.90	10.65	21.55
2400	1/2" thick, clear	17.45	14.50	31.95
2450	Tempered	20.50	14.50	35
2500	Tinted	19.05	14.50	33.55
2600	3/4" thick, clear	24.50	23	47.50
2650	Tempered	28.50	23	51.50
3000	Spandrel glass, panels, 1/4" plate glass insul w/fiberglass, 1" thick	11.65	6.65	18.30
3100	2" thick	13.55	6.65	20.20
3200	Galvanized steel backing, add	4.03		4.03
3300	3/8" plate glass, 1" thick	19.50	6.65	26.15
3400	2" thick	21.50	6.65	28.15
4000	Polycarbonate, masked, clear or colored, 1/8" thick	6.60	4.70	11.30
4100	3/16" thick	7.95	4.84	12.79
4200	1/4" thick	8.80	5.15	13.95
4300	3/8" thick	16.20	5.30	21.50
5000	Facing panel, textured al, 4' x 8' x 5/16" plywood backing, sgl face	2.46	2.64	5.10
5100	Double face	3.80	2.64	6.44
5200	4' x 10' x 5/16" plywood backing, single face	2.68	2.64	5.32
5300	Double face	3.94	2.64	6.58
5400	4' x 12' x 5/16" plywood backing, single face	3.45	2.64	6.09
5500	Sandwich panel, 22 Ga. galv., both sides 2" insulation, enamel exterior	7.85	4.60	12.45
5600	Polyvinylidene floride exterior finish	8.25	4.60	12.85
5700	26 Ga., galv. both sides, 1" insulation, colored 1 side	4.13	4.35	8.48
5800	Colored 2 sides	5.30	4.35	9.65

SHELL

B

Costs are listed for exterior door systems by material, type and size. Prices between sizes listed can be interpolated with reasonable accuracy. Prices are per opening for a complete door system including frame.

B2030 110				Glazed Doors, Steel or Aluminum				
	MATERIAL	TYPE	DOORS	SPECIFICATION	OPENING	COST PER OPNG.		
						MAT.	INST.	TOTAL
5600	St. Stl. & glass	revolving	stock unit	manual oper.	6'-0" x 7'-0"	31,700	7,225	38,925
5650				auto Cntrls.	6'-10" x 7'-0"	44,500	7,700	52,200
5700	Bronze	revolving	stock unit	manual oper.	6'-10" x 7'-0"	36,900	14,600	51,500
5750				auto Cntrls.	6'-10" x 7'-0"	49,700	15,100	64,800
5800	St. Stl. & glass	balanced	standard	economy	3'-0" x 7'-0"	6,850	1,225	8,075
5850				premium	3'-0" x 7'-0"	11,900	1,575	13,475
6300	Alum. & glass	w/o transom	narrow stile	w/panic Hrdwre.	3'-0" x 7'-0"	1,200	795	1,995
6350				dbl. door, Hrdwre.	6'-0" x 7'-0"	2,075	1,300	3,375
6400			wide stile	hdwre.	3'-0" x 7'-0"	1,450	780	2,230
6450				dbl. door, Hdwre.	6'-0" x 7'-0"	2,850	1,550	4,400
6500			full vision	hdwre.	3'-0" x 7'-0"	1,775	1,250	3,025
6550				dbl. door, Hdwre.	6'-0" x 7'-0"	2,525	1,775	4,300
6600			non-standard	hdwre.	3'-0" x 7'-0"	1,025	780	1,805
6650				dbl. door, Hdwre.	6'-0" x 7'-0"	2,050	1,550	3,600
6700			bronze fin.	hdwre.	3'-0" x 7'-0"	1,100	780	1,880
6750				dbl. door, Hrdwre.	6'-0" x 7'-0"	2,200	1,550	3,750
6800			black fin.	hdwre.	3'-0" x 7'-0"	1,800	780	2,580
6850				dbl. door, Hdwre.	6'-0" x 7'-0"	3,625	1,550	5,175
6900		w/transom	narrow stile	hdwre.	3'-0" x 10'-0"	1,475	900	2,375
6950				dbl. door, Hdwre.	6'-0" x 10'-0"	2,050	1,575	3,625
7000			wide stile	hdwre.	3'-0" x 10'-0"	1,675	1,075	2,750
7050				dbl. door, Hdwre.	6'-0" x 10'-0"	2,200	1,875	4,075
7100			full vision	hdwre.	3'-0" x 10'-0"	1,850	1,200	3,050
7150				dbl. door, Hdwre.	6'-0" x 10'-0"	2,350	2,075	4,425
7200			non-standard	hdwre.	3'-0" x 10'-0"	1,100	840	1,940
7250				dbl. door, Hdwre.	6'-0" x 10'-0"	2,175	1,675	3,850
7300			bronze fin.	hdwre.	3'-0" x 10'-0"	1,150	840	1,990
7350				dbl. door, Hdwre.	6'-0" x 10'-0"	2,325	1,675	4,000
7400			black fin.	hdwre.	3'-0" x 10'-0"	1,875	840	2,715
7450				dbl. door, Hdwre.	6'-0" x 10'-0"	3,750	1,675	5,425
7500		revolving	stock design	minimum	6'-10" x 7'-0"	17,100	2,900	20,000
7550				average	6'-0" x 7'-0"	21,100	3,650	24,750
7600				maximum	6'-10" x 7'-0"	27,700	4,850	32,550
7650				min., automatic	6'-10" x 7'-0"	29,900	3,375	33,275

SHELL B

Important: See the Reference Section for critical supporting data - Reference Numbers and City Cost Indexes

B20 Exterior Enclosure

B2030 Exterior Doors

B2030 110	Glazed Doors, Steel or Aluminum

	MATERIAL	TYPE	DOORS	SPECIFICATION	OPENING	COST PER OPNG.		
						MAT.	INST.	TOTAL
7700	Alum & glass			avg., automatic	6'-10" x 7'-0"	33,900	4,125	38,025
7750				max., automatic	6'-10" x 7'-0"	40,500	5,325	45,825
7800		balanced	standard	economy	3'-0" x 7'-0"	5,475	1,225	6,700
7850				premium	3'-0" x 7'-0"	6,800	1,550	8,350
7900		mall front	sliding panels	alum. fin.	16'-0" x 9'-0"	2,550	615	3,165
7950					24'-0" x 9'-0"	3,700	1,150	4,850
8000				bronze fin.	16'-0" x 9'-0"	2,975	715	3,690
8050					24'-0" x 9'-0"	4,300	1,350	5,650
8100			fixed panels	alum. fin.	48'-0" x 9'-0"	6,875	885	7,760
8150				bronze fin.	48'-0" x 9'-0"	8,000	1,025	9,025
8200		sliding entrance	5' x 7' door	electric oper.	12'-0" x 7'-6"	6,500	1,150	7,650
8250		sliding patio	temp. glass	economy	6'-0" x 7'-0"	1,275	213	1,488
8300			temp. glass	economy	12'-0" x 7'-0"	2,225	285	2,510
8350				premium	6'-0" x 7'-0"	1,925	320	2,245
8400					12'-0" x 7'-0"	3,350	430	3,780

SHELL

B

B2030 Exterior Doors

Exterior Door System

The table below lists exterior door systems by material, type and size. Prices between sizes listed can be interpolated with reasonable accuracy. Prices are per opening for a complete door system including frame and required hardware.

Wood doors in this table are designed with wood frames and metal doors with hollow metal frames. Depending upon quality the total material and installation cost of a wood door frame is about the same as a hollow metal frame.

SHELL B

B2030 210 — Wood Doors

	MATERIAL	TYPE	DOORS	SPECIFICATION	OPENING	COST PER OPNG. MAT.	INST.	TOTAL
2350	Birch	solid core	single door	hinged	2'-6" x 6'-8"	960	232	1,192
2400					2'-6" x 7'-0"	965	234	1,199
2450					2'-8" x 7'-0"	970	234	1,204
2500					3'-0" x 7'-0"	980	238	1,218
2550			double door	hinged	2'-6" x 6'-8"	1,775	420	2,195
2600					2'-6" x 7'-0"	1,800	425	2,225
2650					2'-8" x 7'-0"	1,800	425	2,225
2700					3'-0" x 7'-0"	1,825	435	2,260
2750	Wood	combination	storm & screen	hinged	3'-0" x 6'-8"	300	57	357
2800					3'-0" x 7'-0"	330	62.50	392.50
2850		overhead	panels, H.D.	manual oper.	8'-0" x 8'-0"	635	425	1,060
2900					10'-0" x 10'-0"	950	475	1,425
2950					12'-0" x 12'-0"	1,400	570	1,970
3000					14'-0" x 14'-0"	2,000	655	2,655
3050					20'-0" x 16'-0"	4,125	1,325	5,450
3100				electric oper.	8'-0" x 8'-0"	1,350	640	1,990
3150					10'-0" x 10'-0"	1,675	690	2,365
3200					12'-0" x 12'-0"	2,125	785	2,910
3250					14'-0" x 14'-0"	2,725	870	3,595
3300					20'-0" x 16'-0"	4,925	1,750	6,675

B2030 220 — Steel Doors

	MATERIAL	TYPE	DOORS	SPECIFICATION	OPENING	COST PER OPNG. MAT.	INST.	TOTAL
3350	Steel 18 Ga.	hollow metal	1 door w/frame	no label	2'-6" x 7'-0"	965	241	1,206
3400					2'-8" x 7'-0"	965	241	1,206
3450					3'-0" x 7'-0"	965	241	1,206
3500					3'-6" x 7'-0"	1,025	254	1,279
3550					4'-0" x 8'-0"	1,175	255	1,430
3600			2 doors w/frame	no label	5'-0" x 7'-0"	1,850	445	2,295
3650					5'-4" x 7'-0"	1,850	445	2,295
3700					6'-0" x 7'-0"	1,850	445	2,295
3750					7'-0" x 7'-0"	1,975	470	2,445
3800					8'-0" x 8'-0"	2,250	475	2,725
3850			1 door w/frame	"A" label	2'-6" x 7'-0"	1,075	290	1,365

Important: See the Reference Section for critical supporting data - Reference Numbers and City Cost Indexes

B2030 220	Steel Doors

	MATERIAL	TYPE	DOORS	SPECIFICATION	OPENING	COST PER OPNG.		
						MAT.	INST.	TOTAL
3900	Steel, 18 Ga.	hollow metal	1 door w/frame	"A" label	2'-8" x 7'-0"	1,075	294	1,369
3950					3'-0" x 7'-0"	1,075	294	1,369
4000					3'-6" x 7'-0"	1,150	300	1,450
4050					4'-0" x 8'-0"	1,200	315	1,515
4100			2 doors w/frame	"A" label	5'-0" x 7'-0"	2,050	540	2,590
4150					5'-4" x 7'-0"	2,100	545	2,645
4200					6'-0" x 7'-0"	2,100	545	2,645
4250					7'-0" x 7'-0"	2,250	555	2,805
4300					8'-0" x 8'-0"	2,250	560	2,810
4350	Steel 24 Ga.	overhead	sectional	manual oper.	8'-0" x 8'-0"	555	425	980
4400					10'-0" x 10'-0"	765	475	1,240
4450					12'-0" x 12'-0"	1,050	570	1,620
4500					20'-0" x 14'-0"	3,100	1,225	4,325
4550				electric oper.	8'-0" x 8'-0"	1,275	640	1,915
4600					10'-0" x 10'-0"	1,500	690	2,190
4650					12'-0" x 12'-0"	1,775	785	2,560
4700					20'-0" x 14'-0"	3,900	1,650	5,550
4750	Steel	overhead	rolling	manual oper.	8'-0" x 8'-0"	850	685	1,535
4800					10'-0" x 10'-0"	1,125	780	1,905
4850					12'-0" x 12'-0"	1,450	910	2,360
4900					14'-0" x 14'-0"	1,925	1,375	3,300
4950					20'-0" x 12'-0"	2,375	1,225	3,600
5000					20'-0" x 16'-0"	2,725	1,825	4,550
5050				electric oper.	8'-0" x 8'-0"	1,650	905	2,555
5100					10'-0" x 10'-0"	1,925	1,000	2,925
5150					12'-0" x 12'-0"	2,250	1,125	3,375
5200					14'-0" x 14'-0"	2,725	1,600	4,325
5250					20'-0" x 12'-0"	3,175	1,450	4,625
5300					20'-0" x 16'-0"	3,525	2,050	5,575
5350				fire rated	10'-0" x 10'-0"	1,475	995	2,470
5400			rolling grill	manual oper.	10'-0" x 10'-0"	2,150	1,100	3,250
5450					15'-0" x 8'-0"	2,550	1,375	3,925
5500		vertical lift	1 door w/frame	motor operator	16"-0" x 16"-0"	18,000	4,425	22,425
5550					32'-0" x 24'-0"	37,100	2,950	40,050
6000	Aluminum	combination	storm & screen	hinged	3'-0" x 6'-8"	237	61	298
6050					3'-0" x 7'-0"	261	67	328
6100		overhead	rolling grill	manual oper.	12'-0" x 12'-0"	3,450	1,925	5,375
6150				motor oper.	12'-0" x 12'-0"	4,550	2,125	6,675
6200	Alum. & Fbrgls.	overhead	heavy duty	manual oper.	12'-0" x 12'-0"	1,950	570	2,520
6250				electric oper.	12'-0" x 12'-0"	2,675	785	3,460

SHELL

B

Multiple ply roofing is the most popular covering for minimum pitch roofs.

Built Up Ply

B3010 105	Built-Up	COST PER S.F.		
		MAT.	INST.	TOTAL
1200	Asphalt flood coat w/gravel; not incl. insul, flash., nailers			
1300				
1400	Asphalt base sheets & 3 plies #15 asphalt felt, mopped	.54	1.34	1.88
1500	On nailable deck	.59	1.41	2
1600	4 plies #15 asphalt felt, mopped	.76	1.47	2.23
1700	On nailable deck	.69	1.55	2.24
1800	Coated glass base sheet, 2 plies glass (type IV), mopped	.55	1.34	1.89
1900	For 3 plies	.66	1.47	2.13
2000	On nailable deck	.62	1.55	2.17
2300	4 plies glass fiber felt (type IV), mopped	.80	1.47	2.27
2400	On nailable deck	.73	1.55	2.28
2500	Organic base sheet & 3 plies #15 organic felt, mopped	.59	1.50	2.09
2600	On nailable deck	.57	1.55	2.12
2700	4 plies #15 organic felt, mopped	.71	1.34	2.05
2750				
2800	Asphalt flood coat, smooth surface			
2900	Asphalt base sheet & 3 plies #15 asphalt felt, mopped	.58	1.23	1.81
3000	On nailable deck	.54	1.28	1.82
3100	Coated glass fiber base sheet & 2 plies glass fiber felt, mopped	.50	1.18	1.68
3200	On nailable deck	.48	1.23	1.71
3300	For 3 plies, mopped	.61	1.28	1.89
3400	On nailable deck	.57	1.34	1.91
3700	4 plies glass fiber felt (type IV), mopped	.72	1.28	2
3800	On nailable deck	.68	1.34	2.02
3900	Organic base sheet & 3 plies #15 organic felt, mopped	.56	1.23	1.79
4000	On nailable decks	.52	1.28	1.80
4100	4 plies #15 organic felt, mopped	.66	1.34	2
4200	Coal tar pitch with gravel surfacing			
4300	4 plies #15 tarred felt, mopped	1.17	1.41	2.58
4400	3 plies glass fiber felt (type IV), mopped	.96	1.55	2.51
4500	Coated glass fiber base sheets 2 plies glass fiber felt, mopped	.96	1.55	2.51
4600	On nailable decks	.85	1.64	2.49
4800	3 plies glass fiber felt (type IV), mopped	1.33	1.41	2.74
4900	On nailable decks	1.22	1.47	2.69

SHELL B

Important: See the Reference Section for critical supporting data - Reference Numbers and City Cost Indexes

B3010 Roof Coverings

Full Adhered Single Ply Membrane

Preformed Metal

Flat Seam

Batten Seam

Standing Seam

B3010 120	Single Ply Membrane	COST PER S.F.		
		MAT.	INST.	TOTAL
1000	CSPE (Chlorosulfonated polyethylene), 35 mils, fully adhered	1.36	.76	2.12
2000	EPDM (Ethylene propylene diene monomer), 45 mils, fully adhered	.75	.76	1.51
4000	Modified bit., SBS modified, granule surface cap sheet, mopped, 150 mils	.50	1.47	1.97
4500	APP modified, granule surface cap sheet, torched, 180 mils	.55	.99	1.54
6000	Reinforced PVC, 48 mils, loose laid and ballasted with stone	1.04	.39	1.43
6200	Fully adhered with adhesive	1.27	.76	2.03

B3010 130	Preformed Metal Roofing	COST PER S.F.		
		MAT.	INST.	TOTAL
0200	Corrugated roofing, aluminum, mill finish, .0175" thick, .272 P.S.F.	.68	1.38	2.06
0250	.0215 thick, .334 P.S.F.	.88	1.38	2.26

B3010 135	Formed Metal	COST PER S.F.		
		MAT.	INST.	TOTAL
1000	Batten seam, formed copper roofing, 3"min slope, 16 oz., 1.2 P.S.F.	4.53	4.57	9.10
1100	18 oz., 1.35 P.S.F.	5.05	5	10.05
2000	Zinc copper alloy, 3" min slope, .020" thick, .88 P.S.F.	5.80	4.17	9.97
3000	Flat seam, copper, 1/4" min. slope, 16 oz., 1.2 P.S.F.	4.03	4.17	8.20
3100	18 oz., 1.35 P.S.F.	4.53	4.37	8.90
5000	Standing seam, copper, 2-1/2" min. slope, 16 oz., 1.25 P.S.F.	4.33	3.87	8.20
5100	18 oz., 1.40 P.S.F.	4.88	4.17	9.05
6000	Zinc copper alloy, 2-1/2" min. slope, .020" thick, .87 P.S.F.	5.70	4.17	9.87
6100	.032" thick, 1.39 P.S.F.	7.75	4.57	12.32

B3010   Roof Coverings

Shingle and Tile

Shingles and tiles are practical in applications where the roof slope is more than 3-1/2" per foot of rise. Lines 1100 through 6000 list the various materials and the weight per square foot.

Roof Edges

The table below lists the costs for various types of roof perimeter edge treatments.

Roof edge systems include the cost per L.F. for a 2" x 8" treated wood nailer fastened at 4'-0" O.C. and a diagonally cut 4" x 6" treated wood cant.

Roof edge and base flashing are assumed to be made from the same material.

SHELL B

B3010 140	Shingle & Tile	COST PER S.F.		
		MAT.	INST.	TOTAL
1095	Asphalt roofing			
1100	Strip shingles,4" slope,inorganic class A 210-235 lb/Sq.	.38	.85	1.23
1150	Organic, class C, 235-240 lb./sq.	.49	.93	1.42
1200	Premium laminated multi-layered, class A, 260-300 lb./Sq.	.64	1.27	1.91
1545	Metal roofing			
1550	Alum., shingles, colors, 3"min slope, .019"thick, 0.4 PSF	1.85	.93	2.78
1850	Steel, colors, 3" min slope, 26 gauge, 1.0 PSF	2.21	1.91	4.12
2795	Slate roofing			
2800	4" min. slope, shingles, 3/16" thick, 8.0 PSF	6.15	2.38	8.53
3495	Wood roofing			
3500	4" min slope, cedar shingles, 16" x 5", 5" exposure 1.6 PSF	1.87	1.84	3.71
4000	Shakes, 18", 8-1/2" exposure, 2.8 PSF	1.15	2.26	3.41
5095	Tile roofing			
5100	Aluminum, mission, 3" min slope, .019" thick, 0.65 PSF	6.15	1.78	7.93
6000	Clay, Americana, 3" minimum slope, 8 PSF	5.70	2.52	8.22

B3010 Roof Coverings

B3010 320	Roof Deck Rigid Insulation	COST PER S.F.		
		MAT.	INST.	TOTAL
0100	Fiberboard low density, 1/2" thick, R1.39			
0150	1" thick R2.78	.37	.50	.87
0300	1 1/2" thick R4.17	.55	.50	1.05
0350	2" thick R5.56	.75	.50	1.25
0370	Fiberboard high density, 1/2" thick R1.3	.22	.40	.62
0380	1" thick R2.5	.40	.50	.90
0390	1 1/2" thick R3.8	.65	.50	1.15
0410	Fiberglass, 3/4" thick R2.78	.51	.40	.91
0450	15/16" thick R3.70	.67	.40	1.07
0550	1-5/16" thick R5.26	1.16	.40	1.56
0650	2 7/16" thick R10	1.41	.50	1.91
1510	Polyisocyanurate 2#/CF density, 1" thick R7.14	.36	.29	.65
1550	1 1/2" thick R10.87	.39	.32	.71
1600	2" thick R14.29	.50	.37	.87
1650	2 1/2" thick R16.67	.55	.38	.93
1700	3" thick R21.74	.80	.40	1.20
1750	3 1/2" thick R25	.81	.40	1.21
1800	Tapered for drainage	.42	.29	.71
1810	Expanded polystyrene, 1#/CF density, 3/4" thick R2.89	.21	.27	.48
1820	2" thick R7.69	.45	.32	.77
1830	Extruded polystyrene, 15 PSI compressive strength, 1" thick R5	.25	.27	.52
1835	2" thick R10	.39	.32	.71
1840	3" thick R15	.75	.40	1.15
2550	40 PSI compressive strength, 1" thick R5	.39	.27	.66
2600	2" thick R10	.76	.32	1.08
2650	3" thick R15	1.11	.40	1.51
2700	4" thick R20	1.49	.40	1.89
2750	Tapered for drainage	.55	.29	.84
2810	60 PSI compressive strength, 1" thick R5	.46	.28	.74
2850	2" thick R10	.83	.33	1.16
2900	Tapered for drainage	.67	.29	.96
3050	Composites with 2" EPS			
3060	1" Fiberboard	.86	.42	1.28
3070	7/16" oriented strand board	1.02	.50	1.52
3080	1/2" plywood	1.10	.50	1.60
3090	1" perlite	.90	.50	1.40
4000	Composites with 1-1/2" polyisocyanurate			
4010	1" fiberboard	.92	.50	1.42
4020	1" perlite	.97	.47	1.44
4030	7/16" oriented strand board	1.11	.50	1.61

B3010 420			Roof Edges		COST PER L.F.		
	EDGE TYPE	DESCRIPTION	SPECIFICATION	FACE HEIGHT		MAT.	INST.	TOTAL
1000	Aluminum	mill finish	.050" thick	4"		8.70	8.25	16.95
1100				6"		9	8.50	17.50
1300		duranodic	.050" thick	4"		8.55	8.25	16.80
1400				6"		9.25	8.50	17.75
1600		painted	.050" thick	4"		9.15	8.25	17.40
1700				6"		10.05	8.50	18.55
2000	Copper	plain	16 oz.	4"		6.15	8.25	14.40
2100				6"		7	8.50	15.50
2300			20 oz.	4"		6.70	9.40	16.10
2400				6"		7.40	9.05	16.45
2700	Sheet Metal	galvanized	20 Ga.	4"		6.85	9.70	16.55
2800				6"		8.05	9.70	17.75

SHELL

B

325

B30 Roofing

B3010 Roof Coverings

B3010 420 — Roof Edges

	EDGE TYPE	DESCRIPTION	SPECIFICATION	FACE HEIGHT		COST PER L.F.		
						MAT.	INST.	TOTAL
3000	Sheet metal	galvanized	24 Ga.	4"		6.15	8.25	14.40
3100				6"		7	8.25	15.25

B3010 430 — Flashing

	MATERIAL	BACKING	SIDES	SPECIFICATION	QUANTITY	COST PER S.F.		
						MAT.	INST.	TOTAL
0040	Aluminum	none		.019"		.72	2.77	3.49
0050				.032"		1.17	2.77	3.94
0300		fabric	2	.004"		1.02	1.22	2.24
0400		mastic		.004"		1.02	1.22	2.24
0700	Copper	none		16 oz.	<500 lbs.	2.88	3.49	6.37
0800				24 oz.	<500 lbs.	5.15	3.82	8.97
2000	Copper lead	fabric	1	2 oz.		1.55	1.22	2.77
3500	PVC black	none		.010"		.18	1.41	1.59
3700				.030"		.33	1.41	1.74
4200	Neoprene			1/16"		1.63	1.41	3.04
4500	Stainless steel	none		.015"	<500 lbs.	3.48	3.49	6.97
4600	Copper clad				>2000 lbs.	3.36	2.59	5.95
5000	Plain			32 ga.		2.38	2.59	4.97

Important: See the Reference Section for critical supporting data - Reference Numbers and City Cost Indexes

SHELL B

B3010 Roof Coverings

B3010 610 — Gutters

	SECTION	MATERIAL	THICKNESS	SIZE	FINISH	COST PER L.F.		
						MAT.	INST.	TOTAL
0050	Box	aluminum	.027"	5"	enameled	1.20	4.12	5.32
0100					mill	1.36	4.12	5.48
0200			.032"	5"	enameled	1.33	4.12	5.45
0500		copper	16 Oz.	4"	lead coated	8.90	4.12	13.02
0600					mill	5.55	4.12	9.67
1000		steel galv.	28 Ga.	5"	enameled	1.23	4.12	5.35
1200			26 Ga.	5"	mill	1.05	4.12	5.17
1800		vinyl		4"	colors	.94	3.88	4.82
1900				5"	colors	1.10	3.88	4.98
2300		hemlock or fir		4"x5"	treated	8	4.27	12.27
3000	Half round	copper	16 Oz.	4"	lead coated	8.25	4.12	12.37
3100					mill	4.51	4.12	8.63
3600		steel galv.	28 Ga.	5"	enameled	1.23	4.12	5.35
4102		stainless steel		5"	mill	5.50	4.12	9.62
5000		vinyl		4"	white	.75	3.88	4.63

B3010 620 — Downspouts

	MATERIALS	SECTION	SIZE	FINISH	THICKNESS	COST PER V.L.F.		
						MAT.	INST.	TOTAL
0100	Aluminum	rectangular	2"x3"	embossed mill	.020"	.75	2.60	3.35
0150				enameled	.020"	1.07	2.60	3.67
0250			3"x4"	enameled	.024"	1.90	3.53	5.43
0300		round corrugated	3"	enameled	.020"	1.14	2.60	3.74
0350			4"	enameled	.025"	1.64	3.53	5.17
0500	Copper	rectangular corr.	2"x3"	mill	16 Oz.	4.30	2.60	6.90
0600		smooth		mill	16 Oz.	5.25	2.60	7.85
0700		rectangular corr.	3"x4"	mill	16 Oz.	5.45	3.41	8.86
1300	Steel	rectangular corr.	2"x3"	galvanized	28 Ga.	.59	2.60	3.19
1350				epoxy coated	24 Ga.	1.16	2.60	3.76
1400		smooth		galvanized	28 Ga.	.89	2.60	3.49
1450		rectangular corr.	3"x4"	galvanized	28 Ga.	1.65	3.41	5.06
1500				epoxy coated	24 Ga.	2.10	3.41	5.51
1550		smooth		galvanized	28 Ga.	1.35	3.41	4.76
1652		round corrugated	3"	galvanized	28 Ga.	.83	2.60	3.43
1700			4"	galvanized	28 Ga.	1.12	3.41	4.53
1750			5"	galvanized	28 Ga.	1.69	3.81	5.50
2000	Steel pipe	round	4"	black	X.H.	13.05	24.50	37.55
2552	S.S. tubing sch.5	rectangular	3"x4"	mill		26	3.41	29.41

B3010 630 — Gravel Stop

	MATERIALS	SECTION	SIZE	FINISH	THICKNESS	COST PER L.F.		
						MAT.	INST.	TOTAL
5100	Aluminum	extruded	4"	mill	.050"	4.20	3.41	7.61
5200			4"	duranodic	.050"	4.06	3.41	7.47
5300			8"	mill	.050"	5.65	3.96	9.61
5400			8"	duranodic	.050"	5.50	3.96	9.46
6000			12"-2 pc.	duranodic	.050"	6.90	4.95	11.85
6100	Stainless	formed	6"	mill	24 Ga.	7.85	3.67	11.52

B SHELL

B3020 Roof Openings

Roof Hatch

Smoke Hatch

Skylight

B3020 110	Skylights	COST PER S.F.		
		MAT.	INST.	TOTAL
5100	Skylights, plastic domes, insul curbs, nom. size to 10 S.F., single glaze	15.15	10.35	25.50
5200	Double glazing	21	12.75	33.75
5300	10 S.F. to 20 S.F., single glazing	9.25	4.19	13.44
5400	Double glazing	18.30	5.25	23.55
5500	20 S.F. to 30 S.F., single glazing	11.65	3.56	15.21
5600	Double glazing	16.60	4.19	20.79
5700	30 S.F. to 65 S.F., single glazing	16.25	2.71	18.96
5800	Double glazing	12.45	3.56	16.01
6000	Sandwich panels fiberglass, 9-1/16"thick, 2 S.F. to 10 S.F.	17.10	8.25	25.35
6100	10 S.F. to 18 S.F.	15.30	6.25	21.55
6200	2-3/4" thick, 25 S.F. to 40 S.F.	24.50	5.60	30.10
6300	40 S.F. to 70 S.F.	20	5	25

B3020 210	Hatches	COST PER OPNG.		
		MAT.	INST.	TOTAL
0200	Roof hatches, with curb, and 1" fiberglass insulation, 2'-6"x3'-0",aluminum	560	165	725
0300	Galvanized steel 165 lbs.	385	165	550
0400	Primed steel 164 lbs.	455	165	620
0500	2'-6"x4'-6" aluminum curb and cover, 150 lbs.	665	184	849
0600	Galvanized steel 220 lbs.	550	184	734
0650	Primed steel 218 lbs.	695	184	879
0800	2'x6"x8'-0" aluminum curb and cover, 260 lbs.	1,400	251	1,651
0900	Galvanized steel, 360 lbs.	1,050	251	1,301
0950	Primed steel 358 lbs.	960	251	1,211
1200	For plexiglass panels, add to the above	380		380
2100	Smoke hatches, unlabeled not incl. hand winch operator, 2'-6"x3', galv	465	200	665
2200	Plain steel, 160 lbs.	550	200	750
2400	2'-6"x8'-0",galvanized steel, 360 lbs.	1,150	275	1,425
2500	Plain steel, 350 lbs.	1,050	275	1,325
3000	4'-0"x8'-0", double leaf low profile, aluminum cover, 359 lb.	1,575	207	1,782
3100	Galvanized steel 475 lbs.	1,375	207	1,582
3200	High profile, aluminum cover, galvanized curb, 361 lbs.	1,425	207	1,632

For information about Means Estimating Seminars, see yellow pages 12 and 13 in back of book

C INTERIORS

C1010 Partitions

The Concrete Block Partition Systems are defined by weight and type of block, thickness, type of finish and number of sides finished. System components include joint reinforcing on alternate courses and vertical control joints.

C1010 102	Concrete Block Partitions - Regular Weight							
	TYPE	THICKNESS (IN.)	TYPE FINISH	SIDES FINISHED		COST PER S.F.		
						MAT.	INST.	TOTAL
1000	Hollow	4	none	0		1.18	4.68	5.86
1010			gyp. plaster 2 coat	1		1.54	6.85	8.39
1020				2		1.90	8.95	10.85
1200			portland - 3 coat	1		1.43	7.10	8.53
1400			5/8" drywall	1		1.61	6.20	7.81
1500		6	none	0		1.69	5	6.69
1510			gyp. plaster 2 coat	1		2.05	7.15	9.20
1520				2		2.41	9.30	11.71
1700			portland - 3 coat	1		1.94	7.45	9.39
1900			5/8" drywall	1		2.12	6.50	8.62
1910				2		2.55	8	10.55
2000		8	none	0		1.82	5.40	7.22
2010			gyp. plaster 2 coat	1		2.18	7.55	9.73
2020			gyp. plaster 2 coat	2		2.54	9.65	12.19
2200			portland - 3 coat	1		2.07	7.80	9.87
2400			5/8" drywall	1		2.25	6.90	9.15
2410				2		2.68	8.40	11.08
2500		10	none	0		2.53	5.60	8.13
2510			gyp. plaster 2 coat	1		2.89	7.75	10.64
2520				2		3.25	9.90	13.15
2700			portland - 3 coat	1		2.78	8.05	10.83
2900			5/8" drywall	1		2.96	7.10	10.06
2910				2		3.39	8.60	11.99
3000	Solid	2	none	0		1.02	4.63	5.65
3010			gyp. plaster	1		1.38	6.80	8.18
3020				2		1.74	8.90	10.64
3200			portland - 3 coat	1		1.27	7.05	8.32
3400			5/8" drywall	1		1.45	6.15	7.60
3410				2		1.88	7.65	9.53
3500		4	none	0		1.57	4.85	6.42
3510			gyp. plaster	1		2.02	7.05	9.07
3520				2		2.29	9.10	11.39
3700			portland - 3 coat	1		1.82	7.30	9.12
3900			5/8" drywall	1		2	6.35	8.35
3910				2		2.43	7.85	10.28

Important: See the Reference Section for critical supporting data - Reference Numbers and City Cost Indexes

INTERIORS C

C1010 Partitions

C1010 102	Concrete Block Partitions - Regular Weight							

	TYPE	THICKNESS (IN.)	TYPE FINISH	SIDES FINISHED		COST PER S.F.		
						MAT.	INST.	TOTAL
4002	Solid	6	none	0		1.84	5.20	7.04
4010			gyp. plaster	1		2.20	7.35	9.55
4020				2		2.56	9.45	12.01
4200			portland - 3 coat	1		2.09	7.65	9.74
4400			5/8" drywall	1		2.27	6.70	8.97
4410				2		2.70	8.20	10.90

C1010 104	Concrete Block Partitions - Lightweight							

	TYPE	THICKNESS (IN.)	TYPE FINISH	SIDES FINISHED		COST PER S.F.		
						MAT.	INST.	TOTAL
5000	Hollow	4	none	0		1.37	4.58	5.95
5010			gyp. plaster	1		1.73	6.75	8.48
5020				2		2.09	8.85	10.94
5200			portland - 3 coat	1		1.62	7	8.62
5400			5/8" drywall	1		1.80	6.10	7.90
5410				2		2.23	7.60	9.83
5500		6	none	0		1.84	4.90	6.74
5520			gyp. plaster	2		2.56	9.15	11.71
5700			portland - 3 coat	1		2.09	7.35	9.44
5900			5/8" drywall	1		2.27	6.40	8.67
5910				2		2.70	7.90	10.60
6000		8	none	0		2.23	5.25	7.48
6010			gyp. plaster	1		2.59	7.40	9.99
6020				2		2.95	9.50	12.45
6200			portland - 3 coat	1		2.48	7.65	10.13
6400			5/8" drywall	1		2.66	6.75	9.41
6410				2		3.09	8.25	11.34
6500		10	none	0		2.90	5.45	8.35
6510			gyp. plaster	1		3.26	7.60	10.86
6520				2		3.62	9.75	13.37
6700			portland - 3 coat	1		3.15	7.90	11.05
6900			5/8" drywall	1		3.33	6.95	10.28
6910				2		3.76	8.45	12.21
7000	Solid	4	none	0		1.49	4.79	6.28
7010			gyp. plaster	1		1.85	6.95	8.80
7020				2		2.21	9.05	11.26
7200			portland - 3 coat	1		1.74	7.20	8.94
7400			5/8" drywall	1		1.92	6.30	8.22
7410				2		2.35	7.80	10.15
7500		6	none	0		2.22	5.15	7.37
7510			gyp. plaster	1		2.66	7.35	10.01
7520				2		2.94	9.40	12.34
7700			portland - 3 coat	1		2.47	7.60	10.07
7900			5/8" drywall	1		2.65	6.65	9.30
7910				2		3.08	8.15	11.23
8000		8	none	0		2.98	5.55	8.53
8010			gyp. plaster	1		3.34	7.70	11.04
8020				2		3.70	9.80	13.50
8200			portland - 3 coat	1		3.23	7.95	11.18
8400			5/8" drywall	1		3.41	7.05	10.46
8410				2		3.84	8.55	12.39

C INTERIORS

C1010 Partitions

Single Wythe

Structural facing tile
8W series
8" x 16"

Double Wythe

Single Wythe

Structural facing tile
6T series
5-1/3" x 12"

Double Wythe

C1010 120	Tile Partitions	COST PER S.F.		
		MAT.	INST.	TOTAL
1000	8W series 8"x16", 4" thick wall, reinf every 2 courses, glazed 1 side	8.35	5.65	14
1100	Glazed 2 sides	10.60	6	16.60
1200	Glazed 2 sides, using 2 wythes of 2" thick tile	14.60	10.80	25.40
1300	6" thick wall, horizontal reinf every 2 courses, glazed 1 side	11.75	5.90	17.65
1400	Glazed 2 sides, each face different color, 2" and 4" tile	15.65	11.05	26.70
1500	8" thick wall, glazed 2 sides using 2 wythes of 4" thick tile	16.70	11.30	28
1600	10" thick wall, glazed 2 sides using 1 wythe of 4" tile & 1 wythe of 6" tile	20	11.55	31.55
1700	Glazed 2 sides cavity wall, using 2 wythes of 4" thick tile	16.70	11.30	28
1800	12" thick wall, glazed 2 sides using 2 wythes of 6" thick tile	23.50	11.80	35.30
1900	Glazed 2 sides cavity wall, using 2 wythes of 4" thick tile	16.70	11.30	28
2100	6T series 5-1/3"x12" tile, 4" thick, non load bearing glazed one side,	7.75	8.85	16.60
2200	Glazed two sides	13.05	9.95	23
2300	Glazed two sides, using two wythes of 2" thick tile	13.20	17.30	30.50
2400	6" thick, glazed one side	11.95	9.25	21.20
2500	Glazed two sides	14.95	10.50	25.45
2600	Glazed two sides using 2" thick tile and 4" thick tile	14.35	17.50	31.85
2700	8" thick, glazed one side	15.05	10.80	25.85
2800	Glazed two sides using two wythes of 4" thick tile	15.50	17.70	33.20
2900	Glazed two sides using 6" thick tile and 2" thick tile	18.55	17.90	36.45
3000	10" thick cavity wall, glazed two sides using two wythes of 4" tile	15.50	17.70	33.20
3100	12" thick, glazed two sides using 4" thick tile and 8" thick tile	23	19.65	42.65
3200	2" thick facing tile, glazed one side, on 6" concrete block	8.85	10.25	19.10
3300	On 8" concrete block	8.95	10.60	19.55
3400	On 10" concrete block	9.65	10.80	20.45

C1010 Partitions

Gypsum board, single layer each side on wood studs.

Gypsum board, sound deadening board each side, with 1-1/2" insulation on wood studs.

Gypsum board, two layers each side on wood studs.

Gypsum board, single layer each side on metal studs.

Gypsum board, sound deadening board each side on metal studs.

Gypsum board two layers one side, single layer opposite side, with 3-1/2" insulation on metal studs.

INTERIORS

C

C1010 124		Drywall Partitions/Wood Stud Framing					
	FACE LAYER	BASE LAYER	FRAMING	OPPOSITE FACE	INSULATION	COST PER S.F.		
						MAT.	INST.	TOTAL
1200	5/8" FR drywall	none	2 x 4, @ 16" O.C.	same	0	1.01	2.58	3.59
1250				5/8" reg. drywall	0	.99	2.58	3.57
1300				nothing	0	.70	1.72	2.42
1400		1/4" SD gypsum	2 x 4 @ 16" O.C.	same	1-1/2" fiberglass	1.94	3.95	5.89
1450				5/8" FR drywall	1-1/2" fiberglass	1.70	3.48	5.18
1500				nothing	1-1/2" fiberglass	1.39	2.62	4.01
1600		resil. channels	2 x 4 @ 16", O.C.	same	1-1/2" fiberglass	1.76	5	6.76
1650				5/8" FR drywall	1-1/2" fiberglass	1.61	4.01	5.62
1700				nothing	1-1/2" fiberglass	1.30	3.15	4.45
1800		5/8" FR drywall	2 x 4 @ 24" O.C.	same	0	1.48	3.27	4.75
1850				5/8" FR drywall	0	1.20	2.84	4.04
1900				nothing	0	.89	1.98	2.87
2200		5/8" FR drywall	2 rows-2 x 4	same	2" fiberglass	2.48	4.72	7.20
2250			16"O.C.	5/8" FR drywall	2" fiberglass	2.20	4.29	6.49
2300				nothing	2" fiberglass	1.89	3.43	5.32
2400	5/8" WR drywall	none	2 x 4, @ 16" O.C.	same	0	1.03	2.58	3.61
2450				5/8" FR drywall	0	1.02	2.58	3.60
2500				nothing	0	.71	1.72	2.43
2600		5/8" FR drywall	2 x 4, @ 24" O.C.	same	0	1.50	3.27	4.77
2650				5/8" FR drywall	0	1.21	2.84	4.05
2700				nothing	0	.90	1.98	2.88
2800	5/8 VF drywall	none	2 x 4, @ 16" O.C.	same	0	1.83	2.76	4.59
2850				5/8" FR drywall	0	1.42	2.67	4.09
2900				nothing	0	1.11	1.81	2.92

C1010 Partitions

C1010 124	Drywall Partitions/Wood Stud Framing							

	FACE LAYER	BASE LAYER	FRAMING	OPPOSITE FACE	INSULATION	COST PER S.F.		
						MAT.	INST.	TOTAL
3000	5/8 VF drywall	5/8" FR drywall	2 x 4 , 24" O.C.	same	0	2.30	3.45	5.75
3050				5/8" FR drywall	0	1.61	2.93	4.54
3100				nothing	0	1.30	2.07	3.37
3200	1/2" reg drywall	3/8" reg drywall	2 x 4, @ 16" O.C.	same	0	1.37	3.44	4.81
3252				5/8"FR drywall	0	1.19	3.01	4.20
3300				nothing	0	.88	2.15	3.03

C1010 126	Drywall Partitions/Metal Stud Framing							

	FACE LAYER	BASE LAYER	FRAMING	OPPOSITE FACE	INSULATION	COST PER S.F.		
						MAT.	INST.	TOTAL
5200	5/8" FR drywall	none	1-5/8" @ 24" O.C.	same	0	.87	2.28	3.15
5250				5/8" reg. drywall	0	.85	2.28	3.13
5300				nothing	0	.56	1.42	1.98
5400			3-5/8" @ 24" O.C.	same	0	.95	2.30	3.25
5450				5/8" reg. drywall	0	.93	2.30	3.23
5500				nothing	0	.64	1.44	2.08
5600		1/4" SD gypsum	1-5/8" @ 24" O.C.	same	0	1.35	3.22	4.57
5650				5/8" FR drywall	0	1.11	2.75	3.86
5700				nothing	0	.80	1.89	2.69
5800			2-1/2" @ 24" O.C.	same	0	1.38	3.23	4.61
5850				5/8" FR drywall	0	1.14	2.76	3.90
5900				nothing	0	.83	1.90	2.73
6000		5/8" FR drywall	2-1/2" @ 16" O.C.	same	0	1.91	3.67	5.58
6050				5/8" FR drywall	0	1.63	3.24	4.87
6100				nothing	0	1.32	2.38	3.70
6200			3-5/8" @ 24" O.C.	same	0	1.51	3.16	4.67
6250				5/8"FR drywall	3-1/2" fiberglass	1.67	3	4.67
6300				nothing	0	.92	1.87	2.79
6400	5/8" WR drywall	none	1-5/8" @ 24" O.C.	same	0	.89	2.28	3.17
6450				5/8" FR drywall	0	.88	2.28	3.16
6500				nothing	0	.57	1.42	1.99
6600			3-5/8" @ 24" O.C.	same	0	.97	2.30	3.27
6650				5/8" FR drywall	0	.96	2.30	3.26
6700				nothing	0	.65	1.44	2.09
6800		5/8" FR drywall	2-1/2" @ 16" O.C.	same	0	1.93	3.67	5.60
6850				5/8" FR drywall	0	1.64	3.24	4.88
6900				nothing	0	1.33	2.38	3.71
7000			3-5/8" @ 24" O.C.	same	0	1.53	3.16	4.69
7050				5/8"FR drywall	3-1/2" fiberglass	1.68	3	4.68
7100				nothing	0	.93	1.87	2.80
7200	5/8" VF drywall	none	1-5/8" @ 24" O.C.	same	0	1.69	2.46	4.15
7250				5/8" FR drywall	0	1.28	2.37	3.65
7300				nothing	0	.97	1.51	2.48
7400			3-5/8" @ 24" O.C.	same	0	1.77	2.48	4.25
7450				5/8" FR drywall	0	1.36	2.39	3.75
7500				nothing	0	1.05	1.53	2.58
7600		5/8" FR drywall	2-1/2" @ 16" O.C.	same	0	2.73	3.85	6.58
7650				5/8" FR drywall	0	2.04	3.33	5.37
7700				nothing	0	1.73	2.47	4.20
7800			3-5/8" @ 24" O.C.	same	0	2.33	3.34	5.67
7850				5/8"FR drywall	3-1/2" fiberglass	2.08	3.09	5.17
7900				nothing	0	1.33	1.96	3.29

INTERIORS C

C1010 Partitions

C1010 128	Drywall Components	COST PER S.F.		
		MAT.	INST.	TOTAL
0060	Metal studs, 24" O.C. including track, load bearing, 20 gage, 2-1/2"	.73	.80	1.53
0080	3-5/8"	.86	.82	1.68
0100	4"	.78	.83	1.61
0120	6"	1.15	.85	2
0140	Metal studs, 24" O.C. including track, load bearing, 18 gage, 2-1/2"	.73	.80	1.53
0160	3-5/8"	.86	.82	1.68
0180	4"	.78	.83	1.61
0200	6"	1.15	.85	2
0220	16 gage, 2-1/2"	.84	.91	1.75
0240	3-5/8"	1.01	.93	1.94
0260	4"	1.06	.95	2.01
0280	6"	1.33	.97	2.30
0300	Non load bearing, 25 gage, 1-5/8"	.25	.56	.81
0340	3-5/8"	.33	.58	.91
0360	4"	.43	.58	1.01
0380	6"	.50	.59	1.09
0400	20 gage, 2-1/2"	.54	.71	1.25
0420	3-5/8"	.62	.72	1.34
0440	4"	.70	.72	1.42
0460	6"	.88	.74	1.62
0540	Wood studs including blocking, shoe and double top plate, 2"x4", 12"O.C.	.48	1.07	1.55
0560	16" O.C.	.39	.86	1.25
0580	24" O.C.	.30	.69	.99
0600	2"x6", 12" O.C.	.77	1.22	1.99
0620	16" O.C.	.62	.95	1.57
0640	24" O.C.	.47	.74	1.21
0642	Furring one side only, steel channels, 3/4", 12" O.C.	.24	1.59	1.83
0644	16" O.C.	.22	1.41	1.63
0646	24" O.C.	.15	1.07	1.22
0647	1-1/2" , 12" O.C.	.35	1.78	2.13
0648	16" O.C.	.32	1.55	1.87
0649	24" O.C.	.21	1.22	1.43
0650	Wood strips, 1" x 3", on wood, 12" O.C.	.31	.78	1.09
0651	16" O.C.	.23	.59	.82
0652	On masonry, 12" O.C.	.31	.86	1.17
0653	16" O.C.	.23	.65	.88
0654	On concrete, 12" O.C.	.31	1.64	1.95
0655	16" O.C.	.23	1.23	1.46
0665	Gypsum board, one face only, exterior sheathing, 1/2"	.32	.76	1.08
0680	Interior, fire resistant, 1/2"	.24	.43	.67
0700	5/8"	.28	.43	.71
0720	Sound deadening board 1/4"	.24	.47	.71
0740	Standard drywall 3/8"	.22	.43	.65
0760	1/2"	.24	.43	.67
0780	5/8"	.26	.43	.69
0800	Tongue & groove coreboard 1"	.50	1.78	2.28
0820	Water resistant, 1/2"	.24	.43	.67
0840	5/8"	.29	.43	.72
0860	Add for the following:, foil backing	.11		.11
0880	Fiberglass insulation, 3-1/2"	.44	.27	.71
0900	6"	.50	.32	.82
0920	Rigid insulation 1"	.40	.43	.83
0940	Resilient furring @ 16" O.C.	.20	1.34	1.54
0960	Taping and finishing	.03	.43	.46
0980	Texture spray	.04	.47	.51
1000	Thin coat plaster	.08	.53	.61
1040	2"x4" staggered studs 2"x6" plates & blocking	.56	.93	1.49

C INTERIORS

Gypsum plaster and gypsum
lath on metal studs.

Gypsum plaster and diamond
metal lath on metal studs.

Gypsum plaster and gypsum
lath on wood studs.

Gypsum plaster and diamond
metal lath on wood studs.

INTERIORS C

C1010 140	Plaster Partitions/Metal Stud Framing							
	TYPE	FRAMING	LATH	OPPOSITE FACE		COST PER S.F.		
						MAT.	INST.	TOTAL
1000	2 coat gypsum	2-1/2" @ 16"O.C.	3/8" gypsum	same		2.46	6.20	8.66
1010				nothing		1.47	3.59	5.06
1100		3-1/4" @ 24"O.C.	1/2" gypsum	same		2.41	6	8.41
1110				nothing		1.42	3.34	4.76
1500	2 coat vermiculite	2-1/2" @ 16"O.C.	3/8" gypsum	same		2.26	6.70	8.96
1510				nothing		1.38	3.88	5.26
1600		3-1/4" @ 24"O.C.	1/2" gypsum	same		2.21	6.50	8.71
1610				nothing		1.33	3.63	4.96
2000	3 coat gypsum	2-1/2" @ 16"O.C.	3/8" gypsum	same		2.34	6.95	9.29
2010				nothing		1.42	3.98	5.40
2020			3.4lb. diamond	same		2.24	6.95	9.19
2030				nothing		1.36	3.98	5.34
2040			2.75lb. ribbed	same		2.09	6.95	9.04
2050				nothing		1.29	3.98	5.27
2100		3-1/4" @ 24"O.C.	1/2" gypsum	same		2.29	6.75	9.04
2110				nothing		1.37	3.73	5.10
2120			3.4lb. ribbed	same		2.37	6.75	9.12
2130				nothing		1.41	3.73	5.14
3500	3 coat gypsum	2-1/2" @ 16"O.C.	3/8" gypsum	same		2.87	8.90	11.77
3510	W/med. Keenes			nothing		1.69	4.97	6.66
3520			3.4lb. diamond	same		2.77	8.90	11.67
3530				nothing		1.63	4.97	6.60
3540			2.75lb. ribbed	same		2.62	8.90	11.52
3550				nothing		1.56	4.97	6.53
3600		3-1/4" @ 24"O.C.	1/2" gypsum	same		2.82	8.70	11.52
3610				nothing		1.64	4.72	6.36
3620			3.4lb. ribbed	same		2.90	8.70	11.60

Important: See the Reference Section for critical supporting data - Reference Numbers and City Cost Indexes

C10 Interior Construction

C1010 Partitions

C1010 140 — Plaster Partitions/Metal Stud Framing

	TYPE	FRAMING	LATH	OPPOSITE FACE		COST PER S.F. MAT.	INST.	TOTAL
3630	3 coat gypsum	3-1/4" @ 24"O.C.	3.4 lb ribbed	nothing		1.68	4.72	6.40
4000	3 coat gypsum	2-1/2" @ 16"O.C.	3/8" gypsum	same		2.88	9.90	12.78
4010	W/hard Keenes			nothing		1.69	5.50	7.19
4022			3.4 lb. diamond	same		2.78	9.90	12.68
4032				nothing		1.63	5.50	7.13
4040			2.75lb. ribbed	same		2.63	9.90	12.53
4050				nothing		1.56	5.50	7.06
4100		3-1/4" @ 24"O.C.	1/2" gypsum	same		2.83	9.70	12.53
4110				nothing		1.64	5.20	6.84
4120			3.4lb. ribbed	same		2.91	9.70	12.61
4130				nothing		1.68	5.20	6.88

C1010 142 — Plaster Partitions/Wood Stud Framing

	TYPE	FRAMING	LATH	OPPOSITE FACE		COST PER S.F. MAT.	INST.	TOTAL
5000	2 coat gypsum	2"x4" @ 16"O.C.	3/8" gypsum	same		2.47	6.05	8.52
5010				nothing		1.48	3.52	5
5100		2"x4" @ 24"O.C.	1/2" gypsum	same		2.38	5.95	8.33
5110				nothing		1.39	3.38	4.77
5500	2 coat vermiculite	2"x4" @ 16"O.C.	3/8" gypsum	same		2.27	6.55	8.82
5510				nothing		1.39	3.81	5.20
5600		2"x4" @ 24"O.C.	1/2" gypsum	same		2.18	6.45	8.63
5610				nothing		1.30	3.67	4.97
6000	3 coat gypsum	2"x4" @ 16"O.C.	3/8" gypsum	same		2.35	6.80	9.15
6010				nothing		1.43	3.91	5.34
6020			3.4lb. diamond	same		2.25	6.90	9.15
6030				nothing		1.37	3.94	5.31
6040			2.75lb. ribbed	same		2.08	6.90	8.98
6050				nothing		1.29	3.96	5.25
6100		2"x4" @ 24"O.C.	1/2" gypsum	same		2.26	6.70	8.96
6110				nothing		1.34	3.77	5.11
6120			3.4lb. ribbed	same		2.01	6.80	8.81
6130				nothing		1.21	3.80	5.01
7500	3 coat gypsum	2"x4" @ 16"O.C.	3/8" gypsum	same		2.88	8.80	11.68
7510	W/med Keenes			nothing		1.70	4.90	6.60
7520			3.4lb. diamond	same		2.78	8.85	11.63
7530				nothing		1.64	4.93	6.57
7540			2.75lb. ribbed	same		2.61	8.90	11.51
7550				nothing		1.56	4.95	6.51
7600		2"x4" @ 24"O.C.	1/2" gypsum	same		2.79	8.65	11.44
7610				nothing		1.61	4.76	6.37
7620			3.4lb. ribbed	same		2.87	8.85	11.72
7630				nothing		1.65	4.83	6.48
8000	3 coat gypsum	2"x4" @ 16"O.C.	3/8" gypsum	same		2.89	9.75	12.64
8010	W/hard Keenes			nothing		1.70	5.40	7.10
8020			3.4lb. diamond	same		2.79	9.80	12.59
8030				nothing		1.64	5.45	7.09
8040			2.75lb. ribbed	same		2.62	9.85	12.47
8050				nothing		1.56	5.45	7.01
8100		2"x4" @ 24"O.C.	1/2" gypsum	same		2.80	9.65	12.45
8110				nothing		1.61	5.25	6.86
8120			3.4lb. ribbed	same		2.88	9.80	12.68
8130				nothing		1.65	5.35	7

C1010 Partitions

C1010 144	Plaster Partition Components	COST PER S.F.		
		MAT.	INST.	TOTAL
0060	Metal studs, 16" O.C., including track, non load bearing, 25 gage, 1-5/8"	.38	.87	1.25
0080	2-1/2"	.38	.87	1.25
0100	3-1/4"	.45	.89	1.34
0120	3-5/8"	.45	.89	1.34
0140	4"	.58	.90	1.48
0160	6"	.68	.91	1.59
0180	Load bearing, 20 gage, 2-1/2"	.99	1.11	2.10
0200	3-5/8"	1.17	1.13	2.30
0220	4"	1.23	1.16	2.39
0240	6"	1.56	1.17	2.73
0260	16 gage 2-1/2"	1.15	1.26	2.41
0280	3-5/8"	1.39	1.30	2.69
0300	4"	1.46.	1.32	2.78
0320	6"	1.83	1.34	3.17
0340	Wood studs, including blocking, shoe and double plate, 2"x4" , 12" O.C.	.48	1.07	1.55
0360	16" O.C.	.39	.86	1.25
0380	24" O.C.	.30	.69	.99
0400	2"x6" , 12" O.C.	.77	1.22	1.99
0420	16" O.C.	.62	.95	1.57
0440	24" O.C.	.47	.74	1.21
0460	Furring one face only, steel channels, 3/4" , 12" O.C.	.24	1.59	1.83
0480	16" O.C.	.22	1.41	1.63
0500	24" O.C.	.15	1.07	1.22
0520	1-1/2" , 12" O.C.	.35	1.78	2.13
0540	16" O.C.	.32	1.55	1.87
0560	24"O.C.	.21	1.22	1.43
0580	Wood strips 1"x3", on wood., 12" O.C.	.31	.78	1.09
0600	16"O.C.	.23	.59	.82
0620	On masonry, 12" O.C.	.31	.86	1.17
0640	16" O.C.	.23	.65	.88
0660	On concrete, 12" O.C.	.31	1.64	1.95
0680	16" O.C.	.23	1.23	1.46
0700	Gypsum lath. plain or perforated, nailed to studs, 3/8" thick	.50	.49	.99
0720	1/2" thick	.50	.52	1.02
0740	Clipped to studs, 3/8" thick	.50	.55	1.05
0760	1/2" thick	.50	.59	1.09
0780	Metal lath, diamond painted, nailed to wood studs, 2.5 lb.	.35	.49	.84
0800	3.4 lb.	.44	.52	.96
0820	Screwed to steel studs, 2.5 lb.	.35	.52	.87
0840	3.4 lb.	.42	.55	.97
0860	Rib painted, wired to steel, 2.75 lb	.37	.55	.92
0880	3.4 lb	.54	.59	1.13
0900	4.0 lb	.55	.64	1.19
0910				
0920	Gypsum plaster, 2 coats	.38	2.03	2.41
0940	3 coats	.55	2.42	2.97
0960	Perlite or vermiculite plaster, 2 coats	.40	2.32	2.72
0980	3 coats	.69	2.89	3.58
1000	Stucco, 3 coats, 1" thick, on wood framing	.65	4.03	4.68
1020	On masonry	.25	3.19	3.44
1100	Metal base galvanized and painted 2-1/2" high	.66	1.55	2.21

C1010 Partitions

Folding Accordion

Folding Leaf

Movable and Borrow Lites

C1010 205	Partitions	COST PER S.F.		
		MAT.	INST.	TOTAL
0360	Folding accordion, vinyl covered, acoustical, 3 lb. S.F., 17 ft max. hgt	25.50	8.55	34.05
0380	5 lb. per S.F. 27 ft max height	35	9	44
0400	5.5 lb. per S.F., 17 ft. max height	39	9.50	48.50
0420	Commercial, 1.75 lb per S.F., 8 ft. max height	18.20	3.79	21.99
0440	2.0 Lb per S.F., 17 ft. max height	18.80	5.70	24.50
0460	Industrial, 4.0 lb. per S.F. 27 ft. max height	32.50	11.40	43.90
0480	Vinyl clad wood or steel, electric operation 6 psf	45.50	5.35	50.85
0500	Wood, non acoustic, birch or mahogany	23.50	2.85	26.35
0560	Folding leaf, aluminum framed acoustical 12 ft.high.,5.5 lb per S.F.,min.	33	14.25	47.25
0580	Maximum	40	28.50	68.50
0600	6.5 lb. per S.F., minimum	35	14.25	49.25
0620	Maximum	43	28.50	71.50
0640	Steel acoustical, 7.5 per S.F., vinyl faced, minimum	49	14.25	63.25
0660	Maximum	60	28.50	88.50
0680	Wood acoustic type, vinyl faced to 18' high 6 psf, minimum	46	14.25	60.25
0700	Average	55	18.95	73.95
0720	Maximum	71	28.50	99.50
0740	Formica or hardwood faced, minimum	47.50	14.25	61.75
0760	Maximum	50.50	28.50	79
0780	Wood, low acoustical type to 12 ft. high 4.5 psf	34.50	17.05	51.55
0840	Demountable, trackless wall, cork finish, semi acous, 1-5/8"th, min	34	2.63	36.63
0860	Maximum	32	4.49	36.49
0880	Acoustic, 2" thick, minimum	27.50	2.80	30.30
0900	Maximum	47.50	3.79	51.29
0920	In-plant modular office system, w/prehung steel door			
0940	3" thick honeycomb core panels			
0960	12' x 12', 2 wall	9.60	.53	10.13
0970	4 wall	8.90	.70	9.60
0980	16' x 16', 2 wall	10	.37	10.37
0990	4 wall	6.40	.37	6.77
1000	Gypsum, demountable, 3" to 3-3/4" thick x 9' high, vinyl clad	5.35	1.98	7.33
1020	Fabric clad	13.20	2.15	15.35
1040	1.75 system, vinyl clad hardboard, paper honeycomb core panel			
1060	1-3/4" to 2-1/2" thick x 9' high	8.95	1.98	10.93
1080	Unitized gypsum panel system, 2" to 2-1/2" thick x 9' high			
1100	Vinyl clad gypsum	11.45	1.98	13.43
1120	Fabric clad gypsum	18.85	2.15	21
1140	Movable steel walls, modular system			
1160	Unitized panels, 48" wide x 9' high			
1180	Baked enamel, pre-finished	13	1.58	14.58
1200	Fabric clad	18.65	1.69	20.34

C INTERIORS

C1020   Interior Doors

Single Leaf   Sliding Entrance   Rolling Overhead

C1020 102	Special Doors	COST PER OPNG.		
		MAT.	INST.	TOTAL
2500	Single leaf, wood, 3'-0"x7'-0"x1 3/8", birch, solid core	320	160	480
2510	Hollow core	267	153	420
2530	Hollow core, lauan	250	153	403
2540	Louvered pine	370	153	523
2550	Paneled pine	380	153	533
2600	Hollow metal, comm. quality, flush, 3'-0"x7'-0"x1-3/8"	440	168	608
2650	3'-0"x10'-0" openings with panel	720	217	937
2700	Metal fire, comm. quality, 3'-0"x7'-0"x1-3/8"	540	175	715
2800	Kalamein fire, comm. quality, 3'-0"x7'-0"x1-3/4"	415	320	735
3200	Double leaf, wood, hollow core, 2 - 3'-0"x7'-0"x1-3/8"	395	292	687
3300	Hollow metal, comm. quality, B label, 2'-3'-0"x7'-0"x1-3/8"	1,025	370	1,395
3400	6'-0"x10'-0" opening, with panel	1,375	455	1,830
3500	Double swing door system, 12'-0"x7'-0", mill finish	6,325	2,375	8,700
3700	Black finish	6,900	2,450	9,350
3800	Sliding entrance door and system mill finish	7,425	2,025	9,450
3900	Bronze finish	8,125	2,200	10,325
4000	Black finish	8,475	2,275	10,750
4100	Sliding panel mall front, 16'x9' opening, mill finish	2,550	615	3,165
4200	Bronze finish	3,325	800	4,125
4300	Black finish	4,075	985	5,060
4400	24'x9' opening mill finish	3,700	1,150	4,850
4500	Bronze finish	4,800	1,500	6,300
4600	Black finish	5,925	1,850	7,775
4700	48'x9' opening mill finish	6,875	885	7,760
4800	Bronze finish	8,950	1,150	10,100
4900	Black finish	11,000	1,425	12,425
5000	Rolling overhead steel door, manual, 8' x 8' high	850	685	1,535
5100	10' x 10' high	1,125	780	1,905
5200	20' x 10' high	2,300	1,100	3,400
5300	12' x 12' high	1,450	910	2,360
5400	Motor operated, 8' x 8' high	1,650	905	2,555
5500	10' x 10' high	1,925	1,000	2,925
5600	20' x 10' high	3,100	1,325	4,425
5700	12' x 12' high	2,250	1,125	3,375
5800	Roll up grille, aluminum, manual, 10' x 10' high, mill finish	2,400	1,325	3,725
5900	Bronze anodized	3,850	1,325	5,175
6000	Motor operated, 10' x 10' high, mill finish	3,500	1,550	5,050
6100	Bronze anodized	4,950	1,550	6,500
6200	Steel, manual, 10' x 10' high	2,150	1,100	3,250
6300	15' x 8' high	2,550	1,375	3,925
6400	Motor operated, 10' x 10' high	3,250	1,325	4,575
6500	15' x 8' high	3,650	1,600	5,250

Important: See the Reference Section for critical supporting data - Reference Numbers and City Cost Indexes

INTERIORS C

C1030 Fittings

Toilet Units   **Entrance Screens**

Urinal Screens

C1030 110	Toilet Partitions	COST PER UNIT		
		MAT.	INST.	TOTAL
0380	Toilet partitions, cubicles, ceiling hung, marble	1,625	405	2,030
0400	Painted metal	500	213	713
0420	Plastic laminate	650	213	863
0460	Stainless steel	1,300	213	1,513
0480	Handicap addition	355		355
0520	Floor and ceiling anchored, marble	1,775	325	2,100
0540	Painted metal	460	171	631
0560	Plastic laminate	730	171	901
0600	Stainless steel	1,500	171	1,671
0620	Handicap addition	315		315
0660	Floor mounted marble	1,050	270	1,320
0680	Painted metal	515	122	637
0700	Plastic laminate	640	122	762
0740	Stainless steel	1,575	122	1,697
0760	Handicap addition	315		315
0780	Juvenile deduction	43.50		43.50
0820	Floor mounted with handrail marble	995	270	1,265
0840	Painted metal	460	142	602
0860	Plastic laminate	850	142	992
0900	Stainless steel	1,475	142	1,617
0920	Handicap addition	315		315
0960	Wall hung, painted metal	580	122	702
1020	Stainless steel	1,450	122	1,572
1040	Handicap addition	315		315
1080	Entrance screens, floor mounted, 54″ high, marble	655	90	745
1100	Painted metal	207	57	264
1140	Stainless steel	790	57	847
1300	Urinal screens, floor mounted, 24″ wide, laminated plastic	390	107	497
1320	Marble	575	126	701
1340	Painted metal	189	107	296
1380	Stainless steel	660	107	767
1428	Wall mounted wedge type, painted metal	250	85.50	335.50
1460	Stainless steel	540	85.50	625.50
1500	Partitions, shower stall, single wall, painted steel, 2′-8″ x 2′-8″	800	198	998
1510	Fiberglass, 2′-8″ x 2′-8″	650	220	870
1520	Double wall, enameled steel, 2′-8″ x 2′-8″	915	198	1,113
1530	Stainless steel, 2′-8″ x 2′-8″	1,475	198	1,673
1560	Tub enclosure, sliding panels, tempered glass, aluminum frame	340	247	587
1570	Chrome/brass frame-deluxe	970	330	1,300

C INTERIORS

C1030 Fittings

C1030 310	Storage Specialties, EACH	COST EACH		
		MAT.	INST.	TOTAL
0200	Lockers, steel, single tier, 5' to 6' high, per opening, min.	118	35.50	153.50
0210	Maximum	200	41	241
0600	Shelving, metal industrial, braced, 3' wide, 1' deep	21.50	9.35	30.85
0610	2' deep	32.50	9.90	42.40

C1030 510	Identifying/Visual Aid Specialties, EACH	COST EACH		
		MAT.	INST.	TOTAL
0100	Control boards, magnetic, porcelain finish, framed, 24" x 18"	208	107	315
0110	96" x 48"	1,100	171	1,271
0120	Directory boards, outdoor, black plastic, 36" x 24"	755	425	1,180
0130	36" x 36"	870	570	1,440
0140	Indoor, economy, open faced, 18" x 24"	104	122	226
0500	Signs, interior electric exit sign, wall mounted, 6"	109	60.50	169.50
0510	Street, reflective alum., dbl. face, 4 way, w/bracket	116	28.50	144.50
0520	Letters, cast aluminum, 1/2" deep, 4" high	25	23.50	48.50
0530	1" deep, 10" high	34	23.50	57.50
0540	Plaques, cast aluminum, 20" x 30"	895	213	1,108
0550	Cast bronze, 36" x 48"	3,525	425	3,950

C1030 520	Identifying/Visual Aid Specialties, S.F.	COST PER S.F.		
		MAT.	INST.	TOTAL
0100	Bulletin board, cork sheets, no frame, 1/4" thick	3.74	2.94	6.68
0120	Aluminum frame, 1/4" thick, 3' x 5'	8.40	3.58	11.98
0200	Chalkboards, wall hung, alum, frame & chalktrough	11.30	1.89	13.19
0210	Wood frame & chalktrough	10.55	2.03	12.58
0220	Sliding board, one board with back panel	37.50	1.82	39.32
0230	Two boards with back panel	56.50	1.82	58.32
0240	Liquid chalk type, alum. frame & chalktrough	11.30	1.89	13.19
0250	Wood frame & chalktrough	21	1.89	22.89

C1030 710	Bath and Toilet Accessories, EACH	COST EACH		
		MAT.	INST.	TOTAL
0100	Specialties, bathroom accessories, st. steel, curtain rod, 5' long, 1" diam	33.50	33	66.50
0120	Dispenser, towel, surface mounted	45.50	26.50	72
0140	Grab bar, 1-1/4" diam., 12" long	20.50	17.80	38.30
0160	Mirror, framed with shelf, 18" x 24"	125	21.50	146.50
0170	72" x 24"	315	71	386
0180	Toilet tissue dispenser, surface mounted, single roll	11.90	14.25	26.15
0200	Towel bar, 18" long	38.50	18.55	57.05
0300	Medicine cabinets, sliding mirror doors, 20" x 16" x 4-3/4", unlighted	95	61	156
0310	24" x 19" x 8-1/2", lighted	150	85.50	235.50

C1030 730	Bath and Toilet Accessories, L.F.	COST PER L.F.		
		MAT.	INST.	TOTAL
0100	Partitions, hospital curtain, ceiling hung, polyester oxford cloth	15.90	4.16	20.06
0110	Designer oxford cloth	30.50	5.25	35.75

C1030 830	Fabricated Cabinets & Counters, L.F.	COST PER L.F.		
		MAT.	INST.	TOTAL
0110	Household, base, hardwood, one top drawer & one door below x 12" wide	146	34.50	180.50
0120	Four drawer x 24" wide	320	38.50	358.50
0130	Wall, hardwood, 30" high with one door x 12" wide	140	39	179
0140	Two doors x 48" wide	330	46.50	376.50
0150	Counter top-laminated plastic, stock, economy	9.65	14.25	23.90
0160	Custom-square edge, 7/8" thick	12.20	32	44.20

Important: See the Reference Section for critical supporting data - Reference Numbers and City Cost Indexes

C1030 Fittings

C1030 830	Fabricated Cabinets & Counters, L.F.	COST PER L.F.		
		MAT.	INST.	TOTAL
0170	School, counter, wood, 32" high	193	42.50	235.50
0180	Metal, 84" high	298	57	355

C1030 910	Other Fittings, EACH	COST EACH		
		MAT.	INST.	TOTAL
0500	Mail boxes, horizontal, rear loaded, aluminum, 5" x 6" x 15" deep	34.50	12.55	47.05
0510	Front loaded, aluminum, 10" x 12" x 15" deep	128	21.50	149.50
0520	Vertical, front loaded, aluminum, 15" x 5" x 6" deep	29	12.55	41.55
0530	Bronze, duranodic finish	48.50	12.55	61.05
0540	Letter slot, post office	242	53.50	295.50
0550	Mail counter, window, post office, with grille	565	213	778
0700	Turnstiles, one way, 4' arm, 46" diam., manual	295	171	466
0710	Electric	990	710	1,700
0720	3 arm, 5'-5' diam. & 7' high, manual	2,575	855	3,430
0730	Electric	3,325	1,425	4,750

The table below lists the cost per flight for 4'-0" wide stairs. Side walls are not included. Railings are included in the prices.

C2010 110	Stairs	COST PER FLIGHT		
		MAT.	INST.	TOTAL
0470	Stairs, C.I.P. concrete, w/o landing, 12 risers, w/o nosing	1,275	1,800	3,075
0480	With nosing	1,825	2,050	3,875
0550	W/landing, 12 risers, w/o nosing	1,400	2,200	3,600
0560	With nosing	1,925	2,450	4,375
0570	16 risers, w/o nosing	1,750	2,775	4,525
0580	With nosing	2,475	3,125	5,600
0590	20 risers, w/o nosing	2,125	3,325	5,450
0600	With nosing	3,025	3,775	6,800
0610	24 risers, w/o nosing	2,500	3,900	6,400
0620	With nosing	3,575	4,400	7,975
0630	Steel, grate type w/nosing & rails, 12 risers, w/o landing	2,200	925	3,125
0640	With landing	2,875	1,275	4,150
0660	16 risers, with landing	3,600	1,575	5,175
0680	20 risers, with landing	4,325	1,875	6,200
0700	24 risers, with landing	5,075	2,175	7,250
0701				
0710	Cement fill metal pan & picket rail, 12 risers, w/o landing	2,875	925	3,800
0720	With landing	3,775	1,375	5,150
0740	16 risers, with landing	4,725	1,700	6,425
0760	20 risers, with landing	5,700	2,000	7,700
0780	24 risers, with landing	6,650	2,325	8,975
0790	Cast iron tread & pipe rail, 12 risers, w/o landing	2,875	925	3,800
0800	With landing	3,775	1,375	5,150
1120	Wood, prefab box type, oak treads, wood rails 3'-6" wide, 14 risers	1,750	355	2,105
1150	Prefab basement type, oak treads, wood rails 3'-0" wide, 14 risers	845	87.50	932.50

C3010 Wall Finishes

C3010 230	Paint & Covering	COST PER S.F.		
		MAT.	INST.	TOTAL
0060	Painting, interior on plaster and drywall, brushwork, primer & 1 coat	.10	.57	.67
0080	Primer & 2 coats	.15	.75	.90
0100	Primer & 3 coats	.21	.92	1.13
0120	Walls & ceilings, roller work, primer & 1 coat	.10	.38	.48
0140	Primer & 2 coats	.15	.48	.63
0160	Woodwork incl. puttying, brushwork, primer & 1 coat	.10	.82	.92
0180	Primer & 2 coats	.15	1.08	1.23
0200	Primer & 3 coats	.21	1.47	1.68
0260	Cabinets and casework, enamel, primer & 1 coat	.10	.92	1.02
0280	Primer & 2 coats	.15	1.13	1.28
0300	Masonry or concrete, latex, brushwork, primer & 1 coat	.19	.77	.96
0320	Primer & 2 coats	.25	1.10	1.35
0340	Addition for block filler	.12	.94	1.06
0380	Fireproof paints, intumescent, 1/8" thick 3/4 hour	1.87	.75	2.62
0400	3/16" thick 1 hour	4.18	1.13	5.31
0420	7/16" thick 2 hour	5.40	2.63	8.03
0440	1-1/16" thick 3 hour	8.80	5.25	14.05
0500	Gratings, primer & 1 coat	.20	1.15	1.35
0600	Pipes over 12" diameter	.50	3.68	4.18
0700	Structural steel, brushwork, light framing 300-500 S.F./Ton	.06	1.43	1.49
0720	Heavy framing 50-100 S.F./Ton	.06	.71	.77
0740	Spraywork, light framing 300-500 S.F./Ton	.07	.32	.39
0760	Heavy framing 50-100 S.F./Ton	.07	.35	.42
0800	Varnish, interior wood trim, no sanding sealer & 1 coat	.08	.92	1
0820	Hardwood floor, no sanding 2 coats	.14	.19	.33
0840	Wall coatings, acrylic glazed coatings, minimum	.29	.70	.99
0860	Maximum	.59	1.21	1.80
0880	Epoxy coatings, minimum	.36	.70	1.06
0900	Maximum	1.12	2.16	3.28
0940	Exposed epoxy aggregate, troweled on, 1/16" to 1/4" aggregate, minimum	.55	1.57	2.12
0960	Maximum	1.20	2.83	4.03
0980	1/2" to 5/8" aggregate, minimum	1.11	2.83	3.94
1000	Maximum	1.89	4.60	6.49
1020	1" aggregate, minimum	1.93	4.09	6.02
1040	Maximum	2.95	6.70	9.65
1060	Sprayed on, minimum	.52	1.25	1.77
1080	Maximum	.96	2.54	3.50
1100	High build epoxy 50 mil, minimum	.62	.94	1.56
1120	Maximum	1.06	3.87	4.93
1140	Laminated epoxy with fiberglass minimum	.67	1.25	1.92
1160	Maximum	1.19	2.54	3.73
1180	Sprayed perlite or vermiculite 1/16" thick, minimum	.24	.13	.37
1200	Maximum	.68	.58	1.26
1260	Wall coatings, vinyl plastic, minimum	.31	.50	.81
1280	Maximum	.75	1.53	2.28
1300	Urethane on smooth surface, 2 coats, minimum	.23	.32	.55
1320	Maximum	.51	.55	1.06
1340	3 coats, minimum	.31	.44	.75
1360	Maximum	.68	.78	1.46
1380	Ceramic-like glazed coating, cementitious, minimum	.44	.84	1.28
1420	Resin base, minimum	.31	.58	.89
1440	Maximum	.50	1.12	1.62
1460	Wall coverings, aluminum foil	.96	1.33	2.29
1500	Vinyl backing	5.15	1.52	6.67
1520	Cork tiles, 12"x12", light or dark, 3/16" thick	2.81	1.52	4.33
1540	5/16" thick	3.41	1.55	4.96
1560	Basketweave, 1/4" thick	5.25	1.52	6.77
1580	Natural, non-directional, 1/2" thick	5.90	1.52	7.42

INTERIORS

C

C3010 Wall Finishes

C3010 230	Paint & Covering	COST PER S.F.		
		MAT.	INST.	TOTAL
1600	12"x36", granular, 3/16" thick	1.13	.95	2.08
1620	1" thick	1.46	.98	2.44
1640	12"x12", polyurethane coated, 3/16" thick	3.53	1.52	5.05
1660	5/16" thick	5.05	1.55	6.60
1661	Paneling, prefinished plywood, birch	1.33	2.03	3.36
1662	Mahogany, African	2.48	2.13	4.61
1663	Philippine (lauan)	1.07	1.71	2.78
1664	Oak or cherry	3.19	2.13	5.32
1665	Rosewood	4.52	2.67	7.19
1666	Teak	3.19	2.13	5.32
1667	Chestnut	4.71	2.28	6.99
1668	Pecan	2.04	2.13	4.17
1669	Walnut	5.15	2.13	7.28
1670	Wood board, knotty pine, finished	1.79	3.55	5.34
1671	Rough sawn cedar	2.20	3.55	5.75
1672	Redwood	4.76	3.55	8.31
1673	Aromatic cedar	3.73	3.80	7.53
1680	Cork wallpaper, paper backed, natural	2.01	.76	2.77
1700	Color	2.49	.76	3.25
1720	Gypsum based, fabric backed, minimum	.75	.46	1.21
1740	Average	1.10	.51	1.61
1760	Maximum	1.22	.57	1.79
1780	Vinyl wall covering, fabric back, light weight	.64	.57	1.21
1800	Medium weight	.79	.76	1.55
1820	Heavy weight	1.28	.84	2.12
1840	Wall paper, double roll, solid pattern, avg. workmanship	.32	.57	.89
1860	Basic pattern, avg. workmanship	.72	.68	1.40
1880	Basic pattern, quality workmanship	1.68	.84	2.52
1900	Grass cloths with lining paper, minimum	.69	.91	1.60
1920	Maximum	2.23	1.04	3.27
1940	Ceramic tile, thin set, 4-1/4" x 4-1/4"	2.30	3.59	5.89
1942				

C3010 235	Paint Trim	COST PER L.F.		
		MAT.	INST.	TOTAL
2040	Painting, wood trim, to 6" wide, enamel, primer & 1 coat	.10	.46	.56
2060	Primer & 2 coats	.15	.58	.73
2080	Misc. metal brushwork, ladders	.40	4.60	5
2100	Pipes, to 4" dia.	.07	.97	1.04
2120	6" to 8" dia.	.13	1.94	2.07
2140	10" to 12" dia.	.39	2.90	3.29
2160	Railings, 2" pipe	.14	2.30	2.44
2180	Handrail, single	.10	.92	1.02

INTERIORS C

C3020 Floor Finishes

C3020 410	Tile & Covering	COST PER S.F.		
		MAT.	INST.	TOTAL
0060	Carpet tile, nylon, fusion bonded, 18" x 18" or 24" x 24", 24 oz.	3.72	.53	4.25
0080	35 oz.	3.94	.53	4.47
0100	42 oz.	5.15	.53	5.68
0140	Carpet, tufted, nylon, roll goods, 12' wide, 26 oz.	3.66	.61	4.27
0160	36 oz.	5.85	.61	6.46
0180	Woven, wool, 36 oz.	9.20	.61	9.81
0200	42 oz.	9.40	.61	10.01
0220	Padding, add to above, minimum	.48	.28	.76
0240	Maximum	.83	.28	1.11
0260	Composition flooring, acrylic, 1/4" thick	1.40	4.08	5.48
0280	3/8" thick	1.79	4.72	6.51
0300	Epoxy, minimum	2.41	3.15	5.56
0320	Maximum	2.93	4.34	7.27
0340	Epoxy terrazzo, minimum	5.85	4.60	10.45
0360	Maximum	9	6.15	15.15
0380	Mastic, hot laid, 1-1/2" thick, minimum	3.53	3.08	6.61
0400	Maximum	4.53	4.08	8.61
0420	Neoprene 1/4" thick, minimum	3.48	3.90	7.38
0440	Maximum	4.79	4.94	9.73
0460	Polyacrylate with ground granite 1/4", minimum	3.28	2.89	6.17
0480	Maximum	6.35	4.43	10.78
0500	Polyester with colored quart 2 chips 1/16", minimum	3.18	1.99	5.17
0520	Maximum	4.92	3.15	8.07
0540	Polyurethane with vinyl chips, minimum	6.80	1.99	8.79
0560	Maximum	9.75	2.47	12.22
0600	Concrete topping, granolithic concrete, 1/2" thick	.20	3.31	3.51
0620	1" thick	.40	3.40	3.80
0640	2" thick	.79	3.91	4.70
0660	Heavy duty 3/4" thick, minimum	2.62	5.15	7.77
0680	Maximum	2.62	6.10	8.72
0700	For colors, add to above, minimum	.37	1.18	1.55
0720	Maximum	.75	1.31	2.06
0740	Exposed aggregate finish, minimum	4.49	.62	5.11
0760	Maximum	14.20	.83	15.03
0780	Abrasives, .25 P.S.F. add to above, minimum	.28	.46	.74
0800	Maximum	.39	.46	.85
0820	Dust on coloring, add, minimum	.69	.30	.99
0840	Maximum	2.43	.62	3.05
0860	1/2" integral, minimum	.23	3.31	3.54
0880	Maximum	.23	3.31	3.54
0900	Dustproofing, add, minimum	.09	.20	.29
0920	Maximum	.16	.30	.46
0930	Paint	.25	1.10	1.35
0940	Hardeners, metallic add, minimum	.33	.46	.79
0960	Maximum	.99	.67	1.66
0980	Non-metallic, minimum	.20	.46	.66
1000	Maximum	.61	.67	1.28
1020	Integral topping, 3/16" thick	.07	1.96	2.03
1040	1/2" thick	.18	2.06	2.24
1060	3/4" thick	.28	2.30	2.58
1080	1" thick	.37	2.60	2.97
1100	Terrazzo, minimum	2.74	7.05	9.79
1120	Maximum	5.10	15.35	20.45
1280	Resilient, asphalt tile, 1/8" thick on concrete, minimum	1.13	.96	2.09
1300	Maximum	1.24	.96	2.20
1320	On wood, add for felt underlay	.21		.21
1340	Cork tile, minimum	4.13	1.22	5.35
1360	Maximum	9.85	1.22	11.07

C3020 Floor Finishes

C3020 410	Tile & Covering	COST PER S.F.		
		MAT.	INST.	TOTAL
1380	Polyethylene, in rolls, minimum	2.83	1.40	4.23
1400	Maximum	5.50	1.40	6.90
1420	Polyurethane, thermoset, minimum	4.35	3.84	8.19
1440	Maximum	5.15	7.70	12.85
1460	Rubber, sheet goods, minimum	3.89	3.20	7.09
1480	Maximum	6.40	4.27	10.67
1500	Tile, minimum	4.46	.96	5.42
1520	Maximum	7.45	1.40	8.85
1580	Vinyl, composition tile, minimum	.87	.77	1.64
1600	Maximum	2.21	.77	2.98
1620	Tile, minimum	1.94	.77	2.71
1640	Maximum	5.30	.77	6.07
1660	Sheet goods, minimum	2.28	1.54	3.82
1680	Maximum	4.38	1.92	6.30
1720	Tile, ceramic natural clay	4.38	3.73	8.11
1730	Marble, synthetic 12"x12"x5/8"	7.40	11.35	18.75
1740	Porcelain type, minimum	4.70	3.73	8.43
1760	Maximum	6.50	4.40	10.90
1800	Quarry tile, mud set, minimum	3.29	4.87	8.16
1820	Maximum	5.75	6.20	11.95
1840	Thin set, deduct		.98	.98
1860	Terrazzo precast, minimum	4.09	5.15	9.24
1880	Maximum	8.85	5.15	14
1900	Non-slip, minimum	17.45	26	43.45
1920	Maximum	19.25	36	55.25
2020	Wood block, end grain factory type, natural finish, 2" thick	3.91	3.41	7.32
2040	Fir, vertical grain, 1"x4", no finish, minimum	2.52	1.67	4.19
2060	Maximum	2.68	1.67	4.35
2080	Prefinished white oak, prime grade, 2-1/4" wide	6.80	2.51	9.31
2100	3-1/4" wide	8.75	2.31	11.06
2120	Maple strip, sanded and finished, minimum	4.01	3.64	7.65
2140	Maximum	4.74	3.64	8.38
2160	Oak strip, sanded and finished, minimum	4.06	3.64	7.70
2180	Maximum	4.85	3.64	8.49
2200	Parquetry, sanded and finished, minimum	4.29	3.80	8.09
2220	Maximum	6.60	5.40	12
2260	Add for sleepers on concrete, treated, 24" O.C., 1"x2"	.66	2.20	2.86
2280	1"x3"	.83	1.70	2.53
2300	2"x4"	.81	.86	1.67
2340	Underlayment, plywood, 3/8" thick	.96	.57	1.53
2350	1/2" thick	1.17	.59	1.76
2360	5/8" thick	1.34	.61	1.95
2370	3/4" thick	1.54	.66	2.20
2380	Particle board, 3/8" thick	.41	.57	.98
2390	1/2" thick	.43	.59	1.02
2400	5/8" thick	.54	.61	1.15
2410	3/4" thick	.61	.66	1.27
2420	Hardboard, 4' x 4', .215" thick	.43	.57	1

INTERIORS C

C3030 Ceiling Finishes

2 Coats of Plaster on Gypsum Lath on Wood Furring

Fiberglass Board on Exposed Suspended Grid System

Plaster and Metal Lath on Metal Furring

C3030 105 — Plaster Ceilings

	TYPE	LATH	FURRING	SUPPORT		COST PER S.F. MAT.	INST.	TOTAL
2400	2 coat gypsum	3/8" gypsum	1"x3" wood, 16" O.C.	wood		1.21	4.30	5.51
2500	Painted			masonry		1.21	4.38	5.59
2600				concrete		1.21	4.90	6.11
2700	3 coat gypsum	3.4# metal	1"x3" wood, 16" O.C.	wood		1.32	4.60	5.92
2800	Painted			masonry		1.32	4.68	6
2900				concrete		1.32	5.20	6.52
3000	2 coat perlite	3/8" gypsum	1"x3" wood, 16" O.C.	wood		1.23	4.50	5.73
3100	Painted			masonry		1.23	4.58	5.81
3200				concrete		1.23	5.10	6.33
3300	3 coat perlite	3.4# metal	1"x3" wood, 16" O.C.	wood		1.17	4.60	5.77
3400	Painted			masonry		1.17	4.68	5.85
3500				concrete		1.17	5.20	6.37
3600	2 coat gypsum	3/8" gypsum	3/4" CRC, 12" O.C.	1-1/2" CRC, 48"O.C.		1.22	5.15	6.37
3700	Painted		3/4" CRC, 16" O.C.	1-1/2" CRC, 48"O.C.		1.20	4.67	5.87
3800			3/4" CRC, 24" O.C.	1-1/2" CRC, 48"O.C.		1.13	4.27	5.40
3900	2 coat perlite	3/8" gypsum	3/4" CRC, 12" O.C.	1-1/2" CRC, 48"O.C		1.24	5.55	6.79
4000	Painted		3/4" CRC, 16" O.C.	1-1/2" CRC, 48"O.C.		1.22	5.05	6.27
4100			3/4" CRC, 24" O.C.	1-1/2" CRC, 48"O.C.		1.15	4.66	5.81
4200	3 coat gypsum	3.4# metal	3/4" CRC, 12" O.C.	1-1/2" CRC, 36" O.C.		1.58	6.75	8.33
4300	Painted		3/4" CRC, 16" O.C.	1-1/2" CRC, 36" O.C.		1.56	6.25	7.81
4400			3/4" CRC, 24" O.C.	1-1/2" CRC, 36" O.C.		1.49	5.85	7.34
4500	3 coat perlite	3.4# metal	3/4" CRC, 12" O.C.	1-1/2" CRC,36" O.C.		1.72	7.40	9.12
4600	Painted		3/4" CRC, 16" O.C.	1-1/2" CRC, 36" O.C.		1.70	6.90	8.60
4700			3/4" CRC, 24" O.C.	1-1/2" CRC, 36" O.C.		1.63	6.50	8.13

C3030 110 — Drywall Ceilings

	TYPE	FINISH	FURRING	SUPPORT		COST PER S.F. MAT.	INST.	TOTAL
4800	1/2" F.R. drywall	painted and textured	1"x3" wood, 16" O.C.	wood		.65	2.65	3.30
4900				masonry		.65	2.73	3.38
5000				concrete		.65	3.25	3.90
5100	5/8" F.R. drywall	painted and textured	1"x3" wood, 16" O.C.	wood		.68	2.65	3.33
5200				masonry		.68	2.73	3.41
5300				concrete		.68	3.25	3.93
5400	1/2" F.R. drywall	painted and textured	7/8"resil. channels	24" O.C.		.55	2.57	3.12

C INTERIORS

C3030 Ceiling Finishes

C3030 110	Drywall Ceilings					COST PER S.F.		
	TYPE	FINISH	FURRING	SUPPORT		MAT.	INST.	TOTAL
5500	1/2" F.R. drywall		1"x2" wood	stud clips		.69	2.48	3.17
5602		painted	1-5/8" metal studs	24" O.C.		.67	2.29	2.96
5700	5/8" F.R. drywall	painted and textured	1-5/8"metal studs	24" O.C.		.70	2.29	2.99
5702								

C3030 140	Plaster Ceiling Components	COST PER S.F.		
		MAT.	INST.	TOTAL
0060	Plaster, gypsum incl. finish, 2 coats			
0080	3 coats	.55	2.71	3.26
0100	Perlite, incl. finish, 2 coats	.40	2.71	3.11
0120	3 coats	.69	3.36	4.05
0140	Thin coat on drywall	.08	.53	.61
0200	Lath, gypsum, 3/8" thick	.50	.68	1.18
0220	1/2" thick	.50	.71	1.21
0240	5/8" thick	.55	.83	1.38
0260	Metal, diamond, 2.5 lb.	.35	.55	.90
0280	3.4 lb.	.44	.59	1.03
0300	Flat rib, 2.75 lb.	.37	.55	.92
0320	3.4 lb.	.54	.59	1.13
0440	Furring, steel channels, 3/4" galvanized , 12" O.C.	.24	1.78	2.02
0460	16" O.C.	.22	1.29	1.51
0480	24" O.C.	.15	.89	1.04
0500	1-1/2" galvanized , 12" O.C.	.35	1.96	2.31
0520	16" O.C.	.32	1.43	1.75
0540	24" O.C.	.21	.96	1.17
0560	Wood strips, 1"x3", on wood, 12" O.C.	.31	1.22	1.53
0580	16" O.C.	.23	.92	1.15
0600	24" O.C.	.16	.61	.77
0620	On masonry, 12" O.C.	.31	1.33	1.64
0640	16" O.C.	.23	1	1.23
0660	24" O.C.	.16	.67	.83
0680	On concrete, 12" O.C.	.31	2.03	2.34
0700	16" O.C.	.23	1.52	1.75
0720	24" O.C.	.16	1.02	1.18
0940	Paint on plaster or drywall, roller work, primer + 1 coat	.10	.38	.48
0960	Primer + 2 coats	.15	.48	.63

C3030 210	Acoustical Ceilings					COST PER S.F.		
	TYPE	TILE	GRID	SUPPORT		MAT.	INST.	TOTAL
5800	5/8" fiberglass board	24" x 48"	tee	suspended		1.11	1.22	2.33
5900		24" x 24"	tee	suspended		1.24	1.35	2.59
6000	3/4" fiberglass board	24" x 48"	tee	suspended		1.80	1.25	3.05
6100		24" x 24"	tee	suspended		1.93	1.38	3.31
6500	5/8" mineral fiber	12" x 12"	1"x3" wood, 12" O.C.	wood		1.66	1.65	3.31
6600				masonry		1.66	1.76	3.42
6700				concrete		1.66	2.46	4.12
6800	3/4" mineral fiber	12" x 12"	1"x3" wood, 12" O.C.	wood		1.80	1.65	3.45
6900				masonry		1.80	1.65	3.45
7000				concrete		1.80	1.65	3.45
7100	3/4"mineral fiber on	12" x 12"	25 ga. channels	runners		2	2.23	4.23
7102	5/8" F.R. drywall							

Important: See the Reference Section for critical supporting data - Reference Numbers and City Cost Indexes

C3030 Ceiling Finishes

C3030 210		Acoustical Ceilings				COST PER S.F.		
	TYPE	TILE	GRID	SUPPORT		MAT.	INST.	TOTAL
7200 7201 7202	5/8" plastic coated Mineral fiber	12" x 12"		adhesive backed		1.71	.43	2.14
7300 7301 7302	3/4" plastic coated Mineral fiber	12" x 12"		adhesive backed		1.85	.43	2.28
7400 7401 7402	3/4" mineral fiber	12" x 12"	conceal 2" bar & channels	suspended		1.92	2.17	4.09

C3030 240	Acoustical Ceiling Components	COST PER S.F.		
		MAT.	INST.	TOTAL
2480	Ceiling boards, eggcrate, acrylic, 1/2" x 1/2" x 1/2" cubes	1.56	.85	2.41
2500	Polystyrene, 3/8" x 3/8" x 1/2" cubes	1.31	.84	2.15
2520	1/2" x 1/2" x 1/2" cubes	1.76	.85	2.61
2540	Fiberglass boards, plain, 5/8" thick	.56	.68	1.24
2560	3/4" thick	1.25	.71	1.96
2580	Grass cloth faced, 3/4" thick	1.78	.85	2.63
2600	1" thick	1.97	.88	2.85
2620	Luminous panels, prismatic, acrylic	1.90	1.07	2.97
2640	Polystyrene	.97	1.07	2.04
2660	Flat or ribbed, acrylic	3.31	1.07	4.38
2680	Polystyrene	2.27	1.07	3.34
2700	Drop pan, white, acrylic	4.85	1.07	5.92
2720	Polystyrene	4.06	1.07	5.13
2740	Mineral fiber boards, 5/8" thick, standard	.68	.63	1.31
2760	Plastic faced	1.06	1.07	2.13
2780	2 hour rating	.92	.63	1.55
2800	Perforated aluminum sheets, .024 thick, corrugated painted	1.94	.87	2.81
2820	Plain	3.38	.85	4.23
3080	Mineral fiber, plastic coated, 12" x 12" or 12" x 24", 5/8" thick	1.35	.43	1.78
3100	3/4" thick	1.49	.43	1.92
3120	Fire rated, 3/4" thick, plain faced	1.29	.43	1.72
3140	Mylar faced	1.41	.43	1.84
3160	Add for ceiling primer	.13		.13
3180	Add for ceiling cement	.36		.36
3240	Suspension system, furring, 1" x 3" wood 12" O.C.	.31	1.22	1.53
3260	T bar suspension system, 2' x 4' grid	.54	.53	1.07
3280	2' x 2' grid	.67	.66	1.33
3300	Concealed Z bar suspension system 12" module	.48	.82	1.30
3320	Add to above for 1-1/2" carrier channels 4' O.C.	.14	.91	1.05
3340	Add to above for carrier channels for recessed lighting	.25	.93	1.18

For information about Means Estimating Seminars, see yellow pages 12 and 13 in back of book

C INTERIORS

D SERVICES

D1010 Elevators and Lifts

Hydraulic

Traction Geared

Traction Gearless

D1010 110	Hydraulic	COST EACH		
		MAT.	INST.	TOTAL
1300	Pass. elev., 1500 lb., 2 Floors, 100 FPM	36,800	11,100	47,900
1400	5 Floors, 100 FPM	55,000	29,500	84,500
1600	2000 lb., 2 Floors, 100 FPM	37,900	11,100	49,000
1700	5 floors, 100 FPM	56,000	29,500	85,500
1900	2500 lb., 2 Floors, 100 FPM	38,900	11,100	50,000
2000	5 floors, 100 FPM	57,000	29,500	86,500
2200	3000 lb., 2 Floors, 100 FPM	40,700	11,100	51,800
2300	5 floors, 100 FPM	58,500	29,500	88,000
2500	3500 lb., 2 Floors, 100 FPM	42,600	11,100	53,700
2600	5 floors, 100 FPM	60,500	29,500	90,000
2800	4000 lb., 2 Floors, 100 FPM	43,700	11,100	54,800
2900	5 floors, 100 FPM	61,500	29,500	91,000
3100	4500 lb., 2 Floors, 100 FPM	45,200	11,100	56,300
3200	5 floors, 100 FPM	63,000	29,500	92,500
4000	Hospital elevators, 3500 lb., 2 Floors, 100 FPM	55,000	11,100	66,100
4100	5 floors, 100 FPM	81,500	29,500	111,000
4300	4000 lb., 2 Floors, 100 FPM	55,000	11,100	66,100
4400	5 floors, 100 FPM	81,500	29,500	111,000
4600	4500 lb., 2 Floors, 100 FPM	61,000	11,100	72,100
4800	5 floors, 100 FPM	88,000	29,500	117,500
4900	5000 lb., 2 Floors, 100 FPM	64,000	11,100	75,100
5000	5 floors, 100 FPM	90,500	29,500	120,000
6700	Freight elevators (Class "B"), 3000 lb., 2 Floors, 50 FPM	56,500	14,100	70,600
6800	5 floors, 100 FPM	87,500	37,000	124,500
7000	4000 lb., 2 Floors, 50 FPM	59,000	14,100	73,100
7100	5 floors, 100 FPM	90,000	37,000	127,000
7500	10,000 lb., 2 Floors, 50 FPM	73,500	14,100	87,600
7600	5 floors, 100 FPM	104,500	37,000	141,500
8100	20,000 lb., 2 Floors, 50 FPM	89,500	14,100	103,600
8200	5 Floors, 100 FPM	120,500	37,000	157,500

D1010 140	Traction Geared Elevators	COST EACH		
		MAT.	INST.	TOTAL
1300	Passenger, 2000 Lb., 5 floors, 200 FPM	82,500	27,800	110,300
1500	15 floors, 350 FPM	129,000	84,000	213,000
1600	2500 Lb., 5 floors, 200 FPM	85,500	27,800	113,300
1800	15 floors, 350 FPM	132,000	84,000	216,000
1900	3000 Lb., 5 floors, 200 FPM	87,000	27,800	114,800
2100	15 floors, 350 FPM	133,500	84,000	217,500
2200	3500 Lb., 5 floors, 200 FPM	89,000	27,800	116,800
2400	15 floors, 350 FPM	135,500	84,000	219,500
2500	4000 Lb., 5 floors, 200 FPM	89,000	27,800	116,800
2700	15 floors, 350 FPM	136,000	84,000	220,000

SERVICES D

D10 Conveying

D1010 Elevators and Lifts

D1010 140	Traction Geared Elevators	COST EACH		
		MAT.	INST.	TOTAL
2800	4500 Lb., 5 floors, 200 FPM	91,500	27,800	119,300
3000	15 floors, 350 FPM	138,000	84,000	222,000
3100	5000 Lb., 5 floors, 200 FPM	94,000	27,800	121,800
3300	15 floors, 350 FPM	140,500	84,000	224,500
4000	Hospital, 3500 Lb., 5 floors, 200 FPM	101,000	27,800	128,800
4200	15 floors, 350 FPM	174,000	84,000	258,000
4300	4000 Lb., 5 floors, 200 FPM	101,000	27,800	128,800
4500	15 floors, 350 FPM	174,000	84,000	258,000
4600	4500 Lb., 5 floors, 200 FPM	106,500	27,800	134,300
4800	15 floors, 350 FPM	179,500	84,000	263,500
4900	5000 Lb., 5 floors, 200 FPM	108,000	27,800	135,800
5100	15 floors, 350 FPM	181,000	84,000	265,000
6000	Freight, 4000 Lb., 5 floors, 50 FPM class 'B'	90,500	28,800	119,300
6200	15 floors, 200 FPM class 'B'	166,500	100,000	266,500
6300	8000 Lb., 5 floors, 50 FPM class 'B'	108,500	28,800	137,300
6500	15 floors, 200 FPM class 'B'	184,500	100,000	284,500
7000	10,000 Lb., 5 floors, 50 FPM class 'B'	126,000	28,800	154,800
7200	15 floors, 200 FPM class 'B'	338,000	100,000	438,000
8000	20,000 Lb., 5 floors, 50 FPM class 'B'	139,500	28,800	168,300
8200	15 floors, 200 FPM class 'B'	351,500	100,000	451,500

D1010 150	Traction Gearless Elevators	COST EACH		
		MAT.	INST.	TOTAL
1700	Passenger, 2500 Lb., 10 floors, 200 FPM	165,500	72,500	238,000
1900	30 floors, 600 FPM	290,000	185,000	475,000
2000	3000 Lb., 10 floors, 200 FPM	167,000	72,500	239,500
2200	30 floors, 600 FPM	291,500	185,000	476,500
2300	3500 Lb., 10 floors, 200 FPM	169,000	72,500	241,500
2500	30 floors, 600 FPM	293,500	185,000	478,500
2700	50 floors, 800 FPM	383,000	297,500	680,500
2800	4000 lb., 10 floors, 200 FPM	169,500	72,500	242,000
3000	30 floors, 600 FPM	293,500	185,000	478,500
3200	50 floors, 800 FPM	383,500	297,500	681,000
3300	4500 lb., 10 floors, 200 FPM	171,500	72,500	244,000
3500	30 floors, 600 FPM	296,000	185,000	481,000
3700	50 floors, 800 FPM	386,000	297,500	683,500
3800	5000 lb., 10 floors, 200 FPM	174,000	72,500	246,500
4000	30 floors, 600 FPM	298,500	185,000	483,500
4200	50 floors, 800 FPM	388,000	297,500	685,500
6000	Hospital, 3500 Lb., 10 floors, 200 FPM	194,000	72,500	266,500
6200	30 floors, 600 FPM	370,500	185,000	555,500
6400	4000 Lb., 10 floors, 200 FPM	194,000	72,500	266,500
6600	30 floors, 600 FPM	370,500	185,000	555,500
6800	4500 Lb., 10 floors, 200 FPM	199,500	72,500	272,000
7000	30 floors, 600 FPM	376,000	185,000	561,000
7200	5000 Lb., 10 floors, 200 FPM	201,500	72,500	274,000
7400	30 floors, 600 FPM	378,000	185,000	563,000

SERVICES

D

D1090 Other Conveying Systems

D1090 510	Chutes	COST EACH		
		MAT.	INST.	TOTAL
0100	Chutes, linen or refuse incl. sprinklers, galv. steel, 18" diam., per floor	945	283	1,228
0110	Aluminized steel, 36" diam., per floor	1,275	355	1,630
0120	Mail, 8-3/4" x 3-1/2", aluminum & glass, per floor	660	198	858
0130	Bronze or stainless, per floor	1,025	220	1,245

Important: See the Reference Section for critical supporting data - Reference Numbers and City Cost Indexes

SERVICES D

Minimum Plumbing Fixture Requirements

Type of Building or Occupancy (2)	Water Closets (14) (Fixtures per Person)		Urinals (5,10) (Fixtures per Person)		Lavatories (Fixtures per Person)		Bathtubs or Showers (Fixtures per Person)	Drinking Fountains (Fixtures per Person) (3, 13)
	Male	Female	Male	Female	Male	Female		
Assembly Places- Theatres, Auditoriums, Convention Halls, etc.-for permanent employee use	1: 1 - 15 2: 16 - 35 3: 36 - 55 Over 55, add 1 fixture for each additional 40 persons	1: 1 - 15 2: 16 - 35 3: 36 - 55	0: 1 - 9 1: 10 - 50 Add one fixture for each additional 50 males		1 per 40	1 per 40		
Assembly Places- Theatres, Auditoriums, Convention Halls, etc. - for public use	1: 1 - 100 2: 101 - 200 3: 201 - 400 Over 400, add 1 fixture for each additional 500 males and 1 for each additional 125 females	3: 1 - 50 4: 51 - 100 8: 101 - 200 11: 201 - 400	1: 1 - 100 2: 101 - 200 3: 201 - 400 4: 401 - 600 Over 600, add 1 fixture for each additional 300 males		1: 1 - 200 2: 201 - 400 3: 401 - 750 Over 750, add 1 fixture for each additional 500 persons	1: 1 - 200 2: 201 - 400 3: 401 - 750		1: 1 - 150 2: 151 - 400 3: 401 - 750 Over 750, add one fixture for each additional 500 persons
Dormitories (9) School or Labor	1 per 10 Add 1 fixture for each additional 25 males (over 10) and 1 for each additional 20 females (over 8)	1 per 8	1 per 25 Over 150, add 1 fixture for each additional 50 males		1 per 12 Over 12 add 1 fixture for each additional 20 males and 1 for each 15 additional females	1 per 12	1 per 8 For females add 1 bathtub per 30. Over 150, add 1 per 20	1 per 150 (12)
Dormitories- for Staff Use	1: 1 - 15 2: 16 - 35 3: 36 - 55 Over 55, add 1 fixture for each additional 40 persons	1: 1 - 15 3: 16 - 35 4: 36 - 55	1 per 50		1 per 40	1 per 40	1 per 8	
Dwellings: Single Dwelling Multiple Dwelling or Apartment House	1 per dwelling 1 per dwelling or apartment unit				1 per dwelling 1 per dwelling or apartment unit		1 per dwelling 1 per dwelling or apartment unit	
Hospital Waiting rooms	1 per room				1 per room			1 per 150 (12)
Hospitals- for employee use	1: 1 - 15 2: 16 - 35 3: 36 - 55 Over 55, add 1 fixture for each additional 40 persons	1: 1 - 15 3: 16 - 35 4: 36 - 55	0: 1 - 9 1: 10 - 50 Add 1 fixture for each additional 50 males		1 per 40	1 per 40		
Hospitals: Individual Room Ward Room	1 per room 1 per 8 patients				1 per room 1 per 10 patients		1 per room 1 per 20 patients	1 per 150 (12)
Industrial (6) Warehouses Workshops, Foundries and similar establishments- for employee use	1: 1 -10 2: 11 - 25 3: 26 - 50 4: 51 - 75 5: 76 - 100 Over 100, add 1 fixture for each additional 30 persons	1: 1 -10 2: 11 - 25 3: 26 - 50 4: 51 - 75 5: 76 - 100			Up to 100, per 10 persons Over 100, 1 per 15 persons (7, 8)		1 shower for each 15 persons exposed to excessive heat or to skin contamination with poisonous, infectious or irritating material	1 per 150 (12)
Institutional - Other than Hospitals or Penal Institutions (on each occupied floor)	1 per 25	1 per 20	0: 1 - 9 1: 10 - 50 Add 1 fixture for each additional 50 males		1 per 10	1 per 10	1 per 8	1 per 150 (12)
Institutional - Other than Hospitals or Penal Institutions (on each occupied floor)- for employee use	1: 1 - 15 2: 16 - 35 3: 36 - 55 Over 55, add 1 fixture for each additional 40 persons	1: 1 - 15 3: 16 - 35 4: 36 - 55	0: 1 - 9 1: 10 - 50 Add 1 fixture for each additional 50 males		1 per 40	1 per 40	1 per 8	1 per 150 (12)
Office or Public Buildings	1: 1 - 100 2: 101 - 200 3: 201 - 400 Over 400, add 1 fixture for each additional 500 males and 1 for each additional 150 females	3: 1 - 50 4: 51 - 100 8: 101 - 200 11: 201 - 400	1: 1 - 100 2: 101 - 200 3: 201 - 400 4: 401 - 600 Over 600, add 1 fixture for each additional 300 males		1: 1 - 200 2: 201 - 400 3: 401 - 750 Over 750, add 1 fixture for each additional 500 persons	1: 1 - 200 2: 201 - 400 3: 401 - 750		1 per 150 (12)
Office or Public Buildings - for employee use	1: 1 - 15 2: 16 - 35 3: 36 - 55 Over 55, add 1 fixture for each additional 40 persons	1: 1 - 15 3: 16 - 35 4: 36 - 55	0: 1 - 9 1: 10 - 50 Add 1 fixture for each additional 50 males		1 per 40	1 per 40		

Minimum Plumbing Fixture Requirements

Type of Building or Occupancy	Water Closets (14) (Fixtures per Person)		Urinals (5, 10) (Fixtures per Person)		Lavatories (Fixtures per Person)		Bathtubs or Showers (Fixtures per Person)	Drinking Fountains (Fixtures per Person) (3, 13)
	Male	Female	Male	Female	Male	Female		
Penal Institutions - for employee use	1: 1 - 15 2: 16 - 35 3: 36 - 55 Over 55, add 1 fixture for each additional 40 persons	1: 1 - 15 3: 16 - 35 4: 36 - 55	0: 1 - 9 1: 10 - 50 Add 1 fixture for each additional 50 males		1 per 40	1 per 40		1 per 150 (12)
Penal Institutions - for prison use Cell	1 per cell				1 per cell			1 per cellblock floor
Exercise room	1 per exercise room		1 per exercise room		1 per exercise room			1 per exercise room
Restaurants, Pubs and Lounges (11)	1: 1 - 50 2: 51 - 150 3: 151 - 300 Over 300, add 1 fixture for each additional 200 persons	1: 1 - 50 2: 51 - 150 4: 151 - 300	1: 1 - 150 Over 150, add 1 fixture for each additional 150 males		1: 1 - 150 2: 151 - 200 3: 201 - 400 Over 400, add 1 fixture for each additional 400 persons	1: 1 - 150 2: 151 - 200 3: 201 - 400		
Schools - for staff use All Schools	1: 1 - 15 2: 16 - 35 3: 36 - 55 Over 55, add 1 fixture for each additional 40 persons	1: 1 - 15 3: 16 - 35 4: 36 - 55	1 per 50		1 per 40	1 per 40		
Schools - for student use: Nursery	1: 1 - 20 2: 21 - 50 Over 50, add 1 fixture for each additional 50 persons	1: 1 - 20 2: 21 - 50			1: 1 - 25 2: 26 - 50 Over 50, add 1 fixture for each additional 50 persons	1: 1 - 25 2: 26 - 50		1 per 150 (12)
Elementary	1 per 30	1 per 25	1 per 75		1 per 35	1 per 35		1 per 150 (12)
Secondary	1 per 40	1 per 30	1 per 35		1 per 40	1 per 40		1 per 150 (12)
Others (Colleges, Universities, Adult Centers, etc.	1 per 40	1 per 30	1 per 35		1 per 40	1 per 40		1 per 150 (12)
Worship Places Educational and Activities Unit	1 per 150	1 per 75	1 per 150		1 per 2 water closets			1 per 150 (12)
Worship Places Principal Assembly Place	1 per 150	1 per 75	1 per 150		1 per 2 water closets			1 per 150 (12)

Notes:
1. The figures shown are based upon one (1) fixture being the minimum required for the number of persons indicated or any fraction thereof.
2. Building categories not shown on this table shall be considered separately by the Administrative Authority.
3. Drinking fountains shall not be installed in toilet rooms.
4. Laundry trays. One (1) laundry tray or one (1) automatic washer standpipe for each dwelling unit or one (1) laundry trays or one (1) automatic washer standpipes, or combination thereof, for each twelve (12) apartments. Kitchen sinks, one (1) for each dwelling or apartment unit.
5. For each urinal added in excess of the minimum required, one water closet may be deducted. The number of water closets shall not be reduced to less than two-thirds (2/3) of the minimum requirement.
6. As required by ANSI Z4.1-1968, Sanitation in Places of Employment.
7. Where there is exposure to skin contamination with poisonous, infectious, or irritating materials, provide one (1) lavatory for each five (5) persons.
8. Twenty-four (24) lineal inches of wash sink or eighteen (18) inches of a circular basin, when provided with water outlets for such space shall be considered equivalent to one (1) lavatory.
9. Laundry trays, one (1) for each fifty (50) persons. Service sinks, one (1) for each hundred (100) persons.
10. General. In applying this schedule of facilities, consideration shall be given to the accessibility of the fixtures. Conformity purely on a numerical basis may not result in an installation suited to the need of the individual establishment. For example, schools should be provided with toilet facilities on each floor having classrooms.
   a. Surrounding materials, wall and floor space to a point two (2) feet in front of urinal lip and four (4) feet above the floor, and at least two (2) feet to each side of the urinal shall be lined with non-absorbent materials.
   b. Trough urinals shall be prohibited.
11. A restaurant is defined as a business which sells food to be consumed on the premises.
   a. The number of occupants for a drive-in restaurant shall be considered as equal to the number of parking stalls.
   b. Employee toilet facilities shall not to be included in the above restaurant requirements. Hand washing facilities shall be available in the kitchen for employees.
12. Where food is consumed indoors, water stations may be substituted for drinking fountains. Offices, or public buildings for use by more than six (6) persons shall have one (1) drinking fountain for the first one hundred fifty (150) persons and one additional fountain for each three hundred (300) persons thereafter.
13. There shall be a minimum of one (1) drinking fountain per occupied floor in schools, theaters, auditoriums, dormitories, offices of public building.
14. The total number of water closets for females shall be at least equal to the total number of water closets and urinals required for males.

SERVICES D

D2010 Plumbing Fixtures

One Piece Wall Hung Water Closet

Wall Hung Urinal

Side By Side Water Closet Group

Floor Mount Water Closet

Stall Type Urinal

Back to Back Water Closet Group

D2010 110	Water Closet Systems	COST EACH		
		MAT.	INST.	TOTAL
1800	Water closet, vitreous china, elongated			
1840	Tank type, wall hung, one piece	800	520	1,320
1880	Close coupled two piece	895	520	1,415
1920	Floor mount, one piece	735	555	1,290
1960	One piece low profile	625	555	1,180
2000	Two piece close coupled	370	555	925
2040	Bowl only with flush valve			
2080	Wall hung	795	590	1,385
2120	Floor mount	545	565	1,110
2122				

D2010 120	Water Closets, Group	COST EACH		
		MAT.	INST.	TOTAL
1760	Water closets, battery mount, wall hung, side by side, first closet	845	610	1,455
1800	Each additional water closet, add	795	575	1,370
3000	Back to back, first pair of closets	1,500	810	2,310
3100	Each additional pair of closets, back to back	1,450	795	2,245

D2010 210	Urinal Systems	COST EACH		
		MAT.	INST.	TOTAL
2000	Urinal, vitreous china, wall hung	445	585	1,030
2040	Stall type	730	700	1,430

D2010 Plumbing Fixtures

Systems are complete with trim and rough-in (supply, waste and vent) to connect to supply branches and waste mains.

Vanity Top

Wall Hung

Counter Top Single Bowl

Counter Top Double Bowl

D2010 310	Lavatory Systems	COST EACH		
		MAT.	INST.	TOTAL
1560	Lavatory w/trim, vanity top, PE on CI, 20" x 18", Vanity top by others.	325	525	850
1600	19" x 16" oval	242	525	767
1640	18" round	345	525	870
1680	Cultured marble, 19" x 17"	269	525	794
1720	25" x 19"	269	525	794
1760	Stainless, self-rimming, 25" x 22"	360	525	885
1800	17" x 22"	350	525	875
1840	Steel enameled, 20" x 17"	242	540	782
1880	19" round	240	540	780
1920	Vitreous china, 20" x 16"	345	550	895
1960	19" x 16"	345	550	895
2000	22" x 13"	355	550	905
2040	Wall hung, PE on CI, 18" x 15"	610	580	1,190
2080	19" x 17"	610	580	1,190
2120	20" x 18"	575	580	1,155
2160	Vitreous china, 18" x 15"	530	595	1,125
2200	19" x 17"	520	595	1,115
2240	24" x 20"	765	595	1,360
2241				
2243				

D2010 410	Kitchen Sink Systems	COST EACH		
		MAT.	INST.	TOTAL
1720	Kitchen sink w/trim, countertop, PE on CI, 24"x21", single bowl	345	575	920
1760	30" x 21" single bowl	395	575	970
1800	32" x 21" double bowl	440	620	1,060
1840	42" x 21" double bowl	660	625	1,285
1880	Stainless steel, 19" x 18" single bowl	480	575	1,055
1920	25" x 22" single bowl	520	575	1,095
1960	33" x 22" double bowl	710	620	1,330
2000	43" x 22" double bowl	805	625	1,430
2040	44" x 22" triple bowl	1,025	655	1,680
2080	44" x 24" corner double bowl	700	625	1,325
2120	Steel, enameled, 24" x 21" single bowl	237	575	812
2160	32" x 21" double bowl	287	620	907
2240	Raised deck, PE on CI, 32" x 21", dual level, double bowl	415	785	1,200
2280	42" x 21" dual level, triple bowl	810	850	1,660

Important: See the Reference Section for critical supporting data - Reference Numbers and City Cost Indexes

D2010 Plumbing Fixtures

Laboratory Sink

Service Sink

Systems are complete with trim and rough-in (supply, waste and vent) to connect to supply branches and waste mains.

Single Compartment Sink

Double Compartment Sink

D2010 420	Laundry Sink Systems	COST EACH		
		MAT.	INST.	TOTAL
1740	Laundry sink w/trim, PE on CI, black iron frame			
1760	24" x 20", single compartment	495	560	1,055
1840	48" x 21" double compartment	415	610	1,025
1920	Molded stone, on wall, 22" x 21" single compartment	252	560	812
1960	45"x 21" double compartment	410	610	1,020
2040	Plastic, on wall or legs, 18" x 23" single compartment	217	550	767
2080	20" x 24" single compartment	250	550	800
2120	36" x 23" double compartment	288	595	883
2160	40" x 24" double compartment	365	595	960
2200	Stainless steel, countertop, 22" x 17" single compartment	495	560	1,055
2240	19" x 22" single compartment	600	560	1,160
2280	33" x 22" double compartment	600	610	1,210

D2010 430	Laboratory Sink Systems	COST EACH		
		MAT.	INST.	TOTAL
1580	Laboratory sink w/trim, polyethylene, single bowl,			
1600	Double drainboard, 54" x 24" O.D.	1,000	735	1,735
1640	Single drainboard, 47" x 24"O.D.	1,150	735	1,885
1680	70" x 24" O.D.	1,175	735	1,910
1760	Flanged, 14-1/2" x 14-1/2" O.D.	380	660	1,040
1800	18-1/2" x 18-1/2" O.D.	385	660	1,045
1840	23-1/2" x 20-1/2" O.D.	425	660	1,085
1920	Polypropylene, cup sink, oval, 7" x 4" O.D.	237	580	817
1960	10" x 4-1/2" O.D.	231	580	811
1961				

D2010 Plumbing Fixtures

D2010 440	Service Sink Systems	COST EACH		
		MAT.	INST.	TOTAL
4260	Service sink w/trim, PE on CI, corner floor, 28" x 28", w/rim guard	880	740	1,620
4300	Wall hung w/rim guard, 22" x 18"	970	855	1,825
4340	24" x 20"	990	855	1,845
4380	Vitreous china, wall hung 22" x 20"	910	855	1,765
4383				

Important: See the Reference Section for critical supporting data - Reference Numbers and City Cost Indexes

SERVICES D

D2010 Plumbing Fixtures

Recessed Bathtub   Corner Bathtub

Systems are complete with trim and rough-in (supply, waste and vent) to connect to supply branches and waste mains.

Circular Wash Fountain

Semi-Circular Wash Fountain

D2010 510	Bathtub Systems	COST EACH		
		MAT.	INST.	TOTAL
1960	Bathtub, recessed, P.E. on Cl., 42" x 37"	1,125	605	1,730
2000	48" x 42"	1,825	650	2,475
2040	72" x 36"	1,875	725	2,600
2080	Mat bottom, 5' long	635	630	1,265
2120	5'-6" long	1,350	650	2,000
2160	Corner, 48" x 42"	1,825	630	2,455
2200	Formed steel, enameled, 4'-6" long	540	580	1,120
2240	5' long	495	590	1,085

D2010 610	Group Wash Fountain Systems	COST EACH		
		MAT.	INST.	TOTAL
1740	Group wash fountain, precast terrazzo			
1760	Circular, 36" diameter	3,225	945	4,170
1800	54" diameter	3,875	1,025	4,900
1840	Semi-circular, 36" diameter	2,900	945	3,845
1880	54" diameter	3,550	1,025	4,575
1960	Stainless steel, circular, 36" diameter	3,675	880	4,555
2000	54" diameter	4,775	975	5,750
2040	Semi-circular, 36" diameter	3,200	880	4,080
2080	54" diameter	4,150	975	5,125
2160	Thermoplastic, circular, 36" diameter	2,825	715	3,540
2200	54" diameter	3,300	830	4,130
2240	Semi-circular, 36" diameter	2,675	715	3,390
2280	54" diameter	3,125	830	3,955

D2010 Plumbing Fixtures

Square Shower Stall

Corner Angle Shower Stall

Systems are complete with trim and rough-in (supply, waste and vent) to connect to supply branches and waste mains.

Wall Mounted, Low Back

Wall Mounted, No Back

D2010 710	Shower Systems	COST EACH		
		MAT.	INST.	TOTAL
1560	Shower, stall, baked enamel, molded stone receptor, 30" square	490	870	1,360
1600	32" square	465	870	1,335
1640	Terrazzo receptor, 32" square	890	870	1,760
1680	36" square	1,075	920	1,995
1720	36" corner angle	985	920	1,905
1800	Fiberglass one piece, three walls, 32" square	565	800	1,365
1840	36" square	610	800	1,410
1880	Polypropylene, molded stone receptor, 30" square	465	870	1,335
1920	32" square	565	870	1,435
1960	Built-in head, arm, bypass, stops and handles	110	227	337

D2010 810	Drinking Fountain Systems	COST EACH		
		MAT.	INST.	TOTAL
1740	Drinking fountain, one bubbler, wall mounted			
1760	Non recessed			
1800	Bronze, no back	1,175	345	1,520
1840	Cast iron, enameled, low back	565	345	910
1880	Fiberglass, 12" back	680	345	1,025
1920	Stainless steel, no back	980	345	1,325
1960	Semi-recessed, poly marble	815	345	1,160
2040	Stainless steel	930	345	1,275
2080	Vitreous china	690	345	1,035
2120	Full recessed, poly marble	910	345	1,255
2200	Stainless steel	835	345	1,180
2240	Floor mounted, pedestal type, aluminum	575	470	1,045
2320	Bronze	1,375	470	1,845
2360	Stainless steel	1,375	470	1,845

D2010 Plumbing Fixtures

Wall Hung Water Cooler

Floor Mounted Water Cooler

Systems are complete with trim and rough-in (supply, waste and vent) to connect to supply branches and waste mains.

D2010 820	Water Cooler Systems	COST EACH		
		MAT.	INST.	TOTAL
1840	Water cooler, electric, wall hung, 8.2 GPH	685	445	1,130
1880	Dual height, 14.3 GPH	1,000	455	1,455
1920	Wheelchair type, 7.5 G.P.H.	1,650	445	2,095
1960	Semi recessed, 8.1 G.P.H.	990	445	1,435
2000	Full recessed, 8 G.P.H.	1,525	475	2,000
2040	Floor mounted, 14.3 G.P.H.	645	385	1,030
2080	Dual height, 14.3 G.P.H.	1,075	470	1,545
2120	Refrigerated compartment type, 1.5 G.P.H.	1,300	385	1,685

D2010 Plumbing Fixtures

Two Fixture   Three Fixture

*Common wall is with adjacent bathrom

D2010 920	Two Fixture Bathroom, Two Wall Plumbing	COST EACH		
		MAT.	INST.	TOTAL
1180	Bathroom, lavatory & water closet, 2 wall plumbing, stand alone	1,075	1,450	2,525
1200	Share common plumbing wall*	995	1,250	2,245

D2010 922	Two Fixture Bathroom, One Wall Plumbing	COST EACH		
		MAT.	INST.	TOTAL
2220	Bathroom, lavatory & water closet, one wall plumbing, stand alone	1,025	1,300	2,325
2240	Share common plumbing wall*	925	1,100	2,025

D2010 924	Three Fixture Bathroom, One Wall Plumbing	COST EACH		
		MAT.	INST.	TOTAL
1150	Bathroom, three fixture, one wall plumbing			
1160	Lavatory, water closet & bathtub			
1170	Stand alone	1,525	1,675	3,200
1180	Share common plumbing wall *	1,250	1,225	2,475

D2010 926	Three Fixture Bathroom, Two Wall Plumbing	COST EACH		
		MAT.	INST.	TOTAL
2130	Bathroom, three fixture, two wall plumbing			
2140	Lavatory, water closet & bathtub			
2160	Stand alone	1,525	1,700	3,225
2180	Long plumbing wall common *	1,325	1,350	2,675
3610	Lavatory, bathtub & water closet			
3620	Stand alone	1,625	1,950	3,575
3640	Long plumbing wall common *	1,550	1,750	3,300
4660	Water closet, corner bathtub & lavatory			
4680	Stand alone	2,725	1,725	4,450
4700	Long plumbing wall common *	2,575	1,300	3,875
6100	Water closet, stall shower & lavatory			
6120	Stand alone	1,500	2,200	3,700
6140	Long plumbing wall common *	1,425	2,050	3,475
7060	Lavatory, corner stall shower & water closet			
7080	Stand alone	1,900	2,025	3,925
7100	Short plumbing wall common *	1,725	1,450	3,175

SERVICES D

D2010 Plumbing Fixtures

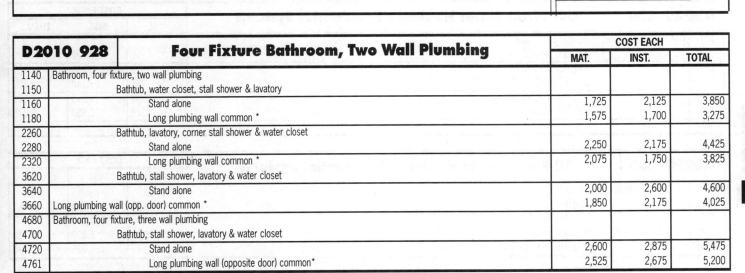

Four Fixture   Five Fixture

*Common wall is with adjacent bathroom

SERVICES

D

D2010 928	Four Fixture Bathroom, Two Wall Plumbing	COST EACH		
		MAT.	INST.	TOTAL
1140	Bathroom, four fixture, two wall plumbing			
1150	Bathtub, water closet, stall shower & lavatory			
1160	Stand alone	1,725	2,125	3,850
1180	Long plumbing wall common *	1,575	1,700	3,275
2260	Bathtub, lavatory, corner stall shower & water closet			
2280	Stand alone	2,250	2,175	4,425
2320	Long plumbing wall common *	2,075	1,750	3,825
3620	Bathtub, stall shower, lavatory & water closet			
3640	Stand alone	2,000	2,600	4,600
3660	Long plumbing wall (opp. door) common *	1,850	2,175	4,025
4680	Bathroom, four fixture, three wall plumbing			
4700	Bathtub, stall shower, lavatory & water closet			
4720	Stand alone	2,600	2,875	5,475
4761	Long plumbing wall (opposite door) common*	2,525	2,675	5,200

D2010 932	Five Fixture Bathroom, Two Wall Plumbing	COST EACH		
		MAT.	INST.	TOTAL
1320	Bathroom, five fixture, two wall plumbing			
1340	Bathtub, water closet, stall shower & two lavatories			
1360	Stand alone	2,575	3,250	5,825
1400	One short plumbing wall common *	2,425	2,825	5,250
2360	Bathroom, five fixture, three wall plumbing			
2380	Water closet, bathtub, two lavatories & stall shower			
2400	Stand alone	3,100	3,300	6,400
2440	One short plumbing wall common *	2,950	2,875	5,825
4080	Bathroom, five fixture, one wall plumbing			
4100	Bathtub, two lavatories, corner stall shower & water closet			
4120	Stand alone	2,925	2,975	5,900
4160	Share common wall *	2,350	2,025	4,375

Installation includes piping and fittings within 10' of heater. Gas heaters require vent piping (not included with these units).

Gas Fired

Oil Fired

D2020 220	Gas Fired Water Heaters - Residential Systems	COST EACH		
		MAT.	INST.	TOTAL
2200	Gas fired water heater, residential, 100° F rise			
2260	30 gallon tank, 32 GPH	705	1,025	1,730
2300	40 gallon tank, 32 GPH	810	1,150	1,960
2340	50 gallon tank, 63 GPH	955	1,150	2,105
2380	75 gallon tank, 63 GPH	1,300	1,275	2,575
2420	100 gallon tank, 63 GPH	1,725	1,350	3,075
2422				

D2020 230	Oil Fired Water Heaters - Residential Systems	COST EACH		
		MAT.	INST.	TOTAL
2200	Oil fired water heater, residential, 100° F rise			
2220	30 gallon tank, 103 GPH	1,100	945	2,045
2260	50 gallon tank, 145 GPH	1,850	1,050	2,900
2300	70 gallon tank, 164 GPH	2,125	1,200	3,325
2340	85 gallon tank, 181 GPH	5,175	1,225	6,400

SERVICES D

D2020 Domestic Water Distribution

Systems below include piping and fittings within 10' of heater. Electric water heaters do not require venting. Gas fired heaters require vent piping (not included in these prices).

Electric

Gas Fired

D2020 240	Electric Water Heaters - Commercial Systems	COST EACH		
		MAT.	INST.	TOTAL
1800	Electric water heater, commercial, 100° F rise			
1820	50 gallon tank, 9 KW 37 GPH	2,975	855	3,830
1860	80 gal, 12 KW 49 GPH	3,775	1,050	4,825
1900	36 KW 147 GPH	5,150	1,150	6,300
1940	120 gal, 36 KW 147 GPH	5,550	1,225	6,775
1980	150 gal, 120 KW 490 GPH	17,300	1,325	18,625
2020	200 gal, 120 KW 490 GPH	18,400	1,350	19,750
2060	250 gal, 150 KW 615 GPH	20,400	1,575	21,975
2100	300 gal, 180 KW 738 GPH	21,900	1,675	23,575
2140	350 gal, 30 KW 123 GPH	16,800	1,800	18,600
2180	180 KW 738 GPH	22,500	1,800	24,300
2220	500 gal, 30 KW 123 GPH	21,600	2,100	23,700
2260	240 KW 984 GPH	30,700	2,100	32,800
2300	700 gal, 30 KW 123 GPH	25,900	2,400	28,300
2340	300 KW 1230 GPH	38,500	2,400	40,900
2380	1000 gal, 60 KW 245 GPH	31,400	3,375	34,775
2420	480 KW 1970 GPH	40,400	3,375	43,775
2460	1500 gal, 60 KW 245 GPH	45,800	4,150	49,950
2500	480 KW 1970 GPH	63,000	4,150	67,150

D2020 250	Gas Fired Water Heaters - Commercial Systems	COST EACH		
		MAT.	INST.	TOTAL
1760	Gas fired water heater, commercial, 100° F rise			
1780	75.5 MBH input, 63 GPH	2,100	1,325	3,425
1820	95 MBH input, 86 GPH	3,350	1,325	4,675
1860	100 MBH input, 91 GPH	3,575	1,375	4,950
1900	115 MBH input, 110 GPH	4,175	1,425	5,600
1980	155 MBH input, 150 GPH	5,000	1,600	6,600
2020	175 MBH input, 168 GPH	5,225	1,700	6,925
2060	200 MBH input, 192 GPH	5,375	1,925	7,300
2100	240 MBH input, 230 GPH	5,775	2,100	7,875
2140	300 MBH input, 278 GPH	7,375	2,375	9,750
2180	390 MBH input, 374 GPH	8,250	2,400	10,650
2220	500 MBH input, 480 GPH	10,100	2,600	12,700
2260	600 MBH input, 576 GPH	11,500	2,800	14,300

SERVICES

D

D2020 Domestic Water Distribution

Oil fired water heater systems include piping and fittings within 10' of heater. Oil fired heaters require vent piping (not included in these systems).

Oil Fired

D2020 260	Oil Fired Water Heaters - Commercial Systems	COST EACH		
		MAT.	INST.	TOTAL
1800	Oil fired water heater, commercial, 100° F rise			
1820	103 MBH output, 116 GPH	1,350	1,150	2,500
1900	134 MBH output, 161 GPH	2,500	1,475	3,975
1940	161 MBH output, 192 GPH	2,750	1,675	4,425
1980	187 MBH output, 224 GPH	3,075	1,950	5,025
2060	262 MBH output, 315 GPH	3,600	2,250	5,850
2100	341 MBH output, 409 GPH	7,600	2,400	10,000
2140	420 MBH output, 504 GPH	8,875	2,600	11,475
2180	525 MBH output, 630 GPH	10,000	2,925	12,925
2220	630 MBH output, 756 GPH	11,600	2,925	14,525
2260	735 MBH output, 880 GPH	12,100	3,375	15,475
2300	840 MBH output, 1000 GPH	12,600	3,375	15,975
2340	1050 MBH output, 1260 GPH	13,600	3,475	17,075
2380	1365 MBH output, 1640 GPH	15,900	3,950	19,850
2420	1680 MBH output, 2000 GPH	17,700	4,975	22,675
2460	2310 MBH output, 2780 GPH	22,700	6,100	28,800
2500	2835 MBH output, 3400 GPH	25,700	6,100	31,800
2540	3150 MBH output, 3780 GPH	28,800	8,400	37,200

D2040 Rain Water Drainage

Design Assumptions: Vertical conductor size is based on a maximum rate of rainfall of 4″ per hour. To convert roof area to other rates multiply "Max. S.F. Roof Area" shown by four and divide the result by desired local rate. The answer is the local roof area that may be handled by the indicated pipe diameter.

Basic cost is for roof drain, 10′ of vertical leader and 10′ of horizontal, plus connection to the main.

Pipe Dia.	Max. S.F. Roof Area	Gallons per Min.
2″	544	23
3″	1610	67
4″	3460	144
5″	6280	261
6″	10,200	424
8″	22,000	913

D2040 210	Roof Drain Systems	COST EACH		
		MAT.	INST.	TOTAL
1880	Roof drain, DWV PVC, 2″ diam., piping, 10′ high	156	495	651
1920	For each additional foot add	1.49	15	16.49
1960	3″ diam., 10′ high	192	575	767
2000	For each additional foot add	2.64	16.70	19.34
2040	4″ diam., 10′ high	225	645	870
2080	For each additional foot add	3.44	18.45	21.89
2120	5″ diam., 10′ high	525	745	1,270
2160	For each additional foot add	5.35	20.50	25.85
2200	6″ diam., 10′ high	570	825	1,395
2240	For each additional foot add	6.55	22.50	29.05
2280	8″ diam., 10′ high	1,350	1,400	2,750
2320	For each additional foot add	11.50	28.50	40
3940	C.I., soil, single hub, service wt., 2″ diam. piping, 10′ high	261	540	801
3980	For each additional foot add	5.10	14.05	19.15
4120	3″ diam., 10′ high	360	585	945
4160	For each additional foot add	6.95	14.75	21.70
4200	4″ diam., 10′ high	425	635	1,060
4240	For each additional foot add	8.95	16.10	25.05
4280	5″ diam., 10′ high	595	705	1,300
4320	For each additional foot add	9.05	18.10	27.15
4360	6″ diam., 10′ high	750	755	1,505
4400	For each additional foot add	15.15	18.85	34
4440	8″ diam., 10′ high	1,625	1,525	3,150
4480	For each additional foot add	23.50	31.50	55
6040	Steel galv. sch 40 threaded, 2″ diam. piping, 10′ high	330	520	850
6080	For each additional foot add	5.35	13.80	19.15
6120	3″ diam., 10′ high	680	750	1,430
6160	For each additional foot add	11.25	20.50	31.75
6200	4″ diam., 10′ high	1,050	965	2,015
6240	For each additional foot add	18.30	24.50	42.80
6280	5″ diam., 10′ high	1,925	1,225	3,150
6320	For each additional foot add	30.50	34	64.50
6360	6″ diam, 10′ high	2,450	1,500	3,950
6400	For each additional foot add	41.50	44.50	86
6440	8″ diam., 10′ high	5,275	2,050	7,325
6480	For each additional foot add	63	51	114

SERVICES

D

D2090 Other Plumbing Systems

D2090 810	Piping - Installed - Unit Costs	COST PER L.F.		
		MAT.	INST.	TOTAL
0840	Cast iron, soil, B & S, service weight, 2" diameter	5.10	14.05	19.15
0860	3" diameter	6.95	14.75	21.70
0880	4" diameter	8.95	16.10	25.05
0900	5" diameter	9.05	18.10	27.15
0920	6" diameter	15.15	18.85	34
0940	8" diameter	23.50	31.50	55
0960	10" diameter	43	34.50	77.50
0980	12" diameter	61	39	100
1040	No hub, 1-1/2" diameter	5.85	12.45	18.30
1060	2" diameter	6	13.20	19.20
1080	3" diameter	7.90	13.80	21.70
1100	4" diameter	10	15.25	25.25
1120	5" diameter	14.40	16.55	30.95
1140	6" diameter	17.95	17.40	35.35
1160	8" diameter	28.50	27	55.50
1180	10" diameter	50.50	30.50	81
1220	Copper tubing, hard temper, solder, type K, 1/2" diameter	1.97	6.30	8.27
1260	3/4" diameter	3.32	6.65	9.97
1280	1" diameter	4.37	7.45	11.82
1300	1-1/4" diameter	5.40	8.75	14.15
1320	1-1/2" diameter	7.15	9.80	16.95
1340	2" diameter	10.95	12.30	23.25
1360	2-1/2" diameter	16.35	14.75	31.10
1380	3" diameter	23	16.40	39.40
1400	4" diameter	38.50	23.50	62
1480	5" diameter	100	27.50	127.50
1500	6" diameter	146	36	182
1520	8" diameter	258	40.50	298.50
1560	Type L, 1/2" diameter	1.65	6.05	7.70
1600	3/4" diameter	2.42	6.45	8.87
1620	1" diameter	3.41	7.20	10.61
1640	1-1/4" diameter	4.68	8.45	13.13
1660	1-1/2" diameter	5.95	9.45	15.40
1680	2" diameter	9.25	11.70	20.95
1700	2-1/2" diameter	14.05	14.25	28.30
1720	3" diameter	18.95	15.80	34.75
1740	4" diameter	32.50	22.50	55
1760	5" diameter	91	26	117
1780	6" diameter	113	34.50	147.50
1800	8" diameter	198	38	236
1840	Type M, 1/2" diameter	1.30	5.85	7.15
1880	3/4" diameter	1.90	6.30	8.20
1900	1" diameter	2.62	7	9.62
1920	1-1/4" diameter	3.83	8.20	12.03
1940	1-1/2" diameter	5.15	9.10	14.25
1960	2" diameter	8.05	11.15	19.20
1980	2-1/2" diameter	12.05	13.80	25.85
2000	3" diameter	16.10	15.25	31.35
2020	4" diameter	29	22	51
2060	6" diameter	113	33	146
2080	8" diameter	186	36	222
2120	Type DWV, 1-1/4" diameter	4.04	8.20	12.24
2160	1-1/2" diameter	5	9.10	14.10
2180	2" diameter	6.70	11.15	17.85
2200	3" diameter	12	15.25	27.25
2220	4" diameter	21	22	43
2240	5" diameter	62.50	24.50	87
2260	6" diameter	89.50	33	122.50

D2090 Other Plumbing Systems

D2090 810	Piping - Installed - Unit Costs	COST PER L.F.		
		MAT.	INST.	TOTAL
2280	8″ diameter	221	36	257
2800	Plastic,PVC, DWV, schedule 40, 1-1/4″ diameter	1.31	11.70	13.01
2820	1-1/2″ diameter	1.32	13.65	14.97
2830	2″ diameter	1.49	15	16.49
2840	3″ diameter	2.64	16.70	19.34
2850	4″ diameter	3.44	18.45	21.89
2890	6″ diameter	6.55	22.50	29.05
3010	Pressure pipe 200 PSI, 1/2″ diameter	1.04	9.10	10.14
3030	3/4″ diameter	1.08	9.65	10.73
3040	1″ diameter	1.16	10.70	11.86
3050	1-1/4″ diameter	1.32	11.70	13.02
3060	1-1/2″ diameter	1.42	13.65	15.07
3070	2″ diameter	1.62	15	16.62
3080	2-1/2″ diameter	2.52	15.80	18.32
3090	3″ diameter	2.89	16.70	19.59
3100	4″ diameter	3.88	18.45	22.33
3110	6″ diameter	7.50	22.50	30
3120	8″ diameter	12.75	28.50	41.25
4000	Steel, schedule 40, threaded, black, 1/2″ diameter	1.41	7.80	9.21
4020	3/4″ diameter	1.69	8.05	9.74
4030	1″ diameter	2.38	9.25	11.63
4040	1-1/4″ diameter	3.02	9.95	12.97
4050	1-1/2″ diameter	3.48	11.05	14.53
4060	2″ diameter	4.55	13.80	18.35
4070	2-1/2″ diameter	7.70	17.70	25.40
4080	3″ diameter	9.95	20.50	30.45
4090	4″ diameter	15.65	24.50	40.15
4100	5″ diameter	28.50	34	62.50
4110	6″ diameter	36	44.50	80.50
4120	8″ diameter	56	51	107
4130	10″ diameter	85.50	60	145.50
4140	12″ diameter	94.50	76.50	171
4200	Galvanized, 1/2″ diameter	1.67	7.80	9.47
4220	3/4″ diameter	1.94	8.05	9.99
4230	1″ diameter	2.71	9.25	11.96
4240	1-1/4″ diameter	3.52	9.95	13.47
4250	1-1/2″ diameter	4.08	11.05	15.13
4260	2″ diameter	5.35	13.80	19.15
4270	2-1/2″ diameter	8.70	17.70	26.40
4280	3″ diameter	11.25	20.50	31.75
4290	4″ diameter	18.30	24.50	42.80
4300	5″ diameter	30.50	34	64.50
4310	6″ diameter	41.50	44.50	86
4320	8″ diameter	63	51	114
4330	10″ diameter	92.50	60	152.50
4340	12″ diameter	127	76.50	203.50
5010	Flanged, black, 1″ diameter	6.35	13.90	20.25
5020	1-1/4″ diameter	7.15	15.20	22.35
5030	1-1/2″ diameter	6.90	16.80	23.70
5040	2″ diameter	8.75	21.50	30.25
5050	2-1/2″ diameter	10.90	27	37.90
5060	3″ diameter	13	30.50	43.50
5070	4″ diameter	21.50	37.50	59
5080	5″ diameter	33.50	46	79.50
5090	6″ diameter	42	58.50	100.50
5100	8″ diameter	62.50	77	139.50
5110	10″ diameter	95	91.50	186.50
5120	12″ diameter	139	105	244

SERVICES

D

D2090 Other Plumbing Systems

D2090 810	Piping - Installed - Unit Costs	COST PER L.F.		
		MAT.	INST.	TOTAL
5720	Grooved joints, black, 3/4" diameter	2.18	6.90	9.08
5730	1" diameter	2.03	7.80	9.83
5740	1-1/4" diameter	2.60	8.45	11.05
5750	1-1/2" diameter	2.43	9.65	12.08
5760	2" diameter	3.63	12.30	15.93
5770	2-1/2" diameter	5.30	15.50	20.80
5900	3" diameter	6.60	17.70	24.30
5910	4" diameter	12.05	19.65	31.70
5920	5" diameter	22	24	46
5930	6" diameter	28	33	61
5940	8" diameter	42	37	79
5950	10" diameter	64	44.50	108.50
5960	12" diameter	87.50	51	138.50

Important: See the Reference Section for critical supporting data - Reference Numbers and City Cost Indexes

D3010 Energy Supply

Forced Hot Water Heating System

Fin Tube Radiation

Terminal Unit Heater

D3010 510	Apartment Building Heating - Fin Tube Radiation	COST PER S.F.		
		MAT.	INST.	TOTAL
1740	Heating systems, fin tube radiation, forced hot water			
1760	1,000 S.F. area, 10,000 C.F. volume	5.38	4.75	10.13
1800	10,000 S.F. area, 100,000 C.F. volume	1.70	2.86	4.56
1840	20,000 S.F. area, 200,000 C.F. volume	1.72	3.22	4.94
1880	30,000 S.F. area, 300,000 C.F. volume	1.61	3.11	4.72

D3010 520	Commercial Building Heating - Fin Tube Radiation	COST PER S.F.		
		MAT.	INST.	TOTAL
1940	Heating systems, fin tube radiation, forced hot water			
1960	1,000 S.F. bldg, one floor	10.60	10.70	21.30
2000	10,000 S.F., 100,000 C.F., total two floors	2.65	4.09	6.74
2040	100,000 S.F., 1,000,000 C.F., total three floors	1.15	1.84	2.99
2080	1,000,000 S.F., 10,000,000 C.F., total five floors	.56	.98	1.54

D3010 530	Commercial Bldg. Heating - Terminal Unit Heaters	COST PER S.F.		
		MAT.	INST.	TOTAL
1860	Heating systems, terminal unit heaters, forced hot water			
1880	1,000 S.F. bldg., one floor	10.51	9.83	20.34
1920	10,000 S.F. bldg., 100,000 C.F. total two floors	2.40	3.39	5.79
1960	100,000 S.F. bldg., 1,000,000 C.F. total three floors	1.19	1.70	2.89
2000	1,000,000 S.F. bldg., 10,000,000 C.F. total five floors	.76	1.10	1.86

SERVICES

D

Small Electric Boiler Systems Considerations:

1. Terminal units are fin tube baseboard radiation rated at 720 BTU/hr with 200° water temperature or 820 BTU/hr steam.
2. Primary use being for residential or smaller supplementary areas, the floor levels are based on 7-1/2" ceiling heights.
3. All distribution piping is copper for boilers through 205 MBH. All piping for larger systems is steel pipe.

Large Electric Boiler System Considerations:

1. Terminal units are all unit heaters of the same size. Quantities are varied to accommodate total requirements.
2. All air is circulated through the heaters a minimum of three times per hour.
3. As the capacities are adequate for commercial use, gross output rating by floor levels are based on 10' ceiling height.
4. All distribution piping is black steel pipe.

D3020 102	Small Heating Systems, Hydronic, Electric Boilers	COST PER S.F.		
		MAT.	INST.	TOTAL
1100	Small heating systems, hydronic, electric boilers			
1120	Steam, 1 floor, 1480 S.F., 61 M.B.H.	4.31	6.10	10.41
1160	3,000 S.F., 123 M.B.H.	3.20	5.35	8.55
1200	5,000 S.F., 205 M.B.H.	2.73	4.92	7.65
1240	2 floors, 12,400 S.F., 512 M.B.H.	2.17	4.89	7.06
1280	3 floors, 24,800 S.F., 1023 M.B.H.	2.24	4.83	7.07
1320	34,750 S.F., 1,433 M.B.H.	2.03	4.70	6.73
1360	Hot water, 1 floor, 1,000 S.F., 41 M.B.H.	7.25	3.41	10.66
1400	2,500 S.F., 103 M.B.H.	4.94	6.10	11.04
1440	2 floors, 4,850 S.F., 205 M.B.H.	4.61	7.35	11.96
1480	3 floors, 9,700 S.F., 410 M.B.H.	4.55	7.55	12.10

D3020 104	Large Heating Systems, Hydronic, Electric Boilers	COST PER S.F.		
		MAT.	INST.	TOTAL
1230	Large heating systems, hydronic, electric boilers			
1240	9,280 S.F., 135 K.W., 461 M.B.H., 1 floor	2.46	2.32	4.78
1280	14,900 S.F., 240 K.W., 820 M.B.H., 2 floors	3.01	3.80	6.81
1320	18,600 S.F., 296 K.W., 1,010 M.B.H., 3 floors	2.93	4.12	7.05
1360	26,100 S.F., 420 K.W., 1,432 M.B.H., 4 floors	2.91	4.04	6.95
1400	39,100 S.F., 666 K.W., 2,273 M.B.H., 4 floors	2.52	3.38	5.90
1440	57,700 S.F., 900 K.W., 3,071 M.B.H., 5 floors	2.37	3.32	5.69
1480	111,700 S.F., 1,800 K.W., 6,148 M.B.H., 6 floors	2.17	2.85	5.02
1520	149,000 S.F., 2,400 K.W., 8,191 M.B.H., 8 floors	2.15	2.84	4.99
1560	223,300 S.F., 3,600 K.W., 12,283 M.B.H., 14 floors	2.26	3.23	5.49

SERVICES D

D3020 Heat Generating Systems

Boiler Selection: The maximum allowable working pressures are limited by ASME "Code for Heating Boilers" to 15 PSI for steam and 160 PSI for hot water heating boilers, with a maximum temperature limitation of 250° F. Hot water boilers are generally rated for a working pressure of 30 PSI. High pressure boilers are governed by the ASME "Code for Power Boilers" which is used almost universally for boilers operating over 15 PSIG. High pressure boilers used for a combination of heating/process loads are usually designed for 150 PSIG.

Boiler ratings are usually indicated as either Gross or Net Output. The Gross Load is equal to the Net Load plus a piping and pickup allowance. When this allowance cannot be determined, divide the gross output rating by 1.25 for a value equal to or greater than the next heat loss requirement of the building.

Table below lists installed cost per boiler and includes insulating jacket, standard controls, burner and safety controls. Costs do not include piping or boiler base pad. Outputs are Gross.

D3020 106	Boilers, Hot Water & Steam	COST EACH		
		MAT.	INST.	TOTAL
0600	Boiler, electric, steel, hot water, 12 K.W., 41 M.B.H.	3,175	1,050	4,225
0620	30 K.W., 103 M.B.H.	3,775	1,150	4,925
0640	60 K.W., 205 M.B.H.	4,650	1,250	5,900
0660	120 K.W., 410 M.B.H.	5,150	1,525	6,675
0680	210 K.W., 716 M.B.H.	5,975	2,275	8,250
0700	510 K.W., 1,739 M.B.H.	15,000	4,225	19,225
0720	720 K.W., 2,452 M.B.H.	18,200	4,775	22,975
0740	1,200 K.W., 4,095 M.B.H.	23,500	5,475	28,975
0760	2,100 K.W., 7,167 M.B.H.	44,900	6,900	51,800
0780	3,600 K.W., 12,283 M.B.H.	69,000	11,600	80,600
0820	Steam, 6 K.W., 20.5 M.B.H.	3,000	1,150	4,150
0840	24 K.W., 81.8 M.B.H.	3,650	1,250	4,900
0860	60 K.W., 205 M.B.H.	5,050	1,375	6,425
0880	150 K.W., 512 M.B.H.	7,300	2,100	9,400
0900	510 K.W., 1,740 M.B.H.	19,300	5,175	24,475
0920	1,080 K.W., 3,685 M.B.H.	27,100	7,450	34,550
0940	2,340 K.W., 7,984 M.B.H.	55,500	11,600	67,100
0980	Gas, cast iron, hot water, 80 M.B.H.	1,700	1,275	2,975
1000	100 M.B.H.	1,925	1,400	3,325
1020	163 M.B.H.	2,375	1,875	4,250
1040	280 M.B.H.	2,900	2,075	4,975
1060	544 M.B.H.	6,875	3,700	10,575
1080	1,088 M.B.H.	7,500	4,700	12,200
1100	2,000 M.B.H.	13,500	7,325	20,825
1120	2,856 M.B.H.	19,200	9,375	28,575
1140	4,720 M.B.H.	54,500	12,900	67,400
1160	6,970 M.B.H.	69,500	21,100	90,600
1180	For steam systems under 2,856 M.B.H., add 8%			
1240	Steel, hot water, 72 M.B.H.	2,225	690	2,915
1260	101 M.B.H.	2,550	765	3,315
1280	132 M.B.H.	2,925	810	3,735
1300	150 M.B.H.	3,375	920	4,295
1320	240 M.B.H.	5,175	1,050	6,225
1340	400 M.B.H.	7,375	1,725	9,100
1360	640 M.B.H.	10,000	2,300	12,300
1380	800 M.B.H.	11,700	2,750	14,450
1400	960 M.B.H.	14,500	3,075	17,575
1420	1,440 M.B.H.	20,200	3,950	24,150

SERVICES

D

D3020 Heat Generating Systems

D3020 106	Boilers, Hot Water & Steam	COST EACH		
		MAT.	INST.	TOTAL
1440	2,400 M.B.H.	32,600	6,900	39,500
1460	3,000 M.B.H.	40,400	9,200	49,600
1520	Oil, cast iron, hot water, 109 M.B.H.	2,125	1,575	3,700
1540	173 M.B.H.	2,925	1,875	4,800
1560	236 M.B.H.	3,200	2,200	5,400
1580	1,084 M.B.H.	7,725	5,000	12,725
1600	1,600 M.B.H.	10,200	7,175	17,375
1620	2,480 M.B.H.	15,100	9,150	24,250
1640	3,550 M.B.H.	19,400	11,000	30,400
1660	Steam systems same price as hot water			
1700	Steel, hot water, 103 M.B.H.	1,400	860	2,260
1720	137 M.B.H.	3,000	1,000	4,000
1740	225 M.B.H.	3,525	1,125	4,650
1760	315 M.B.H.	4,100	1,450	5,550
1780	420 M.B.H.	5,425	1,975	7,400
1800	630 M.B.H.	7,100	2,650	9,750
1820	735 M.B.H.	7,100	2,875	9,975
1840	1,050 M.B.H.	11,300	3,775	15,075
1860	1,365 M.B.H.	13,100	4,275	17,375
1880	1,680 M.B.H.	15,200	4,800	20,000
1900	2,310 M.B.H.	16,700	6,600	23,300
1940	3,150 M.B.H.	21,900	10,600	32,500

SERVICES D

Important: See the Reference Section for critical supporting data - Reference Numbers and City Cost Indexes

D3020 Heat Generating Systems

Unit Heater

Fossil Fuel Boiler System Considerations:

1. Terminal units are horizontal unit heaters. Quantities are varied to accommodate total heat loss per building.
2. Unit heater selection was determined by their capacity to circulate the building volume a minimum of three times per hour in addition to the BTU output.
3. Systems shown are forced hot water. Steam boilers cost slightly more than hot water boilers. However, this is compensated for by the smaller size or fewer terminal units required with steam.
4. Floor levels are based on 10' story heights.
5. MBH requirements are gross boiler output.

D3020 108	Heating Systems, Unit Heaters	COST PER S.F.		
		MAT.	INST.	TOTAL
1260	Heating systems, hydronic, fossil fuel, terminal unit heaters,			
1280	Cast iron boiler, gas, 80 M.B.H., 1,070 S.F. bldg.	6.68	7.29	13.97
1320	163 M.B.H., 2,140 S.F. bldg.	4.40	4.94	9.34
1360	544 M.B.H., 7,250 S.F. bldg.	3.12	3.56	6.68
1400	1,088 M.B.H., 14,500 S.F. bldg.	2.43	3.37	5.80
1440	3,264 M.B.H., 43,500 S.F. bldg.	2.06	2.53	4.59
1480	5,032 M.B.H., 67,100 S.F. bldg.	2.62	2.62	5.24
1520	Oil, 109 M.B.H., 1,420 S.F. bldg.	6.85	6.50	13.35
1560	235 M.B.H., 3,150 S.F. bldg.	4.38	4.70	9.08
1600	940 M.B.H., 12,500 S.F. bldg.	3.14	3.08	6.22
1640	1,600 M.B.H., 21,300 S.F. bldg.	3	2.95	5.95
1680	2,480 M.B.H., 33,100 S.F. bldg.	3.04	2.66	5.70
1720	3,350 M.B.H., 44,500 S.F. bldg.	2.60	2.72	5.32
1800	300 M.B.H., 4,000 S.F. bldg.	4.13	3.32	7.45
1840	2,360 M.B.H., 31,500 S.F. bldg.	2.87	2.72	5.59
1880	Steel boiler, gas, 72 M.B.H., 1,020 S.F. bldg.	6.35	5.40	11.75
1920	240 M.B.H., 3,200 S.F. bldg.	4.30	4.30	8.60
1960	480 M.B.H., 6,400 S.F. bldg.	3.52	3.19	6.71
2000	800 M.B.H., 10,700 S.F. bldg.	3.02	2.84	5.86
2040	1,960 M.B.H., 26,100 S.F. bldg.	2.69	2.57	5.26
2080	3,000 M.B.H., 40,000 S.F. bldg.	2.67	2.61	5.28
2120	Oil, 97 M.B.H., 1,300 S.F. bldg.	5.80	6.15	11.95
2160	315 M.B.H., 4,550 S.F. bldg.	3.68	3.13	6.81
2200	525 M.B.H., 7,000 S.F. bldg.	4.10	3.18	7.28
2240	1,050 M.B.H., 14,000 S.F. bldg.	3.41	3.21	6.62
2280	2,310 M.B.H. 30,800 S.F. bldg.	3	2.79	5.79
2320	3,150 M.B.H., 42,000 S.F. bldg.	2.71	2.82	5.53

SERVICES

D

D3020 Heat Generating Systems

Fin Tube Radiator

Fossil Fuel Boiler System Considerations:

1. Terminal units are commercial steel fin tube radiation. Quantities are varied to accommodate total heat loss per building.
2. Systems shown are forced hot water. Steam boilers cost slightly more than hot water boilers. However, this is compensated for by the smaller size or fewer terminal units required with steam.
3. Floor levels are based on 10' story heights.
4. MBH requirements are gross boiler output.

D3020 110	Heating System, Fin Tube Radiation	COST PER S.F.		
		MAT.	INST.	TOTAL
3230	Heating systems, hydronic, fossil fuel, fin tube radiation			
3240	Cast iron boiler, gas, 80 MBH, 1,070 S.F. bldg.	8.01	10.84	18.85
3280	169 M.B.H., 2,140 S.F. bldg.	4.98	6.90	11.88
3320	544 M.B.H., 7,250 S.F. bldg.	4.11	6	10.11
3360	1,088 M.B.H., 14,500 S.F. bldg.	3.48	5.90	9.38
3400	3,264 M.B.H., 43,500 S.F. bldg.	3.21	5.15	8.36
3440	5,032 M.B.H., 67,100 S.F. bldg.	3.79	5.20	8.99
3480	Oil, 109 M.B.H., 1,420 S.F. bldg.	9.30	11.90	21.20
3520	235 M.B.H., 3,150 S.F. bldg.	5.35	7.10	12.45
3560	940 M.B.H., 12,500 S.F. bldg.	4.23	5.60	9.83
3600	1,600 M.B.H., 21,300 S.F. bldg.	4.17	5.55	9.72
3640	2,480 M.B.H., 33,100 S.F. bldg.	4.22	5.30	9.52
3680	3,350 M.B.H., 44,500 S.F. bldg.	3.77	5.30	9.07
3720	Coal, 148 M.B.H., 1,975 S.F. bldg.	6.35	6.60	12.95
3760	300 M.B.H., 4,000 S.F. bldg.	5.05	5.65	10.70
3800	2,360 M.B.H., 31,500 S.F. bldg.	4	5.25	9.25
3840	Steel boiler, gas, 72 M.B.H., 1,020 S.F. bldg.	8.75	10.35	19.10
3880	240 M.B.H., 3,200 S.F. bldg.	5.35	6.80	12.15
3920	480 M.B.H., 6,400 S.F. bldg.	4.54	5.60	10.14
3960	800 M.B.H., 10,700 S.F. bldg.	4.43	5.95	10.38
4000	1,960 M.B.H., 26,100 S.F. bldg.	3.82	5.15	8.97
4040	3,000 M.B.H., 40,000 S.F. bldg.	3.77	5.15	8.92
4080	Oil, 97 M.B.H., 1,300 S.F. bldg.	6.85	9.80	16.65
4120	315 M.B.H., 4,550 S.F. bldg.	4.60	5.35	9.95
4160	525 M.B.H., 7,000 S.F. bldg.	5.20	5.70	10.90
4200	1,050 M.B.H., 14,000 S.F. bldg.	4.52	5.75	10.27
4240	2,310 M.B.H., 30,800 S.F. bldg.	4.11	5.30	9.41
4280	3,150 M.B.H., 42,000 S.F. bldg.	3.83	5.40	9.23

Important: See the Reference Section for critical supporting data - Reference Numbers and City Cost Indexes

D30 HVAC

D3030 Cooling Generating Systems

Chilled Water Supply & Return Piping

Air Cooled Water Chiller Unit

Roof

Insulate

Return

Fan Coil Unit

Supply

Finish Ceiling

*Cooling requirements would lead to a choice of multiple chillers.

D3030 110	Chilled Water, Air Cooled Condenser Systems	COST PER S.F.		
		MAT.	INST.	TOTAL
1180	Packaged chiller, air cooled, with fan coil unit			
1200	Apartment corridors, 3,000 S.F., 5.50 ton	4.63	6	10.63
1240	6,000 S.F., 11.00 ton	3.80	4.80	8.60
1280	10,000 S.F., 18.33 ton	3.38	3.65	7.03
1320	20,000 S.F., 36.66 ton	2.52	2.64	5.16
1360	40,000 S.F., 73.33 ton	3.41	2.68	6.09
1440	Banks and libraries, 3,000 S.F., 12.50 ton	7.35	7.10	14.45
1480	6,000 S.F., 25.00 ton	6.85	5.75	12.60
1520	10,000 S.F., 41.66 ton	5.10	4.02	9.12
1560	20,000 S.F., 83.33 ton	5.90	3.65	9.55
1600	40,000 S.F., 167 ton*			
1680	Bars and taverns, 3,000 S.F., 33.25 ton	12.70	8.40	21.10
1720	6,000 S.F., 66.50 ton	11.50	6.50	18
1760	10,000 S.F., 110.83 ton	9.80	2.08	11.88
1800	20,000 S.F., 220 ton*			
1840	40,000 S.F., 440 ton*			
1920	Bowling alleys, 3,000 S.F., 17.00 ton	9.40	7.95	17.35
1960	6,000 S.F., 34.00 ton	6.95	5.60	12.55
2000	10,000 S.F., 56.66 ton	6.10	4.05	10.15
2040	20,000 S.F., 113.33 ton	6.70	3.69	10.39
2080	40,000 S.F., 227 ton*			
2160	Department stores, 3,000 S.F., 8.75 ton	6.60	6.70	13.30
2200	6,000 S.F., 17.50 ton	5.15	5.15	10.30
2240	10,000 S.F., 29.17 ton	3.96	3.74	7.70
2280	20,000 S.F., 58.33 ton	3.39	2.77	6.16
2320	40,000 S.F., 116.66 ton	4.29	2.76	7.05
2400	Drug stores, 3,000 S.F., 20.00 ton	10.75	8.20	18.95
2440	6,000 S.F., 40.00 ton	8.20	6.10	14.30
2480	10,000 S.F., 66.66 ton	9	5.25	14.25
2520	20,000 S.F., 133.33 ton	7.95	3.89	11.84
2560	40,000 S.F., 267 ton*			
2640	Factories, 2,000 S.F., 10.00 ton	6.30	6.80	13.10
2680	6,000 S.F., 20.00 ton	5.85	5.45	11.30
2720	10,000 S.F., 33.33 ton	4.34	3.83	8.17
2760	20,000 S.F., 66.66 ton	5.10	3.46	8.56
2800	40,000 S.F., 133.33 ton	4.66	2.83	7.49
2880	Food supermarkets, 3,000 S.F., 8.50 ton	6.50	6.65	13.15
2920	6,000 S.F., 17.00 ton	5.05	5.15	10.20
2960	10,000 S.F., 28.33 ton	3.77	3.63	7.40
3000	20,000 S.F., 56.66 ton	3.27	2.70	5.97
3040	40,000 S.F., 113.33 ton	4.20	2.76	6.96
3120	Medical centers, 3,000 S.F., 7.00 ton	5.75	6.50	12.25

D

SERVICES

D3030 Cooling Generating Systems

D3030 110	Chilled Water, Air Cooled Condenser Systems	COST PER S.F.		
		MAT.	INST.	TOTAL
3160	6,000 S.F., 14.00 ton	4.41	4.97	9.38
3200	10,000 S.F., 23.33 ton	4.04	3.80	7.84
3240	20,000 S.F., 46.66 ton	3.05	2.79	5.84
3280	40,000 S.F., 93.33 ton	3.81	2.71	6.52
3360	Offices, 3,000 S.F., 9.50 ton	5.80	6.60	12.40
3400	6,000 S.F., 19.00 ton	5.65	5.40	11.05
3440	10,000 S.F., 31.66 ton	4.19	3.79	7.98
3480	20,000 S.F., 63.33 ton	5.10	3.53	8.63
3520	40,000 S.F., 126.66 ton	4.58	2.86	7.44
3600	Restaurants, 3,000 S.F., 15.00 ton	8.35	7.35	15.70
3640	6,000 S.F., 30.00 ton	6.50	5.55	12.05
3680	10,000 S.F., 50.00 ton	5.75	4.11	9.86
3720	20,000 S.F., 100.00 ton	6.75	3.86	10.61
3760	40,000 S.F., 200 ton*			
3840	Schools and colleges, 3,000 S.F., 11.50 ton	6.90	6.95	13.85
3880	6,000 S.F., 23.00 ton	6.45	5.65	12.10
3920	10,000 S.F., 38.33 ton	4.80	3.95	8.75
3960	20,000 S.F., 76.66 ton	5.70	3.69	9.39
4000	40,000 S.F., 153 ton*			

Important: See the Reference Section for critical supporting data - Reference Numbers and City Cost Indexes

SERVICES D

D3030 Cooling Generating Systems

Reciprocating Package Chiller
Condenser Water
Cooling Tower
Cooling Tower Water Makeup
Chilled Water Supply & Return Piping
Roof Structure
Finish Ceiling
Insulate
Return
Supply
Fan Coil Unit

*Cooling requirements would lead to a
 choice of multiple chillers

D3030 115	Chilled Water, Cooling Tower Systems	COST PER S.F.		
		MAT.	INST.	TOTAL
1300	Packaged chiller, water cooled, with fan coil unit			
1320	Apartment corridors, 4,000 S.F., 7.33 ton	5.55	5.85	11.40
1360	6,000 S.F., 11.00 ton	4.43	5.05	9.48
1400	10,000 S.F., 18.33 ton	3.17	3.75	6.92
1440	20,000 S.F., 26.66 ton	2.54	2.81	5.35
1480	40,000 S.F., 73.33 ton	3.78	2.88	6.66
1520	60,000 S.F., 110.00 ton	3.74	2.95	6.69
1600	Banks and libraries, 4,000 S.F., 16.66 ton	6.90	6.45	13.35
1640	6,000 S.F., 25.00 ton	5.85	5.70	11.55
1680	10,000 S.F., 41.66 ton	4.83	4.23	9.06
1720	20,000 S.F., 83.33 ton	6.70	4.06	10.76
1760	40,000 S.F., 166.66 ton	6.55	4.98	11.53
1800	60,000 S.F., 250.00 ton	6.50	5.30	11.80
1880	Bars and taverns, 4,000 S.F., 44.33 ton	11.55	8.15	19.70
1920	6,000 S.F., 66.50 ton	14.75	8.45	23.20
1960	10,000 S.F., 110.83 ton	14.10	6.70	20.80
2000	20,000 S.F., 221.66 ton	13.15	7.15	20.30
2040	40,000 S.F., 440 ton*			
2080	60,000 S.F., 660 ton*			
2160	Bowling alleys, 4,000 S.F., 22.66 ton	8.15	7.10	15.25
2200	6,000 S.F., 34.00 ton	7.10	6.25	13.35
2240	10,000 S.F., 56.66 ton	6.80	4.59	11.39
2280	20,000 S.F., 113.33 ton	7.85	4.29	12.14
2320	40,000 S.F., 226.66 ton	7.60	5.15	12.75
2360	60,000 S.F., 340 ton			
2440	Department stores, 4,000 S.F., 11.66 ton	6.35	6.35	12.70
2480	6,000 S.F., 17.50 ton	5.05	5.25	10.30
2520	10,000 S.F., 29.17 ton	3.85	3.92	7.77
2560	20,000 S.F., 58.33 ton	3.52	2.92	6.44
2600	40,000 S.F., 116.66 ton	4.65	3.03	7.68
2640	60,000 S.F., 175.00 ton	5.35	4.93	10.28
2720	Drug stores, 4,000 S.F., 26.66 ton	8.75	7.30	16.05
2760	6,000 S.F., 40.00 ton	7.65	6.25	13.90
2800	10,000 S.F., 66.66 ton	9.40	5.70	15.10
2840	20,000 S.F., 133.33 ton	8.90	4.47	13.37
2880	40,000 S.F., 266.67 ton	8.75	5.60	14.35
2920	60,000 S.F., 400 ton*			
3000	Factories, 4,000 S.F., 13.33 ton	5.90	6.15	12.05
3040	6,000 S.F., 20.00 ton	5.15	5.30	10.45
3080	10,000 S.F., 33.33 ton	4.22	4.04	8.26
3120	20,000 S.F., 66.66 ton	5.25	3.64	8.89
3160	40,000 S.F., 133.33 ton	5.10	3.12	8.22

D3030 Cooling Generating Systems

D3030 115	Chilled Water, Cooling Tower Systems	COST PER S.F.		
		MAT.	INST.	TOTAL
3200	60,000 S.F., 200.00 ton	5.75	5.05	10.80
3280	Food supermarkets, 4,000 S.F., 11.33 ton	6.20	6.30	12.50
3320	6,000 S.F., 17.00 ton	4.54	5.15	9.69
3360	10,000 S.F., 28.33 ton	3.77	3.88	7.65
3400	20,000 S.F., 56.66 ton	3.58	2.92	6.50
3440	40,000 S.F., 113.33 ton	4.60	3.02	7.62
3480	60,000 S.F., 170.00 ton	5.30	4.91	10.21
3560	Medical centers, 4.000 S.F., 9.33 ton	5.30	5.80	11.10
3600	6,000 S.F., 14.00 ton	4.35	5.05	9.40
3640	10,000 S.F., 23.33 ton	3.48	3.77	7.25
3680	20,000 S.F., 46.66 ton	3.15	2.86	6.01
3720	40,000 S.F., 93.33 ton	4.23	2.97	7.20
3760	60,000 S.F., 140.00 ton	4.93	4.88	9.81
3840	Offices, 4,000 S.F., 12.66 ton	5.70	6.10	11.80
3880	6,000 S.F., 19.00 ton	5.10	5.40	10.50
3920	10,000 S.F., 31.66 ton	4.19	4.09	8.28
3960	20,000 S.F., 63.33 ton	5.15	3.66	8.81
4000	40,000 S.F., 126.66 ton	5.55	4.77	10.32
4040	60,000 S.F., 190.00 ton	5.55	5	10.55
4120	Restaurants, 4,000 S.F., 20.00 ton	7.30	6.55	13.85
4160	6,000 S.F., 30.00 ton	6.40	5.85	12.25
4200	10,000 S.F., 50.00 ton	6.25	4.40	10.65
4240	20,000 S.F., 100.00 ton	7.50	4.26	11.76
4280	40,000 S.F., 200.00 ton	6.80	4.84	11.64
4320	60,000 S.F., 300.00 ton	7.25	5.50	12.75
4400	Schools and colleges, 4,000 S.F., 15.33 ton	6.50	6.30	12.80
4440	6,000 S.F., 23.00 ton	5.50	5.60	11.10
4480	10,000 S.F., 38.33 ton	4.53	4.15	8.68
4520	20,000 S.F., 76.66 ton	6.35	4.01	10.36
4560	40,000 S.F., 153.33 ton	6.15	4.88	11.03
4600	60,000 S.F., 230.00 ton	6	5.10	11.10

D3050 Terminal & Package Units

Roof

Roof Top Unit

Return Duct

Fan Coil Unit

Insulated Supply Duct

Return Grille

Supply Diff.

Finish Ceiling

*Above normal capacity

Rooftop, Single Zone System

D3050 150	Rooftop Single Zone Unit Systems	COST PER S.F.		
		MAT.	INST.	TOTAL
1260	Rooftop, single zone, air conditioner			
1280	Apartment corridors, 500 S.F., .92 ton	3.15	2.90	6.05
1320	1,000 S.F., 1.83 ton	3.13	2.87	6
1360	1500 S.F., 2.75 ton	2.44	2.36	4.80
1400	3,000 S.F., 5.50 ton	2.20	2.29	4.49
1440	5,000 S.F., 9.17 ton	2.10	2.09	4.19
1480	10,000 S.F., 18.33 ton	2.17	1.96	4.13
1560	Banks or libraries, 500 S.F., 2.08 ton	7.10	6.50	13.60
1600	1,000 S.F., 4.17 ton	5.55	5.40	10.95
1640	1,500 S.F., 6.25 ton	5	5.20	10.20
1680	3,000 S.F., 12.50 ton	4.77	4.74	9.51
1720	5,000 S.F., 20.80 ton	4.91	4.46	9.37
1760	10,000 S.F., 41.67 ton	4.96	4.44	9.40
1840	Bars and taverns, 500 S.F. 5.54 ton	11.60	8.65	20.25
1880	1,000 S.F., 11.08 ton	10.95	7.45	18.40
1920	1,500 S.F., 16.62 ton	11.40	6.95	18.35
1960	3,000 S.F., 33.25 ton	11.70	6.65	18.35
2000	5,000 S.F., 55.42 ton	10.40	6.65	17.05
2040	10,000 S.F., 110.83 ton*			
2080	Bowling alleys, 500 S.F., 2.83 ton	7.50	7.30	14.80
2120	1,000 S.F., 5.67 ton	6.80	7.05	13.85
2160	1,500 S.F., 8.50 ton	6.50	6.45	12.95
2200	3,000 S.F., 17.00 ton	6.70	6.20	12.90
2240	5,000 S.F., 28.33 ton	6.85	6.05	12.90
2280	10,000 S.F., 56.67 ton	6.20	6.05	12.25
2360	Department stores, 500 S.F., 1.46 ton	4.99	4.58	9.57
2400	1,000 S.F., 2.92 ton	3.88	3.78	7.66
2480	3,000 S.F., 8.75 ton	3.34	3.32	6.66
2520	5,000 S.F., 14.58 ton	3.45	3.18	6.63
2560	10,000 S.F., 29.17 ton	3.52	3.10	6.62
2640	Drug stores, 500 S.F., 3.33 ton	8.85	8.60	17.45
2680	1,000 S.F., 6.67 ton	8	8.30	16.30
2720	1,500 S.F., 10.00 ton	7.65	7.60	15.25
2760	3,000 S.F., 20.00 ton	7.85	7.15	15
2800	5,000 S.F., 33.33 ton	8.05	7.10	15.15
2840	10,000 S.F., 66.67 ton	7.30	7.10	14.40
2920	Factories, 500 S.F., 1.67 ton	5.70	5.25	10.95
3000	1,500 S.F., 5.00 ton	4	4.15	8.15
3040	3,000 S.F., 10.00 ton	3.81	3.79	7.60
3080	5,000 S.F., 16.67 ton	3.94	3.64	7.58
3120	10,000 S.F., 33.33 ton	4.03	3.54	7.57
3200	Food supermarkets, 500 S.F., 1.42 ton	4.84	4.45	9.29

SERVICES

D

D3050 Terminal & Package Units

D3050 150	Rooftop Single Zone Unit Systems	COST PER S.F.		
		MAT.	INST.	TOTAL
3240	1,000 S.F., 2.83 ton	3.73	3.65	7.38
3280	1,500 S.F., 4.25 ton	3.40	3.53	6.93
3320	3,000 S.F., 8.50 ton	3.25	3.23	6.48
3360	5,000 S.F., 14.17 ton	3.35	3.10	6.45
3400	10,000 S.F., 28.33 ton	3.43	3.01	6.44
3480	Medical centers, 500 S.F., 1.17 ton	3.99	3.67	7.66
3520	1,000 S.F., 2.33 ton	3.98	3.66	7.64
3560	1,500 S.F., 3.50 ton	3.10	3.02	6.12
3640	5,000 S.F., 11.67 ton	2.67	2.65	5.32
3680	10,000 S.F., 23.33 ton	2.75	2.50	5.25
3760	Offices, 500 S.F., 1.58 ton	5.40	4.96	10.36
3800	1,000 S.F., 3.17 ton	4.21	4.10	8.31
3840	1,500 S.F., 4.75 ton	3.80	3.95	7.75
3880	3,000 S.F., 9.50 ton	3.62	3.61	7.23
3920	5,000 S.F., 15.83 ton	3.75	3.45	7.20
3960	10,000 S.F., 31.67 ton	3.83	3.37	7.20
4000	Restaurants, 500 S.F., 2.50 ton	8.55	7.85	16.40
4040	1,000 S.F., 5.00 ton	6	6.25	12.25
4080	1,500 S.F., 7.50 ton	5.75	5.70	11.45
4120	3,000 S.F., 15.00 ton	5.90	5.45	11.35
4160	5,000 S.F., 25.00 ton	5.90	5.35	11.25
4200	10,000 S.F., 50.00 ton	5.45	5.30	10.75
4240	Schools and colleges, 500 S.F., 1.92 ton	6.55	6	12.55
4280	1,000 S.F., 3.83 ton	5.10	4.95	10.05
4360	3,000 S.F., 11.50 ton	4.39	4.37	8.76
4400	5,000 S.F., 19.17 ton	4.52	4.11	8.63

SERVICES D

Important: See the Reference Section for critical supporting data - Reference Numbers and City Cost Indexes

D3050 Terminal & Package Units

*Note A: Small single zone unit recommended.

*Note B: A combination of multizone units recommended.

D3050 155	Rooftop Multizone Unit Systems	COST PER S.F.		
		MAT.	INST.	TOTAL
1240	Rooftop, multizone, air conditioner			
1260	Apartment corridors, 1,500 S.F., 2.75 ton. See Note A.			
1280	3,000 S.F., 5.50 ton	12.70	3.70	16.40
1320	10,000 S.F., 18.30 ton	9.35	3.57	12.92
1360	15,000 S.F., 27.50 ton	8.05	3.55	11.60
1400	20,000 S.F., 36.70 ton	8.15	3.57	11.72
1440	25,000 S.F., 45.80 ton	7.05	3.58	10.63
1520	Banks or libraries, 1,500 S.F., 6.25 ton	29	8.45	37.45
1560	3,000 S.F., 12.50 ton	24.50	8.25	32.75
1600	10,000 S.F., 41.67 ton	16	8.15	24.15
1640	15,000 S.F., 62.50 ton	10.45	8.10	18.55
1680	20,000 S.F., 83.33 ton	10.45	8.10	18.55
1720	25,000 S.F., 104.00 ton	9.40	8.05	17.45
1800	Bars and taverns, 1,500 S.F., 16.62 ton	62.50	12.10	74.60
1840	3,000 S.F., 33.24 ton	47	11.70	58.70
1880	10,000 S.F., 110.83 ton	22.50	11.65	34.15
1920	15,000 S.F., 165 ton, See Note B			
1960	20,000 S.F., 220 ton, See Note B			
2000	25,000 S.F., 275 ton, See Note B			
2080	Bowling alleys, 1,500 S.F., 8.50 ton	39	11.45	50.45
2120	3,000 S.F., 17.00 ton	33	11.20	44.20
2160	10,000 S.F., 56.70 ton	22	11.10	33.10
2200	15,000 S.F., 85.00 ton	14.20	11	25.20
2240	20,000 S.F., 113.00 ton	12.80	10.95	23.75
2280	25,000 S.F., 140.00 ton see Note B			
2360	Department stores, 1,500 S.F., 4.37 ton, See Note A.			
2400	3,000 S.F., 8.75 ton	20	5.90	25.90
2440	10,000 S.F., 29.17 ton	12.95	5.65	18.60
2520	20,000 S.F., 58.33 ton	7.30	5.65	12.95
2560	25,000 S.F., 72.92 ton	7.30	5.65	12.95
2640	Drug stores, 1,500 S.F., 10.00 ton	46	13.45	59.45
2680	3,000 S.F., 20.00 ton	34	13	47
2720	10,000 S.F., 66.66 ton	16.75	12.95	29.70
2760	15,000 S.F., 100.00 ton	15.10	12.90	28
2800	20,000 S.F., 135 ton, See Note B			
2840	25,000 S.F., 165 ton, See Note B			
2920	Factories, 1,500 S.F., 5 ton, See Note A			
3000	10,000 S.F., 33.33 ton	14.80	6.50	21.30
3040	15,000 S.F., 50.00 ton	12.80	6.50	19.30
3080	20,000 S.F., 66.66 ton	8.35	6.50	14.85

SERVICES

D

D3050 Terminal & Package Units

D3050 155	Rooftop Multizone Unit Systems	COST PER S.F.		
		MAT.	INST.	TOTAL
3120	25,000 S.F., 83.33 ton	8.35	6.50	14.85
3200	Food supermarkets, 1,500 S.F., 4.25 ton, See Note A			
3240	3,000 S.F., 8.50 ton	19.60	5.75	25.35
3280	10,000 S.F., 28.33 ton	12.45	5.50	17.95
3320	15,000 S.F., 42.50 ton	10.85	5.55	16.40
3360	20,000 S.F., 56.67 ton	7.10	5.50	12.60
3400	25,000 S.F., 70.83 ton	7.10	5.50	12.60
3480	Medical centers, 1,500 S.F., 3.5 ton, See Note A			
3520	3,000 S.F., 7.00 ton	16.15	4.71	20.86
3560	10,000 S.F., 23.33 ton	10.95	4.53	15.48
3600	15,000 S.F., 35.00 ton	10.35	4.53	14.88
3640	20,000 S.F., 46.66 ton	8.95	4.56	13.51
3680	25,000 S.F., 58.33 ton	5.85	4.53	10.38
3760	Offices, 1,500 S.F., 4.75 ton, See Note A			
3800	3,000 S.F., 9.50 ton	22	6.40	28.40
3840	10,000 S.F., 31.66 ton	14.10	6.15	20.25
3920	20,000 S.F., 63.33 ton	7.95	6.15	14.10
3960	25,000 S.F., 79.16 ton	7.95	6.15	14.10
4000	Restaurants, 1,500 S.F., 7.50 ton	34.50	10.10	44.60
4040	3,000 S.F., 15.00 ton	29	9.90	38.90
4080	10,000 S.F., 50.00 ton	19.20	9.80	29
4120	15,000 S.F., 75.00 ton	12.55	9.70	22.25
4160	20,000 S.F., 100.00 ton	11.30	9.70	21
4200	25,000 S.F., 125 ton, See Note B			
4240	Schools and colleges, 1,500 S.F., 5.75 ton	26.50	7.75	34.25
4320	10,000 S.F., 38.33 ton	14.70	7.50	22.20
4360	15,000 S.F., 57.50 ton	9.60	7.45	17.05
4400	20,000 S.F., 76.66 ton	9.60	7.45	17.05
4440	25,000 S.F., 95.83 ton	9.15	7.45	16.60

Important: See the Reference Section for critical supporting data - Reference Numbers and City Cost Indexes

D3050 Terminal & Package Units

Self-Contained Water Cooled System

D3050 160	Self-contained, Water Cooled Unit Systems	COST PER S.F.		
		MAT.	INST.	TOTAL
1280	Self-contained, water cooled unit	2.24	2.32	4.56
1300	Apartment corridors, 500 S.F., .92 ton	2.25	2.30	4.55
1320	1,000 S.F., 1.83 ton	2.24	2.31	4.55
1360	3,000 S.F., 5.50 ton	1.84	1.94	3.78
1400	5,000 S.F., 9.17 ton	1.82	1.81	3.63
1440	10,000 S.F., 18.33 ton	2.77	1.63	4.40
1520	Banks or libraries, 500 S.F., 2.08 ton	4.68	2.31	6.99
1560	1,000 S.F., 4.17 ton	4.20	4.41	8.61
1600	3,000 S.F., 12.50 ton	4.15	4.10	8.25
1640	5,000 S.F., 20.80 ton	6.30	3.68	9.98
1680	10,000 S.F., 41.66 ton	5.45	3.69	9.14
1760	Bars and taverns, 500 S.F., 5.54 ton	10	3.90	13.90
1800	1,000 S.F., 11.08 ton	10.40	7	17.40
1840	3,000 S.F., 33.25 ton	14.45	5.55	20
1880	5,000 S.F., 55.42 ton	13.55	5.80	19.35
1920	10,000 S.F., 110.00 ton	13.20	5.75	18.95
2000	Bowling alleys, 500 S.F., 2.83 ton	6.40	3.15	9.55
2040	1,000 S.F., 5.66 ton	5.70	6	11.70
2080	3,000 S.F., 17.00 ton	8.60	5.05	13.65
2120	5,000 S.F., 28.33 ton	7.70	4.85	12.55
2160	10,000 S.F., 56.66 ton	7.20	4.97	12.17
2200	Department stores, 500 S.F., 1.46 ton	3.29	1.62	4.91
2240	1,000 S.F., 2.92 ton	2.94	3.09	6.03
2280	3,000 S.F., 8.75 ton	2.90	2.88	5.78
2320	5,000 S.F., 14.58 ton	4.41	2.59	7
2360	10,000 S.F., 29.17 ton	3.97	2.50	6.47
2440	Drug stores, 500 S.F., 3.33 ton	7.50	3.70	11.20
2480	1,000 S.F., 6.66 ton	6.70	7.05	13.75
2520	3,000 S.F., 20.00 ton	10.10	5.90	16
2560	5,000 S.F., 33.33 ton	10.60	5.80	16.40
2600	10,000 S.F., 66.66 ton	8.50	5.85	14.35
2680	Factories, 500 S.F., 1.66 ton	3.74	1.85	5.59
2720	1,000 S.F. 3.37 ton	3.39	3.57	6.96
2760	3,000 S.F., 10.00 ton	3.31	3.28	6.59
2800	5,000 S.F., 16.66 ton	5.05	2.95	8
2840	10,000 S.F., 33.33 ton	4.53	2.85	7.38
2920	Food supermarkets, 500 S.F., 1.42 ton	3.19	1.57	4.76
2960	1,000 S.F., 2.83 ton	3.47	3.57	7.04

D SERVICES

D3050 Terminal & Package Units

D3050 160	Self-contained, Water Cooled Unit Systems	COST PER S.F.		
		MAT.	INST.	TOTAL
3000	3,000 S.F., 8.50 ton	2.85	3	5.85
3040	5,000 S.F., 14.17 ton	2.82	2.79	5.61
3080	10,000 S.F., 28.33 ton	3.86	2.42	6.28
3160	Medical centers, 500 S.F., 1.17 ton	2.63	1.30	3.93
3200	1,000 S.F., 2.33 ton	2.86	2.95	5.81
3240	3,000 S.F., 7.00 ton	2.35	2.49	4.84
3280	5,000 S.F., 11.66 ton	2.32	2.30	4.62
3320	10,000 S.F., 23.33 ton	3.53	2.07	5.60
3400	Offices, 500 S.F., 1.58 ton	3.56	1.77	5.33
3440	1,000 S.F., 3.17 ton	3.88	4	7.88
3480	3,000 S.F., 9.50 ton	3.15	3.12	6.27
3520	5,000 S.F., 15.83 ton	4.79	2.81	7.60
3560	10,000 S.F., 31.67 ton	4.31	2.71	7.02
3640	Restaurants, 500 S.F., 2.50 ton	5.65	2.78	8.43
3680	1,000 S.F., 5.00 ton	5.05	5.30	10.35
3720	3,000 S.F., 15.00 ton	4.99	4.93	9.92
3760	5,000 S.F., 25.00 ton	7.55	4.44	11.99
3800	10,000 S.F., 50.00 ton	4.57	4.08	8.65
3880	Schools and colleges, 500 S.F., 1.92 ton	4.31	2.13	6.44
3920	1,000 S.F., 3.83 ton	3.85	4.06	7.91
3960	3,000 S.F., 11.50 ton	3.81	3.77	7.58
4000	5,000 S.F., 19.17 ton	5.80	3.40	9.20
4040	10,000 S.F., 38.33 ton	5	3.40	8.40

Important: See the Reference Section for critical supporting data - Reference Numbers and City Cost Indexes

D3050 Terminal & Package Units

Self-Contained Air Cooled System

D3050 165	Self-contained, Air Cooled Unit Systems	COST PER S.F.		
		MAT.	**INST.**	**TOTAL**
1300	Self-contained, air cooled unit			
1320	Apartment corridors, 500 S.F., .92 ton	3	2.95	5.95
1360	1,000 S.F., 1.83 ton	2.96	2.92	5.88
1400	3,000 S.F., 5.50 ton	2.46	2.73	5.19
1440	5,000 S.F., 9.17 ton	2.14	2.61	4.75
1480	10,000 S.F., 18.33 ton	2.23	2.40	4.63
1560	Banks or libraries, 500 S.F., 2.08 ton	6.30	3.70	10
1600	1,000 S.F., 4.17 ton	5.55	6.20	11.75
1640	3,000 S.F., 12.50 ton	4.87	5.90	10.77
1680	5,000 S.F., 20.80 ton	5.10	5.45	10.55
1720	10,000 S.F., 41.66 ton	4.25	5.35	9.60
1800	Bars and taverns, 500 S.F., 5.54 ton	12.50	8.10	20.60
1840	1,000 S.F., 11.08 ton	12.35	11.85	24.20
1880	3,000 S.F., 33.25 ton	11.10	10.25	21.35
1920	5,000 S.F., 55.42 ton	10.55	10.30	20.85
1960	10,000 S.F., 110.00 ton	10.70	10.25	20.95
2040	Bowling alleys, 500 S.F., 2.83 ton	8.60	5.05	13.65
2080	1,000 S.F., 5.66 ton	7.60	8.45	16.05
2120	3,000 S.F., 17.00 ton	6.90	7.40	14.30
2160	5,000 S.F., 28.33 ton	5.95	7.25	13.20
2200	10,000 S.F., 56.66 ton	5.70	7.25	12.95
2240	Department stores, 500 S.F., 1.46 ton	4.43	2.61	7.04
2280	1,000 S.F., 2.92 ton	3.92	4.36	8.28
2320	3,000 S.F., 8.75 ton	3.39	4.15	7.54
2360	5,000 S.F., 14.58 ton	3.39	4.15	7.54
2400	10,000 S.F., 29.17 ton	3.06	3.73	6.79
2480	Drug stores, 500 S.F., 3.33 ton	10.10	5.95	16.05
2520	1,000 S.F., 6.66 ton	8.95	9.95	18.90
2560	3,000 S.F., 20.00 ton	8.20	8.75	16.95
2600	5,000 S.F., 33.33 ton	7	8.50	15.50
2640	10,000 S.F., 66.66 ton	6.75	8.55	15.30
2720	Factories, 500 S.F., 1.66 ton	5.15	3	8.15
2760	1,000 S.F., 3.33 ton	4.51	4.98	9.49
2800	3,000 S.F., 10.00 ton	3.89	4.73	8.62
2840	5,000 S.F., 16.66 ton	4.04	4.35	8.39
2880	10,000 S.F., 33.33 ton	3.49	4.26	7.75
2960	Food supermarkets, 500 S.F., 1.42 ton	4.30	2.53	6.83
3000	1,000 S.F., 2.83 ton	4.56	4.52	9.08
3040	3,000 S.F., 8.50 ton	3.79	4.23	8.02
3080	5,000 S.F., 14.17 ton	3.29	4.03	7.32
3120	10,000 S.F., 28.33 ton	2.96	3.62	6.58
3200	Medical centers, 500 S.F., 1.17 ton	3.56	2.09	5.65

D3050 Terminal & Package Units

D3050 165	Self-contained, Air Cooled Unit Systems	COST PER S.F.		
		MAT.	INST.	TOTAL
3240	1,000 S.F., 2.33 ton	3.79	3.73	7.52
3280	3,000 S.F., 7.00 ton	3.13	3.49	6.62
3320	5,000 S.F., 16.66 ton	2.72	3.31	6.03
3360	10,000 S.F., 23.33 ton	2.84	3.05	5.89
3440	Offices, 500 S.F., 1.58 ton	4.82	2.83	7.65
3480	1,000 S.F., 3.16 ton	5.15	5.05	10.20
3520	3,000 S.F., 9.50 ton	3.71	4.50	8.21
3560	5,000 S.F., 15.83 ton	3.89	4.14	8.03
3600	10,000 S.F., 31.66 ton	3.34	4.04	7.38
3680	Restaurants, 500 S.F., 2.50 ton	7.55	4.45	12
3720	1,000 S.F., 5.00 ton	6.70	7.50	14.20
3760	3,000 S.F., 15.00 ton	5.85	7.10	12.95
3800	5,000 S.F., 25.00 ton	5.25	6.40	11.65
3840	10,000 S.F., 50.00 ton	5.20	6.35	11.55
3920	Schools and colleges, 500 S.F., 1.92 ton	5.80	3.42	9.22
3960	1,000 S.F., 3.83 ton	5.15	5.75	10.90
4000	3,000 S.F., 11.50 ton	4.47	5.45	9.92
4040	5,000 S.F., 19.17 ton	4.67	5	9.67
4080	10,000 S.F., 38.33 ton	3.92	4.91	8.83

Important: See the Reference Section for critical supporting data - Reference Numbers and City Cost Indexes

D3050 Terminal & Package Units

*Cooling requirements would lead to more than one system.

D3050 170	Split Systems With Air Cooled Condensing Units	COST PER S.F.		
		MAT.	INST.	TOTAL
1260	Split system, air cooled condensing unit			
1280	Apartment corridors, 1,000 S.F., 1.83 ton	1.30	1.83	3.13
1320	2,000 S.F., 3.66 ton	1.09	1.86	2.95
1360	5,000 S.F., 9.17 ton	1.43	2.35	3.78
1400	10,000 S.F., 18.33 ton	1.57	2.56	4.13
1440	20,000 S.F., 36.66 ton	1.65	2.59	4.24
1520	Banks and libraries, 1,000 S.F., 4.17 ton	2.46	4.23	6.69
1560	2,000 S.F., 8.33 ton	3.25	5.30	8.55
1600	5,000 S.F., 20.80 ton	3.58	5.80	9.38
1640	10,000 S.F., 41.66 ton	3.75	5.90	9.65
1680	20,000 S.F., 83.32 ton	4.32	6.10	10.42
1760	Bars and taverns, 1,000 S.F., 11.08 ton	7.70	8.35	16.05
1800	2,000 S.F., 22.16 ton	10.15	9.95	20.10
1840	5,000 S.F., 55.42 ton	8.60	9.30	17.90
1880	10,000 S.F., 110.84 ton	10.40	9.70	20.10
1920	20,000 S.F., 220 ton*			
2000	Bowling alleys, 1,000 S.F., 5.66 ton	3.59	7.90	11.49
2040	2,000 S.F., 11.33 ton	4.43	7.25	11.68
2080	5,000 S.F., 28.33 ton	4.88	7.90	12.78
2120	10,000 S.F., 56.66 ton	5.10	8	13.10
2160	20,000 S.F., 113.32 ton	6.45	8.65	15.10
2320	Department stores, 1,000 S.F., 2.92 ton	1.76	2.89	4.65
2360	2,000 S.F., 5.83 ton	1.85	4.08	5.93
2400	5,000 S.F., 14.58 ton	2.28	3.72	6
2440	10,000 S.F., 29.17 ton	2.50	4.06	6.56
2480	20,000 S.F., 58.33 ton	2.62	4.12	6.74
2560	Drug stores, 1,000 S.F., 6.66 ton	4.23	9.30	13.53
2600	2,000 S.F., 13.32 ton	5.20	8.50	13.70
2640	5,000 S.F., 33.33 ton	6	9.40	15.40
2680	10,000 S.F., 66.66 ton	6.05	9.85	15.90
2720	20,000 S.F., 133.32 ton*			
2800	Factories, 1,000 S.F., 3.33 ton	2.01	3.30	5.31
2840	2,000 S.F., 6.66 ton	2.11	4.66	6.77
2880	5,000 S.F., 16.66 ton	2.86	4.63	7.49
2920	10,000 S.F., 33.33 ton	3	4.70	7.70
2960	20,000 S.F., 66.66 ton	3.01	4.93	7.94
3040	Food supermarkets, 1,000 S.F., 2.83 ton	1.70	2.82	4.52
3080	2,000 S.F., 5.66 ton	1.80	3.97	5.77
3120	5,000 S.F., 14.66 ton	2.22	3.61	5.83
3160	10,000 S.F., 28.33 ton	2.43	3.94	6.37
3200	20,000 S.F., 56.66 ton	2.55	4	6.55
3280	Medical centers, 1,000 S.F., 2.33 ton	1.41	2.28	3.69

SERVICES

D

D3050 Cooling Generating Systems

D3050 170	Split Systems With Air Cooled Condensing Units	COST PER S.F.		
		MAT.	INST.	TOTAL
3320	2,000 S.F., 4.66 ton	1.48	3.26	4.74
3360	5,000 S.F., 11.66 ton	1.82	2.96	4.78
3400	10,000 S.F., 23.33 ton	2	3.24	5.24
3440	20,000 S.F., 46.66 ton	2.10	3.28	5.38
3520	Offices, 1,000 S.F., 3.17 ton	1.91	3.15	5.06
3560	2,000 S.F., 6.33 ton	2.01	4.43	6.44
3600	5,000 S.F., 15.83 ton	2.47	4.04	6.51
3640	10,000 S.F., 31.66 ton	2.72	4.41	7.13
3680	20,000 S.F., 63.32 ton	2.86	4.69	7.55
3760	Restaurants, 1,000 S.F., 5.00 ton	3.17	7	10.17
3800	2,000 S.F., 10.00 ton	3.91	6.40	10.31
3840	5,000 S.F., 25.00 ton	4.30	6.95	11.25
3880	10,000 S.F., 50.00 ton	4.50	7.05	11.55
3920	20,000 S.F., 100.00 ton	5.70	7.65	13.35
4000	Schools and colleges, 1,000 S.F., 3.83 ton	2.27	3.88	6.15
4040	2,000 S.F., 7.66 ton	3	4.89	7.89
4080	5,000 S.F., 19.17 ton	3.29	5.35	8.64
4120	10,000 S.F., 38.33 ton	3.45	5.40	8.85

D3090 Other HVAC Systems/Equip

Vitrified Clay Garage Exhaust System

Dual Exhaust System

D3090 320	Garage Exhaust Systems	COST PER BAY		
		MAT.	INST.	TOTAL
1040	Garage, single exhaust, 3" outlet, cars & light trucks, one bay	1,800	980	2,780
1060	Additional bays up to seven bays	325	138	463
1500	4" outlet, trucks, one bay	1,800	980	2,780
1520	Additional bays up to six bays	335	138	473
1600	5" outlet, diesel trucks, one bay	2,050	980	3,030
1650	Additional single bays up to six	590	160	750
1700	Two adjoining bays	2,050	980	3,030
2000	Dual exhaust, 3" outlets, pair of adjoining bays	2,100	1,075	3,175
2100	Additional pairs of adjoining bays	605	160	765

R10520-120 System Classification

System Classification
Rules for installation of sprinkler systems vary depending on the classification of occupancy falling into one of three categories as follows:

Light Hazard Occupancy
The protection area allotted per sprinkler should not exceed 225 S.F., with the maximum distance between lines and sprinklers on lines being 15'. The sprinklers do not need to be staggered. Branch lines should not exceed eight sprinklers on either side of a cross main. Each large area requiring more than 100 sprinklers and without a sub-dividing partition should be supplied by feed mains or risers sized for ordinary hazard occupancy.
Maximum system area = 52,000 S.F.

Included in this group are:

Churches	Nursing Homes
Clubs	Offices
Educational	Residential
Hospitals	Restaurants
Institutional	Theaters and Auditoriums
Libraries	(except stages and prosceniums)
(except large stack rooms)	Unused Attics
Museums	

Ordinary Hazard Occupancy
The protection area allotted per sprinkler shall not exceed 130 S.F. of noncombustible ceiling and 130 S.F. of combustible ceiling. The maximum allowable distance between sprinkler lines and sprinklers on line is 15'. Sprinklers shall be staggered if the distance between heads exceeds 12'. Branch lines should not exceed eight sprinklers on either side of a cross main.
Maximum system area = 52,000 S.F.

Included in this group are:

Group 1	Group 2
Automotive Parking and Showrooms	Cereal Mills
Bakeries	Chemical Plants—Ordinary
Beverage manufacturing	Confectionery Products
Canneries	Distilleries
Dairy Products Manufacturing/Processing	Dry Cleaners
Electronic Plans	Feed Mills
Glass and Glass Products Manufacturing	Horse Stables
Laundries	Leather Goods Manufacturing
Restaurant Service Areas	Libraries—Large Stack Room Areas
	Machine Shops
	Metal Working
	Mercantile
	Paper and Pulp Mills
	Paper Process Plants
	Piers and Wharves
	Post Offices
	Printing and Publishing
	Repair Garages
	Stages
	Textile Manufacturing
	Tire Manufacturing
	Tobacco Products Manufacturing
	Wood Machining
	Wood Product Assembly

Extra Hazard Occupancy
The protection area allotted per sprinkler shall not exceed 100 S.F. of noncombustible ceiling and 100 S.F. of combustible ceiling. The maximum allowable distance between lines and between sprinklers on lines is 12'. Sprinklers on alternate lines shall be staggered if the distance between sprinklers on lines exceeds 8'. Branch lines should not exceed six sprinklers on either side of a cross main.
Maximum system area:
Design by pipe schedule = 25,000 S.F.
Design by hydraulic calculation = 40,000 S.F.

Included in this group are:

Group 1	Group 2
Aircraft hangars	Asphalt Saturating
Combustible Hydraulic Fluid Use Area	Flammable Liquids Spraying
Die Casting	Flow Coating
Metal Extruding	Manufactured/Modular Home
Plywood/Particle Board Manufacturing	Building Assemblies (where
Printing (inkls with flash points < 100	finished enclosure is present
degrees F	and has combustible interiors)
Rubber Reclaiming, Compounding,	Open Oil Quenching
Drying, Milling, Vulcanizing	Plastics Processing
Saw Mills	Solvent Cleaning
Textile Picking, Opening, Blending,	Varnish and Paint Dipping
Garnetting, Carding, Combing of	
Cotton, Synthetics, Wood Shoddy,	
or Burlap	
Upholstering with Plastic Foams	

SERVICES D

R10520-110 Sprinkler Systems (Automatic)

Sprinkler systems may be classified by type as follows:

1. **Wet Pipe System.** A system employing automatic sprinklers attached to a piping system containing water and connected to a water supply so that water discharges immediately from sprinklers opened by a fire.

2. **Dry Pipe System.** A system employing automatic sprinklers attached to a piping system containing air under pressure, the release of which as from the opening of sprinklers permits the water pressure to open a valve known as a "dry pipe valve". The water then flows into the piping system and out the opened sprinklers.

3. **Pre-Action System.** A system employing automatic sprinklers attached to a piping system containing air that may or may not be under pressure, with a supplemental heat responsive system of generally more sensitive characteristics than the automatic sprinklers themselves, installed in the same areas as the sprinklers; actuation of the heat responsive system, as from a fire, opens a valve which permits water to flow into the sprinkler piping system and to be discharged from any sprinklers which may be open.

4. **Deluge System.** A system employing open sprinklers attached to a piping system connected to a water supply through a valve which is opened by the operation of a heat responsive system installed in the same areas as the sprinklers. When this valve opens, water flows into the piping system and discharges from all sprinklers attached thereto.

5. **Combined Dry Pipe and Pre-Action Sprinkler System.** A system employing automatic sprinklers attached to a piping system containing air under pressure with a supplemental heat responsive system of generally more sensitive characteristics than the automatic sprinklers themselves, installed in the same areas as the sprinklers; operation of the heat responsive system, as from a fire, actuates tripping devices which open dry pipe valves simultaneously and without loss of air pressure in the system. Operation of the heat responsive system also opens approved air exhaust valves at the end of the feed main which facilitates the filling of the system with water which usually precedes the opening of sprinklers. The heat responsive system also serves as an automatic fire alarm system.

6. **Limited Water Supply System.** A system employing automatic sprinklers and conforming to these standards but supplied by a pressure tank of limited capacity.

7. **Chemical Systems.** Systems using halon, carbon dioxide, dry chemical or high expansion foam as selected for special requirements. Agent may extinguish flames by chemically inhibiting flame propagation, suffocate flames by excluding oxygen, interrupting chemical action of oxygen uniting with fuel or sealing and cooling the combustion center.

8. **Firecycle System.** Firecycle is a fixed fire protection sprinkler system utilizing water as its extinguishing agent. It is a time delayed, recycling, preaction type which automatically shuts the water off when heat is reduced below the detector operating temperature and turns the water back on when that temperature is exceeded. The system senses a fire condition through a closed circuit electrical detector system which controls water flow to the fire automatically. Batteries supply up to 90 hour emergency power supply for system operation. The piping system is dry (until water is required) and is monitored with pressurized air. Should any leak in the system piping occur, an alarm will sound, but water will not enter the system until heat is sensed by a firecycle detector.

Area coverage sprinkler systems may be laid out and fed from the supply in any one of several patterns as shown below. It is desirable, if possible, to utilize a central feed and achieve a shorter flow path from the riser to the furthest sprinkler. This permits use of the smallest sizes of pipe possible with resulting savings.

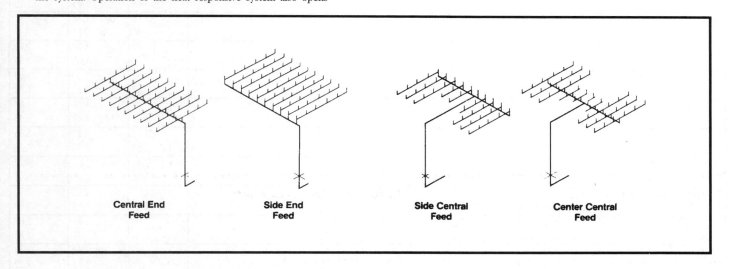

Central End Feed   Side End Feed   Side Central Feed   Center Central Feed

Dry Pipe System: A system employing automatic sprinklers attached to a piping system containing air under pressure, the release of which as from the opening of sprinklers permits the water pressure to open a valve known as a "dry pipe valve". The water then flows into the piping system and out the opened sprinklers.

All areas are assumed to be open.

SERVICES D

D4010 310	Dry Pipe Sprinkler Systems	COST PER S.F.		
		MAT.	INST.	TOTAL
0520	Dry pipe sprinkler systems, steel, black, sch. 40 pipe			
0530	Light hazard, one floor, 500 S.F.	4.93	4.42	9.35
0560	1000 S.F.	2.74	2.55	5.29
0580	2000 S.F.	2.44	2.60	5.04
0600	5000 S.F.	1.24	1.76	3
0620	10,000 S.F.	.84	1.45	2.29
0640	50,000 S.F.	.63	1.28	1.91
0660	Each additional floor, 500 S.F.	.94	2.13	3.07
0680	1000 S.F.	.75	1.74	2.49
0700	2000 S.F.	.75	1.63	2.38
0720	5000 S.F.	.59	1.40	1.99
0740	10,000 S.F.	.53	1.29	1.82
0760	50,000 S.F.	.49	1.14	1.63
1000	Ordinary hazard, one floor, 500 S.F.	4.99	4.47	9.46
1020	1000 S.F.	2.77	2.58	5.35
1040	2000 S.F.	2.53	2.73	5.26
1060	5000 S.F.	1.41	1.87	3.28
1080	10,000 S.F.	1.10	1.91	3.01
1100	50,000 S.F.	1.03	1.87	2.90
1140	Each additional floor, 500 S.F.	1	2.18	3.18
1160	1000 S.F.	.84	1.96	2.80
1180	2000 S.F.	.88	1.81	2.69
1200	5000 S.F.	.80	1.53	2.33
1220	10,000 S.F.	.71	1.51	2.22
1240	50,000 S.F.	.73	1.39	2.12
1500	Extra hazard, one floor, 500 S.F.	6.75	5.55	12.30
1520	1000 S.F.	3.96	3.98	7.94
1540	2000 S.F.	2.74	3.47	6.21
1560	5000 S.F.	1.59	2.62	4.21
1580	10,000 S.F.	1.67	2.51	4.18
1600	50,000 S.F.	1.83	2.42	4.25
1660	Each additional floor, 500 S.F.	1.30	2.70	4
1680	1000 S.F.	1.28	2.57	3.85
1700	2000 S.F.	1.17	2.56	3.73
1720	5000 S.F.	.98	2.25	3.23
1740	10,000 S.F.	1.18	2.07	3.25
1760	50,000 S.F.	1.23	1.99	3.22
2020	Grooved steel, black, sch. 40 pipe, light hazard, one floor, 2000 S.F.	2.32	2.25	4.57
2060	10,000 S.F.	.74	1.24	1.98
2100	Each additional floor, 2000 S.F.	.67	1.33	2

D4010 Sprinklers

D4010 310	Dry Pipe Sprinkler Systems	COST PER S.F.		
		MAT.	INST.	TOTAL
2150	10,000 S.F.	.43	1.08	1.51
2200	Ordinary hazard, one floor, 2000 S.F.	2.39	2.36	4.75
2250	10,000 S.F.	.90	1.60	2.50
2300	Each additional floor, 2000 S.F.	.74	1.44	2.18
2350	10,000 S.F.	.59	1.44	2.03
2400	Extra hazard, one floor, 2000 S.F.	2.60	2.96	5.56
2450	10,000 S.F.	1.31	2.08	3.39
2500	Each additional floor, 2000 S.F.	1.04	2.10	3.14
2550	10,000 S.F.	.91	1.81	2.72
3050	Grooved steel black sch. 10 pipe, light hazard, one floor, 2000 S.F.	2.30	2.23	4.53
3100	10,000 S.F.	.77	1.23	2
3150	Each additional floor, 2000 S.F.	.65	1.31	1.96
3200	10,000 S.F.	.46	1.07	1.53
3250	Ordinary hazard, one floor, 2000 S.F.	2.37	2.36	4.73
3300	10,000 S.F.	.92	1.56	2.48
3350	Each additional floor, 2000 S.F.	.72	1.44	2.16
3400	10,000 S.F.	.61	1.40	2.01
3450	Extra hazard, one floor, 2000 S.F.	2.59	2.95	5.54
3500	10,000 S.F.	1.21	2.04	3.25
3550	Each additional floor, 2000 S.F.	1.03	2.09	3.12
3600	10,000 S.F.	.86	1.79	2.65
4050	Copper tubing, type M, light hazard, one floor, 2000 S.F.	2.48	2.19	4.67
4100	10,000 S.F.	.98	1.24	2.22
4150	Each additional floor, 2000 S.F.	.84	1.30	2.14
4200	10,000 S.F.	.67	1.09	1.76
4250	Ordinary hazard, one floor, 2000 S.F.	2.62	2.45	5.07
4300	10,000 S.F.	1.18	1.48	2.66
4350	Each additional floor, 2000 S.F.	1.12	1.51	2.63
4400	10,000 S.F.	.85	1.30	2.15
4450	Extra hazard, one floor, 2000 S.F.	2.94	2.98	5.92
4500	10,000 S.F.	2.28	2.22	4.50
4550	Each additional floor, 2000 S.F.	1.38	2.12	3.50
4600	10,000 S.F.	1.53	1.93	3.46
5050	Copper tubing, type M, T-drill system, light hazard, one floor			
5060	2000 S.F.	2.49	2.07	4.56
5100	10,000 S.F.	.94	1.04	1.98
5150	Each additional floor, 2000 S.F.	.85	1.18	2.03
5200	10,000 S.F.	.63	.89	1.52
5250	Ordinary hazard, one floor, 2000 S.F.	2.53	2.10	4.63
5300	10,000 S.F.	1.14	1.31	2.45
5350	Each additional floor, 2000 S.F.	.88	1.18	2.06
5400	10,000 S.F.	.78	1.09	1.87
5450	Extra hazard, one floor, 2000 S.F.	2.72	2.48	5.20
5500	10,000 S.F.	1.90	1.64	3.54
5550	Each additional floor, 2000 S.F.	1.16	1.62	2.78
5600	10,000 S.F.	1.15	1.35	2.50

SERVICES

D

Preaction System: A system employing automatic sprinklers attached to a piping system containing air that may or may not be under pressure, with a supplemental heat responsive system of generally more sensitive characteristics than the automatic sprinklers themselves, installed in the same areas as the sprinklers. Actuation of the heat responsive system, as from a fire, opens a valve which permits water to flow into the sprinkler piping system and to be discharged from those sprinklers which were opened by heat from the fire.

All areas are assumed to be opened.

SERVICES D

D4010 350	Preaction Sprinkler Systems	COST PER S.F.		
		MAT.	INST.	TOTAL
0520	Preaction sprinkler systems, steel, black, sch. 40 pipe			
0530	Light hazard, one floor, 500 S.F.	5.05	3.53	8.58
0560	1000 S.F.	2.85	2.61	5.46
0580	2000 S.F.	2.49	2.59	5.08
0600	5000 S.F.	1.29	1.75	3.04
0620	10,000 S.F.	.87	1.44	2.31
0640	50,000 S.F.	.66	1.27	1.93
0660	Each additional floor, 500 S.F.	1.12	1.90	3.02
0680	1000 S.F.	.85	1.74	2.59
0700	2000 S.F.	.85	1.63	2.48
0720	5000 S.F.	.64	1.39	2.03
0740	10,000 S.F.	.56	1.28	1.84
0760	50,000 S.F.	.57	1.18	1.75
1000	Ordinary hazard, one floor, 500 S.F.	5.20	3.83	9.03
1020	1000 S.F.	2.82	2.58	5.40
1040	2000 S.F.	2.70	2.74	5.44
1060	5000 S.F.	1.41	1.86	3.27
1080	10,000 S.F.	1.07	1.90	2.97
1100	50,000 S.F.	.99	1.84	2.83
1140	Each additional floor, 500 S.F.	1.29	2.20	3.49
1160	1000 S.F.	.83	1.77	2.60
1180	2000 S.F.	.80	1.80	2.60
1200	5000 S.F.	.88	1.65	2.53
1220	10,000 S.F.	.76	1.74	2.50
1240	50,000 S.F.	.77	1.62	2.39
1500	Extra hazard, one floor, 500 S.F.	6.90	4.90	11.80
1520	1000 S.F.	3.88	3.63	7.51
1540	2000 S.F.	2.70	3.46	6.16

D4010 Sprinklers

D4010 350	Preaction Sprinkler Systems	COST PER S.F.		
		MAT.	INST.	TOTAL
1560	5000 S.F.	1.65	2.83	4.48
1580	10,000 S.F.	1.64	2.78	4.42
1600	50,000 S.F.	1.77	2.69	4.46
1660	Each additional floor, 500 S.F.	1.50	2.71	4.21
1680	1000 S.F.	1.24	2.55	3.79
1700	2000 S.F.	1.13	2.55	3.68
1720	5000 S.F.	.93	2.26	3.19
1740	10,000 S.F.	1.07	2.09	3.16
1760	50,000 S.F.	1.08	1.96	3.04
2020	Grooved steel, black, sch. 40 pipe, light hazard, one floor, 2000 S.F.	2.42	2.25	4.67
2060	10,000 S.F.	.77	1.23	2
2100	Each additional floor of 2000 S.F.	.77	1.33	2.10
2150	10,000 S.F.	.46	1.07	1.53
2200	Ordinary hazard, one floor, 2000 S.F.	2.44	2.36	4.80
2250	10,000 S.F.	.87	1.59	2.46
2300	Each additional floor, 2000 S.F.	.79	1.44	2.23
2350	10,000 S.F.	.56	1.43	1.99
2400	Extra hazard, one floor, 2000 S.F.	2.56	2.95	5.51
2450	10,000 S.F.	1.21	2.05	3.26
2500	Each additional floor, 2000 S.F.	1	2.09	3.09
2550	10,000 S.F.	.79	1.78	2.57
3050	Grooved steel, black, sch. 10 pipe light hazard, one floor, 2000 S.F.	2.40	2.23	4.63
3100	10,000 S.F.	.80	1.22	2.02
3150	Each additional floor, 2000 S.F.	.75	1.31	2.06
3200	10,000 S.F.	.49	1.06	1.55
3250	Ordinary hazard, one floor, 2000 S.F.	2.39	2.21	4.60
3300	10,000 S.F.	.77	1.54	2.31
3350	Each additional floor, 2000 S.F.	.77	1.44	2.21
3400	10,000 S.F.	.58	1.39	1.97
3450	Extra hazard, one floor, 2000 S.F.	2.55	2.94	5.49
3500	10,000 S.F.	1.09	2.01	3.10
3550	Each additional floor, 2000 S.F.	.99	2.08	3.07
3600	10,000 S.F.	.74	1.76	2.50
4050	Copper tubing, type M, light hazard, one floor, 2000 S.F.	2.59	2.19	4.78
4100	10,000 S.F.	1.01	1.23	2.24
4150	Each additional floor, 2000 S.F.	.95	1.30	2.25
4200	10,000 S.F.	.58	1.07	1.65
4250	Ordinary hazard, one floor, 2000 S.F.	2.67	2.45	5.12
4300	10,000 S.F.	1.15	1.47	2.62
4350	Each additional floor, 2000 S.F.	.95	1.34	2.29
4400	10,000 S.F.	.74	1.19	1.93
4450	Extra hazard, one floor, 2000 S.F.	2.90	2.97	5.87
4500	10,000 S.F.	2.14	2.19	4.33
4550	Each additional floor, 2000 S.F.	1.34	2.11	3.45
4600	10,000 S.F.	1.41	1.90	3.31
5050	Copper tubing, type M, T-drill system, light hazard, one floor			
5060	2000 S.F.	2.59	2.07	4.66
5100	10,000 S.F.	.97	1.03	2
5150	Each additional floor, 2000 S.F.	.95	1.18	2.13
5200	10,000 S.F.	.66	.88	1.54
5250	Ordinary hazard, one floor, 2000 S.F.	2.58	2.10	4.68
5300	10,000 S.F.	1.11	1.30	2.41
5350	Each additional floor, 2000 S.F.	.93	1.19	2.12
5400	10,000 S.F.	.80	1.14	1.94
5450	Extra hazard, one floor, 2000 S.F.	2.68	2.47	5.15
5500	10,000 S.F.	1.76	1.61	3.37
5550	Each additional floor, 2000 S.F.	1.12	1.61	2.73
5600	10,000 S.F.	1.03	1.32	2.35

SERVICES

D

D4010 Sprinklers

Deluge System: A system employing open sprinklers attached to a piping system connected to a water supply through a valve which is opened by the operation of a heat responsive system installed in the same areas as the sprinklers. When this valve opens, water flows into the piping system and discharges from all sprinklers attached thereto.

D4010 370	Deluge Sprinkler Systems	COST PER S.F.		
		MAT.	INST.	TOTAL
0520	Deluge sprinkler systems, steel, black, sch. 40 pipe			
0530	Light hazard, one floor, 500 S.F.	6.10	3.57	9.67
0560	1000 S.F.	3.36	2.51	5.87
0580	2000 S.F.	2.93	2.60	5.53
0600	5000 S.F.	1.47	1.75	3.22
0620	10,000 S.F.	.95	1.44	2.39
0640	50,000 S.F.	.67	1.27	1.94
0660	Each additional floor, 500 S.F.	1.12	1.90	3.02
0680	1000 S.F.	.85	1.74	2.59
0700	2000 S.F.	.85	1.63	2.48
0720	5000 S.F.	.64	1.39	2.03
0740	10,000 S.F.	.56	1.28	1.84
0760	50,000 S.F.	.57	1.18	1.75
1000	Ordinary hazard, one floor, 500 S.F.	6.60	4.08	10.68
1020	1000 S.F.	3.36	2.60	5.96
1040	2000 S.F.	3.14	2.75	5.89
1060	5000 S.F.	1.59	1.86	3.45
1080	10,000 S.F.	1.15	1.90	3.05
1100	50,000 S.F.	1.03	1.88	2.91
1140	Each additional floor, 500 S.F.	1.29	2.20	3.49
1160	1000 S.F.	.83	1.77	2.60
1180	2000 S.F.	.80	1.80	2.60
1200	5000 S.F.	.80	1.52	2.32
1220	10,000 S.F.	.73	1.54	2.27
1240	50,000 S.F.	.73	1.49	2.22
1500	Extra hazard, one floor, 500 S.F.	7.95	4.94	12.89
1520	1000 S.F.	4.64	3.78	8.42
1540	2000 S.F.	3.14	3.47	6.61
1560	5000 S.F.	1.70	2.60	4.30
1580	10,000 S.F.	1.66	2.54	4.20
1600	50,000 S.F.	1.78	2.46	4.24
1660	Each additional floor, 500 S.F.	1.50	2.71	4.21

SERVICES D

D4010 Sprinklers

D4010 370	Deluge Sprinkler Systems	COST PER S.F.		
		MAT.	INST.	TOTAL
1680	1000 S.F.	1.24	2.55	3.79
1700	2000 S.F.	1.13	2.55	3.68
1720	5000 S.F.	.93	2.26	3.19
1740	10,000 S.F.	1.10	2.17	3.27
1760	50,000 S.F.	1.14	2.11	3.25
2000	Grooved steel, black, sch. 40 pipe, light hazard, one floor			
2020	2000 S.F.	2.87	2.26	5.13
2060	10,000 S.F.	.86	1.24	2.10
2100	Each additional floor, 2,000 S.F.	.77	1.33	2.10
2150	10,000 S.F.	.46	1.07	1.53
2200	Ordinary hazard, one floor, 2000 S.F.	2.44	2.36	4.80
2250	10,000 S.F.	.95	1.59	2.54
2300	Each additional floor, 2000 S.F.	.79	1.44	2.23
2350	10,000 S.F.	.56	1.43	1.99
2400	Extra hazard, one floor, 2000 S.F.	3	2.96	5.96
2450	10,000 S.F.	1.30	2.06	3.36
2500	Each additional floor, 2000 S.F.	1	2.09	3.09
2550	10,000 S.F.	.79	1.78	2.57
3000	Grooved steel, black, sch. 10 pipe, light hazard, one floor			
3050	2000 S.F.	2.53	2.14	4.67
3100	10,000 S.F.	.88	1.22	2.10
3150	Each additional floor, 2000 S.F.	.75	1.31	2.06
3200	10,000 S.F.	.49	1.06	1.55
3250	Ordinary hazard, one floor, 2000 S.F.	2.86	2.37	5.23
3300	10,000 S.F.	.85	1.54	2.39
3350	Each additional floor, 2000 S.F.	.77	1.44	2.21
3400	10,000 S.F.	.58	1.39	1.97
3450	Extra hazard, one floor, 2000 S.F.	2.99	2.95	5.94
3500	10,000 S.F.	1.17	2.01	3.18
3550	Each additional floor, 2000 S.F.	.99	2.08	3.07
3600	10,000 S.F.	.74	1.76	2.50
4000	Copper tubing, type M, light hazard, one floor			
4050	2000 S.F.	3.02	2.20	5.22
4100	10,000 S.F.	1.09	1.23	2.32
4150	Each additional floor, 2000 S.F.	.94	1.30	2.24
4200	10,000 S.F.	.58	1.07	1.65
4250	Ordinary hazard, one floor, 2000 S.F.	3.11	2.46	5.57
4300	10,000 S.F.	1.23	1.47	2.70
4350	Each additional floor, 2000 S.F.	.95	1.34	2.29
4400	10,000 S.F.	.74	1.19	1.93
4450	Extra hazard, one floor, 2000 S.F.	3.34	2.98	6.32
4500	10,000 S.F.	2.25	2.20	4.45
4550	Each additional floor, 2000 S.F.	1.34	2.11	3.45
4600	10,000 S.F.	1.41	1.90	3.31
5000	Copper tubing, type M, T-drill system, light hazard, one floor			
5050	2000 S.F.	3.03	2.08	5.11
5100	10,000 S.F.	1.05	1.03	2.08
5150	Each additional floor, 2000 S.F.	.94	1.19	2.13
5200	10,000 S.F.	.66	.88	1.54
5250	Ordinary hazard, one floor, 2000 S.F.	3.02	2.11	5.13
5300	10,000 S.F.	1.19	1.30	2.49
5350	Each additional floor, 2000 S.F.	.93	1.18	2.11
5400	10,000 S.F.	.80	1.14	1.94
5450	Extra hazard, one floor, 2000 S.F.	3.12	2.48	5.60
5500	10,000 S.F.	1.84	1.61	3.45
5550	Each additional floor, 2000 S.F.	1.12	1.61	2.73
5600	10,000 S.F.	1.03	1.32	2.35

SERVICES

D

Firecycle is a fixed fire protection sprinkler system utilizing water as its extinguishing agent. It is a time delayed, recycling, preaction type which automatically shuts the water off when heat is reduced below the detector operating temperature and turns the water back on when that temperature is exceeded.

The system senses a fire condition through a closed circuit electrical detector system which controls water flow to the fire automatically. Batteries supply up to 90 hours emergency power supply for system operation. The piping system is dry (until water is required) and is monitored with pressurized air. Shouldany leak in the system piping occur, an alarm will sound, but water will not enter the system until heat is sensed by a Firecycle detector.

D4010 390	Firecycle Sprinkler Systems	COST PER S.F.		
		MAT.	INST.	TOTAL
0520	Firecycle sprinkler systems, steel black sch. 40 pipe			
0530	Light hazard, one floor, 500 S.F.	21	6.95	27.95
0560	1000 S.F.	10.90	4.32	15.22
0580	2000 S.F.	6.50	3.19	9.69
0600	5000 S.F.	2.89	2	4.89
0620	10,000 S.F.	1.71	1.57	3.28
0640	50,000 S.F.	.85	1.30	2.15
0660	Each additional floor of 500 S.F.	1.16	1.91	3.07
0680	1000 S.F.	.87	1.74	2.61
0700	2000 S.F.	.73	1.62	2.35
0720	5000 S.F.	.65	1.40	2.05
0740	10,000 S.F.	.60	1.29	1.89
0760	50,000 S.F.	.59	1.18	1.77
1000	Ordinary hazard, one floor, 500 S.F.	21.50	7.25	28.75
1020	1000 S.F.	10.85	4.29	15.14
1040	2000 S.F.	6.60	3.32	9.92
1060	5000 S.F.	3.01	2.11	5.12
1080	10,000 S.F.	1.91	2.03	3.94
1100	50,000 S.F.	1.34	2.10	3.44
1140	Each additional floor, 500 S.F.	1.33	2.21	3.54
1160	1000 S.F.	.85	1.77	2.62
1180	2000 S.F.	.91	1.65	2.56
1200	5000 S.F.	.81	1.53	2.34
1220	10,000 S.F.	.72	1.51	2.23
1240	50,000 S.F.	.74	1.43	2.17
1500	Extra hazard, one floor, 500 S.F.	23	8.30	31.30
1520	1000 S.F.	11.80	5.35	17.15
1540	2000 S.F.	6.70	4.05	10.75
1560	5000 S.F.	3.12	2.85	5.97
1580	10,000 S.F.	2.45	2.89	5.34

Important: See the Reference Section for critical supporting data - Reference Numbers and City Cost Indexes

D4010 Sprinklers

D4010 390	Firecycle Sprinkler Systems	COST PER S.F.		
		MAT.	**INST.**	**TOTAL**
1600	50,000 S.F.	2.18	3.18	5.36
1660	Each additional floor, 500 S.F.	1.54	2.72	4.26
1680	1000 S.F.	1.26	2.55	3.81
1700	2000 S.F.	1.15	2.55	3.70
1720	5000 S.F.	.94	2.27	3.21
1740	10,000 S.F.	1.11	2.10	3.21
1760	50,000 S.F.	1.14	2.05	3.19
2020	Grooved steel, black, sch. 40 pipe, light hazard, one floor			
2030	2000 S.F.	6.40	2.84	9.24
2060	10,000 S.F.	1.75	1.90	3.65
2100	Each additional floor, 2000 S.F.	.79	1.33	2.12
2150	10,000 S.F.	.50	1.08	1.58
2200	Ordinary hazard, one floor, 2000 S.F.	6.45	2.95	9.40
2250	10,000 S.F.	1.79	1.82	3.61
2300	Each additional floor, 2000 S.F.	.81	1.44	2.25
2350	10,000 S.F.	.60	1.44	2.04
2400	Extra hazard, one floor, 2000 S.F.	6.55	3.54	10.09
2450	10,000 S.F.	2.02	2.16	4.18
2500	Each additional floor, 2000 S.F.	1.02	2.09	3.11
2550	10,000 S.F.	.83	1.79	2.62
3050	Grooved steel, black, sch. 10 pipe light hazard, one floor,			
3060	2000 S.F.	6.40	2.82	9.22
3100	10,000 S.F.	1.64	1.35	2.99
3150	Each additional floor, 2000 S.F.	.77	1.31	2.08
3200	10,000 S.F.	.53	1.07	1.60
3250	Ordinary hazard, one floor, 2000 S.F.	6.40	2.95	9.35
3300	10,000 S.F.	1.73	1.68	3.41
3350	Each additional floor, 2000 S.F.	.79	1.44	2.23
3400	10,000 S.F.	.62	1.40	2.02
3450	Extra hazard, one floor, 2000 S.F.	6.55	3.53	10.08
3500	10,000 S.F.	1.92	2.12	4.04
3550	Each additional floor, 2000 S.F.	1.01	2.08	3.09
3600	10,000 S.F.	.78	1.77	2.55
4060	Copper tubing, type M, light hazard, one floor, 2000 S.F.	6.60	2.78	9.38
4100	10,000 S.F.	1.85	1.36	3.21
4150	Each additional floor, 2000 S.F.	.96	1.30	2.26
4200	10,000 S.F.	.74	1.09	1.83
4250	Ordinary hazard, one floor, 2000 S.F.	6.65	3.04	9.69
4300	10,000 S.F.	1.99	1.60	3.59
4350	Each additional floor, 2000 S.F.	.97	1.34	2.31
4400	10,000 S.F.	.77	1.18	1.95
4450	Extra hazard, one floor, 2000 S.F.	6.90	3.56	10.46
4500	10,000 S.F.	3.04	2.33	5.37
4550	Each additional floor, 2000 S.F.	1.36	2.11	3.47
4600	10,000 S.F.	1.45	1.91	3.36
5060	Copper tubing, type M, T-drill system, light hazard, one floor 2000 S.F.	6.60	2.66	9.26
5100	10,000 S.F.	1.81	1.16	2.97
5150	Each additional floor, 2000 S.F.	1.06	1.24	2.30
5200	10,000 S.F.	.70	.89	1.59
5250	Ordinary hazard, one floor, 2000 S.F.	6.60	2.69	9.29
5300	10,000 S.F.	1.95	1.43	3.38
5350	Each additional floor, 2000 S.F.	.95	1.18	2.13
5400	10,000 S.F.	.84	1.15	1.99
5450	Extra hazard, one floor, 2000 S.F.	6.70	3.06	9.76
5500	10,000 S.F.	2.59	1.72	4.31
5550	Each additional floor, 2000 S.F.	1.14	1.61	2.75
5600	10,000 S.F.	1.07	1.33	2.40

D SERVICES

Wet Pipe System. A system employing automatic sprinklers attached to a piping system containing water and connected to a water supply so that water discharges immediately from sprinklers opened by heat from a fire.

All areas are assumed to be open.

D4010 410	Wet Pipe Sprinkler Systems	COST PER S.F.		
		MAT.	INST.	TOTAL
0520	Wet pipe sprinkler systems, steel, black, sch. 40 pipe			
0530	Light hazard, one floor, 500 S.F.	1.30	2.25	3.55
0560	1000 S.F.	1.94	2.30	4.24
0580	2000 S.F.	2.09	2.35	4.44
0600	5000 S.F.	1.03	1.65	2.68
0620	10,000 S.F.	.67	1.38	2.05
0640	50,000 S.F.	.50	1.25	1.75
0660	Each additional floor, 500 S.F.	.64	1.91	2.55
0680	1000 S.F.	.61	1.77	2.38
0700	2000 S.F.	.60	1.61	2.21
0720	5000 S.F.	.44	1.38	1.82
0740	10,000 S.F.	.39	1.27	1.66
0760	50,000 S.F.	.33	.98	1.31
1000	Ordinary hazard, one floor, 500 S.F.	1.42	2.42	3.84
1020	1000 S.F.	1.91	2.27	4.18
1040	2000 S.F.	2.17	2.49	4.66
1060	5000 S.F.	1.15	1.76	2.91
1080	10,000 S.F.	.87	1.84	2.71
1100	50,000 S.F.	.81	1.80	2.61
1140	Each additional floor, 500 S.F.	.80	2.16	2.96
1160	1000 S.F.	.58	1.75	2.33
1180	2000 S.F.	.68	1.79	2.47
1200	5000 S.F.	.70	1.67	2.37
1220	10,000 S.F.	.59	1.73	2.32
1240	50,000 S.F.	.61	1.61	2.22
1500	Extra hazard, one floor, 500 S.F.	3.66	3.74	7.40
1520	1000 S.F.	2.44	3.21	5.65
1540	2000 S.F.	2.30	3.32	5.62
1560	5000 S.F.	1.49	2.89	4.38
1580	10,000 S.F.	1.44	2.75	4.19
1600	50,000 S.F.	1.58	2.66	4.24
1660	Each additional floor, 500 S.F.	1.01	2.67	3.68
1680	1000 S.F.	.99	2.53	3.52
1700	2000 S.F.	.88	2.53	3.41
1720	5000 S.F.	.73	2.25	2.98
1740	10,000 S.F.	.90	2.08	2.98
1760	50,000 S.F.	.92	1.99	2.91
2020	Grooved steel, black sch. 40 pipe, light hazard, one floor, 2000 S.F.	2.01	2.01	4.02
2060	10,000 S.F.	.70	1.22	1.92
2100	Each additional floor, 2000 S.F.	.52	1.31	1.83

Important: See the Reference Section for critical supporting data - Reference Numbers and City Cost Indexes

D4010 Sprinklers

D4010 410	Wet Pipe Sprinkler Systems	COST PER S.F.		
		MAT.	INST.	TOTAL
2150	10,000 S.F.	.29	1.06	1.35
2200	Ordinary hazard, one floor, 2000 S.F.	2.03	2.12	4.15
2250	10,000 S.F.	.67	1.53	2.20
2300	Each additional floor, 2000 S.F.	.54	1.42	1.96
2350	10,000 S.F.	.39	1.42	1.81
2400	Extra hazard, one floor, 2000 S.F.	2.15	2.71	4.86
2450	10,000 S.F.	.99	1.99	2.98
2500	Each additional floor, 2000 S.F.	.75	2.07	2.82
2550	10,000 S.F.	.62	1.77	2.39
3050	Grooved steel black sch. 10 pipe, light hazard, one floor, 2000 S.F.	1.99	1.99	3.98
3100	10,000 S.F.	.60	1.16	1.76
3150	Each additional floor, 2000 S.F.	.50	1.29	1.79
3200	10,000 S.F.	.32	1.05	1.37
3250	Ordinary hazard, one floor, 2000 S.F.	2.01	2.12	4.13
3300	10,000 S.F.	.69	1.49	2.18
3350	Each additional floor, 2000 S.F.	.52	1.42	1.94
3400	10,000 S.F.	.41	1.38	1.79
3450	Extra hazard, one floor, 2000 S.F.	2.14	2.70	4.84
3500	10,000 S.F.	.89	1.95	2.84
3550	Each additional floor, 2000 S.F.	.74	2.06	2.80
3600	10,000 S.F.	.57	1.75	2.32
4050	Copper tubing, type M, light hazard, one floor, 2000 S.F.	2.17	1.95	4.12
4100	10,000 S.F.	.81	1.17	1.98
4150	Each additional floor, 2000 S.F.	.69	1.28	1.97
4200	10,000 S.F.	.53	1.07	1.60
4250	Ordinary hazard, one floor, 2000 S.F.	2.26	2.21	4.47
4300	10,000 S.F.	.95	1.41	2.36
4350	Each additional floor, 2000 S.F.	.80	1.44	2.24
4400	10,000 S.F.	.65	1.28	1.93
4450	Extra hazard, one floor, 2000 S.F.	2.49	2.73	5.22
4500	10,000 S.F.	1.94	2.13	4.07
4550	Each additional floor, 2000 S.F.	1.09	2.09	3.18
4600	10,000 S.F.	1.24	1.89	3.13
5050	Copper tubing, type M, T-drill system, light hazard, one floor			
5060	2000 S.F.	2.18	1.83	4.01
5100	10,000 S.F.	.77	.97	1.74
5150	Each additional floor, 2000 S.F.	.70	1.16	1.86
5200	10,000 S.F.	.49	.87	1.36
5250	Ordinary hazard, one floor, 2000 S.F.	2.17	1.86	4.03
5300	10,000 S.F.	.91	1.24	2.15
5350	Each additional floor, 2000 S.F.	.68	1.16	1.84
5400	10,000 S.F.	.63	1.13	1.76
5450	Extra hazard, one floor, 2000 S.F.	2.27	2.23	4.50
5500	10,000 S.F.	1.56	1.55	3.11
5550	Each additional floor, 2000 S.F.	.92	1.63	2.55
5600	10,000 S.F.	.86	1.31	2.17

SERVICES

D

D4020 Standpipes

- Roof
- Roof connections with hose gate valves (for combustible roof)
- Hose connections on each floor (size based on class of service)
- Check Valve
- Siamese inlet connections (for fire department use)

D4020 310	Wet Standpipe Risers, Class I	COST PER FLOOR		
		MAT.	INST.	TOTAL
0550	Wet standpipe risers, Class I, steel black sch. 40, 10' height			
0560	4" diameter pipe, one floor	2,375	2,450	4,825
0580	Additional floors	575	760	1,335
0600	6" diameter pipe, one floor	4,225	4,325	8,550
0620	Additional floors	1,100	1,200	2,300
0640	8" diameter pipe, one floor	6,400	5,225	11,625
0660	Additional floors	1,550	1,475	3,025

D4020 310	Wet Standpipe Risers, Class II	COST PER FLOOR		
		MAT.	INST.	TOTAL
1030	Wet standpipe risers, Class II, steel black sch. 40, 10' height			
1040	2" diameter pipe, one floor	945	885	1,830
1060	Additional floors	310	340	650
1080	2-1/2" diameter pipe, one floor	1,325	1,300	2,625
1100	Additional floors	355	400	755

D4020 310	Wet Standpipe Risers, Class III	COST PER FLOOR		
		MAT.	INST.	TOTAL
1530	Wet standpipe risers, Class III, steel black sch. 40, 10' height			
1540	4" diameter pipe, one floor	2,450	2,450	4,900
1560	Additional floors	485	635	1,120
1580	6" diameter pipe, one floor	4,300	4,325	8,625
1600	Additional floors	1,125	1,200	2,325
1620	8" diameter pipe, one floor	6,475	5,225	11,700
1640	Additional floors	1,575	1,475	3,050

SERVICES D

Roof

Roof connections with hose gate valves (for combustible roof)

Hose connections on each floor (size based on class of service)

Check valve

Siamese inlet connections (for fire department use)

D4020 330	Dry Standpipe Risers, Class I	COST PER FLOOR		
		MAT.	INST.	TOTAL
0530	Dry standpipe riser, Class I, steel black sch. 40, 10' height			
0540	4" diameter pipe, one floor	1,525	2,000	3,525
0560	Additional floors	525	720	1,245
0580	6" diameter pipe, one floor	3,150	3,450	6,600
0600	Additional floors	1,050	1,175	2,225
0620	8" diameter pipe, one floor	4,800	4,200	9,000
0640	Additional floors	1,500	1,450	2,950

D4020 330	Dry Standpipe Risers, Class II	COST PER FLOOR		
		MAT.	INST.	TOTAL
1030	Dry standpipe risers, Class II, steel black sch. 40, 10' height			
1040	2" diameter pipe, one floor	790	920	1,710
1060	Additional floor	264	300	564
1080	2-1/2" diameter pipe, one floor	1,025	1,075	2,100
1100	Additional floors	310	360	670

D4020 330	Dry Standpipe Risers, Class III	COST PER FLOOR		
		MAT.	INST.	TOTAL
1530	Dry standpipe risers, Class III, steel black sch. 40, 10' height			
1540	4" diameter pipe, one floor	1,525	1,950	3,475
1560	Additional floors	450	650	1,100
1580	6" diameter pipe, one floor	3,175	3,450	6,625
1600	Additional floors	1,075	1,175	2,250
1620	8" diameter pipe, one floor	4,850	4,200	9,050
1640	Additional floor	1,550	1,450	3,000

SERVICES

D

D4020 Standpipes

D4020 410	Fire Hose Equipment	COST EACH		
		MAT.	INST.	TOTAL
0100	Adapters, reducing, 1 piece, FxM, hexagon, cast brass, 2-1/2" x 1-1/2"	34.50		34.50
0200	Pin lug, 1-1/2" x 1"	28.50		28.50
0250	3" x 2-1/2"	43		43
0300	For polished chrome, add 75% mat.			
0400	Cabinets, D.S. glass in door, recessed, steel box, not equipped			
0500	Single extinguisher, steel door & frame	68.50	111	179.50
0550	Stainless steel door & frame	142	111	253
0600	Valve, 2-1/2" angle, steel door & frame	76.50	74	150.50
0650	Aluminum door & frame	101	74	175
0700	Stainless steel door & frame	133	74	207
0750	Hose rack assy, 2-1/2" x 1-1/2" valve & 100' hose, steel door & frame	155	148	303
0800	Aluminum door & frame	229	148	377
0850	Stainless steel door & frame	299	148	447
0900	Hose rack assy,& extinguisher,2-1/2"x1-1/2" valve & hose,steel door & frame	169	177	346
0950	Aluminum	292	177	469
1000	Stainless steel	320	177	497
1550	Compressor, air, dry pipe system, automatic, 200 gal., 1/3 H.P.	790	380	1,170
1600	520 gal., 1 H.P.	935	380	1,315
1650	Alarm, electric pressure switch (circuit closer)	163	18.95	181.95
2500	Couplings, hose, rocker lug, cast brass, 1-1/2"	32		32
2550	2-1/2"	42.50		42.50
3000	Escutcheon plate, for angle valves, polished brass, 1-1/2"	12		12
3050	2-1/2"	29		29
3500	Fire pump, electric, w/controller, fittings, relief valve			
3550	4" pump, 30 H.P., 500 G.P.M.	15,600	2,750	18,350
3600	5" pump, 40 H.P., 1000 G.P.M.	22,800	3,125	25,925
3650	5" pump, 100 H.P., 1000 G.P.M.	25,500	3,475	28,975
3700	For jockey pump system, add	2,800	445	3,245
5000	Hose, per linear foot, synthetic jacket, lined,			
5100	300 lb. test, 1-1/2" diameter	1.78	.34	2.12
5150	2-1/2" diameter	2.97	.40	3.37
5200	500 lb. test, 1-1/2" diameter	1.85	.34	2.19
5250	2-1/2" diameter	3.20	.40	3.60
5500	Nozzle, plain stream, polished brass, 1-1/2" x 10"	29.50		29.50
5550	2-1/2" x 15" x 13/16" or 1-1/2"	53		53
5600	Heavy duty combination adjustable fog and straight stream w/handle 1-1/2"	279		279
5650	2-1/2" direct connection	420		420
6000	Rack, for 1-1/2" diameter hose 100 ft. long, steel	36.50	44.50	81
6050	Brass	56	44.50	100.50
6500	Reel, steel, for 50 ft. long 1-1/2" diameter hose	83	63.50	146.50
6550	For 75 ft. long 2-1/2" diameter hose	130	63.50	193.50
7050	Siamese, w/plugs & chains, polished brass, sidewalk, 4" x 2-1/2" x 2-1/2"	345	355	700
7100	6" x 2-1/2" x 2-1/2"	580	445	1,025
7200	Wall type, flush, 4" x 2-1/2" x 2-1/2"	325	177	502
7250	6" x 2-1/2" x 2-1/2"	445	193	638
7300	Projecting, 4" x 2-1/2" x 2-1/2"	300	177	477
7350	6" x 2-1/2" x 2-1/2"	495	193	688
7400	For chrome plate, add 15% mat.			
8000	Valves, angle, wheel handle, 300 Lb., rough brass, 1-1/2"	33	41	74
8050	2-1/2"	56.50	70.50	127
8100	Combination pressure restricting, 1-1/2"	52	41	93
8150	2-1/2"	110	70.50	180.50
8200	Pressure restricting, adjustable, satin brass, 1-1/2"	77.50	41	118.50
8250	2-1/2"	109	70.50	179.50
8300	Hydrolator, vent and drain, rough brass, 1-1/2"	46	41	87
8350	2-1/2"	46	41	87
8400	Cabinet assy, incls. adapter, rack, hose, and nozzle	485	268	753

Important: See the Reference Section for critical supporting data - Reference Numbers and City Cost Indexes

D4090 Other Fire Protection Systems

General: Automatic fire protection (suppression) systems other than water sprinklers may be desired for special environments, high risk areas, isolated locations or unusual hazards. Some typical applications would include:

1. Paint dip tanks
2. Securities vaults
3. Electronic data processing
4. Tape and data storage
5. Transformer rooms
6. Spray booths
7. Petroleum storage
8. High rack storage

Piping and wiring costs are highly variable and are not included.

D4090 910	FM200 Fire Suppression	COST EACH		
		MAT.	INST.	TOTAL
0020	Detectors with brackets			
0040	Fixed temperature heat detector	31	60.50	91.50
0060	Rate of temperature rise detector	37	60.50	97.50
0080	Ion detector (smoke) detector	82.50	78.50	161
0200	Extinguisher agent			
0240	200 lb FM200, container	6,375	221	6,596
0280	75 lb carbon dioxide cylinder	1,050	147	1,197
0320	Dispersion nozzle			
0340	FM200 1-1/2" dispersion nozzle	55	35	90
0380	Carbon dioxide 3" x 5" dispersion nozzle	55	27.50	82.50
0420	Control station			
0440	Single zone control station with batteries	1,400	485	1,885
0470	Multizone (4) control station with batteries	2,700	970	3,670
0500	Electric mechanical release	138	244	382
0550	Manual pull station	49.50	82	131.50
0640	Battery standby power 10" x 10" x 17"	760	121	881
0740	Bell signalling device	54.50	60.50	115

D4090 920	FM200 Systems	COST PER C.F.		
		MAT.	INST.	TOTAL
0820	Average FM200 system, minimum			1.38
0840	Maximum			2.75

Electric Service

Feeder Installation

Switchgear

D5010 120	Electric Service, 3 Phase - 4 Wire	COST EACH		
		MAT.	INST.	TOTAL
0200	Service installation, includes breakers, metering, 20' conduit & wire			
0220	3 phase, 4 wire, 120/208 volts, 60 A	695	750	1,445
0240	100 A	875	900	1,775
0280	200 A	1,225	1,400	2,625
0320	400 A	2,900	2,550	5,450
0360	600 A	5,550	3,450	9,000
0400	800 A	6,925	4,125	11,050
0440	1000 A	8,800	4,750	13,550
0480	1200 A	10,600	4,900	15,500
0520	1600 A	20,700	7,000	27,700
0560	2000 A	23,300	7,975	31,275
0570	Add 25% for 277/480 volt			

D5010 230	Feeder Installation	COST PER L.F.		
		MAT.	INST.	TOTAL
0200	Feeder installation 600 V, including RGS conduit and XHHW wire, 60 A	4.86	9.05	13.91
0240	100 A	8.80	12	20.80
0280	200 A	16.20	18.55	34.75
0320	400 A	32.50	37	69.50
0360	600 A	70	60.50	130.50
0400	800 A	90.50	72	162.50
0440	1000 A	111	92.50	203.50
0480	1200 A	126	95	221
0520	1600 A	181	144	325
0560	2000 A	222	185	407
1200	Branch installation 600 V, including EMT conduit and THW wire, 15 A	.99	4.61	5.60
1240	20 A	.99	4.61	5.60
1280	30 A	1.62	5.65	7.27
1320	50 A	2.70	6.55	9.25
1360	65 A	3.04	6.95	9.99
1400	85 A	4.69	8.20	12.89
1440	100 A	5.45	8.70	14.15
1480	130 A	6.95	9.80	16.75
1520	150 A	8.50	11.20	19.70
1560	200 A	10.20	12.60	22.80

D5010 240	Switchgear	COST EACH		
		MAT.	INST.	TOTAL
0200	Switchgear inst., incl. swbd., panels & circ bkr, 400 A, 120/208volt	3,575	3,050	6,625
0240	600 A	9,600	4,125	13,725

Important: See the Reference Section for critical supporting data - Reference Numbers and City Cost Indexes

SERVICES D

D5010 Electrical Service/Distribution

D5010 240	Switchgear	COST EACH		
		MAT.	INST.	TOTAL
0280	800 A	12,100	5,875	17,975
0320	1200 A	15,100	9,000	24,100
0360	1600 A	20,500	12,600	33,100
0400	2000 A	26,000	16,100	42,100
0410	Add 20% for 277/480 volt			

SERVICES

D

Duplex Wall Receptacle

Undercarpet Receptacle System

D5020 110	Receptacle (by Wattage)	COST PER S.F.		
		MAT.	INST.	TOTAL
0190	Receptacles include plate, box, conduit, wire & transformer when required			
0200	2.5 per 1000 S.F., .3 watts per S.F.	.41	1.09	1.50
0240	With transformer	.45	1.14	1.59
0280	4 per 1000 S.F., .5 watts per S.F.	.47	1.26	1.73
0320	With transformer	.53	1.34	1.87
0360	5 per 1000 S.F., .6 watts per S.F.	.55	1.49	2.04
0400	With transformer	.64	1.60	2.24
0440	8 per 1000 S.F., .9 watts per S.F.	.56	1.66	2.22
0480	With transformer	.68	1.81	2.49
0520	10 per 1000 S.F., 1.2 watts per S.F.	.59	1.80	2.39
0560	With transformer	.78	2.04	2.82
0600	16.5 per 1000 S.F., 2.0 watts per S.F.	.69	2.24	2.93
0640	With transformer	1.02	2.66	3.68
0680	20 per 1000 S.F.,2.4 watts per S.F.	.73	2.43	3.16
0720	With transformer	1.12	2.92	4.04

D5020 115	Receptacles, Floor	COST PER S.F.		
		MAT.	INST.	TOTAL
0200	Receptacle systems, underfloor duct, 5' on center, low density	4.88	2.38	7.26
0240	High density	5.25	3.06	8.31
0280	7' on center, low density	3.84	2.04	5.88
0320	High density	4.20	2.72	6.92
0400	Poke thru fittings, low density	.91	1.10	2.01
0440	High density	1.82	2.16	3.98
0520	Telepoles, using Romex, low density	.81	.71	1.52
0560	High density	1.61	1.44	3.05
0600	Using EMT, low density	.85	.95	1.80
0640	High density	1.71	1.92	3.63
0720	Conduit system with floor boxes, low density	.86	.82	1.68
0760	High density	1.72	1.64	3.36
0840	Undercarpet power system, 3 conductor with 5 conductor feeder, low density	1.30	.30	1.60
0880	High density	2.53	.58	3.11

Important: See the Reference Section for critical supporting data - Reference Numbers and City Cost Indexes

D5020 Lighting and Branch Wiring

Description: Table 9.2-542 includes the cost for switch, plate, box, conduit in slab or EMT exposed and copper wire. Add 20% for exposed conduit.

No power required for switches.

Federal energy guidelines recommend the maximum lighting area controlled per switch shall not exceed 1000 S.F. and that areas over 500 S.F. shall be so controlled that total illumination can be reduced by at least 50%.

D5020 130	Wall Switch by Sq. Ft.	COST PER S.F.		
		MAT.	INST.	TOTAL
0200	Wall switches, 1.0 per 1000 S.F.	.06	.18	.24
0240	1.2 per 1000 S.F.	.06	.19	.25
0280	2.0 per 1000 S.F.	.08	.28	.36
0320	2.5 per 1000 S.F.	.11	.36	.47
0360	5.0 per 1000 S.F.	.26	.76	1.02
0400	10.0 per 1000 S.F.	.51	1.53	2.04

D5020 135	Miscellaneous Power	COST PER S.F.		
		MAT.	INST.	TOTAL
0200	Miscellaneous power, to .5 watts	.03	.08	.11
0240	.8 watts	.05	.12	.17
0280	1 watt	.07	.16	.23
0320	1.2 watts	.07	.19	.26
0360	1.5 watts	.09	.22	.31
0400	1.8 watts	.11	.26	.37
0440	2 watts	.13	.32	.45
0480	2.5 watts	.16	.38	.54
0520	3 watts	.19	.45	.64

D5020 140	Central A. C. Power (by Wattage)	COST PER S.F.		
		MAT.	INST.	TOTAL
0200	Central air conditioning power, 1 watt	.08	.19	.27
0220	2 watts	.09	.21	.30
0240	3 watts	.11	.24	.35
0280	4 watts	.16	.32	.48
0320	6 watts	.28	.46	.74
0360	8 watts	.33	.47	.80
0400	10 watts	.41	.55	.96

SERVICES

D

D5020 Lighting and Branch Wiring

Motor Installation

D5020 145	Motor Installation	COST EACH		
		MAT.	INST.	TOTAL
0200	Motor installation, single phase, 115V, to and including 1/3 HP motor size	525	740	1,265
0240	To and incl. 1 HP motor size	545	740	1,285
0280	To and incl. 2 HP motor size	580	785	1,365
0320	To and incl. 3 HP motor size	645	800	1,445
0360	230V, to and including 1 HP motor size	525	750	1,275
0400	To and incl. 2 HP motor size	550	750	1,300
0440	To and incl. 3 HP motor size	615	805	1,420
0520	Three phase, 200V, to and including 1-1/2 HP motor size	615	825	1,440
0560	To and incl. 3 HP motor size	655	895	1,550
0600	To and incl. 5 HP motor size	695	995	1,690
0640	To and incl. 7-1/2 HP motor size	710	1,025	1,735
0680	To and incl. 10 HP motor size	1,125	1,275	2,400
0720	To and incl. 15 HP motor size	1,450	1,425	2,875
0760	To and incl. 20 HP motor size	1,850	1,625	3,475
0800	To and incl. 25 HP motor size	1,875	1,650	3,525
0840	To and incl. 30 HP motor size	2,875	1,925	4,800
0880	To and incl. 40 HP motor size	3,550	2,300	5,850
0920	To and incl. 50 HP motor size	6,000	2,650	8,650
0960	To and incl. 60 HP motor size	6,175	2,800	8,975
1000	To and incl. 75 HP motor size	8,050	3,200	11,250
1040	To and incl. 100 HP motor size	16,900	3,800	20,700
1080	To and incl. 125 HP motor size	17,100	4,175	21,275
1120	To and incl. 150 HP motor size	20,700	4,900	25,600
1160	To and incl. 200 HP motor size	24,600	5,950	30,550
1240	230V, to and including 1-1/2 HP motor size	580	815	1,395
1280	To and incl. 3 HP motor size	620	885	1,505
1320	To and incl. 5 HP motor size	665	985	1,650
1360	To and incl. 7-1/2 HP motor size	665	985	1,650
1400	To and incl. 10 HP motor size	1,000	1,200	2,200
1440	To and incl. 15 HP motor size	1,150	1,325	2,475
1480	To and incl. 20 HP motor size	1,725	1,600	3,325
1520	To and incl. 25 HP motor size	1,850	1,625	3,475
1560	To and incl. 30 HP motor size	1,875	1,650	3,525
1600	To and incl. 40 HP motor size	3,450	2,225	5,675
1640	To and incl. 50 HP motor size	3,575	2,350	5,925
1680	To and incl. 60 HP motor size	6,000	2,675	8,675
1720	To and incl. 75 HP motor size	7,350	3,025	10,375
1760	To and incl. 100 HP motor size	8,550	3,375	11,925
1800	To and incl. 125 HP motor size	17,000	3,925	20,925
1840	To and incl. 150 HP motor size	18,400	4,450	22,850
1880	To and incl. 200 HP motor size	20,000	4,950	24,950
1960	460V, to and including 2 HP motor size	700	820	1,520
2000	To and incl. 5 HP motor size	740	895	1,635
2040	To and incl. 10 HP motor size	770	985	1,755
2080	To and incl. 15 HP motor size	1,000	1,125	2,125
2120	To and incl. 20 HP motor size	1,025	1,200	2,225

Important: See the Reference Section for critical supporting data - Reference Numbers and City Cost Indexes

D5020 Lighting and Branch Wiring

D5020 145	Motor Installation	COST EACH		
		MAT.	INST.	TOTAL
2160	To and incl. 25 HP motor size	1,125	1,275	2,400
2200	To and incl. 30 HP motor size	1,450	1,375	2,825
2240	To and incl. 40 HP motor size	1,825	1,475	3,300
2280	To and incl. 50 HP motor size	2,050	1,625	3,675
2320	To and incl. 60 HP motor size	3,025	1,925	4,950
2360	To and incl. 75 HP motor size	3,475	2,125	5,600
2400	To and incl. 100 HP motor size	3,775	2,375	6,150
2440	To and incl. 125 HP motor size	6,200	2,675	8,875
2480	To and incl. 150 HP motor size	7,725	3,000	10,725
2520	To and incl. 200 HP motor size	9,050	3,375	12,425
2600	575V, to and including 2 HP motor size	700	820	1,520
2640	To and incl. 5 HP motor size	740	895	1,635
2680	To and incl. 10 HP motor size	770	985	1,755
2720	To and incl. 20 HP motor size	1,000	1,125	2,125
2760	To and incl. 25 HP motor size	1,025	1,200	2,225
2800	To and incl. 30 HP motor size	1,450	1,375	2,825
2840	To and incl. 50 HP motor size	1,525	1,425	2,950
2880	To and incl. 60 HP motor size	3,000	1,900	4,900
2920	To and incl. 75 HP motor size	3,025	1,925	4,950
2960	To and incl. 100 HP motor size	3,475	2,125	5,600
3000	To and incl. 125 HP motor size	6,125	2,625	8,750
3040	To and incl. 150 HP motor size	6,200	2,675	8,875
3080	To and incl. 200 HP motor size	7,850	3,025	10,875

D5020 155	Motor Feeder	COST PER L.F.		
		MAT.	INST.	TOTAL
0200	Motor feeder systems, single phase, feed up to 115V 1HP or 230V 2 HP	2.38	5.75	8.13
0240	115V 2HP, 230V 3HP	2.45	5.80	8.25
0280	115V 3HP	2.61	6.05	8.66
0360	Three phase, feed to 200V 3HP, 230V 5HP, 460V 10HP, 575V 10HP	2.44	6.15	8.59
0440	200V 5HP, 230V 7.5HP, 460V 15HP, 575V 20HP	2.55	6.30	8.85
0520	200V 10HP, 230V 10HP, 460V 30HP, 575V 30HP	2.79	6.65	9.44
0600	200V 15HP, 230V 15HP, 460V 40HP, 575V 50HP	3.54	7.65	11.19
0680	200V 20HP, 230V 25HP, 460V 50HP, 575V 60HP	5.20	9.70	14.90
0760	200V 25HP, 230V 30HP, 460V 60HP, 575V 75HP	5.50	9.85	15.35
0840	200V 30HP	6.45	10.20	16.65
0920	230V 40HP, 460V 75HP, 575V 100HP	7.70	11.10	18.80
1000	200V 40HP	9.05	11.85	20.90
1080	230V 50HP, 460V 100HP, 575V 125HP	9.75	13.10	22.85
1160	200V 50HP, 230V 60HP, 460V 125HP, 575V 150HP	10.95	13.90	24.85
1240	200V 60HP, 460V 150HP	13.45	16.35	29.80
1320	230V 75HP, 575V 200HP	15.30	17	32.30
1400	200V 75HP	16.50	17.35	33.85
1480	230V 100HP, 460V 200HP	25	20.50	45.50
1560	200V 100HP	34	25.50	59.50
1640	230V 125HP	37	27.50	64.50
1720	200V 125HP, 230V 150HP	38	31.50	69.50
1800	200V 150HP	44	34	78
1880	200V 200HP	67.50	50.50	118
1960	230V 200HP	55.50	37.50	93

SERVICES

D

Fluorescent Fixture

Incandescent Fixture

D5020 210	Fluorescent Fixtures (by Wattage)	COST PER S.F.		
		MAT.	INST.	TOTAL
0190	Fluorescent fixtures recess mounted in ceiling			
0200	1 watt per S.F., 20 FC, 5 fixtures per 1000 S.F.	.71	1.42	2.13
0240	2 watts per S.F., 40 FC, 10 fixtures per 1000 S.F.	1.42	2.80	4.22
0280	3 watts per S.F., 60 FC, 15 fixtures per 1000 S.F	2.13	4.22	6.35
0320	4 watts per S.F., 80 FC, 20 fixtures per 1000 S.F.	2.84	5.60	8.44
0400	5 watts per S.F., 100 FC, 25 fixtures per 1000 S.F.	3.55	7	10.55

D5020 216	Incandescent Fixture (by Wattage)	COST PER S.F.		
		MAT.	INST.	TOTAL
0190	Incandescent fixture recess mounted, type A			
0200	1 watt per S.F., 8 FC, 6 fixtures per 1000 S.F.	.84	1.16	2
0240	2 watt per S.F., 16 FC, 12 fixtures per 1000 S.F.	1.67	2.29	3.96
0280	3 watt per S.F., 24 FC, 18 fixtures, per 1000 S.F.	2.52	3.40	5.92
0320	4 watt per S.F., 32 FC, 24 fixtures per 1000 S.F.	3.35	4.56	7.91
0400	5 watt per S.F., 40 FC, 30 fixtures per 1000 S.F.	4.20	5.70	9.90

SERVICES **D**

Important: See the Reference Section for critical supporting data - Reference Numbers and City Cost Indexes

D5020 Lighting and Branch Wiring

High Bay Fixture

Low Bay Fixture

D5020 226	H.I.D. Fixture, High Bay, 16′ (by Wattage)	COST PER S.F.		
		MAT.	INST.	TOTAL
0190	High intensity discharge fixture, 16′ above work plane			
0240	1 watt/S.F., type E, 42 FC, 1 fixture/1000 S.F.	.93	1.20	2.13
0280	Type G, 52 FC, 1 fixture/1000 S.F.	.93	1.20	2.13
0320	Type C, 54 FC, 2 fixture/1000 S.F.	1.07	1.34	2.41
0440	2 watt/S.F., type E, 84 FC, 2 fixture/1000 S.F.	1.88	2.44	4.32
0480	Type G, 105 FC, 2 fixture/1000 S.F.	1.88	2.44	4.32
0520	Type C, 108 FC, 4 fixture/1000 S.F.	2.15	2.67	4.82
0640	3 watt/S.F., type E, 126 FC, 3 fixture/1000 S.F.	2.82	3.64	6.46
0680	Type G, 157 FC, 3 fixture/1000 S.F.	2.82	3.64	6.46
0720	Type C, 162 FC, 6 fixture/1000 S.F.	3.22	4.02	7.24
0840	4 watt/S.F., type E, 168 FC, 4 fixture/1000 S.F.	3.78	4.87	8.65
0880	Type G, 210 FC, 4 fixture/1000 S.F.	3.78	4.87	8.65
0920	Type C, 243 FC, 9 fixture/1000 S.F.	4.65	5.60	10.25
1040	5 watt/S.F., type E, 210 FC, 5 fixture/1000 S.F.	4.70	6.05	10.75
1080	Type G, 262 FC, 5 fixture/1000 S.F.	4.70	6.05	10.75
1120	Type C, 297 FC, 11 fixture/1000 S.F.	5.70	6.95	12.65

D5020 234	H.I.D. Fixture, High Bay, 30′ (by Wattage)	COST PER S.F.		
		MAT.	INST.	TOTAL
0190	High intensity discharge fixture, 30′ above work plane			
0240	1 watt/S.F., type E, 37 FC, 1 fixture/1000 S.F.	1.04	1.51	2.55
0280	Type G, 45 FC., 1 fixture/1000 S.F.	1.04	1.51	2.55
0320	Type F, 50 FC, 1 fixture/1000 S.F.	.87	1.18	2.05
0440	2 watt/S.F., type E, 74 FC, 2 fixtures/1000 S.F.	2.10	3.03	5.13
0480	Type G, 92 FC, 2 fixtures/1000 S.F.	2.10	3.03	5.13
0520	Type F, 100 FC, 2 fixtures/1000 S.F.	1.77	2.43	4.20
0640	3 watt/S.F., type E, 110 FC, 3 fixtures/1000 S.F.	3.15	4.59	7.74
0680	Type G, 138FC, 3 fixtures/1000 S.F.	3.15	4.59	7.74
0720	Type F, 150 FC, 3 fixtures/1000 S.F.	2.65	3.60	6.25
0840	4 watt/S.F., type E, 148 FC, 4 fixtures/1000 S.F.	4.20	6.10	10.30
0880	Type G, 185 FC, 4 fixtures/1000 S.F.	4.20	6.10	10.30
0920	Type F, 200 FC, 4 fixtures/1000 S.F.	3.54	4.83	8.37
1040	5 watt/S.F., type E, 185 FC, 5 fixtures/1000 S.F.	5.25	7.60	12.85
1080	Type G, 230 FC, 5 fixtures/1000 S.F.	5.25	7.60	12.85
1120	Type F, 250 FC, 5 fixtures/1000 S.F.	4.41	6.05	10.46

D5020 238	H.I.D. Fixture, Low Bay, 8′-10′ (by Wattage)	COST PER S.F.		
		MAT.	INST.	TOTAL
0190	High intensity discharge fixture, 8′-10′ above work plane			
0240	1 watt/S.F., type J, 30 FC, 4 fixtures/1000 S.F.	2.26	2.48	4.74
0280	Type K, 29 FC, 5 fixtures/1000 S.F.	2.09	2.22	4.31
0400	2 watt/S.F., type J, 52 FC, 7 fixtures/1000 S.F.	4.02	4.53	8.55
0440	Type K, 63 FC, 11 fixtures/1000 S.F.	4.51	4.68	9.19
0560	3 watt/S.F., type J, 81 FC, 11 fixtures/1000 S.F.	6.20	6.85	13.05

D5020 Lighting and Branch Wiring

D5020 238	**H.I.D. Fixture, Low Bay, 8'-10' (by Wattage)**	COST PER S.F.		
		MAT.	INST.	TOTAL
0600	Type K, 92 FC, 16 fixtures/1000 S.F.	6.60	6.90	13.50
0720	4 watt/S.F., type J, 103 FC, 14 fixtures/1000 S.F.	8.05	9	17.05
0760	Type K, 127 FC, 22 fixtures/1000 S.F.	9	9.35	18.35
0880	5 watt/S.F., type J, 133 FC, 18 fixtures/1000 S.F.	10.25	11.40	21.65
0920	Type K, 155 FC, 27 fixtures/1000 S.F.	11.10	11.55	22.65

D5020 242	**H.I.D. Fixture, Low Bay, 16' (by Wattage)**	COST PER S.F.		
		MAT.	INST.	TOTAL
0190	High intensity discharge fixture, mounted 16' above work plane			
0240	1 watt/S.F., type J, 28 FC, 4 fixt./1000 S.F.	2.39	2.81	5.20
0280	Type K, 27 FC, 5 fixt./1000 S.F.	2.46	3.14	5.60
0400	2 watt/S.F., type J, 48 FC, 7 fixt/1000 S.F.	4.37	5.40	9.77
0440	Type K, 58 FC, 11 fixt/1000 S.F.	5.20	6.45	11.65
0560	3 watt/S.F., type J, 75 FC, 11 fixt/1000 S.F.	6.75	8.25	15
0600	Type K, 85 FC, 16 fixt/1000 S.F.	7.65	9.55	17.20
0720	4 watt/S.F., type J, 95 FC, 14 fixt/1000 S.F.	8.80	10.85	19.65
0760	Type K, 117 FC, 22 fixt/1000 S.F.	10.45	12.85	23.30
0880	5 watt/S.F., type J, 122 FC, 18 fixt/1000 S.F.	11.15	13.65	24.80
0920	Type K, 143 FC, 27 fixt/1000 S.F.	12.90	16	28.90

D50 Electrical

D5030 Communications and Security

D5030 910	Communication & Alarm Systems	COST EACH		
		MAT.	INST.	TOTAL
0200	Communication & alarm systems, includes outlets, boxes, conduit & wire			
0210	Sound system, 6 outlets	5,425	6,125	11,550
0220	12 outlets	7,800	9,800	17,600
0240	30 outlets	13,700	18,500	32,200
0280	100 outlets	41,900	62,000	103,900
0320	Fire detection systems, 12 detectors	2,775	5,100	7,875
0360	25 detectors	4,800	8,625	13,425
0400	50 detectors	9,300	16,900	26,200
0440	100 detectors	16,800	30,500	47,300
0480	Intercom systems, 6 stations	3,175	4,200	7,375
0560	25 stations	11,200	16,100	27,300
0640	100 stations	42,500	59,000	101,500
0680	Master clock systems, 6 rooms	4,475	6,850	11,325
0720	12 rooms	6,900	11,700	18,600
0760	20 rooms	9,150	16,500	25,650
0800	30 rooms	15,600	30,500	46,100
0840	50 rooms	25,300	51,500	76,800
0920	Master TV antenna systems, 6 outlets	2,825	4,325	7,150
0960	12 outlets	5,275	8,075	13,350
1000	30 outlets	10,900	18,700	29,600
1040	100 outlets	36,300	61,000	97,300

Generator System

A: Engine
B: Battery
C: Charger
D: Transfer Switch
E: Muffler

D5090 210	Generators (by kW)	COST PER kW		
		MAT.	INST.	TOTAL
0190	Generator sets, include battery, charger, muffler & transfer switch			
0200	Gas/gasoline operated, 3 phase, 4 wire, 277/480 volt, 7.5 kW	880	218	1,096.67
0240	11.5 kW	815	168	983
0280	20 kW	550	109	659
0320	35 kW	375	71	446
0360	80 kW	283	42.50	325.50
0400	100 kW	248	42	290
0440	125 kW	405	39	444
0480	185 kW	360	30	390
0560	Diesel engine with fuel tank, 30 kW	585	83	668
0600	50 kW	435	65.50	500.50
0640	75 kW	375	52	427
0680	100 kW	315	44	359
0760	150 kW	252	35	287
0840	200 kW	213	28.50	241.50
0880	250 kW	200	23.50	223.50
0960	350 kW	176	19.55	195.55
1040	500 kW	190	15.20	205.20

For information about Means Estimating Seminars, see yellow pages 12 and 13 in back of book

Important: See the Reference Section for critical supporting data - Reference Numbers and City Cost Indexes

E EQUIPMENT & FURNISHINGS

E1010 Commercial Equipment

E1010 110	Security/Vault, EACH	COST EACH		
		MAT.	INST.	TOTAL
0100	Bank equipment, drive up window, drawer & mike, no glazing, economy	4,200	1,100	5,300
0110	Deluxe	8,450	2,175	10,625
0120	Night depository, economy	6,300	1,100	7,400
0130	Deluxe	9,125	2,175	11,300
0140	Pneumatic tube systems, 2 station, standard	23,400	3,675	27,075
0150	Teller, automated, 24 hour, single unit	41,300	3,675	44,975
0160	Teller window, bullet proof glazing, 44" x 60"	2,950	665	3,615
0170	Pass through, painted steel, 72" x 40"	3,000	1,375	4,375
0300	Safe, office type, 1 hr. rating, 34" x 20" x 20"	1,800		1,800
0310	4 hr. rating, 62" x 33" x 20"	7,450		7,450
0320	Data storage, 4 hr. rating, 63" x 44" x 16"	10,000		10,000
0330	Jewelers, 63" x 44" x 16"	16,000		16,000
0340	Money, "B" label, 9" x 14" x 14"	435		435
0350	Tool and torch resistive, 24" x 24" x 20"	7,200	166	7,366
0500	Security gates-scissors type, painted steel, single, 6' high, 5-1/2' wide	149	273	422
0510	Double gate, 7-1/2' high, 14' wide	455	545	1,000

E1010 510	Mercantile Equipment, EACH	COST EACH		
		MAT.	INST.	TOTAL
0015	Barber equipment, chair, hydraulic, economy	460	17.80	477.80
0020	Deluxe	3,025	26.50	3,051.50
0100	Checkout counter, single belt	2,400	66.50	2,466.50
0110	Double belt, power take-away	4,050	74	4,124
0200	Display cases, freestanding, glass and aluminum, 3'-6" x 3' x 1'-0" deep	505	107	612
0220	Wall mounted, glass and aluminum, 3' x 4' x 1'-4" deep	1,250	171	1,421
0320	Frozen food, chest type, 12 ft. long	6,100	269	6,369

E1010 610	Laundry/Dry Cleaning, EACH	COST EACH		
		MAT.	INST.	TOTAL
0100	Laundry equipment, dryers, gas fired, residential, 16 lb. capacity	620	164	784
0110	Commercial, 30 lb. capacity, single	2,825	164	2,989
0120	Dry cleaners, electric, 20 lb. capacity	33,200	4,875	38,075
0130	30 lb. capacity	45,600	6,500	52,100
0140	Ironers, commercial, 120" with canopy, 8 roll	162,500	13,900	176,400
0150	Institutional, 110", single roll	28,800	2,425	31,225
0160	Washers, residential, 4 cycle	710	164	874
0170	Commercial, coin operated, deluxe	3,000	164	3,164

Important: See the Reference Section for critical supporting data - Reference Numbers and City Cost Indexes

EQUIPMENT & FURNISHINGS E

E10 Equipment

E1020 Institutional Equipment

E1020 110	Ecclesiastical Equipment, EACH	COST EACH		
		MAT.	INST.	TOTAL
0090	Church equipment, altar, wood, custom, plain	2,000	305	2,305
0100	Granite, custom, deluxe	25,500	4,050	29,550
0110	Baptistry, fiberglass, economy	3,050	1,100	4,150
0120	Bells & carillons, keyboard operation	16,300	6,700	23,000
0130	Confessional, wood, single, economy	2,550	710	3,260
0140	Double, deluxe	15,900	2,125	18,025
0150	Steeples, translucent fiberglas, 30" square, 15' high	3,950	1,400	5,350

E1020 130	Ecclesiastical Equipment, L.F.	COST PER L.F.		
		MAT.	INST.	TOTAL
0100	Arch. equip., church equip. pews, bench type, hardwood, economy	70.50	21.50	92
0110	Deluxe	140	28.50	168.50

E1020 210	Library Equipment, EACH	COST EACH		
		MAT.	INST.	TOTAL
0110	Library equipment, carrels, metal, economy	245	85.50	330.50
0120	Hardwood, deluxe	900	107	1,007

E1020 230	Library Equipment, L.F.	COST PER L.F.		
		MAT.	INST.	TOTAL
0100	Library equipment, book shelf, metal, single face, 90" high x 10" shelf	135	35.50	170.50
0110	Double face, 90" high x 10" shelf	335	77	412
0120	Charging desk, built-in, with counter, plastic laminate	480	61	541

E1020 310	Theater and Stage Equipment, EACH	COST EACH		
		MAT.	INST.	TOTAL
0200	Movie equipment, changeover, economy	435		435
0210	Film transport, incl. platters and autowind, economy	4,725		4,725
0220	Lamphouses, incl. rectifiers, xenon, 1000W	6,250	243	6,493
0230	4000W	8,850	325	9,175
0240	Projector mechanisms, 35 mm, economy	10,400		10,400
0250	Deluxe	14,300		14,300
0260	Sound systems, incl. amplifier, single, economy	3,100	540	3,640
0270	Dual, Dolby/super sound	15,500	1,225	16,725
0280	Projection screens, wall hung, manual operation, 50 S.F., economy	275	85.50	360.50
0290	Electric operation, 100 S.F., deluxe	1,875	425	2,300

E1020 320	Theater and Stage Equipment, S.F.	COST PER S.F.		
		MAT.	INST.	TOTAL
0090	Movie equipment, projection screens, rigid in wall, acrylic, 1/4" thick	39.50	4.09	43.59
0100	1/2" thick	45.50	6.15	51.65
0110	Stage equipment, curtains, velour, medium weight	7.15	1.42	8.57
0120	Silica based yarn, fireproof	13.75	17.05	30.80
0130	Stages, portable with steps, folding legs, 8" high	23.50		23.50
0140	Telescoping platforms, aluminum, deluxe	35	22	57

E1020 330	Theater and Stage Equipment, L.F.	COST PER L.F.		
		MAT.	INST.	TOTAL
0100	Stage equipment, curtain track, heavy duty	47.50	47.50	95
0110	Lights, border, quartz, colored	156	24.50	180.50

E1020 610	Detention Equipment, EACH	COST EACH		
		MAT.	INST.	TOTAL
0110	Detention equipment, cell front rolling door, 7/8" bars, 5' x 7' high	5,075	1,150	6,225
0120	Cells, prefab., including front, 5' x 7' x 7' deep	8,375	1,525	9,900

EQUIPMENT & FURNISHINGS

E

E1020 Institutional Equipment

E1020 610	Detention Equipment, EACH	COST EACH		
		MAT.	INST.	TOTAL
0130	Doors and frames, 3' x 7', single plate	3,800	580	4,380
0140	Double plate	4,700	580	5,280
0150	Toilet apparatus, incl wash basin	3,125	735	3,860
0160	Visitor cubicle, vision panel, no intercom	2,750	1,150	3,900

E1020 710	Laboratory Equipment, EACH	COST EACH		
		MAT.	INST.	TOTAL
0110	Laboratory equipment, glassware washer, distilled water, economy	6,825	540	7,365
0120	Deluxe	9,000	975	9,975
0140	Radio isotope	12,500		12,500

E1020 720	Laboratory Equipment, S.F.	COST PER S.F.		
		MAT.	INST.	TOTAL
0100	Arch. equip., lab equip., counter tops, acid proof, economy	28	10.40	38.40
0110	Stainless steel	94.50	10.40	104.90

E1020 730	Laboratory Equipment, L.F.	COST PER L.F.		
		MAT.	INST.	TOTAL
0110	Laboratory equipment, cabinets, wall, open	114	42.50	156.50
0120	Base, drawer units	365	47.50	412.50
0130	Fume hoods, not incl. HVAC, economy	570	158	728
0140	Deluxe incl. fixtures	1,350	355	1,705

E1020 810	Medical Equipment, EACH	COST EACH		
		MAT.	INST.	TOTAL
0100	Dental equipment, central suction system, economy	3,175	410	3,585
0110	Compressor-air, deluxe	24,200	865	25,065
0120	Chair, hydraulic, economy	3,975	865	4,840
0130	Deluxe	8,450	1,725	10,175
0140	Drill console with accessories, economy	3,600	271	3,871
0150	Deluxe	10,300	271	10,571
0160	X-ray unit, portable	4,800	108	4,908
0170	Panoramic unit	21,700	725	22,425
0300	Medical equipment, autopsy table, standard	8,025	490	8,515
0310	Deluxe	14,800	820	15,620
0320	Incubators, economy	3,250		3,250
0330	Deluxe	18,200		18,200
0700	Station, scrub-surgical, single, economy	2,350	164	2,514
0710	Dietary, medium, with ice	14,500		14,500
0720	Sterilizers, general purpose, single door, 20" x 20" x 28"	13,800		13,800
0730	Floor loading, double door, 28" x 67" x 52"	184,500		184,500
0740	Surgery tables, standard	12,500	780	13,280
0750	Deluxe	28,400	1,100	29,500
0770	Tables, standard, with base cabinets, economy	1,175	285	1,460
0780	Deluxe	3,500	425	3,925
0790	X-ray, mobile, economy	12,100		12,100
0800	Stationary, deluxe	197,000		197,000

E10 Equipment

E1030 Vehicular Equipment

E1030 110 — Vehicular Service Equipment, EACH

		COST EACH		
		MAT.	INST.	TOTAL
0110	Automotive equipment, compressors, electric, 1-1/2 H.P., std. controls	325	790	1,115
0120	5 H.P., dual controls	2,000	1,175	3,175
0130	Hoists, single post, 4 ton capacity, swivel arms	4,725	2,975	7,700
0140	Dual post, 12 ton capacity, adjustable frame	8,125	7,900	16,025
0150	Lube equipment, 3 reel type, with pumps	7,500	2,375	9,875
0160	Product dispenser, 6 nozzles, w/vapor recovery, not incl. piping, installed	19,800		19,800
0800	Scales, dial type, built in floor, 5 ton capacity, 8' x 6' platform	6,850	2,550	9,400
0810	10 ton capacity, 9' x 7' platform	10,100	3,650	13,750
0820	Truck (including weigh bridge), 20 ton capacity, 24' x 10'	10,500	4,275	14,775

E1030 210 — Parking Control Equipment, EACH

		COST EACH		
		MAT.	INST.	TOTAL
0110	Parking equipment, automatic gates, 8 ft. arm, one way	3,650	880	4,530
0120	Traffic detectors, single treadle	2,325	405	2,730
0130	Booth for attendant, economy	5,750		5,750
0140	Deluxe	24,400		24,400
0150	Ticket printer/dispenser, rate computing	7,800	695	8,495
0160	Key station on pedestal	705	237	942

E1030 310 — Loading Dock Equipment, EACH

		COST EACH		
		MAT.	INST.	TOTAL
0110	Dock bumpers, rubber blocks, 4-1/2" thick, 10" high, 14" long	49	16.40	65.40
0120	6" thick, 20" high, 11" long	134	33	167
0130	Dock boards, H.D., 5' x 5', aluminum, 5000 lb. capacity	1,325		1,325
0140	16,000 lb. capacity	1,375		1,375
0150	Dock levelers, hydraulic, 7' x 8', 10 ton capacity	8,500	1,125	9,625
0160	Dock lifters, platform, 6' x 6', portable, 3000 lb. capacity	8,325		8,325
0170	Dock shelters, truck, scissor arms, economy	1,350	425	1,775
0180	Deluxe	2,000	855	2,855

E EQUIPMENT & FURNISHINGS

E1090 Other Equipment

E1090 210	Solid Waste Handling Equipment, EACH	COST EACH		
		MAT.	INST.	TOTAL
0110	Waste handling, compactors, single bag, 250 lbs./hr., hand fed	7,500	495	7,995
0120	Heavy duty industrial, 5 C.Y. capacity	17,300	2,375	19,675
0130	Incinerator, electric, 100 lbs./hr., economy	9,800	2,400	12,200
0140	Gas, 2000 lbs./hr., deluxe	209,500	18,700	228,200
0150	Shredder, no baling, 35 tons/hr.	259,000		259,000
0160	Incl. baling, 50 tons/day	517,000		517,000

E1090 350	Food Service Equipment, EACH	COST EACH		
		MAT.	INST.	TOTAL
0110	Kitchen equipment, bake oven, single deck	4,650	111	4,761
0120	Broiler, without oven	2,975	111	3,086
0130	Commercial dish washer, semiautomatic, 50 racks/hr.	7,200	680	7,880
0140	Automatic, 275 racks/hr.	20,900	2,925	23,825
0150	Cooler, beverage, reach-in, 6 ft. long	3,925	147	4,072
0160	Food warmer, counter, 1.65 kw	665		665
0170	Fryers, with submerger, single	1,600	126	1,726
0180	Double	1,625	177	1,802
0190	Kettles, steam jacketed, 20 gallons	5,900	204	6,104
0200	Range, restaurant type, burners, 2 ovens and 24" griddle	4,800	147	4,947
0210	Range hood, incl. carbon dioxide system, economy	3,550	295	3,845
0220	Deluxe	26,600	885	27,485

E1090 360	Food Service Equipment, S.F.	COST PER S.F.		
		MAT.	INST.	TOTAL
0110	Refrigerators, prefab, walk-in, 7'-6" high, 6' x 6'	113	15.60	128.60
0120	12' x 20'	71	7.80	78.80

E1090 410	Residential Equipment, EACH	COST EACH		
		MAT.	INST.	TOTAL
0110	Arch. equip., appliances, range, cook top, 4 burner, economy	204	81	285
0120	Built in, single oven 30" wide, economy	490	81	571
0130	Standing, single oven-21" wide, economy	272	66.50	338.50
0140	Double oven-30" wide, deluxe	1,700	66.50	1,766.50
0150	Compactor, residential, economy	455	85.50	540.50
0160	Deluxe	515	142	657
0170	Dish washer, built-in, 2 cycles, economy	272	244	516
0180	4 or more cycles, deluxe	595	490	1,085
0190	Garbage disposer, sink type, economy	46.50	97.50	144
0200	Deluxe	160	97.50	257.50
0210	Refrigerator, no frost, 10 to 12 C.F., economy	485	66.50	551.50
0220	21 to 29 C.F., deluxe	2,275	222	2,497

E1090 610	School Equipment, EACH	COST EACH		
		MAT.	INST.	TOTAL
0110	School equipment, basketball backstops, wall mounted, wood, fixed	1,125	745	1,870
0120	Suspended type, electrically operated	3,925	1,450	5,375
0130	Bleachers-telescoping, manual operation, 15 tier, economy (per seat)	74.50	26.50	101
0140	Power operation, 30 tier, deluxe (per seat)	283	47.50	330.50
0150	Weight lifting gym, universal, economy	2,700	665	3,365
0160	Deluxe	14,900	1,325	16,225
0170	Scoreboards, basketball, 1 side, economy	2,150	660	2,810
0180	4 sides, deluxe	24,700	9,125	33,825
0800	Vocational shop equipment, benches, metal	405	171	576
0810	Wood	430	171	601
0820	Dust collector, not incl. ductwork, 6' diam.	3,000	450	3,450
0830	Planer, 13" x 6"	1,350	213	1,563

E1090 Other Equipment

E1090 620	School Equipment, S.F.	COST PER S.F.		
		MAT.	INST.	TOTAL
0110	School equipment, gym mats, naugahyde cover, 2" thick	3.70		3.70
0120	Wrestling, 1" thick, heavy duty	4.95		4.95

E1090 810	Athletic, Recreational, and Therapeutic Equipment, EACH	COST EACH		
		MAT.	INST.	TOTAL
0110	Sauna, prefabricated, incl. heater and controls, 7' high, 6' x 4'	3,875	650	4,525
0120	10' x 12'	9,525	1,425	10,950
0130	Heaters, wall mounted, to 200 C.F.	505		505
0140	Floor standing, to 1000 C.F., 12500 W	1,650	162	1,812
0610	Shooting range incl. bullet traps, controls, separators, ceilings, economy	24,200	5,600	29,800
0620	Deluxe	44,000	9,425	53,425
0650	Sport court, squash, regulation, in existing building, economy			16,000
0660	Deluxe			29,900
0670	Racketball, regulation, in existing building, economy	2,125	8,075	10,200
0680	Deluxe	28,400	16,100	44,500
0700	Swimming pool equipment, diving stand, stainless steel, 1 meter	4,000	315	4,315
0710	3 meter	5,425	2,125	7,550
0720	Diving boards, 16 ft. long, aluminum	2,775	315	3,090
0730	Fiberglass	2,225	315	2,540
0740	Filter system, sand, incl. pump, 6000 gal./hr.	1,225	545	1,770
0750	Lights, underwater, 12 volt with transformer, 300W	159	1,225	1,384
0760	Slides, fiberglass with aluminum handrails & ladder, 6' high, straight	2,425	535	2,960
0780	12' high, straight with platform	2,225	710	2,935

E1090 820	Athletic, Recreational, and Therapeutic Equipment, S.F.	COST PER S.F.		
		MAT.	INST.	TOTAL
0110	Swimming pools, residential, vinyl liner, metal sides	9.75	5.70	15.45
0120	Concrete sides	11.65	10.50	22.15
0130	Gunite shell, plaster finish, 350 S.F.	19.35	21.50	40.85
0140	800 S.F.	15.55	12.65	28.20
0150	Motel, gunite shell, plaster finish	24	27.50	51.50
0160	Municipal, gunite shell, tile finish, formed gutters	100	47	147

EQUIPMENT & FURNISHINGS

E

E20 Furnishings

E2010 Fixed Furnishings

E2010 310	Window Treatment, EACH	COST EACH		
		MAT.	INST.	TOTAL
0110	Furnishings, blinds, exterior, aluminum, louvered, 1'-4" wide x 3'-0" long	46	42.50	88.50
0120	1'-4" wide x 6'-8" long	92.50	47.50	140
0130	Hemlock, solid raised, 1'-4" wide x 3'-0" long	60.50	42.50	103
0140	1'-4" wide x 6'-9" long	114	47.50	161.50
0150	Polystyrene, louvered, 1'-3" wide x 3'-3" long	44	42.50	86.50
0160	1'-3" wide x 6'-8" long	95.50	47.50	143
0200	Interior, wood folding panels, louvered, 7" x 20" (per pair)	46.50	25	71.50
0210	18" x 40" (per pair)	144	25	169

E2010 320	Window Treatment, S.F.	COST PER S.F.		
		MAT.	INST.	TOTAL
0110	Furnishings, blinds-interior, venetian-aluminum, stock, 2" slats, economy	1.76	.72	2.48
0120	Custom, 1" slats, deluxe	7.15	.97	8.12
0130	Vertical, PVC or cloth, T&B track, economy	8.60	.93	9.53
0140	Deluxe	10.45	1.07	11.52
0150	Draperies, unlined, economy	2.46		2.46
0160	Lightproof, deluxe	7.40		7.40
0510	Shades, mylar, wood roller, single layer, non-reflective	6.10	.62	6.72
0520	Metal roller, triple layer, heat reflective	10.30	.62	10.92
0530	Vinyl, light weight, 4 ga.	.45	.62	1.07
0540	Heavyweight, 6 ga.	1.43	.62	2.05
0550	Vinyl coated cotton, lightproof decorator shades	1.91	.62	2.53
0560	Woven aluminum, 3/8" thick, light and fireproof	4.71	1.22	5.93

E2010 420	Fixed Floor Grilles and Mats, S.F.	COST PER S.F.		
		MAT.	INST.	TOTAL
0110	Floor mats, recessed, inlaid black rubber, 3/8" thick, solid	23	2.14	25.14
0120	Colors, 1/2" thick, perforated	32.50	2.14	34.64
0130	Link-including nosings, steel-galvanized, 3/8" thick	8.80	2.14	10.94
0140	Vinyl, in colors	20.50	2.14	22.64

E2010 510	Fixed Multiple Seating, EACH	COST EACH		
		MAT.	INST.	TOTAL
0110	Seating, painted steel, upholstered, economy	124	24.50	148.50
0120	Deluxe	400	30.50	430.50
0400	Seating, lecture hall, pedestal type, economy	140	39	179
0410	Deluxe	485	59	544
0500	Auditorium chair, veneer construction	154	39	193
0510	Fully upholstered, spring seat	195	39	234

EQUIPMENT & FURNISHINGS E

Important: See the Reference Section for critical supporting data - Reference Numbers and City Cost Indexes

E2020 Moveable Furnishings

E2020 210	Furnishings/EACH	COST EACH		
		MAT.	INST.	TOTAL
0200	Hospital furniture, beds, manual, economy	1,000		1,000
0210	Deluxe	2,750		2,750
0220	All electric, economy	1,475		1,475
0230	Deluxe	4,025		4,025
0240	Patient wall systems, no utilities, economy, per room	1,000		1,000
0250	Deluxe, per room	1,850		1,850
0300	Hotel furnishings, standard room set, economy, per room	1,675		1,675
0310	Deluxe, per room	8,675		8,675
0500	Office furniture, standard employee set, economy, per person	345		345
0510	Deluxe, per person	2,300		2,300
0550	Posts, portable, pedestrian traffic control, economy	103		103
0560	Deluxe	315		315
0700	Restaurant furniture, booth, molded plastic, stub wall and 2 seats, economy	315	213	528
0710	Deluxe	1,050	285	1,335
0720	Upholstered seats, foursome, single-economy	870	85.50	955.50
0730	Foursome, double-deluxe	1,575	142	1,717

E2020 220	Furniture and Accessories, L.F.	COST PER L.F.		
		MAT.	INST.	TOTAL
0210	Dormitory furniture, desk top (built-in),laminated plastc, 24"deep, economy	27.50	17.05	44.55
0220	30" deep, deluxe	154	21.50	175.50
0230	Dressing unit, built-in, economy	198	71	269
0240	Deluxe	595	107	702
0310	Furnishings, cabinets, hospital, base, laminated plastic	232	85.50	317.50
0320	Stainless steel	320	85.50	405.50
0330	Counter top, laminated plastic, no backsplash	40	21.50	61.50
0340	Stainless steel	111	21.50	132.50
0350	Nurses station, door type, laminated plastic	269	85.50	354.50
0360	Stainless steel	340	85.50	425.50
0710	Restaurant furniture, bars, built-in, back bar	171	85.50	256.50
0720	Front bar	235	85.50	320.50
0910	Wardrobes & coatracks, standing, steel, single pedestal, 30" x 18" x 63"	71		71
0920	Double face rack, 39" x 26" x 70"	64.50		64.50
0930	Wall mounted rack, steel frame & shelves, 12" x 15" x 26"	59.50	6.05	65.55
0940	12" x 15" x 50"	33.50	3.18	36.68

For information about Means Estimating Seminars, see yellow pages 12 and 13 in back of book

EQUIPMENT & FURNISHINGS

E

F SPECIAL CONSTRUCTION

F1010 Special Structures

F1010 120	Air-Supported Structures, S.F.	COST PER S.F.		
		MAT.	INST.	TOTAL
0110	Air supported struc., polyester vinyl fabric, 24oz., warehouse, 5000 S.F.	17.90	.27	18.17
0120	50,000 S.F.	8.20	.21	8.41
0130	Tennis, 7,200 S.F.	16.70	.22	16.92
0140	24,000 S.F.	11.35	.22	11.57
0150	Woven polyethylene, 6 oz., shelter, 3,000 S.F.	7.05	.44	7.49
0160	24,000 S.F.	6	.22	6.22
0170	Teflon coated fiberglass, stadium cover, economy	42.50	.12	42.62
0180	Deluxe	50	.16	50.16
0190	Air supported storage tank covers, reinf. vinyl fabric, 12 oz., 400 S.F.	16.15	.38	16.53
0200	18,000 S.F.	4.96	.34	5.30

F1010 210	Pre-Engineered Structures, EACH	COST EACH		
		MAT.	INST.	TOTAL
0600	Radio towers, guyed, 40 lb. section, 50' high, 70 MPH basic wind speed	2,450	1,100	3,550
0610	90 lb. section, 400' high, wind load 70 MPH basic wind speed	31,700	11,700	43,400
0620	Self supporting, 60' high, 70 MPH basic wind speed	5,100	2,050	7,150
0630	190' high, wind load 90 MPH basic wind speed	30,400	8,225	38,625
0700	Shelters, aluminum frame, acrylic glazing, 8' high, 3' x 9'	2,800	960	3,760
0710	9' x 12'	4,850	1,500	6,350

F1010 320	Other Special Structures, S.F.	COST PER S.F.		
		MAT.	INST.	TOTAL
0110	Swimming pool enclosure, transluscent, freestanding, economy	13.10	4.27	17.37
0120	Deluxe	192	12.20	204.20
0510	Tension structures, steel frame, polyester vinyl fabric, 12,000 S.F.	10.50	1.93	12.43
0520	20,800 S.F.	10.75	1.75	12.50

F1010 330	Special Structures, EACH	COST EACH		
		MAT.	INST.	TOTAL
0110	Kiosks, round, 5' diam., 1/4" fiberglass wall, 8' high	6,175		6,175
0120	Rectangular, 5' x 9', 1" insulated dbl. wall fiberglass, 7'-6" high	10,700		10,700
0200	Silos, concrete stave, industrial, 12' diam., 35' high	19,900	17,700	37,600
0210	25' diam., 75' high	65,500	38,900	104,400
0220	Steel prefab, 30,000 gal., painted, economy	19,000	4,400	23,400
0230	Epoxy-lined, deluxe	39,300	8,825	48,125

F1010 340	Special Structures, S.F.	COST PER S.F.		
		MAT.	INST.	TOTAL
0110	Comfort stations, prefab, mobile on steel frame, economy	40		40
0120	Permanent on concrete slab, deluxe	227	39.50	266.50
0210	Domes, bulk storage, wood framing, wood decking, 50' diam.	42.50	1.40	43.90
0220	116' diam.	25.50	1.62	27.12
0230	Steel framing, metal decking, 150' diam.	30.50	9.10	39.60
0240	400' diam.	25	6.95	31.95
0250	Geodesic, wood framing, wood panels, 30' diam.	30.50	1.54	32.04
0260	60' diam.	23	1	24
0270	Aluminum framing, acrylic panels, 40' diam.			93.50
0280	Aluminum panels, 400' diam.			33
0310	Garden house, prefab, wood, shell only, 48 S.F.	21	4.27	25.27
0320	200 S.F.	38	17.80	55.80
0410	Greenhouse, shell-stock, residential, lean-to, 8'-6" long x 3'-10" wide	38.50	25	63.50
0420	Freestanding, 8'-6" long x 13'-6" wide	30	7.90	37.90
0430	Commercial-truss frame, under 2000 S.F., deluxe			43
0440	Over 5,000 S.F., economy			26
0450	Institutional-rigid frame, under 500 S.F., deluxe			112
0460	Over 2,000 S.F., economy			44

Important: See the Reference Section for critical supporting data - Reference Numbers and City Cost Indexes

SPECIAL CONSTRUCTION F

F1010 Special Structures

F1010 340	Special Structures, S.F.	COST PER S.F.		
		MAT.	INST.	TOTAL
0510	Hangar, prefab, galv. roof and walls, bottom rolling doors, economy	10.75	3.61	14.36
0520	Electric bifolding doors, deluxe	11.70	4.71	16.41

F1020 Integrated Construction

F1020 110	Integrated Construction, EACH	COST EACH		
		MAT.	INST.	TOTAL
0110	Integrated ceilings, radiant electric, 2' x 4' panel, manila finish	96.50	19.40	115.90
0120	ABS plastic finish	112	37.50	149.50

F1020 120	Integrated Construction, S.F.	COST PER S.F.		
		MAT.	INST.	TOTAL
0110	Integrated ceilings, Luminaire, suspended, 5' x 5' modules, 50% lighted	4.32	10.20	14.52
0120	100% lighted	6.45	18.35	24.80
0130	Dimensionaire, 2' x 4' module tile system, no air bar	2.61	1.71	4.32
0140	With air bar, deluxe	3.83	1.71	5.54
0220	Pedestal access floors, stl pnls, no stringers, vinyl covering, >6000 S.F.	15.15	1.90	17.05
0230	With stringers, high pressure laminate covering, under 6000 S.F.	19.40	2.13	21.53
0240	Carpet covering, under 6000 S.F.	21.50	2.13	23.63
0250	Aluminum panels, no stringers, no covering	32.50	1.71	34.21

F1020 250	Special Purpose Room, EACH	COST EACH		
		MAT.	INST.	TOTAL
0110	Portable booth, acoustical, 27 db 1000 hz., 15 S.F. floor	3,325		3,325
0120	55 S.F. flr.	6,825		6,825

F1020 260	Special Purpose Room, S.F.	COST PER S.F.		
		MAT.	INST.	TOTAL
0110	Anechoic chambers, 7' high, 100 cps cutoff, 25 S.F.			2,175
0120	200 cps cutoff, 100 S.F.			1,125
0130	Audiometric rooms, under 500 S.F.	48.50	17.40	65.90
0140	Over 500 S.F.	46	14.25	60.25
0300	Darkrooms, shell, not including door, 240 S.F., 8' high	40	7.10	47.10
0310	64 S.F., 12' high	92.50	13.35	105.85
0510	Music practice room, modular, perforated steel, under 500 S.F.	29.50	12.20	41.70
0520	Over 500 S.F.	25	10.65	35.65

F1020 330	Special Construction, L.F.	COST PER L.F.		
		MAT.	INST.	TOTAL
0110	Spec. const., air curtains, shipping & receiving, 8'high x 5'wide, economy	345	99	444
0120	20' high x 8' wide, heated, deluxe	1,425	247	1,672
0130	Customer entrance, 10' high x 5' wide, economy	335	99	434
0140	12' high x 4' wide, heated, deluxe	560	124	684

F1030 Special Construction Systems

F1030 120	Sound, Vibration, and Seismic Construction, S.F.	COST PER S.F.		
		MAT.	INST.	TOTAL
0020	Special construction, acoustical, enclosure, 4" thick, 8 psf panels	30.50	17.80	48.30
0030	Reverb chamber, 4" thick, parallel walls	43	21.50	64.50
0110	Sound absorbing panels, 2'-6" x 8', painted metal	11.35	5.95	17.30
0120	Vinyl faced	8.90	5.35	14.25
0130	Flexible transparent curtain, clear	6.85	6.90	13.75
0140	With absorbing foam, 75% coverage	9.55	6.90	16.45
0150	Strip entrance, 2/3 overlap	5.45	11	16.45
0160	Full overlap	5.75	12.90	18.65
0200	Audio masking system, plenum mounted, over 10,000 S.F.	.67	.22	.89
0210	Ceiling mounted, under 5,000 S.F.	1.28	.40	1.68

F1030 210	Radiation Protection, EACH	COST EACH		
		MAT.	INST.	TOTAL
0110	Shielding, lead x-ray protection, radiography room, 1/16" lead, economy	6,300	2,975	9,275
0120	Deluxe	7,875	4,975	12,850
0210	Deep therapy x-ray, 1/4" lead economy	19,300	9,325	28,625
0220	Deluxe	25,000	12,400	37,400

F1030 220	Radiation Protection, S.F.	COST PER S.F.		
		MAT.	INST.	TOTAL
0110	Shielding, lead, gypsum board, 5/8" thick, 1/16" lead	5.60	4.99	10.59
0120	1/8" lead	11.25	5.70	16.95
0130	Lath, 1/16" thick	5.70	5.50	11.20
0140	1/8" thick	11.65	6.20	17.85
0150	Radio frequency, copper, prefab screen type, economy	26.50	4.74	31.24
0160	Deluxe	34	5.90	39.90

F1030 910	Other Special Construction Systems, EACH	COST EACH		
		MAT.	INST.	TOTAL
0110	Disappearing stairways, folding, pine, 8'-6" ceiling	105	107	212
0200	Fire escape, galvanized steel, 8'-0" to 10'-6" ceiling	1,325	855	2,180
0220	Automatic electric, wood, 8' to 9' ceiling	6,875	855	7,730
0300	Fireplace prefabricated, freestanding or wall hung, economy	1,250	330	1,580
0310	Deluxe	3,700	475	4,175
0320	Woodburning stoves, cast iron, economy	775	655	1,430
0330	Deluxe	2,375	1,075	3,450

SPECIAL CONSTRUCTION

F

F1040 Special Facilities

F1040 210	Ice Rinks, EACH	COAT EACH		
		MAT.	INST.	TOTAL
0100	Ice skating rink, 85' x 200', 55° system, 5 mos., 100 ton			545,500
0110	90° system, 12 mos., 135 ton			605,500
0120	Dash boards, acrylic screens, polyethylene coated plywood	126,500	28,900	155,400
0130	Fiberglass and aluminum construction	176,000	28,900	204,900

F1040 510	Liquid & Gas Storage Tanks, EACH	COST EACH		
		MAT.	INST.	TOTAL
0100	Tanks, steel, ground level, 100,000 gal.			141,500
0110	10,000,000 gal.			2,784,500
0120	Elevated water, 50,000 gal.		147,500	205,500
0130	1,000,000 gal.		690,000	962,000
0150	Cypress wood, ground level, 3,000 gal.	7,475	8,500	15,975
0160	Redwood, ground level, 45,000 gal.	56,500	23,000	79,500

SPECIAL CONSTRUCTION F

F1040 Special Facilities

F1040 910	Special Construction, EACH	COST EACH		
		MAT.	INST.	TOTAL
0110	Special construction, bowling alley incl. pinsetter, scorer etc., economy	38,600	8,525	47,125
0120	Deluxe	50,500	9,475	59,975
0130	For automatic scorer, economy, add	7,000		7,000
0140	Deluxe, add	7,925		7,925
0200	Chimney, metal, high temp. steel jacket, factory lining, 24" diameter	10,200	3,850	14,050
0210	60" diameter	74,500	16,700	91,200
0220	Poured concrete, brick lining, 10' diam., 200' high			1,230,000
0230	20' diam., 500' high			5,400,000
0240	Radial brick, 3'-6" I.D., 75' high			206,500
0250	7' I.D., 175' high			617,000
0300	Control tower, modular, 12' x 10', incl. instrumentation, economy			510,000
0310	Deluxe			710,500
0400	Garage costs, residential, prefab, wood, single car economy	3,500	855	4,355
0410	Two car deluxe	10,500	1,700	12,200
0500	Hangars, prefab, galv. steel, bottom rolling doors, economy (per plane)	11,100	4,025	15,125
0510	Electrical bi-folding doors, deluxe (per plane)	15,500	5,500	21,000

For information about Means Estimating Seminars, see yellow pages 12 and 13 in back of book

SPECIAL CONSTRUCTION

F

G BUILDING SITEWORK

G10 Site Preparation

G1030 Site Earthwork

G1030 805	Trenching	COST PER L.F.		
		MAT.	INST.	TOTAL
1310	Trenching, backhoe, 0 to 1 slope, 2' wide, 2' deep, 3/8 C.Y. bucket	.93	1.70	2.63
1330	4' deep, 3/8 C.Y. bucket	1.36	3.28	4.64
1360	10' deep, 1 C.Y. bucket	3.01	6.30	9.31
1400	4' wide, 2' deep, 3/8 C.Y. bucket	1.98	3.42	5.40
1420	4' deep, 1/2 C.Y. bucket	2.79	5.10	7.89
1450	10' deep, 1 C.Y. bucket	6.25	12.40	18.65
1480	18' deep, 2-1/2 C.Y. bucket	9.30	19.50	28.80
1520	6' wide, 6' deep, 5/8 C.Y. bucket	7.25	10.75	18
1540	10' deep, 1 C.Y. bucket	9.50	15.50	25
1570	20' deep, 3-1/2 C.Y. bucket	14.90	25	39.90
1640	8' wide, 12' deep, 1-1/4 C.Y. bucket	12.40	16.80	29.20
1680	24' deep, 3-1/2 C.Y. bucket	33	56	89
1730	10' wide, 20' deep, 3-1/2 C.Y. bucket	27	42	69
1740	24' deep, 3-1/2 C.Y. bucket	42	70	112
3500	1 to 1 slope, 2' wide, 2' deep, 3/8 C.Y. bucket	1.48	2.55	4.03
3540	4' deep, 3/8 C.Y. bucket	2.92	6.70	9.62
3600	10' deep, 1 C.Y. bucket	13.90	27	40.90
3800	4' wide, 2' deep, 3/8 C.Y. bucket	2.52	4.35	6.87
3840	4' deep, 1/2 C.Y. bucket	4.45	7	11.45
3900	10' deep, 1 C.Y. bucket	15.15	28	43.15
4030	6' wide, 6' deep, 5/8 C.Y. bucket	10.45	14.40	24.85
4050	10' deep, 1 C.Y. bucket	18.45	32	50.45
4080	20' deep, 3-1/2 C.Y. bucket	46.50	85	131.50
4500	8' wide, 12' deep, 1-1/4 C.Y. bucket	23	32.50	55.50
4650	24' deep, 3-1/2 C.Y. bucket	96	176	272
4800	10' wide, 20' deep, 3-1/2 C.Y. bucket	58.50	98.50	157
4850	24' deep, 3-1/2 C.Y. bucket	104	187	291

G1030 815	Pipe Bedding	COST PER L.F.		
		MAT.	INST.	TOTAL
1440	Pipe bedding, side slope 0 to 1, 1' wide, pipe size 6" diameter	.30	.49	.79
1460	2' wide, pipe size 8" diameter	.67	1.09	1.76
1500	Pipe size 12" diameter	.71	1.15	1.86
1600	4' wide, pipe size 20" diameter	1.76	2.83	4.59
1660	Pipe size 30" diameter	1.87	3.01	4.88
1680	6' wide, pipe size 32" diameter	3.26	5.25	8.51
1740	8' wide, pipe size 60" diameter	5.45	8.80	14.25
1780	12' wide, pipe size 84" diameter	10.70	17.25	27.95

Important: See the Reference Section for critical supporting data - Reference Numbers and City Cost Indexes

G20 Site Improvements

G2010 Roadways

G2010 210	Roadway Pavement	COST PER L.F.		
		MAT.	INST.	TOTAL
1500	Roadways, bituminous conc. paving, 2-1/2" thick, 20' wide	42	46.50	88.50
1580	30' wide	52.50	52.50	105
1800	3" thick, 20' wide	44.50	45.50	90
1880	30' wide	56	53	109
2100	4" thick, 20' wide	49.50	45.50	95
2180	30' wide	64	53	117
2400	5" thick, 20' wide	58.50	46	104.50
2480	30' wide	77	54	131
3000	6" thick, 20' wide	79.50	47.50	127
3080	30' wide	109	56	165
3300	8" thick 20' wide	98.50	47.50	146
3380	30' wide	137	56.50	193.50
3600	12" thick 20' wide	121	51	172
3700	32' wide	180	63	243

G2020 Parking Lots

G2020 210	Parking Lots	COST PER CAR		
		MAT.	INST.	TOTAL
1500	Parking lot, 90° angle parking, 3″ bituminous paving, 6″ gravel base	380	245	625
1540	10″ gravel base	450	283	733
1560	4″ bituminous paving, 6″ gravel base	460	245	705
1600	10″ gravel base	530	283	813
1620	6″ bituminous paving, 6″ gravel base	590	255	845
1660	10″ gravel base	665	293	958
1800	60° angle parking, 3″ bituminous paving, 6″ gravel base	435	269	704
1840	10″ gravel base	520	315	835
1860	4″ bituminous paving, 6″ gravel base	530	268	798
1900	10″ gravel base	615	310	925
1920	6″ bituminous paving, 6″ gravel base	685	279	964
1960	10″ gravel base	765	325	1,090
2200	45° angle parking, 3″ bituminous paving, 6″ gravel base	490	291	781
2240	10″ gravel base	590	340	930
2260	4″ bituminous paving, 6″ gravel base	600	290	890
2300	10″ gravel base	695	340	1,035
2320	6″ bituminous paving, 6″ gravel base	775	305	1,080
2360	10″ gravel base	870	355	1,225

Important: See the Reference Section for critical supporting data - Reference Numbers and City Cost Indexes

G20 Site Improvements

G2040 Site Development

G2040 810	Flagpoles	COST EACH		
		MAT.	INST.	TOTAL
0110	Flagpoles, on grade, aluminum, tapered, 20' high	905	470	1,375
0120	70' high	8,325	1,175	9,500
0130	Fiberglass, tapered, 23' high	1,175	470	1,645
0140	59' high	6,900	1,050	7,950
0150	Concrete, internal halyard, 20' high	1,475	380	1,855
0160	100' high	14,800	950	15,750

G2040 950	Other Site Development, EACH	COST EACH		
		MAT.	INST.	TOTAL
0110	Grandstands, permanent, closed deck, steel, economy (per seat)			33
0120	Deluxe (per seat)			88
0130	Composite design, economy (per seat)			82.50
0140	Deluxe (per seat)			154

G BUILDING SITEWORK

G3030 Storm Sewer

G3030 210	Manholes & Catch Basins	COST PER EACH		
		MAT.	INST.	TOTAL
1920	Manhole/catch basin, brick, 4' I.D. riser, 4' deep	800	1,325	2,125
1980	10' deep	1,850	3,075	4,925
3200	Block, 4' I.D. riser, 4' deep	730	1,050	1,780
3260	10' deep	1,375	2,525	3,900
4620	Concrete, cast-in-place, 4' I.D. riser, 4' deep	940	1,800	2,740
4680	10' deep	2,100	4,375	6,475
5820	Concrete, precast, 4' I.D. riser, 4' deep	1,200	965	2,165
5880	10' deep	2,275	2,250	4,525
6200	6' I.D. riser, 4' deep	1,875	1,375	3,250
6260	10' deep	3,750	3,250	7,000

Important: See the Reference Section for critical supporting data - Reference Numbers and City Cost Indexes

Light Pole

G4020 210	Light Pole (Installed)	COST EACH		
		MAT.	INST.	TOTAL
0200	Light pole, aluminum, 20' high, 1 arm bracket	900	860	1,760
0240	2 arm brackets	1,000	860	1,860
0280	3 arm brackets	1,100	890	1,990
0320	4 arm brackets	1,200	890	2,090
0360	30' high, 1 arm bracket	1,625	1,100	2,725
0400	2 arm brackets	1,725	1,100	2,825
0440	3 arm brackets	1,825	1,125	2,950
0480	4 arm brackets	1,925	1,125	3,050
0680	40' high, 1 arm bracket	1,975	1,450	3,425
0720	2 arm brackets	2,075	1,450	3,525
0760	3 arm brackets	2,175	1,500	3,675
0800	4 arm brackets	2,275	1,500	3,775
0840	Steel, 20' high, 1 arm bracket	1,100	920	2,020
0880	2 arm brackets	1,200	920	2,120
0920	3 arm brackets	1,200	950	2,150
0960	4 arm brackets	1,300	950	2,250
1000	30' high, 1 arm bracket	1,275	1,150	2,425
1040	2 arm brackets	1,350	1,150	2,500
1080	3 arm brackets	1,375	1,175	2,550
1120	4 arm brackets	1,475	1,175	2,650
1320	40' high, 1 arm bracket	1,650	1,550	3,200
1360	2 arm brackets	1,750	1,550	3,300
1400	3 arm brackets	1,750	1,600	3,350
1440	4 arm brackets	1,850	1,600	3,450

For information about Means Estimating Seminars, see yellow pages 12 and 13 in back of book

BUILDING SITEWORK

G

Reference Section

All the reference information is in one section making it easy to find what you need to know . . . and easy to use the book on a daily basis. This section is visually identified by a vertical gray bar on the edge of pages.

In this Reference Section, we've included General Conditions; Historical Cost Indexes for cost comparisons over time; Location Factors; a Glossary; and an explanation of all Abbreviations used in the book.

Table of Contents

General Conditions...................................... 450
Location Factors 451
Historical Cost Indexes.............................. 457
Glossary.. 465
Abbreviations 469
Forms 472

449

General Conditions, Overhead & Profit

The total building costs in the Commercial/Industrial/Institutional section include a 10% allowance for general conditions and a 15% allowance for the general contractor's overhead and profit and contingencies.

General contractor overhead includes indirect costs such as permits, Workers' Compensation, insurances, supervision and bonding fees. Overhead will vary with the size of project, the contractor's operating procedures and location. Profits will vary with economic activity and local conditions.

Contingencies provide for unforeseen construction difficulties which include material shortages and weather. In all situations, the appraiser should give consideration to possible adjustment of the 25% factor used in developing the Commercial/Industrial/Institutional models.

Architectural Fees

Tabulated below are typical percentage fees by project size, for good professional architectural service. Fees may vary from those listed depending upon degree of design difficulty and economic conditions in any particular area.

Rates can be interpolated horizontally and vertically. Various portions of the same project requiring different rates should be adjusted proportionately. For alterations, add 50% to the fee for the first $500,000 of project cost and add 25% to the fee for project cost over $500,000.

Architectural fees tabulated below include Engineering fees.

Insurance Exclusions

Many insurance companies exclude from coverage such items as architect's fees, excavation, foundations below grade, underground piping and site preparation. Since exclusions vary among insurance companies, it is recommended that for greatest accuracy each exclusion be priced separately using the unit-in-place section.

As a rule of thumb, exclusions can be calculated at 9% of total building cost plus the appropriate allowance for architect's fees.

Building Types	Total Project Size in Thousands of Dollars						
	100	250	500	1,000	5,000	10,000	50,000
Factories, garages, warehouses, repetitive housing	9.0%	8.0%	7.0%	6.2%	5.3%	4.9%	4.5%
Apartments, banks, schools, libraries, offices, municipal buildings	12.2	12.3	9.2	8.0	7.0	6.6	6.2
Churches, hospitals, homes, laboratories, museums, research	15.0	13.6	12.7	11.9	9.5	8.8	8.0
Memorials, monumental work, decorative furnishings	—	16.0	14.5	13.1	10.0	9.0	8.3

Location Factors

Costs shown in *Means cost data publications* are based on National Averages for materials and installation. To adjust these costs to a specific location, simply multiply the base cost by the factor for that city. The data is arranged alphabetically by state and postal zip code numbers. For a city not listed, use the factor for a nearby city with similar economic characteristics.

STATE/ZIP	CITY	Residential	Commercial
ALABAMA			
350-352	Birmingham	.87	.87
354	Tuscaloosa	.73	.78
355	Jasper	.71	.77
356	Decatur	.77	.79
357-358	Huntsville	.84	.86
359	Gadsden	.73	.80
360-361	Montgomery	.76	.80
362	Anniston	.68	.74
363	Dothan	.75	.76
364	Evergreen	.71	.76
365-366	Mobile	.80	.82
367	Selma	.72	.77
368	Phenix City	.73	.79
369	Butler	.71	.76
ALASKA			
995-996	Anchorage	1.26	1.25
997	Fairbanks	1.28	1.24
998	Juneau	1.26	1.25
999	Ketchikan	1.27	1.30
ARIZONA			
850,853	Phoenix	.87	.88
852	Mesa/Tempe	.84	.84
855	Globe	.80	.82
856-857	Tucson	.84	.85
859	Show Low	.83	.84
860	Flagstaff	.85	.86
863	Prescott	.82	.83
864	Kingman	.81	.83
865	Chambers	.81	.81
ARKANSAS			
716	Pine Bluff	.75	.80
717	Camden	.65	.69
718	Texarkana	.70	.72
719	Hot Springs	.64	.69
720-722	Little Rock	.81	.81
723	West Memphis	.74	.77
724	Jonesboro	.73	.77
725	Batesville	.71	.74
726	Harrison	.72	.74
727	Fayetteville	.68	.72
728	Russellville	.70	.73
729	Fort Smith	.76	.78
CALIFORNIA			
900-902	Los Angeles	1.06	1.07
903-905	Inglewood	1.04	1.03
906-908	Long Beach	1.03	1.04
910-912	Pasadena	1.04	1.04
913-916	Van Nuys	1.07	1.06
917-918	Alhambra	1.08	1.05
919-921	San Diego	1.04	1.04
922	Palm Springs	1.02	1.03
923-924	San Bernardino	1.04	1.02
925	Riverside	1.06	1.06
926-927	Santa Ana	1.04	1.03
928	Anaheim	1.07	1.07
930	Oxnard	1.07	1.06
931	Santa Barbara	1.07	1.06
932-933	Bakersfield	1.04	1.04
934	San Luis Obispo	1.07	1.05
935	Mojave	1.05	1.02
936-938	Fresno	1.10	1.08
939	Salinas	1.12	1.10
940-941	San Francisco	1.21	1.22
942,956-958	Sacramento	1.11	1.10
943	Palo Alto	1.16	1.15
944	San Mateo	1.19	1.16
945	Vallejo	1.13	1.12
946	Oakland	1.18	1.17
947	Berkeley	1.21	1.16
948	Richmond	1.22	1.16
949	San Rafael	1.20	1.16
950	Santa Cruz	1.14	1.13
951	San Jose	1.18	1.17
952	Stockton	1.09	1.07
953	Modesto	1.08	1.08

STATE/ZIP	CITY	Residential	Commercial
CALIFORNIA (CONT'D)			
954	Santa Rosa	1.14	1.12
955	Eureka	1.09	1.05
959	Marysville	1.10	1.09
960	Redding	1.10	1.08
961	Susanville	1.11	1.08
COLORADO			
800-802	Denver	.96	.96
803	Boulder	.94	.93
804	Golden	.92	.94
805	Fort Collins	.91	.93
806	Greeley	.80	.86
807	Fort Morgan	.94	.93
808-809	Colorado Springs	.92	.94
810	Pueblo	.93	.93
811	Alamosa	.90	.92
812	Salida	.91	.92
813	Durango	.93	.92
814	Montrose	.88	.91
815	Grand Junction	.93	.92
816	Glenwood Springs	.92	.94
CONNECTICUT			
060	New Britain	1.06	1.06
061	Hartford	1.05	1.06
062	Willimantic	1.06	1.06
063	New London	1.05	1.04
064	Meriden	1.05	1.05
065	New Haven	1.06	1.07
066	Bridgeport	1.06	1.06
067	Waterbury	1.06	1.06
068	Norwalk	1.06	1.06
069	Stamford	1.07	1.08
D.C.			
200-205	Washington	.93	.97
DELAWARE			
197	Newark	1.00	1.03
198	Wilmington	1.01	1.02
199	Dover	1.00	1.03
FLORIDA			
320,322	Jacksonville	.79	.81
321	Daytona Beach	.86	.88
323	Tallahassee	.72	.76
324	Panama City	.67	.72
325	Pensacola	.75	.81
326,344	Gainesville	.77	.81
327-328,347	Orlando	.85	.86
329	Melbourne	.87	.91
330-332,340	Miami	.83	.87
333	Fort Lauderdale	.84	.86
334,349	West Palm Beach	.84	.84
335-336,346	Tampa	.87	.87
337	St. Petersburg	.77	.81
338	Lakeland	.83	.87
339,341	Fort Myers	.81	.83
342	Sarasota	.85	.85
GEORGIA			
300-303,399	Atlanta	.89	.90
304	Statesboro	.67	.73
305	Gainesville	.74	.78
306	Athens	.74	.80
307	Dalton	.70	.76
308-309	Augusta	.76	.79
310-312	Macon	.78	.79
313-314	Savannah	.79	.80
315	Waycross	.71	.77
316	Valdosta	.71	.75
317	Albany	.75	.79
318-319	Columbus	.80	.81
HAWAII			
967	Hilo	1.22	1.20
968	Honolulu	1.23	1.23

STATE/ZIP	CITY	Residential	Commercial
STATES & POSS.			
969	Guam	1.62	1.31
IDAHO			
832	Pocatello	.88	.90
833	Twin Falls	.72	.76
834	Idaho Falls	.72	.79
835	Lewiston	.97	.97
836-837	Boise	.89	.91
838	Coeur d'Alene	.84	.84
ILLINOIS			
600-603	North Suburban	1.09	1.07
604	Joliet	1.11	1.07
605	South Suburban	1.09	1.07
606-608	Chicago	1.15	1.12
609	Kankakee	1.00	1.00
610-611	Rockford	1.03	1.03
612	Rock Island	.96	.96
613	La Salle	1.01	.99
614	Galesburg	.98	.98
615-616	Peoria	1.01	1.00
617	Bloomington	.96	.98
618-619	Champaign	.98	1.00
620-622	East St. Louis	.97	.97
623	Quincy	.97	.96
624	Effingham	.98	.96
625	Decatur	.97	.97
626-627	Springfield	.98	.98
628	Centralia	.96	.95
629	Carbondale	.96	.95
INDIANA			
460	Anderson	.91	.90
461-462	Indianapolis	.96	.94
463-464	Gary	1.00	.97
465-466	South Bend	.91	.89
467-468	Fort Wayne	.91	.89
469	Kokomo	.92	.88
470	Lawrenceburg	.86	.87
471	New Albany	.85	.85
472	Columbus	.92	.89
473	Muncie	.91	.89
474	Bloomington	.94	.90
475	Washington	.90	.90
476-477	Evansville	.90	.91
478	Terre Haute	.90	.91
479	Lafayette	.91	.88
IOWA			
500-503,509	Des Moines	.92	.91
504	Mason City	.77	.81
505	Fort Dodge	.76	.80
506-507	Waterloo	.80	.82
508	Creston	.81	.83
510-511	Sioux City	.86	.87
512	Sibley	.73	.77
513	Spencer	.75	.77
514	Carroll	.75	.78
515	Council Bluffs	.81	.88
516	Shenandoah	.74	.76
520	Dubuque	.84	.88
521	Decorah	.77	.78
522-524	Cedar Rapids	.93	.92
525	Ottumwa	.83	.85
526	Burlington	.87	.86
527-528	Davenport	.98	.96
KANSAS			
660-662	Kansas City	.95	.96
664-666	Topeka	.79	.85
667	Fort Scott	.85	.84
668	Emporia	.72	.80
669	Belleville	.74	.79
670-672	Wichita	.81	.86
673	Independence	.75	.78
674	Salina	.73	.80
675	Hutchinson	.68	.75
676	Hays	.74	.79
677	Colby	.76	.79
678	Dodge City	.73	.80
679	Liberal	.68	.74
KENTUCKY			
400-402	Louisville	.93	.92
403-405	Lexington	.84	.83

STATE/ZIP	CITY	Residential	Commercial
KENTUCKY (CONT'D)			
406	Frankfort	.82	.84
407-409	Corbin	.67	.71
410	Covington	.93	.94
411-412	Ashland	.94	.95
413-414	Campton	.68	.72
415-416	Pikeville	.77	.80
417-418	Hazard	.67	.71
420	Paducah	.89	.91
421-422	Bowling Green	.89	.90
423	Owensboro	.82	.85
424	Henderson	.91	.91
425-426	Somerset	.67	.71
427	Elizabethtown	.86	.87
LOUISIANA			
700-701	New Orleans	.85	.86
703	Thibodaux	.80	.83
704	Hammond	.78	.81
705	Lafayette	.77	.78
706	Lake Charles	.79	.81
707-708	Baton Rouge	.78	.79
710-711	Shreveport	.78	.79
712	Monroe	.73	.78
713-714	Alexandria	.73	.77
MAINE			
039	Kittery	.80	.85
040-041	Portland	.88	.89
042	Lewiston	.88	.89
043	Augusta	.83	.86
044	Bangor	.86	.88
045	Bath	.81	.86
046	Machias	.82	.85
047	Houlton	.86	.87
048	Rockland	.80	.84
049	Waterville	.75	.83
MARYLAND			
206	Waldorf	.84	.87
207-208	College Park	.84	.90
209	Silver Spring	.85	.89
210-212	Baltimore	.89	.93
214	Annapolis	.85	.90
215	Cumberland	.86	.89
216	Easton	.68	.74
217	Hagerstown	.85	.89
218	Salisbury	.75	.78
219	Elkton	.81	.82
MASSACHUSETTS			
010-011	Springfield	1.05	1.03
012	Pittsfield	1.02	1.01
013	Greenfield	1.01	1.01
014	Fitchburg	1.08	1.06
015-016	Worcester	1.11	1.09
017	Framingham	1.12	1.08
018	Lowell	1.13	1.11
019	Lawrence	1.12	1.10
020-022, 024	Boston	1.19	1.15
023	Brockton	1.11	1.08
025	Buzzards Bay	1.10	1.05
026	Hyannis	1.09	1.07
027	New Bedford	1.11	1.08
MICHIGAN			
480,483	Royal Oak	1.03	1.00
481	Ann Arbor	1.03	1.01
482	Detroit	1.10	1.07
484-485	Flint	.97	.97
486	Saginaw	.94	.95
487	Bay City	.95	.95
488-489	Lansing	.96	.96
490	Battle Creek	.93	.92
491	Kalamazoo	.92	.91
492	Jackson	.94	.94
493,495	Grand Rapids	.83	.84
494	Muskegon	.89	.90
496	Traverse City	.81	.85
497	Gaylord	.84	.87
498-499	Iron Mountain	.90	.92
MINNESOTA			
550-551	Saint Paul	1.13	1.10
553-555	Minneapolis	1.18	1.13

Location Factors

STATE/ZIP	CITY	Residential	Commercial
MINNESOTA (CONT'D)			
556-558	Duluth	1.10	1.04
559	Rochester	1.04	1.02
560	Mankato	1.00	.99
561	Windom	.84	.89
562	Willmar	.85	.91
563	St. Cloud	1.07	1.07
564	Brainerd	.96	.98
565	Detroit Lakes	.97	.99
566	Bemidji	.94	.98
567	Thief River Falls	.92	.95
MISSISSIPPI			
386	Clarksdale	.61	.67
387	Greenville	.68	.75
388	Tupelo	.64	.71
389	Greenwood	.65	.69
390-392	Jackson	.73	.75
393	Meridian	.66	.74
394	Laurel	.62	.69
395	Biloxi	.75	.79
396	McComb	.74	.77
397	Columbus	.64	.70
MISSOURI			
630-631	St. Louis	1.01	1.02
633	Bowling Green	.90	.93
634	Hannibal	.87	.90
635	Kirksville	.80	.88
636	Flat River	.93	.95
637	Cape Girardeau	.86	.93
638	Sikeston	.83	.90
639	Poplar Bluff	.83	.90
640-641	Kansas City	1.02	1.03
644-645	St. Joseph	.93	.95
646	Chillicothe	.85	.85
647	Harrisonville	.94	.97
648	Joplin	.82	.85
650-651	Jefferson City	.88	.92
652	Columbia	.87	.93
653	Sedalia	.85	.91
654-655	Rolla	.88	.87
656-658	Springfield	.84	.87
MONTANA			
590-591	Billings	.88	.90
592	Wolf Point	.84	.87
593	Miles City	.86	.86
594	Great Falls	.89	.90
595	Havre	.81	.87
596	Helena	.88	.89
597	Butte	.83	.87
598	Missoula	.83	.86
599	Kalispell	.82	.85
NEBRASKA			
680-681	Omaha	.89	.89
683-685	Lincoln	.78	.84
686	Columbus	.69	.76
687	Norfolk	.77	.82
688	Grand Island	.77	.83
689	Hastings	.76	.80
690	Mccook	.69	.75
691	North Platte	.75	.80
692	Valentine	.66	.72
693	Alliance	.65	.71
NEVADA			
889-891	Las Vegas	1.00	1.02
893	Ely	.90	.91
894-895	Reno	.97	.97
897	Carson City	.97	.97
898	Elko	.97	.93
NEW HAMPSHIRE			
030	Nashua	.90	.93
031	Manchester	.90	.93
032-033	Concord	.86	.91
034	Keene	.73	.77
035	Littleton	.81	.82
036	Charleston	.70	.75
037	Claremont	.72	.75
038	Portsmouth	.84	.89

STATE/ZIP	CITY	Residential	Commercial
NEW JERSEY			
070-071	Newark	1.12	1.11
072	Elizabeth	1.15	1.10
073	Jersey City	1.12	1.10
074-075	Paterson	1.12	1.11
076	Hackensack	1.11	1.10
077	Long Branch	1.12	1.09
078	Dover	1.12	1.10
079	Summit	1.12	1.09
080,083	Vineland	1.09	1.07
081	Camden	1.10	1.07
082,084	Atlantic City	1.13	1.06
085-086	Trenton	1.11	1.09
087	Point Pleasant	1.10	1.08
088-089	New Brunswick	1.12	1.09
NEW MEXICO			
870-872	Albuquerque	.85	.89
873	Gallup	.85	.89
874	Farmington	.85	.89
875	Santa Fe	.85	.88
877	Las Vegas	.85	.88
878	Socorro	.85	.88
879	Truth/Consequences	.84	.86
880	Las Cruces	.83	.84
881	Clovis	.85	.87
882	Roswell	.85	.88
883	Carrizozo	.85	.89
884	Tucumcari	.86	.88
NEW YORK			
100-102	New York	1.36	1.32
103	Staten Island	1.27	1.26
104	Bronx	1.29	1.25
105	Mount Vernon	1.15	1.16
106	White Plains	1.19	1.16
107	Yonkers	1.21	1.19
108	New Rochelle	1.21	1.17
109	Suffern	1.13	1.11
110	Queens	1.27	1.26
111	Long Island City	1.30	1.27
112	Brooklyn	1.32	1.28
113	Flushing	1.29	1.27
114	Jamaica	1.29	1.26
115,117,118	Hicksville	1.19	1.21
116	Far Rockaway	1.28	1.27
119	Riverhead	1.20	1.21
120-122	Albany	.95	.96
123	Schenectady	.96	.96
124	Kingston	1.02	1.07
125-126	Poughkeepsie	1.06	1.09
127	Monticello	1.03	1.06
128	Glens Falls	.88	.92
129	Plattsburgh	.92	.91
130-132	Syracuse	.96	.95
133-135	Utica	.93	.93
136	Watertown	.91	.94
137-139	Binghamton	.92	.92
140-142	Buffalo	1.06	1.02
143	Niagara Falls	1.01	1.00
144-146	Rochester	.98	.98
147	Jamestown	.88	.92
148-149	Elmira	.87	.91
NORTH CAROLINA			
270,272-274	Greensboro	.73	.76
271	Winston-Salem	.73	.76
275-276	Raleigh	.74	.76
277	Durham	.73	.76
278	Rocky Mount	.64	.69
279	Elizabeth City	.61	.70
280	Gastonia	.74	.75
281-282	Charlotte	.74	.75
283	Fayetteville	.71	.76
284	Wilmington	.72	.75
285	Kinston	.62	.68
286	Hickory	.62	.68
287-288	Asheville	.72	.75
289	Murphy	.66	.68
NORTH DAKOTA			
580-581	Fargo	.80	.85
582	Grand Forks	.77	.82
583	Devils Lake	.80	.82
584	Jamestown	.75	.79
585	Bismarck	.80	.85

STATE/ZIP	CITY	Residential	Commercial
NORTH DAKOTA (CONT'D)			
586	Dickinson	.77	.83
587	Minot	.80	.86
588	Williston	.77	.82
OHIO			
430-432	Columbus	.96	.95
433	Marion	.93	.92
434-436	Toledo	1.01	.99
437-438	Zanesville	.91	.90
439	Steubenville	.96	.95
440	Lorain	1.01	.97
441	Cleveland	1.02	1.01
442-443	Akron	.99	.98
444-445	Youngstown	.96	.96
446-447	Canton	.94	.93
448-449	Mansfield	.95	.94
450	Hamilton	.96	.93
451-452	Cincinnati	.96	.94
453-454	Dayton	.92	.90
455	Springfield	.94	.91
456	Chillicothe	.96	.93
457	Athens	.90	.90
458	Lima	.92	.93
OKLAHOMA			
730-731	Oklahoma City	.80	.82
734	Ardmore	.78	.80
735	Lawton	.81	.82
736	Clinton	.77	.80
737	Enid	.77	.80
738	Woodward	.76	.80
739	Guymon	.67	.68
740-741	Tulsa	.79	.80
743	Miami	.82	.81
744	Muskogee	.72	.72
745	Mcalester	.74	.76
746	Ponca City	.77	.79
747	Durant	.76	.79
748	Shawnee	.76	.79
749	Poteau	.77	.80
OREGON			
970-972	Portland	1.02	1.04
973	Salem	1.01	1.03
974	Eugene	1.01	1.02
975	Medford	.99	1.02
976	Klamath Falls	1.00	1.02
977	Bend	1.02	1.02
978	Pendleton	.99	.99
979	Vale	.99	.94
PENNSYLVANIA			
150-152	Pittsburgh	.99	1.00
153	Washington	.94	.98
154	Uniontown	.91	.97
155	Bedford	.89	.94
156	Greensburg	.95	.97
157	Indiana	.91	.95
158	Dubois	.90	.96
159	Johnstown	.90	.96
160	Butler	.93	.97
161	New Castle	.93	.96
162	Kittanning	.94	.98
163	Oil City	.89	.94
164-165	Erie	.97	.95
166	Altoona	.89	.94
167	Bradford	.89	.95
168	State College	.92	.95
169	Wellsboro	.88	.92
170-171	Harrisburg	.94	.96
172	Chambersburg	.89	.92
173-174	York	.88	.93
175-176	Lancaster	.91	.92
177	Williamsport	.85	.88
178	Sunbury	.89	.92
179	Pottsville	.89	.92
180	Lehigh Valley	.98	.99
181	Allentown	1.01	.99
182	Hazleton	.89	.94
183	Stroudsburg	.92	.96
184-185	Scranton	.95	.96
186-187	Wilkes-Barre	.91	.94
188	Montrose	.89	.94
189	Doylestown	1.04	1.05

STATE/ZIP	CITY	Residential	Commercial
PENNSYLVANIA (CONT'D)			
190-191	Philadelphia	1.14	1.14
193	Westchester	1.07	1.06
194	Norristown	1.03	1.06
195-196	Reading	.96	.99
PUERTO RICO			
009	San Juan	.84	.87
RHODE ISLAND			
028	Newport	1.08	1.06
029	Providence	1.08	1.06
SOUTH CAROLINA			
290-292	Columbia	.73	.75
293	Spartanburg	.71	.73
294	Charleston	.72	.75
295	Florence	.66	.73
296	Greenville	.70	.73
297	Rock Hill	.65	.67
298	Aiken	.83	.84
299	Beaufort	.67	.70
SOUTH DAKOTA			
570-571	Sioux Falls	.77	.80
572	Watertown	.73	.77
573	Mitchell	.75	.76
574	Aberdeen	.76	.79
575	Pierre	.76	.78
576	Mobridge	.73	.77
577	Rapid City	.76	.78
TENNESSEE			
370-372	Nashville	.85	.87
373-374	Chattanooga	.77	.79
375,380-381	Memphis	.85	.88
376	Johnson City	.72	.80
377-379	Knoxville	.75	.79
382	Mckenzie	.70	.75
383	Jackson	.71	.78
384	Columbia	.72	.78
385	Cookeville	.68	.76
TEXAS			
750	Mckinney	.75	.80
751	Waxahackie	.76	.80
752-753	Dallas	.83	.84
754	Greenville	.69	.73
755	Texarkana	.74	.77
756	Longview	.68	.73
757	Tyler	.74	.79
758	Palestine	.67	.72
759	Lufkin	.72	.74
760-761	Fort Worth	.83	.81
762	Denton	.77	.77
763	Wichita Falls	.79	.79
764	Eastland	.72	.72
765	Temple	.75	.75
766-767	Waco	.77	.79
768	Brownwood	.68	.71
769	San Angelo	.72	.75
770-772	Houston	.85	.87
773	Huntsville	.69	.72
774	Wharton	.70	.75
775	Galveston	.83	.85
776-777	Beaumont	.82	.82
778	Bryan	.74	.81
779	Victoria	.74	.76
780	Laredo	.73	.77
781-782	San Antonio	.80	.82
783-784	Corpus Christi	.77	.78
785	Mc Allen	.75	.75
786-787	Austin	.79	.80
788	Del Rio	.66	.69
789	Giddings	.70	.71
790-791	Amarillo	.78	.80
792	Childress	.76	.77
793-794	Lubbock	.76	.79
795-796	Abilene	.75	.78
797	Midland	.76	.77
798-799,885	El Paso	.75	.77
UTAH			
840-841	Salt Lake City	.83	.88
842,844	Ogden	.81	.86

Location Factors

STATE/ZIP	CITY	Residential	Commercial
UTAH (CONT'D)			
843	Logan	.82	.87
845	Price	.72	.78
846-847	Provo	.83	.87
VERMONT			
050	White River Jct.	.73	.76
051	Bellows Falls	.75	.76
052	Bennington	.74	.76
053	Brattleboro	.75	.77
054	Burlington	.80	.85
056	Montpelier	.82	.83
057	Rutland	.81	.85
058	St. Johnsbury	.75	.77
059	Guildhall	.74	.76
VIRGINIA			
220-221	Fairfax	.85	.91
222	Arlington	.87	.91
223	Alexandria	.90	.93
224-225	Fredericksburg	.75	.84
226	Winchester	.71	.79
227	Culpeper	.77	.79
228	Harrisonburg	.67	.76
229	Charlottesville	.72	.82
230-232	Richmond	.81	.85
233-235	Norfolk	.81	.84
236	Newport News	.79	.83
237	Portsmouth	.77	.83
238	Petersburg	.78	.84
239	Farmville	.68	.73
240-241	Roanoke	.72	.77
242	Bristol	.67	.76
243	Pulaski	.66	.73
244	Staunton	.68	.75
245	Lynchburg	.69	.77
246	Grundy	.67	.74
WASHINGTON			
980-981,987	Seattle	1.01	1.05
982	Everett	1.03	1.01
983-984	Tacoma	1.00	1.02
985	Olympia	1.00	1.01
986	Vancouver	.97	1.02
988	Wenatchee	.90	.95
989	Yakima	.95	.98
990-992	Spokane	1.00	.96
993	Richland	.97	.96
994	Clarkston	.96	.95
WEST VIRGINIA			
247-248	Bluefield	.88	.89
249	Lewisburg	.89	.92
250-253	Charleston	.97	.96
254	Martinsburg	.84	.90
255-257	Huntington	.96	.96
258-259	Beckley	.90	.93
260	Wheeling	.92	.97
261	Parkersburg	.91	.95
262	Buckhannon	.91	.96
263-264	Clarksburg	.90	.96
265	Morgantown	.91	.96
266	Gassaway	.92	.95
267	Romney	.86	.93
268	Petersburg	.88	.94
WISCONSIN			
530,532	Milwaukee	1.05	1.00
531	Kenosha	1.03	.98
534	Racine	1.01	.98
535	Beloit	.99	.97
537	Madison	.99	.96
538	Lancaster	.97	.94
539	Portage	.96	.94
540	New Richmond	.98	.95
541-543	Green Bay	1.01	.96
544	Wausau	.95	.93
545	Rhinelander	.95	.95
546	La Crosse	.94	.94
547	Eau Claire	.98	.96
548	Superior	.97	.95
549	Oshkosh	.95	.93
WYOMING			
820	Cheyenne	.75	.78
821	Yellowstone Nat. Pk.	.70	.75

STATE/ZIP	CITY	Residential	Commercial
WYOMING (CONT'D)			
822	Wheatland	.71	.75
823	Rawlins	.69	.74
824	Worland	.69	.74
825	Riverton	.70	.75
826	Casper	.76	.81
827	Newcastle	.68	.73
828	Sheridan	.73	.78
829-831	Rock Springs	.73	.76
CANADIAN FACTORS (reflect Canadian currency)			
ALBERTA			
	Calgary	1.06	1.05
	Edmonton	1.05	1.06
	Fort McMurray	1.03	1.01
	Lethbridge	1.04	1.01
	Lloydminster	1.03	1.01
	Medicine Hat	1.04	1.01
	Red Deer	1.04	1.01
BRITISH COLUMBIA			
	Kamloops	1.01	1.03
	Prince George	1.01	1.03
	Vancouver	1.08	1.09
	Victoria	1.01	1.04
MANITOBA			
	Brandon	1.00	.96
	Portage la Prairie	1.00	.96
	Winnipeg	1.01	1.00
NEW BRUNSWICK			
	Bathurst	.91	.91
	Dalhousie	.91	.91
	Fredericton	.99	.94
	Moncton	.91	.92
	Newcastle	.91	.91
	Saint John	.99	.95
NEWFOUNDLAND			
	Corner Brook	.92	.94
	St. John's	.93	.94
NORTHWEST TERRITORIES			
	Yellowknife	.98	.99
NOVA SCOTIA			
	Dartmouth	.93	.96
	Halifax	.93	.97
	New Glasgow	.93	.95
	Sydney	.91	.93
	Yarmouth	.93	.95
ONTARIO			
	Barrie	1.10	1.05
	Brantford	1.12	1.06
	Cornwall	1.11	1.05
	Hamilton	1.12	1.09
	Kingston	1.11	1.05
	Kitchener	1.06	1.02
	London	1.10	1.07
	North Bay	1.08	1.04
	Oshawa	1.10	1.05
	Ottawa	1.12	1.06
	Owen Sound	1.09	1.04
	Peterborough	1.09	1.04
	Sarnia	1.12	1.06
	St. Catharines	1.05	1.01
	Sudbury	1.03	1.01
	Thunder Bay	1.08	1.02
	Toronto	1.15	1.11
	Windsor	1.09	1.03
PRINCE EDWARD ISLAND			
	Charlottetown	.88	.91
	Summerside	.88	.91
QUEBEC			
	Cap-de-la-Madeleine	1.11	1.00
	Charlesbourg	1.11	1.00
	Chicoutimi	1.11	1.00
	Gatineau	1.10	1.00
	Laval	1.11	1.00
	Montreal	1.11	1.02

STATE/ZIP	CITY	Residential	Commercial
QUEBEC (CONT'D)			
	Quebec	1.13	1.03
	Sherbrooke	1.10	1.00
	Trois Rivieres	1.11	1.00
SASKATCHEWAN			
	Moose Jaw	.91	.92
	Prince Albert	.90	.91
	Regina	.92	.93
	Saskatoon	.91	.92
YUKON			
	Whitehorse	.90	.91

Historical Cost Indexes

The following tables are the estimated Historical Cost Indexes based on a 30-city national average with a base of 100 on January 1, 1993.

The indexes may be used to:

1. Estimate and compare construction costs for different years in the same city.
2. Estimate and compare construction costs in different cities for the same year.
3. Estimate and compare construction costs in different cities for different years.
4. Compare construction trends in any city with the national average.

EXAMPLES

1. Estimate and compare construction costs for different years in the same city.

   A. To estimate the construction cost of a building in Lexington, KY in 1970, knowing that it cost $915,000 in 2005.

   $$\text{Index Lexington, KY in 1970} = 26.9$$
   $$\text{Index Lexington, KY in 2005} = 121.9$$

   $$\frac{\text{Index 1970}}{\text{Index 2005}} \quad x \quad \text{Cost 2005} \quad = \quad \text{Cost 1970}$$

   $$\frac{26.9}{121.9} \quad x \quad \$915,000 \quad = \quad \$202,000$$

   Construction Cost in Lexington, KY in 1970 = $202,000

   B. To estimate the current construction cost of a building in Boston, MA that was built in 1980 for $900,000.

   $$\text{Index Boston, MA in 1980} = 64.0$$
   $$\text{Index Boston, MA in 2005} = 170.0$$

   $$\frac{\text{Index 2005}}{\text{Index 1980}} \quad x \quad \text{Cost 1980} \quad = \quad \text{Cost 2005}$$

   $$\frac{170.0}{64.0} \quad x \quad \$900,000 \quad = \quad \$2,390,500$$

   Construction Cost in Boston in 2005 = $2,390,500

2. Estimate and compare construction costs in different cities for the same year.

   To compare the construction cost of a building in Topeka, KS in 2005 with the known cost of $800,000 in Baltimore, MD in 2005

   $$\text{Index Topeka, KS in 2005} = 125.6$$
   $$\text{Index Baltimore, MD in 2005} = 136.6$$

   $$\frac{\text{Index Topeka}}{\text{Index Baltimore}} \quad x \quad \text{Cost Baltimore} \quad = \quad \text{Cost Topeka}$$

   $$\frac{125.6}{136.6} \quad x \quad \$800,000 \quad = \quad \$735,500$$

   Construction Cost in Topeka in 2005 = $735,500

3. Estimate and compare construction costs in different cities for different years.

   To compare the construction cost of a building in Detroit, MI in 2005 with the known construction cost of $5,000,000 for the same building in San Francisco, CA in 1980.

   $$\text{Index Detroit, MI in 2005} = 157.5$$
   $$\text{Index San Francisco, CA in 1980} = 75.2$$

   $$\frac{\text{Index Detroit 2005}}{\text{Index San Francisco 1980}} \quad x \quad \frac{\text{Cost San Francisco}}{1980} \quad = \quad \frac{\text{Cost}}{\text{Detroit 2005}}$$

   $$\frac{157.5}{75.2} \quad x \quad \$5,000,000 \quad = \quad \$10,472,000$$

   Construction Cost in Detroit in 2005 = $10,472,000

4. Compare construction trends in any city with the national average.

   To compare the construction cost in Las Vegas, NV from 1975 to 2005 with the increase in the National Average during the same time period.

Index Las Vegas, NV for 1975	= 42.8	For 2005 =	149.4
Index 30 City Average for 1975	= 43.7	For 2005 =	148.5

   A. National Average escalation From 1975 to 2005

   $$= \frac{\text{Index — 30 City 2005}}{\text{Index — 30 City 1975}}$$

   $$= \frac{148.5}{43.7}$$

   National Average escalation From 1975 to 2005 = 3.40 or increased by 240%

   B. Escalation for Las Vegas, NV From 1975 to 2005

   $$= \frac{\text{Index Las Vegas, NV 2005}}{\text{Index Las Vegas, NV 1975}}$$

   $$= \frac{149.4}{42.8}$$

   Las Vegas escalation 1975 — 2005 = 3.49 or increased by 249%

   Conclusion: Construction costs in Las Vegas are higher than National average costs and increased at a greater rate from 1975 to 2005 than the National Average.

Year	National 30 City Average	Birmingham	Huntsville	Mobile	Montgomery	Tuscaloosa	Anchorage	Phoenix	Tuscon	Fort Smith	Little Rock	Anaheim	Bakersfield	Fresno	Los Angeles	Oxnard
		Alabama					**Alaska**	**Arizona**		**Arkansas**		**California**				
Jan 2005	148.5E	128.7E	126.5E	120.2E	117.2E	115.2E	184.3E	129.3E	124.7E	115.2E	119.6E	157.0E	153.5E	158.8E	157.4E	156.7E
2004	132.8	115.9	112.9	107.0	104.9	102.9	167.0	116.5	113.6	103.8	108.6	142.1	139.5	143.8	142.0	140.9
2003	129.7	113.1	110.7	104.8	102.6	100.8	163.5	113.9	110.6	101.8	105.8	139.4	137.2	142.1	139.6	138.7
2002	126.7	110.0	103.4	103.6	101.7	99.7	159.5	113.3	110.2	100.4	102.1	136.5	133.4	136.0	136.4	136.3
2001	122.2	106.0	100.5	100.6	98.3	95.9	152.6	109.0	106.4	97.3	98.7	132.5	129.3	132.8	132.4	132.4
2000	118.9	104.1	98.9	99.1	94.2	94.6	148.3	106.9	104.9	94.4	95.7	129.4	125.2	129.4	129.9	129.7
1999	116.6	101.2	97.4	97.7	92.5	92.8	145.9	105.5	103.3	93.1	94.4	127.9	123.6	126.9	128.7	128.2
1998	113.6	96.2	94.0	94.8	90.3	89.8	143.8	102.1	101.0	90.6	91.4	125.2	120.3	123.9	125.8	124.7
1997	111.5	94.6	92.4	93.3	88.8	88.3	142.0	101.8	100.7	89.3	90.1	124.0	119.1	122.3	124.6	123.5
1996	108.9	90.9	91.4	92.3	87.5	86.5	140.4	98.3	97.4	86.5	86.7	121.7	117.2	120.2	122.4	121.5
1995	105.6	87.8	88.0	88.8	84.5	83.5	138.0	96.1	95.5	85.0	85.6	120.1	115.7	117.5	120.9	120.0
1994	103.0	85.9	86.1	86.8	82.6	81.7	134.0	93.7	93.1	82.4	83.9	118.1	113.6	114.5	119.0	117.5
1993	100.0	82.8	82.7	86.1	82.1	78.7	132.0	90.9	90.9	80.9	82.4	115.0	111.3	112.9	115.6	115.4
1992	97.9	81.7	81.5	84.9	80.9	77.6	128.6	88.8	89.4	79.7	81.2	113.5	108.1	110.3	113.7	113.7
1991	95.7	80.5	78.9	83.9	79.8	76.6	127.4	88.1	88.5	78.3	80.0	111.0	105.8	108.2	110.9	111.4
1990	93.2	79.4	77.6	82.7	78.4	75.2	125.8	86.4	87.0	77.1	77.9	107.7	102.7	103.3	107.5	107.4
1989	91.0	77.5	76.1	81.1	76.8	73.7	123.4	85.3	85.6	75.6	76.4	105.6	101.0	101.7	105.3	105.4
1988	88.5	75.7	74.7	79.7	75.4	72.3	121.4	84.4	84.3	74.2	75.0	102.7	97.6	99.2	102.6	102.3
1987	85.7	74.0	73.8	77.9	74.5	71.7	119.3	79.7	80.6	72.2	72.8	100.9	96.1	98.0	100.0	101.1
1986	83.7	72.8	71.7	76.7	72.3	69.4	117.3	78.9	78.8	70.7	72.3	99.2	94.2	93.8	97.8	100.4
1985	81.8	71.1	70.5	72.7	70.9	67.9	116.0	78.1	77.7	69.6	71.0	95.4	92.2	92.6	94.6	96.6
1984	80.6	69.7	69.0	75.2	69.4	66.3	113.8	79.4	79.2	69.1	70.3	93.4	89.8	90.6	92.1	94.0
1983	78.2	67.4	68.8	72.9	69.4	65.5	104.5	77.5	77.9	66.8	68.7	90.9	88.8	88.3	89.7	91.8
1980	60.7	55.2	54.1	56.8	56.8	54.0	91.4	63.7	62.4	53.1	55.8	68.7	69.4	68.7	67.4	69.9
1975	43.7	40.0	40.9	41.8	40.1	37.8	57.3	44.5	45.2	39.0	38.7	47.6	46.6	47.7	48.3	47.0
1970	27.8	24.1	25.1	25.8	25.4	24.2	43.0	27.2	28.5	24.3	22.3	31.0	30.8	31.1	29.0	31.0
1965	21.5	19.6	19.3	19.6	19.5	18.7	34.9	21.8	22.0	18.7	18.5	23.9	23.7	24.0	22.7	23.9
1960	19.5	17.9	17.5	17.8	17.7	16.9	31.7	19.9	20.0	17.0	16.8	21.7	21.5	21.8	20.6	21.7
1955	16.3	14.8	14.7	14.9	14.9	14.2	26.6	16.7	16.7	14.3	14.5	18.2	18.1	18.3	17.3	18.2
1950	13.5	12.2	12.1	12.3	12.3	11.7	21.9	13.8	13.8	11.8	11.6	15.1	15.0	15.1	14.3	15.1
1945	8.6	7.8	7.7	7.8	7.8	7.5	14.0	8.8	8.8	7.5	7.4	9.6	9.5	9.6	9.1	9.6
1940	6.6	6.0	6.0	6.1	6.0	5.8	10.8	6.8	6.8	5.8	5.7	7.4	7.4	7.4	7.0	7.4

Year	National 30 City Average	Riverside	Sacramento	San Diego	San Francisco	Santa Barbara	Stockton	Vallejo	Colorado Springs	Denver	Pueblo	Bridge-Port	Bristol	Hartford	New Britain	New Haven
		California							**Colorado**			**Connecticut**				
Jan 2005	148.5E	156.1E	161.6E	153.8E	180.1E	156.4E	158.0E	165.5E	138.5E	141.1E	136.5E	156.7E	156.0E	156.1E	155.7E	157.3E
2004	132.8	141.0	147.2	139.0	163.6	141.4	144.2	148.6	125.6	127.2	123.5	143.6	142.9	142.4	142.7	144.3
2003	129.7	138.7	144.9	136.6	162.1	139.2	141.8	146.4	123.1	123.7	120.7	141.3	140.2	140.3	140.0	141.6
2002	126.7	135.3	138.4	134.2	157.9	135.9	136.9	143.8	118.7	121.3	116.7	133.8	133.6	133.1	133.3	134.8
2001	122.2	131.4	135.5	129.7	151.8	131.5	134.1	140.2	113.3	117.2	113.1	128.6	128.5	128.5	128.3	128.6
2000	118.9	128.0	131.5	127.1	146.9	128.7	130.7	137.1	109.8	111.8	108.8	122.7	122.6	122.9	122.4	122.7
1999	116.6	126.5	129.4	124.9	145.1	127.2	127.9	135.3	107.0	109.1	107.1	121.1	121.3	121.2	121.1	121.3
1998	113.6	123.7	125.9	121.3	141.9	123.7	124.7	132.5	103.3	106.5	103.7	119.1	119.4	120.0	119.7	120.0
1997	111.5	122.6	124.7	120.3	139.2	122.4	123.3	130.6	101.1	104.4	102.0	119.2	119.5	119.9	119.7	120.0
1996	108.9	120.6	122.4	118.4	136.8	120.5	121.4	128.1	98.5	101.4	99.7	117.4	117.6	117.9	117.8	118.1
1995	105.6	119.2	119.5	115.4	133.8	119.0	119.0	122.5	96.1	98.9	96.8	116.0	116.5	116.9	116.3	116.5
1994	103.0	116.6	114.8	113.4	131.4	116.4	115.5	119.7	94.2	95.9	94.6	114.3	115.0	115.3	114.7	114.8
1993	100.0	114.3	112.5	111.3	129.6	114.3	115.2	117.5	92.1	93.8	92.6	108.7	107.8	108.4	106.3	106.1
1992	97.9	111.8	110.8	109.5	127.9	112.2	112.7	115.2	90.6	91.5	90.9	106.7	106.0	106.6	104.5	104.4
1991	95.7	109.4	108.5	107.7	125.8	110.0	111.0	113.4	88.9	90.4	89.4	97.6	97.3	98.0	97.3	97.9
1990	93.2	107.0	104.9	105.6	121.8	106.4	105.4	111.4	86.9	88.8	88.8	96.3	95.9	96.6	95.9	96.5
1989	91.0	105.2	103.1	103.6	119.2	104.6	103.7	109.8	85.6	88.7	87.4	94.5	94.2	94.9	94.2	94.6
1988	88.5	102.6	100.5	101.0	115.2	101.4	101.2	107.2	83.8	87.0	85.6	92.8	92.6	93.3	92.6	93.0
1987	85.7	100.3	98.7	99.0	111.0	99.3	99.6	103.2	81.6	84.8	83.8	92.4	93.0	93.3	93.0	92.6
1986	83.7	98.6	94.4	96.9	108.0	97.0	96.1	99.1	81.9	83.1	82.7	88.5	88.7	89.4	88.6	88.8
1985	81.8	94.8	92.2	94.2	106.2	93.2	93.7	97.0	79.6	81.0	80.4	86.1	86.2	87.2	86.1	86.3
1984	80.6	91.8	93.8	92.0	102.8	90.6	91.9	95.6	79.8	84.1	80.5	83.5	83.4	84.4	83.0	83.2
1983	78.2	89.4	92.0	90.3	101.0	88.8	91.0	93.0	77.3	80.5	77.4	79.6	79.9	81.1	79.5	79.3
1980	60.7	68.7	71.3	68.1	75.2	71.1	71.2	71.9	60.7	60.9	59.5	61.5	60.7	61.9	60.7	61.4
1975	43.7	47.3	49.1	47.7	49.8	46.1	47.6	46.5	43.1	42.7	42.5	45.2	45.5	46.0	45.2	46.0
1970	27.8	31.0	32.1	30.7	31.6	31.3	31.8	31.8	27.6	26.1	27.3	29.2	28.2	29.6	28.2	29.3
1965	21.5	23.9	24.8	24.0	23.7	24.1	24.5	24.5	21.3	20.9	21.0	22.4	21.7	22.6	21.7	23.2
1960	19.5	21.7	22.5	21.7	21.5	21.9	22.3	22.2	19.3	19.0	19.1	20.0	19.8	20.0	19.8	20.0
1955	16.3	18.2	18.9	18.2	18.0	18.4	18.7	18.6	16.2	15.9	16.0	16.8	16.6	16.8	16.6	16.8
1950	13.5	15.0	15.6	15.0	14.9	15.2	15.4	15.4	13.4	13.2	13.2	13.9	13.7	13.8	13.7	13.9
1945	8.6	9.6	10.0	9.6	9.5	9.7	9.9	9.8	8.5	8.4	8.4	8.8	8.7	8.8	8.7	8.9
1940	6.6	7.4	7.7	7.4	7.3	7.5	7.6	7.6	6.6	6.5	6.5	6.8	6.8	6.8	6.8	6.8

Historical Cost Indexes

Year	National 30 City Average	Connecticut Norwalk	Connecticut Stamford	Connecticut Waterbury	Delaware Wilmington	D.C. Washington	Florida Fort Lauderdale	Florida Jacksonville	Florida Miami	Florida Orlando	Florida Tallahassee	Florida Tampa	Georgia Albany	Georgia Atlanta	Georgia Columbus	Georgia Macon
Jan 2005	148.5E	156.8E	159.7E	156.3E	150.7E	142.8E	127.0E	119.9E	128.1E	126.7E	111.4E	128.7E	116.2E	132.3E	119.5E	116.5E
2004	132.8	143.6	146.5	143.2	136.1	126.8	114.8	107.8	115.6	113.2	100.6	116.6	102.8	119.6	102.0	105.0
2003	129.7	140.6	145.1	140.8	133.0	124.6	109.0	105.6	110.5	107.8	98.0	103.5	100.5	115.8	99.1	102.7
2002	126.7	133.5	136.2	133.9	129.4	120.3	107.1	104.6	107.4	106.7	97.3	102.8	99.7	113.0	98.3	101.9
2001	122.2	128.2	131.5	128.8	124.8	115.9	104.3	100.8	104.6	103.8	94.8	100.3	96.4	109.1	95.5	99.0
2000	118.9	121.5	126.4	123.2	117.4	113.8	102.4	99.0	101.8	101.2	93.6	98.9	94.9	106.1	94.1	97.0
1999	116.6	120.0	122.0	121.2	116.6	111.5	101.4	98.0	100.8	100.1	92.5	97.9	93.5	102.9	92.8	95.8
1998	113.6	118.8	120.8	120.1	112.3	109.6	99.4	96.2	99.0	98.4	90.9	96.3	91.4	100.8	90.4	93.3
1997	111.5	118.9	120.9	120.2	110.7	106.4	98.6	95.4	98.2	97.2	89.9	95.5	89.9	98.5	88.7	91.6
1996	108.9	117.0	119.0	118.4	108.6	105.4	95.9	92.6	95.9	95.1	88.0	93.6	86.6	94.3	83.9	88.3
1995	105.6	115.5	117.9	117.3	106.1	102.3	94.0	90.9	93.7	93.5	86.5	92.2	85.0	92.0	82.6	86.8
1994	103.0	113.9	116.4	115.8	105.0	99.6	92.2	88.9	91.8	91.5	84.5	90.2	82.3	89.6	80.6	83.7
1993	100.0	108.8	110.6	104.8	101.5	96.3	87.4	86.1	87.1	88.5	82.1	87.7	79.5	85.7	77.8	80.9
1992	97.9	107.2	109.0	103.1	100.3	94.7	85.7	84.0	85.3	87.1	80.8	86.2	78.2	84.3	76.5	79.6
1991	95.7	100.6	103.2	96.5	94.5	92.9	85.1	82.8	85.2	85.5	79.7	86.3	76.3	82.6	75.4	78.4
1990	93.2	96.3	98.9	95.1	92.5	90.4	83.9	81.1	84.0	82.9	78.4	85.0	75.0	80.4	74.0	76.9
1989	91.0	94.4	97.0	93.4	89.5	87.5	82.4	79.7	82.5	81.6	76.9	83.5	73.4	78.6	72.4	75.3
1988	88.5	92.3	94.9	91.8	87.7	85.0	80.9	78.0	81.0	80.1	75.4	81.9	71.8	76.9	70.8	73.6
1987	85.7	92.0	92.7	92.5	85.1	82.1	78.8	76.0	77.9	76.6	74.4	79.4	72.2	73.6	70.3	72.3
1986	83.7	88.2	89.7	87.8	83.8	80.8	78.9	75.0	79.9	76.2	72.3	79.1	68.5	72.0	67.5	69.8
1985	81.8	85.3	86.8	85.6	81.1	78.8	76.7	73.4	78.3	73.9	70.6	77.3	66.9	70.3	66.1	68.1
1984	80.6	82.6	83.7	82.5	79.7	79.1	73.8	72.6	75.6	73.0	69.6	76.5	65.9	68.6	65.2	67.5
1983	78.2	78.5	79.8	78.7	76.3	76.0	71.5	70.4	72.9	71.1	67.6	73.6	65.6	68.6	64.7	65.3
1980	60.7	60.7	60.9	62.3	58.5	59.6	55.3	55.8	56.5	56.7	53.5	57.2	51.7	54.0	51.1	51.3
1975	43.7	44.7	45.0	46.3	42.9	43.7	42.1	40.3	43.2	41.5	38.1	41.3	37.5	38.4	36.2	36.5
1970	27.8	28.1	28.2	28.9	27.0	26.3	25.7	22.8	27.0	26.2	24.5	24.2	23.8	25.2	22.8	23.4
1965	21.5	21.6	21.7	22.2	20.9	21.8	19.8	17.4	19.3	20.2	18.9	18.6	18.3	19.8	17.6	18.0
1960	19.5	19.7	19.7	20.2	18.9	19.4	18.0	15.8	17.6	18.3	17.2	16.9	16.7	17.1	16.0	16.4
1955	16.3	16.5	16.5	17.0	15.9	16.3	15.1	13.2	14.7	15.4	14.4	14.1	14.0	14.4	13.4	13.7
1950	13.5	13.6	13.7	14.0	13.1	13.4	12.5	11.0	12.2	12.7	11.9	11.7	11.5	11.9	11.0	11.3
1945	8.6	8.7	8.7	8.9	8.4	8.6	7.9	7.0	7.8	8.1	7.6	7.5	7.4	7.6	7.0	7.2
1940	6.6	6.7	6.7	6.9	6.4	6.6	6.1	5.4	6.0	6.2	5.8	5.7	5.7	5.8	5.5	5.6

Year	National 30 City Average	Georgia Savannah	Hawaii Honolulu	Idaho Boise	Idaho Pocatello	Illinois Chicago	Illinois Decatur	Illinois Joliet	Illinois Peoria	Illinois Rockford	Illinois Springfield	Indiana Anderson	Indiana Evansville	Indiana Fort Wayne	Indiana Gary	Indiana Indianapolis
Jan 2005	148.5E	117.5E	181.4E	134.4E	133.0E	164.4E	143.5E	158.1E	147.4E	151.5E	144.0E	131.8E	134.7E	130.4E	143.1E	138.8E
2004	132.8	105.6	161.2	122.0	120.5	149.3	129.3	145.1	133.9	138.1	130.3	120.9	123.5	120.1	132.2	125.0
2003	129.7	103.0	159.4	120.2	118.7	146.2	127.4	143.5	132.4	137.3	128.4	119.7	121.4	118.2	131.4	122.5
2002	126.7	102.0	157.2	118.3	117.1	141.2	123.8	138.5	129.6	132.9	125.3	117.5	119.0	116.4	129.1	120.5
2001	122.2	99.0	150.0	114.3	113.4	135.8	120.1	133.7	124.3	127.8	119.8	113.4	115.6	112.1	123.4	116.4
2000	118.9	97.5	144.8	112.9	112.1	131.2	115.1	124.6	119.0	122.2	116.2	109.8	111.5	108.4	117.8	113.2
1999	116.6	96.0	143.0	110.2	109.6	129.6	113.0	122.6	116.4	120.7	113.8	107.1	109.3	106.8	112.8	110.6
1998	113.6	93.7	140.4	107.4	107.0	125.2	110.1	119.8	113.8	115.5	111.1	105.0	107.2	104.5	111.1	108.0
1997	111.5	92.0	139.8	104.6	104.7	121.3	107.8	117.5	111.5	113.1	108.9	101.9	104.4	101.8	110.3	105.2
1996	108.9	88.6	134.5	102.2	102.1	118.8	106.6	116.2	109.3	111.5	106.5	100.0	102.1	99.9	107.5	102.7
1995	105.6	87.4	130.3	99.5	98.2	114.2	98.5	110.5	102.3	103.6	98.1	96.4	97.2	95.0	100.7	100.1
1994	103.0	85.3	124.0	94.8	95.0	111.3	97.3	108.9	100.9	102.2	97.0	93.6	95.8	93.6	99.1	97.1
1993	100.0	82.0	122.0	92.2	92.1	107.6	95.6	106.8	98.9	99.6	95.2	91.2	94.3	91.5	96.7	93.9
1992	97.9	80.8	120.0	91.0	91.0	104.3	94.4	104.2	97.3	98.2	94.0	89.5	92.9	89.9	95.0	91.5
1991	95.7	79.5	106.1	89.5	89.4	100.9	92.3	100.0	95.9	95.8	91.5	87.8	91.4	88.3	93.3	89.1
1990	93.2	77.9	102.8	88.2	88.1	98.4	90.9	98.4	93.7	94.0	90.1	84.6	89.3	83.4	88.4	87.1
1989	91.0	76.0	102.8	86.6	86.5	93.7	89.4	92.8	91.6	92.1	88.6	82.3	87.8	81.7	86.5	85.1
1988	88.5	74.5	101.1	83.9	83.7	90.6	87.5	89.8	88.5	87.9	86.8	80.8	85.5	80.0	84.6	83.1
1987	85.7	72.0	99.1	81.4	81.5	86.6	84.4	86.7	86.3	85.2	85.0	78.8	82.4	78.1	81.7	80.4
1986	83.7	70.8	97.5	80.6	80.3	84.4	83.5	85.3	85.1	84.5	83.4	77.0	80.9	76.5	79.6	78.6
1985	81.8	68.9	94.7	78.0	78.0	82.4	81.9	83.4	83.7	83.0	81.5	75.2	79.8	75.0	77.8	77.1
1984	80.6	67.9	90.7	76.6	76.8	80.2	80.4	82.2	83.6	80.6	80.1	73.5	77.4	73.5	77.3	75.9
1983	78.2	66.3	87.2	76.0	75.8	79.0	78.8	80.6	81.5	78.9	78.8	70.9	75.0	71.2	75.2	74.0
1980	60.7	52.2	68.9	60.3	59.5	62.8	62.3	63.4	64.5	61.6	61.1	56.5	59.0	56.7	59.8	57.9
1975	43.7	36.9	44.6	40.8	40.5	45.7	43.1	44.5	44.7	42.7	42.4	39.5	41.7	39.9	41.9	40.6
1970	27.8	21.0	30.4	26.7	26.6	29.1	28.0	28.6	29.0	27.5	27.6	25.4	26.4	25.5	27.1	26.2
1965	21.5	16.4	21.8	20.6	20.5	22.7	21.5	22.1	22.4	21.2	21.3	19.5	20.4	19.7	20.8	20.7
1960	19.5	14.9	19.8	18.7	18.6	20.2	19.6	20.0	20.3	19.2	19.3	17.7	18.7	17.9	18.9	18.4
1955	16.3	12.5	16.6	15.7	15.6	16.9	16.4	16.8	17.0	16.1	16.2	14.9	15.7	15.0	15.9	15.5
1950	13.5	10.3	13.7	13.0	12.9	14.0	13.6	13.9	14.0	13.3	13.4	12.3	12.9	12.4	13.1	12.8
1945	8.6	6.6	8.8	8.3	8.2	8.9	8.6	8.9	9.0	8.5	8.6	7.8	8.3	7.9	8.4	8.1
1940	6.6	5.1	6.8	6.4	6.3	6.9	6.7	6.8	6.9	6.5	6.6	6.0	6.4	6.1	6.5	6.3

Historical Cost Indexes

Year	National 30 City Average	Indiana Muncie	South Bend	Terre Haute	Cedar Rapids	Daven-port	Des Moines	Sioux City	Water-loo	Topeka	Wichita	Lexing-ton	Louis-ville	Baton Rouge	Lake Charles	New Orleans
Jan 2005	148.5E	131.4E	131.7E	134.2E	134.8E	140.9E	133.7E	128.1E	120.0E	125.6E	126.1E	121.9E	135.2E	116.0E	118.8E	127.2E
2004	132.8	120.0	120.6	123.0	122.7	128.6	121.6	116.2	108.3	112.6	113.4	110.1	120.3	105.6	109.0	115.3
2003	129.7	118.7	119.2	121.7	120.9	126.4	120.2	114.6	105.9	109.4	111.3	107.8	118.7	103.3	106.1	112.4
2002	126.7	116.7	117.1	119.9	116.0	122.1	116.7	111.4	103.7	107.8	109.6	106.2	116.4	102.5	105.0	110.6
2001	122.2	112.8	111.6	115.5	112.5	117.5	113.2	107.7	100.6	104.3	105.2	103.3	112.7	99.4	101.3	104.6
2000	118.9	109.1	106.9	110.8	108.8	112.6	108.9	99.0	98.1	101.0	101.1	101.4	109.3	97.8	99.7	102.1
1999	116.6	105.7	103.7	107.9	104.1	109.2	107.6	96.7	96.4	98.7	99.3	99.7	106.6	96.1	97.6	99.5
1998	113.6	103.6	102.3	106.0	102.5	106.8	103.8	95.1	94.8	97.4	97.4	97.4	101.5	94.7	96.1	97.7
1997	111.5	101.3	101.2	104.6	101.2	105.6	102.5	93.8	93.5	96.3	96.2	96.1	99.9	93.3	96.8	96.2
1996	108.9	99.4	99.1	102.1	99.1	102.0	99.1	90.9	90.7	93.6	93.8	94.4	98.3	91.5	94.9	94.3
1995	105.6	95.6	94.6	97.2	95.4	96.6	94.4	88.2	88.9	91.1	91.0	91.6	94.6	89.6	92.7	91.6
1994	103.0	93.2	93.0	95.8	93.0	92.3	92.4	85.9	86.8	89.4	88.1	89.6	92.2	87.7	89.9	89.5
1993	100.0	91.0	90.7	94.4	91.1	90.5	90.7	84.1	85.2	87.4	86.2	87.3	89.4	86.4	88.5	87.8
1992	97.9	89.4	89.3	93.1	89.8	89.3	89.4	82.9	83.8	86.2	85.0	85.8	88.0	85.2	87.3	86.6
1991	95.7	87.7	87.5	91.7	88.5	88.0	88.4	81.8	81.6	84.4	83.9	84.1	84.4	83.2	85.3	85.8
1990	93.2	83.9	85.1	89.1	87.1	86.5	86.9	80.4	80.3	83.2	82.6	82.8	82.6	82.0	84.1	84.5
1989	91.0	81.9	83.5	86.7	85.3	84.9	85.3	79.0	78.8	81.6	81.2	81.3	80.1	80.6	82.7	83.4
1988	88.5	80.4	82.1	84.9	83.4	83.1	83.6	76.8	77.2	80.1	79.5	79.4	78.2	79.0	81.1	82.0
1987	85.7	78.5	79.6	83.0	82.4	81.1	81.2	75.4	75.6	78.9	78.4	77.3	76.3	77.1	78.0	81.3
1986	83.7	76.9	77.5	81.1	79.5	80.0	79.8	73.8	73.7	76.9	76.2	76.5	75.3	76.1	78.1	79.2
1985	81.8	75.2	76.0	79.2	75.9	77.6	77.1	72.1	72.3	75.0	74.7	74.5	74.5	74.9	78.5	78.2
1984	80.6	73.6	75.1	78.0	80.3	77.8	76.6	74.4	73.7	75.1	74.6	75.6	73.7	77.3	78.3	77.5
1983	78.2	71.2	74.5	76.1	79.6	77.0	75.5	74.6	73.0	73.2	72.4	74.9	74.1	75.1	76.6	74.7
1980	60.7	56.1	58.3	59.7	62.7	59.6	61.8	59.3	57.7	58.9	58.0	59.3	59.8	59.1	60.0	57.2
1975	43.7	39.2	40.3	41.9	43.1	41.0	43.1	41.0	40.0	42.0	43.1	42.7	42.5	40.4	40.6	41.5
1970	27.8	25.3	26.2	27.0	28.2	26.9	27.6	26.6	25.9	27.0	25.5	26.9	25.9	25.0	26.7	27.2
1965	21.5	19.5	20.2	20.9	21.7	20.8	21.7	20.5	20.0	20.8	19.6	20.7	20.3	19.4	20.6	20.4
1960	19.5	17.8	18.3	18.9	19.7	18.8	19.5	18.6	18.2	18.9	17.8	18.8	18.4	17.6	18.7	18.5
1955	16.3	14.9	15.4	15.9	16.6	15.8	16.3	15.6	15.2	15.8	15.0	15.8	15.4	14.8	15.7	15.6
1950	13.5	12.3	12.7	13.1	13.7	13.1	13.5	12.9	12.6	13.1	12.4	13.0	12.8	12.2	13.0	12.8
1945	8.6	7.8	8.1	8.4	8.7	8.3	8.6	8.2	8.0	8.3	7.9	8.3	8.1	7.8	8.3	8.2
1940	6.6	6.0	6.3	6.5	6.7	6.4	6.6	6.3	6.2	6.4	6.1	6.4	6.3	6.0	6.4	6.3

Year	National 30 City Average	Shreve-port	Lewis-ton	Portland	Balti-more	Boston	Brockton	Fall River	Law-rence	Lowell	New Bedford	Pitts-field	Spring-field	Wor-cester	Ann Arbor	Dear-born
Jan 2005	148.5E	116.1E	131.2E	130.9E	136.6E	170.0E	158.8E	159.7E	162.4E	163.1E	159.7E	148.2E	151.3E	160.0E	148.7E	156.9E
2004	132.8	105.7	119.8	119.4	121.4	154.1	144.4	143.6	146.4	146.7	143.6	131.9	136.5	143.2	135.9	142.1
2003	129.7	103.9	117.7	117.3	118.1	150.2	139.6	140.6	143.3	143.6	140.5	128.9	133.8	139.0	134.1	139.0
2002	126.7	102.1	117.5	117.1	115.6	145.6	136.0	135.0	136.9	137.3	134.9	124.7	128.3	134.9	131.4	134.3
2001	122.2	98.3	114.6	114.3	111.7	140.9	132.1	131.4	133.1	132.9	131.3	120.3	124.5	130.1	126.8	129.8
2000	118.9	96.2	105.7	105.4	107.7	138.9	129.4	128.4	129.7	130.3	128.3	116.7	121.0	127.3	124.3	125.8
1999	116.6	94.8	105.0	104.7	106.4	136.2	126.7	126.3	127.3	127.4	126.2	115.0	118.8	123.8	117.5	122.7
1998	113.6	92.0	102.8	102.5	104.1	132.8	125.0	124.7	125.0	125.3	124.6	114.2	117.1	122.6	116.0	119.7
1997	111.5	90.1	101.7	101.4	102.2	132.1	124.4	124.7	125.1	125.6	124.7	114.6	117.4	122.9	113.5	117.9
1996	108.9	88.5	99.5	99.2	99.6	128.7	121.6	122.2	121.7	122.1	122.2	112.7	115.0	120.2	112.9	116.2
1995	105.6	86.7	96.8	96.5	96.1	128.6	119.6	117.7	120.5	119.6	117.2	110.7	112.7	114.8	106.1	110.5
1994	103.0	85.0	95.2	94.9	94.4	124.9	114.2	113.7	116.8	116.9	111.6	109.2	111.1	112.9	105.0	109.0
1993	100.0	83.6	93.0	93.0	93.1	121.1	111.7	111.5	114.9	114.3	109.4	107.1	108.9	110.3	102.8	105.8
1992	97.9	82.3	91.7	91.7	90.9	118.0	110.0	109.8	110.9	110.2	107.7	105.7	107.2	108.6	101.5	103.2
1991	95.7	81.6	89.8	89.9	89.1	115.2	105.9	104.7	107.3	105.5	105.1	100.6	103.2	105.7	95.1	98.3
1990	93.2	80.3	88.5	88.6	85.6	110.9	103.7	102.9	105.2	102.8	102.6	98.7	101.4	103.2	93.2	96.5
1989	91.0	78.8	86.7	86.7	83.5	107.2	100.4	99.7	102.9	99.7	99.8	94.7	96.4	99.1	89.6	94.6
1988	88.5	77.0	84.0	84.1	81.2	102.5	96.7	96.4	97.7	96.6	96.4	91.9	92.7	94.4	85.8	91.4
1987	85.7	74.5	81.3	81.4	78.6	97.3	94.5	94.8	94.9	94.8	95.0	89.4	90.5	92.2	87.4	88.7
1986	83.7	74.1	79.0	79.0	75.7	95.1	91.2	90.8	92.5	90.8	90.7	87.1	87.8	89.8	81.7	85.3
1985	81.8	73.6	76.7	77.0	72.7	92.8	88.8	88.7	89.4	88.2	88.6	85.0	85.6	86.7	80.1	82.7
1984	80.6	73.3	75.0	75.1	72.4	88.1	85.5	85.8	86.8	84.1	84.8	82.8	83.5	83.6	78.9	81.1
1983	78.2	72.2	73.1	72.9	70.6	84.0	82.5	82.2	82.7	80.9	81.3	79.5	80.6	81.3	77.7	79.4
1980	60.7	58.7	57.3	58.5	53.6	64.0	63.7	64.1	63.4	62.7	63.1	61.8	62.0	62.3	62.9	64.0
1975	43.7	40.5	41.9	42.1	39.8	46.6	45.8	45.7	46.2	45.7	46.1	45.5	45.8	46.0	44.1	44.9
1970	27.8	26.4	26.5	25.8	25.1	29.2	29.1	29.2	29.1	28.9	29.0	28.6	28.5	28.7	28.5	28.9
1965	21.5	20.3	20.4	19.4	20.2	23.0	22.5	22.5	22.5	22.2	22.4	22.0	22.4	22.1	22.0	22.3
1960	19.5	18.5	18.6	17.6	17.5	20.5	20.4	20.4	20.4	20.2	20.3	20.0	20.1	20.1	20.0	20.2
1955	16.3	15.5	15.6	14.7	14.7	17.2	17.1	17.1	17.1	16.9	17.1	16.8	16.9	16.8	16.7	16.9
1950	13.5	12.8	12.9	12.2	12.1	14.2	14.1	14.1	14.1	14.0	14.1	13.8	14.0	13.9	13.8	14.0
1945	8.6	8.1	8.2	7.8	7.7	9.1	9.0	9.0	9.0	8.9	9.0	8.9	8.9	8.9	8.8	8.9
1940	6.6	6.3	6.3	6.0	6.0	7.0	6.9	7.0	7.0	6.9	6.9	6.8	6.9	6.8	6.8	6.9

Year	National 30 City Average	Michigan Detroit	Flint	Grand Rapids	Kala-mazoo	Lansing	Sagi-naw	Minnesota Duluth	Minne-apolis	Roches-ter	Mississippi Biloxi	Jackson	Missouri Kansas City	St. Joseph	St. Louis	Spring-field
Jan 2005	148.5E	157.5E	142.2E	124.0E	133.8E	142.1E	140.0E	153.0E	165.8E	150.3E	116.6E	110.5E	152.4E	140.4E	150.5E	128.8E
2004	132.8	141.9	129.6	112.8	122.5	129.7	127.6	139.1	150.1	137.0	105.3	99.3	135.1	126.9	135.5	116.3
2003	129.7	138.7	127.9	110.9	120.8	127.7	126.0	136.7	146.5	134.5	101.4	96.7	131.9	124.8	133.3	113.9
2002	126.7	134.2	126.9	107.4	119.9	125.3	124.4	131.9	139.5	130.0	101.0	96.3	128.4	122.7	129.6	112.5
2001	122.2	129.4	122.4	104.3	115.2	120.1	119.7	131.4	136.1	124.9	98.7	93.9	121.8	114.7	125.5	106.7
2000	118.9	125.3	119.8	102.7	111.6	117.6	115.9	124.1	131.1	120.1	97.2	92.9	118.2	111.9	122.4	104.3
1999	116.6	122.6	113.7	100.9	106.1	111.5	110.1	120.3	126.5	117.1	95.7	91.8	114.9	106.1	119.8	101.1
1998	113.6	119.5	112.3	99.7	104.9	110.1	108.7	117.7	124.6	115.5	92.1	89.5	108.2	104.2	115.9	98.9
1997	111.5	117.6	110.8	98.9	104.2	108.7	107.5	115.9	121.9	114.1	90.9	88.3	106.4	102.4	113.2	97.3
1996	108.9	116.0	109.7	93.9	103.5	107.8	106.9	115.0	120.4	113.5	87.1	85.4	103.2	99.9	110.1	94.6
1995	105.6	110.1	104.1	91.1	96.4	98.3	102.3	100.3	111.9	102.5	84.2	83.5	99.7	96.1	106.3	89.5
1994	103.0	108.5	102.9	89.7	95.0	97.1	101.1	99.1	109.3	101.1	82.4	81.7	97.3	92.8	103.2	88.2
1993	100.0	105.4	100.7	87.7	92.7	95.3	98.8	99.8	106.7	99.9	80.0	79.4	94.5	90.9	99.9	86.2
1992	97.9	102.9	99.4	86.4	91.4	94.0	97.4	98.5	105.3	98.6	78.8	78.2	92.9	89.5	98.6	84.9
1991	95.7	97.6	93.6	85.3	90.0	92.1	90.3	96.0	102.9	97.3	77.6	76.9	90.9	88.1	96.4	83.9
1990	93.2	96.0	91.7	84.0	87.4	90.2	88.9	94.3	100.5	95.6	76.4	75.6	89.5	86.8	94.0	82.6
1989	91.0	94.0	90.2	82.4	85.8	88.6	87.4	92.5	97.4	93.7	75.0	74.1	87.3	85.1	91.8	80.9
1988	88.5	90.8	87.8	80.7	84.1	86.3	85.6	90.5	94.6	91.9	73.5	72.9	84.9	82.9	89.0	79.2
1987	85.7	87.8	85.4	79.7	82.6	85.2	84.5	88.5	92.0	90.6	72.4	71.3	81.7	82.7	85.5	78.2
1986	83.7	84.3	82.8	77.5	80.9	82.7	81.5	87.2	89.7	88.4	70.6	69.9	79.9	79.4	83.5	75.6
1985	81.8	81.6	80.7	75.8	79.4	80.0	80.5	85.2	87.9	86.5	69.4	68.5	78.2	77.3	80.8	73.2
1984	80.6	79.4	79.5	74.3	78.7	79.0	79.9	84.9	86.9	86.0	68.6	67.5	77.7	76.7	77.7	72.7
1983	78.2	77.0	77.4	74.1	76.2	77.1	78.0	81.6	81.4	82.6	68.0	67.3	76.2	76.8	76.2	71.6
1980	60.7	62.5	63.1	59.8	61.7	60.3	63.3	63.4	64.1	64.8	54.3	54.3	59.1	60.4	59.7	56.7
1975	43.7	45.8	44.9	41.8	43.7	44.1	44.8	45.2	46.0	45.3	37.9	37.7	42.7	45.4	44.8	41.1
1970	27.8	29.7	28.5	26.7	28.1	27.8	28.7	28.9	29.5	28.9	24.4	21.3	25.3	28.3	28.4	26.0
1965	21.5	22.1	21.9	20.6	21.7	21.5	22.2	22.3	23.4	22.3	18.8	16.4	20.6	21.8	21.8	20.0
1960	19.5	20.1	20.0	18.7	19.7	19.5	20.1	20.3	20.5	20.2	17.1	14.9	19.0	19.8	19.5	18.2
1955	16.3	16.8	16.7	15.7	16.5	16.3	16.9	17.0	17.2	17.0	14.3	12.5	16.0	16.6	16.3	15.2
1950	13.5	13.9	13.8	13.0	13.6	13.5	13.9	14.0	14.2	14.0	11.8	10.3	13.2	13.7	13.5	12.6
1945	8.6	8.9	8.8	8.3	8.7	8.6	8.9	8.9	9.0	8.9	7.6	6.6	8.4	8.8	8.6	8.0
1940	6.6	6.8	6.8	6.4	6.7	6.6	6.9	6.9	7.0	6.9	5.8	5.1	6.5	6.7	6.7	6.2

Year	National 30 City Average	Montana Billings	Great Falls	Nebraska Lincoln	Omaha	Nevada Las Vegas	Reno	New Hampshire Man-chester	Nashua	New Jersey Camden	Jersey City	Newark	Pater-son	Trenton	NM Albu-querque	NY Albany
Jan 2005	148.5E	131.8E	132.6E	124.1E	131.1E	149.4E	143.3E	136.8E	136.5E	157.4E	162.5E	163.7E	163.0E	160.0E	130.5E	141.6E
2004	132.8	118.2	117.9	112.1	119.4	137.1	130.5	124.2	123.9	143.4	145.4	147.2	146.6	146.0	118.3	128.6
2003	129.7	115.8	115.8	110.2	117.3	133.8	128.3	122.4	122.2	142.0	144.3	145.7	145.0	143.0	116.4	126.8
2002	126.7	114.6	114.5	108.2	115.0	131.9	126.4	119.9	119.6	135.9	138.5	140.2	140.1	137.6	114.6	122.6
2001	122.2	117.5	117.7	101.3	111.6	127.8	122.5	116.2	116.2	133.4	136.7	136.9	136.8	136.0	111.4	119.2
2000	118.9	113.7	113.9	98.8	107.0	125.8	118.2	111.9	111.9	128.4	130.5	132.6	132.4	130.6	109.0	116.5
1999	116.6	112.1	112.7	96.6	104.8	121.9	114.4	109.6	109.6	125.3	128.8	131.6	129.5	129.6	106.7	114.6
1998	113.6	109.7	109.1	94.9	101.2	118.1	111.4	110.1	110.0	124.3	127.7	128.9	128.5	127.9	103.8	113.0
1997	111.5	107.4	107.6	93.3	99.5	114.6	109.8	108.6	108.6	121.9	125.1	126.4	126.4	125.2	100.8	110.0
1996	108.9	108.0	107.5	91.4	97.4	111.8	108.9	106.5	106.4	119.1	122.5	122.5	123.8	121.8	98.6	108.3
1995	105.6	104.7	104.9	85.6	93.4	108.5	105.0	100.9	100.8	107.0	112.2	111.9	112.1	111.2	96.3	103.6
1994	103.0	100.2	100.7	84.3	91.0	105.3	102.2	97.8	97.7	105.4	110.6	110.1	110.6	108.2	93.4	102.4
1993	100.0	97.9	97.8	82.3	88.7	102.8	99.9	95.4	95.4	103.7	109.0	108.2	109.0	106.4	90.1	99.8
1992	97.9	95.5	96.4	81.1	87.4	99.4	98.4	90.4	90.4	102.0	107.5	107.0	107.8	103.9	87.5	98.5
1991	95.7	94.2	95.1	80.1	86.4	97.5	95.8	87.8	87.8	95.0	98.5	95.6	100.0	96.6	86.2	96.1
1990	93.2	92.9	93.8	78.8	85.0	96.3	94.5	86.3	86.3	93.0	93.5	93.8	97.2	94.5	84.9	93.2
1989	91.0	91.3	92.2	77.3	83.6	94.8	92.3	84.8	84.8	90.4	91.4	91.8	95.5	91.9	83.3	88.4
1988	88.5	89.5	90.4	75.8	82.0	93.0	90.5	83.2	83.2	87.7	89.5	89.7	92.5	89.5	81.7	86.8
1987	85.7	85.4	87.3	75.3	79.8	91.5	88.3	82.4	82.5	86.1	88.3	88.7	90.3	87.9	78.1	84.5
1986	83.7	86.2	87.3	72.6	78.8	90.1	87.2	79.7	79.7	84.6	86.1	86.7	87.8	86.1	78.9	81.7
1985	81.8	83.9	84.3	71.5	77.5	87.6	85.0	78.1	78.1	81.3	83.3	83.5	84.5	82.8	76.7	79.5
1984	80.6	83.2	82.9	70.9	77.6	85.9	83.3	76.4	75.5	78.2	80.3	80.1	81.3	78.5	75.9	77.3
1983	78.2	80.3	80.0	72.1	77.3	83.0	81.9	73.0	72.2	74.6	76.5	76.7	76.5	74.2	73.1	74.2
1980	60.7	63.9	64.6	58.5	63.5	64.6	63.0	56.6	56.0	58.6	60.6	60.1	60.0	58.9	59.0	59.5
1975	43.7	43.1	43.8	40.9	43.2	42.8	41.9	41.3	40.8	42.3	43.4	44.2	43.9	43.8	40.3	43.9
1970	27.8	28.5	28.9	26.4	26.8	29.4	28.0	26.2	25.6	27.2	27.8	29.0	27.8	27.4	26.4	28.3
1965	21.5	22.0	22.3	20.3	20.6	22.4	21.6	20.6	19.7	20.9	21.4	23.8	21.4	21.3	20.6	22.3
1960	19.5	20.0	20.3	18.5	18.7	20.2	19.6	18.0	17.9	19.0	19.4	19.4	19.4	19.2	18.5	19.3
1955	16.3	16.7	17.0	15.5	15.7	16.9	16.4	15.1	15.0	16.0	16.3	16.3	16.3	16.1	15.6	16.2
1950	13.5	13.9	14.0	12.8	13.0	14.0	13.6	12.5	12.4	13.2	13.5	13.5	13.5	13.3	12.9	13.4
1945	8.6	8.8	9.0	8.1	8.3	8.9	8.7	8.0	7.9	8.4	8.6	8.6	8.6	8.5	8.2	8.5
1940	6.6	6.8	6.9	6.3	6.4	6.9	6.7	6.2	6.1	6.5	6.6	6.6	6.6	6.6	6.3	6.6

Year	National 30 City Average	New York Bing-hamton	Buffalo	New York	Roches-ter	Schen-ectady	Syracuse	Utica	Yonkers	North Carolina Charlotte	Durham	Greens-boro	Raleigh	Winston-Salem	N.Dakota Fargo	Ohio Akron
Jan 2005	148.5E	135.8E	150.4E	195.2E	144.0E	141.2E	140.2E	136.7E	175.7E	110.9E	112.4E	112.4E	112.4E	111.5E	125.2E	144.1E
2004	132.8	123.5	136.1	177.7	132.0	128.4	127.3	124.4	161.8	98.9	100.0	100.1	100.3	99.4	113.0	131.9
2003	129.7	121.8	132.8	173.4	130.3	126.7	124.7	121.7	160.0	96.2	97.5	97.5	97.8	96.8	110.6	129.6
2002	126.7	119.0	128.5	170.1	127.1	123.0	121.8	118.4	154.8	94.8	96.1	96.1	96.3	95.5	106.3	127.2
2001	122.2	116.0	125.2	164.4	123.1	120.1	118.6	115.3	151.4	91.5	92.9	92.9	93.3	92.3	103.1	123.5
2000	118.9	112.4	122.3	159.2	120.0	117.5	115.1	112.4	144.8	90.3	91.6	91.6	91.9	91.0	97.9	117.8
1999	116.6	108.6	120.2	155.9	116.8	114.7	113.7	108.5	140.6	89.3	90.2	90.3	90.5	90.1	96.8	116.0
1998	113.6	109.0	119.1	154.4	117.2	114.2	113.6	108.6	141.4	88.2	89.1	89.2	89.4	89.0	95.2	113.1
1997	111.5	107.2	115.7	150.3	115.2	111.2	110.6	107.0	138.8	86.8	87.7	87.8	87.9	87.6	93.5	110.6
1996	108.9	105.5	114.0	148.0	113.3	109.5	108.2	105.2	137.2	85.0	85.9	86.0	86.1	85.8	91.7	107.9
1995	105.6	99.3	110.1	140.7	106.6	104.6	104.0	101.7	129.4	81.8	82.6	82.6	82.7	82.6	88.4	103.4
1994	103.0	98.1	107.2	137.1	105.2	103.5	102.3	96.5	128.1	80.3	81.1	81.0	81.2	81.1	87.0	102.2
1993	100.0	95.7	102.2	133.3	102.0	100.8	99.6	94.1	126.3	78.2	78.9	78.9	78.9	78.8	85.8	100.3
1992	97.9	93.7	100.1	128.2	99.3	99.3	98.1	90.5	123.9	77.1	77.7	77.8	77.8	77.6	83.1	98.0
1991	95.7	89.5	96.8	124.4	96.0	96.7	95.5	88.7	121.5	76.0	76.6	76.8	76.7	76.6	82.6	96.8
1990	93.2	87.0	94.1	118.1	94.6	93.9	91.0	85.6	111.4	74.8	75.4	75.5	75.5	75.3	81.3	94.6
1989	91.0	85.4	91.8	114.8	92.6	89.6	88.4	84.1	102.1	73.3	74.0	74.0	74.0	73.8	79.7	92.9
1988	88.5	83.6	89.4	106.5	87.9	87.7	86.5	82.4	99.9	71.5	72.2	72.2	72.2	72.0	78.2	91.1
1987	85.7	82.4	85.8	102.6	85.8	85.3	85.1	82.1	98.9	70.6	71.2	71.2	71.1	71.0	77.4	90.1
1986	83.7	80.2	84.4	99.8	83.7	82.4	83.3	79.2	96.6	68.4	69.1	69.1	69.1	68.9	75.4	87.7
1985	81.8	77.5	83.2	94.9	81.6	80.1	81.0	77.0	92.8	66.9	67.6	67.7	67.6	67.5	73.4	86.8
1984	80.6	75.4	81.3	91.1	80.6	78.5	79.0	76.1	89.5	66.3	66.6	66.8	66.2	66.0	72.2	82.8
1983	78.2	73.2	78.2	86.5	77.1	75.3	76.5	74.5	85.4	64.6	65.1	65.7	64.9	64.6	70.5	79.4
1980	60.7	58.0	60.6	66.0	60.5	60.3	61.6	58.5	65.9	51.1	52.2	52.5	51.7	50.8	57.4	62.3
1975	43.7	42.5	45.1	49.5	44.8	43.7	44.8	42.7	47.1	36.1	37.0	37.0	37.3	36.1	39.3	44.7
1970	27.8	27.0	28.9	32.8	29.2	27.8	28.5	26.9	30.0	20.9	23.7	23.6	23.4	23.0	25.8	28.3
1965	21.5	20.8	22.2	25.5	22.8	21.4	21.9	20.7	23.1	16.0	18.2	18.2	18.0	17.7	19.9	21.8
1960	19.5	18.9	19.9	21.5	19.7	19.5	19.9	18.8	21.0	14.4	16.6	16.6	16.4	16.1	18.1	19.9
1955	16.3	15.8	16.7	18.1	16.5	16.3	16.7	15.8	17.6	12.1	13.9	13.9	13.7	13.5	15.2	16.6
1950	13.5	13.1	13.8	14.9	13.6	13.5	13.8	13.0	14.5	10.0	11.5	11.5	11.4	11.2	12.5	13.7
1945	8.6	8.3	8.8	9.5	8.7	8.6	8.8	8.3	9.3	6.4	7.3	7.3	7.2	7.1	8.0	8.8
1940	6.6	6.4	6.8	7.4	6.7	6.7	6.8	6.4	7.2	4.9	5.6	5.6	5.6	5.5	6.2	6.8

Year	National 30 City Average	Ohio Canton	Cincin-nati	Cleve-land	Colum-bus	Dayton	Lorain	Spring-field	Toledo	Youngs-town	Oklahoma Lawton	Oklahoma City	Tulsa	Oregon Eugene	Port-land	PA Allen-town
Jan 2005	148.5E	137.4E	138.0E	149.0E	139.2E	132.7E	143.4E	133.9E	145.4E	141.5E	121.1E	121.1E	118.4E	150.1E	152.7E	146.5E
2004	132.8	125.6	124.3	135.6	126.2	121.1	131.5	122.0	132.8	129.3	109.3	109.4	107.3	136.8	137.8	133.9
2003	129.7	123.7	121.3	132.7	123.7	119.4	126.1	120.0	130.4	126.2	107.5	107.8	105.0	134.2	135.9	130.3
2002	126.7	120.9	119.2	130.0	120.9	117.5	124.3	118.0	128.4	123.6	106.3	106.0	104.1	132.2	133.9	128.1
2001	122.2	117.7	115.9	125.9	117.1	114.0	120.4	114.7	123.8	119.9	102.0	102.1	99.7	129.8	131.2	123.5
2000	118.9	112.7	110.1	121.3	112.5	109.0	115.0	109.2	115.7	114.1	98.7	98.9	97.5	126.3	127.4	119.9
1999	116.6	110.6	107.9	118.7	109.5	107.1	112.0	106.4	113.6	111.9	97.5	97.4	96.2	120.9	124.3	117.8
1998	113.6	108.5	105.0	114.8	106.8	104.5	109.4	104.1	111.2	109.0	94.3	94.5	94.5	120.0	122.2	115.7
1997	111.5	106.4	102.9	112.9	104.7	102.6	107.4	102.2	108.5	107.2	92.9	93.2	93.0	118.1	119.6	113.9
1996	108.9	103.2	100.2	110.1	101.1	98.8	103.8	98.4	105.1	104.1	90.9	91.3	91.2	114.4	116.1	112.0
1995	105.6	98.8	97.1	106.4	99.1	94.6	97.1	92.0	100.6	100.1	85.3	88.0	89.0	112.2	114.3	108.3
1994	103.0	97.7	95.0	104.8	95.4	93.0	96.0	90.1	99.3	98.9	84.0	86.5	87.6	107.6	109.2	106.4
1993	100.0	95.9	92.3	101.9	93.9	90.6	93.9	87.7	98.3	97.2	81.3	83.9	84.8	107.2	108.8	103.6
1992	97.9	94.5	90.6	98.7	92.6	89.2	92.2	86.4	97.1	96.0	80.1	82.7	83.4	101.1	102.6	101.6
1991	95.7	93.4	88.6	97.2	90.6	87.7	91.1	84.9	94.2	92.7	78.3	80.7	81.4	99.5	101.1	98.9
1990	93.2	92.1	86.7	95.1	88.1	85.9	89.4	83.4	92.9	91.8	77.1	79.9	80.0	98.0	99.4	95.0
1989	91.0	90.3	84.6	93.6	85.8	82.6	87.9	81.0	91.4	88.4	75.8	78.5	78.4	95.2	97.0	92.4
1988	88.5	89.1	82.9	92.1	83.9	81.0	85.2	79.5	89.6	86.9	74.3	77.0	76.8	93.5	95.3	89.8
1987	85.7	87.7	80.9	89.6	81.5	79.5	86.9	78.0	86.1	85.4	72.6	74.4	74.7	91.4	91.9	86.9
1986	83.7	84.9	78.8	87.3	79.7	78.0	81.6	76.1	84.6	83.8	71.6	74.1	73.9	90.0	91.7	84.7
1985	81.8	83.7	78.2	86.2	77.7	76.3	80.7	74.3	84.7	82.7	71.2	73.9	73.8	88.7	90.1	81.9
1984	80.6	81.3	77.4	82.5	76.1	75.9	79.5	73.9	83.5	80.3	70.6	72.9	72.8	89.5	89.3	78.4
1983	78.2	76.9	74.9	77.9	74.5	71.6	77.2	71.9	80.4	77.3	69.1	71.5	72.2	89.0	89.2	75.0
1980	60.7	60.6	59.8	61.0	58.6	56.6	60.8	56.2	64.0	61.5	52.2	55.5	57.2	68.9	68.4	58.7
1975	43.7	44.0	43.7	44.6	42.2	40.5	42.7	40.0	44.9	44.5	38.6	38.6	39.2	44.6	44.9	42.3
1970	27.8	27.8	28.2	29.6	26.7	27.0	27.5	25.4	29.1	29.6	24.1	23.4	24.1	30.4	30.0	27.3
1965	21.5	21.5	20.7	21.1	20.2	20.0	21.1	19.6	21.3	21.0	18.5	17.8	20.6	23.4	22.6	21.0
1960	19.5	19.5	19.3	19.6	18.7	18.1	18.1	17.8	19.4	19.4	16.9	16.1	18.1	21.2	21.1	19.1
1955	16.3	16.3	16.2	16.5	15.7	15.2	16.1	14.9	16.2	16.0	14.1	13.5	15.1	17.8	17.7	16.0
1950	13.5	13.5	13.3	13.6	13.0	12.5	13.3	12.3	13.4	13.2	11.7	11.2	12.5	14.7	14.6	13.2
1945	8.6	8.6	8.5	8.7	8.3	8.0	8.5	7.8	8.6	8.4	7.4	7.1	8.0	9.4	9.3	8.4
1940	6.6	6.7	6.5	6.7	6.4	6.2	6.5	6.1	6.6	6.5	5.7	5.5	6.1	7.2	7.2	6.5

Year	National 30 City Average	Pennsylvania Erie	Harrisburg	Philadelphia	Pittsburgh	Reading	Scranton	RI Providence	South Carolina Charleston	Columbia	South Dakota Rapid City	Sioux Falls	Tennessee Chattanooga	Knoxville	Memphis	Nashville
Jan 2005	148.5E	140.2E	140.6E	167.4E	147.2E	145.2E	141.7E	155.8E	110.6E	110.0E	115.0E	118.4E	116.7E	115.9E	129.4E	128.3E
2004	132.8	127.0	126.2	148.4	133.1	128.7	129.0	138.8	98.9	98.4	103.9	107.3	105.0	104.5	115.6	115.8
2003	129.7	124.5	123.8	145.5	130.5	126.3	126.0	135.9	96.1	95.7	101.3	104.0	102.8	102.3	111.1	110.9
2002	126.7	122.1	121.1	142.0	127.8	123.4	124.2	131.1	94.8	93.9	100.2	103.2	102.2	101.2	107.6	109.2
2001	122.2	118.9	118.4	136.9	124.1	120.4	121.1	127.8	92.1	91.3	96.8	99.7	98.2	96.5	102.6	104.3
2000	118.9	114.6	114.0	132.1	120.9	115.9	117.7	122.8	89.9	89.1	94.2	98.0	96.9	95.0	100.8	100.8
1999	116.6	113.8	112.8	129.8	119.6	114.8	116.3	121.6	89.0	88.1	92.4	95.8	95.9	93.4	99.7	98.6
1998	113.6	109.8	110.5	126.6	117.1	112.5	113.7	120.3	88.0	87.2	90.7	93.6	94.9	92.4	97.2	96.8
1997	111.5	108.5	108.8	123.3	113.9	110.8	112.3	118.9	86.5	85.6	89.3	92.1	93.6	90.8	96.1	94.9
1996	108.9	106.8	105.7	120.3	110.8	108.7	109.9	117.2	84.9	84.1	87.2	90.0	91.6	88.8	94.0	91.9
1995	105.6	99.8	100.6	117.1	106.3	103.6	103.8	111.1	82.7	82.2	84.0	84.7	89.2	86.0	91.2	87.6
1994	103.0	98.2	99.4	115.2	103.7	102.2	102.6	109.6	81.2	80.6	82.7	83.1	87.4	84.1	89.0	84.8
1993	100.0	94.9	96.6	107.4	99.0	98.6	99.8	108.2	78.4	77.9	81.2	81.9	85.1	82.3	86.8	81.9
1992	97.9	92.1	94.8	105.3	96.5	96.7	97.8	106.9	77.1	76.8	80.0	80.7	83.6	77.7	85.4	80.7
1991	95.7	90.5	92.5	101.7	93.2	94.1	94.8	96.1	75.0	75.8	78.7	79.7	81.9	76.6	83.0	79.1
1990	93.2	88.5	89.5	98.5	91.2	90.8	91.3	94.1	73.7	74.5	77.2	78.4	79.9	75.1	81.3	77.1
1989	91.0	86.6	86.5	94.2	89.4	87.9	88.8	92.4	72.1	72.8	75.7	77.0	78.5	73.7	81.2	74.8
1988	88.5	84.9	84.0	89.8	87.7	85.3	86.4	89.1	70.5	71.3	74.2	75.5	77.0	72.1	79.7	73.1
1987	85.7	82.6	81.8	86.8	85.6	82.9	83.3	86.3	70.2	70.2	73.6	74.9	74.4	70.2	77.5	71.0
1986	83.7	81.5	79.6	84.9	83.6	79.7	81.3	85.2	67.4	68.1	71.0	72.3	74.2	69.2	75.3	68.1
1985	81.8	79.6	77.2	82.2	81.5	77.7	80.0	83.0	65.9	66.9	69.7	71.2	72.5	67.7	74.3	66.7
1984	80.6	78.3	75.4	79.0	78.9	75.9	78.8	80.3	64.3	65.5	68.7	70.4	71.4	66.4	73.3	66.3
1983	78.2	76.3	72.2	74.4	75.5	74.0	75.0	76.3	65.3	65.2	67.7	69.7	69.3	64.1	72.8	65.1
1980	60.7	59.4	56.9	58.7	61.3	57.7	58.4	59.2	50.6	51.1	54.8	57.1	55.0	51.6	55.9	53.1
1975	43.7	43.4	42.2	44.5	44.5	43.6	41.9	42.7	35.0	36.4	37.7	39.6	39.1	37.3	40.7	37.0
1970	27.8	27.5	25.8	27.5	28.7	27.1	27.0	27.3	22.8	22.9	24.8	25.8	24.9	22.0	23.0	22.8
1965	21.5	21.1	20.1	21.7	22.4	20.9	20.8	21.9	17.6	17.6	19.1	19.9	19.2	17.1	18.3	17.4
1960	19.5	19.1	18.6	19.4	19.7	19.0	18.9	19.1	16.0	16.0	17.3	18.1	17.4	15.5	16.6	15.8
1955	16.3	16.1	15.6	16.3	16.5	15.9	15.9	16.0	13.4	13.4	14.5	15.1	14.6	13.0	13.9	13.3
1950	13.5	13.3	12.9	13.5	13.6	13.2	13.1	13.2	11.1	11.1	12.0	12.5	12.1	10.7	11.5	10.9
1945	8.6	8.5	8.2	8.6	8.7	8.4	8.3	8.4	7.1	7.1	7.7	8.0	7.7	6.8	7.3	7.0
▼ 1940	6.6	6.5	6.3	6.6	6.7	6.4	6.5	6.5	5.4	5.5	5.9	6.2	6.0	5.3	5.6	5.4

Year	National 30 City Average	Texas Abilene	Amarillo	Austin	Beaumont	Corpus Christi	Dallas	El Paso	Fort Worth	Houston	Lubbock	Odessa	San Antonio	Waco	Wichita Falls	Utah Ogden
Jan 2005	148.5E	114.3E	117.8E	117.7E	120.2E	114.7E	124.2E	112.9E	119.9E	128.4E	116.2E	111.1E	121.3E	116.3E	116.5E	126.4E
2004	132.8	102.9	106.3	105.7	108.5	102.9	112.0	101.5	108.4	115.9	104.7	99.9	108.4	104.8	105.1	115.2
2003	129.7	100.5	104.4	103.9	106.1	100.4	109.3	99.5	105.5	113.4	102.3	97.6	104.8	102.7	103.0	112.7
2002	126.7	99.9	101.9	102.4	104.5	98.9	107.9	98.7	104.6	111.5	100.5	95.9	103.7	100.5	101.4	110.8
2001	122.2	93.4	98.4	99.8	102.3	96.6	103.8	95.5	100.9	107.8	97.6	93.4	100.5	97.8	97.3	107.7
2000	118.9	93.4	98.1	99.1	101.6	96.9	102.7	92.4	99.9	106.0	97.6	93.4	99.4	97.3	96.9	104.6
1999	116.6	91.8	94.5	96.0	99.7	94.0	101.0	90.7	97.6	104.6	96.0	92.1	98.0	94.8	95.5	103.3
1998	113.6	89.8	92.7	94.2	97.9	91.8	97.9	88.4	94.5	101.3	93.3	90.1	94.8	92.7	92.9	98.5
1997	111.5	88.4	91.3	92.8	96.8	90.3	96.1	87.0	93.3	100.1	91.9	88.8	93.4	91.4	91.5	96.1
1996	108.9	86.8	89.6	90.9	95.3	88.5	94.1	86.7	91.5	97.9	90.2	87.1	92.3	89.7	89.9	94.1
1995	105.6	85.2	87.4	89.3	93.7	87.4	91.4	85.2	89.5	95.9	88.4	85.6	88.9	86.4	86.8	92.2
1994	103.0	83.3	85.5	87.0	91.8	84.6	89.6	82.2	87.4	93.4	87.0	83.9	87.0	84.9	85.4	89.4
1993	100.0	81.4	83.4	84.9	90.0	82.8	87.8	80.2	85.5	91.1	84.8	81.9	85.0	83.0	83.5	87.1
1992	97.9	80.3	82.3	83.8	88.9	81.6	86.2	79.0	84.1	89.8	83.6	80.7	83.9	81.8	82.4	85.1
1991	95.7	79.2	81.1	82.8	87.8	80.6	85.9	78.0	83.3	87.9	82.7	79.8	83.3	80.6	81.5	84.1
1990	93.2	78.0	80.1	81.3	86.5	79.3	84.5	76.7	82.1	85.4	81.5	78.6	80.7	79.6	80.3	83.4
1989	91.0	76.6	78.7	79.7	85.1	77.8	82.5	75.2	80.6	84.0	80.0	77.0	78.7	78.1	78.7	81.9
1988	88.5	75.0	77.2	78.3	83.9	76.3	81.1	73.6	79.8	82.5	78.4	75.5	77.1	76.5	77.4	80.4
1987	85.7	75.9	76.4	76.0	81.3	75.6	79.9	72.2	77.7	81.3	77.3	74.6	75.7	75.2	77.4	79.1
1986	83.7	72.5	74.1	75.3	80.5	73.5	78.9	70.7	76.7	80.3	75.4	72.4	74.3	72.8	74.5	77.3
1985	81.8	71.1	72.5	74.5	79.3	72.3	77.6	69.4	75.1	79.6	74.0	71.2	73.9	71.7	73.3	75.2
1984	80.6	70.5	71.6	72.8	78.7	71.6	79.8	68.3	77.1	80.7	72.7	69.8	73.1	71.3	72.4	74.9
1983	78.2	68.3	70.6	71.0	75.0	70.2	77.8	66.3	74.9	79.8	71.0	69.0	71.8	69.4	70.1	76.1
1980	60.7	53.4	55.2	54.5	57.6	54.5	57.9	53.1	57.0	59.4	55.6	57.2	55.0	54.9	55.4	62.2
1975	43.7	37.6	39.0	39.0	39.6	38.1	40.7	38.0	40.4	41.2	38.9	37.9	39.0	38.6	38.0	40.0
1970	27.8	24.5	24.9	24.9	25.7	24.5	25.5	23.7	25.9	25.4	25.1	24.6	23.3	24.8	24.5	26.8
1965	21.5	18.9	19.2	19.2	19.9	18.9	19.9	19.0	19.9	20.0	19.4	19.0	18.5	19.2	18.9	20.6
1960	19.5	17.1	17.4	17.4	18.1	17.1	18.2	17.0	18.1	18.2	17.6	17.3	16.8	17.4	17.2	18.8
1955	16.3	14.4	14.6	14.6	15.1	14.4	15.3	14.3	15.2	15.2	14.8	14.5	14.1	14.6	14.4	15.7
1950	13.5	11.9	12.1	12.1	12.5	11.9	12.6	11.8	12.5	12.6	12.2	12.0	11.6	12.1	11.9	13.0
1945	8.6	7.6	7.7	7.7	8.0	7.6	8.0	7.5	8.0	8.0	7.8	7.6	7.4	7.7	7.6	8.3
▼ 1940	6.6	5.9	5.9	5.9	6.1	5.8	6.2	5.8	6.2	6.2	6.0	5.9	5.7	5.9	5.8	6.4

Historical Cost Indexes

Year	National 30 City Average	Utah Salt Lake City	Vermont Burlington	Vermont Rutland	Virginia Alexandria	Virginia Newport News	Virginia Norfolk	Virginia Richmond	Virginia Roanoke	Washington Seattle	Washington Spokane	Washington Tacoma	West Virginia Charleston	West Virginia Huntington	Wisconsin Green Bay	Wisconsin Kenosha
Jan 2005	148.5E	129.8E	125.3E	124.4E	136.6E	122.4E	123.7E	124.4E	112.6E	154.7E	141.5E	150.5E	141.2E	140.6E	140.8E	144.3E
2004	132.8	117.8	113.0	112.3	121.5	108.9	110.2	110.9	99.7	138.0	127.6	134.5	124.8	125.6	128.9	132.4
2003	129.7	116.0	110.7	110.1	119.5	104.8	106.2	108.6	97.0	134.9	125.9	133.0	123.3	123.6	127.4	130.9
2002	126.7	113.7	109.0	108.4	115.1	102.9	104.1	106.6	95.2	132.7	123.9	131.4	121.2	120.9	123.0	127.3
2001	122.2	109.1	105.7	105.2	110.8	99.8	100.3	102.9	92.1	127.9	120.3	125.7	114.6	117.5	119.1	123.6
2000	118.9	106.5	98.9	98.3	108.1	96.5	97.6	100.2	90.7	124.6	118.3	122.9	111.5	114.4	114.6	119.1
1999	116.6	104.5	98.2	97.7	106.1	95.6	96.5	98.8	89.8	123.3	116.7	121.6	110.6	113.4	112.1	115.8
1998	113.6	99.5	97.8	97.3	104.1	93.7	93.9	97.0	88.3	119.4	114.3	118.3	106.7	109.0	109.5	112.9
1997	111.5	97.2	96.6	96.3	101.2	91.6	91.7	92.9	86.9	118.1	111.7	117.2	105.3	107.7	105.6	109.1
1996	108.9	94.9	95.1	94.8	99.7	90.2	90.4	91.6	85.5	115.2	109.2	114.3	103.1	104.8	103.8	106.4
1995	105.6	93.1	91.1	90.8	96.3	86.0	86.4	87.8	82.8	113.7	107.4	112.8	95.8	97.2	97.6	97.9
1994	103.0	90.2	89.5	89.3	93.9	84.6	84.8	86.3	81.4	109.9	104.0	108.3	94.3	95.3	96.3	96.2
1993	100.0	87.9	87.6	87.6	91.6	82.9	83.0	84.3	79.5	107.3	103.9	106.7	92.6	93.5	94.0	94.3
1992	97.9	86.0	86.1	86.1	90.1	81.0	81.6	82.0	78.3	105.1	101.4	103.7	91.4	92.3	92.0	92.1
1991	95.7	84.9	84.2	84.2	88.2	77.6	77.9	79.8	77.3	102.2	100.0	102.2	89.7	88.6	88.6	89.8
1990	93.2	84.3	83.0	82.9	86.1	76.3	76.7	77.6	76.1	100.1	98.5	100.5	86.1	86.8	86.7	87.8
1989	91.0	82.8	81.4	81.3	83.4	74.9	75.2	76.0	74.2	96.1	96.8	98.4	84.5	84.8	84.3	85.6
1988	88.5	81.3	79.7	79.7	81.0	73.4	73.7	74.1	72.3	94.2	95.0	96.6	82.8	83.1	82.0	83.0
1987	85.7	79.8	79.0	79.0	77.9	71.0	71.7	72.7	70.2	91.9	92.4	92.9	81.1	81.3	80.2	81.7
1986	83.7	78.1	76.4	76.4	76.8	70.3	70.5	71.0	67.8	90.5	91.9	92.8	79.9	80.2	77.8	79.3
1985	81.8	75.9	74.8	74.9	75.1	68.7	68.8	69.5	67.2	88.3	89.0	91.2	77.7	77.7	76.7	77.4
1984	80.6	75.2	73.7	73.7	75.0	67.9	68.3	68.8	66.2	89.2	88.1	90.0	75.4	75.8	75.0	75.1
1983	78.2	74.5	70.9	71.5	72.9	65.9	66.5	68.9	66.0	87.6	86.9	88.8	73.8	74.4	73.4	73.2
1980	60.7	57.0	55.3	58.3	57.3	52.5	52.4	54.3	51.3	67.9	66.3	66.7	57.7	58.3	58.6	58.3
1975	43.7	40.1	41.8	43.9	41.7	37.2	36.9	37.1	37.1	44.9	44.4	44.5	41.0	40.0	40.9	40.5
1970	27.8	26.1	25.4	26.8	26.2	23.9	21.5	22.0	23.7	28.8	29.3	29.6	26.1	25.8	26.4	26.5
1965	21.5	20.0	19.8	20.6	20.2	18.4	17.1	17.2	18.3	22.4	22.5	22.8	20.1	19.9	20.3	20.4
1960	19.5	18.4	18.0	18.8	18.4	16.7	15.4	15.6	16.6	20.4	20.8	20.8	18.3	18.1	18.4	18.6
1955	16.3	15.4	15.1	15.7	15.4	14.0	12.9	13.1	13.9	17.1	17.4	17.4	15.4	15.2	15.5	15.6
1950	13.5	12.7	12.4	13.0	12.7	11.6	10.7	10.8	11.5	14.1	14.4	14.4	12.7	12.5	12.8	12.9
1945	8.6	8.1	7.9	8.3	8.1	7.4	6.8	6.9	7.3	9.0	9.2	9.2	8.1	8.0	8.1	8.2
1940	6.6	6.3	6.1	6.4	6.2	5.7	5.3	5.3	5.7	7.0	7.1	7.1	6.2	6.2	6.3	6.3

Year	National 30 City Average	Wisconsin Madison	Wisconsin Milwaukee	Wisconsin Racine	Wyoming Cheyenne	Canada Calgary	Canada Edmonton	Canada Hamilton	Canada London	Canada Montreal	Canada Ottawa	Canada Quebec	Canada Toronto	Canada Vancouver	Canada Winnipeg
Jan 2005	148.5E	141.6E	147.4E	145.0E	115.0E	155.1E	156.4E	160.9E	156.8E	150.1E	156.0E	151.0E	163.6E	160.0E	147.4E
2004	132.8	129.7	134.2	132.7	104.7	138.8	139.4	142.6	140.4	134.5	141.0	135.7	146.0	141.7	129.7
2003	129.7	128.4	131.1	131.5	102.6	133.6	134.2	139.1	137.0	130.2	137.4	131.4	142.3	137.0	124.4
2002	126.7	124.5	128.2	126.5	101.7	122.5	122.2	136.3	134.1	127.2	134.9	128.4	139.8	134.6	121.4
2001	122.2	120.6	123.9	123.3	99.0	117.5	117.4	131.5	129.1	124.4	130.2	125.5	134.7	130.2	117.2
2000	118.9	116.6	120.5	118.7	98.1	115.9	115.8	130.0	127.5	122.8	128.8	124.1	133.1	128.4	115.6
1999	116.6	115.9	117.4	115.5	96.9	115.3	115.2	128.1	125.6	120.8	126.8	121.9	131.2	127.1	115.2
1998	113.6	110.8	113.4	112.7	95.4	112.5	112.4	126.3	123.9	119.0	124.7	119.6	128.5	123.8	113.7
1997	111.5	106.1	110.1	109.3	93.2	110.7	110.6	124.4	121.9	114.6	122.8	115.4	125.7	121.9	111.4
1996	108.9	104.4	107.1	106.5	91.1	109.1	109.0	122.6	120.2	112.9	121.0	113.6	123.9	119.0	109.7
1995	105.6	96.5	103.9	97.8	87.6	107.4	107.4	119.9	117.5	110.8	118.2	111.5	121.6	116.2	107.6
1994	103.0	94.5	100.6	96.1	85.4	106.7	106.6	116.5	114.2	109.5	115.0	110.2	117.8	115.1	105.5
1993	100.0	91.3	96.7	93.8	82.9	104.7	104.5	113.7	111.9	106.9	112.0	107.0	114.9	109.2	102.8
1992	97.9	89.2	93.9	91.7	81.7	103.4	103.2	112.4	110.7	104.2	110.8	103.6	113.7	108.0	101.6
1991	95.7	86.2	91.6	89.3	80.3	102.1	102.0	108.2	106.7	101.8	106.9	100.4	109.0	106.8	98.6
1990	93.2	84.3	88.9	87.3	79.1	98.0	97.1	103.8	101.4	99.0	102.7	96.8	104.6	103.2	95.1
1989	91.0	81.8	86.4	85.0	77.8	95.6	94.7	98.2	96.5	94.9	98.5	92.5	98.5	97.6	92.9
1988	88.5	79.9	84.1	82.4	76.3	93.9	93.0	95.4	93.2	90.6	94.0	88.3	94.8	95.9	90.0
1987	85.7	78.1	81.1	81.1	76.7	90.2	89.3	89.7	90.1	87.2	89.6	85.0	90.4	94.2	87.1
1986	83.7	76.1	78.9	78.6	73.4	91.2	90.1	88.9	87.9	84.8	87.7	82.4	89.3	93.2	85.3
1985	81.8	74.3	77.4	77.0	72.3	90.2	89.1	85.8	84.9	82.4	83.9	79.9	86.5	89.8	83.4
1984	80.6	72.5	76.3	74.7	73.8	89.6	88.1	85.1	84.1	81.6	82.9	80.0	85.2	89.2	82.5
1983	78.2	71.9	74.0	73.0	73.8	84.6	82.8	80.1	79.1	76.4	77.6	75.8	79.9	83.3	77.9
1980	60.7	56.8	58.8	58.1	56.9	64.9	63.3	63.7	61.9	59.2	60.9	59.3	60.9	65.0	61.7
1975	43.7	40.7	43.3	40.7	40.6	42.2	41.6	42.9	41.6	39.7	41.5	39.0	42.2	42.4	39.2
1970	27.8	26.5	29.4	26.5	26.0	28.9	28.6	28.5	27.8	25.6	27.6	26.0	25.6	26.0	23.1
1965	21.5	20.6	21.8	20.4	20.0	22.3	22.0	22.0	21.4	18.7	21.2	20.1	19.4	20.5	17.5
1960	19.5	18.1	19.0	18.6	18.2	20.2	20.0	20.0	19.5	17.0	19.3	18.2	17.6	18.6	15.8
1955	16.3	15.2	15.9	15.6	15.2	17.0	16.8	16.7	16.3	14.3	16.2	15.3	14.8	15.5	13.3
1950	13.5	12.5	13.2	12.9	12.6	14.0	13.9	13.8	13.5	11.8	13.4	12.6	12.2	12.8	10.9
1945	8.6	8.0	8.4	8.2	8.0	9.0	8.8	8.8	8.6	7.5	8.5	8.0	7.8	8.2	7.0
1940	6.6	6.2	6.5	6.3	6.2	6.9	6.8	6.8	6.6	5.8	6.6	6.2	6.0	6.3	5.4

Glossary

Accent lighting–Fixtures or directional beams of light arranged to bring attention to an object or area.

Acoustical material–A material fabricated for the sole purpose of absorbing sound.

Addition–An expansion to an existing structure generally in the form of a room, floor(s) or wing(s). An increase in the floor area or volume of a structure.

Aggregate–Materials such as sand, gravel, stone, vermiculite, perlite and fly ash (slag) which are essential components in the production of concrete, mortar or plaster.

Air conditioning system–An air treatment to control the temperature, humidity and cleanliness of air and to provide for its distribution throughout the structure.

Air curtain–A stream of air directed downward to prevent the loss of hot or cool air from the structure and inhibit the entrance of dust and insects. Generally installed at loading platforms.

Anodized aluminum–Aluminum treated by an electrolytic process to produce an oxide film that is corrosion resistant.

Arcade–A covered passageway between buildings often with shops and offices on one or both sides.

Ashlar–A square-cut building stone or a wall constructed of cut stone.

Asphalt paper–A sheet paper material either coated or saturated with asphalt to increase its strength and resistance to water.

Asphalt shingles–Shingles manufactured from saturated roofing felts, coated with asphalt and topped with mineral granules to prevent weathering.

Assessment ratio–The ratio between the market value and assessed valuation of a property.

Back-up–That portion of a masonry wall behind the exterior facing–usually load bearing.

Balcony–A platform projecting from a building either supported from below or cantilevered. Usually protected by railing.

Bay–Structural component consisting of beams and columns occurring consistently throughout the structure.

Beam–A horizontal structural framing member that transfers loads to vertical members (columns) or bearing walls.

Bid–An offer to perform work described in a contract at a specified price.

Board foot–A unit or measure equal in volume to a board one foot long, one foot wide and one inch thick.

Booster pump–A supplemental pump installed to increase or maintain adequate pressure in the system.

Bowstring roof–A roof supported by trusses fabricated in the shape of a bow and tied together by a straight member.

Brick veneer–A facing of brick laid against, but not bonded to, a wall.

Bridging–A method of bracing joists for stiffness, stability and load distribution.

Broom finish–A method of finishing concrete by lightly dragging a broom over freshly placed concrete.

Built-up roofing–Installed on flat or extremely low-pitched roofs, a roof covering composed of plies or laminations of saturated roofing felts alternated with layers of coal tar pitch or asphalt and surfaced with a layer of gravel or slag in a thick coat of asphalt and finished with a capping sheet.

Caisson–A watertight chamber to expedite work on foundations or structure below water level.

Cantilever–A structural member supported only at one end.

Carport–An automobile shelter having one or more sides open to the weather.

Casement window–A vertical opening window having one side fixed.

Caulking–A resilient compound generally having a silicone or rubber base used to prevent infiltration of water or outside air.

Cavity wall–An exterior wall usually of masonry having an inner and outer wall separated by a continuous air space for thermal insulation.

Cellular concrete–A lightweight concrete consisting of cement mixed with gas-producing materials which, in turn, forms bubbles resulting in good insulating qualities.

Chattel mortgage–A secured interest in a property as collateral for payment of a note.

Clapboard–A wood siding used as exterior covering in frame construction. It is applied horizontally with grain running lengthwise. The thickest section of the board is on the bottom.

Clean room–An environmentally controlled room usually found in medical facilities or precision manufacturing spaces where it is essential to eliminate dust, lint or pathogens.

Close studding–A method of construction whereby studs are spaced close together and the intervening spaces are plastered.

Cluster housing–A closely grouped series of houses resulting in a high density land use.

Cofferdam–A watertight enclosure used for foundation construction in waterfront areas. Water is pumped out of the cofferdam allowing free access to the work area.

Collar joint–Joint between the collar beam and roof rafters.

Column–A vertical structural member supporting horizontal members (beams) along the direction of its longitudinal axis.

Combination door (windows)–Door (windows) having interchangeable screens and glass for seasonal use.

Common area–Spaces either inside or outside the building designated for use by the occupant of the building but not the general public.

Common wall–A wall used jointly by two dwelling units.

Compound wall–A wall constructed of more than one material.

Concrete–A composite material consisting of sand, coarse aggregate (gravel, stone or slag), cement and water that when mixed and allowed to harden forms a hard stone-like material.

Concrete floor hardener–A mixture of chemicals applied to the surface of concrete to produce a dense, wear-resistant bearing surface.

Conduit–A tube or pipe used to protect electric wiring.

Coping–The top cover or capping on a wall.

Craneway–Steel or concrete column and beam supports and rails on which a crane travels.

Curtain wall–A non-bearing exterior wall not supported by beams or girders of the steel frame.

Dampproofing–Coating of a surface to prevent the passage of water or moisture.

Dead load–Total weight of all structural components plus permanently attached fixtures and equipment.

Decibel–Unit of acoustical measurement.

Decorative block–A concrete masonry unit having a special treatment on its face.

Deed restriction–A restriction on the use of a property as set forth in the deed.

Depreciation–A loss in property value caused by physical aging, functional or economic obsolescence.

Direct heating–Heating of spaces by means of exposed heated surfaces (stove, fire, radiators, etc.).

Distributed load–A load that acts evenly over a structural member.

Dock bumper–A resilient material attached to a loading dock to absorb the impact of trucks backing in.

Dome–A curved roof shape spanning an area.

Dormer–A structure projecting from a sloping roof.

Double-hung window–A window having two vertical sliding sashes–one covering the upper section and one covering the lower.

Downspout–A vertical pipe for carrying rain water from the roof to the ground.

Drain tile–A hollow tile used to drain water-soaked soil.

Dressed size–In lumber about 3/8" - 1/2" less than nominal size after planing and sawing.

Drop panel–The depressed surface on the bottom side of a flat concrete slab which surrounds a column.

Dry pipe sprinkler system–A sprinkler system that is activated with water only in case of fire. Used in areas susceptible to freezing or to avoid the hazards of leaking pipes.

Drywall–An interior wall constructed of gypsum board, plywood or wood paneling. No water is required for application.

Duct–Pipe for transmitting warm or cold air. A pipe containing electrical cable or wires.

Dwelling–A structure designed as living quarters for one or more families.

Easement–A right of way or free access over land owned by another.

Eave–The lower edge of a sloping roof that overhangs the side wall of the structure.

Economic rent–That rent on a property sufficient to pay all operating costs exclusive of services and utilities.

Economic life–The term during which a structure is expected to be profitable. Generally shorter than the physical life of the structure.

Economic obsolescence–Loss in value due to unfavorable economic influences occurring from outside the structure itself.

Effective age–The age a structure appears to be based on observed physical condition determined by degree of maintenance and repair.

Elevator–A mechanism for vertical transport of personnel or freight equipped with car or platform.

Ell–A secondary wing or addition to a structure at right angles to the main structure.

Eminent domain–The right of the state to take private property for public use.

Envelope–The shape of a building indicating volume.

Equity–Value of an owner's interest calculated by subtracting outstanding mortgages and expenses from the value of the property.

Escheat–The assumption of ownership by the state, of property whose owner cannot be determined.

Facade–The exterior face of a building sometimes decorated with elaborate detail.

Face brick–Brick manufactured to present an attractive appearance.

Facing block–A concrete masonry unit having a decorative exterior finish.

Feasibility study–A detailed evaluation of a proposed project to determine its financial potential.

Felt paper–Paper sheathing used on walls as an insulator and to prevent infiltration of moisture.

Fenestration–Pertaining to the density and arrangement of windows in a structure.

Fiberboard–A building material composed of wood fiber compressed with a binding agent, produced in sheet form.

Fiberglass–Fine spun filaments of glass processed into various densities to produce thermal and acoustical insulation.

Field house–A long structure used for indoor athletic events.

Finish floor–The top or wearing surface of the floor system.

Fireproofing–The use of fire-resistant materials for the protection of structural members to ensure structural integrity in the event of fire.

Fire stop–A material or member used to seal an opening to prevent the spread of fire.

Flashing–A thin, impervious material such as copper or sheet metal used to prevent air or water penetration.

Float finish–A concrete finish accomplished by using a flat tool with a handle on the back.

Floor drain–An opening installed in a floor for removing excess water into a plumbing system.

Floor load–The live load the floor system has been designed for and which may be applied safely.

Flue–A heat-resistant enclosed passage in a chimney to remove gaseous products of combustion from a fireplace or boiler.

Footing–The lowest portion of the foundation wall that transmits the load directly to the soil.

Foundation–Below grade wall system that supports the structure.

Foyer–A portion of the structure that serves as a transitional space between the interior and exterior.

Frontage–That portion of a lot that is placed facing a street, public way or body of water.

Frost action–The effects of freezing and thawing on materials and the resultant structural damage.

Functional obsolescence–An inadequacy caused by outmoded design, dated construction materials or over or undersized areas, all of which cause excessive operating costs.

Furring strips–Wood or metal channels used for attaching gypsum or metal lath to masonry walls as a finish or for leveling purposes.

Gambrel roof (mansard)–A roof system having two pitches on each side.

Garden apartment–Two or three story apartment building with common outside areas.

General contractor–The prime contractor responsible for work on the construction site.

Girder–A principal horizontal supporting member.

Girt–A horizontal framing member for adding rigid support to columns which also acts as a support for sheathing or siding.

Grade beam–That portion of the foundation system that directly supports the exterior walls of the building.

Gross area–The total enclosed floor area of a building.

Ground area–The area computed by the exterior dimensions of the structure.

Ground floor–The floor of a building in closest proximity to the ground.

Grout–A mortar containing a high water content used to fill joints and cavities in masonry work.

Gutter–A wood or sheet metal conduit set along the building eaves used to channel rainwater to leaders or downspouts.

Gunite–A concrete mix placed pneumatically.

Gypsum–Hydrated calcium sulphate used as both a retarder in Portland cement and as a major ingredient in plaster of Paris. Used in sheets as a substitute for plaster.

Hall–A passageway providing access to various parts of a building–a large room for entertainment or assembly.

Hangar–A structure for the storage or repair of aircraft.

Head room–The distance between the top of the finished floor to the bottom of the finished ceiling.

Hearth–The floor of a fireplace and the adjacent area of fireproof material.

Heat pump–A mechanical device for providing either heating or air conditioning.

Hip roof–A roof whose four sides meet at a common point with no gabled ends.

I-beam–A structural member having a cross-section resembling the letter "I".

Insulation–Material used to reduce the effects of heat, cold or sound.

Jack rafter–An unusually short rafter generally found in hip roofs.

Jalousie–Adjustable glass louvers which pivot simultaneously in a common frame.

Jamb–The vertical member on either side of a door frame or window frame.

Joist–Parallel beams of timber, concrete or steel used to support floor and ceiling.

Junction box–A box that protects splices or joints in electrical wiring.

Kalamein door–Solid core wood doors clad with galvanized sheet metal.

Kip–A unit of weight equal to 1,000 pounds.

Lally column–A concrete-filled steel pipe used as a vertical support.

Laminated beam–A beam built up by gluing together several pieces of timber.

Landing–A platform between flights of stairs.

Lath–Strips of wood or metal used as a base for plaster.

Lead-lined door, sheetrock–Doors or sheetrock internally lined with sheet lead to provide protection from X-ray radiation.

Lean-to–A small shed or building addition with a single pitched roof attached to the exterior wall of the main building.

Lintel–A horizontal framing member used to carry a load over a wall opening.

Live load–The moving or movable load on a structure composed of furnishings, equipment or personnel weight.

Load bearing partition–A partition that supports a load in addition to its own weight.

Loading dock leveler–An adjustable platform for handling off-on loading of trucks.

Loft building–A commercial/industrial type building containing large, open unpartitioned floor areas.

Louver window–A window composed of a series of sloping, overlapping blades or slats that may be adjusted to admit varying degrees of air or light.

Main beam–The principal load bearing beam used to transmit loads directly to columns.

Mansard roof–A roof having a double pitch on all four sides, the lower level having the steeper pitch.

Mansion–An extremely large and imposing residence.

Masonry–The utilization of brick, stone, concrete or block for walls and other building components.

Mastic–A sealant or adhesive compound generally used as either a binding agent or floor finish.

Membrane fireproofing–A coating of metal lath and plaster to provide resistance to fire and heat.

Mesh reinforcement–An arrangement of wire tied or welded at their intersection to provide strength and resistance to cracking.

Metal lath–A diamond-shaped metallic base for plaster.

Mezzanine–A low story situated between two main floors.

Mill construction–A heavy timber construction that achieves fire resistance by using large wood structural members, noncombustible bearing and non-bearing walls and omitting concealed spaces under floors and roof.

Mixed occupancy–Two or more classes of occupancy in a single structure.

Molded brick–A specially shaped brick used for decorative purposes.

Monolithic concrete–Concrete poured in a continuous process so there are no joints.

Movable partition–A non-load bearing demountable partition that can be relocated and can be either ceiling height or partial height.

Moving walk–A continually moving horizontal passenger carrying device.

Multi-zone system–An air conditioning system that is capable of handling several individual zones simultaneously.

Net floor area–The usable, occupied area of a structure excluding stairwells, elevator shafts and wall thicknesses.

Non-combustible construction–Construction in which the walls, partitions and the structural members are of non-combustible materials.

Nurses call system–An electrically operated system for use by patients or personnel for summoning a nurse.

Occupancy rate–The number of persons per room, per dwelling unit, etc.

One-way joist connection–A framing system for floors and roofs in a concrete building consisting of a series of parallel joists supported by girders between columns.

Open web joist–A lightweight prefabricated metal chord truss.

Parapet–A low guarding wall at the point of a drop. That portion of the exterior wall that extends above the roof line.

Parging–A thin coat of mortar applied to a masonry wall used primarily for waterproofing purposes.

Parquet–Inlaid wood flooring set in a simple design.

Peaked roof–A roof of two or more slopes that rises to a peak.

Penthouse–A structure occupying usually half the roof area and used to house elevator, HVAC equipment, or other mechanical or electrical systems.

Pier–A column designed to support a concentrated load.

Pilaster–A column usually formed of the same material and integral with, but projecting from, a wall.

Pile–A concrete, wood or steel column usually less than 2′ in diameter which is driven or otherwise introduced into the soil to carry a vertical load or provide vertical support.

Pitched roof–A roof having one or more surfaces with a pitch greater than 10°.

Plenum–The space between the suspended ceiling and the floor above.

Plywood–Structural wood made of three or more layers of veneer and bonded together with glue.

Porch–A structure attached to a building to provide shelter for an entranceway or to serve as a semi-enclosed space.

Post and beam framing–A structural framing system where beams rest on posts rather than bearing walls.

Post-tensioned concrete–Concrete that has the reinforcing tendons tensioned after the concrete has set.

Pre-cast concrete–Concrete structural components fabricated at a location other than in-place.

Pre-stressed concrete–Concrete that has the reinforcing tendons tensioned prior to the concrete setting.

Purlins–A horizontal structural member supporting the roof deck and resting on the trusses, girders, beams or rafters.

Rafters–Structural members supporting the roof deck and covering.

Resilient flooring–A manufactured interior floor covering material in either sheet or tile form, that is resilient.

Ridge–The horizontal line at the junction of two sloping roof edges.

Rigid frame–A structural framing system in which all columns and beams are rigidly connected; there are no hinged joints.

Roof drain–A drain designed to accept rainwater and discharge it into a leader or downspout.

Rough floor–A layer of boards or plywood nailed to the floor joists that serves as a base for the finished floor.

Sanitary sewer–A sewer line designed to carry only liquid or waterborne waste from the structure to a central treatment plant.

Sawtooth roof–Roof shape found primarily in industrial roofs that creates the appearance of the teeth of a saw.

Semi-rigid frame–A structural system wherein the columns and beams are so attached that there is some flexibility at the joints.

Septic tank–A covered tank in which waste matter is decomposed by natural bacterial action.

Service elevator–A combination passenger and freight elevator.

Sheathing–The first covering of exterior studs by boards, plywood or particle board.

Shoring–Temporary bracing for structural components during construction.

Span–The horizontal distance between supports.

Specification–A detailed list of materials and requirements for construction of a building.

Storm sewer–A system of pipes used to carry rainwater or surface waters.

Stucco–A cement plaster used to cover exterior wall surfaces usually applied over a wood or metal lath base.

Stud–A vertical wooden structural component.

Subfloor–A system of boards, plywood or particle board laid over the floor joists to form a base.

Superstructure–That portion of the structure above the foundation or ground level.

Terrazzo–A durable floor finish made of small chips of colored stone or marble, embedded in concrete, then polished to a high sheen.

Tile–A thin piece of fired clay, stone or concrete used for floor, roof or wall finishes.

Unit cost–The cost per unit of measurement.

Vault–A room especially designed for storage.

Veneer–A thin surface layer covering a base of common material.

Vent–An opening serving as an outlet for air.

Wainscot–The lower portion of an interior wall whose surface differs from that of the upper wall.

Wall bearing construction–A structural system where the weight of the floors and roof are carried directly by the masonry walls rather than the structural framing system.

Waterproofing–Any of a number of materials applied to various surfaces to prevent the infiltration of water.

Weatherstrip–A thin strip of metal, wood or felt used to cover the joint between the door or window sash and the jamb, casing or sill to keep out air, dust, rain, etc.

Wing–A building section or addition projecting from the main structure.

X-ray protection–Lead encased in sheetrock or plaster to prevent the escape of radiation.

Abbreviation	Meaning
A	Area Square Feet; Ampere
ABS	Acrylonitrile Butadiene Stryrene; Asbestos Bonded Steel
A.C.	Alternating Current; Air-Conditioning; Asbestos Cement; Plywood Grade A & C
A.C.I.	American Concrete Institute
AD	Plywood, Grade A & D
Addit.	Additional
Adj.	Adjustable
af	Audio-frequency
A.G.A.	American Gas Association
Agg.	Aggregate
A.H.	Ampere Hours
A hr.	Ampere-hour
A.H.U.	Air Handling Unit
A.I.A.	American Institute of Architects
AIC	Ampere Interrupting Capacity
Allow.	Allowance
alt.	Altitude
Alum.	Aluminum
a.m.	Ante Meridiem
Amp.	Ampere
Anod.	Anodized
Approx.	Approximate
Apt.	Apartment
Asb.	Asbestos
A.S.B.C.	American Standard Building Code
Asbe.	Asbestos Worker
A.S.H.R.A.E.	American Society of Heating, Refrig. & AC Engineers
A.S.M.E.	American Society of Mechanical Engineers
A.S.T.M.	American Society for Testing and Materials
Attchmt.	Attachment
Avg.	Average
A.W.G.	American Wire Gauge
AWWA	American Water Works Assoc.
Bbl.	Barrel
B&B	Grade B and Better; Balled & Burlapped
B.&S.	Bell and Spigot
B.&W.	Black and White
b.c.c.	Body-centered Cubic
B.C.Y.	Bank Cubic Yards
BE	Bevel End
B.F.	Board Feet
Bg. cem.	Bag of Cement
BHP	Boiler Horsepower; Brake Horsepower
B.I.	Black Iron
Bit.; Bitum.	Bituminous
Bk.	Backed
Bkrs.	Breakers
Bldg.	Building
Blk.	Block
Bm.	Beam
Boil.	Boilermaker
B.P.M.	Blows per Minute
BR	Bedroom
Brg.	Bearing
Brhe.	Bricklayer Helper
Bric.	Bricklayer
Brk.	Brick
Brng.	Bearing
Brs.	Brass
Brz.	Bronze
Bsn.	Basin
Btr.	Better
BTU	British Thermal Unit
BTUH	BTU per Hour
B.U.R.	Built-up Roofing
BX	Interlocked Armored Cable
c	Conductivity, Copper Sweat
C	Hundred; Centigrade
C/C	Center to Center, Cedar on Cedar
Cab.	Cabinet
Cair.	Air Tool Laborer
Calc	Calculated
Cap.	Capacity
Carp.	Carpenter
C.B.	Circuit Breaker
C.C.A.	Chromate Copper Arsenate
C.C.F.	Hundred Cubic Feet
cd	Candela
cd/sf	Candela per Square Foot
CD	Grade of Plywood Face & Back
CDX	Plywood, Grade C & D, exterior glue
Cefi.	Cement Finisher
Cem.	Cement
CF	Hundred Feet
C.F.	Cubic Feet
CFM	Cubic Feet per Minute
c.g.	Center of Gravity
CHW	Chilled Water; Commercial Hot Water
C.I.	Cast Iron
C.I.P.	Cast in Place
Circ.	Circuit
C.L.	Carload Lot
Clab.	Common Laborer
Clam	Common maintenance laborer
C.L.F.	Hundred Linear Feet
CLF	Current Limiting Fuse
CLP	Cross Linked Polyethylene
cm	Centimeter
CMP	Corr. Metal Pipe
C.M.U.	Concrete Masonry Unit
CN	Change Notice
Col.	Column
CO_2	Carbon Dioxide
Comb.	Combination
Compr.	Compressor
Conc.	Concrete
Cont.	Continuous; Continued
Corr.	Corrugated
Cos	Cosine
Cot	Cotangent
Cov.	Cover
C/P	Cedar on Paneling
CPA	Control Point Adjustment
Cplg.	Coupling
C.P.M.	Critical Path Method
CPVC	Chlorinated Polyvinyl Chloride
C.Pr.	Hundred Pair
CRC	Cold Rolled Channel
Creos.	Creosote
Crpt.	Carpet & Linoleum Layer
CRT	Cathode-ray Tube
CS	Carbon Steel, Constant Shear Bar Joist
Csc	Cosecant
C.S.F.	Hundred Square Feet
CSI	Construction Specifications Institute
C.T.	Current Transformer
CTS	Copper Tube Size
Cu	Copper, Cubic
Cu. Ft.	Cubic Foot
cw	Continuous Wave
C.W.	Cool White; Cold Water
Cwt.	100 Pounds
C.W.X.	Cool White Deluxe
C.Y.	Cubic Yard (27 cubic feet)
C.Y./Hr.	Cubic Yard per Hour
Cyl.	Cylinder
d	Penny (nail size)
D	Deep; Depth; Discharge
Dis.;Disch.	Discharge
Db.	Decibel
Dbl.	Double
DC	Direct Current
DDC	Direct Digital Control
Demob.	Demobilization
d.f.u.	Drainage Fixture Units
D.H.	Double Hung
DHW	Domestic Hot Water
Diag.	Diagonal
Diam.	Diameter
Distrib.	Distribution
Dk.	Deck
D.L.	Dead Load; Diesel
DLH	Deep Long Span Bar Joist
Do.	Ditto
Dp.	Depth
D.P.S.T.	Double Pole, Single Throw
Dr.	Driver
Drink.	Drinking
D.S.	Double Strength
D.S.A.	Double Strength A Grade
D.S.B.	Double Strength B Grade
Dty.	Duty
DWV	Drain Waste Vent
DX	Deluxe White, Direct Expansion
dyn	Dyne
e	Eccentricity
E	Equipment Only; East
Ea.	Each
E.B.	Encased Burial
Econ.	Economy
E.C.Y	Embankment Cubic Yards
EDP	Electronic Data Processing
EIFS	Exterior Insulation Finish System
E.D.R.	Equiv. Direct Radiation
Eq.	Equation
Elec.	Electrician; Electrical
Elev.	Elevator; Elevating
EMT	Electrical Metallic Conduit; Thin Wall Conduit
Eng.	Engine, Engineered
EPDM	Ethylene Propylene Diene Monomer
EPS	Expanded Polystyrene
Eqhv.	Equip. Oper., Heavy
Eqlt.	Equip. Oper., Light
Eqmd.	Equip. Oper., Medium
Eqmm.	Equip. Oper., Master Mechanic
Eqol.	Equip. Oper., Oilers
Equip.	Equipment
ERW	Electric Resistance Welded
E.S.	Energy Saver
Est.	Estimated
esu	Electrostatic Units
E.W.	Each Way
EWT	Entering Water Temperature
Excav.	Excavation
Exp.	Expansion, Exposure
Ext.	Exterior
Extru.	Extrusion
f.	Fiber stress
F	Fahrenheit; Female; Fill
Fab.	Fabricated
FBGS	Fiberglass
F.C.	Footcandles
f.c.c.	Face-centered Cubic
f'c.	Compressive Stress in Concrete; Extreme Compressive Stress
F.E.	Front End
FEP	Fluorinated Ethylene Propylene (Teflon)
F.G.	Flat Grain
F.H.A.	Federal Housing Administration
Fig.	Figure
Fin.	Finished
Fixt.	Fixture
Fl. Oz.	Fluid Ounces
Flr.	Floor
F.M.	Frequency Modulation; Factory Mutual
Fmg.	Framing
Fndtn.	Foundation

Fori.	Foreman, Inside	I.W.	Indirect Waste	M.C.F.	Thousand Cubic Feet
Foro.	Foreman, Outside	J	Joule	M.C.F.M.	Thousand Cubic Feet per Minute
Fount.	Fountain	J.I.C.	Joint Industrial Council	M.C.M.	Thousand Circular Mils
FPM	Feet per Minute	K	Thousand; Thousand Pounds;	M.C.P.	Motor Circuit Protector
FPT	Female Pipe Thread		Heavy Wall Copper Tubing, Kelvin	MD	Medium Duty
Fr.	Frame	K.A.H.	Thousand Amp. Hours	M.D.O.	Medium Density Overlaid
F.R.	Fire Rating	KCMIL	Thousand Circular Mils	Med.	Medium
FRK	Foil Reinforced Kraft	KD	Knock Down	MF	Thousand Feet
FRP	Fiberglass Reinforced Plastic	K.D.A.T.	Kiln Dried After Treatment	M.F.B.M.	Thousand Feet Board Measure
FS	Forged Steel	kg	Kilogram	Mfg.	Manufacturing
FSC	Cast Body; Cast Switch Box	kG	Kilogauss	Mfrs.	Manufacturers
Ft.	Foot; Feet	kgf	Kilogram Force	mg	Milligram
Ftng.	Fitting	kHz	Kilohertz	MGD	Million Gallons per Day
Ftg.	Footing	Kip.	1000 Pounds	MGPH	Thousand Gallons per Hour
Ft. Lb.	Foot Pound	KJ	Kiljoule	MH, M.H.	Manhole; Metal Halide; Man-Hour
Furn.	Furniture	K.L.	Effective Length Factor	MHz	Megahertz
FVNR	Full Voltage Non-Reversing	K.L.F.	Kips per Linear Foot	Mi.	Mile
FXM	Female by Male	Km	Kilometer	MI	Malleable Iron; Mineral Insulated
Fy.	Minimum Yield Stress of Steel	K.S.F.	Kips per Square Foot	mm	Millimeter
g	Gram	K.S.I.	Kips per Square Inch	Mill.	Millwright
G	Gauss	kV	Kilovolt	Min., min.	Minimum, minute
Ga.	Gauge	kVA	Kilovolt Ampere	Misc.	Miscellaneous
Gal.	Gallon	K.V.A.R.	Kilovar (Reactance)	ml	Milliliter, Mainline
Gal./Min.	Gallon per Minute	KW	Kilowatt	M.L.F.	Thousand Linear Feet
Galv.	Galvanized	KWh	Kilowatt-hour	Mo.	Month
Gen.	General	L	Labor Only; Length; Long;	Mobil.	Mobilization
G.F.I.	Ground Fault Interrupter		Medium Wall Copper Tubing	Mog.	Mogul Base
Glaz.	Glazier	Lab.	Labor	MPH	Miles per Hour
GPD	Gallons per Day	lat	Latitude	MPT	Male Pipe Thread
GPH	Gallons per Hour	Lath.	Lather	MRT	Mile Round Trip
GPM	Gallons per Minute	Lav.	Lavatory	ms	Millisecond
GR	Grade	lb.; #	Pound	M.S.F.	Thousand Square Feet
Gran.	Granular	L.B.	Load Bearing; L Conduit Body	Mstz.	Mosaic & Terrazzo Worker
Grnd.	Ground	L. & E.	Labor & Equipment	M.S.Y.	Thousand Square Yards
H	High; High Strength Bar Joist;	lb./hr.	Pounds per Hour	Mtd.	Mounted
	Henry	lb./L.F.	Pounds per Linear Foot	Mthe.	Mosaic & Terrazzo Helper
H.C.	High Capacity	lbf/sq.in.	Pound-force per Square Inch	Mtng.	Mounting
H.D.	Heavy Duty; High Density	L.C.L.	Less than Carload Lot	Mult.	Multi; Multiply
H.D.O.	High Density Overlaid	L.C.Y.	Loose Cubic Yard	M.V.A.	Million Volt Amperes
Hdr.	Header	Ld.	Load	M.V.A.R.	Million Volt Amperes Reactance
Hdwe.	Hardware	LE	Lead Equivalent	MV	Megavolt
Help.	Helper Average	LED	Light Emitting Diode	MW	Megawatt
HEPA	High Efficiency Particulate Air	L.F.	Linear Foot	MXM	Male by Male
	Filter	Lg.	Long; Length; Large	MYD	Thousand Yards
Hg	Mercury	L & H	Light and Heat	N	Natural; North
HIC	High Interrupting Capacity	LH	Long Span Bar Joist	nA	Nanoampere
HM	Hollow Metal	L.H.	Labor Hours	NA	Not Available; Not Applicable
H.O.	High Output	L.L.	Live Load	N.B.C.	National Building Code
Horiz.	Horizontal	L.L.D.	Lamp Lumen Depreciation	NC	Normally Closed
H.P.	Horsepower; High Pressure	lm	Lumen	N.E.M.A.	National Electrical Manufacturers
H.P.F.	High Power Factor	lm/sf	Lumen per Square Foot		Assoc.
Hr.	Hour	lm/W	Lumen per Watt	NEHB	Bolted Circuit Breaker to 600V.
Hrs./Day	Hours per Day	L.O.A.	Length Over All	N.L.B.	Non-Load-Bearing
HSC	High Short Circuit	log	Logarithm	NM	Non-Metallic Cable
Ht.	Height	L-O-L	Lateralolet	nm	Nanometer
Htg.	Heating	L.P.	Liquefied Petroleum; Low Pressure	No.	Number
Htrs.	Heaters	L.P.F.	Low Power Factor	NO	Normally Open
HVAC	Heating, Ventilation & Air-	LR	Long Radius	N.O.C.	Not Otherwise Classified
	Conditioning	L.S.	Lump Sum	Nose.	Nosing
Hvy.	Heavy	Lt.	Light	N.P.T.	National Pipe Thread
HW	Hot Water	Lt. Ga.	Light Gauge	NQOD	Combination Plug-on/Bolt on
Hyd.;Hydr.	Hydraulic	L.T.L.	Less than Truckload Lot		Circuit Breaker to 240V.
Hz.	Hertz (cycles)	Lt. Wt.	Lightweight	N.R.C.	Noise Reduction Coefficient
I.	Moment of Inertia	L.V.	Low Voltage	N.R.S.	Non Rising Stem
I.C.	Interrupting Capacity	M	Thousand; Material; Male;	ns	Nanosecond
ID	Inside Diameter		Light Wall Copper Tubing	nW	Nanowatt
I.D.	Inside Dimension; Identification	M²CA	Meters Squared Contact Area	OB	Opposing Blade
I.F.	Inside Frosted	m/hr; M.H.	Man-hour	OC	On Center
I.M.C.	Intermediate Metal Conduit	mA	Milliampere	OD	Outside Diameter
In.	Inch	Mach.	Machine	O.D.	Outside Dimension
Incan.	Incandescent	Mag. Str.	Magnetic Starter	ODS	Overhead Distribution System
Incl.	Included; Including	Maint.	Maintenance	O.G.	Ogee
Int.	Interior	Marb.	Marble Setter	O.H.	Overhead
Inst.	Installation	Mat; Mat'l.	Material	O&P	Overhead and Profit
Insul.	Insulation/Insulated	Max.	Maximum	Oper.	Operator
I.P.	Iron Pipe	MBF	Thousand Board Feet	Opng.	Opening
I.P.S.	Iron Pipe Size	MBH	Thousand BTU's per hr.	Orna.	Ornamental
I.P.T.	Iron Pipe Threaded	MC	Metal Clad Cable	OSB	Oriented Strand Board

O.S.&Y.	Outside Screw and Yoke	Rsr	Riser	Tilf.	Tile Layer, Floor
Ovhd.	Overhead	RT	Round Trip	Tilh.	Tile Layer, Helper
OWG	Oil, Water or Gas	S.	Suction; Single Entrance; South	THHN	Nylon Jacketed Wire
Oz.	Ounce	SCFM	Standard Cubic Feet per Minute	THW.	Insulated Strand Wire
P.	Pole; Applied Load; Projection	Scaf.	Scaffold	THWN;	Nylon Jacketed Wire
p.	Page	Sch.; Sched.	Schedule	T.L.	Truckload
Pape.	Paperhanger	S.C.R.	Modular Brick	T.M.	Track Mounted
P.A.P.R.	Powered Air Purifying Respirator	S.D.	Sound Deadening	Tot.	Total
PAR	Parabolic Reflector	S.D.R.	Standard Dimension Ratio	T-O-L	Threadolet
Pc., Pcs.	Piece, Pieces	S.E.	Surfaced Edge	T.S.	Trigger Start
P.C.	Portland Cement; Power Connector	Sel.	Select	Tr.	Trade
P.C.F.	Pounds per Cubic Foot	S.E.R.; S.E.U.	Service Entrance Cable	Transf.	Transformer
P.C.M.	Phase Contract Microscopy	S.F.	Square Foot	Trhv.	Truck Driver, Heavy
P.E.	Professional Engineer;	S.F.C.A.	Square Foot Contact Area	Trlr	Trailer
	Porcelain Enamel;	S.F. Flr.	Square Foot of Floor	Trlt.	Truck Driver, Light
	Polyethylene; Plain End	S.F.G.	Square Foot of Ground	TV	Television
Perf.	Perforated	S.F. Hor.	Square Foot Horizontal	T.W.	Thermoplastic Water Resistant
Ph.	Phase	S.F.R.	Square Feet of Radiation		Wire
P.I.	Pressure Injected	S.F. Shlf.	Square Foot of Shelf	UCI	Uniform Construction Index
Pile.	Pile Driver	S4S	Surface 4 Sides	UF	Underground Feeder
Pkg.	Package	Shee.	Sheet Metal Worker	UGND	Underground Feeder
Pl.	Plate	Sin.	Sine	U.H.F.	Ultra High Frequency
Plah.	Plasterer Helper	Skwk.	Skilled Worker	U.L.	Underwriters Laboratory
Plas.	Plasterer	SL	Saran Lined	Unfin.	Unfinished
Pluh.	Plumbers Helper	S.L.	Slimline	URD	Underground Residential
Plum.	Plumber	Sldr.	Solder		Distribution
Ply.	Plywood	SLH	Super Long Span Bar Joist	US	United States
p.m.	Post Meridiem	S.N.	Solid Neutral	USP	United States Primed
Pntd.	Painted	S-O-L	Socketolet	UTP	Unshielded Twisted Pair
Pord.	Painter, Ordinary	sp	Standpipe	V	Volt
pp	Pages	S.P.	Static Pressure; Single Pole; Self-	V.A.	Volt Amperes
PP; PPL	Polypropylene		Propelled	V.C.T.	Vinyl Composition Tile
P.P.M.	Parts per Million	Spri.	Sprinkler Installer	VAV	Variable Air Volume
Pr.	Pair	spwg	Static Pressure Water Gauge	VC	Veneer Core
P.E.S.B.	Pre-engineered Steel Building	S.P.D.T.	Single Pole, Double Throw	Vent.	Ventilation
Prefab.	Prefabricated	SPF	Spruce Pine Fir	Vert.	Vertical
Prefin.	Prefinished	S.P.S.T.	Single Pole, Single Throw	V.F.	Vinyl Faced
Prop.	Propelled	SPT	Standard Pipe Thread	V.G.	Vertical Grain
PSF; psf	Pounds per Square Foot	Sq.	Square; 100 Square Feet	V.H.F.	Very High Frequency
PSI; psi	Pounds per Square Inch	Sq. Hd.	Square Head	VHO	Very High Output
PSIG	Pounds per Square Inch Gauge	Sq. In.	Square Inch	Vib.	Vibrating
PSP	Plastic Sewer Pipe	S.S.	Single Strength; Stainless Steel	V.L.F.	Vertical Linear Foot
Pspr.	Painter, Spray	S.S.B.	Single Strength B Grade	Vol.	Volume
Psst.	Painter, Structural Steel	sst	Stainless Steel	VRP	Vinyl Reinforced Polyester
P.T.	Potential Transformer	Sswk.	Structural Steel Worker	W	Wire; Watt; Wide; West
P. & T.	Pressure & Temperature	Sswl.	Structural Steel Welder	w/	With
Ptd.	Painted	St.; Stl.	Steel	W.C.	Water Column; Water Closet
Ptns.	Partitions	S.T.C.	Sound Transmission Coefficient	W.F.	Wide Flange
Pu	Ultimate Load	Std.	Standard	W.G.	Water Gauge
PVC	Polyvinyl Chloride	STK	Select Tight Knot	Wldg.	Welding
Pvmt.	Pavement	STP	Standard Temperature & Pressure	W. Mile	Wire Mile
Pwr.	Power	Stpi.	Steamfitter, Pipefitter	W-O-L	Weldolet
Q	Quantity Heat Flow	Str.	Strength; Starter; Straight	W.R.	Water Resistant
Quan.; Qty.	Quantity	Strd.	Stranded	Wrck.	Wrecker
Q.C.	Quick Coupling	Struct.	Structural	W.S.P.	Water, Steam, Petroleum
r	Radius of Gyration	Sty.	Story	WT., Wt.	Weight
R	Resistance	Subj.	Subject	WWF	Welded Wire Fabric
R.C.P.	Reinforced Concrete Pipe	Subs.	Subcontractors	XFER	Transfer
Rect.	Rectangle	Surf.	Surface	XFMR	Transformer
Reg.	Regular	Sw.	Switch	XHD	Extra Heavy Duty
Reinf.	Reinforced	Swbd.	Switchboard	XHHW; XLPE	Cross-Linked Polyethylene Wire
Req'd.	Required	S.Y.	Square Yard		Insulation
Res.	Resistant	Syn.	Synthetic	XLP	Cross-linked Polyethylene
Resi.	Residential	S.Y.P.	Southern Yellow Pine	Y	Wye
Rgh.	Rough	Sys.	System	yd	Yard
RGS	Rigid Galvanized Steel	t.	Thickness	yr	Year
R.H.W.	Rubber, Heat & Water Resistant;	T	Temperature; Ton	Δ	Delta
	Residential Hot Water	Tan	Tangent	%	Percent
rms	Root Mean Square	T.C.	Terra Cotta	~	Approximately
Rnd.	Round	T & C	Threaded and Coupled	Ø	Phase
Rodm.	Rodman	T.D.	Temperature Difference	@	At
Rofc.	Roofer, Composition	T.E.M.	Transmission Electron Microscopy	#	Pound; Number
Rofp.	Roofer, Precast	TFE	Tetrafluoroethylene (Teflon)	<	Less Than
Rohe.	Roofer Helpers (Composition)	T. & G.	Tongue & Groove;	>	Greater Than
Rots.	Roofer, Tile & Slate		Tar & Gravel		
R.O.W.	Right of Way	Th.; Thk.	Thick		
RPM	Revolutions per Minute	Thn.	Thin		
R.S.	Rapid Start	Thrded	Threaded		

RESIDENTIAL
COST ESTIMATE

OWNER'S NAME: _____   APPRAISER: _____

RESIDENCE ADDRESS: _____   PROJECT: _____

CITY, STATE, ZIP CODE: _____   DATE: _____

CLASS OF CONSTRUCTION	RESIDENCE TYPE	CONFIGURATION	EXTERIOR WALL SYSTEM
☐ ECONOMY	☐ 1 STORY	☐ DETACHED	☐ WOOD SIDING—WOOD FRAME
☐ AVERAGE	☐ 1-1/2 STORY	☐ TOWN/ROW HOUSE	☐ BRICK VENEER—WOOD FRAME
☐ CUSTOM	☐ 2 STORY	☐ SEMI-DETACHED	☐ STUCCO ON WOOD FRAME
☐ LUXURY	☐ 2-1/2 STORY		☐ PAINTED CONCRETE BLOCK
	☐ 3 STORY	OCCUPANCY	☐ SOLID MASONRY (AVERAGE & CUSTOM)
	☐ BI-LEVEL	☐ ONE STORY	☐ STONE VENEER—WOOD FRAME
	☐ TRI-LEVEL	☐ TWO FAMILY	☐ SOLID BRICK (LUXURY)
		☐ THREE FAMILY	☐ SOLID STONE (LUXURY)
		☐ OTHER _____	

*LIVING AREA (Main Building)

First Level _____ S.F.
Second Level _____ S.F.
Third Level _____ S.F.
Total _____ S.F.

*LIVING AREA (Wing or Ell) (   )

First Level _____ S.F.
Second Level _____ S.F.
Third Level _____ S.F.
Total _____ S.F.

*LIVING AREA (Wing or Ell) (   )

First Level _____ S.F.
Second Level _____ S.F.
Third Level _____ S.F.
Total _____ S.F.

*Basement Area is not part of living area.

MAIN BUILDING	COSTS PER S.F. LIVING AREA
Cost per Square Foot of Living Area, from Page _____	$
Basement Addition: _____ % Finished, _____ % Unfinished	+
Roof Cover Adjustment: _____ Type, Page _____ (Add or Deduct)	( )
Central Air Conditioning: ☐ Separate Ducts ☐ Heating Ducts, Page _____	+
Heating System Adjustment: _____ Type, Page _____ (Add or Deduct)	( )
Main Building: Adjusted Cost per S.F. of Living Area	$

MAIN BUILDING TOTAL COST $ _____ /S.F. × _____ S.F. × _____ = $ _____
Cost per S.F. Living Area   Living Area   Town/Row House Multiplier (Use 1 for Detached)   TOTAL COST

WING OR ELL ( ) _____ STORY	COSTS PER S.F. LIVING AREA
Cost per Square Foot of Living Area, from Page _____	$
Basement Addition: _____ % Finished, _____ % Unfinished	+
Roof Cover Adjustment: _____ Type, Page _____ (Add or Deduct)	( )
Central Air Conditioning: ☐ Separate Ducts ☐ Heating Ducts, Page _____	+
Heating System Adjustment: _____ Type, Page _____ (Add or Deduct)	( )
Wing or Ell ( ): Adjusted Cost per S.F. of Living Area	$

WING OR ELL (   ) TOTAL COST $ _____ /S.F. × _____ S.F. × = $ _____
Cost per S.F. Living Area   Living Area   TOTAL COST

WING OR ELL ( ) _____ STORY	COSTS PER S.F. LIVING AREA
Cost per Square Foot of Living Area, from Page _____	$
Basement Addition: _____ % Finished, _____ % Unfinished	+
Roof Cover Adjustment: _____ Type, Page _____ (Add or Deduct)	( )
Central Air Conditioning: ☐ Separate Ducts ☐ Heating Ducts, Page _____	+
Heating System Adjustment: _____ Type, Page _____ (Add or Deduct)	( )
Wing or Ell ( ): Adjusted Cost per S.F. of Living Area	$

WING OR ELL (   ) TOTAL COST $ _____ /S.F. × _____ S.F. × = $ _____
Cost per S.F. Living Area   Living Area   TOTAL COST

TOTAL THIS PAGE   [_____]

Page 1 of 2

**RESIDENTIAL
COST ESTIMATE**

		QUANTITY	UNIT COST	
Total Page 1				$
Additional Bathrooms: _____ Full, _____ Half				
Finished Attic: _____ Ft. x _____ Ft.		S.F.		+
Breezeway: ☐ Open ☐ Enclosed _____ Ft. x _____ Ft.		S.F.		+
Covered Porch: ☐ Open ☐ Enclosed _____ Ft. x _____ Ft.		S.F.		+
Fireplace: ☐ Interior Chimney ☐ Exterior Chimney ☐ No. of Flues ☐ Additional Fireplaces				+
Appliances:				+
Kitchen Cabinets Adjustment: (+/–)				
☐ Garage ☐ Carport: _____ Car(s) Description _____ (+/–)				
Miscellaneous:				+

ADJUSTED TOTAL BUILDING COST $ _____

REPLACEMENT COST	
ADJUSTED TOTAL BUILDING COST	$ _____
Site Improvements	
(A) Paving & Sidewalks	$ _____
(B) Landscaping	$ _____
(C) Fences	$ _____
(D) Swimming Pool	$ _____
(E) Miscellaneous	$ _____
TOTAL	$ _____
Location Factor	x _____
Location Replacement Cost	$ _____
Depreciation	–$ _____
LOCAL DEPRECIATED COST	$ _____

INSURANCE COST	
ADJUSTED TOTAL BUILDING COST	$ _____
Insurance Exclusions	
(A) Footings, Site Work, Underground Piping	–$ _____
(B) Architects' Fees	–$ _____
Total Building Cost Less Exclusion	$ _____
Location Factor	x _____
LOCAL INSURABLE REPLACEMENT COST	$ _____

SKETCH AND ADDITIONAL CALCULATIONS

CII APPRAISAL

5. DATE: _____

1. SUBJECT PROPERTY: _____
6. APPRAISER: _____

2. BUILDING: _____

3. ADDRESS: _____

4. BUILDING USE: _____
7. YEAR BUILT: _____

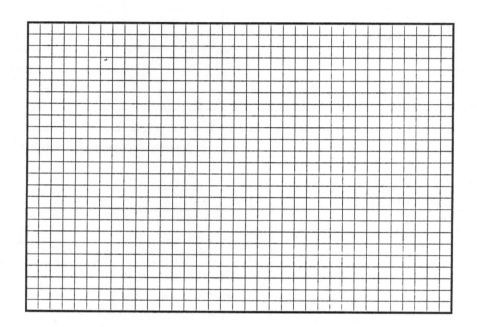

8. EXTERIOR WALL CONSTRUCTION: _____

9. FRAME: _____

10. GROUND FLOOR AREA: _____ S.F. 11. GROSS FLOOR AREA (EXCL. BASEMENT): _____ S.F.

12. NUMBER OF STORIES: _____ 13. STORY HEIGHT: _____

14. PERIMETER: _____ L.F. 15. BASEMENT AREA: _____ S.F.

16. GENERAL COMMENTS: _____

_____

_____

_____

_____

_____

_____

_____

FORMS

	NO.	DESCRIPTION			UNIT	UNIT COST	NEW SF COST	MODEL SF COST	+/- CHANGE
A	**SUBSTRUCTURE**								
A	1010	Standard Foundations	Bay size:		S.F. Gnd.				
A	1030	Slab on Grade	Material:	Thickness:	S.F. Slab				
A	2010	Basement Excavation	Depth:	Area:	S.F. Gnd.				
A	2020	Basement Walls			L.F. Walls				
B	**SHELL**								
	B10 Superstructure								
B	1010	Floor Construction	Elevated floors:		S.F. Floor				
					S.F. Floor				
B	1020	Roof Construction			S.F. Roof				
	B20 Exterior Enclosure								
B	2010	Exterior walls	Material:	Thickness: % of wall	S.F. Walls				
			Material:	Thickness: % of wall	S.F. Walls				
B	2020	Exterior Windows	Type:	% of wall	S.F. Wind.				
			Type:	% of wall Each	S.F. Wind.				
B	2030	Exterior Doors	Type:	Number:	Each				
			Type:	Number:	Each				
	B30 Roofing								
B	3010	Roof Coverings	Material:		S.F. Roof				
			Material:		S.F. Roof				
B	3020	Roof Openings			S.F. Opng.				
C	**INTERIORS**								
C	1010	Partitions:	Material:	Density:	S.F. Part.				
			Material:	Density:					
C	1020	Interior Doors	Type:	Number:	Each				
C	1030	Fittings			Each				
C	2010	Stair Construction			Flight				
C	3010	Wall Finishes	Material:	% of Wall	S.F. Walls				
			Material:		S.F. Walls				
C	3020	Floor Finishes	Material:		S.F. Floor				
			Material:		S.F. Floor				
			Material:		S.F. Floor				
C	3030	Ceiling Finishes	Material:		S.F. Ceil.				

	NO.	SYSTEM/COMPONENT	DESCRIPTION	UNIT	UNIT COST	NEW S.F. COST	MODEL S.F. COST	+/- CHANGE
D		**SERVICES**						
		D10 Conveying						
D	1010	Elevators & Lifts	Type: Capacity: Stops:	Each				
D	1020	Escalators & Moving Walks	Type:	Each				
		D20 Plumbing						
D	2010	Plumbing		Each				
D	2020	Domestic Water Distribution		S.F. Floor				
D	2040	Rain Water Drainage		S.F. Roof				
		D30 HVAC						
D	3010	Energy Supply		Each				
D	3020	Heat Generating Systems	Type:	S.F. Floor				
D	3030	Cooling Generating Systems	Type:	S.F. Floor				
D	3090	Other HVAC Sys. & Equipment		Each				
		D40 Fire Protection						
D	4010	Sprinklers		S.F. Floor				
D	4020	Standpipes		S.F. Floor				
		D50 Electrical						
D	5010	Electrical Service/Distribution		S.F. Floor				
D	5020	Lighting & Branch Wiring		S.F. Floor				
D	5030	Communications & Security		S.F. Floor				
D	5090	Other Electrical Systems		S.F. Floor				
E		**EQUIPMENT & FURNISHINGS**						
E	1010	Commercial Equipment		Each				
E	1020	Institutional Equipment		Each				
E	1030	Vehicular Equipment		Each				
E	1090	Other Equipment		Each				
F		**SPECIAL CONSTRUCTION**						
F	1020	Integrated Construction		S.F.				
F	1040	Special Facilities		S.F.				
G		**BUILDING SITEWORK**						

ITEM		Total Change
17	Model 020 total ____ $_____	
18	Adjusted S.F. cost $_____ item 17 +/- changes	
19	Building area - from item 11 _____ S.F. x adjusted S.F. cost..............	$...
20	Basement area - from item 15 _____ S.F. x S.F. cost $	$..
21	Base building sub-total - item 20 + item 19	$.....
22	Miscellaneous addition (quality, etc.)	$......
23	Sub-total - item 22 + 21	$.......
24	General conditions -25 % of item 23	$.....
25	Sub-total - item 24 + item 23....................	$......
26	Architects fees _____ % of item 25	$
27	Sub-total - item 26 + item 27	$......
28	Location modifier	x....
29	Local replacement cost - item 28 x item 27	$.....
30	Depreciation _____ % of item 29	$
31	Depreciated local replacement cost - item 29 less item 30	$....
32	Exclusions	$.......
33	Net depreciated replacement cost - item 31 less item 32	$....

Base Cost per square foot floor area: *(from square foot table)*

Specify source:

Page: _____ , Model # _____ , Area _____ S.F.,
Exterior wall _____ , Frame _____

Adjustments for exterior wall variation:

Size Adjustment:
   (Interpolate)

Height Adjustment: _____ + _____ = _____

Adjusted Base Cost per square foot:

Building Cost $ _____ x _____ = _____
                  Adjusted Base Cost   *Floor Area*
                  per square foot

Basement Cost $ _____ x _____ = _____
                  Basement Cost   *Basement Area*

Lump Sum Additions

TOTAL BUILDING COST *(Sum of above costs)* _____

Modifications: *(complexity, workmanship, size)* +/- _____ % _____

Location Modifier: City _____ Date _____ x _____

   Local cost of replacement _____

Less depreciation: _____ - _____

   Local cost of replacement less depreciation $ _____

485

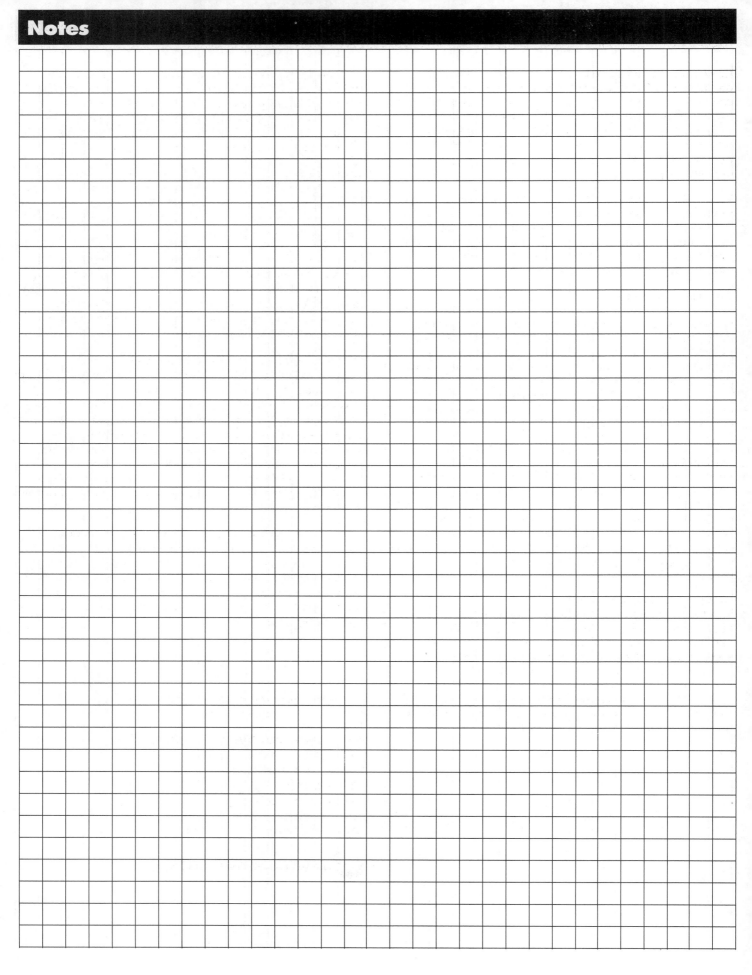

Reed Construction Data, Inc.

Reed Construction Data, Inc., a leading worldwide provider of total construction information solutions, is comprised of three main product groups designed specifically to help construction professionals advance their businesses with timely, accurate and actionable project, product, and cost data. Reed Construction Data is a division of Reed Business Information, a member of the Reed Elsevier plc group of companies.

The *Project, Product, and Cost & Estimating* divisions offer a variety of innovative products and services designed for the full spectrum of design, construction, and manufacturing professionals. Through it's *International* companies, Reed Construction Data's reputation for quality construction market data is growing worldwide.

Cost Information
RSMeans, the undisputed market leader and authority on construction costs, publishes current cost and estimating information in annual cost books and on the CostWorks CD-ROM. RSMeans furnishes the construction industry with a rich library of complementary reference books and a series of professional seminars that are designed to sharpen professional skills and maximize the effective use of cost estimating and management tools. RSMeans also provides construction cost consulting for Owners, Manufacturers, Designers, and Contractors.

Project Data
Reed Construction Data provides complete, accurate and relevant project information through all stages of construction. Customers are supplied industry data through leads, project reports, contact lists, plans and specifications surveys, market penetration analyses and sales evaluation reports. Any of these products can pinpoint a county, look at a state, or cover the country. Data is delivered via paper, e-mail, CD-ROM or the Internet.

Building Product Information
The First Source suite of products is the only integrated building product information system offered to the commercial construction industry for comparing and specifying building products. These print and online resources include *First Source*, CSI's SPEC-DATA™, CSI's MANU-SPEC™, First Source CAD, and Manufacturer Catalogs. Written by industry professionals and organized using CSI's MasterFormat™, construction professionals use this information to make better design decisions.

FirstSourceONL.com combines Reed Construction Data's project, product and cost data with news and information from Reed Business Information's *Building Design & Construction* and *Consulting-Specifying Engineer,* this industry-focused site offers easy and unlimited access to vital information for all construction professionals.

International
BIMSA/Mexico provides construction project news, product information, cost-data, seminars and consulting services to construction professionals in Mexico. Its subsidiary, PRISMA, provides job costing software.

Byggfakta Scandinavia AB, founded in 1936, is the parent company for the leaders of customized construction market data for Denmark, Estonia, Finland, Norway and Sweden. Each company fully covers the local construction market and provides information across several platforms including subscription, ad-hoc basis, electronically and on paper.

Reed Construction Data Canada serves the Canadian construction market with reliable and comprehensive project and product information services that cover all facets of construction. Core services include: *BuildSource, BuildSpec, BuildSelect,* product selection and specification tools available in print and on the Internet; Building Reports, a national construction project lead service; CanaData, statistical and forecasting information; *Daily Commercial News,* a construction newspaper reporting on news and projects in Ontario; and *Journal of Commerce,* reporting news in British Columbia and Alberta.

Cordell Building Information Services, with its complete range of project and cost and estimating services, is Australia's specialist in the construction information industry. Cordell provides in-depth and historical information on all aspects of construction projects and estimation, including several customized reports, construction and sales leads, and detailed cost information among others.

For more information, please visit our Web site at www.reedconstructiondata.com.

Reed Construction Data, Inc., Corporate Office
30 Technology Parkway South
Norcross, GA 30092-2912
(800) 322-6996
(800) 895-8661 (fax)
info@reedbusiness.com
www.reedconstructiondata.com

Means Project Cost Report

By filling out and returning the Project Description, you can receive a discount of $20.00 off any one of the Means products advertised in the following pages. The cost information required includes all items marked (✔) except those where no costs occurred. The sum of all major items should equal the Total Project Cost.

$20.00 Discount per product for each report you submit.

DISCOUNT PRODUCTS AVAILABLE—FOR U.S. CUSTOMERS ONLY—STRICTLY CONFIDENTIAL

Project Description (No remodeling projects, please.)

✔ Type Building _____

✔ Location _____

Capacity _____

✔ Frame _____

✔ Exterior _____

✔ Basement: full ☐ partial ☐ none ☐ crawl ☐

✔ Height in Stories _____

✔ Total Floor Area _____

Ground Floor Area _____

✔ Volume in C.F. _____

% Air Conditioned _____ Tons _____

Comments _____

Owner _____

Architect _____

General Contractor _____

✔ Bid Date _____

Typical Bay Size _____

✔ Labor Force: _____ % Union _____ % Non-Union

✔ Project Description (Circle one number in each line)
  1. Economy 2. Average 3. Custom 4. Luxury
  1. Square 2. Rectangular 3. Irregular 4. Very Irregular

	✔ **Total Project Cost**		$	
A	✔ **General Conditions**		$	
B	✔ **Site Work**		$	
BS	Site Clearing & Improvement			
BE	Excavation	(	C.Y.)	
BF	Caissons & Piling	(	L.F.)	
BU	Site Utilities			
BP	Roads & Walks Exterior Paving	(	S.Y.)	
C	✔ **Concrete**		$	
C	Cast in Place	(	C.Y.)	
CP	Precast	(	S.F.)	
D	✔ **Masonry**		$	
DB	Brick	(	M)	
DC	Block	(	M)	
DT	Tile	(	S.F.)	
DS	Stone	(	S.F.)	
E	✔ **Metals**		$	
ES	Structural Steel	(	Tons)	
EM	Misc. & Ornamental Metals			
F	✔ **Wood & Plastics**		$	
FR	Rough Carpentry	(	MBF)	
FF	Finish Carpentry			
FM	Architectural Millwork			
G	✔ **Thermal & Moisture Protection**		$	
GW	Waterproofing-Dampproofing	(	S.F.)	
GN	Insulation	(	S.F.)	
GR	Roofing & Flashing	(	S.F.)	
GM	Metal Siding/Curtain Wall	(	S.F.)	
H	✔ **Doors and Windows**		$	
HD	Doors	(	Ea.)	
HW	Windows	(	S.F.)	
HH	Finish Hardware			
HG	Glass & Glazing	(	S.F.)	
HS	Storefronts	(	S.F.)	

J	✔ **Finishes**			$
JL	Lath & Plaster	(	S.Y.)	
JD	Drywall	(	S.F.)	
JM	Tile & Marble	(	S.F.)	
JT	Terrazzo	(	S.F.)	
JA	Acoustical Treatment	(	S.F.)	
JC	Carpet	(	S.Y.)	
JF	Hard Surface Flooring	(	S.F.)	
JP	Painting & Wall Covering	(	S.F.)	
K	✔ **Specialties**			$
KB	Bathroom Partitions & Access.	(	S.F.)	
KF	Other Partitions	(	S.F.)	
KL	Lockers	(	Ea.)	
L	✔ **Equipment**			$
LK	Kitchen			
LS	School			
LO	Other			
M	✔ **Furnishings**			$
MW	Window Treatment			
MS	Seating	(	Ea.)	
N	✔ **Special Construction**			$
NA	Acoustical	(	S.F.)	
NB	Prefab. Bldgs.	(	S.F.)	
NO	Other			
P	✔ **Conveying Systems**			$
PE	Elevators	(	Ea.)	
PS	Escalators	(	Ea.)	
PM	Material Handling			
Q	✔ **Mechanical**			$
QP	Plumbing	(No. of fixtures	)	
QS	Fire Protection (Sprinklers)			
QF	Fire Protection (Hose Standpipes)			
QB	Heating, Ventilating & A.C.			
QH	Heating & Ventilating	(BTU Output	)	
QA	Air Conditioning	(	Tons)	
R	✔ **Electrical**			$
RL	Lighting	(	S.F.)	
RP	Power Service			
RD	Power Distribution			
RA	Alarms			
RG	Special Systems			
S	✔ **Mech./Elec. Combined**			$

Product Name _____

Product Number _____

Your Name _____

Title _____

Company _____

☐ Company

☐ Home Street Address _____

City, State, Zip _____

☐ Please send _____ forms.

Please specify the Means product you wish to receive. Complete the address information as requested and return this form with your check (product cost less $20.00) to address below.

RSMeans Company, Inc.,
Square Foot Costs Department
P.O. Box 800
Kingston, MA 02364-9988

For more information
visit Means Web Site
at www.rsmeans.com

Reed Construction Data/RSMeans ... a tradition of excellence in Construction Cost Information and Services since 1942.

Table of Contents

Annual Cost Guides, Page 2
Reference Books, Page 6
Seminars, Page 12
Electronic Data, Page 14
New Titles, Page 15
Order Form, Page 16

Book Selection Guide

The following table provides definitive information on the content of each cost data publication. The number of lines of data provided in each unit price or assemblies division, as well as the number of reference tables and crews is listed for each book. The presence of other elements such as an historical cost index, city cost indexes, square foot models or cross-referenced index is also indicated. You can use the table to help select the Means' book that has the quantity and type of information you most need in your work.

Unit Cost Divisions	Building Construction Costs	Mechanical	Electrical	Repair & Remodel.	Square Foot	Site Work Landsc.	Assemblies	Interior	Concrete Masonry	Open Shop	Heavy Construc.	Light Commercial	Facil. Construc.	Plumbing	Western Construction Costs	Residential
1	1104	822	907	1013		1045		847	1014	1102	1051	752	1528	874	1102	710
2	2505	1468	460	1450		7794		547	1387	2467	4978	692	4305	1584	2490	772
3	1401	113	100	726		1259		201	1782	1396	1404	229	1312	82	1399	248
4	835	18	0	647		657		574	1056	811	594	407	1058	0	821	334
5	1820	239	193	957		768		927	691	1788	1023	805	1802	316	1802	752
6	1489	82	78	1475		452		1406	318	1481	608	1626	1596	47	1826	1757
7	1264	158	74	1241		468		489	408	1264	349	956	1318	168	1264	752
8	1839	28	0	1907		313		1673	645	1821	51	1227	2034	0	1841	1193
9	1595	47	0	1424		233		1688	360	1547	159	1337	1809	47	1586	1216
10	861	47	25	498		193		700	170	863	0	394	906	233	861	217
11	1018	322	169	502		137		813	44	925	107	218	1173	291	925	104
12	307	0	0	47		210		1416	27	298	0	62	1435	0	298	64
13	1153	1005	375	494		388		884	75	1136	279	484	1783	928	1119	193
14	345	36	0	258		36		292	0	344	30	12	343	35	343	6
15	1987	13138	636	1779		1569		1174	59	1996	1766	1217	10865	9695	2017	826
16	1277	470	10063	1002		739		1108	55	1265	760	1081	9834	415	1226	552
17	427	354	427	0		0		0	0	427	0	0	427	356	427	0
Totals	21227	18347	13507	15420		16261		14739	8091	20961	13159	11499	43528	15071	21347	9696

Assembly Divisions	Building Construction Costs	Mechanical	Electrical	Repair & Remodel.	Square Foot	Site Work Landsc.	Assemblies	Interior	Concrete Masonry	Open Shop	Heavy Construc.	Light Commercial	Facil. Construc.	Plumbing	Western Construction Costs	Asm Div	Residential
A		19	0	192	150	540	612	0	550		542	149	24	0		1	374
B		0	0	809	2480	0	5590	333	1914		0	2024	144	0		2	217
C		0	0	635	862	0	1220	1568	145		0	767	238	0		3	588
D		1031	780	693	1823	0	2439	753	0		0	1310	1027	896		4	871
E		0	0	85	255	0	292	5	0		0	255	5	0		5	393
F		0	0	0	123	0	126	0	0		0	123	3	0		6	358
G		465	172	332	111	1856	584	0	482		432	110	113	560		7	299
																8	760
																9	80
																10	0
																11	0
																12	0
Totals		1515	952	2746	5804	2396	10863	2659	3091		974	4738	1554	1456			3940

Reference Section	Building Construction Costs	Mechanical	Electrical	Repair & Remodel.	Square Foot	Site Work Landsc.	Assemblies	Interior	Concrete Masonry	Open Shop	Heavy Construc.	Light Commercial	Facil. Construc.	Plumbing	Western Construction Costs	Residential
Tables	130	45	85	69	4	80	219	46	71	130	61	57	104	51	131	42
Models					102							43				32
Crews	410	410	410	391		410		410	410	393	410	393	391	410	410	393
City Cost Indexes	yes	yes	yes	yes	yes	yes	yes	yes	yes	yes	yes	yes	yes	yes	yes	yes
Historical Cost Indexes	yes	yes	yes	yes	yes	yes	yes	yes	yes	yes	yes	yes	yes	yes	yes	no

Annual Cost Guides

For more information
visit Means Web Site
at www.rsmeans.com

For more information
visit Means Web Site
at www.rsmeans.com

Means Building Construction Cost Data 2005

Available in Both Softbound and Looseleaf Editions

The "Bible" of the industry comes in the standard softcover edition or the looseleaf edition.

Many customers enjoy the convenience and flexibility of the looseleaf binder, which increases the usefulness of *Means Building Construction Cost Data 2005* by making it easy to add and remove pages. You can insert your own cost information pages, so everything is in one place. Copying pages for faxing is easier also. Whichever edition you prefer, softbound or the convenient looseleaf edition, you'll be eligible to receive *The Change Notice* FREE. Current subscribers receive *The Change Notice* via e-mail.

$115.95 per copy, Softbound
Catalog No. 60015

$144.95 per copy, Looseleaf
Catalog No. 61015

Means Building Construction Cost Data 2005

Offers you unchallenged unit price reliability in an easy-to-use arrangement. Whether used for complete, finished estimates or for periodic checks, it supplies more cost facts better and faster than any comparable source. Over 23,000 unit prices for 2005. The City Cost Indexes cover over 930 areas, for indexing to any project location in North America. Order and get *The Change Notice* FREE. You'll have year-long access to the Means Estimating **HOTLINE** FREE with your subscription. Expert assistance when using Means data is just a phone call away.

$115.95 per copy
Over 700 pages, illustrated, available Oct. 2004
Catalog No. 60015

Means Building Construction Cost Data 2005

Metric Version

The Federal Government has stated that all federal construction projects must now use metric documentation. The *Metric Version* of *Means Building Construction Cost Data 2005* is presented in metric measurements covering all construction areas. Don't miss out on these billion dollar opportunities. Make the switch to metric today.

$115.95 per copy
Over 700 pages, illus., available Dec. 2004
Catalog No. 63015

For more information
visit Means Web Site
at www.rsmeans.com

Annual Cost Guides

Means Mechanical Cost Data 2005

- HVAC • Controls

Total unit and systems price guidance for mechanical construction...materials, parts, fittings, and complete labor cost information. Includes prices for piping, heating, air conditioning, ventilation, and all related construction.

Plus new 2005 unit costs for:
- Over 2500 installed HVAC/controls assemblies
- "On Site" Location Factors for over 930 cities and towns in the U.S. and Canada
- Crews, labor and equipment

$115.95 per copy
Over 600 pages, illustrated, available Oct. 2004
Catalog No. 60025

Means Plumbing Cost Data 2005

Comprehensive unit prices and assemblies for plumbing, irrigation systems, commercial and residential fire protection, point-of-use water heaters, and the latest approved materials. This publication and its companion, *Means Mechanical Cost Data*, provide full-range cost estimating coverage for all the mechanical trades.

$115.95 per copy
Over 550 pages, illustrated, available Oct. 2004
Catalog No. 60215

Means Electrical Cost Data 2005

Pricing information for every part of electrical cost planning: More than 15,000 unit and systems costs with design tables; clear specifications and drawings; engineering guides and illustrated estimating procedures; complete labor-hour and materials costs for better scheduling and procurement; the latest electrical products and construction methods.
- A Variety of Special Electrical Systems including Cathodic Protection
- Costs for maintenance, demolition, HVAC/ mechanical, specialties, equipment, and more

$115.95 per copy
Over 450 pages, illustrated, available Oct. 2004
Catalog No. 60035

Means Electrical Change Order Cost Data 2005

You are provided with electrical unit prices exclusively for pricing change orders based on the recent, direct experience of contractors and suppliers. Analyze and check your own change order estimates against the experience others have had doing the same work. It also covers productivity analysis and change order cost justifications. With useful information for calculating the effects of change orders and dealing with their administration.

$115.95 per copy
Over 450 pages, available Oct. 2004
Catalog No. 60235

Means Facilities Maintenance & Repair Cost Data 2005

Published in a looseleaf format, *Means Facilities Maintenance & Repair Cost Data* gives you a complete system to manage and plan your facility repair and maintenance costs and budget efficiently. Guidelines for auditing a facility and developing an annual maintenance plan. Budgeting is included, along with reference tables on cost and management and information on frequency and productivity of maintenance operations.

The only nationally recognized source of maintenance and repair costs. Developed in cooperation with the Army Corps of Engineers.

$252.95 per copy
Over 600 pages, illustrated, available Dec. 2004
Catalog No. 60305

Means Square Foot Costs 2005

It's Accurate and Easy To Use!

- **Updated 2005 price information,** based on nationwide figures from suppliers, estimators, labor experts and contractors.
- "How-to-Use" Sections, with **clear examples** of commercial, residential, industrial, and institutional structures.
- Realistic graphics, offering true-to-life illustrations of building projects.
- Extensive information on using square foot cost data, including **sample estimates** and **alternate pricing methods.**

$126.95 per copy
Over 450 pages, illustrated, available Nov. 2004
Catalog No. 60055

For more information
visit Means Web Site
at www.rsmeans.com

Annual Cost Guides

Means Repair & Remodeling Cost Data 2005

Commercial/Residential

You can use this valuable tool to estimate commercial and residential renovation and remodeling.

Includes: New costs for hundreds of unique methods, materials and conditions that only come up in repair and remodeling. PLUS:

- Unit costs for over 16,000 construction components
- Installed costs for over 90 assemblies
- Costs for 300+ construction crews
- Over 930 "On Site" localization factors for the U.S. and Canada.

$99.95 per copy
Over 650 pages, illustrated, available Nov. 2004
Catalog No. 60045

Means Facilities Construction Cost Data 2005

For the maintenance and construction of commercial, industrial, municipal, and institutional properties. Costs are shown for new and remodeling construction and are broken down into materials, labor, equipment, overhead, and profit. Special emphasis is given to sections on mechanical, electrical, furnishings, site work, building maintenance, finish work, and demolition. More than 45,000 unit costs, plus assemblies and reference sections are included.

$279.95 per copy
Over 1200 pages, illustrated, available Nov. 2004
Catalog No. 60205

Means Residential Cost Data 2005

Contains square foot costs for 30 basic home models with the look of today, plus hundreds of custom additions and modifications you can quote right off the page. With costs for the 100 residential systems you're most likely to use in the year ahead. Complete with blank estimating forms, sample estimates and step-by-step instructions.

Means Light Commercial Cost Data 2005

Specifically addresses the light commercial market, which is an increasingly specialized niche in the industry. Aids you, the owner/designer/contractor, in preparing all types of estimates, from budgets to detailed bids. Includes new advances in methods and materials. Assemblies section allows you to evaluate alternatives in the early stages of design/planning.

Over 13,000 unit costs for 2005 ensure you have the prices you need...when you need them.

$99.95 per copy
Over 600 pages, illustrated, available Oct. 2004
Catalog No. 60175

$99.95 per copy
Over 650 pages, illustrated, available Nov. 2004
Catalog No. 60185

Means Assemblies Cost Data 2005

Means Assemblies Cost Data 2005 takes the guesswork out of preliminary or conceptual estimates. Now you don't have to try to calculate the assembled cost by working up individual components costs. We've done all the work for you.

Presents detailed illustrations, descriptions, specifications and costs for every conceivable building assembly—240 types in all—arranged in the easy-to-use UNIFORMAT II system. Each illustrated "assembled" cost includes a complete grouping of materials and associated installation costs including the installing contractor's overhead and profit.

$189.95 per copy
Over 600 pages, illustrated, available Oct. 2004
Catalog No. 60065

Means Site Work & Landscape Cost Data 2005

Means Site Work & Landscape Cost Data 2005 is organized to assist you in all your estimating needs. Hundreds of fact-filled pages help you make accurate cost estimates efficiently.

Updated for 2005!

- Demolition features—including ceilings, doors, electrical, flooring, HVAC, millwork, plumbing, roofing, walls and windows
- State-of-the-art segmental retaining walls
- Flywheel trenching costs and details
- Updated Wells section
- Landscape materials, flowers, shrubs and trees

$115.95 per copy
Over 600 pages, illustrated, available Nov. 2004
Catalog No. 60285

For more information
visit Means Web Site
at www.rsmeans.com

Annual Cost Guides

Means Open Shop Building Construction Cost Data 2005

The latest costs for accurate budgeting and estimating of new commercial and residential construction... renovation work... change orders... cost engineering. *Means Open Shop BCCD* will assist you to...
• Develop benchmark prices for change orders
• Plug gaps in preliminary estimates, budgets
• Estimate complex projects
• Substantiate invoices on contracts
• Price ADA-related renovations

$115.95 per copy
Over 700 pages, illustrated, available Dec. 2004
Catalog No. 60155

Means Building Construction Cost Data 2005
Western Edition

This regional edition provides more precise cost information for western North America. Labor rates are based on union rates from 13 western states and western Canada. Included are western practices and materials not found in our national edition: tilt-up concrete walls, glu-lam structural systems, specialized timber construction, seismic restraints, landscape and irrigation systems.

$115.95 per copy
Over 650 pages, illustrated, available Dec. 2004
Catalog No. 60225

Means Construction Cost Indexes 2005

Who knows what 2005 holds? What materials and labor costs will change unexpectedly? By how much?
• Breakdowns for 316 major cities.
• National averages for 30 key cities.
• Expanded five major city indexes.
• Historical construction cost indexes.

$252.95 per year
$63.00 individual quarters
Catalog No. 60145 A,B,C,D

Means Concrete & Masonry Cost Data 2005

Provides you with cost facts for virtually all concrete/masonry estimating needs, from complicated formwork to various sizes and face finishes of brick and block, all in great detail. The comprehensive unit cost section contains more than 8,500 selected entries. Also contains an assemblies cost section, and a detailed reference section which supplements the cost data.

$105.95 per copy
Over 450 pages, illustrated, available Dec. 2004
Catalog No. 60115

Means Heavy Construction Cost Data 2005

A comprehensive guide to heavy construction costs. Includes costs for highly specialized projects such as tunnels, dams, highways, airports, and waterways. Information on different labor rates, equipment, and material costs is included. Has unit price costs, systems costs, and numerous reference tables for costs and design. Valuable not only to contractors and civil engineers, but also to government agencies and city/town engineers.

$115.95 per copy
Over 450 pages, illustrated, available Nov. 2004
Catalog No. 60165

Means Heavy Construction Cost Data 2005
Metric Version
Make sure you have the Means industry standard metric costs for the federal, state, municipal and private marketplace. With thousands of up-to-date metric unit prices in tables by CSI standard divisions. Supplies you with assemblies costs using the metric standard for reliable cost projections in the design stage of your project. Helps you determine sizes, material amounts, and has tips for handling metric estimates.

$115.95 per copy
Over 450 pages, illustrated, available Dec. 2004
Catalog No. 63165

Means Interior Cost Data 2005

Provides you with prices and guidance needed to make accurate interior work estimates. Contains costs on materials, equipment, hardware, custom installations, furnishings, labor costs . . . every cost factor for new and remodel commercial and industrial interior construction, including updated information on office furnishings, plus more than 50 reference tables. For contractors, facility managers, owners.

$115.95 per copy
Over 600 pages, illustrated, available Oct. 2004
Catalog No. 60095

Means Labor Rates for the Construction Industry 2005

Complete information for estimating labor costs, making comparisons and negotiating wage rates by trade for over 300 cities (United States and Canada). With 46 construction trades listed by local union number in each city, and historical wage rates included for comparison. No similar book is available through the trade.

Each city chart lists the county and is alphabetically arranged with handy visual flip tabs for quick reference.

$253.95 per copy
Over 300 pages, available Dec. 2004
Catalog No. 60125

For more information
visit Means Web Site
at www.rsmeans.com

Reference Books

Building Security: Strategies & Costs
By David Owen

Unauthorized systems access, terrorism, hurricanes and tornadoes, sabotage, vandalism, fire, explosions, and other threats… All are considerations as building owners, facility managers, and design and construction professionals seek ways to meet security needs.

This comprehensive resource will help you evaluate your facility's security needs — and design and budget for the materials and devices needed to fulfill them. The text and cost data will help you to:

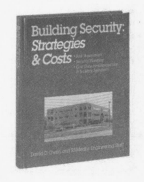

- Identify threats, probability of occurrence, and the potential losses— to determine and address your real vulnerabilities.
- Perform a detailed risk assessment of an existing facility, and prioritize and budget for security enhancement.
- Evaluate and price security systems and construction solutions, so you can make cost-effective choices.

Includes over 130 pages of Means Cost Data for installation of security systems and materials, plus a review of more than 50 security devices and construction solutions—how they work, and how they compare.

$89.95 per copy
Over 400 pages, Hardcover
Catalog No. 37339

Green Building: Project Planning & Cost Estimating
By RSMeans and Contributing Authors

Written by a team of leading experts in sustainable design, this new book is a complete guide to planning and estimating green building projects, a growing trend in building design and construction – commercial, industrial, institutional and residential. It explains:

- All the different criteria for "green-ness"
- What criteria your building needs to meet to get a LEED, Energy Star, or other recognized rating for green buildings
- How the project team works differently on a green versus a traditional building project
- How to select and specify green products
- How to evaluate the cost and value of green products versus conventional ones — not only for their first (installation) cost, but their cost over time (in maintenance and operation).

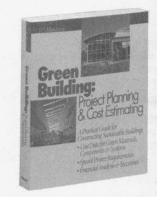

Features an extensive Green Building Cost Data section, which details the available products, how they are specified, and how much they cost.

$89.95 per copy
350 pages, illustrated, Hardcover
Catalog No. 67338

Life Cycle Costing for Facilities
By Alphonse Dell'Isola and Dr. Steven Kirk

Guidance for achieving higher quality design and construction projects at lower costs!

Facility designers and owners are frustrated with cost-cutting efforts that yield the cheapest product, but sacrifice quality. Life Cycle Costing, properly done, enables them to achieve both—high quality, incorporating innovative design, and costs that meet their budgets.

The authors show how LCC can work for a broad variety of projects – from several types of buildings, to roads and bridges, to HVAC and electrical upgrades, to materials and equipment procurement. Case studies include:

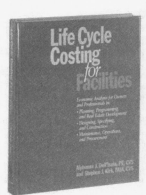

- Health care and nursing facilities
- College campus and high schools
- Office buildings, courthouses, and banks
- Exterior walls, elevators, lighting, HVAC, and more

The book's extensive cost section provides maintenance and replacement costs for facility elements.

$99.95 per copy
450 pages, Hardcover
Catalog No. 67341

For more information
visit Means Web Site
at www.rsmeans.com

Reference Books

Value Engineering: Practical Applications

. . . For Design, Construction, Maintenance & Operations
By Alphonse Dell'Isola, PE

A tool for immediate application—for engineers, architects, facility managers, owners, and contractors. Includes: making the case for VE—the management briefing, integrating VE into planning and budgeting, conducting life cycle costing, integrating VE into the design process, using VE methodology in design review and consultant selection, case studies, VE workbook, and a life cycle costing program on disk.

$79.95 per copy
Over 450 pages, illustrated, Hardcover
Catalog No. 67319

Facilities Operations & Engineering Reference

By the Association for Facilities Engineering and RSMeans

An all-in-one technical referance for planning and managing facility projects and solving day-to-day operations problems. Selected as the official Certified Plant Engineer reference, this handbook covers financial analysis, maintenance, HVAC and energy efficiency, and more.

$109.95 per copy
Over 700 pages, illustrated, Hardcover
Catalog No. 67318

The Building Professional's Guide to Contract Documents

3rd Edition
By Waller S. Poage, AIA, CSI, CVS

This comprehensive treatment of Contract Documents is an important reference for owners, design professionals, contractors, and students.
• Structure your Documents for Maximum Efficiency
• Effectively communicate construction requirements to all concerned
• Understand the Roles and Responsibilities of Construction Professionals
• Improve Methods of Project Delivery
$64.95 per copy, 400 pages
Diagrams and construction forms, Hardcover
Catalog No. 67261A

HVAC: Design Criteria, Options, Selection

Expanded 2nd Edition

By William H. Rowe III, AIA, PE
Includes Indoor Air Quality, CFC Removal, Energy Efficient Systems, and Special Systems by Building Type. Helps you solve a wide range of HVAC system design and selection problems effectively and economically. Gives you clear explanations of the latest ASHRAE standards.

$84.95 per copy
Over 600 pages, illustrated, Hardcover
Catalog No. 67306

Cost Planning & Estimating for Facilities Maintenance

In this unique book, a team of facilities management authorities shares their expertise at:
• Evaluating and budgeting maintenance operations
• Maintaining & repairing key building components
• Applying *Means Facilities Maintenance & Repair Cost Data* to your estimating

Covers special maintenance requirements of the 10 major building types.

$89.95 per copy
Over 475 pages, Hardcover
Catalog No. 67314

Facilities Maintenance Management

By Gregory H. Magee, PE
Now you can get successful management methods and techniques for all aspects of facilities maintenance. This comprehensive reference explains and demonstrates successful management techniques for all aspects of maintenance, repair, and improvements for buildings, machinery, equipment, and grounds. Plus, guidance for outsourcing and managing internal staffs.

$86.95 per copy
Over 280 pages with illustrations, Hardcover
Catalog No. 67249

Builder's Essentials: Advanced Framing Methods

By Scot Simpson

A highly illustrated, "framer-friendly" approach to advanced framing elements. Provides expert, but easy-to-interpret, instruction for laying out and framing complex walls, roofs, and stairs, and special requirements for earthquake and hurricane protection. Also helps bring framers up to date on the latest building code changes, and provides tips on the lead framer's role and responsibilities, how to prepare for a job, and how to get the crew started.

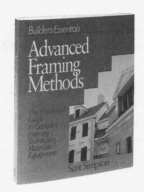

$24.95 per copy
250 pages, illustrated, Softcover
Catalog No. 67330

Reference Books

**For more information
visit Means Web Site
at www.rsmeans.com**

Interior Home Improvement Costs,

New 9th Edition

Estimates for the most popular remodeling and repair projects—from small, do-it-yourself jobs—to major renovations and new construction. Includes: Kitchens & Baths; New Living Space from Your Attic, Basement or Garage; New Floors, Paint & Wallpaper; Tearing Out or Building New Walls; Closets, Stairs & Fireplaces; New Energy-Saving Improvements, Home Theatres, and More!

$24.95 per copy
250 pages, illustrated, Softcover
Catalog No. 67308E

Exterior Home Improvement Costs

New 9th Edition

Estimates for the most popular remodeling and repair projects—from small, do-it-yourself jobs, to major renovations and new construction. Includes: Curb Appeal Projects—Landscaping, Patios, Porches, Driveways and Walkways; New Windows and Doors; Decks, Greenhouses, and Sunrooms; Room Additions and Garages; Roofing, Siding, and Painting; "Green" Improvements to Save Energy & Water

$24.95 per copy
Over 275 pages, illustrated, softcover
Catalog No. 67309E

Builder's Essentials: Plan Reading & Material Takeoff

By Wayne J. DelPico

For Residential and Light Commercial Construction

A valuable tool for understanding plans and specs, and accurately calculating material quantities.
Step-by-step instructions and takeoff procedures based on a full set of working drawings.

$35.95 per copy
Over 420 pages, Softcover
Catalog No. 67307

Planning & Managing Interior Projects

2nd Edition

By Carol E. Farren, CFM

Addresses changes in technology and business, guiding you through commercial design and construction from initial client meetings to post-project administration. Includes: evaluating space requirements, alternative work models, telecommunications and data management, and environmental issues.

$69.95 per copy
Over 400 pages, illustrated, Hardcover
Catalog No. 67245A

Builder's Essentials: Best Business Practices for Builders & Remodelers:
An Easy-to-Use Checklist System
By Thomas N. Frisby

A comprehensive guide covering all aspects of running a construction business, with more than 40 user–friendly checklists. This book provides expert guidance on: increasing your revenue and keeping more of your profit; planning for long-term growth; keeping good employees and managing subcontractors.

$29.95 per copy
Over 220 pages, Softcover
Catalog No. 67329

Builder's Essentials: Framing & Rough Carpentry 2nd Edition

By Scot Simpson

A complete training manual for apprentice and experienced carpenters. Develop and improve your skills with "framer-friendly," easy-to-follow instructions, and step-by-step illustrations. Learn proven techniques for framing walls, floors, roofs, stairs, doors, and windows. Updated guidance on standards, building codes, safety requirements, and more. Also available in Spanish!

$24.95 per copy
Over 150 pages, Softcover
Catalog No. 67298A   Spanish Catalog No. 67298AS

How to Estimate with Means Data & CostWorks

By RSMeans and Saleh Mubarak
New 2nd Edition!

Learn estimating techniques using Means cost data. Includes an instructional version of Means CostWorks CD–ROM with Sample Building Plans.
The step-by-step guide takes you through all the major construction items. Over 300 sample estimating problems are included.

$59.95 per copy
Over 190 pages, Softcover
Catalog No. 67324A

Unit Price Estimating Methods

New 3rd Edition

This new edition includes up-to-date cost data and estimating examples, updated to reflect changes to the CSI numbering system and new features of Means cost data. It describes the most productive, universally accepted ways to estimate, and uses checklists and forms to illustrate shortcuts and timesavers. A model estimate demonstrates procedures. A new chapter explores computer estimating alternatives.

$59.95 per copy
Over 350 pages, illustrated, Hardcover
Catalog No. 67303A

Means Landscape Estimating Methods

4th Edition

By Sylvia H. Fee

This revised edition offers expert guidance for preparing accurate estimates for new landscape construction and grounds maintenance. Includes a complete project estimate featuring the latest equipment and methods, and **two chapters on Life Cycle Costing, and Landscape Maintenance Estimating.**

$62.95 per copy
Over 300 pages, illustrated, Hardcover
Catalog No. 67295B

Means Environmental Remediation Estimating Methods, 2nd Edition

By Richard R. Rast

Guidelines for estimating 50 standard remediation technologies. Use it to prepare preliminary budgets, develop estimates, compare costs and solutions, estimate liability, review quotes, negotiate settlements.

A valuable support tool for *Means Environmental Remediation Unit Price* and *Assemblies* books.

$99.95 per copy
Over 750 pages, illustrated, Hardcover
Catalog No. 64777A

For more information
visit Means Web Site
at www.rsmeans.com

Reference Books

Means Illustrated Construction Dictionary, Condensed Edition, 2nd Edition
By RSMeans

Recognized in the industry as the best resource of its kind, this has been further enhanced with updates to existing terms and the addition of hundreds of new terms and illustrations . . . in keeping with the most recent developments in the industry.

The best portable dictionary for office or field use—an essential tool for contractors, architects, insurance and real estate personnel, homeowners, and anyone who needs quick, clear definitions for construction terms.

$59.95 per copy
Over 500 pages
Catalog No. 67282A

Means Repair and Remodeling Estimating
New 4th Edition
By Edward B. Wetherill & RSMeans

Focuses on the unique problems of estimating renovations of existing structures. It helps you determine the true costs of remodeling through careful evaluation of architectural details and a site visit.

New section on disaster restoration costs.

$69.95 per copy
Over 450 pages, illustrated, Hardcover
Catalog No. 67265B

Facilities Planning & Relocation
New, lower price and user-friendly format.
By David D. Owen

A complete system for planning space needs and managing relocations. Includes step-by-step manual, over 50 forms, and extensive reference section on materials and furnishings.

$89.95 per copy
Over 450 pages, Softcover
Catalog No. 67301

Means Square Foot & Assemblies Estimating Methods
3rd Edition!

Develop realistic Square Foot and Assemblies Costs for budgeting and construction funding. The new edition features updated guidance on square foot and assemblies estimating using UNIFORMAT II. An essential reference for anyone who performs conceptual estimates.

$69.95 per copy
Over 300 pages, illustrated, Hardcover
Catalog No. 67145B

Means Electrical Estimating Methods 3rd Edition

Expanded new edition includes sample estimates and cost information in keeping with the latest version of the CSI MasterFormat and UNIFORMAT II. Complete coverage of Fiber Optic and Uninterruptible Power Supply electrical systems, broken down by components and explained in detail. Includes a new chapter on computerized estimating methods. A practical companion to *Means Electrical Cost Data*.

$64.95 per copy
Over 325 pages, Hardcover
Catalog No. 67230A

Means Mechanical Estimating Methods 3rd Edition

This guide assists you in making a review of plans, specs, and bid packages with suggestions for takeoff procedures, listings, substitutions and pre-bid scheduling. Includes suggestions for budgeting labor and equipment usage. Compares materials and construction methods to allow you to select the best option.

$64.95 per copy
Over 350 pages, illustrated, Hardcover
Catalog No. 67294A

Means Spanish/English Construction Dictionary
By RSMeans, The International Conference of Building Officials (ICBO), and Rolf Jensen & Associates (RJA)

Designed to facilitate communication among Spanish- and English-speaking construction personnel—improving performance and job-site safety. Features the most common words and phrases used in the construction industry, with easy-to-follow pronunciations. Includes extensive building systems and tools illustrations.

$22.95 per copy
250 pages, illustrated, Softcover
Catalog No. 67327

Project Scheduling & Management for Construction
New 3rd Edition
By David R. Pierce, Jr.

A comprehensive yet easy-to-follow guide to construction project scheduling and control—from vital project management principles through the latest scheduling, tracking, and controlling techniques. The author is a leading authority on scheduling with years of field and teaching experience at leading academic institutions. Spend a few hours with this book and come away with a solid understanding of this essential management topic.

$64.95 per copy
Over 300 pages, illustrated, Hardcover
Catalog No. 67247B

**For more information
visit Means Web Site
at www.rsmeans.com**

Concrete Repair and Maintenance Illustrated
By Peter H. Emmons
$69.95 per copy
Catalog No. 67146

Superintending for Contractors:
How to Bring Jobs in On-Time, On-Budget
By Paul J. Cook
$35.95 per copy
Catalog No. 67233

HVAC Systems Evaluation
By Harold R. Colen, PE
$84.95 per copy
Catalog No. 67281

Basics for Builders: How to Survive and Prosper in Construction
By Thomas N. Frisby
$34.95 per copy
Catalog No. 67273

Successful Interior Projects Through Effective Contract Documents
By Joel Downey & Patricia K. Gilbert
Now $24.98 per copy, limited quantity
Catalog No. 67313

Building Spec Homes Profitably
By Kenneth V. Johnson
$29.95 per copy
Catalog No. 67312

Estimating for Contractors
How to Make Estimates that Win Jobs
By Paul J. Cook
$35.95 per copy
Catalog No. 67160

Successful Estimating Methods:
From Concept to Bid
By John D. Bledsoe, PhD, PE
$32.48 per copy
Catalog No. 67287

Total Productive Facilities Management
By Richard W. Sievert, Jr.
$29.98 per copy
Over 270 pages, illustrated, Hardcover
Catalog No. 67321

Means Productivity Standards for Construction
Expanded Edition (*Formerly Man-Hour Standards*)
$49.98 per copy
Over 800 pages, Hardcover
Catalog No. 67236A

Means Plumbing Estimating Methods, 3rd Edition
By Joseph Galeno and Sheldon Greene
Now $29.98 per copy
Catalog No. 67283B

For more information
visit Means Web Site
at www.rsmeans.com

Reference Books

Preventive Maintenance Guidelines for School Facilities
By John C. Maciha

A complete PM program for K-12 schools that ensures sustained security, safety, property integrity, user satisfaction, and reasonable ongoing expenditures.

Includes schedules for weekly, monthly, semiannual, and annual maintenance with hard copy and electronic forms.

$149.95 per copy
Over 225 pages, Hardcover
Catalog No. 67326

Preventive Maintenance for Higher Education Facilities
By Applied Management Engineering, Inc.

An easy-to-use system to help facilities professionals establish the value of PM, and to develop and budget for an appropriate PM program for their college or university. Features interactive campus building models typical of those found in different-sized higher education facilities, and PM checklists linked to each piece of equipment or system in hard copy and electronic format.

$149.95 per copy
150 pages, Hardcover
Catalog No. 67337

Historic Preservation: Project Planning & Estimating
By Swanke Hayden Connell Architects

Expert guidance on managing historic restoration, rehabilitation, and preservation building projects and determining and controlling their costs. Includes:

- How to determine whether a structure qualifies as historic
- Where to obtain funding and other assistance
- How to evaluate and repair more than 75 historic building materials

$99.95 per copy
Over 675 pages, Hardcover
Catalog No. 67323

Means Illustrated Construction Dictionary, 3rd Edition

Long regarded as the Industry's finest, the Means *Illustrated Construction Dictionary* is now even better.
With the addition of over 1,000 new terms and hundreds of new illustrations, it is the clear choice for the most comprehensive and current information. **The companion CD-ROM that comes with this new edition adds many extra features: larger graphics, expanded definitions, and links to both CSI MasterFormat numbers and product information.**

$99.95 per copy
Over 790 pages, Illustrated, Hardcover
Catalog No. 67292A

Designing & Building with the IBC, 2nd Edition
By Rolf Jensen & Associates, Inc.

This updated comprehensive guide helps building professionals make the transition to the 2003 *International Building Code®*. Includes a side-by-side code comparison of the IBC 2003 to the IBC 2000 and the three primary model codes, a quick-find index, and professional code commentary. With illustrations, abbreviations key, and an extensive Resource section.

$99.95 per copy
Over 875 pages
Catalog No. 67328A

Residential & Light Commercial Construction Standards, 2nd Edition
By RSMeans and Contributing Authors

New, updated second edition of this unique collection of industry standards that define quality construction. For contractors, subcontractors, owners, developers, architects, engineers, attorneys, and insurance personnel, this book provides authoritative requirements and recommendations compiled from the nation's leading professional associations, industry publications, and building code organizations.

$59.95 per copy
600 pages, illustrated, Softcover
Catalog No. 67322A

Means Estimating Handbook, 2nd Edition
By RSMeans

Updated new Second Edition answers virtually any estimating technical question - all organized by CSI MasterFormat. This comprehensive reference covers the full spectrum of technical data required to estimate construction costs. The book includes information on sizing, productivity, equipment requirements, code-mandated specifications, design standards and engineering factors.

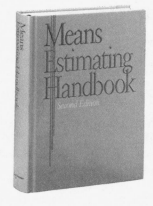

$99.95 per copy
Over 900 pages, Hardcover
Catalog No. 67276A

For more information
visit Means Web Site
at www.rsmeans.com

Seminars

Means CostWorks Training

This one-day seminar course has been designed with the intention of assisting both new and existing users to become more familiar with CostWorks program. The class is broken into two unique sections: (1) A one-half day presentation on the function of each icon; and each student will be shown how to use the software to develop a cost estimate. (2) Hands-on estimating exercises that will ensure that each student thoroughly understands how to use CostWorks. You must bring your own laptop computer to this course.

CostWorks Benefits/Features:
• Estimate in your own spreadsheet format
• Power of Means National Database
• Database automatically regionalized
• Save time with keyword searches
• Save time by establishing common estimate items in "Bookmark" files
• Customize your spreadsheet template
• Hot key to Product Manufacturers' listings and specs
• Merge capability for networking environments
• View crews and assembly components
• AutoSave capability
• Enhanced sorting capability

Unit Price Estimating

This interactive two-day seminar teaches attendees how to interpret project information and process it into final, detailed estimates with the greatest accuracy level.

The single most important credential an estimator can take to the job is the ability to visualize construction in the mind's eye, and thereby estimate accurately.

Some Of What You'll Learn:
• Interpreting the design in terms of cost
• The most detailed, time tested methodology for accurate "pricing"
• Key cost drivers—material, labor, equipment, staging and subcontracts
• Understanding direct and indirect costs for accurate job cost accounting and change order management

Who Should Attend: Corporate and government estimators and purchasers, architects, engineers…and others needing to produce accurate project estimates.

Square Foot and Assemblies Cost Estimating

This two-day course teaches attendees how to quickly deliver accurate square foot estimates using limited budget and design information.

Some Of What You'll Learn:
• How square foot costing gets the estimate done faster
• Taking advantage of a "systems" or "assemblies" format
• The Means "building assemblies/square foot cost approach"
• How to create a very reliable preliminary and systems estimate using bare-bones design information

Who Should Attend: Facilities managers, facilities engineers, estimators, planners, developers, construction finance professionals…and others needing to make quick, accurate construction cost estimates at commercial, government, educational and medical facilities.

Repair and Remodeling Estimating

This two-day seminar emphasizes all the underlying considerations unique to repair/remodeling estimating and presents the correct methods for generating accurate, reliable R&R project costs using the unit price and assemblies methods.

Some Of What You'll Learn:
• Estimating considerations—like labor-hours, building code compliance, working within existing structures, purchasing materials in smaller quantities, unforeseen deficiencies
• Identifies problems and provides solutions to estimating building alterations
• Rules for factoring in minimum labor costs, accurate productivity estimates and allowances for project contingencies
• R&R estimating examples are calculated using unite prices and assemblies data

Who Should Attend: Facilities managers, plant engineers, architects, contractors, estimators, builders…and others who are concerned with the proper preparation and/or evaluation of repair and remodeling estimates.

Mechanical and Electrical Estimating

This two-day course teaches attendees how to prepare more accurate and complete mechanical/electrical estimates, avoiding the pitfalls of omission and double-counting, while understanding the composition and rationale within the Means Mechanical/Electrical database.

Some Of What You'll Learn:
• The unique way mechanical and electrical systems are interrelated
• M&E estimates, conceptual, planning, budgeting and bidding stages
• Order of magnitude, square foot, assemblies and unit price estimating
• Comparative cost analysis of equipment and design alternatives

Who Should Attend: Architects, engineers, facilities managers, mechanical and electrical contractors…and others needing a highly reliable method for developing, understanding and evaluating mechanical and electrical contracts.

Plan Reading and Material Takeoff

This two-day program teaches attendees to read and understand construction documents and to use them in the preparation of material takeoffs.

Some of What You'll Learn:
• Skills necessary to read and understand typical contract documents—blueprints and specifications
• Details and symbols used by architects and engineers
• Construction specifications' importance in conjunction with blueprints
• Accurate takeoff of construction materials and industry-accepted takeoff methods

Who Should Attend: Facilities managers, construction supervisors, office managers…and other responsible for the execution and administration of a construction project including government, medical, commercial, educational or retail facilities.

Facilities Maintenance and Repair Estimating

This two-day course teaches attendees how to plan, budget, and estimate the cost of ongoing and preventive maintenance and repair for existing buildings and grounds.

Some Of What You'll Learn:
• The most financially favorable maintenance, repair and replacement scheduling and estimating
• Auditing and value engineering facilities
• Preventive planning and facilities upgrading
• Determining both in-house and contract-out service costs; annual, asset-protecting M&R plan

Who Should Attend: Facility managers, maintenance supervisors, buildings and grounds superintendents, plant managers, planners, estimators…and others involved in facilities planning and budgeting.

Scheduling and Project Management

This two-day course teaches attendees the most current and proven scheduling and management techniques needed to bring projects in on time and on budget.

Some Of What You'll Learn:
• Crucial phases of planning and scheduling
• How to establish project priorities, develop realistic schedules and management techniques
• Critical Path and Precedence Methods
• Special emphasis on cost control

Who Should Attend: Construction project managers, supervisors, engineers, estimators, contractors…and others who want to improve their project planning, scheduling and management skills.

Advanced Project Management

This two-day seminar will teach you how to effectively manage and control the entire design-build process and allow you to take home tangible skills that will be immediately applicable on existing projects.

Some Of What You'll Learn:
• Value engineering, bonding, fast-tracking and bid package creation
• How estimates and schedules can be integrated to provide advanced project management tools
• Cost engineering, quality control, productivity measurement and improvement
• Front loading a project and predicting its cash flow

Who Should Attend: Owners, project managers, architectural and engineering managers, construction managers, contractors…and anyone else who is responsible for the timely design and completion of construction projects.

Seminars

For more information
visit Means Web Site
at www.rsmeans.com

2005 Means Seminar Schedule

Location	Dates
Las Vegas, NV	March
Washington, DC	April
Phoenix, AZ	April
Denver, CO	May
San Francisco, CA	June
Philadelphia, PA	June
Washington, DC	September
Dallas, TX	September
Las Vegas, NV	October
Orlando, FL	November
Atlantic City, NJ	November
San Diego, CA	December

Note: Call for exact dates and details.

Registration Information

Register Early... Save up to $100! Register 30 days before the start date of a seminar and save $100 off your total fee. *Note: This discount can be applied only once per order. It cannot be applied to team discount registrations or any other special offer.*

How to Register Register by phone today! Means toll-free number for making reservations is: **1-800-334-3509.**

Individual Seminar Registration Fee $895. Individual CostWorks Training Registration Fee $349. To register by mail, complete the registration form and return with your full fee to: Seminar Division, Reed Construction Data, RSMeans Seminars, 63 Smiths Lane, Kingston, MA 02364.

Federal Government Pricing All Federal Government employees save 25% off regular seminar price. Other promotional discounts cannot be combined with Federal Government discount.

Team Discount Program Two to four seminar registrations: Call for pricing.

Multiple Course Discounts When signing up for two or more courses, call for pricing.

Refund Policy Cancellations will be accepted up to ten days prior to the seminar start. There are no refunds for cancellations received later than ten working days prior to the first day of the seminar. A $150 processing fee will be applied for all cancellations. Written notice of cancellation is required . Substitutions can be made at anytime before the session starts. **No-shows are subject to the full seminar fee.**

AACE Approved Courses The RSMeans Construction Estimating and Management Seminars described and offered to you here have each been approved for 14 hours (1.4 recertification credits) of credit by the AACE International Certification Board toward meeting the continuing education requirements for re-certification as a Certified Cost Engineer/Certified Cost Consultant.

AIA Continuing Education We are registered with the AIA Continuing Education System (AIA/CES) and are committed to developing quality learning activities in accordance with the CES criteria. RSMeans seminars meet the AIA/CES criteria for Quality Level 2. AIA members will receive (14) learning units (LUs) for each two-day RSMeans Course.

NASBA CPE Sponsor Credits We are part of the National Registry of CPE Sponsors. Attendees may be eligable for (16) CPE credits.

Daily Course Schedule The first day of each seminar session begins at 8:30 A.M. and ends at 4:30 P.M. The second day is 8:00 A.M.–4:00 P.M. Participants are urged to bring a hand-held calculator since many actual problems will be worked out in each session.

Continental Breakfast Your registration includes the cost of a continental breakfast, a morning coffee break, and an afternoon break. These informal segments will allow you to discuss topics of mutual interest with other members of the seminar. (You are free to make your own lunch and dinner arrangements.)

Hotel/Transportation Arrangements RSMeans has arranged to hold a block of rooms at each hotel hosting a seminar. To take advantage of special group rates when making your reservation, be sure to mention that you are attending the Means Seminar. You are, of course, free to stay at the lodging place of your choice. (**Hotel reservations and transportation arrangements should be made directly by seminar attendees.**)

Important Class sizes are limited, so please register as soon as possible.

Note: Pricing subject to change.

Registration Form Call 1-800-334-3509 to register or FAX 1-800-632-6732. Visit our Web site www.rsmeans.com

Please register the following people for the Means Construction Seminars as shown here. Full payment or deposit is enclosed, and we understand that we must make our own hotel reservations if overnight stays are necessary.

☐ Full payment of $ _____ enclosed.

☐ Bill me

Name of Registrant(s)
(To appear on certificate of completion)

_____

_____

_____

P.O. #: _____

GOVERNMENT AGENCIES MUST SUPPLY PURCHASE ORDER NUMBER

Firm Name _____

Address _____

City/State/Zip _____

Telephone No. fax No. _____

E-mail Address _____

Charge our registration(s) to: ☐ MasterCard ☐ VISA ☐ American Express ☐ Discover

Account No. _____ Exp. Date _____

Cardholder's Signature _____

Seminar Name   City   Dates

_____

_____

Please mail check to: Seminar Division, Reed Construction Data, RSMeans Seminars, 63 Smiths Lane, P.O.Box 800, Kingston, MA 02364 USA

MeansData™

CONSTRUCTION COSTS FOR SOFTWARE APPLICATIONS

Your construction estimating software is only as good as your cost data.

A proven construction cost database is a mandatory part of any estimating package. The following list of softwa providers can offer you MeansData™ as an added feature for their estimating systems. See the table below for w types of products and services they offer (match their numbers). Visit online at **www.rsmeans.com/demosource/** for more information and free demos. Or call their numbers listed below.

1. **3D International**
713-871-7000
venegas@3di.com

2. **4Clicks-Solutions, LLC**
719-574-7721
mbrown@4clicks-solutions.com

3. **Aepco, Inc.**
301-670-4642
blueworks@aepco.com

4. **Applied Flow Technology**
800-589-4943
info@aft.com

5. **ArenaSoft Estimating**
888-370-8806
info@arenasoft.com

6. **ARES Corporation**
925-299-6700
sales@arescorporation.com

7. **BSD - Building Systems Design, Inc.**
888-273-7638
bsd@bsdsoftlink.com

8. **CMS - Computerized Micro Solutions**
800-255-7407
cms@proest.com

9. **Corecon Technologies, Inc.**
714-895-7222
sales@corecon.com

10. **CorVet Systems**
301-622-9069
sales@corvetsys.com

11. **Discover Software**
727-559-0161
sales@discoversoftware.com

12. **Estimating Systems, Inc.**
800-967-8572
esipulsar@adelphia.net

13. **MAESTRO Estimator Schwaab Technology Solutions, Inc.**
281-578-3039
Stefan@schwaabtech.com

14. **Magellan K-12**
936-447-1744
sam.wilson@magellan-K12.com

15. **MC² - Management Computer**
800-225-5622
vkeys@mc2-ice.com

16. **Maximus Asset Solutions**
800-659-9001
assetsolutions@maximus.com

17. **Prime Time**
610-964-8200
dick@primetime.com

18. **Prism Computer Corp.**
800-774-7622
famis@prismcc.com

19. **Quest Solutions, Inc.**
800-452-2342
info@questsolutions.com

20. **RIB Software (Americas), Inc.**
800-945-7093
san@rib-software.com

21. **Shaw Beneco Enterprises, Inc.**
877-719-4748
inquire@beneco.com

22. **Timberline Software Corp.**
800-628-6583
product.info@timberline.com

23. **TMA SYSTEMS, Inc.**
800-862-1130
sales@tmasys.com

24. **US Cost, Inc.**
800-372-4003
sales@uscost.com

25. **Vanderweil Facility Advisors**
617-451-5100
info@VFA.com

26. **Vertigraph, Inc.**
800-989-4243
info-request@vertigraph.com

27. **WinEstimator, Inc.**
800-950-2374
sales@winest.com

TYPE	1	2	3	4	5	6	7	8	9	10	11	12	13	14	15	16	17	18	19	20	21	22	23	24	25	26	27
BID					●		●	●		●	●				●		●		●	●	●	●				●	●
Estimating		●			●	●	●	●	●	●	●	●	●		●		●		●	●		●				●	●
DOC/JOC/SABER		●			●		●			●	●		●			●					●	●					●
IDIQ		●								●	●		●			●					●						●
Asset Mgmt.											●			●		●	●	●							●		●
Facility Mgmt.	●		●								●				●	●	●	●		●			●	●			
Project Mgmt.	●	●				●			●	●	●				●		●		●	●	●						
TAKE-OFF					●			●	●		●				●				●	●	●	●		●		●	●
EARTHWORK								●			●				●		●	●								●	
Pipe Flow				●							●																
HVAC/Plumbing					●			●		●	●																
Roofing					●			●		●	●																
Design	●				●						●						●			●	●						●
Other Offers/Links:																											
Accounting/HR		●			●						●								●	●	●						
Scheduling					●						●		●						●	●				●		●	●
CAD											●			●							●						●
PDA																					●			●			●
Lt. Versions		●						●									●				●						
Consulting	●	●			●			●			●				●	●	●		●					●			●
Training		●			●		●	●	●	●	●				●	●	●	●		●	●	●	●	●	●	●	●

Reseller applications now being accepted. Call Carol Polio Ext. 5107.

FOR MORE INFORMATION
CALL 1-800-448-8182, EXT. 5107 OR FAX 1-800-632-6732

For more information
visit Means Web Site
at www.rsmeans.com

New Titles

Builder's Essentials: Estimating Building Costs for the Residential & Light Commercial Contractor

By Wayne J. DelPico

Step-by-step estimating methods for residential and light commercial contractors. Includes a detailed look at every construction specialty—explaining all the components, takeoff units, and labor needed for well-organized, complete estimates covers:

Correctly interpreting plans and specifications.

Developing accurate and complete labor and material costs.

Understanding direct and indirect overhead costs... and accounting for time-sensitive costs.

Using historical cost data to generate new project budgets.

Plus hard-to-find, professional guidance on what to consider so you can allocate the right amount for profit and contingencies.

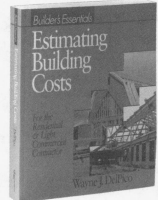

$29.95 per copy
Over 400 pages, illustrated, softcover
Catalog No. 67343

Building & Renovating Schools

This all-inclusive guide covers every step of the school construction process—from initial planning, needs assessment, and design, right through moving into the new facility. A must-have resource for anyone concerned with new school construction or renovation, including architects and engineers, contractors and project managers, facility managers, school administrators and school board members, building committees, community leaders, and anyone else who wants to ensure that the project meets the schools' needs in a cost-effective, timely manner. With square foot cost models for elementary, middle, and high school facilities and real-life case studies of recently completed school projects.

The contributors to this book – architects, construction project managers, contractors, and estimators who specialize in school construction – provide start-to-finish, expert guidance on the process.

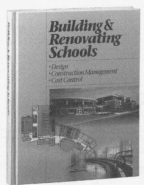

$99.95 per copy
Over 425 pages
Catalog No. 67342

Means ADA Compliance Pricing Guide, New Second Edition

By Adaptive Environments and RSMeans

Completely updated and revised to the new 2004 Americans with Disabilities Act Accessible Guidelines, this book features more than 70 of the most commonly needed modifications for ADA compliance—their design requirements, suggestions, and final cost. Projects range from installing ramps and walkways, widening doorways and entryways, and installing and refitting elevators, to relocating light switches and signage, and remodeling bathrooms and kitchens. Also provided are:

Detailed cost estimates for budgeting modification projects, including estimates for each of 260 alternates.

An assembly estimate for every project, with detailed cost breakdown including materials, labor hours, and contractor's overhead.

3,000 Additional ADA compliance-related unit cost line items.

Costs that are easily adjusted to over 900 cities and towns.

Over 350 pages
$79.99 per copy
Catalog No. 67310A

2005 Order Form

ORDER TOLL FREE 1-800-334-3509
OR FAX 1-800-632-6732.

Qty.	Book No.	COST ESTIMATING BOOKS	Unit Price	Total
	60065	Assemblies Cost Data 2005	$189.95	
	60015	Building Construction Cost Data 2005	115.95	
	61015	Building Const. Cost Data–Looseleaf Ed. 2005	144.95	
	63015	Building Const. Cost Data–Metric Version 2005	115.95	
	60225	Building Const. Cost Data–Western Ed. 2005	115.95	
	60115	Concrete & Masonry Cost Data 2005	105.95	
	60145	Construction Cost Indexes 2005	251.95	
	60145A	Construction Cost Index–January 2005	63.00	
	60145B	Construction Cost Index–April 2005	63.00	
	60145C	Construction Cost Index–July 2005	63.00	
	60145D	Construction Cost Index–October 2005	63.00	
	60345	Contr. Pricing Guide: Resid. R & R Costs 2005	39.95	
	60335	Contr. Pricing Guide: Resid. Detailed 2005	39.95	
	60325	Contr. Pricing Guide: Resid. Sq. Ft. 2005	39.95	
	64025	ECHOS Assemblies Cost Book 2005	189.95	
	64015	ECHOS Unit Cost Book 2005	126.95	
	54005	ECHOS (Combo set of both books)	251.95	
	60235	Electrical Change Order Cost Data 2005	115.95	
	60035	Electrical Cost Data 2005	115.95	
	60205	Facilities Construction Cost Data 2005	279.95	
	60305	Facilities Maintenance & Repair Cost Data 2005	252.95	
	60165	Heavy Construction Cost Data 2005	115.95	
	63165	Heavy Const. Cost Data–Metric Version 2005	115.95	
	60095	Interior Cost Data 2005	115.95	
	60125	Labor Rates for the Const. Industry 2005	253.95	
	60185	Light Commercial Cost Data 2005	99.95	
	60025	Mechanical Cost Data 2005	115.95	
	60155	Open Shop Building Const. Cost Data 2005	115.95	
	60215	Plumbing Cost Data 2005	115.95	
	60045	Repair and Remodeling Cost Data 2005	99.95	
	60175	Residential Cost Data 2005	99.95	
	60285	Site Work & Landscape Cost Data 2005	115.95	
	60055	Square Foot Costs 2005	126.95	
		REFERENCE BOOKS		
	67147A	ADA in Practice	59.98	
	67310A	ADA Compliance Pricing Guide, 2nd Ed.	79.95	
	67273	Basics for Builders: How to Survive and Prosper	34.95	
	67330	Bldrs Essentials: Adv. Framing Methods	24.95	
	67329	Bldrs Essentials: Best Bus. Practices for Bldrs	29.95	
	67298A	Bldrs Essentials: Framing/Carpentry 2nd Ed.	24.95	
	67298AS	Bldrs Essentials: Framing/Carpentry Spanish	24.95	
	67307	Bldrs Essentials: Plan Reading & Takeoff	35.95	
	67261A	Bldg. Prof. Guide to Contract Documents 3rd Ed.	64.95	
	67342	Building & Renovating Schools	99.95	
	67339	Building Security: Strategies & Costs	89.95	
	67312	Building Spec Homes Profitably	29.95	
	67146	Concrete Repair & Maintenance Illustrated	69.95	
	67314	Cost Planning & Est. for Facil. Maint.	89.95	
	67317A	Cyberplaces: The Internet Guide 2nd Ed.	59.95	
	67328A	Designing & Building with the IBC, 2nd Ed.	99.95	
	67230B	Electrical Estimating Methods 3rd Ed.	64.95	

Qty.	Book No.	REFERENCE BOOKS (Cont.)	Unit Price	Total
	64777A	Environmental Remediation Est. Methods 2nd Ed.	$ 99.95	
	67160	Estimating for Contractors	35.95	
	67276A	Estimating Handbook 2nd Ed.	99.95	
	67249	Facilities Maintenance Management	86.95	
	67246	Facilities Maintenance Standards	79.95	
	67318	Facilities Operations & Engineering Reference	109.95	
	67301	Facilities Planning & Relocation	89.95	
	67231	Forms for Building Const. Professional	47.48	
	67260	Fundamentals of the Construction Process	34.98	
	67323	Historic Preservation: Proj. Planning & Est.	99.95	
	67308E	Home Improvement Costs–Int. Projects 9th Ed.	29.95	
	67309E	Home Improvement Costs–Ext. Projects 9th Ed.	29.95	
	67324A	How to Est. w/Means Data & CostWorks 2nd Ed.	59.95	
	67304	How to Estimate with Metric Units	9.98	
	67306	HVAC: Design Criteria, Options, Select. 2nd Ed.	84.95	
	67281	HVAC Systems Evaluation	84.95	
	67282A	Illustrated Construction Dictionary, Condensed	59.95	
	67292A	Illustrated Construction Dictionary, w/CD-ROM	99.95	
	67295B	Landscape Estimating 4th Ed.	62.95	
	67341	Life Cycle Costing	99.95	
	67302	Managing Construction Purchasing	19.98	
	67294A	Mechanical Estimating 3rd Ed.	64.95	
	67245A	Planning and Managing Interior Projects 2nd Ed.	69.95	
	67283A	Plumbing Estimating Methods 2nd Ed.	59.95	
	67326	Preventive Maint. Guidelines for School Facil.	149.95	
	67337	Preventive Maint. for Higher Education Facilities	149.95	
	67236A	Productivity Standards for Constr.–3rd Ed.	49.98	
	67247B	Project Scheduling & Management for Constr. 3rd Ed.	64.95	
	67265B	Repair & Remodeling Estimating 4th Ed.	69.95	
	67322A	Resi. & Light Commercial Const. Stds. 2nd Ed.	59.95	
	67327	Spanish/English Construction Dictionary	22.95	
	67145A	Sq. Ft. & Assem. Estimating Methods 3rd Ed.	69.95	
	67287	Successful Estimating Methods	32.48	
	67313	Successful Interior Projects	24.98	
	67233	Superintending for Contractors	35.95	
	67321	Total Productive Facilities Management	29.98	
	67284	Understanding Building Automation Systems	29.98	
	67303A	Unit Price Estimating Methods 3rd Ed.	59.95	
	67319	Value Engineering: Practical Applications	79.95	

MA residents add 5% state sales tax	
Shipping & Handling**	
Total (U.S. Funds)*	

Prices are subject to change and are for U.S. delivery only. *Canadian customers may call for current prices. **Shipping & handling charges: Add 7% of total order for check and credit card payments. Add 9% of total order for invoiced orders.

Send Order To: **ADDV-1001**

Name (Please Print) _____

Company _____

☐ **Company**
☐ **Home**   Address _____

_____

City/State/Zip _____

Phone # _____ P.O. # _____

(Must accompany all orders being billed)

Mail To: **RSMeans** P.O. Box 800, Kingston, MA 02364-0800